AAAR

THE RIVERS HANDBOOK

The Rivers Handbook

HYDROLOGICAL AND ECOLOGICAL PRINCIPLES

EDITED BY

PETER CALOW
DSc, PhD, CBiol, FIBiol
Department of Animal and Plant Sciences
The University of Sheffield

AND

GEOFFREY E. PETTS
BSc, PhD
School of Geography
The University of Birmingham

IN TWO VOLUMES
VOLUME TWO

OXFORD
BLACKWELL SCIENTIFIC PUBLICATIONS
LONDON EDINBURGH BOSTON
MELBOURNE PARIS BERLIN VIENNA

Editorial Offices:
Osney Mead, Oxford OX2 0EL
25 John Street, London WC1N 2BL
23 Ainslie Place, Edinburgh EH3 6AJ
238 Main Street, Cambridge
Massachusetts 02142, USA
54 University Street, Carlton
Victoria 3053, Australia

Other Editorial Offices:
Librairie Arnette SA
1, rue de Lille
75007 Paris
France

Blackwell Wissenschafts-Verlag GmbH
Düsseldorfer Str. 38
D-10707 Berlin
Germany

Blackwell MZV
Feldgasse 13
A-1238 Wien
Austria

First published in 1994

Set by Excel Typesetters, Hong Kong
Printed and bound in Great Britain
at the University Press, Cambridge

DISTRIBUTORS

Marston Book Services Ltd
PO Box 87
Oxford OX2 0DT
(*Orders*: Tel: 0865 791155
Fax: 0865 791927
Telex: 837515)

USA
Blackwell Scientific Publications, Inc.
238 Main Street
Cambridge, MA 02142
(*Orders*: Tel: 800 759-6102
617 876-7000)

Canada
Oxford University Press
70 Wynford Drive
Don Mills
Ontario M3C 1J9
(*Orders*: Tel: 416 441-2941)

Australia
Blackwell Scientific Publications Pty Ltd
54 University Street
Carlton, Victoria 3053
(*Orders*: Tel: 03 347-5552)

A catalogue record for this title
is available from the British Library

ISBN 0-632-029854

Library of Congress
Cataloging-in-Publication Data

The Rivers handbook: hydrological
and ecological principles.
Includes bibliographical references
and indexes.
1. Rivers. 2. Stream ecology.
I. Calow, Peter II. Petts, Geoffrey E.
GB1205.R57 1994 551.48′3

Contents

List of Contributors

P. D. ARMITAGE *Institute of Freshwater Ecology, The River Laboratory, East Stoke, Wareham, Dorset BH20 6BB, UK*

N. W. ARNELL *Institute of Hydrology, Crowmarsh Gifford, Wallingford, Oxfordshire OX10 8BB, UK*

R. BETTESS *Hydraulics Research, Howbery Park, Wallingford, Oxfordshire OX10 8BA, UK*

A. BROOKES *National Rivers Authority Thames Region, Kings Meadow House, Kings Meadow Road, Reading, Berkshire RG1 8DQ, UK*

P. CROCKETT *National Rivers Authority Severn Trent Region, 550 Streetsbrook Road, Solihull B91 1QT, UK*

B. R. DAVIES *Freshwater Research Unit, Department of Zoology, University of Cape Town, Cape Town, South Africa*

A. J. DOBBS *WRc plc, Henley Road, Medmenham, Marlow, Buckinghamshire SL7 2HD, UK*

D. P. DODGE *Ontario Ministry of Natural Resources, Great Lakes Branch, PO Box 5000, Maple, Toronto, Ontario L6A 1S9, Canada*

N. P. FAWTHROP *National Rivers Authority Anglian Region, Kingfisher House, Goldhay Way, Orton Goldhay, Peterborough PE2 5ZR, UK*

G. M. GÍSLASON *Institute of Biology, University of Iceland, Grensasvegur 12, IS-108 Reykjavik, Iceland*

J. A. GORE *Center For Environmental Research and Service, The Troy State University System, Troy, Alabama 36082, USA*

R. D. HEY *School of Environmental Sciences, University of East Anglia, Norwich, Norfolk NR4 7TJ, UK*

A. R. G. LARGE *Department of Geography, University of Technology, Loughborough, Leicestershire LE11 3TU, UK*

P. LARSEN *Institute of Hydraulic Structures and Agricultural Engineering, University of Karlsruhe, Kaiserstrasse 12, D-76128 Karlsruhe, Germany*

C. C. MACK *Ontario Ministry of Natural Resources, Great Lakes Branch, PO Box 5000, Maple, Ontario L6A 1S9, Canada*

I. MADDOCK *Department of Geography, University of Technology, Loughborough, Leics LE11 3TU, UK*

J. A. MAWDSLEY *National Rivers Authority Northumbria–Yorkshire Region, Eldon House, Regent Centre, Gosforth, Newcastle upon Tyne NE3 3UD, UK*

J. L. METCALFE-SMITH *Rivers Research Branch, National Water Research Institute, PO BOX 5050, 867 Lakeshore Road, Burlington, Ontario L7R 4A6, Canada*

A. M. MILNER *Environmental and Natural Resources Institute, University of Alaska, 707 A Street, Anchorage, Alaska 99501, USA*

J. H. O'KEEFFE *Institute For Water Research, Rhodes University, Grahamstown 6140, South Africa*

W. PARR *WRc plc, Henley Road, Medmenham, Marlow, Buckinghamshire SL7 2HD, UK*

G. E. PETTS *School of Geography, University of Birmingham, Edgbaston, Birmingham BI5 2TT, UK*

J. L. RASMUSSEN *USDI Fish and Wildlife Service, Large River Fisheries Coordination Office, US Fish and Wildlife Service, 608 East Cherry, Columbia, Missouri 65201, USA*

C. B. STALNAKER *USDI National Biological Survey, 4512 McMurry Avenue, Fort Collins, Colorado 80525-3400, USA*

R. A. SWEETING *National Rivers Authority Thames Region, 6th Floor, Reading Bridge House, Reading Bridge, Reading RG1 8DQ, UK*

M. C. THOMS *Department of Geography, University of Sydney, New South Wales, Australia*

A. J. UNDERWOOD *Institute of Marine Ecology, Zoology Building A08, University of Sydney, New South Wales 2006, Australia*

P. M. WADE *International Centre of Landscape Ecology, Department of Geography, University of Technology, Loughborough, Leicestershire LE11 3TU, UK*

K. F. WALKER *River Murray Laboratory, Department of Zoology, University of Adelaide, South Australia 5005, Australia*

T. F. ZABEL *WRC plc, Henley Road, Medmenham, Marlow, Buckinghamshire SL7 2HD, UK*

Preface

Putting principles into practice is perhaps the ultimate arbiter of sound science. On the other hand, nature can be managed without science – by trial-and-error and folklaw. Yet as medicine, agriculture and, indeed, river management itself have amply demonstrated, the application of sound science can make the practice considerably more reliable and effective.

The Rivers Handbook has been constructed in this spirit: 'to provide a synthesis of scientific knowledge on rivers and to provide the basis for achieving scientifically sound and environmentally sensitive management' (Preface to Vol. 1). The first volume explored the principles; seeking systematically not only to describe physical, chemical and biological features of rivers but also to provide a critical appraisal of the approaches for their investigation and the models available for synthesis, prediction and refining our understanding.

This volume puts principles into practice. It has five major parts. The first reminds readers of the basics (Chapter 1) and then describes, in general terms, the sensitivity of river systems to environmental perturbation from pollution, hydrological and river channel change. Part of the role of science is to make these observations not only quantitative but good descriptors of the key features of complex systems. This is the business of Part 2 which reviews problems associated with biological sampling and water-quality monitoring. Part 3 moves into the predictive mode, examining the application of mathematical models with respect to the effects of physical change on the physical properties of rivers and the effects of physical and chemical changes on the biota. Part 4 focuses on the tools and approaches that are available for solving, or mitigating, particular problems, i.e. integrating environmentally sensitive tools and approaches into contemporary river management practices. Finally, Part 5 provides case studies not only from the temperate region–where, as in Vol. 1, the emphasis has been–but from cold environments and dry lands to emphasize where the future needs are likely to be.

Volume 2 completes the *Handbook*. It will not, of course, be the last word on the subject; both principles and practice will develop. The principles will sharpen and the practice will become increasingly sophisticated and effective. What we hope is that this book will have acted as more than just a signpost along the way of these developments. It has benefited, we are sure, from the effort of the individual authors, from the caring attention of our publishers and from the much appreciated work of our support teams. To all these we offer our thanks. As always, of course, in the final analysis we, the editors, accept the blame for any remaining blemishes.

Peter Calow
Geoffrey E. Petts

Abbreviations and Symbols

These are arranged in alphabetical order with Roman first, followed by Greek. One letter can be used for several different terms; we have not attempted to rationalize these since it would have been counter to common usage and potentially would have caused more confusion than it avoided. Because of the widely different ways in which terms with common letters are used, there is little danger of misunderstanding.

A	Area
ABF	Aquatic baseflow
ADF	Average daily flow
AOX	Adsorbable organic halogen
AQC	Analytical quality control
ASPT	Average score per taxon
B	Channel width
BACI	Before–after/control–impact
BATNEEC	Best available techniques (technology) not entailing excessive costs
BBI	Belgian biotic index
BCT	Best conventional technology
BEOL	Biologically effective organic loading
BI	Biotic index
BMWP	Biological Monitoring Working Party
BOD	Biochemical oxygen demand
BPEO	Best practicable environmental option
CCI	Community comparison index
COD	Chemical oxygen demand
CPOM	Coarse particulate organic matter
CSO	Combined sewer outflow
D	Sediment diameter
D	Mean water depth (also d)
D_{35}	Sediment diameter for which 35% of the sample is finer
D_{gr}	Dimensionless sediment diameter (Ackers and White sediment transport theory)
d	Depth of flow (also D)
DO	Dissolved oxygen
DOM	Dissolved organic matter
DWF	Dry-weather flow
EBI	Extended biotic index
EI	Ecotoxicological index
EPT	Ephemeroptera/Plecoptera/Trichoptera
EQO	Environmental quality objective
EQS	Environmental quality standard
F_{gr}	Sediment mobility (Ackers and White sediment transport theory)
Fr	Froude number
FBI	Family-level biotic index
FFG	Functional feeding group
FPOM	Fine particulate organic matter
G	Volume rate of sediment transport per unit width
G_{gr}	Dimensionless sediment transport rate (Ackers and White sediment transport theory)
g	Acceleration due to gravity
H'	Shannon–Wiener index value
h	Frequency of occurrence
HSC	Habitat suitability criteria
IB	Indice biotique
IBI	Index of biotic integrity
IBT	Inter-basin transfer
ICI	Invertebrate community index
IDH	Intermediate disturbance hypothesis

IFIM	Instream flow incremental methodology
IHACRES	Identification of unit hydrographs and component flows from rainfall, evaporation and streamflow data
LC_{50}	Concentration lethal to 50% of those exposed
LoD	Limit of detection
MAC	Maximum admissible concentration
MAP	Mean annual precipitation
MAR	Mean annual runoff
MMF	Minimum maintained flow
n	Manning's roughness
NOEC	No observable effect concentration
NWS	Net withdrawals from storage
PCB	Polychlorinated biphenyl
PHABSIM	Physical habitat simulation
POM	Particulate organic matter
Q	Discharge
Q95	95th percentile discharge
Q_s	Sediment load
QSAR	Quantity structure–activity relation
R_a	Annual reliability
R_m	Time-based reliability
R_v	Volume reliability
r	Shear stress
RCC	River continuum concept
RCS	River conservation system
Re	Reynolds number
Re^*	Boundary Reynolds number
RIVPACS	River invertebrate prediction and classification system
S	Saprobic index
S	Stream slope
S_f	Friction slope
S_{R1}	Equilibrium regime slope for single channel
S_{R2}, S_{R3}, etc.	Equilibrium regime slope for two, three (etc.) channels
S_v	Valley slope
S_y	Bed slope
s	Saprobic value
s	Specific gravity of sediment
SDC	Serial discontinuity concept
SRP	Soluble reactive phosphate
SS	Suspended solids
SU	Systematic unit
t	Time
TBC	Toxicity-based consent
TBI	Trent biotic index
TQ	Tolerance quotient
TUH	T hour unit hydrograph
TWINSPAN	Two-way indicator species analysis
U	Mean current velocity (also V)
U^*	Shear velocity (also V^*)
UES	Uniform emission standard
V	Velocity (also U)
V^*	Shear velocity (also U^*)
w	Stream width
WQO	Water quality objective
WUA	Weighted usable area
X	Sediment concentration by weight
x	Horizontal distance
z	Bed level
γ	Specific weight of fluid
δ	Thickness of viscous sublayer
τ	Shear stress
ν	Kinematic viscosity
Ω	Bagnold power

Part 1: Perturbations and Biological Impacts

Rivers may be viewed as arteries of the catchment ecosystem. Many of the problems for river management today relate to a complex sequence of landuse changes, water resource developments and industrial expansion which have altered the pattern of runoff, the quality of river discharge, and the size distribution and load of sediment transported. Volume 2 of the *Handbook* begins here with a detailed consideration of the sensitivity of river systems to environmental perturbations. The chapters seek not only to describe the changes that have occurred within river ecosystems but also to elucidate the complex interactions between the physical, chemical and biological components that occur in space and time. Part 1 begins with a brief review of the fundamental scientific knowledge of river systems required to underpin environmentally-sensitive river management. Then, the three primary groups of impacts are discussed: pollution (Chapter 2), hydrological change (Chapter 3) and river channel change (Chapter 4). Each chapter describes the ways in which the impacts have affected biota and ecosystem functioning, and identify the processes and pathways that must be managed if ecosystem degradation is to be avoided, or rehabilitation achieved. However, the effect of an impact is measured not only by the magnitude of change induced within the river ecosystem but also by the ability of the ecosystem to recover. Thus, this part of the *Handbook* concludes (Chapter 5) with a discussion of the mechanisms by which biotic communities recover from disturbance.

1: Rivers: Dynamic Components of Catchment Ecosystems

G. E. PETTS

1.1 INTRODUCTION

Throughout the world, river systems have been dramatically altered by human action. These changes have been induced directly by dams and reservoirs (Petts 1984) and channelization (Brookes 1988), and indirectly by land use developments throughout the drainage basins (Lauga *et al* 1986). Some species of both flora and fauna have disappeared; exotic species have invaded; the functional characteristics of the river systems have been disrupted, and there has been a reduction in landscape quality and loss of wilderness areas.

Political concern for river systems has arisen mainly in response to pollution problems, but increasingly there is arising a broader concern for ecosystem sustainability in the face of both socio-economic development and climatic change (e.g. Roberts 1993). The need to develop rivers and their water resources continues to intensify, especially in the developing countries where the existing needs for hydro-electricity and irrigation, the increasing needs of the rapidly growing populations, and the likely increase in per capita resource demands, mean that major river developments will continue (e.g. Petts 1990c). Particular problems are associated with dryland developments (Beaumont 1989; see also Chapter 25) where one solution to water resource problems is the development of large-scale inter-basin transfers.

Over the past two decades, increasing efforts have been directed to the application of scientific principles to the development of environmentally-sensitive approaches for managing rivers (e.g. Gore & Petts 1989; Boon *et al* 1992). The following chapters in this volume provide critical state-of-the-art reviews of our scientific understanding of the way in which rivers are affected by human activity and of the approaches currently available to diagnose and mitigate problems, and to rehabilitate damaged systems. An up-to-date scientific background for river management is provided in Volume 1 of *The Rivers Handbook*; this chapter summarizes the key points that underpin approaches for developing environmentally-sensitive water projects.

The scientific basis for modern approaches to river management

The quantitative and process-based studies of rivers that provide the basis for environmentally-sensitive water management are modern developments. Truly integrated studies of fluvial environments are even more recent. A revolution in scientific endeavour occurred in the 1950s with the increasing application of analytical and experimental techniques, the quantitative description of systems, and the application of physical, chemical and biological principles. Modern approaches to the study of rivers are founded in two important works: *Fluvial Processes in Geomorphology* by L.B. Leopold, M.G. Wolman and J.P. Miller (Leopold *et al* 1964), and *The Ecology of Running Waters* by H.B.N. Hynes (1970). It is only in the last decade, however, that hydrological, geomorphological and biological research have become integrated to establish a new understanding for lotic systems (e.g. Resh *et al* 1988; Stalnaker *et al* 1989; Statzner *et al* 1988) and their land-water ecotones (Naiman & Decamps 1990).

An acceptance that human intervention within drainage basins can transform, and usually degrade, river ecosystems is also a feature of the recent period. George P. Marsh (1864) – the mighty prophet of modern conservation – had demonstrated a remarkable awareness of the short-term and potential long-term consequences of human activity on the environment. These consequences were clearly illustrated in *Man's Role in Changing the Face of the Earth* edited by Thomas and published in 1956. Thirty years after Thomas's book, *The Ecology of River Systems* edited by B.R. Davies and K.F. Walker (1986) demonstrated that virtually all the world's rivers had been very much changed by human impacts well before any serious scientific studies had been undertaken; a paucity of field data had inhibited detailed analyses.

The first major work focusing on human impacts on river ecology was published in 1972. *River Ecology and Man*, edited by R.T. Oglesby, C.A. Carlson and J.A. McCann (Oglesby *et al* 1972), evolved in response to a new, and rapidly growing, insistence that river quality – in terms of nature conservation, aesthetics and recreation – be weighed along with consumptive and ecologically destructive uses. Despite such advances, scientific development remained divorced from practical river management and decision-making. In his closing summary, J.W. Leonard noted (p. 459): 'The specialist can no longer responsibly fall back on the Philistine attitude that he will do his thing and if they are not intelligent enough to make use of his golden discoveries, well and good! but if not it's *their* hard luck.'

Throughout the 1970s and 1980s, isolationist and reductionist scientific philosophies constrained the development of new scientifically-based approaches for river development and management. The failure of scientists to communicate their findings effectively to managers and decision-makers has also contributed to progressive environmental degradation. Nevertheless, significant advances were being made especially on the ecological effects of river regulation (Ward & Stanford 1979; Lillehammer & Saltveit 1984; Craig & Kemper 1987). The growth of research in this field is reflected by the foundation in 1987 of a dedicated interdisciplinary journal, *Regulated Rivers*, which seeks to integrate scientific research and scientifically-based management. Much recent work has focused on the reconstruction of former river conditions (Petts *et al* 1989a), 'a difficult, but essential, exercise if we are to attempt to restore, or at least minimize or turn to our advantage, our impact on large rivers' (Hynes 1986, p. viii).

The synthesis of recent research presented in Volume 1 emphasizes that five principles govern our understanding of the functional characteristics of rivers. Rivers are: (i) three-dimensional systems; (ii) driven by hydrology and fluvial geomorphology; (iii) structured by food webs; (iv) characterized by spiralling processes; and (v) dependent upon change – changing flows, moving sediments and shifting channels. These five principles are elaborated below.

1.2 THE NATURE OF RIVERS

The palimpsestic view of the . . . system is that it is composed of a nested hierarchy of subsystems each having different levels of sensitivity and recovery, the whole being subject to a temporal stream of input (i.e. process) changes. [Chorley *et al* (1984, p. 9)]

From a biological point of view, flowing water has a number of advantages over still water, despite the stress exerted on biota: it is constantly mixed by turbulence providing nutrients, exchange of respiratory gases, and removal of wastes. Some species of aquatic macrophyte, for example, are highly adapted to the improved metabolic conditions of flowing water. Producers and consumers must be efficient at nutrient uptake, especially the sessile biota which cannot pursue their food items, because of the rapid downstream transport of nutrients. Flowing water is also fundamental for the downstream and lateral dispersal of attached and suspended algae, macrophytes, and invertebrates despite the high mortality caused by mechanical stress. Floods cause hydraulic disturbance but this is important in determining the composition of biotic communities within the channel, the riparian zone and the floodplain.

Classifications

Classification systems are necessary for both scientists and managers so that information about rivers can be organized, but for several reasons (Naiman *et al* 1992) most attention remains focused on conceptual and regional approaches. The principal components of the abiotic milieu of river ecosystems are hydrology, temperature and channel morphology – factors that at the regional scale reflect climate and geology. However, drainage basin morphometry and valley morphology are also important.

The world's rivers have adjusted to a sequence of climatic changes, and the related variations of vegetation cover and sea level, that have occurred over the Holocene; i.e. over a period of at least 10 000–12 000 years. Many river valleys contain evidence of pre-Holocene and even pre-Quaternary conditions, and valley form is seen to exert an important influence on the fluvial system.

On the global scale, Falkenmark and Chapman (1989) defined three primary 'climatic' regions (cold, temperate and warm) each subdivided into either humid or dry. Climate and relief explain the global variation of hydrological regimes (Beckinsale 1969). Climate and geology form the basis of a global classification of water quality (Gibbs 1970) which separates Ca-HCO_3 rock-dominated rivers from those dominated by Na-Cl with low (precipitation-dominated) or high (evaporation-dominated) conductivity. Lotspeich (1980) proposed the use of climate and geology as essential controls for the classification of catchments and their river systems but with soil and vegetation as secondary factors. Thus, regional water resource and fisheries management in the USA has utilized an 'ecoregion' approach to delineate large areas ($>1000\,km^2$) based upon climate, physiography and vegetation (Bailey 1978; Rohm *et al* 1987). Some researchers have attempted process-based classifications; streamflow patterns, for example, have been related to lotic community structure for different regions of the USA (Poff & Ward 1989). However, Naiman *et al* (1992) suggest that disturbance regimes, especially floods, droughts and channel changes, should be incorporated into such regional classification schemes.

A useful way to approach studies of individual rivers is to utilize a hierarchical classification

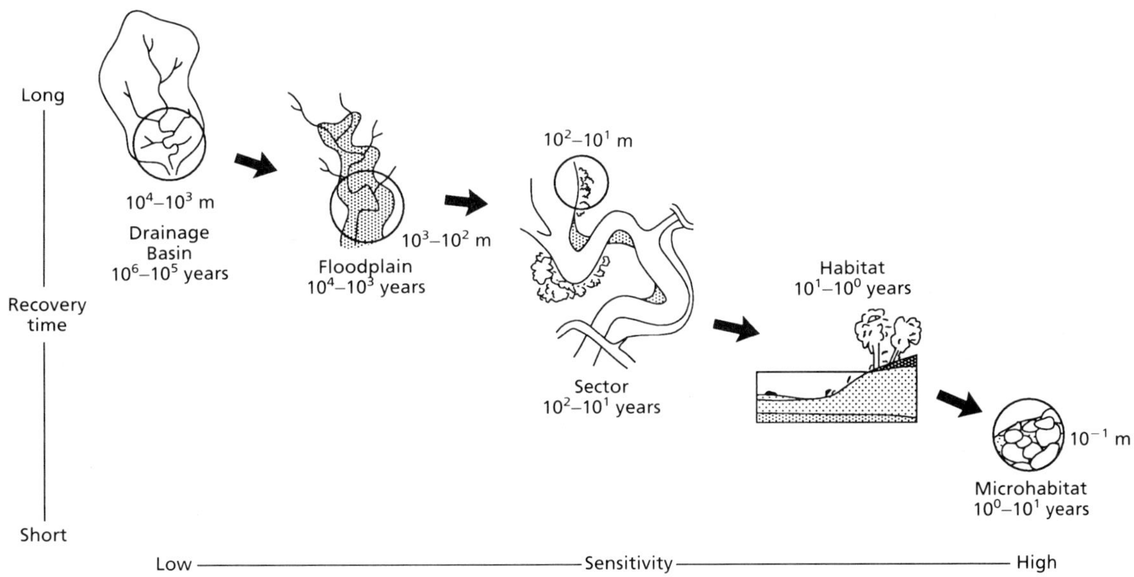

Fig. 1.1 Hierarchical organization of a river system in relation to sensitivity to disturbance and recovery time (after Frissell *et al* 1986).

incorporating spatially nested levels of resolution, such as that proposed by Frissell *et al* (1986) (Fig. 1.1). A river may be viewed as a series of reaches or sectors each receiving and discharging water, sediments, organic matter and nutrients. Individual sectors may be defined by tributary confluences, by changes of valley and channel morphology, or by changes of riparian vegetation; variables reflecting changes of hydrological, water quality and disturbance regimes. Habitats include pools and riffles, and backwater, marginal and main flow sites. Microhabitats relate, for example, to different hydraulic or substrate characteristics. Such a system defines not only structural components but also recognizes that the relative importance of factors controlling the short- and long-term behaviour of streams changes with spatial scale, following Schumm and Lichty (1965). With regard to disturbance, for example, microhabitats are seen to be most susceptible but they also have the shortest recovery time.

Within a single drainage basin the physical, chemical and biological characteristics of a river vary from headwaters to mouth. Although 'drainage area' and 'distance from source' have been used as scale variables for studies of river systems, it is the drainage network that focuses the delivery of water and material in transport from the catchment onto the main river. The number of headwater, fingertip, tributaries (the stream magnitude: Shreve 1967) and the number of confluences (the stream frequency: NERC 1975) relate closely to streamflow but most ecological studies have employed stream ordering techniques as an objective, and widely applicable, classification system. The most commonly used system is that proposed by Strahler (1952) modified after Horton (1945). This convention (see Fig. 1.2a) designates all headwater streams, terminated by the first confluence, as first-order streams; two first-order streams join to form a second-order stream; the confluence of two second-order streams produces a third-order river, and so on. As well as providing an index of scale, ordering also gives a measure of position within the stream network and it has been correlated with both physical and biological features of streams.

For rivers, classification is complicated by both longitudinal and lateral linkages which vary in different ways and to different degrees over different time-scales. This dynamic connectivity is fundamental for sustaining river ecosystems. Classification implies the existence of relatively distinct boundaries which may be identified by a set of discrete variables (Naiman *et al* 1992) but Arnell (1989) emphasizes the difficulties of classifying continuous processes. Thus, it has been argued (see Vol. 1, Chapter 2) that differences between basins represent a continuum of types and that simple classifications may give way to an approach which recognizes the uniqueness of each basin using large hydrological and thematic databases. The use of large database systems would allow the integration of information representing all the different perspectives on fluvial hydrosystems to provide a comprehensive classification scheme.

Systems characterized by longitudinal, lateral and vertical fluxes

Traditional views consider a river as a *longitudinal* gradient or sequence of interlinked zones (Illies & Botosaneanu 1963). The classic high-gradient upland stream will small channel, cold temperatures and highly oxygenated water, dominated by fastwater habitats, contrasts with low-gradient sectors of large floodplain rivers with a diversity of channel forms and floodplain water bodies. In general, large rivers tend to be characterized by more regular and predictable variations of the abiotic variables than headwater streams, and the greater range of habitats and food sources is associated with greater fish diversity and a wide range of trophic adaptations (Welcomme 1977). However, such downstream variations are more than just a structural sequence; they represent a longitudinal continuum dominated by downstream transfers of energy and matter (Vannote *et al* 1980).

With increasing size of the drainage basin, declining gradient and reducing altitude, progressive changes occur in discharge, channel morphology and temperature regime. From a hydrological and geomorphological point of view a river is seen as the artery of each drainage basin,

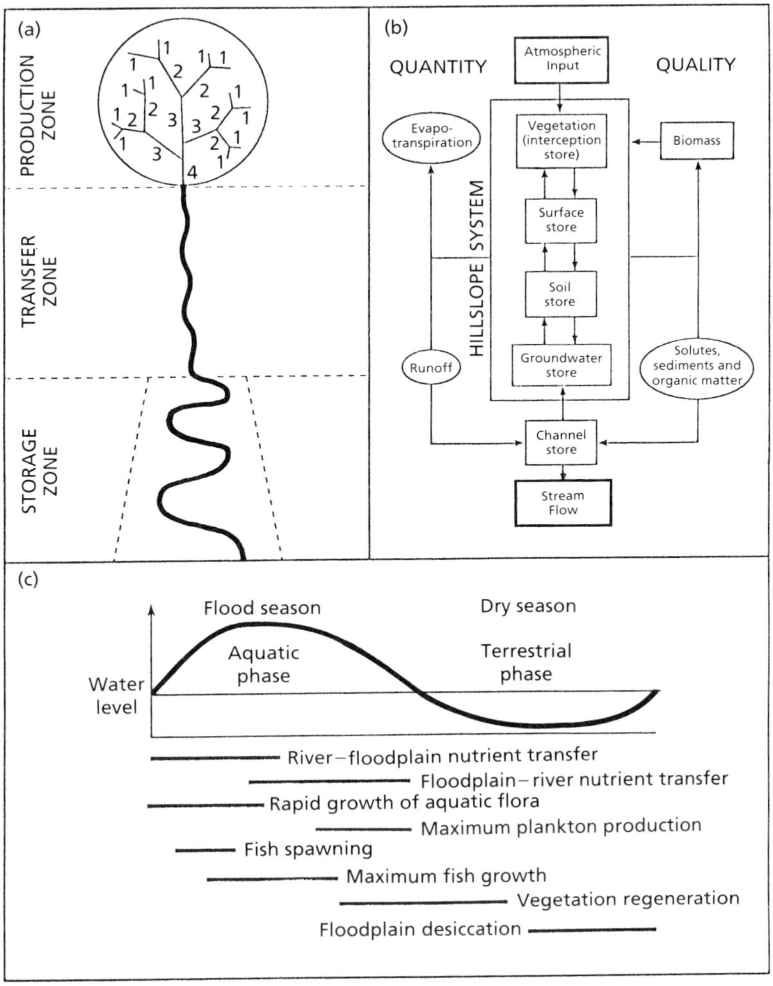

Fig. 1.2 Land-water interactions within fluvial hydrosystems. (a) The three primary zones (after Schumm 1977). (b) Hillslope controls on flow quantity and quality within headwater streams. (c) The influence of the floodpulse within the storage zone (based on Ward 1989).

and large basins can be viewed as a nested hierarchy of basins of different size. Ecosystem-level processes in downstream areas are linked to those in upstream catchments (Minshall *et al* 1985a), and Hynes (1970) described river ecosystems as largely heterotrophic, being dependent upon upstream sources for energy supplies. The simplified model of a river defined by Schumm (1977) recognizes three zones (Fig. 1.2a): headwater catchments (up to about 4th order), a transfer zone, and a downstream depositional or storage zone. This division fits well with the river continuum concept (RCC) (Vannote *et al* 1980), which describes the change in relative abundance of invertebrate functional groups along a river from headwater to mouth. The RCC defines three types of channel: litter-dominated, usually headwater, channels characterized by shredders and collectors; relatively wide and shallow mid-sectors where light and nutrients favour benthic algae production, characterized by scrapers; and lower sectors where high levels of fine particulate organic matter from upstream or from floodplain inputs favour collectors. Predators tend to have similar relative abundances in all sectors.

Herein, it is argued that within fluvial hydrosystems physical, chemical and biological processes are dominated by *longitudinal* (downslope) fluxes over long time-scales (10^3 years), but over a shorter timescale (<100 years) the dominance of longitudinal processes is restricted to zones 1 and 2; in zone 3 lowland floodplain systems, ecosystem dynamics are dominated by *lateral* exchanges and influenced by *vertical* fluxes.

Throughout fluvial hydrosystems important interactions take place across the land–water boundary. These interactions are manifest in a number of ways: organic matter inputs; nutrient fluxes; predation by terrestrial birds and mammals, etc. In zone 1 the variation of river flows, water quality and sediment transport is closely related to hillslope processes (Fig. 1.2b). The hydrological processes and pathways not only determine streamflow pattern (see Vol. 1, Chapter 1) but also influence the relative importance of the different biogeochemical processes that operate at all levels within the hillslope system (see Vol. 1, Chapters 3 and 4).

Many headwater streams rise in forest zones where extensive shading by riparian vegetation limits photosynthesis and supplies large amounts of allochthonous matter in the form of leaf fall. The importance of allochthonous organic material for the trophodynamics of streams has been demonstrated (see Vol. 1, Chapter 14), but the influence of large woody debris on channel morphology and habitat diversity is also well established (see Vol. 1, Chapter 6). Woody debris can play a particularly important role in the ecological development of new streams (Milner 1987; Sidle & Milner 1989) and this may have importance for river restoration.

In the depositional, floodplain zone the dynamics of the river are dominated by *lateral* exchanges between the channel and floodplain, and by *vertical* exchanges between the channel and the underlying alluvial aquifer. Floodplains comprise a range of forms, including channels having varying degrees of connectivity with the main river, and sedimentary units including sand bars, levees and backswamps. Typically a mosaic of patches of different type and age can be identified (see Plate 21.1), although the relative abundance of different patch types, the complexity of the mosaic and the arrangement of patches relate to the channel pattern (meandering, braided, anastomosed) and the degree of channel mobility (see Vol. 1, Chapter 7). Thus, along the Amazon, Salo *et al* (1986) demonstrated that channel migration is the principal process explaining the shifting successional forest mosaic.

The functioning of floodplain-river systems has been described by the 'flood-pulse' concept (Fig. 1.2c) formulated by Junk *et al* (1989) and applied to river fisheries by Bayley (1991). This emphasizes the importance of flood regime upon the ecological dynamics of aquatic and terrestrial biota. Welcomme (1992) considers that elucidation of the role of littoral and riparian ecotones (Naiman & Decamps 1990) has far-reaching implications for the management of rivers, particularly focusing on the need to restore the complex carbon, nutrient and biotic interactions. However, for the development of appropriate management, knowledge is still required concerning the typologies of floodplains and the factors regulating their relative productivity (Welcomme 1992).

The importance of the *vertical* dimension has been demonstrated only recently (Stanford & Ward 1988). Many rivers have formed large alluvial plains that record the history of fluvial processes, especially during the last 15 000 years (Starkel *et al* 1991), a period dominated by rising sea level. The sandy-gravel sediments that fill buried valleys form important riverine aquifers that are closely connected to surface-water bodies – channels and floodplain lakes – losing and gaining flow under different hydrological conditions. Groundwater habitats are colonized by various micro-, meio- and macro-organisms. The seasonally variable influence of groundwater into surface water-bodies also exerts an important influence on their faunas (see special issues of *Regulated Rivers* 6.4 (1991) and 7.1 (1992)).

Gibert *et al* (1990) conceived of a vertical ecotone at the interface between various surface water resource patches and the groundwater. Within the alluvial plain there is a diverse range of vertical ecotones: springs located on the border of the plain and a range of interstitial habitats located under stagnant floodplain lakes, under temporary water bodies, and under the main channel and its tributaries. The vertical ecotone acts as a physical and biological filter. Organisms can move actively or passively into the sediments, for example in response to drought or floods (Marmonier & Creuze des Chatelliers 1991); fine sediments and organic matter with associated nutrients, toxic metals, and micropollutants can accumulate within gravel channel-bed sediments as a result of infiltration (Petts *et al* 1989b), and alluvial aquifers can become contaminated by pollution from surface waters (Danielopol & Marmonier 1992).

Along large rivers, flow and water quality reflect the mixing of water from the different headwater catchments, the routing of flows through the channel network, and channel and floodplain storage. Large basins display a range of weather conditions, lithologies and land uses. Whilst headwater streams are highly sensitive to catchment changes such effects may be buffered downstream due to mixing of runoff from unaffected or differently affected catchments. Alternatively, downstream the cumulative effects of a variety of changes may give rise to significant ecological impacts. Downstream dispersion, such as invertebrate drift, which may be behavioural or accidental, plays a major role in influencing both population expansion through colonization of new habitats and community recovery following disturbance. However, because the degree of isolation increases from small streams to large rivers, Fuchs and Statzner (1990) argue that recovery of the benthic macroinvertebrate fauna may be faster in small streams than in large rivers, in which biological response may lag behind physical and water-quality recovery, a lag measured in tens of years. This is discussed further in Chapter 5.

The simple models of downstream variations in river characteristics provide the building blocks for interpreting real rivers, which are influenced by different climates, geologies, vegetation types and land uses. Variations also reflect complex geomorpholgical histories associated with changes of climate and landforming processes during the Pleistocene and Holocene. Indeed, one criticism of the continuum concept has been that along many rivers there are marked discontinuities related to geological controls, tributary confluences and artificial influences, such as dams. The effect of downstream changes in the effectiveness of disturbances on the structure and function of river ecosystems requires further investigation.

The role of hydrology and geomorphology

The importance of fluvial geomorphology as a template for river ecology was recognized by Hynes (1970, 1975). Lotic ecosystems are in large part determined by a range of hydrological, hydraulic and morphological variables: discharge (annual mean, seasonal regime, short-term variability); hydraulics (flow depth, velocity, shear stress); and channel morphology (width, depth, slope, pattern and substrate characteristics). The importance of channel morphology and local hydraulic variations for biota have been elaborated by Statzner and Higler (1986) and Statzner *et al* (1988).

At the catchment scale, the hydrological regime is an important determinant of ecosystem structure and function. Flow variability during the year is related mainly to climatic factors but catchment geology and the areas of lakes and floodplains within the catchment are also influential. The predictability of extreme flows (floods and droughts) and the magnitude and duration of these events, are important factors in determining biological communities. Flow variability is naturally regulated along a river, particularly as a result of floodplain storage. Large tropical rivers with naturally regulated and climatically predictable flood regimes are associated with a highly adapted biota (Welcomme 1977).

At any point along a river the channel morphology is adjusted to the supplies of water and sediment from upstream, modified by local conditions (see Chapter 12; see also Vol. 1, Chapter 6). Our current knowledge of river channel morphodynamics is founded in the regime theory developed by engineers at the turn of the century, and as advocated in fluvial geomorphology by Lane (1955). Lane conceptualized the relationship of discharge and sediment for a stable channel balance as:

$$Q_s D \propto QS$$

where Q_s = sediment load, Q = stream discharge, D = sediment size and S = stream slope. An increase in the fluid discharge, due, say, to urban development or climate change, or an increase in slope following channelization, will have repercussions for sediment transport, potentially causing channel degradation. On the other hand, an increase in the size or quantity of sediment input to a river, perhaps consequent upon catchment disturbance for construction, may cause channel aggradation with dramatic effects upon channel slope.

Any change in the supply of sediment to the river or a change in the fluid discharge, either in magnitude or distribution through time, will lead to adjustments of channel morphology, which, locally at least, may be rapid (see Chapter 4). A balance between forcing function, process and channel form, however, does not imply static conditions. The morphology of a channel represents a quasi-equilibrium form maintained by inter-adjustments of the different channel dimensions including: width, depth, slope, wetted perimeter, hydraulic radius, bed form, boundary roughness and sinuosity. The quasi-equilibrium channel represents the most probable state given the recent history of flows and sediment loads, and depending upon local constraints such as bank material characteristics, bed sediment, valley width, etc. (Langbein & Leopold 1964, p. 793). Bedrock outcrops or boulder rapids that inhibit channel slope adjustment, and valley narrowing, terraces or even riparian woods, limit lateral movement. Studies of river channel changes in response to climatic variations, neotectonic activity and sea-level changes, and human impacts have demonstrated the complex nature of channel adjustments (Schumm 1977).

Channel morphology is the product of the whole series of competent discharges experienced and a dominant discharge, which would form or maintain the average geometric properties of a mobile-bed channel, has proved impossible to define. Nevertheless, the discharge at bankfull stage is considered to be a useful representative discharge which is reasonably close to the dominant discharge on many rivers. Along most sectors the channel dimensions will show a degree of seasonal and annual variability, particularly in terms of substrate, bedform and cross-sectional shape. Variability also occurs at the scale of individual flood events which can cause erosion and deposition of both channel bed and banks. Nevertheless, channel forms present predictable geometrical characteristics (see also Table 4.3) and these can provide scale frameworks for ecological analyses.

Flowing water provides a wide variety of habitats reflecting the planform and cross-sectional variation of hydraulic conditions (Fig. 1.3a) within the channel. The pattern of velocity variation in space and time has a strong influence on biota, especially benthic invertebrates and fluvial algae. Three general velocity zones may be defined: near the bed over riffles, pools and channel margins. Although flow velocities through pools and riffles may be similar during high flows (Fig. 1.3b), when higher near-bed velocity at the pool mid-point may be locally higher than at the riffle crest (Clifford & Richards 1992), marked differences develop during receding flows. On the downstream face of riffles, relatively high flow velocities are maintained even under low-flow conditions sustaining favourable diffusion gradients for dissolved gases and nutrients. Channel asymmetry plays a particularly important role in maintaining zones of high flow velocities with declining discharge (Fig. 1.3c). Low velocities in pools and lateral 'dead zones' may have different flora and fauna not only because of the advantage of lower shear stresses but also because of the less favourable diffusion gradients. For many species survival of some life stage, at least, is assisted by the presence of refuges along the channel margins; the latter also provide important areas for the development of planktonic algae, zooplankton and fish fry.

Systems structured by food webs

That rivers and streams are part of catchment-scale ecosystems is demonstrated at the community level by running water food webs (see Vol. 1, Chapter 14). The role of food webs is a key topic also because of the importance of river communities in both environmental management and assessment. Food web structure obviously relates to the nature of biological interactions such as competition and predation but it also relates to environmental characteristics (Lawton 1989). Habitat diversity,

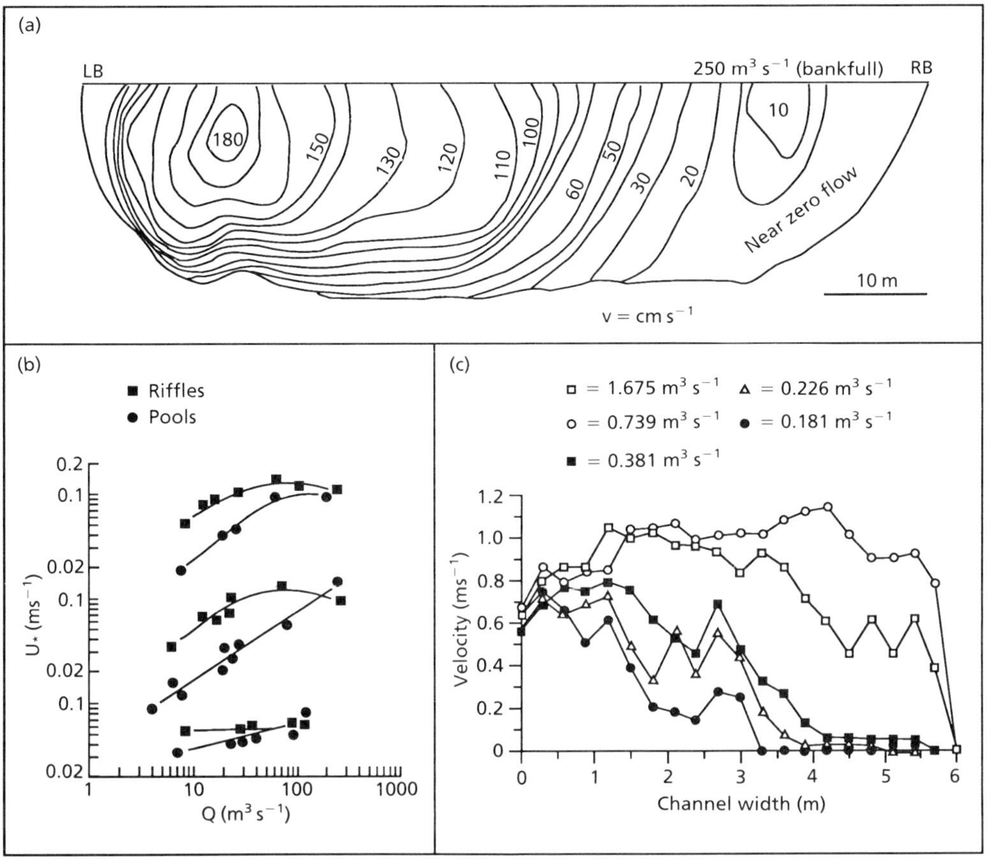

Fig. 1.3 Instream variations of flow velocity. (a) Cross-sectional velocity variations close to bankfull discharge at a riffle on the River Severn, UK (after Bevan & Carling 1992, by permission of John Wiley & Sons, Ltd). Note the region of slow flow with a high residence time near the right bank (RB). (b) Convergence of mean velocity (*U*) with increasing discharge (*Q*) in pools and riffles on the River Severn, UK (after Carling 1991). (c) Cross-sectional variation of velocity at 0.6 depth at various discharges within an asymmetric riffle on the Blackbrook, Leicestershire.

especially the complexity of refugia along the channel margins and on and within the channel bed has an important influence on the shape of the food web. However, hydraulic physical disturbances tend to weaken species interactions rather than alter the basic food web structure.

Primary production can be important in most running waters, including small, shaded streams (Stock & Ward 1989), although autotrophy tends to increase with stream order (Minshall *et al* 1983). Rooted aquatic angiosperms, their algal epiphytes, and the algal elements of biofilms on substrate sediments all contribute to autochthonous primary production, but the grazing of living macrophyte tissue is apparently negligible, although grazing of algae can be important in stony streams. Rivers and streams also receive a continuous input of energy from allochthonous sources. In headwater streams dissolved organic matter can be supplied by surface runoff, interflow through soils and groundwater (Foster & Grieve 1982). In floodplain rivers, following inundation, water returning to the channel contains high levels of dissolved organic carbon (Degenes *et al* 1991). Dissolved organic matter (DOM) can be processed in streams and rivers following conversion to a particulate form by physical-chemical processes or adsorption onto detritus, or it can be

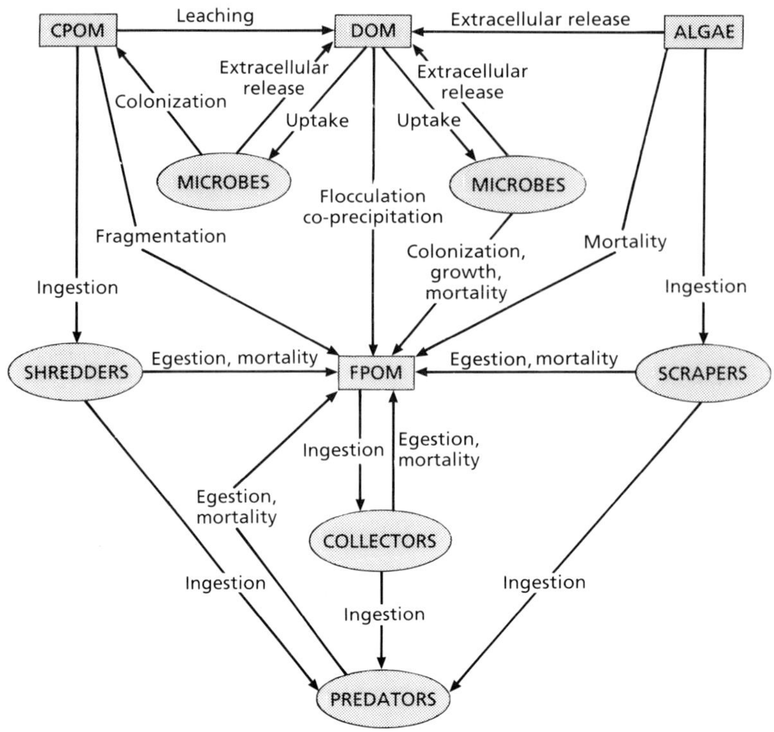

Fig. 1.4 The main interrelationships between nutrient resource pools, microbe state variables and invertebrate functional feeding groups (after Cummins 1992, by permission of John Wiley & Sons, Ltd). CPOM = coarse particulate organic matter (>1 mm); FPOM = fine particulate organic matter (1 mm–0.45 mm); DOM = dissolved organic matter; algae = primary producers.

assimilated by bacteria, some fungi and invertebrates. However, it is the detritus from both allochthonous (wood, leaves, etc.) and autochthonous (dead aquatic animals and plants) sources that is quantitatively the most important fuel for running-water food webs. The importance of detritus is that it comprises not only decomposing organic matter but also an assemblage of living organisms, such as fungi, bacteria, micro-invertebrates and algal cells (Anderson & Cargill 1987).

Detritivorous animals are important in processing organic matter, but micro-organisms play the pivotal role in stream ecosystem dynamics (Fig. 1.4). The breakdown of organic detritus in streams is related to the efficiency of heterotrophic micro-organisms (bacteria and fungi). Fungi, especially aquatic hyphomycetes, play a vital role in the decomposition of coarse particulate organic matter, particularly deciduous leaves, in streams, and in the functioning of allochthonous-based systems. Benthic bacteria associated with biofilms appear to be most important in the processing of fine particles and dissolved organic matter. Slimes on stones, comprising fungi, bacteria, algae and fine particulate organic matter, are important sites for processing of organic material and an important food source for detritivorous insects. The classification of macroinvertebrates into feeding groups – shredders, collectors, scrapers, piercers and predators – has application not only to the investigation of ecosystem properties but also to assessing human impacts. The invertebrates have received particular attention not least because they link the micro-organisms that dominate nutrient cycling on the one hand and fish, often the focus of interest in river management, on the other.

There are a range of predators in streams and rivers, including a variety of invertebrate taxa, fish, amphibia, and streamside birds and mammals. Predators in running waters may not be affected by the population dynamics of individual prey taxa but predation can exert a top-down control on prey. The role of predation by streamside birds and mammals in river ecosystem dynamics is yet to be fully evaluated.

Spiralling, delivery and retention characteristics

River ecosystems are characterized not only by downstream *transfers* but also by important *storages*. Model rivers may be viewed as a series of storages with different transport velocities, retention times and turnover rates. For example, sediment yield to the oceans is but a small fraction of that eroded from hillslopes. Some of the eroded material will be deposited further down the slope or at the foot of the slope, but along a river there are many depositional landforms all of which represent storages: bars, fans, floodplains and deltas. There are also many different sedimentary environments (Happ 1971), some retaining large amounts of organic material, such as cut-off channels and backswamps. That part of the gross erosion or sediment mobilization within a catchment represented by the sediment yield at the catchment outlet is defined by the sediment delivery ratio (Walling 1983). Values of the delivery ratio below 10% are common and an example of a catchment sediment budget is shown in Fig. 1.5a.

Some floodplain sediments together with their organic deposits can remain stable for thousands or tens of thousands of years. Throughout the temperate zone, valley aggradation during the period of Holocene sea-level rise has resulted in the long-term storage of thicknesses of sands, gravels, alluvium and organic deposits. At a smaller scale, woody debris jams, macrophyte beds, and the roots and overhanging branches of riparian vegetation are important retention features for leaf litter and sediment. To varying degrees, instream deposits are periodically flushed from the system following decomposition of debris jams and/or by high floods. Debris jams can retain a substantial proportion of the organic matter in channels, and typical residence times vary from a few years to a few hundred years (Gregory 1992).

The passage of water from hillslopes within headwater catchments to the mouth of the river during a flood involves lags over short timescales. With rising discharge, water enters each successive reach of a channel, part enters storage within that channel such that there is a net reduction in the rate of increase of discharge downstream. Through floodplain sectors, this attenuation is exacerbated by storage on floodplains and by losses through the channel bed into the alluvial aquifer. Flood peak attenuation along the River Tees, UK, is shown in Fig. 1.5b. Attenuation is at a minimum and nearly zero at just above bankfull (at about $375\,m^3\,s^{-1}$) and increases for both lower discharges contained within the channel and also very considerably at higher, floodplain flows (Archer 1989). During storage, water-quality changes take place by interactions between water, sediment, detritus and biota. At flows below bankfull, dead zones separated from the main flow by localized lines of lateral shear can be created by variations of channel shape and vegetation characteristics. Tracer experiments undertaken on the Lower Severn, UK (Bevan & Carling 1992) show typical distributions (Fig. 1.5c). Retardation of the tracer suggests the existence of large-scale flow structures within the reach with exchange across internal shear zones into volumes of slow-moving water (see Fig. 1.3a and Vol. 1, Plate 5.1).

Within any reach of channel a proportion of the downstream flux of nutrients will be taken up by the biota and flow through the food web – the components of which may also be seen as storages – or returned to the water. Within a river metabolic activity depends upon two primary carbon sources, autochthonous matter from primary production and allochthonous carbon from riparian, floodplain or terrestrial sources. At the base of the food web the primary producers and microbial consumers assimilate inorganic nitrogen and phosphorus from the river water; carbon is respired as dissolved inorganic carbon whilst inorganic N and P are recycled to the algae and microbes (see Vol. 1, Part 3).

In running-water systems, the downstream displacement gives rise to nutrient spiralling (see Vol. 1, Chapter 18). Elwood *et al* (1983) demonstrated seasonal changes in the uptake length and residence time of phosphorus in a 120-m reach of Walker Branch, Tennessee, USA. The shortest uptake lengths occurred in late autumn,

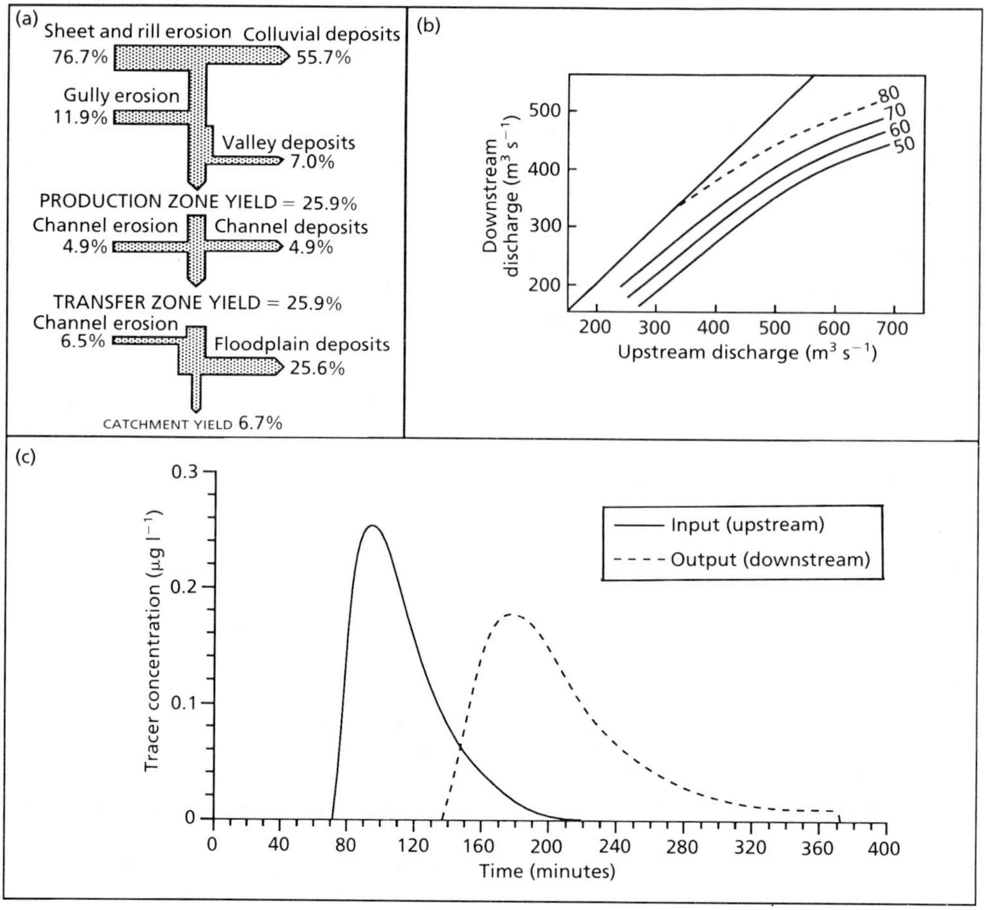

Fig. 1.5 Retention characteristics in river systems. (a) Sediment budget for Coon Creek, Wisconsin, for the period 1938–75 (after Trimble 1981). Note the expected balance between erosion and deposition within the transfer zone has been assumed here. (b) Floodpeak attenuation along a 35-km reach of the River Tees, UK, indicated by the relationship between upstream and downstream discharges with the ratio of 12-hour mean flow to the peak flow as a parameter (after Archer 1989). (c) Examples of Rhodamine WT tracer data for a 1-km pool-riffle reach of the River Severn, UK, illustrating the effect of dead zones on flow and transport processes (after Bevan & Carling 1992, by permission of John Wiley & Sons, Ltd).

associated with maximum amounts of detritus and surface area for microbial colonization. Residence times varied from 16 days in summer to >200 days in winter. Invertebrate collectors can markedly influence the retention process by consuming fine particulate matter. Indeed, Wallace *et al* (1982) suggested that a reduction of lotic insect fauna may decrease shredding and other breakdown activity, subsequent utilization, and downstream transport of organic matter. Moreover, the terrestrial–aquatic ecotones play a major role in the retention of nutrients during transport downstream (Pinay *et al* 1990). The amount of utilization depends on the frequency of spirals, which relates to the rate of cycling and degree of storage, or retentiveness, within the system in relation to the downstream flux rate. A primary need for river management is research on the lateral transfers of organic matter through the terrestrial–aquatic interface (Welcomme 1992).

Systems characterized by change

Traditional river management has been based on the assumption that rivers are in equilibrium with their external environment but the dynamic interactions of the large numbers of variables involved over a range of different time-scales means that such an assumption is probably invalid. Only a few spring-fed streams (Minshall *et al* 1985b) and perhaps some artificially regulated streams may fit a disturbance-free equilibrium model dominated by biotic interactions. For biota, rivers and streams are hazardous environments characterized by stresses and disturbances: erosion and abrasion; siltation and burial; and desiccation, as well as extremes of water quality (temperature, dissolved oxygen, pH, toxic metals). The roles of disturbance in stream ecology (Resh *et al* 1988) and in the structuring of upland (Wissmar & Swanson 1990) and lowland (Salo 1990) land-water ecotones have been elaborated.

A disturbance cannot be defined solely in terms of event magnitude and frequency. The severity of impact will also relate to the timing of the disturbance in relation to stability thresholds, physical and biological, and the effectiveness of recovery processes (see Chapters 5 & 6). Furthermore, two groups of disturbances should be distinguished: relatively frequent and low-magnitude disturbances that are part of the normal regime and more or less predictable; and relatively rare events that are outside the predictable range. In reality these groups are the two end members of a continuum because the frequency of low-magnitude disturbances influences the rate of system recovery to unpredictable disturbance by extreme events.

Over a period of 1–10 years rivers are characterized by changing flows, variable rates of particulate matter transport, variable water quality and changing channel forms. Biota respond to these changing forces, and this in turn generates dynamic biological interactions. Over short time-scales most river systems may be viewed as transient states. Within such systems, an 'equilibrium' should not be defined as a constant environment. River ecosystems characterized by strong recovery processes should be conceived as *quasi-equilibrium* systems within which the system components vary about some average condition over intermediate timescales (10^2 years) (Fig. 1.6).

The impact of an event in disturbing different components of a river ecosystem varies widely, depending on the system component under consideration (such as benthic algae, marginal macrophytes, or floodplain forest). The persistence of a change induced by a disturbance relates

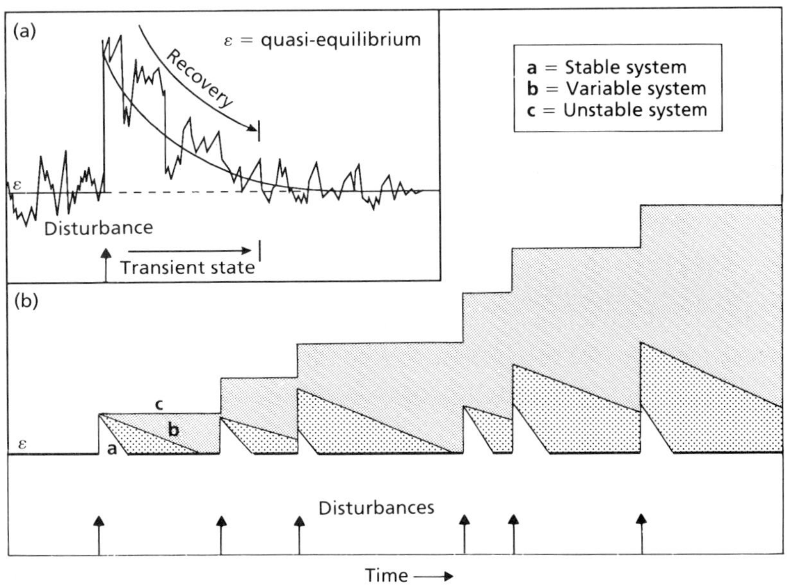

Fig. 1.6 The interactions of disturbances and recovery processes in rivers. (a) A system parameter in quasi-equilibrium is seen to vary about some long-term average condition; when disturbed the system recovers but the duration of transient states during the recovery period is dependent upon the effectiveness of the recovery processes. (b) A series of disturbances has different effects; contrasting systems can have (a) highly effective, (b) intermediate and (c) weak recovery processes.

to the effectiveness of the recovery processes (Fig. 1.6b). For example, the morphology of semi-arid streams may experience progressive change because of the weak physical recovery processes to flood-induced erosion (Wolman & Gerson 1978). In contrast, the insect species of these streams recover quickly following catastrophic wash-out by floods because they have short life cycles and continuous emergence, ensuring rapid recolonization by adults from adjacent riparian areas (Fisher *et al* 1982).

In rivers and streams, disturbance plays a critical role in organizing communities and ecosystems (Townsend 1989; Reice *et al* 1990) and this has important management implications, especially with regard to the long-term effects of human impacts (Petts 1980, 1987). For example, the diversity of vegetation patches on floodplains is related to the rejuvenation of successions associated with channel erosion and deposition so that the patch mosaic reflects age structure and patch type (Petts 1990a,b). Thus, along the River Rhône the floodplain patch mosaic reflects the varying morphological stability of the different channel sectors (Roux *et al* 1989; see also Plate 21.1). In the rapidly shifting braided sectors, frequent disturbance inhibits long plant successions and pioneer populations dominate. Less active sectors, in contrast, are characterized by hardwood forests. River regulation increased the recurrence interval between disturbing events and this led to marked changes of floodplain patch characteristics (Bravard *et al* 1986; Amoros *et al* 1987), pioneer communities and early successional stages being most severely affected. This example illustrates the sensitivity of fluvial ecosystems to changes in the disturbance regime, but it is likely that human impacts also change the type and effectiveness of the recovery processes so that system response may involve new successional directions. In some cases, transient states may persist for 100 years or more after an impact (Petts 1987).

1.3 A MANAGEMENT PERSPECTIVE

River management may be defined as the *maintenance* of some artificial condition (Mellquist 1992). At one end of the spectrum of actions, management involves the redistribution of channel bed sediment to improve fish passage; weed cutting for flood control; the 'protection' of a river bank to prevent erosion; embanking to prevent floodplain inundation; or the removal of riparian vegetation to improve access. At the other end, river management involves the diversion of water for power generation, domestic supply or irrigation, the complete alteration of a river's hydrological and water-quality regimes by large-scale impoundments, or the control of pollution. Ideal river management involves resource utilization without deterioration of the natural basis (Mellquist 1992), a concept promoted in the 'Brundtland Report' (Brundtland 1987). However, in many regions management choices are limited by the historic legacy, and river restoration and rehabilitation measures are required to improve already severely degraded environments.

The historical legacy

In the industrialized countries, major water-quality problems have arisen over a period of more than 100 years, generally in the following order: faecal pollution, organic pollution, salinization, metal pollution, eutrophication, contamination by organic micropollutants, and acidification (Meybeck & Helmer 1989). Severe pollution problems on European rivers began around 1850 and peaked in the 1960s and 70s. The biogeochemistry of a typical 'industrial' river, the Rhine (Kempe *et al* 1991), shows an increase in carbon dioxide pressure and biochemical oxygen demand (BOD) until 1972. Carbon dioxide pressure peaked at more than 6000 ppmv p_{CO_2} (pressures 20 times that of ambient air) indicating very high rates of respiration of organic matter from sewage and industrial wastes. Since then improved sewage treatment has decreased the input of labile organics to the river. The loads of some heavy metals such as cadmium have also declined but the historic legacy of high loadings is reflected in high concentrations of heavy metals remaining in some river-bed (Thoms 1987) and floodplain (Macklin & Klimek 1992) sediments. Nutrient inputs have continued to rise or are being maintained at high levels. The annual load of PO_4-P at Lobith on the Rhine rose from

8000 t year^{-1} in 1963 to 25 000 t year^{-1} in 1978 and this is paralleled by increasing concentrations of NO_3-N (Kempe *et al* 1991). Pollution problems have attracted considerable attention (see Chapter 2) but they were superimposed upon longer term problems arising from catchment land-use changes, river engineering works and water resource development schemes.

Throughout Europe, rivers have been influenced by a long history of catchment alteration and channel engineering (Petts *et al* 1989a). Palaeoenvironmental reconstructions have demonstrated human impacts, beginning with floodplain siltation induced by deforestation, extending over a period of about 4000 years (Starkel *et al* 1991). Progressive aggradation and increasing flood levels, resulting from expanding agriculture and catchment disturbance, during the 18th and 19th centuries were major factors influencing the intensification of river regulation works. Such works to improve navigation, to utilize water power, and to drain floodplain lands were common by the end of the 17th century (Petts *et al* 1989a) but the era of the mega-project opened in the second half of the 19th century (Cosgrove & Petts 1990). The struggle of humans to tame natural rivers, with large labour forces working long hours under harsh conditions, and entrepreneurs motivated by the desire for economic growth was the popular pioneering vision. During the mid-19th century, extensive areas of floodplain were drained following channel straightening and realignment. The regulation of the Tisza River provides a dramatic example: its course was shortened by 340 km and 12.5 × 10^6 ha of floodplain were drained.

In both Europe and North America the scale of river works increased dramatically from 1930 to the mid-1960s. The 1930s were notable for two technological achievements. The first was the construction of massive individual projects. Beginning with the 221-m-high Hoover Dam and its huge reservoir, Lake Mead, on the Colorado river, large dams and reservoirs became widespread. By the end of the 1930s the great multipurpose dam had come to symbolize social advancement and technological prowess. In 1989, worldwide there were 28 large dams over 200 m high. The second notable advance of the 1930s was the efficient application of engineering and technology to problems of resource development and hazard control, symbolized by the fully integrated development of entire river basins. This was first achieved in the Tennessee Valley, USA, and subsequently pursued in many basins throughout the world, notably the Volga, USSR, and in the Snowy Mountains scheme of Australia (White 1977).

Today, most rivers throughout the developed world are intensely regulated, many are hydraulically optimal but ecologically poor, often eutrophic and organically enriched, and their once ecologically diverse and productive corridors have been eliminated. The biological impacts of hydrological and geomorphological change are elaborated in Chapters 3 and 4 respectively. Since 1970, the focus of river development has moved away from Europe and North America, to the less developed and developing countries. The need for food and energy security in the face of rapid population growth is the dominant driving force. For example, in the Himalayas, the exploitable hydro-power potential at this time is estimated at 42 000 MW from 66 sites; the current level of exploitation is only 161 MW (Sharma 1989). In dry regions, intermittent drought and population growth imply an increasing need for water security in agriculture as well as increasing water needs for domestic and industrial supply (see Chapter 25). Falkenmark (1989) has suggested that the problem of water supply in Africa might be solved by inter-basin transfers from the Congo. Long-distance water transfers have also been proposed to meet the expanding dryland developments in USA, including Los Angeles, Tucson and Salt Lake City (Beaumont 1989).

Prospect

The case for environmental conservation is embodied within the aims of the World Conservation Strategy (IUCN 1980): to maintain essential ecological processes and life-support systems; to preserve genetic diversity; and to ensure *sustainable* utilization of species and ecosystems. Given the historic legacy that affects many rivers, Boon (1992) argues that the case for conservation involves five general management options along a spectrum of decreasing conservation value: preservation, limitation, mitigation, restoration and dereliction.

All levels of conservation require management; controls — and the administration of these controls — on the wide range of human activities that influence river ecosystems. These include:

- point-source pollution controls (especially implementation of effluent discharge standards);
- land-use regulation (include the definition of 'protection zones' for groundwater and the use of buffer zones to minimize diffuse-source pollution);
- water allocation to meet seasonally-variable instream and riparian flow needs;
- channel and floodplain management to sustain morphological diversity;
- controls on human use of a river for fishing, recreation, etc.;
- controls on biota to prevent overpopulation of certain species, including deliberate or accidental introductions and biological invasions.

These management options may be organized in a decision-making framework that focuses on achieving maximum resource development while providing for ecological conservation (Petts 1989, 1990a). At the first level, options for modifying river projects (ideally but not exclusively at the design stage) should be considered with the specific aim of maintaining the natural structural and biological dynamics of the affected fluvial hydrosystem. Flow manipulation and channel management are key concerns. The economic and environmental effectiveness of first-order management proposals should then be evaluated in relation to second-order options (direct controls on biota and on human uses).

As noted by Mellquist (1992), a decision to do nothing to a river is an important option in management. In some cases this may require non-use of a natural resource, such as a wild river, implicitly accepting that non-use is an acceptable form of use. The protection of a river from hydropower development, for example, may be seen as a trade-off, the preservation of one being acceptable because of the intensive development of another (e.g. Ledec & Goodland 1988). However, the preservation versus dereliction debate should be rehearsed only when options for limitation, mitigation and restoration have been fully evaluated.

In reality the attainment of the conservation aims is constrained by the goals of socio-economic development. It is necessary therefore to make a pragmatic choice of the best practicable environmental option (Royal Commission 1976; Petts 1990a). Importantly, such an approach recognizes the time-limited nature of decisions, flexibility in management being important so that the 'best' option can be revised once new knowledge is available. The following chapters describe scientifically-based approaches for assessing environmental quality, diagnosing problems, and monitoring environmental change, degradation and recovery. A set of secondary regulation measures (Kitson 1984) are discussed, including tools to *limit* impacts of river and catchment development; to *mitigate* the effects of river regulation, abstraction or waste disposal where such actions are accepted as necessary for socio-economic development; and to *restore* or *rehabilitate* degraded rivers by stimulating and enhancing natural recovery processes.

REFERENCES

Amoros C, Roux AL, Reygrobellet JL, Bravard JP, Pautou G. (1987) A method for applied ecological studies of fluvial hydrosystems. *Regulated Rivers* **1**: 17–36. [1.2]

Anderson NH, Cargill AS. (1987) Nutritional ecology of aquatic detritivorous insects. In: Slansky F, Rodriguez JR (eds) *Nutritional Ecology of Insects, Mites, Spiders and Related Invertebrates*, pp. 903–25. John Wiley & Sons, New York. [1.2]

Archer DR. (1989) Flood wave attenuation due to channel and floodplain storage and effects on flood frequency. In: Bevan K, Carling P (eds) *Floods: Hydrological, Sedimentological and Geomorphological Implications*, pp. 37–46. John Wiley & Sons, Chichester. [1.2]

Arnell NW. (1989) Humid temperate sloping land. In: Falkenmark M, Chapman T (eds) *Comparative Hydrology*. UNESCO, Paris. [1.2]

Bailey RG. (1978) *Description of the Ecoregions of the United States*. Intermontane Region, United States Forest Service, Ogden, Utah. [1.2]

Bayley PB. (1991) The flood-pulse advantage and the restoration of river-floodplain systems. *Regulated Rivers* **6**: 75–86. [1.2]

Beaumont P. (1989) *Environmental Management and Development in Drylands*. 505 pp. Routledge, London. [1.1, 1.3]

Beckinsale RP. (1969) River regimes. In: Chorley RJ (ed) *Water, Earth and Man*, pp. 455–72. Methuen, London. [1.2]

Bevan K, Carling P. (1992) Velocities, roughness and dispersion in the lowland River Severn. In: Carling PA, Petts GE (eds) *Lowland Floodplain Rivers*, pp. 71–94. John Wiley & Sons, Chichester. [1.2]

Boon PJ. (1992) Essential elements in the case for river conservation. In: Boon PJ, Calow P, Petts GE (eds) *River Conservation and Management*, pp. 11–34. John Wiley and Sons, Chichester. [1.3]

Bravard JP, Amoros C, Pautou G. (1986) Impact of civil engineering works on the successions of communities in a fluvial system. A methodological approach applied to a section of the Upper Rhône River, France. *Oikos* **47**: 92–111. [1.2]

Brookes A. (1988) *Channelized Rivers*. John Wiley & Sons, Chichester. [1.1]

Brundtland GH. (1987) *Our Common Future*. Report of the World Commission on Environment and Development. Oxford University Press, Oxford. [1.3]

Carling PA. (1991) An appraisal of the velocity-reversal hypothesis for stable pool-riffle sequences in the River Severn, England. *Earth Surface Processes and Landforms* **16**: 19–32. [1.2]

Chorley RJ, Schumm SA, Sugden DE. (1984) *Geomorphology*. Methuen, London. [1.2]

Clifford NJ, Richards KS. (1992) The reversal hypothesis and the maintenance of riffle-pool sequences: a review and field appraisal. In: Carling PA, Petts GE (eds) *Lowland Floodplain Rivers*, pp. 43–70. John Wiley & Sons, Chichester. [1.2]

Cosgrove DE, Petts GE. (eds) (1990) *Water, Engineering and Landscape*. Belhaven, London. [1.3]

Craig JF, Kemper JB. (eds) (1987) *Regulated Streams: Advances in Ecology*. Plenum Press, New York. [1.1]

Cummins KW. (1992) Catchment characteristics and river ecosystems. In: Boon PJ, Calow P, Petts GE (eds) *River Conservation and Management*, pp. 125–35. John Wiley & Sons, Chichester. [1.2]

Danielopol DL, Marmonier P. (1992) Aspects of research on groundwater along the Rhône, Rhine and Danube. *Regulated Rivers* **7**: 5–16. [1.2]

Davies BR, Walker KF. (eds) (1986) *The Ecology of River Systems*. Dr W. Junk, Dordrecht. [1.1]

Degenes ET, Kempe S, Richey JE. (eds) (1991) *Biogeochemistry of Major World Rivers*. SCOPE 42. John Wiley & Sons, Chichester. [1.2]

Elwood JW, Newbold JD, O'Neill RV, Van Winkle W. (1983) Resource spiralling: an operational paradigm for analysing lotic ecosystems. In: Fontaine TD, Bartell SM (eds) *Dynamics of Lotic Ecosystems*, pp. 3–27. Ann Arbor Science, Ann Arbor, Michigan. [1.2]

Falkenmark M. (1989) The massive water scarcity now threatening Africa – why isn't it being addressed? *Ambio* **18**: 112–18. [1.3]

Falkenmark M, Chapman T. (1989) *Comparative Hydrology*. UNESCO, Paris. [1.2]

Fisher SG, Gray LJ, Grimm NB, Busch DE. (1982) Temporal succession in a desert stream ecosystem following flash flooding. *Ecological Monographs* **52**: 93–110. [1.2]

Foster IDL, Grieve IC. (1982) Short term flucuations in dissolved organic matter concentrations in streamflow draining a forested watershed and their relation to catchment budget. *Earth Surface Processes and Landforms* **7**: 417–25. [1.2]

Frissel CA, Wiss WJ, Warren CE, Huxley MD. (1986) A hierarchical framework for stream classification: viewing streams in a watershed context. *Environmental Management* **10**: 199–214. [1.2]

Fuchs U, Statzner B. (1990) Timescales for the recovery potential of river communities after restoration: lessons to be learned from smaller streams. *Regulated Rivers* **5**: 77–88. [1.2]

Gibbs RJ. (1970) Mechanisms controlling world water chemistry *Science* **170**: 1088–90. [1.2]

Gibert J, Dole-Olivier M-J, Marmonier P, Vervier P. (1990) Surface water–groundwater ecotones. In: Naiman RJ, Decamps H (eds) *The Ecology and Management of Aquatic-Terrestrial Ecotones*, pp. 199–226. UNESCO, Paris; Parthenon, Carnforth, UK. [1.2]

Gore JA, Petts GE. (eds) (1989) *Alternatives in Regulated River Management*. CRC Press, Boca Raton, Florida. [1.1]

Gregory KJ. (1992) Vegetation and river channel process interactions. In: Boon PJ, Calow P, Petts GE (eds) *River Conservation and Management*, pp. 255–70. John Wiley & Sons, Chichester. [1.2]

Happ SC. (1971) Genetic classification of valley sediment deposits. *American Society of Civil Engineers Journal of the Hydraulics Division* **97**: 43–53. [1.2]

Horton RE. (1945) Erosional development of streams and their drainage basins: hydrophysical approach to quantitative geomorphology. *Geological Society of America Bulletin* **56**: 275–370. [1.2]

Hynes HBN. (1970) *The Ecology of Running Waters*. University of Toronto Press, Toronto. [1.1, 1.2]

Hynes HBN. (1975) The stream and its valley. *Verhandlungen der Internationalen Vereingung fur theoretische und angewandte Limnologie* **19**: 1–15. [1.2]

Hynes HBN. (1986) Foreword. In: Davies BR, Walker KF (eds) *The Ecology of River Systems*, pp. VII–X. Dr W. Junk, Dordrecht. [1.1]

Illies J, Botosaneanu L. (1963) Problèmes et mèthodes de la classification et de la zonation ècologique des eaux courantes considerèes surtout du point de rue faunistique. *Mitteilungen der Internationalen*

Vereinigung für theoretische und angewandte Limnologie **12**: 1–57. [1.2]

IUCN. (1980) *World Conservation Strategy*. International Union for Conservation of Nature and Natural Resources, Gland, Switzerland. [1.3]

Junk WJ, Bayley PB, Sparks RE. (1989) The flood-pulse concept in river-floodplain systems. In: Dodge DP (ed) *Proceedings of the International Large River Symposium (LARS)*, pp. 110–27. Canadian Special Publication of Fisheries and Aquatic Sciences 106. [1.2]

Kempe S, Pettine M, Cauwet G. (1991) Biogeochemistry of European Rivers. In: Degenes ET, Kempe S, Richey JE (eds) *Biogeochemistry of Major World Rivers*. SCOPE 42, pp. 169–212. John Wiley & Sons, Chichester. [1.3]

Kitson T. (1984) *Regulated River Basins: a Review of Hydrological Aspects for Operational Management*. UNESCO, Paris. [1.3]

Lane EW. (1955) The importance of fluvial morphology in hydraulic engineering. *American Society of Civil Engineers Proceedings* **81**: 1–17. [1.2]

Langbein WB, Leopold LB. (1964) River meanders – theory of minimum variance. *Professional Paper, United States Geological Survey* 282-B, USGS, Washington, DC. [1.2]

Lauga J, Decamps H, Holland MM. (eds) (1986) *Land use Impacts on Aquatic Ecosystems*. Proceedings of the MAB-UNESCO and PIREN-CNRS Workshop. CRDP, Toulouse. [1.1]

Lawton JH. (1989) Food webs. In: Cherrett JM (ed) *Ecological Concepts*, pp. 43–78. Blackwell Scientific Publications, Oxford. [1.2]

Ledec G, Goodland AC. (1988) *Wildlands: their Protection and Management in Economic Development*. World Bank, Washington, DC. [1.3]

Leopold LB, Wolman MG, Miller JP. (1964) *Fluvial Processes in Geomorphology*. W.H. Freeman, San Francisco. [1.1]

Lillehammer A, Saltveit SJ. (eds) (1984) *Regulated Rivers*. Universiteitsforlaget, Oslo. [1.1]

Lotspeich FB. (1980) Watersheds as the basic ecosystem: this conceptual framework provides a basis for a natural classification system. *Natural Resources Bulletin* **16**: 581–6. [1.2]

Macklin MG, Klimek K. (1992) Dispersal, storage and transformation of metal-contaminated alluvium in the upper Vistula basin, southwest Poland. *Applied Geography* **12**: 7–30. [1.3]

Marmonier P, Creuze des Chatelliers M. (1991) Effects of spates on interstitial assemblages of the upper Rhône river. Importance of spatial heterogeneity. *Hydrobiologia* **210**: 243–51. [1.2]

Marsh GP. (1864) *Man and Nature*. Revised edn (1965); Lowenthal D (ed). Balknap Press of Harvard University Press, Cambridge, Mass., USA. [1.1]

Mellquist P. (1992) River management – objectives and applications. In: Boon PJ, Calow P, Petts GE (eds) *River Conservation and Management*, pp. 1–10. John Wiley & Sons, Chichester. [1.3]

Meybeck M, Helmer R. (1989) The quality of rivers: from pristine stage to global pollution. *Palaeogeography, Palaeolimnology, Palaeoecology (Global and Planetary Change Section)* **75**: 283–309. [1.3]

Milner AM. (1987) Colonization and ecological development of new streams in Glacier Bay National Park, Alaska. *Freshwater Biology* **18**: 53–70. [1.2]

Minshall GW, Petersen RC, Cummins KW, Bott TL, Sedell JR, Cushing CE, Vannote RL. (1983) Interbiome comparison of stream ecosystem dynamics. *Ecological Monographs* **53**: 1–25. [1.2]

Minshall GW, Cummins KW, Peterson RC, Cushing CE, Burns DA, Sedell JR, Vannote RL. (1985a) Developments in stream ecosystem theory. *Canadian Journal of Fisheries and Aquatic Sciences* **42**: 1045–55. [1.2]

Minshall GW, Petersen RC, Nimz CF. (1985b) Species richness in streams of different size from the same drainage basin. *American Naturalist* **125**: 16–38. [1.2]

Naiman RJ, Decamps H. (eds) (1990) *The Ecology and Management of Aquatic-Terrestrial Ecotones*. UNESCO, Paris; Parthenon, Carnforth, UK. [1.1, 1.2]

Naiman RJ, Lonzarich DG, Beechie TJ, Ralph SC. (1992) General principles of classification and the assessment of conservation potential in rivers. In: Boon PJ, Calow P, Petts GE (eds) *River Conservation and Management*, pp. 93–124. John Wiley & Sons, Chichester. [1.2]

NERC. (1975) *The Flood Studies Report* (5 vols). Natural Environment Research Council, Swindon, UK. [1.2]

Oglesby RT, Carlson CA, McCann JA. (eds) (1972) *River Ecology and Man*. Academic Press, New York. [1.1]

Petts GE. (1980) Long-term consequences of upstream impoundment. *Environmental Conservation* **7**: 325–32. [1.2]

Petts GE. (1984) *Impounded Rivers*. John Wiley & Sons, Chichester. [1.1]

Petts GE. (1987) Time-scales for ecological change in regulated rivers. In: Craig JF, Kemper JB (eds) *Regulated Streams — Advances in Ecology*, pp. 257–66. Plenum, New York. [1.2]

Petts GE. (1989) Perspectives for ecological management of regulated rivers. In: Gore JA, Petts GE (eds) *Alternatives in Regulated River Management*, pp. 3–24. CRC Press, Boca Raton, Florida. [1.3]

Petts GE. (1990a) The role of ecotones in aquatic landscape management. In: Naiman RJ, Decamps H (eds) *The Ecology and Management of Aquatic-Terrestrial Ecotones*, pp. 227–60. UNESCO, Paris; Parthenon, Carnforth, UK. [1.2, 1.3]

Petts GE. (1990b) Forested river corridors: a lost resource. In: Cosgrove DE, Petts GE (eds) *Water,*

Engineering and Landscape, pp. 12–34. Belhaven, London. [1.2]

Petts GE. (1990c) Regulation of large rivers: problems and possibilities for environmentally-sound river development in South America. *Interciencia* **15**: 388–410. [1.1]

Petts GE, Moller H, Roux AL. (eds) (1989a) *Historical Change of Large Alluvial Rivers: Western Europe*. John Wiley & Sons, Chichester. [1.1, 1.3]

Petts GE, Thoms MT, Brittan K, Atkin B. (1989b) A freeze-coring technique applied to pollution by fine sediments in gravel-bed rivers. *The Science of the Total Environment* **84**: 259–72. [1.2]

Pinay G, Decamps H, Chauvet E, Fustec E. (1990) Functions of ecotones in fluvial systems. In: Naiman RJ, Decamps H (eds) *The Ecology and Management of Aquatic-Terrestrial Ecotones*, pp. 141–70. UNESCO, Paris; Parthenon, Carnforth, UK. [1.2]

Poff NL, Ward JV. (1989) Implications of streamflow variability and predictability for lotic community structure: a regional analysis for streamflow patterns. *Canadian Journal of Fisheries and Aquatic Sciences* **48**: 1805–18. [1.2]

Reice SR, Wissmar RC, Naiman RJ. (1990) Disturbance regimes, resilience, and recovery of animal communities and habitats in lotic ecosystems. *Environmental Management* **14**: 647–60. [1.2]

Resh VH, Brown AV, Covich AP, Gurtz ME, Li HW, Minshall GW, Reice SR, Sheldon AL, Wallace JB, Wissmar RC. (1988) The role of disturbance in stream ecology. *Journal of the North American Benthological Society* **7**: 433–55. [1.1, 1.2]

Roberts CN (ed.) (1993) *The Changing Global Environment*. Blackwell Publishers, Oxford. [1.1]

Rohm CM, Giese JW, Bennett CG. (1987) Evaluation of an aquatic ecoregion classification of streams in Arkansas. *Journal of Freshwater Ecology* **4**: 127–39. [1.2]

Roux AL, Bravard J-P, Amoros C, Pautou G. (1989) Ecological changes of the French Upper Rhône River since 1750. In: Petts GE, Moller H, Roux AL (eds) *Historical Change of Large Alluvial Rivers: Western Europe*, pp. 323–50. John Wiley & Sons, Chichester. [1.2]

Royal Commission. (1976) *Air Pollution Control: an Integrated Approach*. 5th Report of the Royal Commission on Environmental Pollution. Her Majesty's Stationery Office, London. [1.3]

Salo J. (1990) External processes influencing origin and maintenance of inland water-land ecotones. In: Naiman RJ, Decamps H (eds) *The Ecology and Management of Aquatic-Terrestrial Ecotones*, pp. 37–64. UNESCO, Paris; Parthenon, Carnforth, UK. [1.2]

Salo J, Kalliola R, Hakkinen I, Makinen Y, Niemela P, Puhakka M, Coley PD. (1986) River dynamics and the diversity of Amazon lowland forest. *Nature* **322**: 254–8. [1.2]

Schumm SA. (1977) *The Fluvial System*. John Wiley, New York. [1.2]

Schumm SA, Lichty RW. (1965) Time, space and causality in geomorphology. *American Journal of Science* **263**: 110–19. [1.2]

Sharma CK. (1989) Nepal's hydro schemes: progress and plans *Water, Power and Dam Construction* March, pp. 22–3. [1.3]

Shreve RL. (1967) Infinite topologically random channel networks. *Journal of Geology* **75**: 178–86. [1.2]

Sidle RC, Milner AM. (1989) Stream development in Glacier Bay National Park, Alaska, U.S.A. *Arctic and Alpine Research* **21**: 350–63. [1.2]

Stalnaker CB, Milhous RT, Bovee K. (1989) Hydrology and hydraulics applied to fishery management in large rivers. In: Dodge DP (ed) *Proceedings of the International Large River Symposium (LARS)*. Canadian Special Publication of Fisheries and Aquatic Sciences No. 106, pp. 13–30. [1.1]

Stanford JA, Ward JV. (1988) The hyporheic habitat of river ecosystems. *Nature* **355**: 64–6. [1.2]

Starkel L. (1983) The reflection of hydrologic changes in the fluvial environment of the temperate zone during the last 15 000 years. In: Gregory KJ (ed) *Background to Palaeohydrology*, pp. 213–35. John Wiley & Sons, Chichester.

Starkel L, Gregory KJ, Thornes JB. (eds) (1991) *Temperate Palaeohydrology*. John Wiley & Sons, Chichester. [1.2, 1.3]

Statzner B, Higler B. (1986) Stream hydraulics as a major determinant of benthic invertebrate zonation patterns. *Freshwater Biology* **16**: 127–39. [1.2]

Statzner B, Gore JA, Resh VH. (1988) Hydraulic stream ecology: observed patterns and potential applications. *Journal of the North American Benthological Society* **7**: 307–60. [1.1, 1.2]

Stock MS, Ward AK. (1989) Establishment of a bedrock epilithic community in a small stream: microbial (algal and bacterial) metabolism and physical structure. *Canadian Journal of Fisheries and Aquatic Sciences* **46**: 1874–83. [1.2]

Strahler AN. (1952) Quantitative analysis of watershed geomorphology, *American Geophysical Union Transactions* **38**: 913–20. [1.2]

Thoms MT. (1987) Channel sedimentation within the urbanized River Tame, UK. *Regulated Rivers* **1**: 229–46. [1.3]

Townsend CR. (1989) The patch dynamics concept of stream community ecology. *Journal of the North American Benthological Society* **8**: 36050. [1.2]

Trimble SW. (1981) Changes in sediment storage in the Coon Creek Basin, driftless area, Wisconsin, 1853–1975. *Science* **214**: 181–3. [1.2]

Vannote RL, Minshall GW, Cummins KW, Sedell JR,

Cushing CE. (1980) The river continuum concept. *Canadian Journal of Fisheries and Aquatic Sciences* **37**: 130–7. [1.2]

Wallace JB, Webster JR, Cuffney TF. (1982) Stream detritus dynamics: regulation by invertebrate consumers *Oecologia* **53**: 197–200. [1.2]

Walling DE. (1983) The sediment delivery problem. *Journal of Hydrology* **65**: 209–37. [1.2]

Ward JV. (1989) The four-dimensional nature of lotic ecosystems. *Journal of the North American Benthological Society* **8**: 2–8. [1.2]

Ward JV, Stanford JA. (eds) (1979) *The Ecology of Regulated Streams*. Plenum, New York. [1.1]

Welcomme RL. (1977) *Fisheries Ecology of Floodplain Rivers*. Longman, London. [1.2]

Welcomme RL. (1992) River conservation – future prospects. In: Boon PJ, Calow P, Petts GE (eds) *River Conservation and Management*, pp. 453–62. John Wiley & Sons, Chichester. [1.2]

White GF. (1977) *Environmental Effects of Complex River Development*. Westview, Boulder, Colorado. [1.3]

Wissmar RC, Swanson FJ. (1990) Landscape disturbances and lotic ecotones. In: Naiman RJ, Decamps H (eds) *The Ecology and Management of Aquatic-Terrestrial Ecotones*, pp. 65–89. UNESCO, Paris; Parthenon, Carnforth, UK. [1.2]

Wolman MG, Gerson R. (1978) Relative scales of time and effectiveness of climate in watershed geomorphology. *Earth Surface Proceses* **3**: 189–208. [1.2]

2: River Pollution

R. A. SWEETING

2.1 THE NATURE OF POLLUTION

Pollution is a term applied to any environmental state or manifestation which is harmful or unpleasant to life, resulting from failure to achieve or maintain control over the chemical, physical or biological consequences or side-effects of human scientific, industrial and social habits (Collocott & Dobson 1974). Since the advent of our industrial society we have regarded it as our prerogative to pollute the environment, whether aquatic, aerial or terrestrial. Historically, pollution of the aquatic environment was recognized in Biblical times. In Great Britain it was regarded as an environmental problem as long ago as the 13th century, when laws were passed prohibiting the charcoal burners of Staines in Middlesex from washing their charcoal in the River Thames for fear of befouling the river downstream. Many of the major rivers of Europe, such as the Elbe, Rhine and Danube, and of the USA, including the Ohio, Connecticut, and the Hudson, experienced marked pollution problems in the late 19th and early 20th centuries.

These problems were originally those of gross contamination and subsequent deoxygenation by untreated sewage. High organic suspended solids and high ammonia levels, with high biochemical oxygen demand (BOD) were characteristic of such problems. Reduction in fish species, invertebrate species and macrophyte species was followed by a total disappearance of fish, survival of two or three pollution-tolerant invertebrate groups and massive encroachment by one or two species of emergent macrophytes, or dense algal blooms (Klein 1962).

The problems were gross, the solutions were essentially simple. In the UK they were solved, in theory if not in practice, by the resolution of a series of Royal Commissions on Pollution. Royal Commissions in the UK had their equivalents in other developed countries with the result that most developed countries today have in place effluent controls for sewage-based discharges.

River uses

One of the major corollaries of effluent control policies is that acceptance by society that rivers are suitable environments for getting rid of wastes. The fact that they flow downstream means that, unlike still freshwaters and, to a lesser extent, coastal waters, any pollutant entering them is washed away from its point of entry. When humans were ecological rather than global creatures and lived in relatively small numbers in small communities, rivers were able to cope with the biological loads that were discharged into them. Increasingly large communities, increasing densities of populations and increasing use of resources by developed countries have meant that not only are effluent controls necessary, but effluent treatment to limit the effects of discharges is also accepted as a necessity to maintain a reasonable water quality. A reasonable water quality is important in order that we can benefit from the rivers by a series of 'uses'. In the UK a set of uses have been proposed (Department of the Environment 1992) as follows.

1 Fisheries Ecosystem: six classes suggested from salmonid to no fish.

2 Abstraction for Drinking Water Supply: related to EC directives.

3 Agricultural Abstraction: sets of standards for livestock watering and for irrigation.

4 Industrial Abstraction: suitable standards are being considered.

5 Special Ecosystem: for sites where unusual fauna and/or flora are present.

6 Watersports: chemical and bacteriological standards related to human health risk.

Each of these 'uses' influences the quality of the river water. A complex series of rules, laws and directives in the UK and Europe attempts to control the effects of these uses (Haigh 1987). At present, rivers (and other waters) are an indispensable receptacle for wastes, but this 'use' differs from those listed above in that quality criteria are applied to the effluents, so that the final quality of the receiving river enables it to be suitable for specific purposes. In reality the balance between effluent quality, quantity and river use has been achieved by a combination of historic precedents and economic constraints. The 'use'-related quality objective system is an attempt to plan this process.

Chemical measurements

Pollutants can be measured in two ways: firstly by chemical analysis, and secondly by their effect on the biota present in and around the river. This dichotomy of approach has been one of the greatest strengths and simultaneously one of the greatest weaknesses in the assessment of pollution. Measurement by chemical analysis is the most obvious method of assessing pollution and has the great advantage that it can be applied directly to effluents. It can be applied to simple chemicals such as metals, salts and dissolved gases as well as complex organic compounds such as chlorinated hydrocarbons (Meybeck *et al* 1989). Measurements of some other parameters can also be carried out on effluents relatively easily, e.g. temperature, turbidity, colour and pH. Controls of discharges are relatively simple to operate when applied to these parameters.

However, knowing the nature of a pollutant in an effluent is only the first of a series of requirements. Measurement of the amount of effluent entering a river provides a total loading of that material. Most effluents vary in quantity as well as quality so total load may be very different at different times of the day, of the week or seasons of the year. The river flow will also vary throughout the year: in a chalk stream it may vary by up to 10 times between the smallest (September) and largest flows (March); clay catchments and upland catchments may have variations of flow of the order of 2000-fold over a year. Calculation of dilution factors is rendered more complex by such variation and often the worst situation is used to define consent limits, i.e. maximum effluent concentration and maximum permitted effluent quantity under dry weather flow conditions of the river.

Variations in quantities of chemicals in effluents may create difficulties in assessing pollutant loading of rivers. This is particularly important when dealing with substances which bioaccumulate, as sampling programmes have to be designed to determine total input over defined time periods. Many discharges to rivers are intermittent (e.g. storm overflows) and produce peaks of chemical concentration in the river downstream. As the chemical moves downstream the short burst becomes attenuated, so that a lower, but longer-lasting presence is detected. The concentrations of pollutants in the river will be lower the further downstream it is sampled away from the discharge. Consequently the sensitivity of the analytical technique needs to be much greater with river water than with effluents. The presence of a particular pollutant in a river indicates an input upstream, but does not pinpoint its source.

The consent limit therefore provides the ceiling level of pollutant (concentration multiplied by quantity of effluent) allowed into a river by the regulatory authority. It does not necessarily equate with a 'no observable effect concentration' (NOEC) in the receiving river.

Ecotoxicological effects

Historically, traditional chemical analyses have enabled the recognition and measurement of relatively large amounts of relatively few substances. Over the last 50 years analytical techniques have improved in sensitivity and now can be used to detect thousands of polluting substances. Modern technology uses tens of thousands of different compounds, most of which find their way to the local river via sewage works, industrial discharge or diffuse input (e.g. agriculture). It is not possible to have analytical suites to monitor for the presence of all of these materials, so their effect on biological systems is often used. This is particularly helpful in assessing the toxicity of complex effluents where many similar organic compounds may be released into the discharge. Simple toxicity tests involve the addition of an effluent to certain volumes of water to produce a series of dilutions to which fish, invertebrates or bacteria are introduced. Until recently the end-point of toxicity tests was death of the test organisms (Cairns & Dickson 1973). Over the last 10 years progress has been made towards measurement of sublethal end-points. These can be used for effluents, diluted effluents or river-waters and whereas previously they were all carried out in the laboratory, increasingly the tests are carried out *in situ* (i.e. in the receiving river). The use of rainbow trout (*Oncorhynchus mykiss*) in water intakes is a good illustration of this at the vertebrate level. Monitoring is carried out on respiratory rhythm and heart beat. A pronounced change in either, or both, of these activity patterns elicits an alarm system and further investigations, both biological and chemical (Cairns *et al* 1976).

Invertebrate toxicity tests using the crustaceans *Daphnia magna* or *Gammarus pulex* are frequently used, the former for lethality or oxygen consumption, the latter in a 'scope for growth' situation. 'Scope for growth' involves the calculation of an organism's available resources for growth under specified circumstances. It can be represented by the equation:

$$P = C - F - U - R$$

where P = scope for growth, C = energy consumed, F = energy lost as faeces, U = energy excreted and R = energy metabolized.

The amount of energy available for growth (and reproduction) after all metabolic requirements have been met, will decrease as environmental stressors (e.g. pollution concentrations) increase (Crane & Maltby 1991). This effect on an individual organism, whether eukaryote or prokaryote, mirrors the effects on individuals in the field. Different individuals will react slightly differently to a particular level of pollutants, so that in any population there will be variation (usually normally distributed) in effect. Different species will respond both at a different level of pollutant and in a different way. This is described below.

2.2 ECOLOGICAL IMPACTS OF POLLUTANTS

In a population subjected to a pollutant challenge, there will be intrinsic factors that determine the severity of damage to individuals within it. Age, sex, general condition and sexual condition will influence the capacity of the individual to avoid or cope with the pollutant. In the case of mobile organisms it may involve swimming down a concentration gradient, slowing feeding or moving into the flow and increasing its rate of drift. In non-mobile organisms the production of detoxifying enzymes may be increased (Beverley *et al* 1991) or, as in bacteria, alternative metabolic pathways may be used (Stewart 1990).

A riverine community, because of the different niches occupied by component organisms, will be disproportionately damaged in some areas by extrinsic factors as well. For example, with benthic macroinvertebrate communities, competition between individuals in populations may force the smaller or less fit members into the substrate. In the event of pollution of the river, these individuals may survive to recolonize the more favourable parts of the substrate, replacing those organisms lost. Enhanced detachment and increased drift can occur as a result of pesticide application to streams, resulting in changes in insect distribution (Muirhead-Thomson 1987).

The immediate environment of organisms in the water flow also influences their survival and here the influence of dead zones becomes apparent (see Vol. 1, Plate 5.2 & pp. 112, 210). These zones of relatively static water within rivers may have retention times many times greater than that of the main river. Their stability depends on the nature of the river bed and banks, and the speed of flow. In low-flow summer conditions retention times of water in these dead zones can be up to several weeks. Absorption of pollutants is slower, but so is release downstream. Fish and macroinvertebrates may use them as refuges from chemical assaults under certain conditions of flow. Impaired flows at the edges of rivers will act in similar ways and enable some organisms to escape from high concentrations of toxic materials flowing downstream.

There is a large element of chance in pollution incidents in that at any one time a different group of organisms may be affected. For example, low dissolved oxygen is the most common cause of fish mortality in rivers. It can result from a large number of organic materials entering rivers (e.g. pulp, sewage, silage, milk, urea, ammonia) and depressing the available oxygen. If it coincides with migratory fish presence in that section of the river, then considerable inroads can be made into those populations. The same occurrence days or weeks later may have little effect on those populations. At other times, coarse fish fry or eggs may be severely affected. Such intermittent pollution episodes result in biological communities of an apparently unbalanced nature being present for significant periods of time. The ecological picture presented as a result of intermittent discharges is consequently different from that presented by continuous discharges.

Physical effects

The chemical composition of polluting materials has been the subject of many investigations over the last 150 years. Less attention has been paid to physical pollution studies. Studies in this area have concentrated on temperature, pressure or the nature of suspended or settleable solids.

In the UK, the number of thermal discharges to rivers is over 2000. Thermal discharges (e.g. power station effluents, cooling waters from chemicals manufacture) have direct and indirect effects. The direct temperature effect of increased productivity as a result of raised metabolic rates is illustrated in the middle sections of the River Trent in Nottinghamshire with its numerous coal-fired power-station effluents. Less well documented are temperature influences in the lower River Thames as a result of major sewage works inputs. For organisms well within their thermal range, minor temperature changes are relatively unimportant, but for organisms at the edge of their range this additional environmental stress may decide between species survival or extinction. The return of salmon to the River Thames may be jeopardized by relatively small increases in temperature. In historic times the burbot (*Lota lota*) disappeared from the UK. The gradual warming of Northern Europe over the last 500 years meant that the low temperatures required for incubation of burbot eggs was no longer guaranteed at the edges of its range (which included the UK). Conversely thermal inputs to the St Helen's Canal in Merseyside ensured survival of a population of introduced guppies for several decades. Rapid changes in temperature are not tolerated by poikilotherms as their isoenzyme complement at any one temperature will need to be replaced as variation of more than a few degrees centigrade occurs, a process which may take several days or even weeks.

Indirect temperature effects are well documented in relation to dissolved oxygen content of water and ammonia toxicity. Pressure effects associated with dams, pumped water and changes in air pressure have been the subject of considerable work in the USA. Rapid changes in concentrations of nitrogen (producing symptoms similar to divers' 'bends') with major haemorrhaging of the systemic circulation are indicated by bleeding from the gills and fins. Seismic surveying in lowland Britain in the 1980s produced several instances of swimbladder implosion in communities of freshwater fish: such instances indicate the need to consider pollution in a much broader sense than convention indicates.

Suspended solids are referred to frequently in pollution investigations and yet the majority of problems arise from settleable solids. They are all included in the effects of 'suspended solids'. A major component of sewage is suspended solids with an organic fraction onto which bacteria are attached. The turbidity introduced by these solids prohibits photosynthesis, as does their deposition on the leaves of aquatic macrophytes and on benthic algae. For filter-feeders in the water column and on the substrate the solids may, depending on their size and composition, act as food or interfere with the filtration process. Some solids clog gills of fish and invertebrates leading to asphyxiation. Deposition of silt on gravel can smother diatoms, benthic algae, macroinvertebrates and fish eggs. If the deposits are large then the macrophyte assemblages may be radically altered. In the USA fishery scientists classify their waters as turbid ($>100\,mg\,l^{-1}$ solids), intermediate ($25-100\,mg\,l^{-1}$ solids) or clear ($<25\,mg\,l^{-1}$ solids) according to the mean value of suspended solids. Although a crude classification, it enables distinctions to be made in the predicted faunal composition both in the water column and on the substrate.

The nature of the suspended solids is also important in determining damage. Angular, inorganic material (from shotblasting, china-clay mining, stone-working) produces physical abrasion besides inducing clogging-effects (Alabaster & Lloyd 1980). Formerly sediments were thought to be of minor importance, but investigations into sediment-associated phosphorus in eutrophication indicates that they are of major importance in determining nutrient availability in the water column. Other pollutants are also known to become readily associated with sediments, and as the physical and chemical changes take place in sedimented particles, degradation, solubilization and release of toxic materials to the aqueous phase can occur (Dickson *et al* 1987).

Physical characteristics of effluents other than those described above are also of fundamental importance. Many materials are suspended or dissolved in organic solvents. The nature of these organic solvents is normally lipophilic, i.e. cell-membrane-dissolving or exoskeleton-affecting. Damage caused by such solvents, because of their detergent effect, is irreversible and often greater than the solute or suspension that they carry.

Pesticide applications and veterinary treatments are often geared to most effective contact with the target organisms. By their very design they may specifically affect the aquatic flora or fauna (e.g. high adsorbed particulate suspension or finely divided emulsion).

Biological pollutants

Pollution studies have been centred on chemical analysis and biological effects. The major contributor to river pollution in most developed countries is still the domestic component of sewage. There is a scientific discipline geared to measuring BOD, dissolved oxygen and ammonia in all its forms. Statistical derivations and algorithms to determine percentile concentrations on any number of samples analysed are available in order to help the discharge consent setting process.

The bacterial component of effluents like sewage (and silage) is considerable and normally expressed as coliforms, *Escherichia coli* or faecal *Streptococcus* numbers per 100 ml of receiving water. In Europe the methodology is standardized (Department of Environment 1983) and geared towards two major European Directives – the Drinking Water Quality Directive and the Bathing Beach Directive. Neither of these reports considers the effects on the environment of a massive influx of foreign bacteria other than from a human-use point of view. Coliforms are primarily Gram-negative non-sporing bacteria of the family Enterobacteriaceae. Their importance is that many of them are natural inhabitants of the gut of mammals (and birds, reptiles, amphibians and fish), some are parasitic and several are markedly pathogenic, so that their presence in water indicates a potential risk to human health. *Escherichia coli* (*E. coli*) and *Streptococcus faecalis* (faecal *Streptococcus*) are two of the most common of these coliforms in human and mammalian guts and faeces. Coliforms derived from mammals normally die at temperatures less than 37°C and their rate of death depends on a series of physical and chemical environmental factors (e.g. light intensity, temperature, organic material present). However, many coliforms, although capable of survival in conditions simulating mammalian intestinal conditions, derive from plant material (e.g. *Erwinia*, *Proteus*, *Leuconostoc*). Their presence in large numbers (crude sewage may contain 10^6–10^8 coliforms per ml) influences the microbial ecology of the receiving water, so much so that heterotrophic, sewage-derived bacteria far outnumber the indigenous autotrophic–heterotrophic mix of the riverine flora. In more eutrophic waters the natural balance would favour heterotrophs; in mesotrophic and oligotrophic environments, autotrophs would predominate. The influence of this bacterial input to riverine ecology is both fundamental and massive and yet it remains one of the least investigated components of sewage pollution.

The input of viruses derived from human and domestic animal and plant sources is largely unknown. The Bathing Waters Directive refers to the presence of viruses to provide a legislative framework for immersion sports, but other implications of viral ecology as a result of organic inputs are poorly understood.

In recent years, following outbreaks of giardiasis and cryptosporidiosis in human populations using surface waters at their potable water source, there has been considerable interest in protozoan pathogens (cf. the Badenoch Report; Badenoch 1990). With these parasites, as with viruses and bacteria, there is considerable debate as to whether they should be treated as environmental contaminants and removed from any discharge to the environment or whether they should be regarded as part of the normal riverine biota and eliminated during the process of treatment for drinking-water purposes. There are cost implications for both routes, but the latter provides no environmental benefit.

Pollutant pathways

There are three major routes by which pollutants can enter living organisms – with their food, via their respiration and by contact. The major route of entry depends on the physical and chemical characteristics of the pollutant and the level of organization of the assaulted animal or plant. Some materials have immediate effect on the individual while others bioaccumulate and their effects may not be realized for a considerable time. Chlorine is a good example of the former: in water it can exist as free chlorine or combined with ammoniacal compounds as chloramines. The proportion of combined chlorine depends on the concentration of ammonia in the receiving water. Free chlorine, acting as an oxidizing agent, will kill fish such as rainbow trout in 24 hours at 12°C and at concentrations of $0.3\,mg\,l^{-1}$ (Alabaster & Lloyd 1980). Similar concentrations of the chloramines kill trout over several days. Temperature and pH have little effect on this toxicity. Once contact with the fish epithelium has been achieved, a marked pathology results which is irreversible. Acute poisoning with metals such as copper, zinc and cadmium and with phenols is seen to have similar, direct, rapid effects.

Some compounds are absorbed directly from the water and accumulated. Again metals are good examples of these. Zinc is accumulated in gills, liver, kidney and bone of fish and elevated levels produce a reduction in egg numbers. However, mechanisms exist in the fish whereby zinc can be lost through the gills, in the faeces and via deposition with eggs. Zinc is a trace-element essential for life (being involved in the manufacture of the enzyme carbonic anhydrase) but occurs naturally in some geological formations at levels which initially may seem detrimental to fish. However, under such circumstances fish become acclimated to these levels and have an enhanced detoxification capacity. Absorption of zinc from the surrounding water is influenced by pH and water hardness: calcium competes with zinc for binding sites on the epithelial cells (Lloyd 1992). It is also influenced by the form the metal takes in the water – simple salts will be readily absorbed whereas complexes containing zinc may not be bioavailable, but may eventually become locked up in sediments. Copper is similar in many respects to zinc in its toxicity for fish, but has traditionally been used as a molluscicide, bactericide, algicide and general herbicide. Copper serves as a good illustration of obvious multiple environmental effects with both immediate, direct and long-term effects.

Many synthetic compounds are accumulated by living organisms, e.g. organochlorines, polychlorinated biphenyls (PCBs), organotin compounds. Here the mechanism of toxicity may be different, longer-term and bear little relationship to the acute toxicity data produced for that compound. Because these synthetic compounds are long-lasting there is the opportunity for them to be reworked in the food web and reach concentration factors of several hundred thousand near the top end of the trophic chain. Little is known about their accumulation, degradation mechanism or rate of movement through the various trophic levels. Concern over synthetic compounds is so great in areas like the lower Rhine that programmes of sediment removal have been embarked upon on a major scale. Polychlorinated biphenyls form a major part of the argument for this work, because of their persistence and extreme toxicity under some circumstances. They are a group of 209 related compounds which accumulate in fatty tissues and which were manufactured between 1929 and 1971. Their primary use was in electrical transformers and condensers, but also in paints, printing inks and oils. About 2 million tons have been produced worldwide. It is estimated that 30% of this is cycling through the biota, mainly freshwater. Elevated levels in fish, like eels, with a high fat content present a problem for predators such as herons, otters and humans, where the accumulation will continue. Fish-eating mammals are particularly at risk as during the production of milk by the mother, high PCB concentrations fed to the offspring are undiluted by other food.

During the metabolism of pollutants by the aquatic fauna and flora some degradation to other compounds takes place, or a change in the physical characteristics of the original material may result. Under some circumstances these changes may compound the toxic effect. The use of lead shot by anglers in the UK was for many years suspected as the cause of lead poisoning in swans. Lead as metal lumps is not available for biological uptake. However, as the swans used shot along with gravel from the river bed to help grind up food in their gizzards, metallic lead was ground down. This resulted in significant solubilization which took place in the acid part of the gut. The lead was then absorbed. Following the cessation of use of lead shot by anglers the population of swans on rivers in southern England increased dramatically over 2 years (Lievesley & Perrins 1991).

Such transformations of apparently innocuous materials, whether by physical, chemical or biological mechanisms, render suspect the 'no effect' level proposed in some pollution control policies.

Eutrophication

A whole series of human activities has resulted in the enrichment of rivers in many parts of the world. In the UK, particularly in the south and east, various activities have accelerated eutrophication. Changes in the physical environment such as widening and dredging with consequent slowing of flow and the elimination of bankside trees in order to provide access for leisure activities, commercial activities or river maintenance are predisposing factors in the encouragement of eutrophication. Canalized or 'improved' rivers allow greater light input to the river surface and riverside environment. The very act of canalization increases the available river surface area, thereby increasing light availability further. Eutrophication and the opposite process, oligotrophication, are naturally occurring processes with a long time-scale. The accelerated eutrophication caused by human activity is a rapid process seen over the last 50 years.

Sewage effluent, industrial discharges, farm slurries and agricultural run-off are all major sources of plant nutrients, notably nitrates and phosphates. In rivers in their natural state the limiting factors to plant growth vary according to topography, geology and geography (Haslam 1990). There is consequently a natural variation in plant community composition throughout the rivers of the UK. By increasing light, increasing nitrate, but primarily by addition of phosphate an enhanced plant growth results. This growth may be of relatively small numbers of submerged or emergent macrophytes (*Elodea canadensis*, *Potamogeton crispus*, *Nuphar lutea*), floating-leaved plants (*Lemna minor*), or diatoms (*Stephanodiscus hantzschii*, *Asterionella formosa*) green algae (*Ankistrodesmus*) or blue-green algae (*Microcystis*). The particular species that predominate will depend on the original diversity present, but essentially the effect is one of species impoverishment with a large biomass of very few species. Habitat provision afforded by plant communities is a major influence on the diversity of macroinvertebrates, so that enriched rivers become impoverished in their macroinvertebrate fauna, not only because of direct influences of water quality, but also by indirect influences. Such changes inevitably control the composition of the fish community, which tends towards the limited, cyprinid-dominated coarse fisheries of lowland Britain where two or three species – roach (*Rutilus rutilus*), bream (*Abramis brama*) and dace (*Leuciscus leuciscus*) – make up over 60% of the fish biomass. The appearance of blue-green algal blooms is of major importance in still waters, where their presence may preclude some of the uses referred to in Section 2.1, but the appearance of *Aphanizomenon* in bloom proportions in the lower Thames in late 1990 is of potentially greater significance. As water is abstracted for London's drinking water from the lower Thames, this use may create the requirement to reverse eutrophication.

Algal communities in many nutrient-enriched lowland rivers extend through the whole water column and are only limited by their own competition for light. Very few submerged macrophytes can withstand this shading (the exception being *Nuphar lutea*), so the impoverishment of the riverine ecosystem is magnified. Changes can be observed both in the water column and in the benthic communities: these are sequential so that methods of assessment of eutrophication using algae or diatoms have been developed by aquatic biologists (Whitton *et al* 1991).

Acidification

In the context of aquatic pollution, the decrease in pH of freshwaters in Europe and North America in recent historic times is encompassed by the term acidification. Simple chemistry informs us that it is a shift to the right of the following equation:

$$[H_2O] \leftrightharpoons [H^+] + [OH^-]$$

with pH being defined as $-\log_{10}[H^+]$. Naturally there is variation in the pH of freshwaters depending on the geological influences that can buffer the acidity of natural rainfall (pH of 5.5). The dissolution of calcium carbonate in areas of chalk, limestone, calciferous sandstone and calcium-rich alluvia produces waters of neutral or slightly alkaline pH, which, because of their buffering, are also pH-stable. The pH values of freshwaters in geologically-older and other non-calcium-bearing areas are acid and far more liable to be changed by other factors: the acidification produced by respiration and decomposition of plant material in already slightly acid waters contrasts with the increase in pH that accompanies active photosynthesis as carbon dioxide is absorbed from the water. Faunal and floral communities in both terrestrial and aquatic environments have evolved to cope with local levels of acidity and in more acid areas with the pH variation that naturally occurs.

In the last 150 years, human activity has produced a marked change in many of these poorly buffered waters, primarily by the production of acid rain, a term first coined by Smith in 1872. Rain becomes acid rain by the addition of the products of combustion of fossil fuels – primarily the anions of sulphate, nitrate, ammonia and other micropollutants of organic origin. Immediately a climatological factor has to be incorporated into acid rain – if hilly areas of high rainfall, mountainous and/or more northerly areas of high snowfall, or areas susceptible to extensive fogs and mists coincide with calcium-poor geology, then acidification results. However, according to NERC (1984), dry deposition accounts for two-thirds of the total acid load in the UK, so the formerly simple acid rain theory is not quite as simple as earlier theorists believed. Other causes of acidification in freshwaters are upland coniferous afforestation (Nillson *et al.* 1982) heathland regeneration (Rosenquist 1978), opencast mining as well as post-glacial natural acidification (Pennington 1984).

An increase in acidity produces a change in the solubility of all ionic materials and may, by chemical reaction, induce a change in the solubility of many non-ionic or weakly ionic materials. One of the major effects of the acidity increase is the dissolution of metal ions. Lentic environments are affected more readily than lotic environments by acidification, because of the more finite resources that they possess to counteract the acidification process.

The effects of acidification are the gradual removal of fish species until only the most tolerant are left, the elimination of the crustaceans and the disappearance of molluscs. Depending on the original fauna, the change may be slight or catastrophic. The floral composition is also markedly altered and many macrophytes will disappear altogether, being deleteriously affected by the acidity or by the increasing concentration of metal ions in the water. The metals most commonly affected by pH changes from 7 down to 4 are aluminium, copper, mercury, lead and iron, all of which show significant speciation over this range. With aluminium, for example, the most toxic forms to fish are found in the pH range 5.2–5.8.

Attempts have been made in acid-stressed waters to counteract the increase in acidity by the addition of lime, chalk or limestone. For fisheries, the recovery is apparent within a few years but for large parts of the biota the change to a calcium-rich environment precludes their recolonization. The only effective long-term solution to freshwater acidification is the reduction in emissions of fossil fuel gases, primarily oxides of sulphur and nitrogen. Industrialized countries in the northern hemisphere have agreed to this over the next decade. Changes in forestry practice are available to forestry managers which would decrease the acid enhancement due to these activities. There is some concern that the use of fertilizers on agricultural land also leads to a more acid agricultural run-off directly to watercourses or indirectly via ammonia volatilization – major changes in intensive agriculture are required before this is counteracted.

2.3 PROSPECT

The pollution that characterized the 19th and first half of the 20th centuries was essentially sewage or heavy-industry related. The technology and controls available in the second half of the 20th century are sufficient to check the worst excesses of these effluents. However, the proportion of our rivers that are effluent-derived continues to increase as does the number and complexity of compounds disgorged by our society. Our understanding of riverine ecology has moved from almost nothing to simple cognisance at the same time. So now we are able to demonstrate the complexity of the ecosystem and see the ecological ripples that result from dropping an organochlorine or nutrient stone. The complexity of legislation required to cope with this situation has created a subject in itself. Inevitably emission standards will get stricter and major polluters will have to shoulder an ever-increasing financial penalty. In the 21st century the axiom that effluent discharge to rivers is an inevitable consequence of society may be seriously challenged. The technology that enables us to take suspended solids, micropollutants and bacteria out of effluents is already past the drawing board, so that we may be in sight of having cleaner rivers than for 500 years.

REFERENCES

Alabaster JS, Lloyd R. (1980) *Water Quality Criteria for Freshwater Fish.* Butterworths, London. [2.2]

Badenoch, J. (1990) *Report of the Group of Experts on Cryptosporidium in Water Supplies.* HMSO, London. [2.2]

Beverley M, Garrood AC, Craven M, Johnson I. (1991) Elevation of glutathione S transferase in response to xenobiotic and general stress in molluscs. SETAC-Europe Proceedings. Inaugural meeting, University of Sheffield (unpublished). [2.2]

Cairns J, Dickson KL. (eds) (1973) *Biological Methods for the Assessment of Water Quality.* American Society for Testing and Materials, Philadelphia. [2.1]

Cairns J, Dickson KL, Westlake GL. (eds) (1976) *Biological Monitoring of Water and Effluent Quality.* American Society for Testing and Materials, Philadelphia. [2.1]

Collocott TC, Dobson AB. (eds) (1974) *Chambers Science and Technology Dictionary.* Chambers, Edinburgh. [2.1]

Crane M, Maltby L. (1991) The lethal and sublethal responses of *Gammarus pulex* to stress: sensitivity and sources of variation in an *in situ* bioassay. *Environmental Toxicology and Chemistry.* **10**: 1331–9. [2.1]

Department of Environment. (1983) *Reports on Public Health and Medical Subjects No. 71: The Bacteriological Examination of Drinking Water Supplies 1982.* HMSO, London. [2.2]

Department of Environment. (1992) *River Quality.* HMSO, London. [2.1]

Dickson KL, Maki AW, Brungs WA. (eds) (1987) *Fate and Effects of Sediment Bound Chemicals in Aquatic System.* SETAC Special Publications Series. Pergamon Press, Oxford. [2.2]

Haigh N. (1987) *EEC Environmental Policy and Britain,* 2nd edn. Longman, Harlow, Essex. [2.1]

Haslam SM. (1990) *River Pollution – An Ecological Perspective.* Belhaven Press, London. [2.2]

Klein L. (1962) *River Pollution 2. Causes and Effects.* Butterworths, London. [2.1]

Lievesley P, Perrins CM. (1991) *Swans and Lead Poisoning. Report for the year 1990.* Edward Grey Institute, Oxford. [2.2]

Lloyd R. (1992) *Pollution and Freshwater Fish.* Fishing News Books, Oxford. [2.2]

Meybeck M, Chapman DV, Helmer R. (1989) *Global Freshwater Quality.* Blackwell Publishers, Oxford. [2.1]

Muirhead-Thomson RC. (1987) *Pesticide Impact on Stream Fauna with Special Reference to Macroinvertebrates.* Cambridge University Press. [2.2]

NERC (Natural Environment Research Council) (1984) *Acid Rain.* The Environment Committee: House of Commons. [2.2]

Nillson ST, Miller HG, Miller JD. (1982) Forest growth as a possible cause of soil and water acidification: An examination of the concepts. *Oikos* **39**: 40–9. [2.2]

Pennington WR. (1984) Long term natural acidification of upland sites in Cumbria. *Report Freshwater Biological Association* **52**: 28–46. [2.2]

Rosenquist IT. (1978) Acid precipitation and other possible sources for acidification of rivers and lakes. *Science of the Total Environment* **10**: 39–49. [2.2]

Stewart GSAB. (1990) *In vivo* bioluminescence: New potentials for microbiology. *Letters in Applied Microbiology* **10**: 1–8. [2.2]

Whitton BA, Rott E, Friedrich G. (eds) (1991) *Use of Algae for Monitoring Rivers. Proceedings of an International Symposium, Düsseldorf, 1991.* University of Innsbruck. [2.2]

3: Hydrological Change

J. A. GORE

3.1 INTRODUCTION

The prevailing interactions of velocity, depth and substrate reflect the hydrological integrity of a river channel (Dingman 1984). It is not unreasonable to assume that the structure and function of most aquatic communities is tied to the stability or predictability of hydrological patterns and instream hydraulic conditions (Statzner & Higler 1986). Only recently have lotic ecologists begun to examine the potential impact of the change in hydrological and hydraulic conditions on the instream flora and fauna. The manner in which these hydraulic factors interact influences the distribution of aquatic biota along the length of a lotic system and within a given reach of that system. Although it might appear that adaptation and/or preference for certain ranges of hydraulic conditions are the physical challenges faced by lotic biota, it can be demonstrated (and considered from a management perspective) that increased genetic variability (and species vigour?) accompany a physically variable stream environment (Robinson *et al* 1992).

The basic processes by which hydrological and hydraulic conditions might influence biota have been reviewed by Newbury (1984) and Davis (1986). In introducing the concept of 'hydraulic stream ecology', Statzner *et al* (1988) emphasized the particular attention that must be paid to the complex hydraulic interactions which occur near the substrate or shear zone and their influence on benthic organisms. Although the interactions with complex hydraulic conditions may not be as well defined for those organisms which exist within the water column and can more actively escape turbulent conditions, Heede and Rinne (1990) also view the analysis of hydrological and hydraulic conditions to be critical for effective management of fisheries. In this chapter, the potential effects of hydrological change will be reviewed and it will be necessary to establish some basic hydraulic reference points to understand the magnitude of changes brought about by anthropomorphic alterations of runoff pattern, and discharge regime.

3.2 HYDROLOGICAL EFFECTS

Transition zones

As water flows from headwaters to mouth, changes in slope and channel shape, associated with changes of the sediment load, have the effect of altering the manner in which that flow body dissipates energy. Flow patterns, the distribution and frequency of occurrence of various depth, velocity and substrate combinations, can be expected to change with distance from the headwaters. In general, headwater streams have relatively shallow slopes and gentle flows, but relatively close to the source they undergo a shift to high-velocity flows as the slope dramatically increases. As the river nears the foothills or lowland areas another transitional zone is encountered in which flows are decelerated and the river begins to braid and deposit sediment loads. Farther downstream, with declining slope the river begins to form large sinuous meanders; this is a zone of diverse flow patterns as the river begins to cut backwater areas and form oxbow lakes. Statzner and Higler (1986) proposed that an index of hydraulic stress, laminar sublayer thickness and slope, could be used to describe a 'typical' pristine

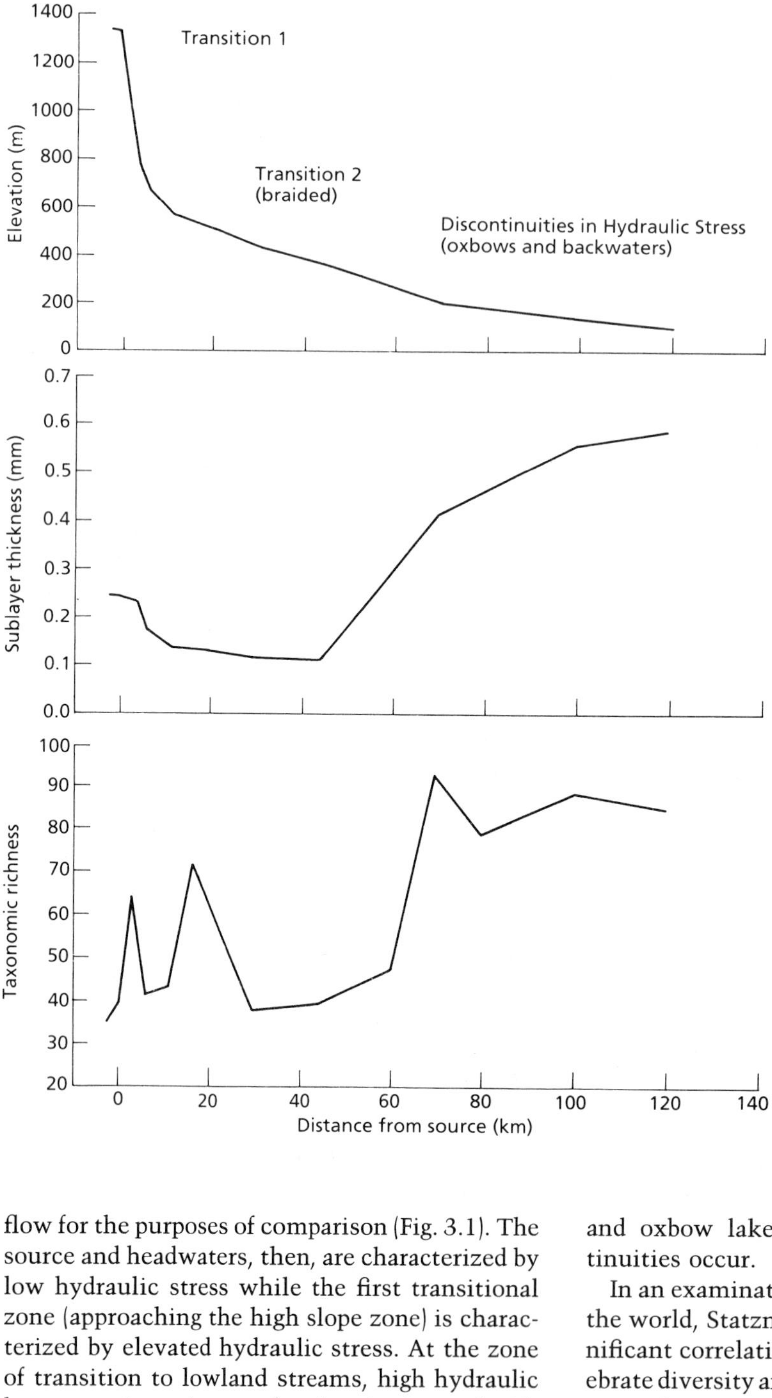

Fig. 3.1 Proposed relationship between hydraulic heterogeneity, complex hydraulics along the substrate (as quantified by sublayer thickness), and diversity of aquatic organisms (taxonomic richness) along the length of a typical river (after Statzner & Higler 1986).

flow for the purposes of comparison (Fig. 3.1). The source and headwaters, then, are characterized by low hydraulic stress while the first transitional zone (approaching the high slope zone) is characterized by elevated hydraulic stress. At the zone of transition to lowland streams, high hydraulic heterogeneity exists as the river begins to braid, but this zone is characterized by lower hydraulic stress. As the river meanders and forms backwater and oxbow lakes, numerous hydraulic discontinuities occur.

In an examination of 14 stream systems around the world, Statzner and Higler (1986) found a significant correlation between high benthic invertebrate diversity and the location of the transitional zones between low and high hydraulic stress. This is not surprising as the availability of substantially different hydraulic microhabitats would increase in these transitional zones and a diverse fauna, existing at the limits of their hydraulic tolerances, could be expected. Indeed, Statzner and Higler suggest that in the zones between the transitions, communities are more stable in their structure, and resilience to disturbance might be different from those communities at the transition zones. This general conclusion is echoed by other studies outside North America, where zonational patterns could not be attributed equally to patterns of particulate organic carbon use as in the river continuum concept (RCC; Vannote *et al* 1980). In Australian streams, for example, temporal and spatial flexibility of the fauna is more attributable to variability in streamflow (Lake *et al* 1986) while King (1981) found that changes in physico-chemical conditions (including water velocity) correlated well with changes in faunal distributions along the length of a short South African coastal river.

The relationship between changes in hydraulic condition and benthic invertebrates extends to the interstitial meiofauna as well. Ward and Voelz (1990) found that substrate condition and localized hydraulic conditions are more important than elevational factors in determining meiofaunal distributions. The areas of transition between meiofaunal communities coincided with the zones of hydraulic transition as predicted by Statzner and Higler (1986).

Longitudinal transition in fish communities is well documented (Sheldon 1968; Jenkins & Freeman 1972; Evans & Noble 1979; Platania 1991). Again, most of the patterns of species replacement or change in community structure are related to identifiable physical habitats related to site-specific hydraulic conditions. Chisholm and Hubert (1986) found that brook trout densities varied as a function of gradient and stream width-to-depth ratios with three identifiable zones of population density, varying from highest densities in low-gradient streams and lowest densities in highest-gradient areas. Platts (1979) also identified an abrupt transition in salmonid-dominated communities at the first zone of transition in hydraulic stress in low-order streams of Idaho. Statzner (1987b) has predicted that zonational changes occur at major hydrological breaks (usually at geomorphic changes in slope). In these areas where a high diversity of hydraulic conditions occurs, diversity of biota has been found to be at its highest.

The points of hydrological change along rivers usually coincide with optimum points of placement for hydraulic control structures, particularly impoundments. Thus, these ecotonal areas, which are reasonably fragile in structure (Naiman *et al* 1988), are often the first communities to be altered by anthropogenic changes which influence discharge regimes. Not only does the physico-chemical nature of release water from impounded waters in these areas influence the structure and function of downstream communities, but alterations in flow patterns influence the distribution of substrate particles and the pattern and distribution of flow which, in turn, create different patterns of microhabitat availability. These discontinuities can have the effect of 'resetting' community structure to mimic those of areas farther upstream where physico-chemical conditions are similar (Fig. 3.2) (Ward & Stanford 1983). The ability of the lotic biota to recover from these changes in physical environment is an index of the ultimate severity of the disturbance.

Superimposed upon these longitudinal and within-reach changes in hydraulic conditions are changes in seasonal hydrograph, short-term events such as floods, and potentially long-term 'events' such as the effects of human intervention upon the flow regime (i.e. afforestation, impoundments, diversions and abstractions) which have the potential to alter temperature patterns and the availability of particulate organic food for lotic biota.

Life-cycle requirements

The life cycles of lotic biota appear to be regulated by two major physical factors: temperature and hydraulic conditions. The impacts of temperature change on aquatic insects is well documented (Hynes 1970). Temperature and day-length seem to be the regulating features which synchronize hatching, maturation of larvae, emergence and mating of adults and, in some cases, egg-laying behaviour.

Not only are there certain temperature optima

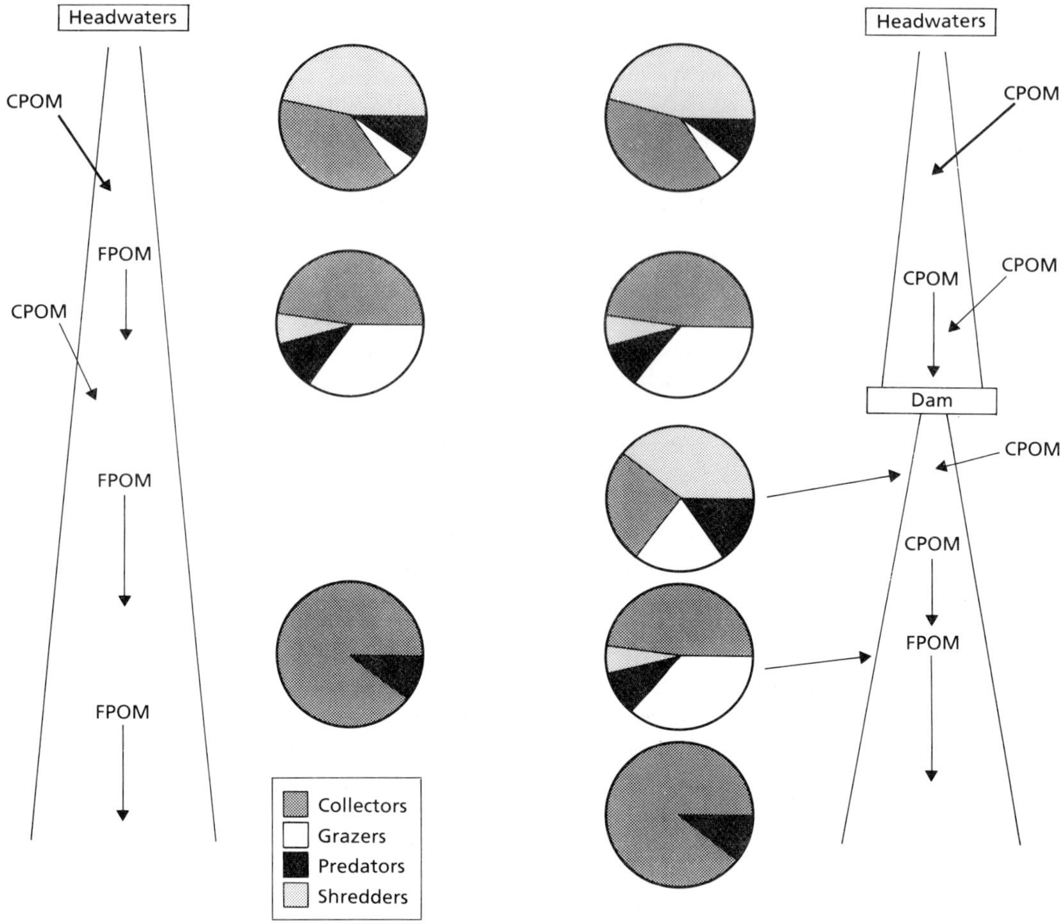

Fig. 3.2 Comparison of the composition of functional feeding goups at the upper, middle and lower reaches of a natural river system (left) and an impounded system (right) where the impacts of the dam include increased proportions of coarse particulate organic matter (CPOM) and less fine particulate organic matter (FPOM) in the tailwater. Increased clarity causes increased periphyton growth for grazers while hypolimnetic releases decrease the number of degree-days in the tailwater by as much as half that in the natural system. (After Stanford & Ward 1984.)

for the hatching of various eggs but a diel cycle also appears to enhance development (Humpesch 1978). Indeed, it appears that greater daily temperature changes reduce the degree-hours required for hatching of some species of aquatic insects (Sweeney & Schnack 1977). Although there is some debate about the importance of this diel temperature cycle, alteration of thermal regimes by change in hydrological condition could result in reduced success of many macroinvertebrate species.

Thermal cues also regulate the ability of aquatic invertebrates to enter a period of quiescence or rapid development prior to emergence as adults. In general, rising temperatures induce diapase and declining temperatures terminate that period of dormancy (Ward 1992). These responses preadapt a species to survive drought periods and to colonize temporary waters. In temperate-zone rivers, rising temperatures in the spring cue growth and maturation among larval and nymphal aquatic insects which have overwintered. In combination with day-length, the summation of degree-days synchronizes the emergence of adults. Declining autumnal temperatures may also cue the emergence of the more rapidly growing summer communities of benthic invertebrates (Ward 1992). Artificial disruption of thermal patterns and/or day-length cues has been shown to disrupt emergence patterns and reduce population success (Ward & Stanford 1982). Higher than normal discharge during snowmelt periods may depress downstream water temperatures and delay emergence (Canton & Ward 1981).

The combined influences of temperature and hydraulic preferences have also been demonstrated for lotic fish species, particularly benthic species, where distinct thermal tolerances correlate well to preferences for pool (higher temperature) or riffle (lower temperature) habitats (Hill & Matthews 1980; Ingersoll & Claussen 1984).

It can easily be demonstrated that any shift in hydrological pattern that leads to an alteration of the established thermal regime of a lotic ecosystem will ultimately lead to a dramatic change in the composition and survival of lotic biota. These hydrological alterations are discussed below.

Statzner *et al* (1988) have summarized the results of many examinations of the relationships between the distribution of lotic biota and the heterogeneity of flows within a stream reach. There is little doubt that a strong interaction between body morphology and hydrodynamic stresses controls the ability of lotic species to forage and maintain position within a preferred habitat. The relationship between flow conditions and life-cycle requirements is not as clear. However, there is recent evidence to indicate that flow patterns also control life-cycle success. Dudley and Anderson (1987) demonstrated that receding water level cued pupation for the tipulid, *Lipsothrix nigrilinea*. In this case, discharge was the governing factor rather than the hydraulic conditions. Pupation by the larvae could be delayed for a year or longer if water levels did not decline.

As growth leads to an alteration of both morphology and cross-sectional area of the organisms exposed to various hydrodynamic conditions, it could be expected that the organisms must migrate to areas of suitable hydraulic fitness in order to continue to forage and mature. Gersabeck and Merritt (1979) calculated distinct velocity preferenda for different instars of larval blackflies, while Gore *et al* (1986) demonstrated significantly different depth distributions for the final instars of the caddisfly, *Hydropsyche angustipennis*. Although insufficient physiological data exist to verify, Gore and Bryant (1990) have speculated that the apparent short-term preference of egg-bearing female crayfish for high-velocity riffles is a result of aeration requirements for successful incubation. Statzner *et al* (1988) concluded that the distribution of the various life stages of the hemipteran, *Aphelocheirus aestivalis*, was controlled by the ability of each instar to respond to changes in the distribution of hydraulic stresses across the substrate. Thus, early instars required low-stress areas (pool and moderate runs) while adults lived almost exclusively in high-velocity riffles. In stream reaches where a high diversity of depth and velocity combinations existed, there also existed a high diversity and density of instars and adults. Diversity of velocities and depths in stream reaches is directly related to discharge *and* to diversity of fish and macroinvertebrate species (Gore *et al* 1992b). Artificial discharge regimes which promote more homogeneous flow patterns might be expected to sustain lower diversities of lotic species and reduce or eliminate some species which are cued to certain hydraulic conditions.

3.3 DISRUPTION OF HYDROLOGICAL PATTERNS

Predictable hydrographic patterns are disrupted by human activity, which either impounds mainstem flow or diverts that flow for human consumption leaving a dramatically reduced flow and altered flow regime in the natural channel of the river. For the most part, significant alterations in physical and chemical conditions also result from impoundment for flood control and irrigation storage, impoundment for hydropower operation (especially hydropeaking), irrigation diversion or abstraction, channelization and afforestation.

Lakes as a hydrological influence

In comparison to the age of most lotic ecosystems, lakes represent a temporary (albeit lengthy) disruption to the flow patterns of riverine habitats. As standing bodies of water, lakes dramatically alter the physical, chemical and biological composition of the newly flowing water in the lake outlet. Since these disruptions constitute a natural geomorphological occurrence, the changes in hydrological and biological conditions can serve as a baseline for comparison to artificial disruptions (impoundments) which mimic lakes.

Standing water lacks the kinetic energy to maintain a large proportion of suspended solids. As a result, lakes act as settling basins for the suspended load carried by the input stream. The receiving stream which is fed primarily by epilimnetic water tends to be sediment poor. The mass and depth of the lake often lead to thermal and chemical stratification of the lake. With stratification, nutrients carried into the lake are concentrated in the hypolimnion (Wetzel 1983). The outlet stream, then, is nutrient poor with the exception of short periods of time when pulses of nutrients enter the receiving stream during mixis. Lentic systems are also characterized by a distinct community of planktonic organisms, which also appear as seston in the outlet streams. The pulses of seston occur on diel cycles as a result of vertical migration of zooplankton to the epilimnion, as seasonal pulses during particular phases of life cycles of planktonic and nectonic species, and as organic particulates generated by biotic activity within the epilimnion (Cushing 1964; Wotton 1982; Perry & Sheldon 1986). Epilimnetic waters are usually warmer than natural stream systems (Mackay 1979). These physical and biological changes are most often expressed in distinctive changes in the biota of the lake outlet ecosystem characterized by highly productive communities of filter-feeders (Richardson & Mackay 1991).

The containment of water within a lake has a tendency to ameliorate the effects of heavy precipitation and runoff events in the catchment. This buffering affect may stabilize discharges in the outlet stream and favour species which rely on constant hydraulic conditions for filter-feeding, particularly blackflies (Morin & Peters 1988). Finally, the mouth of the lake outlet is most often wide and shallow; this tends to reduce turbulence in the outflow water, making flow more laminar (Carlsson *et al* 1977) while making a greater portion of the water column and its sestonic load available to filtering macroinvertebrates (Morin *et al* 1988). Dense populations of filter-feeding blackflies and caddis flies are favoured.

The outflows of most lake ecosystems tends to have high gradients. The high gradient and the sediment-hungry nature of the outflow water combine to wash away fine particulates in the immediate outflow leaving larger substrate particles for attachment by filter-feeders. In addition, the sediment retention of the lake means that the outlet water does not carry a scouring suspended load and allows the increase of dense mats of attached periphyton (Simons 1979; Valett & Stanford 1987).

The more stable thermal regimes in outlet waters of lakes tend to result in fewer species in the community with dominance by filter feeders (Ward & Stanford 1982). Prolonged and elevated summer temperatures may promote higher growth rates while reduced ice-cover in temperate streams may also reduce mechanical damage to filtering species (Valett & Stanford 1987).

The abundance of filter-feeders displays a characteristic decline with distance from the lake outlet. Although increased predation may eliminate some of the filter-feeders (Sheldon & Oswood 1977), the reduction in density is apparently attributable to the rapid depletion of the seston by the very dense populations of filter-feeders near the outlet (Oswood 1979; Perry & Sheldon 1986). The recovery of downstream communities to resemble those of adjacent or upstream communities seems to be a function of the time for the stream restore the physical heterogeneity that accompanies thermal variation, hydrographic variation from spates and variation in suspended load, emphasizing the importance of physical habitat structure on the structure and dynamics of lotic communities (Robinson & Minshall 1990).

Deep-release impoundments

Although reservoirs mimic the dynamics of most lakes in their abilities to trap suspended material, their stratification patterns and development of planktonic fauna, the structure which impounds the water differs significantly from the shallow and wide mouth of a lake outlet. Most impoundment structures are placed for the purpose of flood control, irrigation storage and/or hydroelectricity generation. In all of these cases, storage or development of a hydraulic head necessitates the controlled release of waters from the base of the dam structure. Thus, release waters are derived from the hypolimnion of the reservoir behind the dam. The effect of this hypolimnetic or deep-release can alter the composition of the tailwater fauna for a considerable distance downstream. The impacts of a cold yet thermally constant, nutrient-rich but sediment-hungry discharge are well documented (Ward & Stanford 1979; Lillehammer & Saltveit 1984; Petts 1984; Craig & Kemper 1987).

In general, the effects of hypolimnetic releases

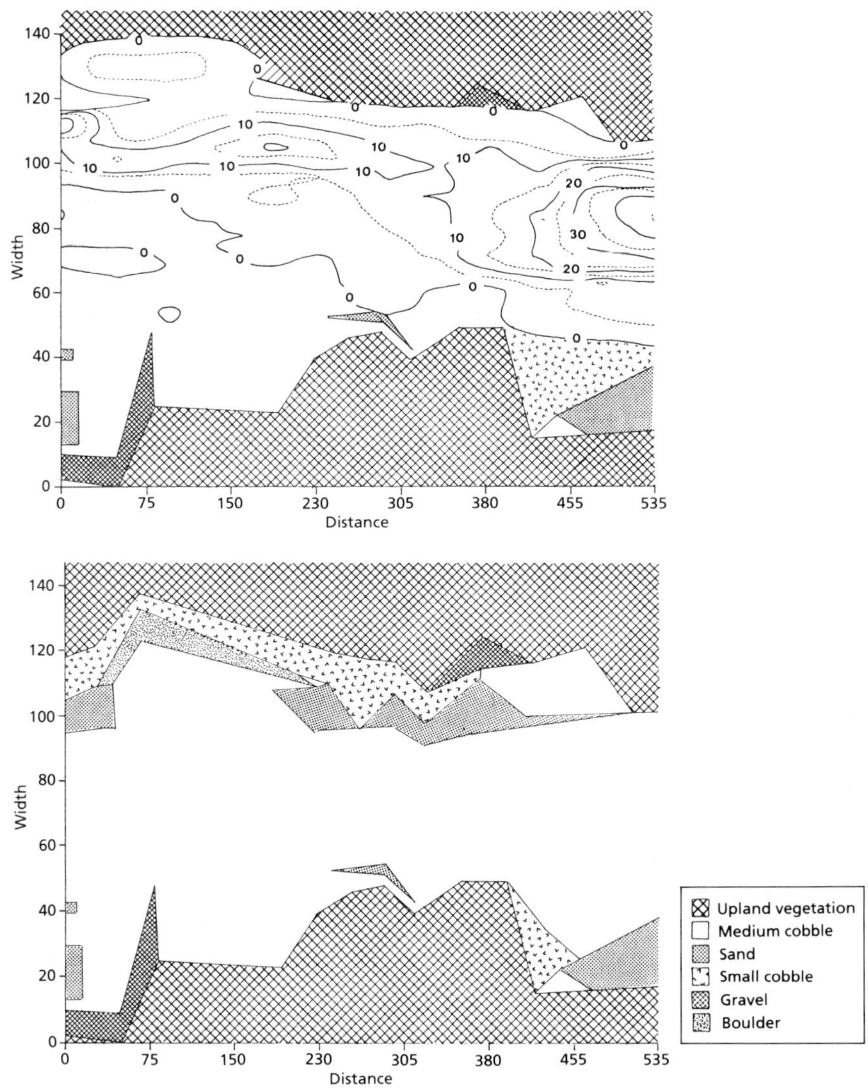

Fig. 3.3 Substrate and flow maps of a site on the Caney Fork River (Tennessee, USA) approximately 500 m downstream of a peaking hydropower facility, showing the degraded channel shape with a straight *thalweg* and substrate dominated by medium cobble (after Gore *et al* 1990).

are to improverish the downstream macroinvertebrate communities by the action of thermal constancy which eliminates the thermal cues needed to complete life cycles (as described above). Nutrient-rich and sediment-poor waters allow the abundant growth of periphyton mats, which further eliminate macroinvertebrate habitats. Fish communities, again due to altered thermal regime and reduced food availability, are also impoverished. The zone of recovery, i.e. the length of stream in which restoration of thermal variability occurs, can extend to distances of over 25 km.

Because the reservoir traps most of the inflowing sediment the released water entrains bed sediments and erodes the river channel, resulting in a channelized system with relatively homogeneous flow patterns (Fig. 3.3). Degradation of the

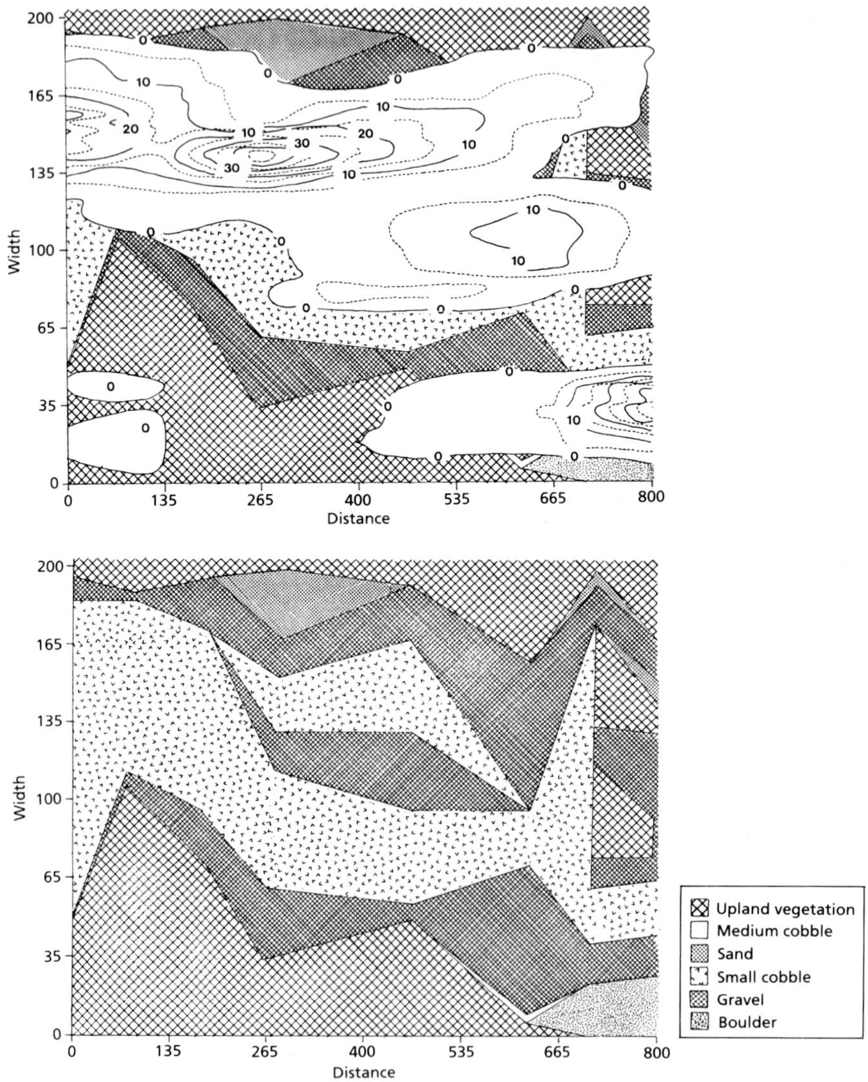

Fig. 3.4 Substrate and flow maps of a site on the Caney Fork River (Tennessee, USA) proximately 5 km downstream of a peaking hydropower facility, showing the aggraded channel with a braided appearance, split *thalweg*, and substrate dominated by sand and gravel (after Gore *et al* 1990).

channel continues until armouring by medium and large cobbles occurs.

In many instances, degradation reduces the water table in the floodplain. Floodplains themselves serve as refugia for some aquatic organisms and as a critical habitat for the completion of life cycles of many micro-organisms which, in turn, serve as food for the lotic communities. Reduced floodplain inundation could reduce this reserve of food for the lotic community, especially juvenile fish, which often spend the first few weeks of their lives in these floodplain areas (Boulton & Lloyd 1992). Combined with water-table reductions, decreased bank stability often results in the alteration of the composition and use of the riparian community while the clear water itself induces the increase in growth of algae (Gore 1977).

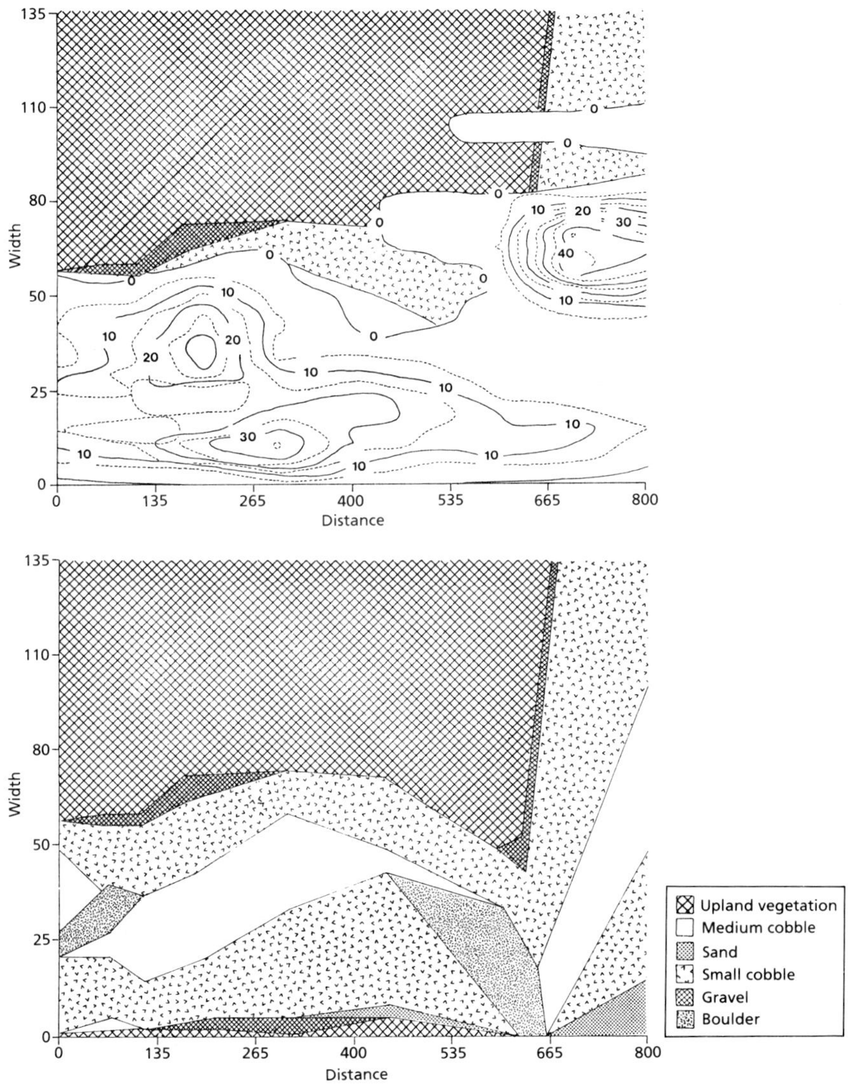

Fig. 3.5 Substrate and flow maps of a site on the Caney Fork River (Tennessee, USA) approximately 15 km downstream of a peaking hydropower facility, showing a minimally impacted section of channel with high flow diversity, heterogeneous substrate and sinuous internal meander (after Gore *et al* 1990).

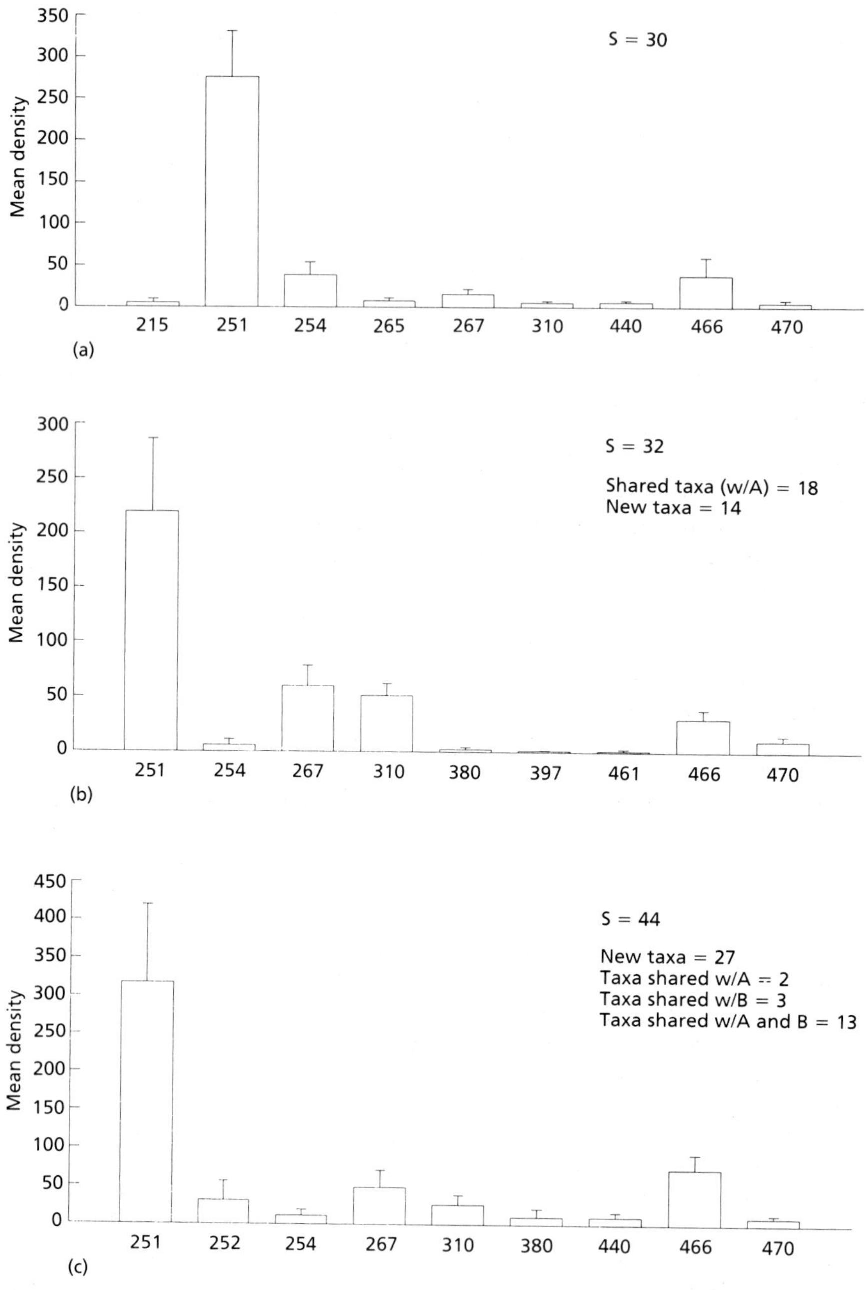

Fig. 3.6 Distribution and density of the 10 most common/dominant midge (Chironomidae) taxa at each of three stations downstream of a hydropower facility (Caney Fork, Tennessee, USA). (a) Zone of degradation. (b) Zone of aggradation/deposition. (c) Historical channel. Taxa listed are: 215 = *Cardiocladius obscurus*; 251 = *Cricotopus tremulus*; 252 = *Cricotopus trifascia*; 254 = *Cricotopus bicinctus*; 265 = *Dicrotendipes* sp.; 267 = *Dicrotendipes nervosa*; 310 = *Microtendipes* sp.; 380 = *Phaenopsectra* sp.; 397 = *Potthastia longimanus*; 440 = *Rheotanytarsus* sp.; 461 = *Stictochironomus devinctus*; 466 = *Synorthocladius semivirens*; 470 = *Tanytarsus* sp. (Data from Gore & Pennington, personal communication, with permission.)

Farther downstream, a zone of aggradation can occur where finer sediments from the degraded tailwaters are redeposited, forming many small islands (Figs 3.4 & 3.5). Fuller discussion of geomorphological changes is presented in Chapter 4.

Aside from the variety of thermal and chemical changes which may alter the composition of downstream communities – see reviews by Ward and Stanford (1979) and Petts (1984) – the changes in flow patterns and substrate composition will alter the amounts and kinds of physical habitat available. Gore and Pennington (1988) reported significant changes in chironomid species in the zones of degradation and aggradation related, in part, to the dramatic changes in dominant substrate when compared to the historical channel (Fig. 3.6). More recently, Grzybkowska *et al* (1990) found that densities of dominant macroinvertebrates, mostly chironomids, were highest along the banks of tailwaters and declined abruptly towards mid-river. In general, higher species richness occurred in depositional areas (Gaschignard 1984). Although not mentioned by these authors, it is not surprising that densities would be reduced near mid-river, the much higher velocities resulting in increased shear across the substrate and decreased suitability of hydraulic habitat.

Release schedules

Although it is difficult to separate the effects of changes in physico-chemical conditions from those of changes in flow pattern, it can be assumed that loss of preferred hydraulic habitat is partially responsible for the impoverished lotic communities which exist in the tailwaters of most impoundments. Of particular importance is the increased shear that peaking hydropower imparts on downstream communities. Release schedules have a dramatic effect on community composition. Unless releases of water are constant, the wetted perimeter of the channel is continually changing. Depending upon the habitat preferences of the species and the duration of time that the substrate is wetted, there is a continual cycle of invasion and stranding by various species of benthic invertebrates and fish (Corrarino & Brusven 1983; Perry & Perry 1986). The effect of reducing discharges also imparts on community composition by increasing the amount of downstream drift of benthic species. Most often increased drift has been reported to accompany flow reductions (Perry & Perry 1986). Gore (1977) and Corrarino and Brusven (1983) reported catastrophic drift of invertebrates when discharges declined rapidly. Gore attributed this to elimination of hydraulic habitat requirements of nearly all benthic species as discharge declined. Peaking hydropower further exacerbates the problems of adjusting to a predictable flow regime. Peaking releases occur as human demand dictates (when electricity consumption is higher than normal distribution loads). As a result the hydrograph of peaking tailwaters is flashy and significantly different from the input stream (Gore *et al* 1989). Although these tailwaters may have relatively stable beds and channel morphology after years of operation, analysis has revealed that habitat availability varies according to the release schedule, with less hydraulic habitat available as discharge increases (Gore *et al* 1989). The response of lotic biota to these changes in discharge is not clearly understood and it is not known if short-term loss of the preferred hydraulic conditions in the habitat is sufficient to alter population success or community dynamics. Gore *et al* (1993) have predicted that shear forces decline rapidly as water surface elevation increases and that the habitat disturbance occurs only for a brief time during the rising limb of each peaking wave. However, a considerable amount of behavioural observation is needed to verify this prediction.

The sudden increase in discharge in tailwater areas can be equivalent to floods with yearly recurrence in the historical channel. This often leads to daily increases in drift of macroinvertebrates as shear dislodges foraging and resting individuals. Increases in drift are usually associated with initial discharge surges (Matter *et al* 1983; Irvine 1985). Although drift has been reported to decline within an hour of the initial surge (Layzer *et al* 1989), Matter *et al* (1983) calculated that as much as 14% of standing crop in tailwaters can be eliminated each month by the drift initiated through the increased shear resulting from peaking operations. The affect of sudden flow fluctuations on tailwater fish communities is not as clear; however, DeJalon *et al* (1988) reported reductions in trout production as a result of the decline in the food source, the benthic macroinvertebrates lost to drift and other reservoir-related phenomena. Both fish and macroinvertebrates shift their physical habitat requirements as spawning or changes in water temperature dominate fish life-histories or changes in size and state of maturity alter habitat preferences of macroinvertebrates. Thus, the magnitude of drift and the success of recruitment to fish populations can also vary in tailwaters affected by hydropeaking operations (Campbell & Neuner 1985; Layzer *et al* 1989). The loss of habitat heterogeneity (i.e. decline in usable habitat) and the effects of unpredictable shear across the substrate serve to aid the impoverishment of tailwater fauna, regardless of the physicochemical conditions which arise as a result of the release structure.

Land use practices

In many cases, demand for water to irrigate croplands or the removal of forest or riparian vegetation can significantly alter the hydraulic conditions within the channel. Irrigation withdrawal, on the one hand, reduces mean discharge within the channel, while return water from poorly managed croplands, in addition to being chemically altered, may result in unusually higher flows in downstream receiving stream channels.

In the case of logging practices such as clearcutting, removal of the dominant vegetation results in reduced water retention by the watershed itself and dramatic increases in surface runoff and the volume of water being transported by the river channel. Although the majority of studies on the effects of clearcutting and other logging practices have concentrated on the consequences of the alteration of sediment transport, thermal regime and availability of particulate organic material as a food source for biota in the effected lotic ecosystem (see, for example, Wallace & Gurtz 1986; Webster *et al* 1990), a few studies have examined concurrent alterations in the hydraulic and hydrological conditions of the river which alter hydraulic habitat for the biota.

The availability of woody debris in low-order streams determines not only the quantity of coarse particulate organic matter available to drive community dynamics, but also the hydraulic characteristics of the channel. Bisson *et al* (1987) have shown that removal of trees and deprivation of the channel of the woody debris contributed by downed trees alters the ability of the channel to maintain pool habitats. In high-gradient streams (10–15%), an increase in average discharge resulting from clearcutting can cause scouring of the channel bed and a coarsening of the substrate, while low-gradient streams tend to be accumulators of sediment. Reaches with coarser substrate provide more diverse microhabitats and higher densities and diversities of macroinvertebrates and fish (Murphy & Hall 1981). Foraging efficiency of trout predators can increase in pools in logged catchments since sedimentation tends to occlude refugia on the substrate and higher overland flow results in increased drift densities of food sources (Wilzbach *et al* 1986). These relationships seem to be related to the amount of logging done in the watershed and the ability of the remaining vegetation to alter surface flow rates. With at least 25% of the forest remaining after logging activities, little significant impact on invertebrate communities could be ascertained in a series of low-order streams in Oregon (Carlson *et al* 1990).

Abstractions and diversions to provide water for irrigation, hydropower and public consumption usually result in significant decreases in water volume flowing through the river channel. Armitage and Petts (1992) have demonstrated that combinations of change in discharge, baseflow and substrate are the factors which govern numbers of species, numbers of families and total abundance of benthic species in stream channels affected by abstraction. Although it is possible that the heterogeneity of flow conditions may remain the same as in prewithdrawal conditions, hydraulic patterns shift to lower velocities and depths. Thus, one might expect that the preferred habitat conditions for high-velocity, riffle-dwelling species and deep-pool-dwelling species might be substantially reduced. In addition, the wetted area available to provide hydraulic cover and overhead cover for edge-dwelling species will also be reduced and, despite the fact that the same amount of edge area exists under reduced flows, the amount of habitat for edge dwellers will also be reduced. Alteration of the benthic community begins as soon as abstraction causes flows to fall below the natural low-stage conditions (Jäger *et al* 1985). Densities and biomass of macroinvertebrates have been reported to be reduced by as much as 75% (Heger & Moog 1986). Initial loss of invertebrates is immediate and attributable to drift, as an escape response from less than favourable physical habitat conditions (Gore 1977). However, species which are adapted to life on intermittently wetted substrates may persist (Bickerton *et al* 1993). Continued low flows do not provide habitat criteria favourable to recolonization (Gore & Milner 1990). Thus, the food support for fish communities does not exist and fish numbers, biomass and diversity also decline. Survival is dependent upon the number of refugia available and contribution of colonists from upstream drift (Armitage 1987).

The effects of reduced flows in a channel reach beyond the biota of the stream channel itself as the flow provides the supply for bank storage, hence riparian growth. In areas where reduced flows prevailed (downstream of hydropower facilities), reduced soil moisture caused lower growth rates and declining abundance of riparian plants, especially of juveniles (Stromberg & Patten 1990). Smith *et al* (1991) predicted that the elimination of floods and high flows could lead to the selective mortality of juvenile plants and the reduction of riparian strip width.

3.4 HYDRAULIC REFERENCE POINTS

In general, there are five major hydraulic conditions which most affect the distribution and ecological success of lotic biota. These are velocity profile, suspended load, bedload movement, water column effects such as turbulence, and substratum interactions (near-bed hydraulics). Singly, or in combination, the changes in these instream conditions can alter the distribution of biota and disrupt community structure. Within a stream reach the distribution of aquatic biota is governed by the interactions of these hydraulic conditions upon the morphology and behaviour of the individual organisms. Full discussions of suspended sediment transport (Chapter 3), and instream hydraulics and sediment transport (Chapter 5) are given in Volume 1.

Complex hydraulics

For any organisms living upon or near the substrate of a river or stream, the interactions of depth and velocity with the profile of the substrate particles is of critical importance to the range of potential microhabitats available (Statzner *et al* 1988). Table 3.1 summarizes some of the critical hydraulic conditions which a benthic organism must encounter. Although studies of flow in flumes have indicated the existence of a non-moving boundary layer (up to 2 mm in thickness) (Nowell & Jumars 1984) and lotic scientists have indicated that morphological and behavioural

Table 3.1 Instream hydraulic characteristics which have been shown to influence the distribution of riverine biota

Simple hydraulic characteristics:	
D:	mean depth (cm)
k:	substrate characteristic (cm)
S:	slope of water surface
w:	stream width (cm)
ρ:	density of water ($g\,cm^{-3}$)
g:	acceleration due to gravity ($cm\,s^{-2}$)
Q:	discharge ($m^3\,s^{-1}$)
U:	mean current velocity ($cm\,s^{-1}$)
ν:	kinematic viscosity ($cm^2\,s^{-1}$)
Complex hydraulic characteristics relevant to describe flow for an area within a stream reach or for an entire stream reach:	
Fr:	Froude number (dimensionless); $Fr > 1$ (shooting) $Fr < 1$ (tranquil)
Re:	Reynolds number (dimensionless); drag characteristics
Re^*:	boundary Reynolds number (dimensionless)
U^*:	shear velocity ($cm\,s^{-1}$)
δ':	thickness of viscous sublayer (cm)
τ:	shear stress ($dyne\,cm^{-2}$)

Formulae:
$Fr = U/(gD)^{0.5}$, $Re = UD/\nu$, $Re^*_1 = U^*_1 k/\nu$, $Re^*_2 = U^*_2 k/\nu$, $U^*_1 = (\tau/\rho)^{0.5}$, $U^*_2 = U/5.75.\log_{10}(12\,D/k)$, $\delta'_1 = 11.5\nu/U^*_1$, $\delta'_2 = 11.5\nu/U^*_2$, $\tau = gSD$.

adaptations are directed towards existence within this boundary layer (Hynes 1970), recent investigations have determined that very few benthic organisms are sufficiently streamlined to take advantage of this boundary layer (see discussions below and Statzner 1987a). Indeed, a true boundary layer rarely exists under natural flow conditions because of the complex profile of the substrate. Thus, aquatic organisms are restricted to those combinations of velocity, depth and substrate which allow morphological and/or behavioural resistance to flow to be exceeded by energetic gains from foraging in these areas. That is, conditions such as shear stress and the thickness of the slower-moving laminations of water which constitute the viscous sublayer are the major determinants of the distribution of most benthic organisms within a given stream reach. Davis and Barmuta (1989) classify near-bed flows as hydraulically smooth or hydraulically rough – chaotic, wake interference, isolated roughness or skimming flows. Each of these flow types is associated with recognizable assemblages of benthic biota. Hydraulic fitness of swimming organisms, and by implication, fish community structure, is also affected by the distribution of turbulent zones (Scarnecchia 1988).

Any hydrological change which leads to increases in shear velocities or shear stresses or reduction in the thickness of the viscous sublayer will reduce the availability of adequate microhabitats to some species while increasing it for others and, presumably, alter the abundances of individuals in the community. For example, Ciborowski and Craig (1989) found that increases in velocity resulted in increased drift and significant changes in aggregations of larval blackflies. Indeed, the position of the nearest upstream organisms influenced the hydraulic conditions of the downstream individual and its ability to feed. In turn, community functioning and food web dynamics may be changed to the extent that some species will be locally eliminated from the community.

Near-bed conditions are further changed with changes in substrate composition or distribution since increased hydraulic roughness increases the rate of sediment deposition (Harvey & Watson 1986). Increased sedimentation rates have been shown to significantly alter the composition of benthic communities as interstitial microhabitats are eliminated (increased predation success; Brusven & Rose 1981) and scour or deposition eliminates primary production (Brusven & Prather 1974; Chutter 1969).

3.5 INTERACTIONS OF FLOW AND MORPHOLOGY

Ambühl (1959) indicated that most benthic invertebrates were sufficiently streamlined to avoid the shear conditions along the substrate and could, therefore, forage across the substratum with relative impunity. This notion was generally accepted by most lotic ecologists and contributed to the search for other land–water interactions which might serve as a major template to the distribution of aquatic biota (both within reaches and along the length of river ecosystems). However, in the past decade a number of scientists have examined the interactions of complex hydraulics and the responses of biota to changes in these conditions. Gore (1983) demonstrated a significant correlation between shell length (as an indicator of exposure to sheer velocities) and velocity distributions among pleurocerid snails. Using laser doppler anemometry, Statzner and Holm (1982, 1989) and Statzner (1987a, 1988) demonstrated that benthic invertebrates do not exhibit the streamlining conditions which Ambühl reported. Indeed, Reynolds numbers (the interaction between frontal area exposed to current, the length of the body, and the velocity and viscosity of the water) were sufficiently high to predict that in order to compensate for the forces of pressure drag and friction drag, and to maintain an energetic gain from foraging activity along the substrate, most benthic organisms were quite limited in the range of conditions in which they could exist. High drag coefficients may be of some advantage to organisms swimming in still waters, where the edges of the prothorax or elytra may aid predators in making fast turns and stops (Nachtigall & Bilo 1965). However, resistance to flow in lotic ecosystems implies that body shape (a compromise between hemispherical and ovoid) and behavioural responses to changes in flow restrict the ability of lotic organisms to move freely within and between stream reaches by limiting energy gains to occupancy of a narrow range of flow microhabitats. Indeed, because of the changes in surface area: volume ratios during the growth of an individual, many species (mayflies, for example) must move to higher-velocity microhabitats as they grow in order to satisfy oxygen requirements (Kovalak 1978). Resistance to entry into the drift increases as much as 1000-fold during the growth of benthic invertebrates (Waringer 1989) and is a function of increase in size and alteration in cross-section profile presented to the direction of flow (Dussart 1987). Gore (1983) suggested that exposure of shells to higher-velocity lamina may result in the erosion or decollation of the top spires of lotic pleurocerid snails. This decollation can result in further bacterial and fungal deterioration of the shells (Burch 1982). Thus, increases in average velocity in a channel might be expected to differentially affect cohorts of the same benthic population. Statzner *et al* (1988) used these hydraulic parameters to demonstrate that the density and distribution of specific cohorts of benthic organisms could be predicted if complex hydraulic conditions such as Reynolds velocity, laminar sublayer thickness, and even Froude number were known at the point of sampling.

The effect of flow is not limited to the interactions of complex hydraulics at the level of the substrate upon benthic invertebrates. Brewer and Parker (1990) determined that the upslope distribution of macrophyte species was determined by the break force upon the stems. The tensile strength was different for all species and the force required to break the stem varied as a power function of the cross-sectional area. Thus, thick-stemmed species are the only species occurring in high-gradient streams, while lowland rivers and large-order rivers tend to present more flow microhabitats for a variety of macrophyte species. Similarly, Scarnecchia (1988) used the fineness ratio (length/maximum diameter) of Webb (1975) to demonstrate that homogeneity of flows in a channelized river tended to reduce the diversity of fish communities. Where an *Fr* value of 4.5 gives minimum drag for fish, stream reaches of high uniform flows are characterized by fish with this more highly streamlined characteristic while species with *Fr* values less than 4 are virtually eliminated from these areas.

3.6 EFFECTS OF FLOW ON BEHAVIOUR AND PHYSIOLOGY

As hydraulic conditions change within a stream channel, the effects upon stream biota are either direct mechanical damage and removal of the organisms from the preferred microhabitat or changes in the behaviour and physiological conditions of the biota. These changes in behaviour and physiology may mean a net loss or reduction in energy for growth and reproduction or the inability to breed or forage in appropriate habitats.

Although Pfeifer and McDiffett (1975) indicated the potential for large increases in nutrient uptake with increased velocity, a minimum current of about $5\,cm\,s^{-1}$ across the surface of periphyton beds has been shown to stimulate increased primary production in lotic ecosystems (Lock & John 1979). The influence of changes in current velocity extend to composition of periphyton communities themselves. Peterson (1986, 1987) found that reduced mean daily flows supported higher diatom biomass in partially exposed areas, with a higher diversity of taxa including unattached colonial species. Those species inhabiting slow current areas were less resistant to desiccation, probably due to slow nutrient renewal.

Among fish species, changes in hydraulic conditions may differentially affect certain cohorts or functional groups within the community. These changes are related to energetic balances tuned to body morphology (see above) and physiological abilities. Among fish, morphological adaptations which minimize exposure to shear and enhance preference for low velocities (and low oxygen consumption) can be predicted by an energy-cost hypothesis (Facey & Grossman 1992). Although it is well known that fish can rapidly adjust their swimming speeds and positions in streams with changes in current velocity (Jones 1968; Hanson & Li 1978), Trump and Leggett (1980) were able to demonstrate that the most efficient swimming conditions (particularly during migration) were conditions in which current velocities can be rapidly detected and are not rapidly varying. Under suboptimal swimming conditions, energetic losses may lead to decreased spawning success or condition factor. In the same manner Gibson (1983) has noted that the change from aggressive to schooling behaviour for migration is not only triggered by changes in temperature in the stream ecosystem but is also tied to changes in buoyancy associated with changes in current velocity.

Among benthic species, shifts in behaviour and/or physiological demand according to hydraulic changes are also common. At minimum, many species of lotic insects have been shown to increase respiratory activity with decreases in current velocity (Feldmeth 1970). Kovalak (1976) found larvae of the trichopteran, *Glossosoma*, to occupy up-stream faces of rocks during times of higher temperatures and lower dissolved oxygen concentrations. Similarly, mayfly nymphs change position on the surface of substrate particles in relation to current velocity and dissolved oxygen (DO) concentrations. As DO decreases, nymphs take positions in higher current velocities, usually exposing themselves to greater shear and mechanical removal (as drift) or greater predation (Wiley & Kohler 1980). At least some species of crayfish exhibit the same apparent high-risk activity (exposure to predation) in order to oxygenate egg clusters (Gore & Bryant 1990). With increases in stream velocity, many crayfish species can alter body posture to counteract the effects of drag, maintain position, and take advantage of increased oxygen in these areas (Maude & Williams 1983).

Drift

The phenomenon of downstream drift of benthic species has been attributed to a variety of factors including negative phototactic responses, escape from introduced toxic substances or increased sediment load, escape from predation, interference competition as density increases, and as a response to a deteriorating microhabitat. An excellent review of the phenomenon of drift is provided by Brittain and Eikeland (1988). Although the influence of some of the factors is controversial, there is little doubt that changes in velocity pattern result in increased drift from areas in which the hydraulic habitat is no longer suitable. Varying discharges, particularly as a result of hydropower operation, are known to initiate catastrophic drift among stream invertebrates and larval fish (Cushman 1985). However, it is not certain if the rate of change of discharge is the primary influencing factor; releases which simulate large floods do not seem to initiate greater drift densities than gradual increases in flow (Irvine & Henriques 1984). However, it does appear that initial changes in flow initiate more drift than subsequent similar flow changes (Irvine 1985), which seems to indicate that areas with unpredictable flow patterns are continually occupied by invertebrates at lower densities than the hydraulic habitat might support under more predictable or stable flow conditions (Ciborowski & Clifford 1983). The addition of sediment to increased flows further increases drift density as scour tends to remove some individuals maintaining position in adequate flow habitats (Ciborowski *et al* 1977). The importance of the hydraulic habitat to the success of benthic species is underlined by the observations that drift has also been seen to increase during conditions of decreasing flow. Catastrophic drift is usually initiated at some discharge greater than conditions in which the majority of the substrate is dewatered indicating that lower velocity tolerances or preferences can also be exceeded (Gore 1977; Corrarino & Brusven 1983). In either case, increased or decreased discharges which promote hydraulic conditions exceeding the preferences of the benthic species (and/or increase sediment loads to the flow; see Culp *et al* 1986) can result in a net loss of benthic species from a given stream reach.

3.7 MANAGEMENT/MITIGATION IMPLICATIONS

It should be apparent that maintenance of a functioning stream ecosystem is, in part or whole, a function of the hydraulic heterogeneity of the river channel. Statzner *et al* (1988) and Heede and Rinne (1990) have demonstrated the close relationship between stream biota and the hydraulic and hydrological conditions which define the physical habitat of most lotic biota. Stream managers should, then, focus attention on the development of techniques and release schemes which provide for currently existing hydraulic conditions within the channel or, in the case of degraded systems, provide new conditions which increase hydraulic heterogeneity to allow the recolonization of the greatest numbers and kinds of biota. Obviously, this does not imply that high diversity will yield a stable stream community, but the greatest variety of physical conditions should allow the fastest rate of colonization and recovery from disturbance as at least *some* habitat will be available to local species as colonizers. A period of dynamic adjustment between competitors, predators and prey will likely adjust the community structure to the proportions and conditions of available hydraulic habitat.

Two management techniques are currently available to ensure maintenance of adequate hydraulic habitats. One method, instream flow prediction and managed discharge regimes, is an *a priori* approach; the other method, channel engineering and restoration, provides an *a posteriori* way of increasing or altering the available hydraulic habitat.

Instream flow techniques encompass a body of field measurements and computer techniques which combine hydrological and hydraulic conditions in the channel with biological criteria for those same parameters in order to predict gain or loss of habitat under new flow regimes or to formulate recommendations to ensure that adequate minimum flows be maintained for retention of the target organisms of management interest. These techniques are described in this volume as well as being summarized in great detail by Wesche and Rechard (1980) and Gustard *et al* (1987). Gore and Nestler (1988) point out that the body of instream flow models currently used for flow recommendations in North America and Europe should be viewed as management models which allow the assessment of economic and biological benefits when designing release schedules rather than an exacting model to predict the ecological consequences of changes in flows. Despite the lack of ecological prediction, these techniques provide a useful tool to stream managers to ascertain which flows will substantially alter habitat availability and provide a 'window' of parameters in which hydrological conditions for maintaining a reasonably stable downstream ecosystem can be produced.

Channel engineering is the construction, after channel disturbance such as diversion and/or channelization, of artificial meanders, embankments and obstacles to increase hydraulic heterogeneity in the channel by reconstructing pool/riffle sequences and providing a high diversity of substrate particles. The construction of these structures is examined in detail in other chapters in this volume as well as in texts by Gore (1985), Brookes (1988) and Gore and Petts (1989). Gore (1985) has suggested that combining instream flow techniques with channel engineering may improve the design of restoration projects and allow a cost/benefit analysis prior to construction of various structures such as weirs and wings. Traditionally, the use of habitat restoration features has been restricted to low-order, high-gradient streams dominated by salmonid fisheries or forage fish. However, recent instream flow analyses suggest that reregulation weirs in large rivers can provide increases in hydraulic heterogeneity and increase habitat availability for target fish (Martin *et al* 1985). Indeed, Gore *et al* (1992b) provide a list of techniques and restoration structures aimed at increasing top width, restoring internal meander, and providing dampening of peaking hydropower waves as methods to establish predictably available habitat for fish and benthos in the receiving streams of large-order rivers.

In both cases, the result is some amount of control of the hydrological and hydraulic conditions which influence the availability of physical habitats for lotic biota and a means to conserve the integrity of lotic ecosystems.

REFERENCES

Ambühl H. (1959) Die Bedeutung der Strömung als ökologischer Faktor. *Schweizerische Zeitschrift für Hydrobiologie* **21**: 133–264. [3.5]

Armitage PD. (1987) The classification of tailwater sites receiving residual flows from upland reservoirs in Great Britain, using macroinvertebrate data. In: Craig JF, Kemper JB (eds) *Regulated Streams. Advances in Ecology*, pp. 131–44. Plenum Press, New York. [3.3]

Armitage PD, Petts GE. (1992) Biotic score and prediction to assess the effects of water abstractions on river macroinvertebrates for conservation purposes. *Aquatic Conservation* **2**: 1–17. [3.3]

Bickerton M, Petts G, Armitage P, Castella E. (1993) Assessing the ecological effects of groundwater abstraction on chalk streams: three examples from eastern England. *Regulated Rivers* **2**: 121–34. [3.3]

Bisson PA, Bilby RE, Bryant MD, Andrew CD, Grette GB, House RA, Murphy ML, Koski KV, Sedell JR. (1987) Large woody debris in forested streams in the Pacific Northwest: past, present, and future. In: Salo EO, Cundy TW (eds) *Proceedings of the Symposium on Streamside Management: Forestry and Fishery Interactions*, pp. 143–231. University of Washington Press, Seattle, WA. [3.3]

Boulton AJ, Lloyd LN. (1992) Flooding frequency and invertebrate emergence from dry floodplain sediments on the River Murray, Australia. *Regulated Rivers* **7**: 137–51. [3.3]

Brewer CA, Parker M. (1990) Adaptations of macrophytes to life in moving water: upslope limits and mechanical properties of stems. *Hydrobiologia* **194**: 133–42. [3.5]

Brittain JE, Eikeland TJ. (1988) Invertebrate drift – a review. *Hydrobiologia* **166**: 77–93. [3.6]

Brookes A. (1988) *Channelized Rivers. Perspectives for Environmental Management*. John Wiley & Sons, Chichester. [3.7]

Brusven MA, Prather KV. (1974) Influence of stream sediments on distribution of macrobenthos. *Journal of the Entomological Society of British Columbia* **71**: 24–32. [3.4]

Brusven MA, Rose ST. (1981) Influence of substrate composition and suspended sediment on insect predation by the torrent sculpin, *Cottus rhotheus*. *Canadian Journal of Fisheries and Aquatic Sciences* **38**: 1444–8. [3.4]

Burch JB. (1982) *Freshwater Snails (Mollusca: Gastropoda) of North America*. EPA-600/3-82-026, US Environmental Protection Agency, Cincinnati, Ohio. [3.5]

Campbell RF, Neuner JH. (1985) Seasonal and diurnal shifts in habitat utilized by resident rainbow trout in western Washington Cascade Mountain streams. In: Olson FW, White RG, Hamre RH (eds) *Symposium on Small Hydropower and Fisheries*, pp. 39–48. American Fisheries Society, Bethesda, MD. [3.3]

Canton SP, Ward JV. (1981) Emergence of Trichoptera from Trout Creek, Colorado, USA. In: Moretti GP (ed) *Proceedings of the Third International Symposium on Trichoptera*, pp. 39–45. W. Junk, The Hague, Netherlands. [3.2]

Carlson JY, Andrus CW, Froehlich HA. (1990) Woody debris, channel features, and macroinvertebrates of streams with logged and undisturbed riparian timber in northeastern Oregon, USA. *Canadian Journal of Fisheries and Aquatic Sciences* **47**: 1103–11. [3.3]

Carlsson M, Nilsson LM, Svensson B, Ulfstrand S, Wotton RS. (1977) Lacustrine seston and other factors influencing the blackflies (Diptera: Simuliidae) inhabiting lake outlets in Swedish Lapland. *Oikos* **29**: 229–38. [3.3]

Chisholm IM, Hubert WA. (1986) Influence of stream gradient on standing stock of brook trout in the Snowy Range, Wyoming. *Northwest Science* **60**: 137–9. [3.2]

Chutter FM. (1969) The effects of silt and sand on the invertebrate fauna of streams and rivers. *Hydrobiologia* **34**: 57–76. [3.4]

Ciborowski JJH, Clifford HF. (1983) Life histories, microdistribution and drift of two mayfly (Ephemeroptera) species in the Pembina River, Alberta, Canada. *Holarctic Ecology* **6**: 3–10. [3.6]

Ciborowski JJH, Craig DA. (1989) Factors influencing dispersion of larval black flies (Diptera: Simuliidae): effects of current velocity and food concentration. *Canadian Journal of Fisheries and Aquatic Sciences* **46**: 1329–41. [3.4]

Ciborowski JJH, Pointing PJ, Corkum LD. (1977) The effect of current velocity and sediment on the drift of the mayfly *Ephemerella subvaria* McDunnough. *Freshwater Biology* **7**: 567–72. [3.6]

Corrarino CA, Brusven MA. (1983) The effects of reduced stream discharge on insect drift and stranding of near shore insects. *Freshwater Invertebrate Biology* **2**: 88–98. [3.3, 3.6]

Craig JF, Kemper JB. (eds) (1987) *Regulated Streams. Advances in Ecology*. Plenum Press, New York. [3.3]

Culp JM, Wrona FJ, Davies RW. (1986) Response of stream benthos and drift to fine sediment deposition versus transport. *Canadian Journal of Zoology* **64**: 1345–51. [3.6]

Cushing CE. (1964) Plankton and water chemistry in the Montreal River lake–stream system, Saskatchewan. *Ecology* **45**: 306–13. [3.3]

Cushman RM. (1985) Review of ecological effects of rapidly varying flows downstream from hydroelectric facilities. *North American Journal of Fisheries Management* **5**: 330–9. [3.6]

Davis JA. (1986) Boundary layers, flow microenvironments and stream Benthos. In: De Deckker P, Williams WD (eds) *Limnology in Australia*, pp. 293–312. CSIRO, Melbourne. [3.1]

Davis JA, Barmuta LA. (1989) An ecologically useful classification of mean and near-bed flows in streams and rivers. *Freshwater Biology* **21**: 271–82. [3.4]

DeJalon DG, Montes C, Barcelo E, Casado C, Menes F. (1988) Effects of hydroelectric scheme on fluvial ecosystems within the Spanish Pyrenees. *Regulated Rivers* **2**: 479–91. [3.3]

Dingman SL. (1984) *Fluvial Hydrology*. W.H. Freeman, San Francisco. [3.1]

Dudley TL, Anderson HH. (1987) The biology and life

cycles of *Lipsothrix* spp. (Diptera: Tipulidae) inhabiting wood in western Oregon streams. *Freshwater Biology* **17**: 437–51. [3.2]

Dussart GBJ. (1987) Effects of water flow on the detachment of some aquatic pulmonate gastropods. *American Malacological Bulletin* **5**: 65–72. [3.5]

Evans JW, Noble BL. (1979) The longitudinal distribution of fishes in an east Texas stream. *American Midland Naturalist* **101**: 333–43. [3.2]

Facey DE, Grossman GD. (1992) The relationship between water velocity, energetic costs, and microhabitat in four North American stream fishes. *Hydrobiologia* **239**: 1–6. [3.6]

Feldmeth CR. (1970) The respiratory energetics of two species of stream caddis fly larvae in relation to water flow. *Comparative Biochemistry and Physiology* **32**: 193–202. [3.6]

Gaschignard O. (1984) Impact d'un crue sur les macroinvertébrés benthiques d'un bras du Rhône. *Verhandlungen Internationalen Vereinigung für theoretische und angewandte Limnologie* **22**: 1997–2001. [3.3]

Gersabeck EF Jr, Merritt RW. (1979) The effect of physical factors on the colonization and relocation behavior of immature black flies (Diptera: Simuliidae). *Environmental Entomology* **8**: 34–9. [3.2]

Gibson RJ. (1983) Water velocity as a factor in the change from aggressive to schooling behavior and subsequent migration of Atlantic salmon smolt (*Salmo salar*). *Naturaliste Canadien* **110**: 143–8. [3.6]

Gore JA. (1977) Reservoir manipulations and benthic macroinvertebrates in a prairie river. *Hydrobiologia* **55**: 113–23. [3.3, 3.6]

Gore JA. (1983) Considerations of size related flow preferences among macroinvertebrates used in instream flow studies. In: Shuval HI (ed) *Developments in Ecology and Environmental Quality*, vol. II, pp. 389–98. Balaban Int. Publ., Jerusalem. [3.5]

Gore JA. (ed) (1985) *The Restoration of Rivers and Streams*. Butterworth, Boston. [3.7]

Gore JA, Bryant RM Jr. (1990) Temporal shifts in physical habitat of the crayfish, *Orconectes neglectus* (Faxon). *Hydrobiologia* **199**: 131–42. [3.2, 3.6]

Gore JA, Milner AM. (1990) Island biogeographic theory: can it be used to predict lotic recovery rates? *Environmental Management* **14**: 737–53. [3.3]

Gore JA, Nestler JM. (1988) Instream flow studies in perspective. *Regulated Rivers* **2**: 93–101. [3.7]

Gore JA, Pennington W. (1988) Changes in larval chironomid habitat with distance from peaking hydropower operations. Annual Meeting of the North American Benthological Society, Tuscaloosa, AL. [3.3]

Gore JA, Petts GE. (eds) (1989) *Alternatives in Regulated River Management*. CRC Press, Boca Raton, FL. [3.7]

Gore JA, Statzner B, Resh VH. (1986) Physical habitat characteristics and microdistribution of final instars of *Hydropsyche angustipennis* (Curtis). *Fifth International Symposium on Trichoptera, Lyon, France*. W. Junk, The Hague, Netherlands. [3.2]

Gore JA, Nestler JM, Layzer JB. (1989) Instream flow predictions and management options for biota affected by peaking-power hydroelectric operations. *Regulated Rivers* **3**: 35–48. [3.3]

Gore JA, Nestler JM, Layzer JB. (1990) Habitat factors in tailwaters with emphasis on peaking hydropower. Technical Report EL-90-2, US Army Corps of Engineers, Waterways Experiment Station, Vicksburg, MS. [3.3]

Gore JA, Crawford DJ, Nestler JM. (1992a) Habitat restoration and enhancement techniques associated with large order, warmwater rivers influenced by peaking hydropower. Technical Report, US Army Engineers, Waterways Experiment Station, Vicksburg, MS (in press).

Gore JA, Layzer JB, Russell IA. (1992b) Non-traditional applications of instream flow techniques for conserving habitat of biota in the Sabie River of southern Africa. In: Boon PJ, Petts GE, Calow P (eds) *River Conservation and Management*, pp. 161–77. John Wiley, New York. [3.2, 3.7]

Gore JA, Niemela S, Resh VH, Statzner B. (1993) Near-substrate hydraulic conditions under artificial floods from peaking hydropower operation: disturbance intensity and duration. *Regulated Rivers* (in press) [3.3]

Grzybkowska M, Hejduk J, Zieliński P. (1990) Seasonal dynamics and production of Chironomidae in a large lowland river upstream and downstream from a new reservoir in Central Poland. *Archiv für Hydrobiologie* **119**: 439–55. [3.3]

Gustard A, Cole G, Marshall D, Bayliss A. (1987) A study of compensation flows in the U.K. Report No. 99, Institute of Hydrology, Oxfordshire, UK. [3.7]

Hanson CH, Li HW. (1978) A research program to examine fish behavior in response to hydraulic flow fields – development of biological design criteria for proposed water diversions. Completion Report, Project C-7679, OWRT, US Department of Interior. [36]

Harvey MD, Watson CC. (1986) Fluvial processes and morphological thresholds in incised channel restoration. *Water Resources Bulletin* **22**: 359–68. [3.4]

Heede BH, Rinne JN. (1990) Hydrodynamic and fluvial morphologic processes: implications for fisheries management and research. *North American Journal of Fisheries Management* **10**: 249–68. [3.1, 3.7]

Heger H, Moog O. (1986) Der Einflu von Wasserableitungen auf das Benthos des Landeckbaches in Osttirol (Österreich). *Berichte Naturwissenschaftlich-Medizinischen Vereins in Innsbruck* **73**: 199–214. [3.3]

Hill LG, Matthews WJ. (1980) Temperature selection by

the darters *Etheostoma spectabile* and *Etheostoma radiosum* (Pisces: Percidae). *American Midland Naturalist* **104**: 412–15. [3.2]

Humpesch U. (1978) Preliminary notes on the effect of temperature and light-condition on the time of hatching in some Heptageniidae (Ephemeroptera). *Verhandlungen der Internationalen Vereinigung für theoretische und angewandte Limnologie* **20**: 2605–11. [3.2]

Hynes HBN. (1970) *The Ecology of Running Waters*. University of Toronto Press, Toronto. [3.2, 3.4]

Ingersoll CG, Claussen DJ. (1984) Temperature selection and critical thermal maxima of the fantial darter, *Etheostoma flabellare*, and johnny darter, *E. nigrum*, related to habitat and season. *Environmental Biology of Fishes* **11**: 131–8. [3.2]

Irvine JR. (1985) Effects of successive flow perturbations on stream invertebrates. *Canadian Journal of Fisheries and Aquatic Sciences* **42**: 1922–7. [3.3, 3.6]

Irvine JR, Henriques PR. (1984) A preliminary investigation on effects of fluctuating flows on invertebrates of the Hawea River, a large regulated river in New Zealand. *New Zealand Journal of Marine and Freshwater Research* **18**: 283–90. [3.6]

Jäger P, Kawecka B, Margreiter-Kownacka M. (1985) Zur Methodik der Untersuchungen der Auswirkungen des Wasserentzuges in Restwasserstrecken auf die Benthosbiozönosen (Fallbeispiel: Radurschlbach): *Österreichische Wasserwirtschaft* (Jahrgang 37) 7/8: 190–202. [3.3]

Jenkins RE, Freeman CA. (1972) Longitudinal distribution and habitat of the fishes of Mason Creek, an upper Roanoke River drainage tributary, Virginia. *Virginia Journal of Science* **23**: 194–202. [3.2]

Jones FRH. (1968) *Fish Migration*. St. Martins, New York. [3.6]

King JM. (1981) The distribution of invertebrate communities in a small South African river. *Hydrobiologia* **83**: 43–65. [3.2]

Kovalak WP. (1976) Seasonal and diel changes in the positioning of *Glossosoma nigrior* Banks (Trichoptera: Glossosomatidae) on artificial substrates. *Canadian Journal of Zoology* **54**: 1585–94. [3.6]

Kovalak WP. (1978) Relationships between size of stream insects and current velocity. *Canadian Journal of Zoology* **56**: 178–86. [3.5]

Lake PS, Barmuta LA, Boulton AJ, Campbell IC, St. Clair RM. (1986) Australian streams and Northern Hemisphere stream ecology: comparisons and problems. *Proceedings of the Ecological Society of Australia* **14**: 61–82. [3.2]

Layzer JB, Nehus TJ, Pennington W, Gore JA, Nestler JM. (1989) Seasonal variation in the composition of drift below a peaking hydroelectric project. *Regulated Rivers* **3**: 29–34. [3.3]

Lillehammer A, Saltveit SJ. (eds) (1984) *Regulated Rivers*. Universitetsforlaget AS, Oslo. [3.3]

Lock MA, John PH. (1979) The effect of flow patterns on uptake of phosphorus by river periphyton. *Limnology and Oceanography* **24**: 376–83. [3.6]

Mackay RJ. (1979) Life history patterns of some species of *Hydropsyche* (Trichoptera: Hydropsychidae) in southern Ontario. *Canadian Journal of Zoology* **57**: 963–75. [3.3]

Martin JL, Curtis LT, Nestler JM. (1985) Effects of flow alterations on trout habitat in the Cumberland River below Wolf Creek Dam. Misc. Paper E-86-11. U.S. Army Engineer, Waterways Experiment Station, Vicksburg, MS. [3.7]

Matter WJ, Hudson PL, Saul GE. (1983) Invertebrate drift and particulate organic material transport in the Savannah River below Lake Hartwell during a peak power generation cycle. In: Fontaine TD, Bartell SM (eds) *Dynamics of Lotic Ecosystems*, pp. 357–369. Ann Arbor Press, Michigan. [3.3]

Maude SH, Williams DD. (1983) Behavior of crayfish in water currents: hydrodynamics of eight species with reference to their distribution patterns in southern Ontario. *Canadian Journal of Fisheries and Aquatic Sciences* **40**: 68–77. [3.6]

Morin A, Peters RH. (1988) Effect of microhabitat features, seston quality, and periphyton abundance of over-wintering black fly larvae in southern Québec. *Limnology and Oceanography* **33**: 431–46. [3.3]

Morin A, Back C, Chalifour A, Boisvert J, Peters RH. (1988) Effect of black fly ingestion and assimilation on seston transport in a Quebec lake outlet. *Canadian Journal of Fisheries and Aquatic Sciences* **45**: 705–14. [3.3]

Murphy ML, Hall JD. (1981) Varied effects of clear-cut logging on predators and their habitat in small streams of the Cascade Mountains, Oregon. *Canadian Journal of Fisheries and Aquatic Sciences* **38**: 137–45. [3.3]

Nachtigall W, Bilo D. (1965) Die Strömungsmechanik des *Dytiscus*-Rumpfes. *Zeitschrift fur Vergleichende Physiologic* **50**: 371–401. [3.5]

Naiman RJ, Décamps H, Pastor J, Johnston CA. (1988) The potential importance of boundaries to fluvial ecosystems. *Journal of the North American Benthological Society* 7: 289–306. [3.2]

Newbury RW. (1984) Hydrologic determinants of aquatic insect habitats. In: Resh VH, Rosenberg DM (eds) *The Ecology of Aquatic Insects*, pp. 323–75. Praeger, New York. [3.1]

Nowell ARM, Jumars PA. (1984) Flow environments of aquatic benthos. *Annual Review of Ecology and Systematics* **15**: 303–28. [3.4]

Oswood MW. (1979) Abundance patterns of filter-feeding caddisflies (Trichoptera: Hydropsychidae) and seston in a Montana (U.S.A.) lake outlet. *Hydrobiologia* **63**: 177–83. [3.3]

Perry SA, Perry WB. (1986) Effects of experimental flow regulation on invertebrate drift and stranding in the Flathead and Kootenai Rivers, Montana, USA. *Hydrobiologia* **134**: 171–82. [3.3]

Perry SA, Sheldon AL. (1986) Effects of exported seston on aquatic insect faunal similarity and species richness in lake outlet streams in Montana, USA. *Hydrobiologia* **137**: 65–77. [3.3]

Peterson CG. (1986) Effects of discharge reduction on diatom colonization below a large hydroelectric dam. *Journal of the North American Benthological Society* **5**: 278–89. [3.6]

Peterson CG. (1987) Influences of flow regime on development and desiccation response of lotic diatom communities. *Ecology* **68**: 946–54. [3.6]

Petts GE. (1984) *Impounded Rivers. Perspectives for Ecological Management*. John Wiley & Sons, Chichester. [3.3]

Pfeifer RF, McDiffett WF. (1975) Some factors affecting primary productivity of stream riffle communities. *Archive für Hydrobiologie* **75**: 306–17. [3.6]

Platania SP. (1991) Fishes of the Rio Chama and upper Rio Grande, New Mexico, with preliminary comments on their longitudinal distribution. *Southwestern Naturalist* **36**: 186–93. [3.2]

Platts WS. (1979) Relationships among stream order, fish populations, and aquatic geomorphology in an Idaho river drainage. *Fisheries* **4**: 5–9. [3.2]

Richardson JS, Mackay RJ. (1991) Lake outlets and the distribution of filter feeders: an assessment of hypotheses. *Oikos* **62**: 370–80. [3.3]

Robinson CT, Minshall GW. (1990) Longitudinal development of macroinvertebrate communities below oligotrophic lake outlets. *Great Basin Naturalist* **50**: 303–11. [3.3]

Robinson CT, Reed LM, Minshall GW. (1992) Influence of flow regime on life history, production, and genetic structure of *Baetis tricaudatus* (Ephemeroptera) and *Hesperoperla pacifica* (Plecoptera). *Journal of the North American Benthological Society* **11**: 278–89. [3.1]

Scarnecchia DL. (1988) The importance of streamlining in influencing fish community structure in channelized and unchannelized reaches of a prairie stream. *Regulated Rivers* **5**: 155–66. [3.4, 3.5]

Sheldon AL. (1968) Species diversity and longitudinal succession in stream fishes. *Ecology* **49**: 193–8. [3.2]

Sheldon AL, Oswood MW. (1977) Blackfly (Diptera: Simuliidae) abundance in a lake outlet: test of a predictive model. *Hydrobiologia* **56**: 113–20. [3.3]

Simons DB. (1979) Effects of stream regulation on channel morphology. In: Ward JV, Stanford JA (eds) *The Ecology of Regulated Streams*, pp. 95–111. Plenum Press, New York. [3.3]

Smith SD, Wellington AB, Nachlinger JL, Fox CA. (1991) Functional responses of riparian vegetation to streamflow diversion in the eastern Sierra Nevada. *Ecological Applications* **1**: 89–97. [3.3]

Stanford JA, Ward JV. (1984) The effects of regulation on the limnology of the Gunnison River: A North American case history. In: Lillehammer A, Saltveit SJ (eds) *Regulated Rivers*, pp. 467–80. Universitetsforlaget AS, Oslo. [3.2]

Statzner B. (1987a) Ökologische Bedeutung der sohlennahen Strömungsgeschwindigkeit für benthische Wirbellose in Flie gewässern. Habil. Thesis, University of Karlsruhe. [3.4, 3.5]

Statzner (1987b) Characteristics of lotic ecosystems and consequences for future research direction. In: Schulze E-D, Zwölfer H (eds) *Potentials and Limitations of Ecosystem Analysis*, pp. 365–90. Ecol. Stud. 61, Springer-Verlag, Berlin. [3.2]

Statzner B. (1988) Growth and Reynolds number of lotic macroinvertebrates: a problem for adaptation of shape to drag. *Oikos* **51**: 84–7. [3.5]

Statzner B, Higler B. (1986) Stream hydraulics as a major determinant of benthic invertebrate zonation patterns. *Freshwater Biology* **16**: 127–39. [3.1, 3.2]

Statzner B, Holm TF. (1982) Morphological adaptations of benthic invertebrates to stream flow – an old question studied by means of a new technique (Laser Doppler Anemometry). *Oecologia* **53**: 290–2. [3.5]

Statzner B, Holm TF. (1989) Morphological adaptation of shape to flow: Microcurrents around lotic macroinvertebrates with known Reynolds numbers at quasinatural flow conditions. *Oecologia* **78**: 145–57. [3.5]

Statzner B, Gore JA, Resh VH. (1988) Hydraulic stream ecology: observed patterns and potential applications. *Journal of the North American Benthological Society* **7**: 307–60. [3.1, 3.2, 3.4, 3.5, 3.7]

Stromberg JC, Patten DC. (1990) Riparian vegetation instream flow requirements: a case study from a diverted stream in the eastern Sierra Nevada, California, USA. *Environmental Management* **14**: 185–94. [3.3]

Sweeney BW, Schnack JA. (1977) Egg development, growth, and metabolism of *Sigara alternata* (Say) (Hemiptera: Corixidae) in fluctuating thermal environments. *Ecology* **58**: 265–77. [3.2]

Trump CL, Leggett WC. (1980) Optimum swimming speeds in fish: the problem of currents. *Canadian Journal of Fisheries and Aquatic Sciences* **37**: 1086–92. [3.6]

Valett HM, Stanford JA. (1987) Food quality and hydropsychid caddisfly density in a lake outlet stream in Glacier National Park, Montana, USA. *Canadian Journal of Fisheries and Aquatic Sciences* **44**: 77–82. [3.3]

Vannote RL, Minshall GW, Cummins KW, Sedell JR, Cushing CE. (1980) The river continuum concept. *Canadian Journal of Fisheries and Aquatic Sciences* **37**: 130–7. [3.2]

Wallace JB, Gurtz ME. (1986) Response of *Baetis* mayflies (Ephemeroptera) to catchment logging. *American Midland Naturalist* **115**: 25–41. [3.3]

Ward JV. (1992) *Aquatic Insect Ecology. 1. Biology and Habitat*. John Wiley, New York. [3.2]

Ward JV, Stanford JA. (1979) Ecological factors controlling stream zoobenthos with emphasis on thermal modification of regulated streams. In: Ward JV, Stanford JA (eds) *The Ecology of Regulated Streams*, pp. 35–56. Plenum Press, New York. [3.3]

Ward JV, Stanford JA. (1982) Thermal responses in the evolutionary ecology of aquatic insects. *Annual Review of Entomology* **27**: 97–117. [3.2, 3.3]

Ward JV, Stanford JA. (1983) The serial discontinuity concept of lotic ecosystems. In: Fontaine TD III, Bartell SM (eds) *Dynamics of Lotic Ecosystems*, pp. 29–42. Ann Arbor Press, Ann Arbor, Michigan. [3.2]

Ward JV, Voelz NJ. (1990) Gradient analysis of interstitial meiofauna along a longitudinal stream profile. *Stygologia* **5**: 93–9. [3.2]

Waringer JA. (1989) Resistance of cased caddis larva to accidental entry into the drift: the contribution of active and passive elements. *Freshwater Biology* **21**: 411–20. [3.5]

Webb PW. (1975) Hydrodynamics and energetics of fish populations. *Fisheries Research Board of Canada Bulletin* No. 190. [3.5]

Webster JR, Golladay SW, Benfield EF, D'Angelo DJ, Peters GT. (1990) Effects of forest disturbance on particulate organic matter budgets of small streams. *Journal of the North American Benthological Society* **9**: 120–40. [3.3]

Wesche TA, Rechard PA. (1980) A summary of instream flow methods for fisheries and related research needs. USDA, Forest Service, Eisenhower Consortium Bulletin No. 9. [3.7]

Wetzel RG. (1983) *Limnology*, 2nd edn. Saunders College Publ., Philadelphia. [3.3]

Wiley MJ, Kohler SL. (1980) Positioning changes of mayfly nymphs due to behavioral regulation of oxygen consumption. *Canadian Journal of Zoology* **58**: 618–22. [3.6]

Wilzbach MA, Cummins KW, Hall JD. (1986) Influence of habitat manipulations on interactions between cutthroat trout and invertebrate drift. *Ecology* **67**: 898–911. [3.3]

Wotton RS. (1982) Does the surface film of a lake provide a source of food for animals living in lake outlets? *Limnology and Oceanography* **27**: 959–60. [3.3]

4: River Channel Change

A. BROOKES

4.1 INTRODUCTION

The morphology of a river channel is formed by the movement of water and sediment in relation to the material locally available in the bed and banks (see Vol. 1, Chapter 6). Hydraulic and morphological variability through space and time determine the different habitats found both within a given river channel and also in the adjacent riparian and floodplain zones. Channels change in a variety of ways through the processes of erosion and deposition, processes that are affected by catchment characteristics. Surface runoff and sediment load may be expected to change in response to land use alterations such as deforestation, agriculture, grazing, urbanization and other influences. However, a change in the discharge and sediment load rarely produces an immediate response but instead initiates a change, or sequence of changes, which may extend over a long period of time. Geomorphological studies have concentrated mainly on understanding the effects of climatic change over the period 10 000–15 000 years (Gregory *et al* 1987) or human impacts over the last 50 years (e.g. Gregory 1977a). Relatively little attention has been given to changes over the historic timescale of 100–500 years, a period of intensifying human impact (Petts 1989a).

Channel changes resulting from the historic development of land and water resources have undoubtedly been extensive (see Chapter 1). Over the last 2000 years, and particularly during the last 300 years, human impacts have had an increasing influence on drainage basins and their watercourses. Table 4.1 lists some of the major human impacts on river channels, which continue today. At present, dams over 15 m high are being completed throughout the world at a rate of about 500 per year. By the year 2000 it is estimated that 60% of the total streamflow in the world will be regulated (Petts 1989a). Understanding the ways in which river channels have changed through historical time may be important for ecologically-sound management. It may be desirable for conservation purposes to restore a river channel to a condition existing prior to development. It may also be necessary for the purposes of environmental impact assessment to determine the magnitude and rate of change arising from a variety of impacts. Also, analyses of historical channel changes should be an integral part of the concept and design of alternative management strategies (see Part 5).

Alterations of channel size and shape, and substrate characteristics, have been shown to have major impacts on benthic invertebrates (e.g. Keefer & Maughan 1985; Petts & Greenwood 1985), fish (e.g. Milhous 1982) and aquatic macrophytes (e.g. Brookes 1987). Changes of river channel mobility and pattern significantly influence instream and floodplain habitats. This chapter outlines some of the main causes and types of channel change and the biological impacts which may arise from such perturbations. The final section identifies the problems that need to be addressed further if ecologically-sound management techniques are to be explored further and improved.

4.2 MORPHOLOGICAL VARIABILITY

It is important to appreciate that hydraulic factors such as depth, slope and velocity of flow directly

Table 4.1 Human influences on channel change

Indirect change	
Land use change:	
	Deforestation
	Afforestation
	Agricultural (e.g. conversion of grazing to arable)
	Urbanization
	Mining
Land drainage:	
	Agricultural drainage
	Surface water sewers
Direct change	
Regulation:	
	Impoundment of water
	Water diversions (e.g. for irrigation)
Channel management:	
	Gravel extraction
	Straightening
	Flood control
	Bank erosion protection
	Dredging

produce bank erosion and sediment transport. These factors tend to determine the cross-sectional shape, pool-riffle formation and meander shape of alluvial river channels. A natural channel is neither straight nor uniform. At the catchment scale, its size increases downstream as tributaries join and increase the flow; as width and depth increase downstream the slope decreases; the size of the sediment load on the bed decreases, typically with cobbles and gravels in the middle reaches and sand or silt farther downstream. Even within a single reach a natural channel is non-uniform: the bed undulates as a series of pools (topographically low areas) and riffles (high areas); the flow varies considerably over short distances and through time over a range from high to low discharge; the bed material of a pool is typically fine-grained sand, whilst a riffle has a concentration of larger rock sizes (gravels).

Pools and riffles result from the interaction of flowing water and mobile sediments. At very high discharges the flow tends to converge at pools, causing scour, whilst the flow is divergent at riffles, corresponding to deposition. In contrast, at low flows a pool is characterized by relatively deep, slow-moving water, and a riffle by fast-flow and a steep water-surface gradient. The cross-section is often symmetrical at a riffle. A meander bend typically has an area of deposition on the inside and an area of erosion at the outer bank. Deposition of material produces a point bar adjacent to a pool, forming an asymmetrical channel shape. The shapes of natural river channels may be complicated further by a variety of bar forms, either attached to the bank or formed mid-channel. A classification of more complex morphological types by Kellerhalls *et al* (1976), includes bars, channel patterns (or planforms) and islands which may form in braided river channels.

Although a natural river channel is formed and maintained by the flow it carries, it is never of sufficient size to contain all discharges. Larger floods invariably overflow the normal channel banks onto the adjacent floodplain, on average about once per year, and this process is important in building a floodplain through deposition of sediment. It is also natural for alluvial channels with sufficiently powerful streams to erode laterally at various rates across their floodplains.

This hydraulic and morphological variability determines different ecological habitats in various parts of a river channel and adjacent floodplain. At low discharges there are a variety of conditions for feeding, breeding and cover, ranging from slow, deep flow in pools to fast, shallow water on riffles. Sufficient water depth is maintained during dry periods to support aquatic life. Habitat diversity and fish diversity are positively associated (Wesche 1985).

Pools, riffles and bars are composed of different bedload materials which provide environments for a variety of benthic organisms. Substrate size, heterogeneity, frequency of turnover of sediment, and the rates of erosional or depositional processes are all thought to determine invertebrate diversity and abundance (Petts 1984). In particular the size of substrate has been related by numerous investigators to the standing crops of benthic invertebrates. According to Pennak and Van Gerpen (1947) benthic invertebrates decrease in number in the series rubble, bedrock, gravel and sand. A similar decrease in the series rubble, coarse gravel, sand, fine gravel and silt was reported by Kimble and Wesche (1975). Larger

Table 4.2 Key geomorphological factors influencing river ecology over different time-scales

Factor	Typical time-scale of change	Method used to determine change of river morphology	Knowledge of biological response over time	Knowledge of biological response using space as a substitute for time
Cross-sectional shape (e.g. bank collapse)	Hours, days, months, years (e.g. during extreme flood events)	Direct observation and measurement	Limited	Fair
Cross-sectional size	As above	As above	Limited	Fair
Pool and riffle sequence	As above	As above	Limited	Fair
Point bars	As above	As above	Limited	Some
Islands	As above	As above	Some	Fair
Substrate (including sedimentation)	As above	As above	Good	Good
Pattern (e.g. lateral migration)	100–50 years	Historical sources such as maps and documents	Limited	Fair
Pattern (e.g. braided to meandering)	250–15 000 years +	Sedimentary evidence; dating techniques (floral and faunal evidence; artifactual remains)	Not applicable	Some

substrate provides insects with a firm surface to hold on to and also provides some protection from the force of the current.

At higher discharges shelter areas are available as protection from higher water velocities. Overhanging banks cut on the outside of a bend adjacent to a pool may provide natural cover for fish. Overbank flooding and sedimentation are important processes affecting floodplain wetlands and wetland vegetation. Wetlands may be important for waterfowl, amphibians and reptiles and some mammals. Cutoff channels may support a number of breeding, feeding and nesting habitats.

Table 4.2 summarizes some key geomorphological factors of natural alluvial river channels which are important for biological and aesthetic diversity. Typical time-scales over which these factors have been observed to change are also shown. Whilst there are many studies which have examined the biological response of changed substrate, for example as a result of sedimentation, there is little knowledge of the impacts in changes of other factors. By contrast, a greater amount of knowledge of biological response is evident from the method of substituting space for time, which is sometimes referred to as the ergodic hypothesis (Table 4.2). Local changes within a single river are often inferred by comparison of biological populations between rivers or between 'natural' and 'changed' sections along individual rivers. The differences between the actual measured values and the expected ones form the basis for the estimation of the nature and amount of change arising from human activities such as land-use change, channelization or dam construction (but see Chapter 6).

4.3 PRINCIPLES OF CHANNEL EQUILIBRIUM AND RESPONSE

A river channel transports both water and sediment and the flow is typically non-uniform, unsteady and turbulent. Under conditions of equilibrium, channels tend to be morphologically stable, transporting the water and sediment load imposed from the catchment upstream without enlarging or aggrading. *It is necessary to understand that a channel which is described as stable may naturally erode its bed and banks.* This erosion may be a relatively slow process which may only be pronounced after a major flood discharge. As the channel migrates across the floodplain it maintains a stable average form. There are a whole range of channel types with different degrees of activity (see Vol. 1, Chapters 6 & 7).

Channels that maintain a stable average form are described by geomorphologists as graded or in equilibrium, whereas engineers usually refer to them as 'in regime' or stable. It is only if different bank materials are encountered during migration, or a new discharge regime is imposed, that the channel becomes unstable. The equilibrium concept implies that stable width, depth, slope (and probably planform) can be expressed as functions of the controlling variables: discharge, sediment supply and channel bed and bank materials (Fig. 4.1). Fluctuations of channel form about an equilibrium condition are termed autogenic changes. Where the sequence of water flow and sediment transported by rivers is changed, then alterations in processes operating within the river channel occur as a direct result. These adjustments are referred to as allogenic changes.

Although a set of determinate equations fully describing stream behaviour have not yet been developed, research has succeeded in defining some of the functional relationships between stream variables. Used with caution, such relationships can suggest the types of channel adjustment to be expected from changes in the upstream catchment or from change in one or more stream characteristics. Some of the most useful relationships which can be used in planning and design are listed in Table 4.3, together with examples of the types of impact which cause river channels to change. The following discussion concentrates on those changes induced by human impact, which has undoubtedly been the strongest influence over the past 5000 years, rather than the effects of climatic change.

The types of channel change which may occur range from the sedimentary bedforms (e.g. pools and riffles), through the cross-section of the river channel, the river channel planform and the drainage network (Gregory 1976). For each of these

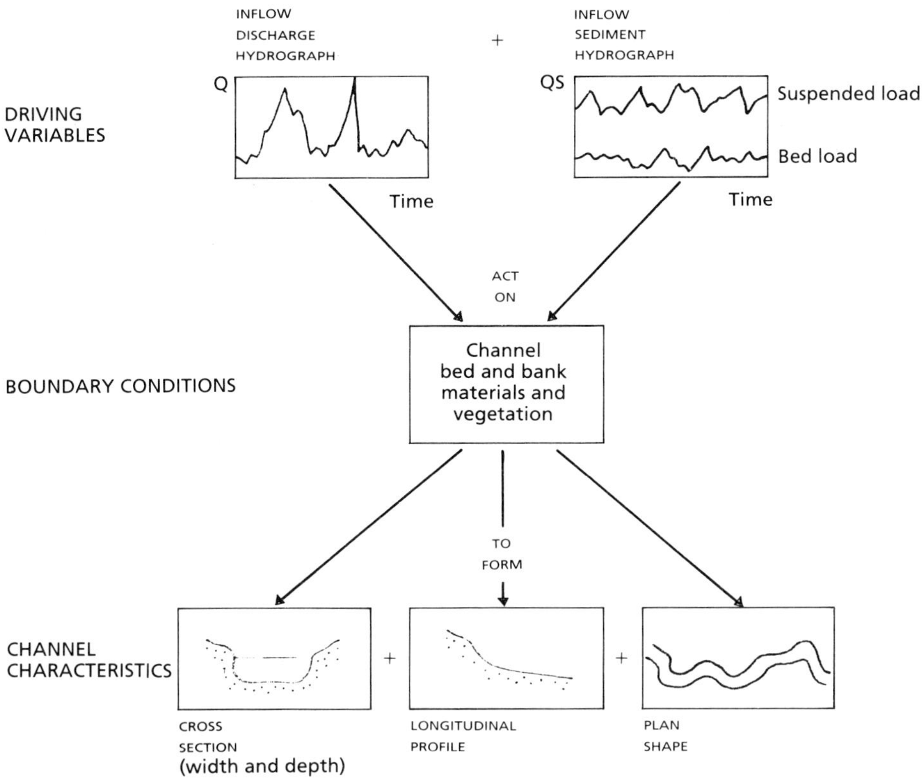

Fig. 4.1 Equilibrium concept for erodible channels.

scales there can be adjustments of size, shape and composition. Examples are given in Table 4.4.

4.4 LAND USE CHANGES AND SEDIMENT YIELD

Changes in the drainage basin may alter rainfall–runoff and sediment yield relationships and may have an indirect effect on channel characteristics. The most extensive changes affecting streams are land use changes attributable to agriculture, forestry, mining, grazing and urbanization. Deforestation for the purpose of grazing or cultivation began in Europe more than 3000 years ago. A major impact of deforestation is accelerated erosion on hillslopes, together with gully erosion and increased supply of sediment to the streams. Runoff characteristics are also affected by the conversion of forest to farming. Indeed, any change that eliminates or reduces vegetative cover is likely to increase sediment discharge proportionately more than the water discharge. Mining (Graf 1979), forest fires (White & Wells 1979) and construction activities (Wolman 1967) are all examples of this type of disturbance. The conversion of forest to farmland in Wisconsin in the USA began in about 1820, leading to much higher sediment loads, generally causing channel-bed aggradation and overbank deposition on floodplains to depths in excess of 1 m (Meade 1976). Channels have tended to become wider, shallower and less sinuous and much of the eroded sediment is still stored in the channel and floodplains. Conversely, afforestation, improved land management or conservation practices reduce soil erosion and have the opposite effect.

In the Southern Piedmont of the USA rapid expansion of intensive crop agriculture from about 1800 until 1920 increased runoff and sediment yield (Trimble 1974). Equation 8 (Table 4.3)

Table 4.3 Functional relationships (after Schumm 1969)

Equations		Examples of change
$Qs^{+}, Qw^{++} \approx S^{-}, d_{50}{}^{+}, D^{+}, W^{+}$	(1)	Long-term effect of urbanization. Increased frequency and magnitude of discharge. Channel erosion (increasing width and depth)
$Qs^{0}, Qw^{+} \approx S^{-}, d_{50}{}^{+}, D^{+}, W^{+}$	(2)	
$Qs^{-}, Qw^{+} \approx S^{-}, d_{50}{}^{+}, D^{+}, W^{-}$ $Qs^{--}, Qw^{-} \approx S^{-}, d_{50}{}^{+}, D^{+}, W^{\pm}$	(3) (4)	Intensification of vegetation cover through afforestation and improved land management reduces sediment loads
$Qs^{-}, Qw^{+} \approx S^{-}, d_{50}{}^{+}, D^{+}, W^{\pm}$	(5)	Diversion of water to a river
$Qs^{+}, Qw^{+} \approx S^{\pm}, d_{50}{}^{\pm}, D^{\pm}, W^{+}$ $Qs^{-}, Qw^{-} \approx S^{\pm}, d_{50}{}^{\pm}, D^{\pm}, W^{-}$	(6) (7)	Commensurate changes of water and sediment discharge with unpredictable changes of slope, flow depth and bed material
$Qs^{++}, Qw^{+} \approx S^{+}, d_{50}{}^{-}, D^{-}, W^{+}$	(8)	Land use change from forest to crop production. Sediment discharge increasing more rapidly than water discharge. Bed changes from gravel to sand, wider shallower channels
$Qs^{+}, Qw^{0} \approx S^{+}, d_{50}{}^{-}, D^{-}, W^{+}$	(9)	
$Qs^{0}, Qw^{-} \approx S^{+}, d_{50}{}^{-}, D^{-}, W^{-}$ $Qs^{-}, Qw^{--} \approx S^{+}, d_{50}{}^{-}, D^{-}, W^{-}$	(10) (11)	Extraction of water from river resulting in narrower stream
$Qs^{+}, Qw^{-} \approx S^{+}, d_{50}{}^{-}, D^{-}, W^{\pm}$	(12)	Increased water and sediment discharge

D = flow depth; d_{50} = median particle size of bedmaterial; Qs = bedload (expressed as a percentage of total load); Qw = water discharge (e.g. mean annual discharge); S = channel slope; W = flow width.
0 indicates no change; +/− indicates increase or decrease respectively; ++/−− indicates a change of considerable magnitude.

Table 4.4 Potential river channel adjustments (after Gregory 1976)

Potential adjustments of	Fluvial landform: River channel cross-section	River channel pattern	Drainage network
Size	*Increase or decrease of river channel capacity* Erosion of bed and banks can produce a larger channel which maintains the same shape. Sedimentation can produce a smaller channel which maintains the same shape	*Increase or decrease of size of pattern* Increase or decrease of meander wavelength whilst preserving the same planform shape	*Increase or decrease of network extent and density* Extension of channels or shrinkage of perennial, intermittent and ephemeral streams
Shape	*Adjustment of shape* Width/depth ratio may be increased or decreased	*Alteration of shape of pattern* A change from regular to irregular meanders	*Drainage pattern changed in shape* Inclusion of new stream channels after deforestation
Composition	*Change in channel sediments* Alteration of grain size of sediments in bed and banks possibly accompanied by development of berms or bars	*Planform metamorphosis* Change from single to multi-thread channel or converse	*Network composition changed* The replacement of channels with no definite stream channel by a clearly defined channel

describes the situation, with sediment discharge (Qs) increasing more than water discharge (Qw). This predicts increased slope, decreased median sediment size, decreased depth of flow and increased flow width. The resulting channels are wider and shallower. These predictions conform to the findings of Trimble (1974) who noted that the composition of channel beds changed from gravel to sand. The increased frequency of flooding arising from the channel aggradation created extensive swamp and marshy areas.

After the 1920s improved conservation practices and reforestation caused drastic reductions in Qs and a moderate increase in Qw (equation 3). Trimble (1974) found that the streams began to erode their accumulated sediments, and channels became entrenched and narrower. Erosion of sandy sediments uncovered coarser bed gravels.

Changes such as these in turn may trigger adjustments in stream systems and associated biological and physical systems. One of the major adverse effects of inorganic sediments on aquatic ecosystems relates to sediment covering the substrate (Hynes 1973). The impacts on fish and aquatic invertebrates of sediments arising from land use changes such as deforestation have been reviewed by Cordone and Kelley (1961). The effects of finer sediments can be direct, by causing abrasive injuries to external organs such as gills, protective mucous coverings and fins, or by smothering nests and eggs. Indirect effects can range from elimination of a preferred food source to elimination of preferred reproductive habitats for fish and aquatic invertebrates. The problem of excess fine sediment within a river channel may be temporary and with the completion of a particular land disturbance activity, there may be a tendency for reduced sediment loads and the gradual removal of deposited sediment.

4.5 URBANIZATION

Developing an area for housing or other urban purposes increases the area of low or zero infiltration capacity and increases the speed of water transmission in channels or surface water sewers (Leopold 1968; Hammer 1972; Dunne & Leopold 1978). Increased discharges arising from paved urban areas often cause enlargement of natural channels downstream in both temperate (e.g. Graf 1975; Richards & Wood 1977) and tropical environments (Whitlow & Gregory 1989). There may also be changes of substrate composition (Thoms 1987).

The impact of urbanization is complicated by the fact that the channel degradation itself, together with erosion of the catchment surface during construction, can cause large amounts of sediment to be released initially to the channel, which must subsequently be carried out of the system (Fig. 4.2). Wolman and Schick (1967) note that sediment which had accumulated within a channel was still present 7 years after construction activities had ended. It is only after the phase of temporary aggradation that the channel itself begins to enlarge in response to increased runoff from paved areas. The rate of enlargement also depends on the type, size and location of the urban area within the catchment.

As soon as a catchment is completely urbanized or development is stable, it is possible that the stream may become adjusted to the changed hydrological state (Fig. 4.2). The immediate results of increased storm runoff may be

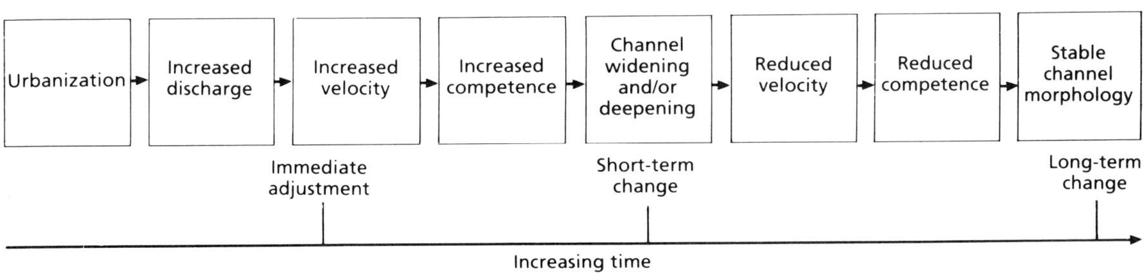

Fig. 4.2 Impact of urbanization (after Morisawa 1985).

increased velocities, leading to erosion of the channel bed and/or banks (see equation 1, Table 4.3). However, negative feedback may eventually occur, leading to a new equilibrium between the changed hydrological state, sediment load and morphology of the channel (Fig. 4.2). Associated with changes in the size of the channel there may be changes in channel shape, and below urban areas it has been suggested that width may change more easily than depth because of the armouring of the channel bed.

For the Avondale stream basin in Harare, Zimbabwe, Whitlow and Gregory (1989) showed how downstream reaches had widened by an average of 1.7 times over a period of 12 years, involving average rates of bank erosion of 0.33 m per year. This was in response to increased peak discharges, estimated to be two times former

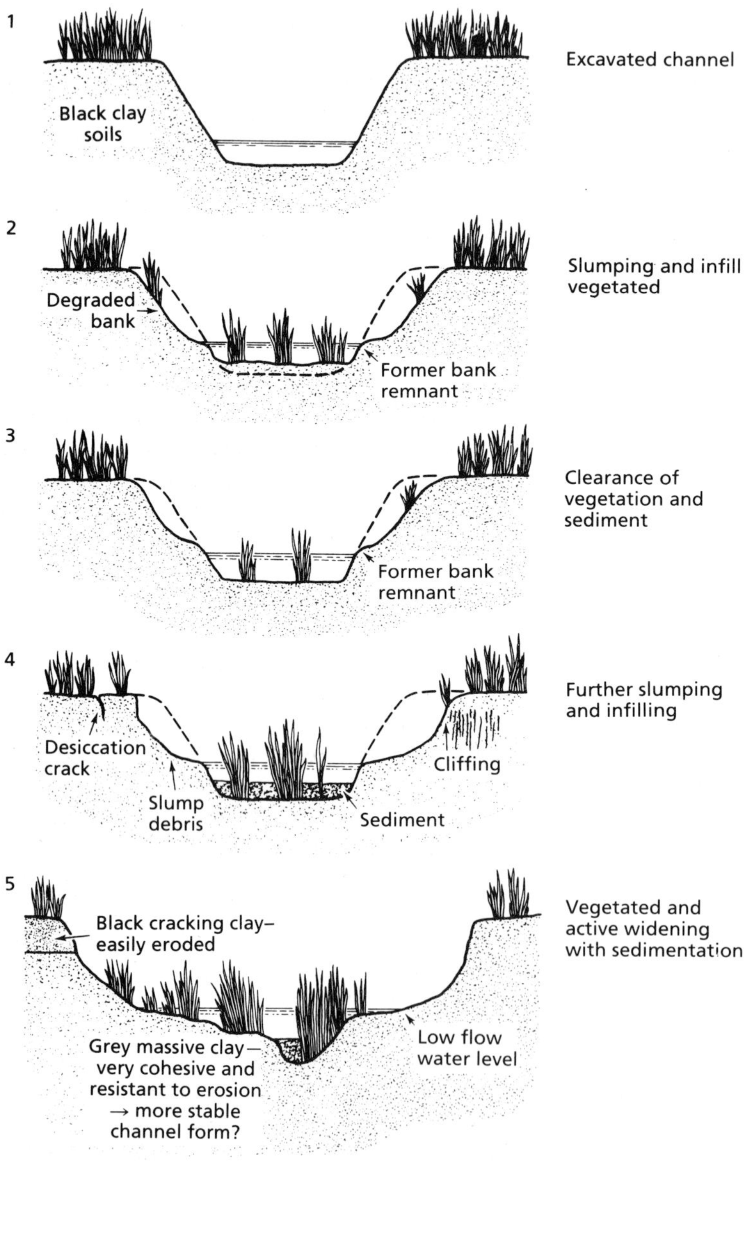

Fig. 4.3 Sequence of channel change in the lower part of the urbanized Avondale stream, Zimbabwe (from Whitlow & Gregory 1989, by permission of John Wiley & Sons).

values. Figure 4.3 depicts the sequence of change in the lower reaches. Essentially the original excavated channel has been modified progressively by human activity and natural processes to produce a broad, irregular channel which is still in a phase of active widening.

Studies of urban river sediments are limited, particularly those concerned with the textural characteristics of the channel bed sediments (Douglas 1985). A study of the River Tame, within and below the Birmingham conurbation, UK, provided information on the structural properties of the river bed (Thoms 1987). The study suggested that whilst the accumulation of fine matrix sediments arising from construction has an immediate impact, in the longer term, as a consequence of the changed hydrological regime, the river not only reduces this impact but also establishes a substrate containing less fine sediment. This impact can influence relatively long reaches. The ecological effects of such changes are little known as yet (Thoms 1987; Gilbert 1989). In general, and in unpolluted water, density of invertebrates tends to fall off as erosion increases (Percival & Whitehead 1929). The unusual patterns of erosion and siltation in urban channels coupled with pollution combine together in a variety of ways to reduce invertebrate numbers (Gilbert 1989).

4.6 ADJUSTMENTS OF RIVER CHANNEL CROSS-SECTIONS BELOW DAMS AND RESERVOIRS

Construction of the Thomson Dam, Australia, caused siltation of the channel substrate for a distance of 25 km downstream (Davey *et al* 1987). However, following construction the magnitude and frequency of peak discharges is often decreased below dams because of the effect of water storage. The sediment load becomes trapped in the reservoir. The channel below a dam typically scours and this has to be allowed for in the design of a dam to prevent subsequent undermining and failure (Komura & Simons 1967). Because reservoirs trap up to 95% of the bed load and suspended sediment carried by the river, this encourages scour immediately below the dam. Thus below Glen Canyon dam on the Colorado River, which was constructed between 1956 and 1963, approximately 10 million m^3 of sediment was scoured from below the dam (Pemberton 1976). However, the bed subsequently became armoured and stabilized between 1965 and 1975.

Williams and Wolman (1984) document changes that have occurred within alluvial river channels downstream of 21 dams in the USA over a period of up to 70 years. The reduction in flood peaks for the affected rivers ranged from 3% to 91%, with an average of 39% of the pre-dam values. Erosion of the channel bed had been experienced at all sites, with a maximum value of 7 m occurring immediately below the dam and decreasing further downstream. The response of channel width to dam closure was found to be more variable. Widths decreased, increased or remained constant depending on the site characteristics. These findings for dams in the USA are supported by information obtained from other geographical regions (see Petts 1984).

The rate of channel change below dams varies but studies suggest that the majority of bed degradation occurs within the first 10–15% of the period of adjustment. Bed degradation can continue and migrate downstream for several decades after dam closure at rates of up to 50 km per year. However, variations in the rate of migration of the zone of bed degradation are dependent on numerous factors, including the nature of the bed and bank materials and the discharge regime. Degradation below a dam may provide increasing sediment load in a downstream direction (Meade & Trimble 1974).

Deposition within regulated rivers may also occur because the regulation of high-magnitude floods artificially slows sediment transport, whilst the sediment supplied by tributaries is unaffected or possibly increased (Petts 1984). Coarse sediments introduced by tributary streams with relatively high gradients are therefore deposited in the main channel, which invariably has a lower slope and wide channel. An extensively documented study is that of the Colorado River below Glen Canyon Dam, where deposition has occurred at every major tributary confluence for distances of up to 24 km and beyond (Howard & Dolan 1981).

Benthic invertebrates are influenced by a

number of factors which are likely to change immediately upon dam closure. However, channel morphology and the composition and stability of the substrate are known to change slowly and it can be anticipated that changes in the flora and fauna are likely to occur over a period of many years. Sedimentation in the Thomson River, southeastern Australia, during construction of a dam reduced the total numbers of macroinvertebrate species and the density of species and individuals (Doeg *et al* 1987). Immediately below the Navajo Dam, Upper Colorado River, USA, densities of invertebrates were observed to increase during a 5-year period following closure of the dam, from 828 m^{-2} to 6727 m^{-2} (Mullan *et al* 1976). This change may be due to improved bed stability following degradation and channel armouring. Increased benthos observed immediately below many dams may be a consequence of improved stability and removal of finer material (Pearson *et al* 1968; Petts 1984). However Voelz and Ward (1990) observed that species which feed on coarse sedimentary detritus (shredders) were absent from below a dam on the Blue River, Colorado.

Further downstream the invertebrate communities may relate to either erosional or sedimentational processes. Thus, 13 km below the Navajo Dam the density of invertebrates was 6727 m^{-2} and the dominant species were of the Ephemeroptera, Diptera, Hydropsychidae, Gastropoda and Amphipoda, all typical of er-

Table 4.5 Channel change and invertebrate distributions within the Afon Rheidol, below Nant-y-Moch Reservoir, Wales, UK (after Petts & Greenwood 1985)

Reach	Control channel	1	2	3	4
Number of samples	10	8	17	18	13
Mean substrate particle size (mm)	54	75	40	13	Gravel and boulders
Proportion less than 2 mm (%)	18.1	6.6	11.3	30.1	19.6
Channel width (m)	14	8	16	22	14
Substrate stability	Unstable	Stable	Unstable	Unstable	Stable
Proportion of sites with moss (%)	0	10	18	0	63
Average number of individuals (per m^2)	336	402	177	313	437
Number of species	10	13	12	21	23
Species diversity	0.917	1.239	1.833	2.250	2.298
Diptera (no. m^{-2})	275	309	91	76	153
Chironomidae	258	116	45	50	114
Simuliidae	16	189	42	2	19
Dicranota sp.	0	4	4	20	7
Ephemeroptera (no. m^{-2})	10	5	21	11	14
Plecoptera (no. m^{-2})	6.5	41.5	29	48	73
Diura bicaudata	5	1.5	5	0	3
Protonemura meyeri	0	7	13	3	41
Leuctra hippopus	1.5	33	11	45	29
Trichoptera (no. m^{-2})	39	41	42	116	163
Polycentropus sp.	1	8	+	+	8
Rhyacophilidae	18	10	5	7	28
Cased caddis	20	23	36	89	116
Oligochaeta (no. m^{-2})	4	2	12	31	8

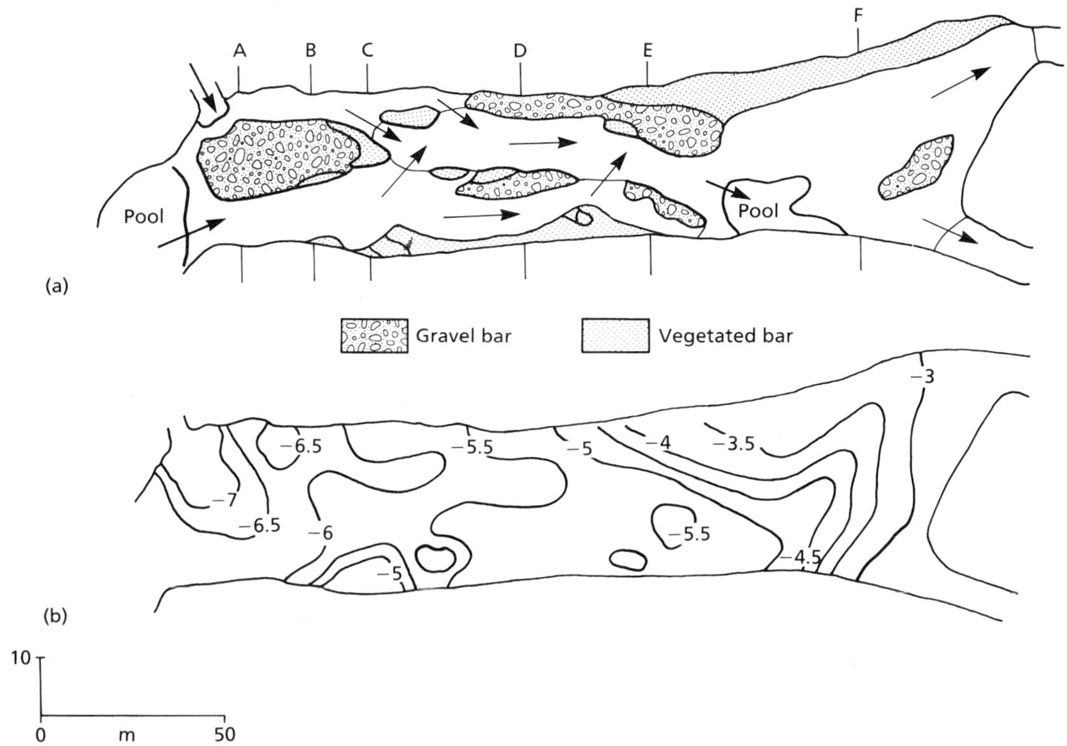

Fig. 4.4 Sedimentation within the River Rheidol below Nant-y-Moch Dam, UK: morphology (a) and mean particle size variations (phi values) of the surface layer (from Petts 1984, by permission of John Wiley & Sons).

osional situations (Mullan *et al* 1976). At a distance of 30 km downstream sedimentation processes were predominant, the increased turbidity reducing the benthic species to between 32 and 192 m^{-2}. These processes also eliminated a trout fishery. Petts (1988) found localized siltation of the substrate below two dams in the UK, particularly below tributary confluences.

The effect of channel changes on benthic macroinvertebrates has been investigated below the Nant-y-Moch reservoir, on the River Rheidol, Wales, UK (Petts & Greenwood 1985). Channel changes had been caused by sedimentation arising from point-source inputs of sediment to the mainstream from tributary streams. Four distinct successive morphological and sedimentological units were identified downstream from the tributary confluence. The invertebrate distributions associated with the channel changes are shown in Table 4.5. The first reach immediately below the confluence junction produced reduced channel dimensions and a coarse, stable substrate (Fig. 4.4). Below this reach (Reach 2) the channel had experienced less change and was braided with an unstable substrate. The third reach was found to be characterized by progressive sedimentation, whilst the fourth reach had an existing substrate stabilized by flow regulation, and mosses and algae covering much of the channel bed. This reach had the largest number of benthic species and greatest species diversity. It is anticipated that reach 1 will extend downstream as the processes of sedimentation change the channel morphology. Reaches 2 and 3 will shift downstream, whilst reach 4 will be progressively eliminated. Changes in the macroinvertebrate populations may be anticipated in accordance with this change, which may take many years.

4.7 DIRECT INFLUENCES

Direct modification of river channels includes dam and reservoir construction, channelization and irrigation diversions. There may be changes which occur as a consequence of a change at a specific location or along a length of river.

Channelization

Since channelization involves manipulation of one or more of the dependent hydraulic variables of slope, depth, width and roughness, then feedback effects may be initiated which proceed to promote a new state of equilibrium (Brookes 1988). Channelization may induce instability not only in the improved channel reach but also upstream and/or downstream of the managed reach. Perhaps the most dramatic adjustments occur in response to slope changes associated with channel straightening or regrading, or to extended bottom widths. For the East and West Prairie rivers in Alberta, Canada, Parker and Andres (1976) found that channelization of a meandering stream increased the slope by providing a shorter channel path. This increase of slope enabled the transport of more sediment than was supplied at the upstream end of the channelized reach and the difference was obtained from the bed, causing degradation which progressed upstream as a nickpoint. An excess of load was then supplied to the downstream part of the channelized reach and because the flatter natural reach downstream could not transport this sedi-

Table 4.6 Examples of the morphological consequences of river channelization

			Morphological effects			
Area applied	Dates	Extent	Downstream	Within	Upstream	Source
Willow R, Iowa	1906–20			440% capacity increase between 1920 and 1958	Series of nickpoints	Daniels (1960); Ruhe (1970)
Blackwater R, Missouri	1910			1000% increase of cross-sectional area		Emerson (1971)
East and West Prairie rivers, Canada	1953–71		Deposition of sediment	500% increase of cross-sectional area	4 m of degradation (1964–74)	Parker & Andres (1976)
Tisza R, Hungary	1850s	32% reduction of length	No effect – it enters the Danube	Bed lowered by a maximum of 2.3 m		Szilagy (1932)
Mississippi R, Baton Rouge	1929–47	35% reduction of length	No effect due to lack of time since completion			Lane (1947)
Colorado R, Laguna Dam		9.6 km length reduction	Insufficient sediment to raise bed level	Bed lowered	Lowering limited by dam located upstream	Lane (1947)
Peabody R, New Hampshire				Reach lowered by a maximum of 3.6 m; width increase of 400%		Yearke (1971)
103 streams in North America	1960–70	Lengths of 0.07–4.2 km		Degradation at only 17 sites; bank stability poor at 7 sites		Brice (1981)

ment it was deposited on the bed. Degradation within the channelized reach caused bank collapse. A similar process operated on the Willow River in Iowa where a 440% increase of capacity was achieved in 38 years following channelization (Daniels 1960) and in the Blackwater River, Missouri, where the cross-sectional area increased by a maximum of 1000% over a 60-year period (Emerson 1971). Table 4.6 provides some further examples of the morphological consequences of artificial straightening.

Adjustments of channel morphology have been noted within resectioned reaches and the problems of overwide channels have been exemplified for part of a flood alleviation scheme constructed on the River Tame in England (Nixon 1966). The channel reverted to its original capacity in less than 30 years in the absence of maintenance; this is due to the enlarged channel being in equilibrium with the design flood flow but out of equilibrium with the normal range of flows which predominate for most of the time. The low flows deposit sediment because of reduced velocities in the overwide reach and these deposits may become stabilized to form more permanent morphological features.

Enlarged cross-sections in the lower reaches of projects sometime function as sediment traps. Flow depths and channel sizes may be reduced as the channels fill with sediment. In the lower reaches of streams the problem may be compounded by low slopes and backwater effects; for instance, within 10 years of the completion of a flood control channel on the San Lorenzo River in California, 350 000 m^3 of sediment had been deposited (Griggs & Paris 1982).

Channelization may have both direct and indirect effects on aquatic habitat and associated organisms (Brookes 1988). Traditional channelization practices may dramatically change the channel morphology to the detriment of the biology of watercourses. In North America the number of published works on the effects of channelization on the flora and fauna of rivers far exceeds those concerned solely with the physical effects. Figure 4.5 compares the channel morphology and hydrology of a natural stream with the morphology of a channel changed by channelization. Channelization can change the original dimensions and shape of a channel, the slope and the planform, changing a heterogeneous system into a homogeneous one (McClellan 1974; Neuhold 1981). The most commonly stated reason for a change in fish populations in channelized rivers is the loss of a natural pool-riffle sequence which provides a variety of flow conditions suitable as cover for fish and for the organisms on which fish feed. Shelter areas are required at high flows to protect fish from abnormally high velocities; these conditions are absent in a channelized river, where a meandering stream, with an abundance of long pools separated by short riffles, may be converted into a straight stream composed mainly of run habitat. Alteration of the width and depth variables in a channel may create shallow and unnatural hydraulic conditions which create an unsuitable habitat for fish and may present topographical difficulties for fish migration.

Disruption of the substrate can have significant impacts on benthic invertebrates, and in an extreme case natural bed materials may be replaced by bedrock (Huggins & Moss 1974). An unstable substrate following channelization may lead to excessive sediment loads, as may other forms of morphological adjustment, and this is detrimental to the fish populations (Graesser 1979). The effect of a changed substrate may be complex; Schmal and Sanders (1978) found a marked seasonal trend with low invertebrate populations coincident with high spring flows and an unstable substrate, and high populations related to a stable substrate at other flows.

The processes of biological recovery take time (see Chapter 5), and may be naturally or artificially induced. Tarplee *et al* (1971), in a study of North Carolina's coastal plains' streams, suggested that species diversity of fish may recover in 15 years provided no further alterations of the stream bed occur. On a resectioned part of the Chariton River in Missouri it was found that after 30 years the channel had tended to revert from the uniform width and depth of the constructed channel back to natural conditions with meanders and a corresponding improvement in fish populations (Congdon 1971). Further examples of the biological impacts of channelization are given in Table 4.7.

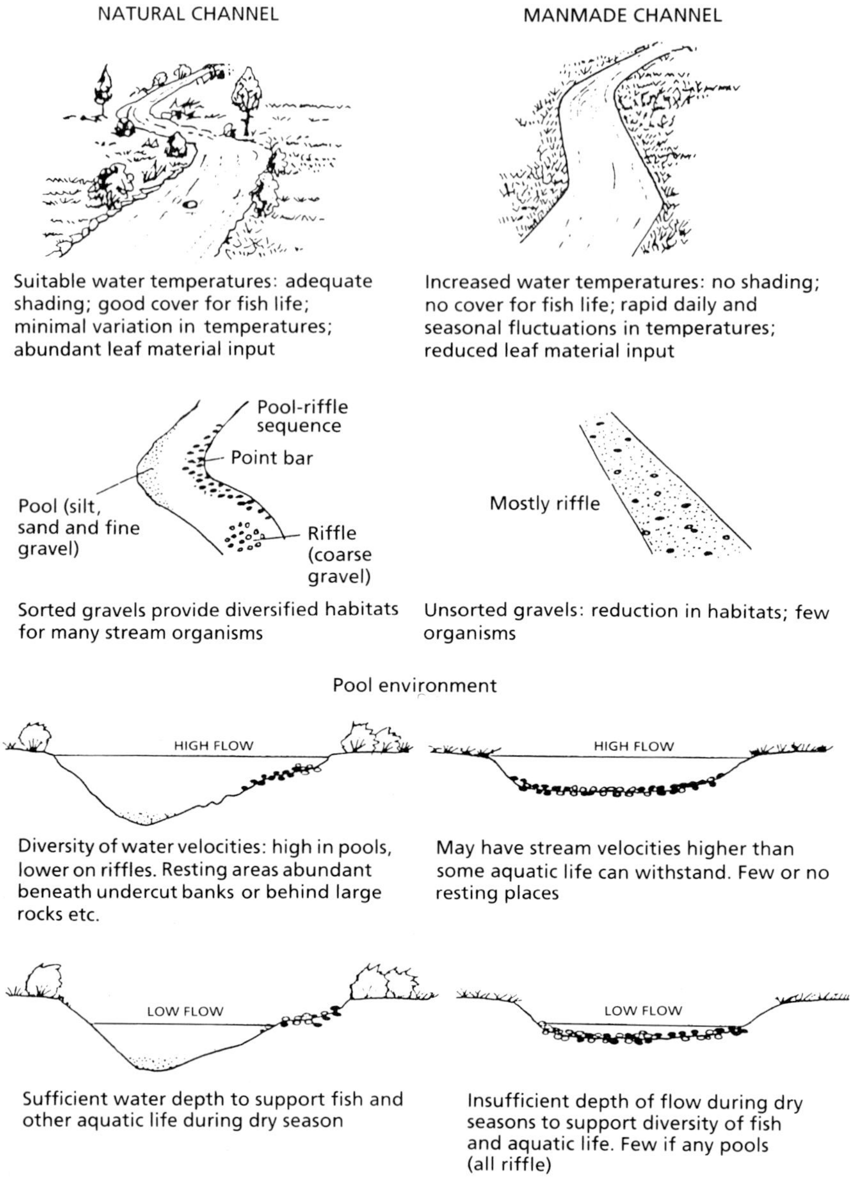

Fig. 4.5 Comparison of the channel morphology and hydrology of a natural stream with a channelized watercourse (based on Corning 1975, by permission of Virginia Department of Game and Inland Fisheries).

Channelization of the Twentymile Creek in northern Mississippi took place at various dates between 1910 and 1966 for drainage and flood control purposes (Shields & Hoover 1991). Straightening and channel enlargement caused a nickpoint to progress upstream, for about 14 km, doubling the channel cross-section over a period of 12 years, causing degradation and bank failure. However, between 1982 and 1986 grade control structures were installed to check the slope and

Table 4.7 Examples of the ecological impacts of river channelization

Location	Type	Date	Variables affected	Reduced parameters	Recovery	Source
Missouri R, Nebraska		1950s	Brush and piles and pools eliminated	Benthic area reduced by 67%; standing crop of drift reduced 88%	Study represents 15 years after channelization	Morris *et al* (1968)
Little Sioux R, Iowa			Lack of suitable substrate	Fewer macroinvertebrates but more drift organisms		Hansen & Muncy (1971)
Buena Vista Marsh, Portage Co, Wisconsin	Dredging	Various	Seasonal effect; substrate only unstable at high flow; at other times vegetation/silt	High invertebrate populations when substrate stable; vegetation and silt favours snails/ midges. Elimination of stonefiles		Schmal & Sanders (1978)
Mud Creek, Douglas Creek, Kanas	Enlarged	1971	Destruction of pool-riffle sequence	Number of fish and species reduced; biomass reduced by 82% and diversity by 49%	2 years after	Huggins & Moss (1974)
Rush Creek, Modoc County, California	Realigned	1969	Destruction of pool-riffle sequence	Trout biomass reduced by 86%; total number by 36% and total biomass by 69%	5 years after construction	Moyle (1976)
Missouri River	Various channel works		Lack of niches (especially pools)	Annual catch reduced by 47%; harvest rate by 48% and standing crop by 63%		Groen & Schmulbach (1978)

prevent further degradation. Rock riprap and vegetation were used to protect stream banks. Engineering stabilization has been shown to have a positive effect on a degraded channel. Bank stability enabled riparian vegetation to increase and fish assemblages near to the grade control structures were shown to be more diverse than at other sites, partly due to an improved substrate (Shields & Hoover 1991).

4.8 PLANFORM AND NETWORK CHANGES

The discussion of changes arising from specific impacts (Sections 4.3 to 4.7) has concentrated mainly on changes of channel cross-section and composition of the substrate. Two areas of study which have received relatively little attention are changes of the planform and network characteristics of river systems. Two types of change of channel planform have been distinguished (Lewin 1977), namely autogenic, which are inherent in the river regime, and allogenic, which occur in response to changes in the river system, including the influence of human activities. A detailed study in England and Wales by Hooke and Redmond (1989) revealed that almost 35% of channels which drain upland areas had shown some change of planform between 1870 and 1950 and that at least some of these changes were either due to the direct (channelization) or indirect (urbanization, deforestation) influences of humans. With many studies it has been difficult to separate the influence of human interference from other variables (e.g. Hooke, 1977). A study along the Dunajee River in southern Poland revealed that the channel had become more sinuous and branching over the period 1787 to 1955 (Klimek & Trafas 1972). This change of shape was attributed to deforestation and the increase of potato cultivation in the 19th century because this allowed for high rates of runoff.

Ecological changes resulting from such planform changes have received even less attention. A recent study of the French Upper Rhône River has elucidated ecological changes which have occurred since 1750 (Roux *et al* 1989). This was achieved by using historical maps, floristic and

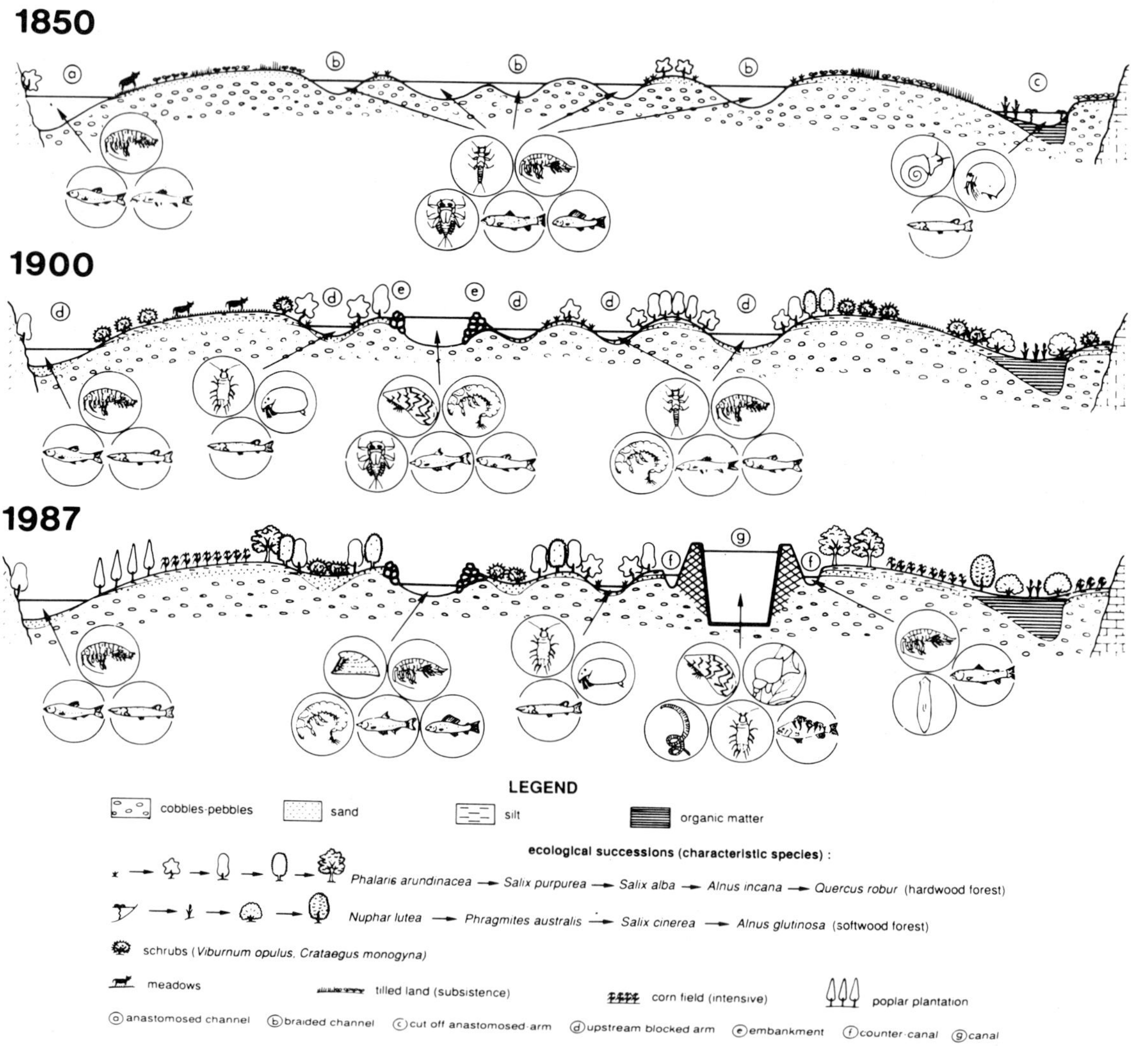

Fig. 4.6 Transect of part of the Upper Rhône River showing channel changes between the years 1850, 1900 and 1987. Certain animals are plotted to illustrate similarities and differences between aquatic communities (from Roux *et al* 1989, by permission of John Wiley & Sons).

faunistic evidence from relic channels and documentary sources. Figure 4.6 is a typical transect which shows the numerous braided and anastomosed channels in 1850 before civil engineering works; in 1900 after the construction of embankments; and in 1987 following canalization and regulation of the river for hydro-electric power generation in the 1980s. The corresponding ecological changes are complex. The short-term impacts of embanking in the last century increased the structural and functional diversity of the ecosystems in the braided sections. The embankments provided new biotopes and the new hydrological and sedimentological processes allowed the development of several ecological sequences which were previously inhibited. The diversity further increased with the canalization and regulation undertaken in the 1980s. However, it is anticipated that the diversity of habitats and communities will eventually decrease due to the lowering of the water table and the reduction of the influence of hydrological and sedimentological processes on the floodplain.

Changes of the drainage network are also widespread as a consequence of both direct and indirect human intervention. New drainage ditches have been dug (e.g. Marshall *et al* 1978; Ovenden & Gregory 1980) and land use changes have probably led to erosion of new channels or rills. For example, Eyles (1977) showed that in the Southern Tablelands of New South Wales, Australia, tree removal and grazing pressure caused dramatic changes between 1840 and 1950. A sequence was observed from a chain of scour ponds, to discontinuous gully, to incised channel and eventually to a permanent stream.

4.9 TIME DIMENSION

A particular problem is the complexity of ecological adjustments to channel changes. Many of the ecological studies may represent transient states and not final equilibria (Petts 1987). For example, changes caused by regulation may be conceptualized as a hierarchy of responses (Petts 1980). First-order impacts occur at the time of dam closure, or shortly afterwards, and affect the transfer of energy and material into and within the downstream river. Second-order impacts are the channel changes and changes of the floodplain dynamics and primary production arising from the local effects of the first-order impacts (Fig. 4.7). Third-order impacts on benthic invertebrates, fish and floodplain fauna result from a combination of all the first- and second-order impacts as well as from biotic interactions between populations. The complete adjustment of biological populations must be preceded by the adjustment of abiotic factors, of which channel change can take a considerable length of time, even hundreds of years.

For channelized rivers the period between the commencement and the attainment of a new state of equilibrium, the relaxation time, has been found to vary. Values have been given ranging from 9 to 15 years for three Mississippi Rivers (Schumm *et al* 1984) to as much as 1000 years elsewhere (Brookes 1988). Recovery of fish populations is dependent on an improvement of habitat, which may be offset by maintenance of channelized reaches. Studies in North America have shown that substantial recovery of fish populations may occur in periods of between 30 and 86 years (Brookes 1988).

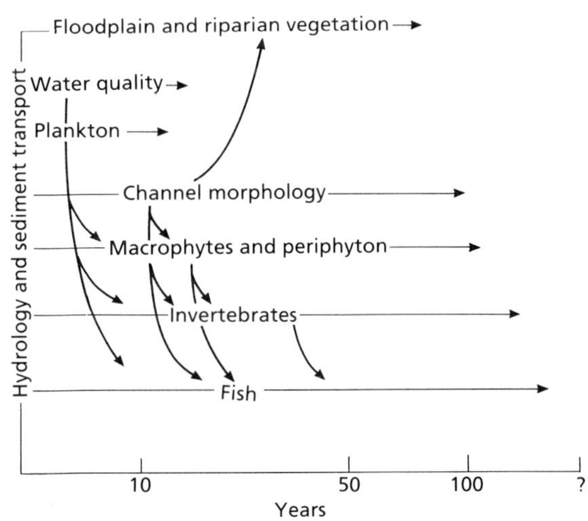

Fig. 4.7 Interrelationships among the major variables and appropriate time-scales for the consideration of adjustment to river regulation (after Petts 1989b, by permission of CRC Press).

4.10 THRESHOLDS AND COMPLEX RESPONSE

In addition to the time sequence for change it is also important to note that a particular change, as evidenced by erosion or deposition, may only occur after a critical threshold level has been exceeded (Schumm 1973). For example, large floods may be responsible for triggering a sequence of change. Perhaps best known are the threshold velocities required to entrain sediment of a certain size. Thresholds can either be extrinsic or intrinsic. Extrinsic thresholds exist within a system but change will not occur without the influence of an external variable. Intrinsic thresholds may be exceeded where the external variables remain relatively constant, yet progressive change within the system renders it unstable and change occurs.

Responses to change may be instantaneous, progressive, delayed or complex. The problems of complex response were outlined by Schumm (1973) who described how two river basins can react in different ways, or at differing rates, to the same stimulus for change. The response is likely to be complex, with spatial and temporal lags. For example, a period of forest clearance may be replaced by conservation measures, and yet it may take a considerable amount of time for the wave of sediment arising from the forest clearance to work through the system.

It is further apparent that the interpretation of river channel change must consider the interrelationships between the channel cross-section, channel pattern and drainage network (Gregory 1976). Dumaresq Creek in New South Wales has undergone an increase in the number of stream channels through extension of the drainage network, reduced channel sizes following reservoir construction and enlargement due to urban runoff (Gregory 1977b).

4.11 PROSPECT

This chapter has illustrated a variety of channel adjustments arising from changes in the discharge and sediment loads of rivers. Channel changes may extend over periods of up to several hundred years. Because many of the responses of a river cannot be predicted by theory alone, an important means for evaluating impact is through documented case studies. The size, shape and composition of channels may change (Table 4.4). Most biological studies have concentrated on the impacts arising from a change of composition of the channel substrate; few have examined the impact of changes of channel size or shape (cf. Fig. 4.3) on the ecology of a watercourse. The precise biological impacts of different types of channel change have yet to be fully quantified. The loss of lotic habitat as a result of straightening, or floodplain modification, is another area which has received relatively little attention. Such studies are important so that knowledge can be extended and improved management strategies developed and implemented.

Discussion of biological impacts is difficult without understanding the intricate relationships between the biotic and abiotic elements of the stream ecosystem. Both abiotic and biotic factors vary seasonally and geographically. If changes caused by regulation are conceptualized as a hierarchy of responses (Petts 1980), then the third-order impacts on benthic invertebrates, fish and floodplain fauna result from a combination of all the first- and second-order impacts as well as from biotic interactions between populations (see Section 4.9). However, this is often complicated because channel changes along a specific length of river may be the result of more than one impact. For example, urbanization often includes channelization to accommodate anticipated increases of peak discharge. In other circumstances channelization may have been undertaken on channels affected by increased sedimentation arising from land use changes in the upstream catchment. Examples of the complex nature of channel changes arising from a variety of impacts have been documented by Brookes and Gregory (1988). Channel changes are further complicated by the presence of thresholds, both intrinsic and extrinsic.

Substantial efforts by geomorphologists have elucidated changes occurring in the 10–50-year time period. However, relatively few studies have so far identified the significance of changes occurring over longer time spans of, say, up to 500 years, the period of intensifying human impact in temperate zones. The lack of study is partly a result of the absence of documentary sources, including maps. However, archaeology is likely to be invaluable as a means of extending our knowledge of this period, and hence the likely ecological repercussions. Through reconstruction of the past it will be possible to determine the rate and magnitude of channel change in response to single and composite impacts. It should also enable improved prediction of the effects of more recent impacts and the development of ecologically-sound management tools (Brookes 1991).

Understanding how river channels have changed is vital if restoration strategies which are both morphologically stable and ecologically desirable are to be developed further.

ACKNOWLEDGEMENTS

The views expressed in this paper are those of the author and not necessarily those of his employer, the National Rivers Authority (Thames Region). Permission to publish this paper is gratefully appreciated.

REFERENCES

Brice JC. (1981) *Stability of Relocated Stream Channels.* Technical Report No. FHWA/RD-80/158, Federal Highways Administration, US Department of Transportation, Washington, DC. [4.7]

Brookes A. (1987) Recovery and adjustment of aquatic vegetation within channelization works in England and Wales. *Journal of Environmental Management* **24**: 365–82. [4.1]

Brookes A. (1988) *Channelized Rivers: Perspectives for Environmental Management.* John Wiley & Sons, Chichester. [4.7, 4.9]

Brookes A. (1991) Geomorphology. In: Gardiner JL (ed) *River Projects and Conservation: A Manual for Holistic Appraisal.* pp. 57–66. John Wiley & Sons, Chichester. [4.11]

Brookes A, Gregory KJ. (1988) Channelization, river engineering and geomorphology. In: Hooke JM (ed) *Geomorphology in Environmental Planning,* pp. 145–67. John Wiley & Sons, Chichester. [4.11]

Congdon JC. (1971) Fish populations of channelized and unchannelized sections of the Chariton River, Missouri. In: Schneberger E, Funk JL (eds) *Stream Channelization: A Symposium,* pp. 52–83. Special Publication No. 2, North Central Division, American Fish Society, Bethesda, Maryland. [4.7]

Cordone AJ, Kelley DW. (1961) The influence of inorganic sediment on aquatic life of streams. *California Fish and Game* **47**: 189–228. [4.4]

Corning RV. (1975) Channelization: shortcut to nowhere. *Virginia Wildlife* **6**: 8. [4.7]

Daniels RB. (1960) Entrenchment of the Willow Creek Drainage Ditch, Harrison County, Iowa, *American Journal of Science.* **258**: 161–76. [4.7]

Davey GW, Doeg TJ, Blyth JD. (1987) Changes in benthic sediment in the Thomson River, Victoria, during construction of the Thomson Dam. *Regulated Rivers* **1**: 71–85. [4.6]

Doeg TJ, Davey GW, Blyth JD. (1987) Response of the aquatic macroinvertebrate communities to dam construction on the Thomson River, Southeastern Australia. *Regulated Rivers* **1**: 195–209. [4.6]

Douglas I. (1985) Urban sedimentology. *Progress in Physical Geography* **9**: 255–80. [4.5]

Dunne T, Leopold LB. (1978) *Water in Environmental Planning.* W.H. Freeman, San Francisco. [4.5]

Emerson JW. (1971) Channelization: a case study. *Science* **173**: 325–6. [4.7]

Eyles RJ. (1977) Birchams Creek: the transition from a chain of ponds to a gully. *Australian Geographical Studies* **15**: 146–57. [4.8]

Gilbert OL. (1989) *The Ecology of Urban Habitats.* Chapman and Hall, London. [4.5]

Graesser NWC. (1979) How land improvement can damage Scottish salmon fisheries. *Salmon and Trout Magazine* **215**: 39–43. [4.7]

Graf WL. (1975) The impact of suburbanization on fluvial geomorphology. *Water Resources Research* **11**: 690–2. [4.5]

Graf WL. (1979) Mining and channel response. *Annals of the Association of American Geographers* **69**: 262–75. [4.4]

Gregory KJ. (1976) Changing river basins. *Geographical Journal* **142**: 237–47. [4.3, 4.10]

Gregory KJ. (1977a) The context of river channel changes. In: Gregory KJ (ed) *River Channel Changes,* pp. 1–12. John Wiley & Sons, Chichester. [4.1]

Gregory KJ. (1977b) Channel and network metamorphosis in northern New South Wales. In: Gregory KJ (ed) *River Channel Changes,* pp. 389–410. John Wiley & Sons, Chichester. [4.10]

Gregory KJ, Lewin J, Thornes JB. (1987) *Palaeohydrology in Practice: a River Basin Analysis.* John Wiley & Sons, Chichester. [4.1]

Griggs GB, Paris L. (1982) Flood control failure: San Lorenzo River, California. *Environmental Management* **6**: 407–19. [4.7]

Groen CL, Schmulbach JC. (1978) The sport fishery of the unchannelized Middle Missouri River. *Transactions of the American Fisheries Society* **107**: 412–18. [4.7]

Hammer TR. (1972) Stream channel enlargement due to urbanization. *Water Resources Research* **8**: 1530–40. [4.5]

Hansen DR, Muncy RJ. (1971) *Effects of Stream Channelization on Fish and Bottom Fauna in the Little Sioux River. Iowa.* Iowa State Water Resources Institute, Ames, Iowa. [4.7]

Hooke JM. (1977) The distribution and nature of changes in river channel pattern. In: Gregory KJ (ed) *River Channel Changes*, pp. 265–80. John Wiley & Sons, Chichester. [4.8]

Hooke JM, Redmond CE. (1989) River channel changes in England and Wales. *Journal of the Institution of Water and Environmental Management* **3**: 328–35. [4.8]

Howard AD, Dolan R. (1981) Geomorphology of the Colorado River in the Grand Canyon. *Journal of Geology* **89**: 307–21. [4.6]

Huggins DG, Moss RE. (1974) Fish population structure in altered and unaltered areas of a small Kansas stream. *Transactions of the Kansas Academy of Science* **77**: 18–30. [4.7]

Hynes HB. (1973) The effects of sedimentation on the biota in running water. In *Proceedings of the 9th Canadian Hydrology Symposium: Fluvial Processes and Sedimentation*, pp. 653–62. Inland Waters Directorate, Canadian Department of the Environment, Canada.

Keefer L, Maughan OE. (1985) Effects of headwater impoundment and channelization on invertebrate drift, *Hydrobiologia* **127**: 161. [4.1]

Kellerhalls R, Church M, Bray DI. (1976) Classification and analysis of river processes. *Journal of the Hydraulics Division. American Society of Civil Engineers* **102**: 813–29. [4.2]

Kimble LA, Wesche TA. (1975) Relationships between selected physical parameters and benthic community structure in a small mountain stream. Water Resources Series No. 55. University of Wyoming, Laramie, Wyoming. [4.2]

Klimek K, Trafas K. (1972) Young Holocene changes in the course of the Durajec River in the Beskid Sadecki Mountains (Western Carpathians). *Studia Geomorphologica Carpatho-Balcanica*, **VI**: 85–92. [4.8]

Komura S, Simons DB. (1967) River bed degradation below dams. *Journal of the Hydraulics, American Society of Civil Engineers* **93**: 1–14. [4.6]

Lane EW. (1947) The effects of cutting off bends in rivers. In: *Proceedings of the Third Hydraulics Conference*. Bulletin No. 31, University of Iowa, Iowa. [4.7]

Leopold LB. (1968) Hydrology for urban land planning: a guidebook on the hydrologic effects of urban land use. US Geological Survey Circular No. 554. [4.5]

Lewin J. (1977) Channel pattern changes. In: Gregory KJ (eds) *River Channel Changes*, pp. 167–84. John Wiley & Sons, Chichester. [4.8]

McClellan TJ. (1974) *Ecological Recovery of Realigned Stream Channels: Portland, Oregon.* Technical Report, Federal Highways Administration, US Department of Transportation, Portland, Oregon. [4.7]

Marshall EJP, Wade PM, Clare P. (1978) Land drainage channels in England and Wales. *Geographical Journal* **144**: 254–63. [4.8]

Meade RH. (1976) Sediment problems in the Savannah River Basin, In: Dillman BL, Steep JM (eds) *The Future of the Savannah River*, pp. 105–29. Water Resources Research Institute, Clemson, South Carolina. [4.4]

Meade RH, Trimble SW. (1974) Changes in sediment loads in rivers of the Atlantic drainage of the United States since 1900. In: *Effects of Man on the Interface of the Hydrological Cycle with the Physical Environment*, pp. 99–104. IASH Publication, No. 113. [4.6]

Milhous RT. (1982) Effect of sediment transport and flow regulation on the ecology of gravel-bed rivers. In: Hey RD, Bathurst JC, Thorne CR (eds) *Gravel-Bed Rivers*, pp. 819–92. John Wiley & Sons, Chichester. [4.1]

Morisawa M. (1985) *Rivers: Form and Process.* Longman, London. [4.5]

Morris LA, Langemeirs RM, Russell TR, Witt A. (1968) Effects of main stem impoundments and channelization upon the limnology of the Missouri River, Nebraska. *Transactions of the American Fisheries Society* **97**: 380–8. [4.7]

Moyle PB. (1976) Some effects of channelization on the fisheries and invertebrates of Rush Creek, Modoc County, California. *California Fish and Game* **63**: 179–86. [4.7]

Mullan JW, Starostka VJ, Stone JL, Wiley RW, Wiltzius WJ. (1976) Factors affecting Upper Colorado River Reservoir tailwater trout fisheries. In: Orsborn JF, Allman CE (eds) *Instream Flow Needs*, Vol. II, pp. 405–23. American Fisheries Society, Bethesda, Maryland. [4.6]

Neuhold JH. (1981) Strategy of stream ecosystem recovery. In: Barrett GW, Rosenburg R (eds) *Stress Effects of Natural Ecosystems*. John Wiley & Sons, New York. [4.7]

Nixon M. (1966) Flood regulation and river training. In: Thorn RB (ed) *River Engineering and Water Conservation Works*, pp. 293–7. Butterworths, London. [4.7]

Ovenden JC, Gregory KJ. (1980) The performance of stream networks in Britain. *Earth Surface Processes* **5**: 47–60. [4.8]

Parker G, Andres D. (1976) Detrimental effects of river channelization. In: *Proceedings of Conference: Rivers 1976, American Society of Civil Engineers*, pp. 1248–66. [4.7]

Pearson WD, Kramer RH, Franklin DR. (1968) Macroinvertebrates in the Green River below Flaming

Gorge Dam, 1964–65 & 1967. *Proceedings of the Utah Academy of Science, Arts and Letters*, **45**: 148–67. [4.6]

Pemberton EL. (1976) Channel changes in the Colorado River below Glen Canyon Dam. In: *Proceedings of the Third Interagency Sedimentation Conference*, pp. 5–61 to 5–73. Washington, DC. [4.6]

Pennak RW, Van Gerpen ED. (1947) Bottom fauna production and physical nature of the substrate in a northern Colorado trout stream, *Ecology* **28**: 42–8. [4.2]

Percival E, Whitehead H. (1929) A quantitative study of the fauna of some types of stream bed. *Journal of Ecology* **17**: 282–314. [4.5]

Petts GE. (1980) Long-term consequences of upstream impoundment. *Environmental Conservation* **7**: 325. [4.9, 4.11]

Petts GE. (1984) *Impounded Rivers: Perspectives for Ecological Management*. John Wiley & Sons, Chichester. [4.2, 4.6]

Petts GE. (1987) Time-scales for ecological change in regulated rivers. In: Craig JF, Kemper JB (eds) *Regulated Streams: Advances in Ecology*, pp. 257–66. Plenum Press, New York. [4.9]

Petts GE. (1988) Accumulation of fine sediment within substrate gravels along two regulated rivers, UK. *Regulated Rivers* **2**: 141–53. [4.6]

Petts GE. (1989a) Historical analysis of fluvial hydrosystems. In: Petts GE, Moller H, Roux AL (eds) *Historical Change of Large Alluvial Rivers: Western Europe*, pp. 1–18. John Wiley & Sons, Chichester. [4.1]

Petts GE. (1989b) Perspectives for ecological management of regulated rivers. In: Gore JA, Petts GE (eds) *Alternatives in Regulated River Management*, pp. 3–24. CRC Press, Boca Raton, Florida. [4.9]

Petts GE, Greenwood MG. (1985) Channel changes and invertebrate faunas below Nant-y-Moch Dam, River Rheidol, Wales, UK. *Hydrobiologia* **122**: 65. [4.1, 4.6]

Richards KS, Wood R. (1977) Urbanization, water redistribution, and their effect on channel processes. In: Gregory KJ (eds) *River Channel Changes*, pp. 369–88. John Wiley & Sons, Chichester. [4.5]

Roux AL, Bravard J-P, Amoros C, Pautou G. (1989) Ecological changes of the French Upper Rhône River since 1750. In: Petts GE, Moller H, Roux AL (eds) *Historical Change of Large Alluvial Rivers: Western Europe*, pp. 323–50. John Wiley & Sons, Chichester. [4.8]

Ruhe RV. (1970) Stream regimen and man's manipulation. In: Coates DR (ed) *Environmental Geomorphology*, pp. 9–23. State University of New York, New York. [4.7]

Schmal RN, Sanders DF. (1978) Effects of stream channelization on aquatic macroinvertebrates. Buena Vista Marsh, Portage County, Wisconsin. Report No. FWS/OBS-78/92. Office of Biological Services, Fish and Wildlife Service, US Department of the Interior, Washington, DC. [4.7]

Schumm SA. (1969) River metamorphosis. *Journal of the Hydraulics Division, American Society of Civil Engineers* **95**: 255–73. [4.3]

Schumm SA. (1973) Geomorphic thresholds and complex response of drainage systems. In: Morisawa ME (ed) *Fluvial Geomorphology*, pp. 299–310. State University of New York, Binghamton, New York. [4.10]

Schumm SA, Harvey MD, Watson CC. (1984) *Incised Channels: Morphology, Dynamics and Control*. Water Resources Publications, Littleton, Colorado. [4.9]

Shields FD Jr, Hoover JJ. (1991) Effects of channel restabilization on habitat density, Twentymile Creek, Mississippi. *Regulated Rivers* **6**: 163–82. [4.7]

Szilagy J. (1932) Flood control on the Tisza River. *Military Engineer* **24**: 632. [4.7]

Tarplee WH, Louder DE, Weber AJ. (1971) *Evaluation of the Effects of Channelization on the Fish Populations in North Carolina's Coastal Streams*. North Carolina Wildlife Resources Commission, North Carolina. [4.7]

Thoms MC. (1987) Channel sedimentation within the urbanized River Thame, UK. *Regulated Rivers* **1**: 229–46. [4.5]

Trimble SW. (1974) *Man-induced Soil Erosion on the Southern Piedmont*. Soil Conservation Society of America, Ankeny, Iowa. [4.4]

Voelz NJ, Ward JV. (1990) Macroinvertebrate responses along a complex regulated stream environment gradient. *Regulated Rivers* **5**: 365–74. [4.6]

Wesche TA. (1985) Stream channel modifications and reclamation structures to enhance fish habitat. In: Gore JA (ed) *The Restoration of Streams and Rivers*, pp. 103–63. Butterworths, Boston. [4.2]

White WD, Wells SG. (1979) Forest fire devegetation and drainage basin adjustments in mountain terrain. In: Rhodes DD, Williams GP (eds) *Adjustments to the Fluvial System*, pp. 199–223. Kendall-Hunt, Dubuque, Iowa. [4.4]

Whitlow JR, Gregory KJ. (1989) Changes in urban stream channels in Zimbabwe. *Regulated Rivers* **4**: 27–42. [4.5]

Williams GP, Wolman MG. (1984) Downstream Effects of Dams in Alluvial Rivers. United States Geological Survey Professional Paper 1286. [4.6]

Wolman MG. (1967) A cycle of sedimentation and erosion in urban river channels. *Geografiska Annaler* **49A**: 385–95. [4.4]

Wolman MG, Schick AP. (1967) Effect of construction on fluvial sediment, urban and suburban areas of Maryland. *Water Resources Research* **3**: 451–64. [4.5]

Yearke LW. (1971) River erosion due to channel relocation. *Civil Engineering* **41**: 39–40. [4.7]

5: System Recovery

A. M. MILNER

5.1 INTRODUCTION

This chapter is principally concerned with the ways in which biotic communities in river systems recover following disturbance. Five key aspects of recovery are reviewed: recovery pathways; indicator end-points used to determine recovery; the role of succession in recovery processes; recovery mechanisms of the biota; and factors affecting recovery rates. Some examples of recovery times following specific disturbances are included. In certain cases the extent to which recovery rates may be enhanced through restoration practices is indicated.

A brief set of definitions is presented here with specific reference to fluvial systems, but the reader is also referred to Chapter 6. Some of the principal problems in the field of recovery concern misinterpretation of terms through differing definitions. The term *disturbance* has suffered this problem due to its frequent use in a great variety of contexts (Rykiel 1985; Peters 1988). Disturbance in general terms is considered 'any relatively discrete event in time that disrupts ecosystem, community or population structure and changes resources, substrate availability or the physical environment' (White & Pickett 1985). Sousa's (1984) definition of disturbance is a discrete removal of organisms that are subsequently replaced by reproduction of resident survivors or by immigration. A key problem of defining disturbance by its effects is that disturbances may not be detected if they occur at a small spatial scale of low intensity and the rate of recovery is rapid, even with regular site monitoring (Lake 1990). In rivers, Resh *et al* (1988) narrow the scope of a disturbance to include only those physical and biological events that occur outside the normal range of predictable frequencies, severities or intensities for that system. For example, an annual spate that causes severe scour and bed material transport in a stream system would not constitute a disturbance within this definition (Gore & Milner 1990), unless its timing or magnitude was 'atypical' (Sparks *et al* 1990). The rationale is that riverine communities are adapted to the annual flood as a predictable event and have evolved various behavioural and reproductive strategies to ensure rapid recovery. In contrast, a 50-year flood in the same stream is unpredictable and would constitute a disturbance.

One practical problem with this definition of disturbance is determining what is the 'normal' or 'typical' range of predictable frequencies, severities or intensities; it may be impossible to determine whether an event is a disturbance without a long-term database for the system under examination (Gore *et al* 1990). This definition of disturbance also requires that recovery must be examined against these constant fluctuations, replacements and processes within the biotic communities in response to 'non-disturbance' events, for example an annual flood. Poff (1992) suggests that organisms are not necessarily pre-adapted to predictable environments; potential adaptation to disturbance predictability is itself dependent upon other attributes of the disturbance regime as some components are interactive. Most of the discussion surrounding disturbances in rivers concerns the effects of altered flow regimes. However, variations in discharge represent only one form (albeit one of the most important) of disturbance affecting riverine communities; other events, particularly those of anthropogenic origin, are unlikely to be predictable and hence will constitute a disturbance under any definition.

Recovery is defined by Davis *et al* (1984) as the process of species returning to 'normal' population levels after an impact. Again how is 'normal' defined? In many cases no or negligible pre-disturbance data exist to determine what 'normal' populations are in rivers and, in addition, over various temporal scales, freshwater communities are continually adjusting to biotic and abiotic factors within the ecosystem (see Section 5.2 below). Recovery is probably best viewed as a return to an ecosystem which closely resembles unstressed surrounding areas or source areas (Gore 1985) or to an unstressed condition (Kelly & Harwell 1990). Recovery is thus a natural process and is distinct from restoration, enhancement or rehabilitation.

Restoration is a process that involves management decisions and manipulation to enhance the rate of recovery (after Davis *et al* 1984). Consequently Gore (1985) defines river restoration as 'recovery enhancement' and considers it a technique to enable disturbed river ecosystems to stabilize at a much higher rate than through the natural physical and biological recovery processes of habitat development and colonization. *Enhancement* refers to improving the current state of the ecosystem without reference to its initial state, while *rehabilitation* is probably a mixture of enhancement and restoration (Magnuson *et al* 1980).

Resistance is the ability of a community to initially resist a disturbance while *resilience* refers to the ability of a disturbed community to recover to a state existing before the disturbance (*sensu* Webster *et al* 1975). Relative to other ecosystems riverine communities are considered to have a low resistance to disturbance (Reice 1985; Doeg *et al* 1989a) but a high resilience (Gurtz & Wallace 1984; Lake & Barmuta 1986; Lake *et al* 1989). High resilience is attributed to the relatively short life cycles of riverine species. The more resilient the fauna then the faster will be the recovery rates of biotic communities in rivers.

5.2 RECOVERY PATHWAYS

Figure 5.1 illustrates recovery pathways that may occur following disturbance to a community whereby recovery may result in different endpoints (A or B) rather than a return to the pre-disturbance community. The pre-disturbance community may only result by implementing restoration procedures. Depending upon the timescale necessary for recovery, the structure of the pre-disturbed community may, in any event, have changed through development and successional

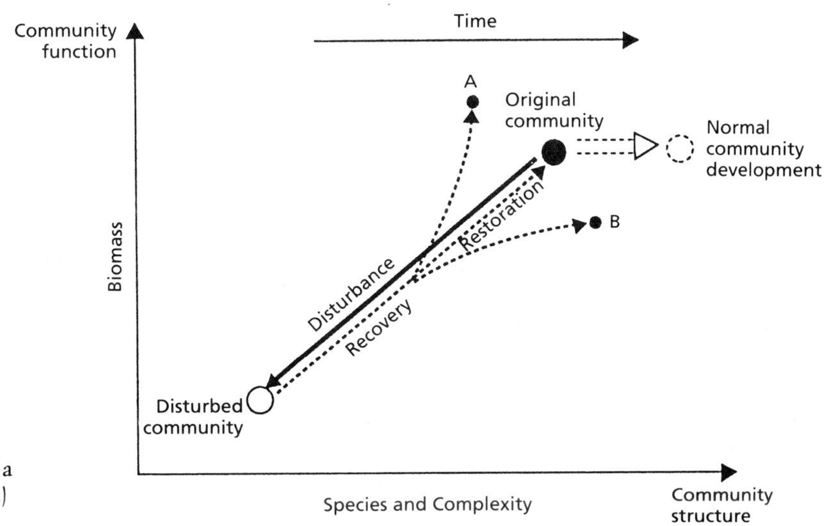

Fig. 5.1 Recovery pathways of a disturbed community (see text) (after Bradshaw 1988).

processes in response to the natural physico-chemical environment and habitat changes. Hence, if a pre-disturbance situation is to be achieved, restoration and recovery may require mimicking the presumed successional stage had the system been permitted to continue normal community processes (Cairns 1990); this is a difficult scenario for a river manager to achieve.

5.3 INDICATOR END-POINTS OF SYSTEM RECOVERY

What indicator or set of indicators constitutes a suitable end-point to recovery that is acceptable to such parties as the river manager, the developer, the polluter and the general public? The establishment of acceptable end-points to recovery is a complex issue and a criterion of immense importance in the management of rivers recovering from disturbances.

In attempting to characterize the initial and long-term recovery patterns of river ecosystems in terms of single, multiple, acute, chronic or cumulative disturbance events, one must select from a number of indicator end-points which reflect some characteristic of the system at some spatio-temporal scale of observation (Kelly & Harwell 1990). The end-points of recovery are potentially different according to the nature of the disturbance (Cairns 1990; Gore *et al* 1990).

The common end-points used in assessing the recovery of river systems are divided into either biotic or abiotic (Fig. 5.2). Biotic recovery is examined principally at the community level. A community can be defined as an assemblage of species populations which occur together in space and time. Although an ecosystem is generally regarded to represent a community together with its abiotic environment, it is impossible to consider a community in isolation from the environment in which it exists (Begon *et al* 1990). At the community level there are both structural and functional end-points to recovery.

Structural end-points cover the make-up of community assemblages in terms of such attributes as density, number of species, species diversity, etc. and typically involve comparison with the pre-disturbance or a reference com-

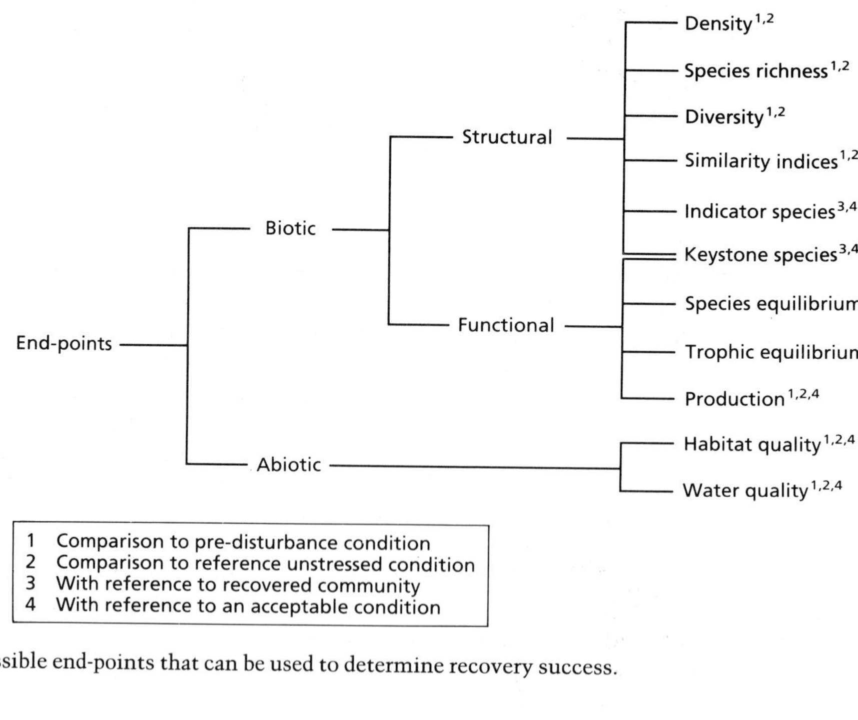

Fig. 5.2 Possible end-points that can be used to determine recovery success.

munity. Similarity indices such as Jaccard's Index (Southwood 1978) can be used to compare the structure of a recovered community with a reference community, as shown by Townsend *et al* (1987) and Gore (1979), who compared the persistence of communities several years apart. Indicator species are individual species selected for their ability to provide an indication of the well-being of the habitat or water quality within a river following recovery, e.g. stoneflies or salmonids. Keystone (or gold star) species can be defined as species which are of special significance to the structure and function of the riverine community and may have important commercial, recreational or conservation value. Examples would be the recovery of recreationally or commercially valuable fish species, such as trout and salmon, or species of conservation value such as the otter. A species could potentially be both an indicator and a keystone, for example salmonid species.

Functional end-points refer to the functioning of the community, for example species equilibrium, and may be evaluated independently of a reference community. It may be that functional end-points are more important when a particular purpose is attempting to be fulfilled (Kelly & Harwell 1900); Another potential indicator of recovery is trophic structure of the macroinvertebrate functional feeding groups (see Vol. 1, Chapter 11 for definitions). Predators are typically the last trophic group to arrive during the recovery process (Gore 1982), and Minshall *et al* (1983) found that trophic structure by functional group was not attained at the end of 3½ years after a flood-induced disturbance. Production is a further possible recovery end-point but has rarely been used in recovery studies due to the intensive sampling and analysis needed for its measurement. However, it is potentially a very important recovery end-point and recent techniques make its calculation considerably easier.

Abiotic end-points to recovery are considered to be such parameters as physical habitat quality or water quality and are assessed independently of the community present. Depending upon the disturbance involved, some examples of physical habitat quality might include percentage of fine sediment among the gravels/cobbles and heterogeneity of the substrate. Water-quality end-points might include such parameters as dissolved oxygen, pH, total suspended sediment or heavy metal concentration and typically reflect performance standards set by a regulatory body.

Most recovery studies have reported density, species richness or a combination of the two through a diversity index as end-points. These structural end-points can, however, indicate the completeness of recovery before it has actually occurred and thus may prove misleading (Shaw & Minshall 1980; Minshall *et al* 1983). Maximum densities may ignore the dominance or species replacement by disturbance-tolerant forms (Gore & Milner 1990). For example, as shown by Krumholz and Minckley (1984), an equilibrium of pollution-tolerant species at densities comparable to pre-disturbance levels can be expected under situations of chronic organic loading (sewage outfalls). Species equilibrium (where rate of immigration of colonizing species is equal to the rate of extinction) frequently lags behind maximum density because of a period of dynamic adjustment before equilibrium. For example Minshall *et al* (1983) reported that following a major flood macroinvertebrates took 400 days to reach maximum densities but between 600 and 650 days to achieve species equilibrium.

However, certain communities may never attain species equilibrium due to constant successional adjustments to abiotic events or disturbances. Some river systems are believed to maximize species diversity by being maintained in a non-equilibrium state reducing competitive hierarchies through intermediate disturbances (Ward & Stanford 1983). Reice (1985) suggests that the intermediate disturbance hypothesis probably does not apply to streams and rivers because competitive hierarchies have not been widely demonstrated. The dynamic equilibrium model (Huston 1979) is proposed by Reice *et al* (1990) as the most appropriate for rivers. This model suggests that the recurrence interval of disturbance events is shorter than the time necessary for competitive or predator–prey interactions to cause the elimination of species. Hence species equilibrium within riverine communities may be an unattainable end-point; this subject creates much debate among river ecologists.

Another important question regarding end-points is whether the structural and functional end-points of a recovered or restored community can be maintained without constant management. Cairns (1990) considers that unless a self-sustaining community based on natural reproduction, succession and adaptation is attained then system recovery has not been achieved.

5.4 RECOVERY AND SUCCESSION

Succession is most simply defined as community changes that occur at a site following a disturbance (after Fisher 1990). Statzner and Higler (1985) suggest that if disturbances cause losses and gains of species, then it would be expected that the biological community in rivers is re-established by means of succession and therefore river systems cannot be viewed independent of time. One of the problems is that successional changes are superimposed and confounded by diel, seasonal and long-term schedules which also produce changes in the community (Fisher 1990).

Primary succession occurs after a disturbance that is so intense that no trace of the previous community exists (Fisher 1990) or where a new channel is formed. Secondary succession can be viewed as community change at sites that have been structured physically and chemically by previous resident assemblages and where organic matter remains to support initial colonizers (after Gore & Milner 1990). It is a rare disturbance in lotic systems that eliminates all organisms from a stretch of river and thus there are a limited number of examples of recovery through primary succession (see Section 5.7 below). Except in the severest of floods and droughts, organisms can obtain refuge in the hyporheic zone and act as a colonizing source within the stream section as pre-disturbance conditions return; hence this is not primary succession (Fisher 1990). The colonization of artificial substrates placed in rivers could be considered an example of primary succession. However, because the scale is so small substrates are likely to be overwhelmed by continued rapid colonization from adjacent communities so that there is no opportunity for distinct successional communities to exist. Therefore, the difference between primary and secondary succession is probably only discernible at scales that limit colonizers. In rivers this scale probably reaches 10^3-10^4 m in length (Fisher 1990). Exceptions to this scale have been demonstrated by McCormick and Stevenson (1991), who found distinct successional algal communities on tiles placed in a stream.

5.5 MECHANISMS OF RECOVERY

Good dispersal abilities, high production rates, short generation times and flexible life history strategies typically enable periphyton species to recover rapidly following disturbances (Steinman & McIntire 1990). Recovery times may be considerably longer if propagule sources upstream from the disturbance area are impacted (Resh *et al* 1988).

Four principal migration sources and mechanisms for stream benthos have been identified: (i) downstream migration or drift; (ii) upstream migration; (iii) vertical migration from within the substrate (the hyporheic zone); and (iv) aerial sources (Williams & Hynes 1976; Williams 1981). Depending upon the disturbance, some or all of these sources may be available in river systems recovering from disturbances. For an artificial stream in Canada where upstream sources of colonizers were present, Williams and Hynes (1976) concluded that the order of importance for recovery in June was drift > aerial sources > movement from the hyporheic zone > upstream movement. Where present, downstream drift is typically the most important recovery mechanism of the benthos (Niemi *et al* 1990) although this is not always the case. Doeg *et al* (1989b) found that for gravel and stones, crawling across the substratum was the most common method of colonization although the scale was small.

For certain major disturbances (e.g. elimination of fauna along an entire channel without upstream sources of drift) aerial colonization represents the major, if not only, source for recovery. Chironomids (Diptera) have a high dispersal ability, being small winged and relatively light (Brundin 1967). Generally most Ephemeroptera (mayflies), Trichoptera (caddisflies), and many

Plecoptera (stoneflies) are considered weak fliers, while Odonata (dragonflies and damselflies), some Coleoptera (beetles) and Hemiptera (bugs) are good fliers (Williams 1981). Even if adult insects are able to invade a drainage there is no certainty that they will colonize the stream. Five years after the Mount St Helens eruption, caddis flies, which were relatively abundant as adults in the impacted drainages, were either absent or at low abundances in the streams (Anderson & Wisseman 1987). Molluscs are typically one of the last taxa to recover following disturbances (Wallace 1990), presumably as a result of poor dispersal mechanisms.

Mechanisms of recovery in fish include redispersal of surviving adults within the watershed and increased recruitment of young through various age classes. In severe disturbances where fish are completely eliminated, fish must colonize from other streams or via the straying of anadromous adults returning to spawn.

5.6 FACTORS AFFECTING RECOVERY RATES

This section overviews the principal factors that influence recovery rates for any particular disturbance. Some attempt is made to prioritize these factors according to their overall influence, beginning with those factors having the most effect on recovery rates. The effects of these factors are summarized in Table 5.1. The scale in this table is arbitrary and relative; a factor of 0 has little effect while a factor of −3 or +3 has the greatest effect, either retarding or enhancing recovery rates respectively.

Physical habitat

Disturbances that impact physical habitat probably exert some of the greatest influences on system recovery. Physical habitat impacts include the introduction of fine sediment onto the river bed from road construction and logging, the removal of large woody debris from rivers, the loss of heterogeneity of the river bed substrate and the lack of diversified flow patterns through channelization and the loss of cover through removal of riparian vegetation These impacts severely influence colonization processes by the biota, frequently resulting in long recovery times. For example, sediment accumulation between the bed cobbles may restrict access to the underside of stones for invertebrates and may eliminate or significantly modify the hyporheic habitat (Campbell & Doeg 1989). For this factor management assistance through restoration may then be needed if recovery rates are not to be greatly retarded.

Recovery of terrestrial and riparian vegetation

River communities are closely linked to terrestrial plant conditions in the surrounding watershed (Cummins & Klug 1979; Vannote *et al* 1980). Hence the recovery of trophic functional groups within the invertebrate community may, in certain disturbances, be long term and dependent upon the recovery of terrestrial vegetation. Fish recovery may also depend on the suitable regrowth of riparian vegetation. For example, redevelopment of riparian habitat was a major factor in the recovery of coho stocks in an impacted drainage following the Mount St Helens eruption in western Washington, USA (Martin *et al* 1986) and of trout populations following a catastrophic debris flow in an Oregon stream (Lamberti *et al* 1991).

Size of disturbed area/upstream sources of drift

Spatial scales are very important in determining recovery rates. The larger the area affected by the disturbance then the greater the distance to colonizing sources, particularly in the case of major disturbances. Source distance effects described by Gore (1979, 1982) are then significant and will determine the immigration rates of colonizing species (see Section 5.7, 'Channel modifications'). Another major factor governing the immigration of invertebrate colonizers following disturbances is whether upstream sources of drift are present. As discussed previously under recovery mechanisms, drift is a major source for colonization and where present significantly increases recovery rates.

Table 5.1 Factors that can potentially influence recovery rates of the biota (the solid bar equals definite + or − to end of that range where the broken bar means variable over that range)

Factor	Influence of factor	Level at which factor influences recovery	Lower recovery rate (−3 −2 −1 0)	Increased recovery rate (0 +1 +2 +3)
Physical habitat	Alteration influences colonization processes	Severe alteration	━━━━━━	
		Minor alteration	━━	
Recovery of terrestrial and riparian vegetation	Affects recovery of certain trophic groups and fish populations	Vegetation growth frequently long-term process	━━━━━━	
Size of disturbed area/ source of drift	Affects immigration of colonizers	Large/no upstream sources of drift	▬ ▬ ▬ ▬ ▬	
		Small/sources of drift present		▬ ▬ ▬ ▬ ▬
Continued disturbances	Arrests succession and recovery	Frequent/large	━━━━━━	
		Infrequent/small	━━	
Previous disturbances	History of events in system	Similar to present disturbance		━━━━━━
		Dissimilar to present disturbance		━━
	No history—relatively stable physical/ chemical environment		━━	
Presence and proximity of refugia	Provides colonizing source	Present/close		▬ ▬ ▬ ▬ ▬
		Absent/far	▬ ▬ ▬ ▬ ▬	
Flushing capacity	Affects persistence of disturbance	High		▬ ▬ ▬ ▬ ▬
		Low	▬ ▬ ▬ ▬ ▬	
Land use/basin characteristics	Affects water quality	Enhanced contaminant input and toxicity	▪ ▬ ▬ ▬	
		Reduced contaminant input and toxicity		▬ ▬ ▬ ▪
Time of year/life cycle stage	Affects availability of colonizers and community resilience	High latitudes — ice-covered season	▪ ▬ ▬ ▬	
		Low latitudes		▬ ▬ ▬ ▬ ▬
Nutrient input and turnover time	Low nutrient inputs and slow turnover reduces system resilience	Slow turnover	▬ ▬	
		Rapid turnover		▬ ▬
Location of disturbance	Lower in river system	Increases diversity of food resources		▬ ▬

Continued disturbances

Continued disturbances will arrest the successional pathways and hold recovery in an early successional stage, thereby increasing recovery times. This continuance of the disturbance parallels the difference between a *pulse disturbance* and a *press disturbance* as defined by Bender *et al* (1984). A pulse disturbance is a disturbance of limited and easily definable duration whereas press disturbances are longer and frequently involve changes in the watershed or the river channel. Examples of pulse disturbances are floods or inputs of non-persistent pollutants while press disturbances include logging, mining, persistent toxic chemicals and stream channelization.

History of previous disturbances

Colonization by biotic communities will probably occur more rapidly in river systems with a history of episodic disturbances due to the potential selection for species within the community that can rapidly recolonize following displacement. Hence a community that experiences regular and frequent floods will likely recover more rapidly to similar type events than a community in a more stable physical environment. It is probable that these communities will also recover more rapidly to dissimilar events such as point- or non-point-source inputs of pollutants. This more rapid recovery is principally due to the adaptive strategies of species within the community for coping with frequent disturbance by selecting for opportunistic species with high vagility, high fecundity and early maturation. Communities living in a more stable physical/chemical environment are likely to be less adapted to respond rapidly when disturbed. For example, unpredictable droughts in normally permanent streams may result in invertebrate recovery rates in excess of 1 year (Ladle & Bass 1981) but in regions with seasonally predictable dewatering, recovery may occur in less than 2 months (Sagar 1983). In addition in many rivers stressed by pollution, the pre-disturbance fauna is made up of pollution-tolerant species which are likely to recover quicker than a diverse pristine fauna.

Presence and proximity of refugia

There are many types of refugia within a river system including localized microhabitats and/or zones within the channel, unique reaches, riparian vegetation, floodplains and groundwaters (Sedell *et al* 1990). In many cases refugia may be inhabited by biota that are different or transient in comparison to the communities that typically occur in the channel. Nevertheless these refugia function as source areas for recolonization of rivers following disturbances that reduce biomass or diversity (Sedell *et al* 1990). If the disturbed river section is far from available refugia, then the vagility of adults, generation times and fecundity of the colonizers assume more importance in dictating recovery rates (Niemi *et al* 1990).

Flushing capacity/persistence of disturbance

The persistence of a disturbance will have a major influence on the ability of a river system to recover. Where flushing capacities (i.e. the ability of a river to flush out contaminants or sediments) are high the persistence of certain disturbances will be reduced and recovery rates enhanced. This persistence is particularly the case with point- and non-point-source chemical pollutants, particularly where the pollutants become incorporated into the sediments, and with disturbances that influence the structural habitat of the river. For example, sediment inputs from highway construction or logging can severely impact benthic organisms and survival of juvenile fish (Everest *et al* 1987). Sand deposition on river substrate has been shown to hinder upstream movements of aquatic insects (Luedtke & Brusven 1976). Deposited sediment may be difficult to remove without substantial flushing flows.

Land use/basin characteristics

Land use within a catchment will influence the transport of contaminants to a river and affects the magnitude and frequency of disturbances. For example, poor management of drainage patterns on agricultural land without riparian vegetation may increase the runoff of sediment and other potential contaminants into the river. This may result in continued disturbances and retard recovery times. Basin geology influences the ability of a watershed to buffer water-quality degradation within a river, which in turn influences the toxicity and persistence of various chemical pollutants and thereby affects recovery rates. For example, metal toxicity will be greater at lower hardness levels, which could contribute to longer recovery rates.

Time of year of disturbance and life cycles of the biota

The time of year that a disturbance occurs may significantly affect the ability of certain species to follow dispersal routes for colonization, especially in the case of aquatic insects. This is particularly evident at higher latitudes where macroinvertebrates have long seasonal life cycles compared to more milder climes where life cycles are frequently aseasonal (Wallace 1990). For example, some multivoltine taxa (e.g. Chironomidae) may pass from the egg through adult stages in under 2 weeks in warm-water streams (Fisher & Gray 1983). At higher latitudes many species are univoltine (one generation a year) while in some taxa (e.g. Plecoptera) certain species have one generation every 2 or 3 years (Oswood *et al*, in press). Hence aerial colonization by aquatic insects following disturbances at these higher latitudes can only occur at certain times during the ice-free season and in certain species with long life cycles potentially only once every 2 or 3 years.

Even in more temperate latitudes, the timing of the disturbance during the year may significantly influence recovery rates. An atypical summer flood in an Irish river caused longer recovery times to summer invertebrate assemblages than a comparable more common flood in the winter (Giller *et al* 1991) while colonization in natural gravel patches has been shown to be more rapid in summer than winter in an Australian river (Doeg *et al* 1989a). The time of the year also determines which life stage of an organism is present following a disturbance; for example, during drought some animals survive as eggs which enhances recovery when suitable conditions return (Morrison 1990).

Thus the sequence of colonization, succession and possible end-point community may be significantly influenced by which species are able to become part of the colonizing pool at the time following the disturbance.

Nutrient input and recycling

Disturbances which affect autotrophic production or the allochthonous inputs to a river will influence recovery rates of benthic organisms and fish (DeAngelis *et al* 1990). Productivity in some streams has been shown to be nutrient limited (e.g. Peterson *et al* 1985) indicating that disturbances which alter the nutrient loading to, and recycling within, a river channel will also influence the system's ability to recover. Where nutrient inputs are low and there is a low turnover and cycling of nutrients, resilience will consequently be low (Webster *et al* 1975) and thus recovery times will be longer (DeAngelis *et al* 1990). In rivers there is the further complicating factor of flow, which tends to move allochthonous nutrients downstream. Webster and Patten (1979) suggested the term 'nutrient spiralling' to describe the coupled process of nutrient cycling and transport, and the 'tightness' of the spirals is an indication of the retention and reutilization of a nutrient. Retention devices, such as fallen trees, will increase the tightness of the spirals, and, if retention efficiency is disturbed, recovery times will likely be increased. Restoration techniques in rivers sometimes involve increasing retention efficiency.

Location of disturbance in stream course and stream order

The farther downstream a disturbance occurs then the greater the upstream source of colonizers (particularly for colonization by drift) and, in light of the river continuum concept (Vannote *et al* 1980), the greater the availability of different types of food resources due to upstream processing (Gore & Milner 1990). Higher stream orders (larger rivers with many tributary inputs) may thus potentially recover more rapidly.

5.7 RECOVERY RATES

Levels of disturbance

From the foregoing, it is evident that many factors influence recovery rates, which consequently can be extremely variable. Recovery rates also depend upon the end-point selected. Nevertheless some general conclusions are drawn here about the ranges of recovery rates for any particular disturbance. Most of the recovery times given here refer to time to attain maximum densities, biomass or taxa richness unless otherwise stated. In the case of macroinvertebrates, 90% of the cases reviewed by Niemi *et al* (1990) indicated recovery times within a year. This section covers only a small percentage of the recovery studies reported, more to provide a range and indicate circumstances where recovery is longer than is typical for any particular disturbance. These ranges are

Table 5.2 Range of recovery times for macroinvertebrates and fish following various disturbances (the broken bar equals tentative range)

Disturbance		Level	Recovery time in years: Macroinvertebrates	Recovery time in years: Fish
Watershed disturbances	Logging	3 and 4	5.0 – >20.0	
	Mining	2, 3 and 4	15.0 – >20.0	
	Fires	3	3.0 – 5.0 (tentative)	
	Volcanic activity	1	5.0 – >20.0 ? (tentative)	5.0 – 10.0
Channel modifications	New channel (no upstream colonizers)	1	10.0 – >20.0	
	New channel (upstream colonizers present)	2	~0.3 – 1.0	2.0 – 20.0
Floods	Adapted fauna	1B and 3	0.1 – 0.4	
	Non-adapted fauna	1A(?) and 3	~0.35 – 3.0	~1.5 – 3.5
Drought	Adapted fauna	2 and 3	0.1 – 0.4	
	Non-adapted fauna	2 and 3	0.8 – 1.0	
Pesticides	Indirect application	2 and 3	1.0 – 3.0	1.0 – ~2.5
	Direct application	2	~0.3 – 5.0	
Chemical spills and toxic pollutants	Persistent	2 and 3	2.0 – 10.0	
	Non-persistent and flushed-out	2 and 3	0.5 – 1.0	1.0 – ~2.5
Organic pollutants		3	0.5 – 1.0	

Scale: 0, 0.2, 0.4, 0.6, 0.8, 1.0, 2.0, 3.0, 4.0, 5.0, 10.0, 15.0, 20.0, >20.0

▬▬ Macroinvertebrates; ••••• Fish

summarized in Table 5.2. The reader is directed to excellent reviews by Yount and Niemi (1990) and Niemi *et al* (1990) for a more detailed treatment of this aspect of recovery.

To aid in determining likely recovery times from disturbances, Gore and Milner (1990) suggest that disturbance first be categorized at a number of different levels.

Level 1A

Disturbance completely destroys communities along the entire stream length leaving no upstream or downstream sources to colonize and may result in a new stream channel. This level of disturbance leads to recovery through primary successional processes. Examples include volcanic deposits and ice movements.

Level 1B

Disturbance completely destroys surface communities along the entire stream length leaving no upstream or downstream sources to colonize; but there remains a remnant hyporheic component or near-surface source of eggs and smaller instars. Such disturbance probably leads to recovery via secondary succession. Examples are floods in low-order streams with no confluence with a downstream source (e.g. some small desert streams and coastal streams that empty directly into the ocean) and some forms of drought.

Level 2A

Disturbance completely destroys communities in a reach of stream but upstream and downstream colonization sources remain, leading to recovery through primary succession. Examples are chemical spills, reclaimed or diverted river channels, and surface mining effects.

Level 2B

As 2A, but the hyporheic zone is not destroyed, leading to secondary succession. Examples are application of insecticides, some chemical spills and some types of dewatering.

Level 3

This level of disturbance results in reduction of species abundance and diversity from pre-disturbance levels in a section of stream but does not completely eliminate the benthos, leading to secondary succession recovery processes. Examples include floods and chronic non-point- and point-source pollutants.

Level 4

Disturbances of this level result in reduction of species abundance and/or diversity or loss of benthos compared to pre-disturbance levels in discrete patches within a stream section but such that proximal patches are unaffected. Recovery is by secondary succession. As previously pointed out, if these patches represent too small a scale, then succession does not take place. Examples are sediment inputs from highway construction and logging.

Watershed disturbances

One of the principal watershed disturbance effects on rivers arises from timber harvest through the effects of clear-cutting, road construction and in certain areas planting (levels 3 and 4). Commercial logging and associated road building may cause a variety of effects on stream communities, including increased sediment deposition (Everest *et al* 1987), increased discharge, increased stream temperatures (Beschta *et al* 1987), alterations in the quantity, quality and timing of allochthonous matter, and increases in autochthonous primary production (Webster *et al* 1983; Duncan & Brusven 1985a,b). Clear-cutting of a watershed is a large-scale, low-frequency, anthropogenic disturbance and the stream biota have not evolved appropriate adaptations to such disturbances (Gurtz & Wallace 1984), hence recovery can be lengthy.

This type of disturbance illustrates the problems of determining recovery rates where cumulative effects influence riverine communities. Cumulative effects result from environmental change due to the accumulation and interaction of localized or small impacts which increases the effect on the resource (Burns 1991). For example, increased sediment delivery to a river channel from logging practices, combined with altered sediment transport and storage downstream due to changed large woody debris inputs, may cause stream-bed aggradation, channel widening and degradation of the physical habitat (Smith 1989). Cumulative effects are not always synergistic in their action and may be antagonistic. Increases in streamflow may moderate the effects of increased stream temperatures that result from the removal of stream vegetation (Chamberlain *et al* 1991). In logged watersheds of the southern Appalachian Mountains, USA, there was a significant shift in energy base to autochthonous production (Webster *et al* 1992) accompanied by an associated shift in benthic invertebrates dominated by scrapers and collectors (Gurtz & Wallace 1984; Wallace *et al* 1988). Wallace and Gurtz (1986) found that the invertebrate taxa (*Baetis* and chironomids) that increased possess short generation times and respond rapidly. A similar increase in primary production was observed in low-order streams of southeast Alaska (Duncan & Brusven 1985a) although no significant alteration in the ratio of functional groups was apparent (Duncan & Brusven 1985b). In addition responses by invertebrates to major watershed disturbances like logging are frequently substrate mediated with more taxa recovering in larger substrates (Gurtz & Wallace 1984).

Thus potentially after the deleterious effects of sediment from timber harvest and road construction have been flushed out, the invertebrate densities and biomass in the clear-cut area could be considered acceptable when compared to the predisturbance condition. However, the trophic structure of the benthic community (particularly the shredders) could take considerably longer to recover and be dependent upon riparian zone development to supply allochthonous leaf input (Haeffner & Wallace 1981; Wallace *et al* 1988). These results again indicate the problems for a river manager in determining what is an acceptable end-point to community recovery.

The increased invertebrate biomass may be considered of value for it will potentially lead to increased growth and abundance of fish. Although this effect has been reported in a number of logged southeastern Alaskan streams, increases in juvenile salmon growth and densities may be nullified by a shortage of essential winter habitat (Heifetz *et al* 1986; Murphy *et al* 1986), leading to increased overwintering mortality. Hence the manager of a river in a watershed disturbed by logging may decide that the best end-point may be a keystone species, for example the abundance and production of salmon. Physical habitat quality may have an extremely long recovery time for river channels in watersheds disturbed by logging, particularly in certain geographical areas. Large woody debris (material >10 cm in diameter and >1 m long – see Murphy & Koski 1989) is an important structural part of salmon habitat in stream orders 1–4 in forests of the Pacific Northwest in the USA. A model developed by Murphy and Koski (1989) predicts that 90 years after logging, a 70% reduction in large woody debris would occur and recovery to pre-logging levels would take at least 250 years. Four case histories of the effects of logging on the response and recovery of stream channel habitat indicated that channel recovery ranged from 5 to 60 years and recovery of riparian vegetation ranged from 100 to 200 years (Sullivan *et al* 1987).

Strip mining (level 2 and 3 disturbances) within a watershed also causes long recovery times of the benthos due to the regrowth of vegetation, with estimates of 20–25 years to reach pre-disturbance densities and species composition (Tolbert & Vaughan 1979).

Fires are an integral part of many ecosystems and constitute a significant watershed disturbance (level 3). Fires alter large woody debris, sediment suspension, nutrient cycling and leaf litter input. Heavy accumulations of fine sediment diminish invertebrate standing crops and cause recovery to be delayed until the sediments are flushed out. Following the 1988 fires in Yellowstone National Park, USA, Minshall *et al* (1989) estimated that benthic invertebrates would take at least 3 years to recover.

The Mount St Helens eruption in May 1980 also provided an opportunity to study the recovery of stream communities after a major disturbance to the watershed. In most watersheds near the blast zone, riparian vegetation was eliminated and ash deposits of a few millimetres to over 5 cm were reported (Cook *et al* 1981). Rushforth *et al* (1986) reported that within 15 months or less of the eruption, algal communities were established in streams throughout the blast zone although due to continued disturbances some of these communities remained in an early successional stage (principally diatoms). Two years after the eruption, few aquatic insect species had reinvaded the denuded streams, and Wilzbach *et al* (1983) suggested that it may be 100 years for complete recovery in some streams due to the instability of channel banks and substrates which lead to continued disturbances at successive high-flow events. Recovery is proceeding more rapidly in watersheds where some or all of the forest remained intact compared to watersheds where the forest was completely destroyed (Hawkins & Sedell 1990). Restoration measures have been employed in a number of drainages to enhance recovery but these have caused a number of controversies (see Franklin *et al* 1988).

Channel modifications

Channel relocation and reconstruction projects are relatively rare but provide an indication of recovery times where a major physical modification of the river channel occurs. These cases are not recovery *per se* as there is no original condition with which to compare but provide information on colonization and community development within a primary successional framework with reference areas used for comparison. These projects represent level 2 disturbances as there is typically an upstream source of colonizers. However, channelization of rivers can lead to long recovery times for both fish and macroinvertebrate communities (Niemi *et al* 1990), particularly where upstream sources are absent and where the physical habitat is impaired.

In a colonization study of a newly formed channel in the Tongue River, Wyoming, USA, maximum densities of macroinvertebrates were obtained in less than 90 days with equilibrium levels reached in about 200 days when source areas were less than 200 m upstream (Gore 1979, 1982). Recovery times increased proportionally according to distance from source areas of colonizers. Macroinvertebrate recovery was dependent upon accumulation of organic detritus and establishment of periphyton populations for the invasion of grazers and detritivores. Predators were the last trophic group to arrive and trophic equilibrium was attained after taxa equilibrium. For a new stream channel associated with the Ponesk Burn in the uplands of Scotland, Doughty and Turner (1991) reported maximum diversity and similarity to a reference site after 80–100 days of invertebrate colonization, and taxa equilibrium at one site was reached after 140–150 days. Trophic equilibrium was reached at this site after 476 days. Equilibrium was not attained at the other two sites. Source distance effects on colonization rates were also observed in this study. Aquatic insects took 5.5 years to achieve similarity to control areas after a channel relocation project in Wisconsin, USA; the long recovery rate being attributed to sandy substrates and lack of coarse particulate organic matter (Narf 1985). In contrast, Malmqvist *et al* (1991) found that invertebrate diversity in a man-made channel (1.5 km long) was similar to that in reference lake-outlet streams within 1 year.

Overall, where channelization reduces substrate heterogeneity and stability, it is likely to reduce the abundance and species richness of the colonizing invertebrate communities (Hortle & Lake 1982). Resident fish populations (mostly Centrachidae) were not established in the new channel of the Tongue River 2 years after colonization had commenced despite the placement of cover structures and boulders to diversify flow patterns and enhance recovery (Gore & Johnson 1980). Similarly in the new channel associated with the Ponesk Burn, brown trout populations had not reached the levels of the reference site after 2 years although recovery was retarded by further disturbances (Doughty & Turner 1991). One of the longest recovery times in the literature relates to a channelized section of the Luxapalila River, Mississippi, USA, where fish populations were considered not to have recovered after 52 years (Arner *et al* 1976).

A recently formed stream following glacial recession permits a unique opportunity to examine the colonization of a new stream channel along its entire length without upstream source areas (equivalent of level I disturbance). After approximately 25 years of development, species equilibrium and maximum diversity have not yet been reached and no non-insect forms have colonized (Milner 1987 and unpublished). However, maximum densities were attained some 14 years earlier. This study illustrates that colonization (and hence recovery) is a long-term process when there are no upstream source areas for drift and refugia areas are distant.

Floods

Floods are rarely of sufficient magnitude to remove the entire stream biota (Minshall *et al* 1983) and so typically fall under level 3 disturbances. Nevertheless floods constitute significant and common disturbances to community structure and function, and recovery times can be relatively long, particularly where there is no or little history of floods in the system. During floods, scour and redistribution of sediment can influence the physical habitat and thereby restrict the ability of macroinvertebrates and attached algae to colonize.

Frequent floods can truncate algal successional sequences and favour well-attached resistant taxa or taxa with recolonization and growth rages high enough to accumulate biomass between floods (Power & Stewart 1987). These authors found that following a flood the filamentous alga *Rhizoclonium* (with specialized basal attachment cells) was reduced four-fold in occurrence whereas *Spirogyra* (usually unbranched and free-floating) decreased by a factor of 20. The distributional pattern of these algae was restored after 26 days. Although filamentous forms of algae are more susceptible to scour during flood events, habitats colonized by certain filamentous forms are frequently characterized by high light and nutrient levels, thereby enhancing faster growth and recovery rates (Steinman & McIntire 1990). In lowland desert streams, gross primary production has been shown to exceed respiration after 5 days following flash flooding (Fisher *et al* 1982).

As mentioned previously, Minshall *et al* (1983) reported recovery times of 650 days for macroinvertebrates to attain species equilibrium after a major flood on the Teton River, Idaho, USA (possibly level 1 leading to primary succession) although maximum densities were reached significantly earlier. After a flood, which had reduced densities to 5% of pre-flood levels in a small Irish river, macroinvertebrates only reached 50% of previous densities after 2 years (Giller *et al* 1991). However, there was no apparent effect on the resident salmonid populations. Molles (1985) reported that biomass and diversity of invertebrates was similar within 1 year of a flash flood on a northern New Mexico stream although differences in community structure remained after 2 years.

In contrast, recovery rates are typically more rapid in rivers where the fauna are adapted to frequent floods. In an unstable, braided river in New Zealand, where flows were highly variable and floods common, numbers and densities of the invertebrate fauna had recovered to pre-flood levels within 132 days following a major flood (Scrimgeour *et al* 1988). In Arizona desert streams in the USA, invertebrate densities recovered within 21 days after flash floods (Fisher *et al* 1982) and within 6 weeks after a 100-year magnitude flood (Collins *et al* 1981).

Matthews (1986) reported fish similarity to have recovered in 8 months following a catastrophic flood in an Arkansas stream (USA), while Meffe and Minckley (1987) showed rapid recovery of fish in desert streams due to behavioural responses during flooding which allowed the native fish to persist. Rainbow trout recolonized a 1-mile stretch of a Californian river within 4 years following the complete elimination of the resident population (Lambert 1988).

Drought

In contrast to floods, drying of stream channels normally occurs gradually and thus many species have evolved life history or behavioural adaptations for rapid recovery when flow returns. Most studies report macroinvertebrate communities recovered within 1 year (e.g. Larimore *et al* 1959; Ladle & Bass 1981, Griswold *et al* 1982; Morrison 1990) and within 3 months where the community has a history of similar droughts (e.g. Harrison 1966; Sagar 1983). Fish were also shown by Larimore *et al* (1959) to have recovered within 1 year.

Pesticides

Pesticides cover a wide range of toxic chemicals used to control or eradicate undesirable organisms and include insecticides and herbicides. Pesticides can affect waterbodies either indirectly from application to or within the watershed or directly to the river. Recovery times can vary from <1 month to >25 years depending upon the magnitude, toxicity and continued use of the pesticide and its overall persistence (Wallace 1990). This disturbance together with chemical discharges and spills probably causes the greatest variability in recovery times. This topic has been reviewed extensively by Muirhead-Thomson (1987).

Indirect forest spraying with pesticides typically results in recovery times for invertebrates of less than 3 years (Hoffman & Drooz 1953; Dimond 1967; Kreutzweiser & Kingsbury 1987). Similar recovery times have been reported for trout and Atlantic salmon fry and parr (Keenleyside 1967; Warner & Fenderson 1962). Pesticide application frequently induces post-spray drift of invertebrates (Morrison 1977), which thereby become susceptible to predation and physical damage (Kreutzweiser & Kingsbury 1987).

Most direct applications of pesticides have been the result of experimental additions. An acute experimental spill with the insecticide parathion showed macroinvertebrate diversity recovered within 3 months (Ghetti & Gorbi 1985), while Morrison (1977), using experimental treatments with rotenone in three Scottish hill streams, documented densities of invertebrate fauna recovered within 1 year. The rotenone treatment caused increased drift and some mortality of invertebrates. Cook and Moore (1969) reported aquatic insect taxa returned to pre-disturbance levels within 3 months (except Plecoptera – 6 months) following rotenone treatment of a Californian stream. Densities and species diversity of invertebrates took 17 months to recover after an intended point-source introduction of DDT into a Montana stream (Schoenthal 1963). However, in a Nigerian stream, macroinvertebrates had recovered within 3 months after an accidental input of the insecticide, Gammalin-20 (Victor & Ogbeidu 1986). Experimental additions of the insecticide methoxychlor to a small headwater stream in western North Carolina eliminated most insect taxa apart from a few long-lived forms. No upstream sources of colonizers were present and although trophic recovery was complete in 2 years, taxonomic recovery required 5 years (Wallace *et al* 1991). This study highlights the apparent stochastic element to recovery depending upon the timing of the treatment and the flight period of the ovipositing adults when no upstream sources of colonizers are present.

Chemical spills and toxic pollutants

This disturbance is similar to pesticides in causing some of the most variable recovery times, depending upon the initial scale of mortality to the river fauna and the continued persistence of the chemical after the spillage has occurred. Chemical spills typically come under level 2 and 3 disturbances depending upon the impact to the faunas. Edwards (1987) and Yount and Niemi (1990) in their review quote recovery times of nearly instantaneous to >10 years for these types of disturbances.

A fire in a chemical storehouse near Basel, Switzerland in November 1986 (the 'Sandoz Accident') caused what was categorized at the time as 'one of the worst chemical spills ever' into the River Rhine (see Capel *et al* 1988). The spill caused a massive mortality of benthic organisms and fish. Nevertheless within 1 year the benthos had recovered to pre-disturbance levels (Capel *et al* 1988) and within 2 years 40 of the 47 indigenous fish species present before the spill had recolonized (Lelek & Kohler 1990). However, the pre-accident benthos was not highly diversified for it was already affected by chronic chemical contamination. This case again illustrates rapid recovery where the fauna is adapted to previous disturbances and stresses and where there is a lack of persistence of the chemicals at the affected areas.

The recovery times of periphyton communities exposed to heavy metals may be longer than other disturbances, particularly where the metals adsorb to the sediments or become sequestered inside the cells of the periphytic organisms (Steinman & McIntire 1990). The potential long-term recovery rates of benthic communities completely eliminated by metal pollutants were shown by Chadwick *et al* (1986) in a headwater stream in Montana that had been affected by copper mining and domestic wastes for over 100 years. Invertebrates did not colonize until 3 years after the water quality improved, and after 10 years only partial recovery had been recorded when compared to reference areas. The slow recovery time was attributed to possible continued leaching of metals from streambanks and the lack of an upstream source of colonizers. The continued input of toxic material from abandoned waste lagoons was cited by Hill *et al* (1975) as the principal reason that 2 years after the cessation of pollutant inputs from a chemical plant, aquatic insects had only partially recovered. Where flushing flows remove the persistence of the chemical and there is no residual toxicity, Smith and Distler (1981) reported benthic communities recovered within 6–9 months after being destroyed by a toxic chemical discharge.

Pollution control measures on the River Ebbw Fawr in Wales, UK, improved the water quality of an effluent from a steelworks, and fish, particularly brown trout, began to colonize rapidly (Turnpenny & Williams 1981). At Sixteenmile Creek in Montana, USA, brook and brown trout populations required 3 years to recover after a large quantity of toxaphene-based cattle dip was dumped into the creek (Workman 1981). Recovery was a function of reproductive success and survival of the fish and because the physical habitat had not been significantly altered. Similar recovery times were reported by Johnston and Cheverie (1980) for brook and brown trout populations following the poisoning of a stream in northern Canada with an accidental spill of the insecticide endosulfan, and by Valdez *et al* (1977) for Dolly Varden charr after a massive spill of drilling muds (an aqueous suspension of oils, phenols, bentonite, lignite and ferrochrome lignosulphonate) in a maritime tundra stream in Alaska. In each of these studies sufficient time was necessary for juvenile trout and the charr to move through three age classes to replace the older fish that had been lost. Faster recovery times of 1 year were documented for fish in streams after kills of warm-water species (Gunning & Berra 1969; Olmstead & Cloutman 1974).

Organic pollutants

In most cases recovery from organic pollutants that are biodegradable is more rapid than recovery from chemical or toxic pollutants as typically the pollutants are less persistent, providing there are no further inputs. Given the short generation times of periphyton and their dispersal and colonization abilities, periphyton recovery should only take weeks or months following the cessation of organic pollutant input (Steinman & McIntire 1990).

In the Big River in Ireland, benthic invertebrates had recovered in similarity to upstream reference areas after 8 months following the cessation of organic wastes input from a factory (Douglas & McCreanor 1990). Within 1 year of the improvement of water quality from gross organic pollution in the River Derwent, UK, Brinkhurst (1965) observed a diverse fauna of invertebrates, including molluscs and oxygen-sensitive taxa. Probably one of the most dramatic cases of recovery following organic pollution occurred in the River Thames estuary in England as documented by Gameson and Wheeler (1977). Although a grossly polluted river from the first half of the 19th century, improvements in sewage works effluents since 1955 have dramatically improved the river and no hydrogen sulphide has been present in the water since 1965. A rapid recovery by fish has followed and by 1967, 62 species had been found which represented more than a single capture.

Restoration

With reference to Table 5.1 the factors which cause a −3 on the recovery scale are candidates for the application of restoration techniques to enhance recovery rates. These include disturbances where the physical habitat is altered, terrestrial and riparian vegetation are significantly affected, there are large disturbed areas far from colonizing sources and where there are continued disturbances arresting recovery.

5.8 CONCLUSIONS

From this overview the following eight key points can be made to summarize the principal conclusions regarding recovery of biotic communities from disturbance that is important to the river manager.

1 The most common forms of anthropogenic disturbances to rivers reported in the literature are DDT insecticide applications in forest sprayings, logging activity, dredging/channelization, and rotenone applications. The most common form of natural disturbance is floods.

2 Recovery end-points most commonly used in studies are total density, total biomass and taxa richness, although they typically occur earlier than other potential end-points such as taxa equilibrium, keystone species recovery, production and physical habitat recovery. The most suitable end-point is system and disturbance specific and the river manager must decide which is the most appropriate under the circumstances. It may not always be in a manager's best interests for a river system to return exactly to the predisturbance community which, in any event with the passing of time, may be difficult to define. Also in many cases of disturbance, no predisturbance data are available and thus reference areas are necessary for comparison.

3 For most disturbances, macroinvertebrate total density, total biomass and taxa richness typically recover within 1 year (see Fig. 5.3). Where upstream sources of drift are absent and refugia are distant, recovery times of macroinvertebrates can be significantly longer. At the order level, time to recovery for the major riverine insect groups are ranked: Diptera < Ephemeroptera < Trichoptera < Plecoptera (see Fig. 5.4).

4 Fish population recovery is generally longer than macroinvertebrates and in many cases

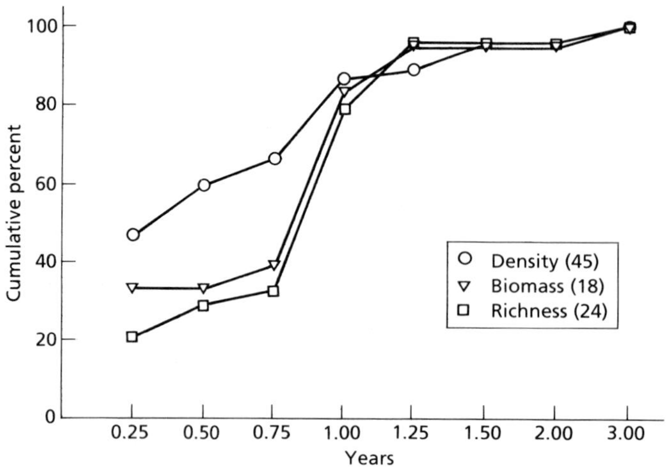

Fig. 5.3 Recovery times of macroinvertebrate total density, total biomass and taxa richness following disturbances (from Niemi *et al* 1990).

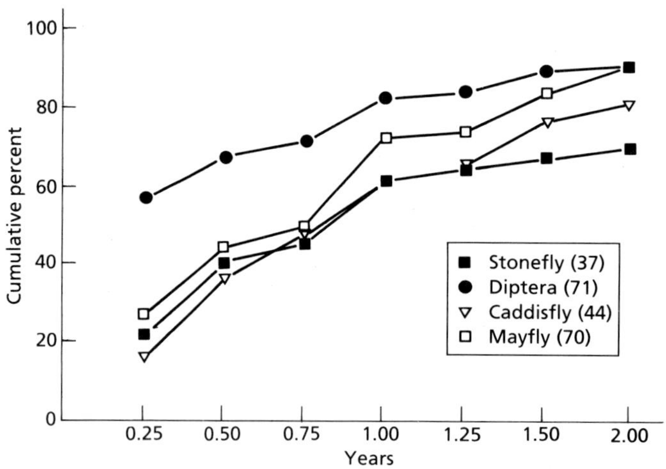

Fig. 5.4 Cumulative percentage over time for recovery end-points for four macroinvertebrate orders (from Niemi *et al* 1990).

requires juveniles to pass through yearly age classes before recovery is complete.

5 Macroinvertebrate communities, with species assemblages that are opportunistic with short life cycles selected as a result of a history of previous disturbances, will recover more rapidly to future similar disturbances when compared to a community in a relatively stable environment. It is also likely the former communities will also recover more rapidly to dissimilar disturbances.

6 Level 3 and 4 disturbances typically result in relatively rapid recovery times compared to level 1 and 2, except where the watershed or the physical habitat is significantly disturbed. Watershed disturbance (e.g. mining and logging) and channel modifications which alter the physical habitat result in the longest recovery times of the biotic communities. These types of disturbances frequently necessitate restoration techniques.

7 Pesticides, chemical spills and toxic pollutant disturbances typically result in the most variable recovery times depending upon the persistence of the toxicant and the flushing capacity of the system. Recovery in a number of systems has been delayed by the continued leaching of toxi-

cants after the initial disturbance had ceased. **8** Recovery from organic pollutants is typically more rapid than chemical and toxic pollutants.

ACKNOWLEDGEMENTS

I am extremely grateful to Ross Doughty, Paul Giller, David Gilvear, Jim Gore, Sam Lake and Bruce Wallace for their constructive comments on an earlier version of this manuscript. Thanks to Geoff Petts and an anonymous reviewer for additional comments and to Maureen Milner for preparing the figures and tables.

REFERENCES

Anderson NH, Wisseman RW. (1987) Recovery of the Trichoptera fauna near Mt. St. Helens five years after the 1980 eruption. In: Bournaud M, Tachet H (eds) *Proceedings of the 5th International Symposium on Trichoptera*, pp. 367–73. W. Junk, Dordrecht. [5.5]

Arner DH, Robinette HR, Frasier JE, Gray MH. (1976) *Effects of Channelization of the Luxapalila River on Fish, Aquatic Invertebrates, Water Quality, and Furbearers*. Office of Biological Sciences, US Department of Interior, Fish and Wildlife Service, Washington, DC. [5.7]

Begon M, Harper JL, Townsend CR. (1990) *Ecology: Individuals, Populations and Communities*, 2nd edn. Blackwell Scientific Publications, Oxford. [5.3]

Bender EA, Case TJ, Gilpin ME. (1984) Perturbation experiments in community ecology: theory and practice. *Ecology* **65**: 1–13. [5.6]

Beschta RL, Bilby RE, Brown GW, Holtby LB, Hofstra TD. (1987) Stream temperature and aquatic habitat: fisheries and forestry interactions. In: Salo EO, Cundy TW (eds) *Streamside Management: Forestry and Fishery Interactions*, pp. 191–232. Forestry Resources Contribution no. 57, University of Washington, Seattle. [5.7]

Bradshaw AD. (1988) Alternative endpoints for reclamation. In: Cairns JJ (ed.) *Rehabilitating Damaged Ecosystems*, Vol. II, pp. 69–85. CRC Press, Boca Raton. [5.2]

Brinkhurst RO. (1965) Observations on the recovery of a British river from gross organic pollution. *Hydrobiologia* **25**: 9–51. [5.7]

Brundin L. (1967) Insects and the problem of austral disjunctive distribution. *Annual Review of Entomology* **12**: 149–68. [5.5]

Burns DC. (1991) Cumulative effects of small modifications to habitat. *Fisheries* **16**: 12–17. [5.7]

Cairns JJ Jr. (1990) Lack of theoretical basis for predicting rate and pathways of recovery. *Environmental Management* **14**: 517–26. [5.2, 5.3]

Campbell IC, Doeg TJ. (1989) Impact of timber harvesting and production on streams: a review. *Australian Journal of Marine and Freshwater Sciences* **40**: 519–39. [5.6]

Capel PD, Giger W, Reichert R, Wanner O. (1988) Accidental input of pesticides into the Rhine River. *Environmental Science and Technology* **22**: 992–7. [5.7]

Chadwick JW, Canton SP, Dent RL. (1986) Recovery of benthic invertebrate communities in Silver Bow Creek, Montana, following improved metal mine wastwater treatment. *Water, Air and Soil Pollution* **28**: 427–38. [5.7]

Chamberlain TW, Harr RD, Everest FH. (1991) Timber harvesting, silviculture, and watershed processes. In: *Influences of Forest and Rangeland Management on Salmonid Habitat*, pp. 181–205. American Fisheries Society Special Publication. American Fisheries Society, Betscheda. [5.7]

Collins JP, Young G, Howell J, Minckley WL. (1981) Impact of flooding in a Sonoran desert stream, including elimination of an endangered fish population (*Poeciliopsis O. occidentadlis*, Poeciliidae). *Southwestern Naturalist* **26**: 415–23. [5.7]

Cook RJ, Barron JC, Rapendick RI, Williams GJ. (1981) Impact on agriculture of the Mount St. Helens eruptions. *Science* **211**: 16–22. [5.7]

Cook SF, Moore RL. (1969) The effect of a rotenone treatment on the insect fauna of a Californian stream. *Transactions of the American Fisheries Society* **98**: 539–44. [5.7]

Cummins KW, Klug MJ. (1979) Feeding ecology of stream invertebrates. *Annual Review of Ecology and Systematics* **10**: 147–72. [5.6]

Davis WP, Hoss DE, Scott GI, Sheridan PF. (1984) Fisheries resource impacts from spills of oil or hazardous substances. In: Cairns J Jr, Buikema AL Jr (eds) *Restoration of Habitats Impacted by Oil Spills*, pp. 157–72. Butterworth, Boston. [5.1]

DeAngelis DL, Mulholland PJ, Elwood JW, Palumbo AV, Steinman AD. (1990) Biochemical cycling constraints on stream ecosytem recovery. *Environmental Management* **14**: 685–97. [5.6]

Dimond J. (1967) Pesticides and stream insects. *Maine Forest Service Bulletin* No. 23, pp. 1–21. [5.7]

Doeg TJ, Lake PS, Marchant R. (1989a) Colonization of experimentally disturbed patches by stream macroinvertebrates in the Acheron River, Victoria. *Australian Journal of Ecology* **14**: 207–20. [5.1, 5.6]

Doeg TJ, Marchant R, Douglas M, Lake PS. (1989b) Experimental colonization of sand, gravel and stones by macroinvertebrates in the Acheron River, southeastern Australia. *Freshwater Biology* **22**: 57–64. [5.5]

Doughty CR, Turner MJ. (1991) *The Ponesk Burn Diversion: Colonization by Benthic Invertebrates and Fish.* Technical Report No. 98. Clyde River Purification Board, East Kilbride. [5.7]

Douglas DJ, McCreanor J. (1990) The Big River, Co. Louth, Ireland: a case study in recovery. *Annals of Limnology* **26**: 73–9. [5.7]

Duncan WFA, Brusven MA. (1985a) Energy dynamics of three low-order southeast Alaska streams: allochthonous processes. *Journal of Freshwater Ecology* **3**: 233–48. [5.7]

Duncan WFA, Brusven MA. (1985b) Energy dynamics of three low-order southeast Alaska streams: autochthonous production. *Journal of Freshwater Ecology* **3**: 155–66. [5.7]

Edwards RW. (1987) Ecological assessment of the degradation and recovery of rivers from pollution. In: Ravera O (ed.) *Ecological Assessment of Environmental Degradation, Pollution and Recovery*, pp. 159–94. Elsevier Press, Amsterdam. [5.7]

Everest FH, Beschta RL, Scrivener JC, Koski KV, Sedell JR, Cederholm CJ. (1987) Fine sediment and salmonid production: a paradox. In: Salo EO, Cundy TW (eds) *Streamside Management: Forestry and Fishery Interactions*, pp. 98–142. Forestry Resources Contribution No. 57, University of Washington, Seattle. [5.6, 5.7]

Fisher SG. (1990) Recovery processes in lotic ecosystems: limits of successional theory. *Environmental Management* **14**: 725–36. [5.4]

Fisher SG, Gray LJ. (1983) Secondary production and organic matter processing by collector macroinvertebrates in a desert stream. *Ecology* **64**: 1217–24. [5.6]

Fisher SG, Gray LG, Grimm NB, Busch DE. (1982) Temporal succession in a desert stream ecosystem following flash flooding. *Ecological Monographs* **52**: 93–110. [5.7]

Franklin JF, Frenzen PM, Swanson FJ. (1988) Re-creation of ecosystems at Mount St. Helens: Contrasts in artificial and natural approaches. In: Cairns J Jr (ed) *Rehabilitating Damaged Ecosystems*, vol. II pp. 1–34. CRC Press, Boca Raton. [5.7]

Gameson ALH, Wheeler A. (1977) Restoration and recovery of the Thames estuary. In: Cairns J Jr, Dickson KL, Herricks EE (eds) *Recovery and Restoration of Damaged Ecosystems*, pp. 72–101. University of Virginia Press, Charlottesville. [5.7]

Ghetti PF, Gorbi G. (1985) Effects of acute pollution on macroinvertebrates in a stream. *Verhandlungen Internationale Vereinigung für Theoretische und Angewandte Limnologie* **22**: 2426–31. [5.7]

Giller PS, Sangpradub N, Twomey H. (1991) Catastrophic flooding and macroinvertebrate community structure. *Verhandlungen Internationale Vereinigung für Theoretische und Angewandte Limnologie* **24**: 1724–9. [5.6, 5.7]

Gore JA. (1979) Patterns of initial benthic recolonization of a reclaimed coal strip-mined river channel. *Canadian Journal of Zoology* **57**: 2429–39. [5.3, 5.6, 5.7]

Gore JA. (1982) Benthic invertebrate colonization: source distance effects on community composition. *Hydrobiologia* **94**: 183–93. [5.3, 5.6, 5.7]

Gore JA. (1985) Mechanisms of colonization and habitat enhancement for benthic macroinvertebrates in restored river channels. In: Gore JA (ed) *The Restoration of Rivers and Streams*, pp. 81–101. Butterworths, Boston. [5.1]

Gore JA, Johnson LS. (1980) *Establishment of Biotic and Hydrologic Stability in a Reclaimed Coal Strip-Mined River Channel.* Institute of Energy and Environment, University of Wyoming, Laramie. [5.7]

Gore JA, Milner AM. (1990) Island biogeographic theory: can it be used to predict lotic recovery rates? *Environmental Management* **14**: 737–54. [5.1, 5.3, 5.4, 5.6, 5.7]

Gore JA, Kelly JR, Yount DJ. (1990) Application of ecological theory to determining recovery potential of disturbed lotic ecosystems: research needs and priorities. *Environmental Management* **14**: 755–62. [5.1, 5.3]

Griswold BL, Edwards CJ, Woods LC III. (1982) Recolonization of Macroinvertebrates and fish in a channelized stream after drought. *Ohio Journal of Science* **82**: 96–102. [5.7]

Gunning GE, Berra TM. (1969) Fish repopulation of experimentally decimated segments in the headwaters of the two streams. *Transactions of the American Fisheries Society* **98**: 305–8. [5.7]

Gurtz ME, Wallace JB. (1984) Substrate-mediated responses of stream invertebrates to disturbance. *Ecology* **65**: 1556–69. [5.1, 5.7]

Haefner JD, Wallace JB. (1981) Shifts in aquatic insect populations in first-order Appalachian stream following a decade of old field succession. *Canadian Journal of Fisheries and Aquatic Sciences* **38**: 353–9. [5.7]

Harrison AD. (1966) Recolonization of a Rhodesian stream after drought. *Archiv für Hydrobiologie* **62**: 405–21. [5.7]

Hawkins CP, Sedell JR. (1990) The role of refugia in the recolonization of streams devastated by the 1980 eruption of Mount St. Helens. *Northwest Science* **64**: 271–4. [5.7]

Heifetz J, Murphy ML, Koski K. (1986) Effects of logging on winter habitat of juvenile salmonids in Alaskan streams. *North American Journal of Fisheries Management* **6**: 52–8. [5.7]

Hill DM, Taylor EA, Saylor CF. (1975) Status of faunal recovery in the north fork Houlston River, Tennessee and Virginia. *Proceedings of the Southeastern Associ-*

ation of Game and Fish Commissioners **28**: 398–413. [5.7]

Hoffman CH, Drooz AT. (1953) Effects of a C-47 airplane application of DDT on fish-food organisms in two Pennsylvania watersheds. *American Midland Naturalist* **50**: 172–88. [5.7]

Hortle KG, Lake PS. (1982) Macroinvertebrate assemblages in channelized and unchannelized sections of the Bunyip River, Victoria. *Australian Journal of Marine and Freshwater Research* **33**: 1071–82. [5.7]

Huston M. (1979) A general hypothesis of species diversity. *American Naturalist* **113**: 81–101. [5.3]

Johnston CE, Cheverie JC. (1980) Repopulation of a coastal stream by brook trout and rainbow trout after Endosulfan poisoning. *Progressive Fish Culturist* **42**: 107–10. [5.7]

Keenleyside MHA. (1967) Effects of forest spraying with DDT in New Brunswick on food of young Atlantic salmon. *Journal of the Fisheries Research Board of Canada* **24**: 807–22. [5.7]

Kelly JR, Harwell MA. (1990) Indicators of ecosystem recovery. *Environmental Management* **14**: 527–45. [5.1, 5.3]

Kreutzweiser DP, Kingsbury PD. (1987) Permethrin treatments in Canadian forests. Part 2: Impact on stream invertebrates. *Pesticide Science* **19**: 49–60. [5.7]

Krumholz LA, Minckley WL. (1984) Changes in a fish population in the upper Ohio River following temporary pollution abatement. *Transactions of the American Fisheries Society* **93**: 1–6. [5.3]

Ladle M, Bass JAB. (1981) The ecology of a small chalk stream and its responses to drying during drought conditions. *Archiv für Hydrobiologie* **90**: 448–66. [5.6, 5.7]

Lake PS. (1990) Disturbing hard and soft bottom communities: A comparison of marine and freshwater environments. *Australian Journal of Ecology* **15**: 477–88. [5.1]

Lake PS, Barmuta LA. (1986) Stream benthic communities: persistent presumptions and current speculations. In: DeDecker P, Williams WD (eds) *Limnology in Australia*, pp. 263–76. Dr W. Junk, Dordrecht. [5.1]

Lake PS, Doeg TJ, Marchant R. (1989) Effects of multiple disturbance on macroinvertebrate communities in the Acheron River, Victoria. *Australian Journal of Ecology* **14**: 507–14. [5.1]

Lambert TR. (1988) Recolonization of a small stream by rainbow trout following a flood event. In: *68th Annual Conference of the Western Association of Wildlife Agencies*, pp. 258–64. Western Division, American Fisheries Society, Albuquerque. [5.7]

Lamberti GA, Gregory SV, Ashkenas LR, Wildman RC, Moore KMS. (1991) Stream ecosystem recovery following a catastrophic debris flow. *Canadian Journal of Fisheries and Aquatic Sciences* **48**: 196–208. [5.6]

Larimore RW, Childers WF, Heckrotte C. (1959) Destruction and re-establishment of stream fish and invertebrates affected by drought. *Transactions of the American Fisheries Society* **88**: 261–85. [5.7]

Lelek A, Kohler C. (1990) Restoration of fish communities of the Rhine River two years after a heavy pollution wave. *Regulated Rivers: Research and Management* **5**: 57–66. [5.7]

Luedtke RJ, Brusven MA. (1976) Effects of sand sedimentation on colonization of stream insects. *Journal of the Fisheries Research Board of Canada* **33**: 1881–6. [5.6]

McCormick PV, Stevenson JR. (1991) Mechanisms of benthic algal succession in lotic environments. *Ecology* **72**: 1835–48. [5.4]

Magnuson JJ, Reiger HA, Christie WJ, Sonzogni WC. (1980) To rehabilitate and restore great lake ecosystems. In: Cairns J Jr (ed) *The Recovery Process in Damaged Ecosystems*, pp. 95–112. Ann Arbor Science Publishers, Ann Arbor. [5.1]

Malmqvist B, Rundle S, Brönmark C, Erlandson A. (1991) Invertebrate colonization of a new, man-made stream in southern Sweden. *Freshwater Biology* **26**: 307–24. [5.7]

Martin DJ, Wasserman LJ, Dale VH. (1986) Influence of riparian vegetation of posteruption survival of coho salmon fingerlings on the westside streams of Mount St. Helens, Washington. *North American Journal of Fisheries Management* **6**: 1–18. [5.6]

Matthews WJ. (1986) Fish faunal structure in an Ozark stream: stability, persistence and a catastrophic flood. *Copeia*, pp. 388–97. [5.7]

Meffe GK, Minckley WL. (1987) Persistence and stability of fish and invertebrate assemblages in a repeatedly disturbed Sonoran Desert stream. *American Midland Naturalist* **117**: 177–91. [5.7]

Milner AM. (1987) Colonization and ecological development of new streams in Glacier Bay National Park, Alaska. *Freshwater Biology* **18**: 53–70. [5.7]

Minshall GW, Andrews DA, Manuel-Faler CY. (1983) Application of island biogeographic theory to streams: macroinvertebrate recolonization of the Teton River, Idaho. In: Barnes JR, Minshall GW (eds) *Stream Ecology: Application and Testing of General Ecological Theory*, pp. 279–97. Plenum Press, New York. [5.3, 5.7]

Minshall GW, Brock JT, Varley JD. (1989) Wildfire and Yellowstone's stream ecosystems. *BioScience* **39**: 707–15. [5.7]

Molles MC Jr. (1985) Recovery of a stream invertebrate community from a flash flood in Tesuque Creek, New Mexico. *The Southwestern Naturalist* **30**: 279–87. [5.7]

Morrison BR. (1977) The effects of rotenone on the invertebrate fauna of three hill streams in Scotland.

Fisheries Management **8**: 128–38. [5.7]
Morrison BR. (1990) Recolonisation of four small streams in central Scotland following drought conditions in 1984. *Hydrobiologia* **208**: 261–67. [5.6, 5.7]
Muirhead-Thomson RC. (1987) *Pesticide Impact on Stream Fauna with Special Reference to Macroinvertebrates.* Cambridge University Press, Cambridge. [5.7]
Murphy ML, Koski K. (1989) Input and depletion of woody debris in Alaska streams and implications for streamside management. *North American Journal of Fishery Management* **9**: 427–36. [5.7]
Murphy ML, Heifetz J, Johnson SW, Koski KV, Thedinga J. (1986) Effects of clear-cut logging with and without buffer strips on juvenile salmonids in Alaskan streams. *Canadian Journal of Fisheries and Aquatic Sciences* **43**: 1521–33. [5.7]
Narf RP. (1985) Aquatic insect colonization and substrate changes in a relocated stream segment. *The Great Lakes Entomologist* **18**: 83–92. [5.7]
Niemi GJ, DeVore P, Detenbeck N, Taylor D, Yount JD, Lima A, Pastor J, Naiman RJ. (1990) Overview of case studies on recovery of aquatic systems from disturbance. *Environmental Management* **14**: 571–88. [5.5, 5.6, 5.7, 5.8]
Olmstead LL, Cloutman DG. (1974) Repopulation after a fish kill in Mud Creek, Washington County, Arkansas, following pesticide pollution. *Transactions of the American Fisheries Society* **103**: 79–87. [5.7]
Oswood MW, Irons JG, Milner AM. (in press) Alaska. In: Cushing CE, Minshall GW, Cummins KW (eds) *River and Stream Ecosystems.* Elsevier Press, Amsterdam. [5.6]
Peters RH. (1988) Some general problems for ecology illustrated by food web theory. *Ecology* **69**: 1673–6. [5.1]
Peterson BJ, Hobbie JE, Hersey AE, Lock MA, Ford TE, Vestal JR, McKinley VL, Hullar MAJ, Miller MC, Ventullo RM, Volk GS. (1985) Transformation of a tundra river from heterotrophy to autotrophy by addition of phosphorus. *Science* **229**: 1383–6. [5.6]
Poff NL. (1992) Why disturbance can be predictable: a perspective on the definition of disturbance in streams. *Journal of the North American Benthological Association* **11**: 86–92. [5.1]
Power ME, Stewart AJ. (1987) Disturbance and recovery of an algal assemblage following flooding in an Oklahoma stream. *American Midland Naturalist* **117**: 333–45. [5.7]
Reice SR. (1985) Experimental disturbance and the maintenance of species diversity in a stream community. *Oecologia* **67**: 90–7. [5.1, 5.3]
Reice SR, Wissmar RC, Naiman RJ. (1990) Disturbance regimes, resilience, and recovery of animal communities and habitats in lotic ecosystems. *Environmental Management* **14**: 647–59. [5.3]
Resh VH, Brown AV, Covich AP, Gurtz ME, Li HW, Minshall GW, Reice SR, Sheldon AL, Wallace JB, Wissmar RC. (1988) The role of disturbance theory in stream ecology. *Journal of the North American Benthological Society* **7**: 433–55. [5.1, 5.5]
Rushforth SR, Squires LE, Cushing CE. (1986) Algal communities of springs and streams in the Mt. St. Helens region, Washington, U.S.A. following the May 1980 eruption. *Journal of Phycology* **22**: 129–37. [5.7]
Rykiel EJ. (1985) Towards a definition of ecological disturbance. *Australian Journal of Ecology* **10**: 361–6. [5.1]
Sagar PM. (1983) Invertebrate recolonization of previously dry channels in the Rakaia River. *New Zealand Journal of Marine and Freshwater Research* **17**: 377–86. [5.6, 5.7]
Schoenthal ND. (1963) Some effects of DDT on cold water fish and fish-food organisms. PhD Thesis. Montana State College, Bozeman. [5.7]
Scrimgeour GJ, Davidson RJ, Davidson JM. (1988) Recovery of benthic macroinvertebrates and epilithic communities following a large flood, in an unstable, braided, New Zealand river. *New Zealand Journal of Marine and Freshwater Research* **22**: 337–44. [5.7]
Sedell JR, Reeves GH, Hauer FR, Stanford JA, Hawkins CP. (1990) Role of refugia in recovery from disturbances: Modern fragmented and disconnected river systems. *Environmental Management* **14**: 711–24. [5.6]
Shaw DW, Minshall GW. (1980) Colonization of an introduced substrate by stream macroinvertebrates. *Oikos* **34**: 259–71. [5.3]
Smith DL, Distler DA. (1981) Recovery of a benthic macroinvertebrate community following a toxic chemical discharge in a sandy plains stream. *Southwestern Naturalist* **25**: 547–51. [5.7]
Smith RD. (1989) Current research investigating channel unit distribution in streams of southeast Alaska. In: Alexander EB (ed) *Watershed 89: Proceedings of a Conference on the Stewardship of Soil, Air and Water Resources.* Alaska Region R-10-MB-77, U.S. Forest Service, Juneau. [5.7]
Sousa WP. (1984) The role of disturbance in natural communities. *Annual Review of Ecology and Systematics* **15**: 353–91. [5.1]
Southwood TR. (1978) *Ecological Methods with Particular Reference to the Study of Insect Populations.* Chapman & Hall, London. [5.3]
Sparks RE, Bayley PB, Kohler SL, Osbourne LL. (1990) Disturbance and recovery of large floodplain rivers. *Environmental Management* **14**: 699–709. [5.1]
Statzner B, Higler B. (1985) Questions and comments on the river continuum concept. *Canadian Journal of Fisheries and Aquatic Sciences* **42**: 1038–44. [5.4]
Steinman AD, McIntire CD. (1990) Recovery of lotic

periphyton communities after disturbance. *Environmental Management* **14**: 589–604. [5.5, 5.7]

Sullivan K, Lisle TE, Dolloff CA. (1987) Stream channels: the link between Forests and fishes. In: Salo EO, Cundy TW (eds) *Streamside Management: Forestry and Fishery Interactions*, pp. 39–97. Forestry Resources Contribution No. 57, University of Washington, Seattle. [5.7]

Tolbert VR, Vaughan GL. (1979) Stripmining as it relates to benthic insect communities and their recovery. *West Virginia Academy of Science* **51**: 168–81. [5.7]

Townsend CR, Hildrew AL, Schofield K. (1987) Persistence of stream invertebrate communities in relation to the environmental variability. *Journal of Animal Ecology* **56**: 597–613. [5.3]

Turnpenny AWH, Williams R. (1981) Factors affecting the recovery of fish populations in an industrial river. *Environmental Pollution (Series A)* **26**: 39–58. [5.7]

Valdez RA, Helm WT, Neuhold JM. (1977) Aquatic ecology. In: Merritt ML, Fuller RG (eds) *The Environment of Amchitka Island*, pp. 287–313. Technical Information Center, ERDA, Washington DC. [5.7]

Vannote RL, Minshall GW, Cummins KW, Sedell JR, Cushing CE. (1980) The river continuum concept. *Canadian Journal of Fisheries and Aquatic Sciences* **37**: 130–7. [5.6]

Victor R, Ogbeidu AE. (1986) Recolonisation of macrobenthic invertebrates in a Nigerian stream after pesticide treatment and associated disruption. *Environmental Pollution (Series A)* **41**: 125–37. [5.7]

Wallace JB. (1990) Recovery of lotic macroinvertebrate communities from disturbance. *Environmental Management* **14**: 605–20. [5.5, 5.6, 5.7]

Wallace JB, Gurtz ME. (1986) Response of *Baetis* mayflies (Ephemeroptera) to catchment logging. *American Midland Naturalist* **115**: 25–41. [5.7]

Wallace JB, Gurtz ME, Smith-Cuffney F. (1988) Long-term comparisons of insect abundances in disturbed and undisturbed Appalachian headwater streams. *Verhandlungen Internationale Vereinigung für Theoretische und Angewandte Limnologie* **23**: 1224–31. [5.7]

Wallace JB, Huryn AD, Lugthart GJ. (1990) Colonization of a headwater stream during three years of seasonal insecticide applications. *Hydrobiologia* **211**: 65–76. [5.5, 5.7]

Ward JV, Stanford JA. (1983) The intermediate disturbance hypothesis: an explanation for biotic diversity patterns in lotic ecosystems. In: Fontaine TD, Bartell SM (eds) *Dynamics of Lotic Ecosystems*, pp. 347–56. Ann Arbor Science Publishers, Ann Arbor. [5.3]

Warner K, Fenderson OC. (1962) Effects of DDT spraying for forest insects on Maine trout streams. *Journal of Wildlife Management* **26**: 86–93. [5.7]

Webster JR, Patten BC. (1979) Effects of watershed perturbation on stream potassium and calcium dynamics. *Ecological Monographs* **49**: 51–72. [5.6]

Webster JR, Waide JB, Patten BC. (1975) Nutrient recycling and the stability of ecosystems. In: Howell FG, Gentry JB, Smith MH (eds) *Mineral Cycling in Southeastern Ecosystems*, pp. 1–27. US Energy Research and Development Administration (CONF-740513). [5.1, 5.6]

Webster JR, Gurtz ME, Hains JJ, Meyer JL, Swank WT, Waide JB, Wallace JB. (1983) Stability of stream ecosystems. In: Barnes JR, Minshall GW (eds) *Stream Ecology: Application and Testing of General Ecological Theory*, pp. 355–95. Plenum Press, New York. [5.7]

Webster JR, Golladay SW, Benfield EF, Meyer JL, Swank WT, Wallace JB. (1992) Catchment disturbance and stream response: An overview of stream research at Coweeta Hydrological Laboratory. In: Boon PJ, Calow P, Petts GE (eds) *River Conservation and Management*, pp. 231–53. John Wiley & Sons, Chichester. [5.7]

White PS, Pickett STA. (1985) Natural disturbance and patch dynamics: an introduction. In: White PS, Pickett STA (eds) *The ecology of natural disturbance and patch dynamics*, pp. 3–13. Academic Press, New York. [5.1]

Williams DD. (1981) Migrations and distributions of stream benthos. In: Lock MA, Williams DD (eds) *Perspectives in Running Water Ecology*, pp. 155–207. Plenum Press, New York. [5.5]

Williams DD, Hynes HBN. (1976) The recolonization mechanisms of stream benthos. *Oikos* **27**: 265–72. [5.5]

Wilzbach P, Dudley TH, Hall JD. (1983) *Recovery Patterns in Stream Communities Impacted by the Mt. St. Helens Eruption*. Water Resources Research Institute Report No. WRRI-83. Oregon State University, Corvallis. [5.7]

Workman DL. (1981) Recovery of rainbow trout and brown trout populations following chemical poisoning in Sixteenmile Creek, Montana. *North American Journal of Fisheries Management* **1**: 144–50. [5.7]

Yount JD, Niemi GJ. (1990) Recovery of lotic communities and ecosystems from disturbance — a narrative review of case studies. *Environmental Management* **14**: 547–69. [5.7]

Part 2: Monitoring Programmes

Part 1 demonstrated the sensitivity of river ecosystems to human impacts. These often take place remote from the river channel itself. There is an obvious need to be able to improve our ability to determine when, where and how changes will take place. Better environmental assessment procedures are required both for management and for the development of improved models to predict ecosystem responses to human impacts. Fundamental to environmental assessment is monitoring. This part begins with a critical review of the problems associated with sampling and testing hypotheses about spatial and temporal patterns in rivers as they relate to human disturbance in natural ecosystems (Chapter 6). It is not the intention here to address all applications of monitoring in river management, since these are detailed in specialist textbooks and monographs. In practice, monitoring has developed particularly with regard to water quality. Thus, in the following two chapters, a focus on water quality is used to illustrate the application of monitoring. Chapter 7 reviews the techniques and approaches involved in water-quality monitoring and Chapter 8 provides a detailed appraisal of the use of macroinvertebrate communities for biological water-quality assessment. Reference should be made as well to Chapter 13 in Volume 1, that also addresses the design of sampling programmes.

6: Spatial and Temporal Problems with Monitoring

A. J. UNDERWOOD

6.1 INTRODUCTION

Despite the enormous and widespread need to be able to identify and, where possible, predict the effects of human disturbances in natural ecosystems, there is still insufficient attention paid to the basic requirements of design of sampling and analysis of quantitative data from field surveys. This is increasingly discussed in the context of experimental ecology (e.g. Underwood 1981, 1986; Hurlbert 1984). It does not yet seem to have been the focus of sufficient attention in the areas of environmental monitoring and assessment of anthropogenic impact, even though there have been some excellent accounts of general procedures (notably Green 1979; Clarke & Green 1988).

Monitoring, as an ongoing exercise in many habitats, is in particular need of justification of procedures and rationales (see Underwood 1989 for examples). The purpose of this review will be to consider some of the problems associated with sampling and testing hypotheses about spatial and temporal patterns in rivers, as they relate to such issues as environmental impacts. Some of the inadequacies of routine monitoring programmes, where there are no clearly articulated logical hypotheses (e.g. Connor & Simberloff 1986; Underwood 1990) will also be discussed. The context of this discussion is devoid of any particular understanding of the nuances of sampling procedures in freshwater systems. There are two justifications for this, given the author's background in marine habitats (which might be excusable if they were thought of as very salty, very wide rivers?). First, these topics are discussed elsewhere in this volume (see Chapters 7 & 8) as are the potential disturbances and impacts (see Chapters 4 & 5). Second, and perhaps more importantly, the issues are general and transcend particularities of habitat, although obviously the organisms to be examined and their associations in assemblages will be very different from one habitat to another. A discussion of the sampling problem in rivers is presented by Norris *et al* in Vol. 1 (Chapter 13).

The general approach used will be to examine deficiencies in certain sorts of sampling and monitoring programmes with a view to increasing the relationships between the data that can be gathered in routine sampling, testable hypotheses and interpretations in relation to causation. The particular emphasis will be to increase the usefulness of sampling to establish causality between an observed disturbance and a measurable biological response (see also Clarke & Green 1988; Underwood & Peterson 1988). For this reason, the focus will be on measurements of abundances of populations of single species, although any other appropriate variable could be examined in the designs discussed (e.g. species diversity, rate of growth of individuals, production per unit area, scope for growth, etc.). Sophisticated multivariate procedures exist for many types of environmental monitoring involving numerous species in an assemblage (e.g. Clarke 1993; Warwick 1993), but these do not extend to the complex scenarios and designs that are often needed to detect impacts in very patchy and variable habitats (Clarke 1993).

Types of disturbance

Broadly, there are three types of human disturbances to natural habitats. These are sometimes classed as 'press' and 'pulse' disturbances (Bender *et al* 1984) and 'catastrophes'. Although there are some problems with this type of trichotomy (see below), it is useful to think about the nature, time-course and possible responses of populations to the various types of anthropogenic influences. *Pulse disturbances* are acute, short-term episodes of disturbance, for example an accidental discharge of chemicals through an outfall would, if it caused any biological effect, create a temporary pulse of change in a population. The disturbance is removed quickly and the response of the system is short-term, although a short-term change may itself cause long-term consequences. In contrast, a *press disturbance* is a sustained or chronic interference with a natural population. For example, building a weir and altering the rate of flow of water is a long-term environmental change that may have an impact on downstream populations because of changed physical features of the habitat. In this case, there would be long-term and usually non-recoverable changes in the populations. Finally, *catastrophes* are major, often planned, destructions of habitat. In these cases, there is no possibility that the organisms could recover – their habitat is removed. This is an easily predicted outcome.

There are some difficulties because, for example, a pulse disturbance may actually cause long-term effects. For example, there may be a pulse that removes predators temporarily from an area. Their prey then have a temporary escape from predation and remain in large numbers through a critical period of growth so that when the pulse disturbance is gone and the predators can recolonize from elsewhere, they cannot consume the prey that escaped them during or immediately after the disturbance. As a result, there can be long-term changes in dominance in the assemblage in such a habitat (Dayton 1971; Connell 1975).

As another problem, there are many cases where environmental disturbances will involve both types of perturbation. Thus, building a weir will cause a pulse disturbance due to disruption of sediments and, often, the introduction of chemicals during the construction phase. After it is built, there is a permanent press on the flow of water. Both may cause effects on populations and the interpretation of subsequent changes in the abundances of organisms may only be interpretable in the light of knowledge of the interaction between short-term and long-term disturbances.

A further difficulty lies in the definition of time-scales for the organisms concerned. For example, the presence of a mine may affect local rivers during its operation (Underwood 1991). If the mine is active for, say, 40 years, that will represent a press for many populations with shorter turnover times. For any longer-lived species, the closure of the mine after 40 years means that it was short-term and therefore a pulse. Obviously, the time-course of either type of disturbance is intimately related to the life cycle and longevity of the potentially affected organisms (Frank 1981; Connell & Sousa 1983). This makes the dichotomy between the two types of disturbance less useful than if there were no ambiguities.

Nevertheless, it is instructive to think in terms of the different types of disturbance. Where there is a pulse, environmental management requires that the time-course and rate of change of the population be considered (Bender *et al* 1984). The effect on abundance will depend on the inertia and resilience of the populations being affected (Underwood 1989; see later). The crucial issues for conservation of species will be their capacity to tolerate short-term disruptions to their habitat and their capacity to recolonize once the disturbance has ended.

Where press disturbances are happening, the major issue will be the changed abundances in the longer term, i.e. the new long-run averages of the population after the disturbance starts. Management must then focus on long-term sustenance of the population, for example to ensure that it retains the minimal numbers necessary to continue to breed.

Where there are planned or unfortunate catastrophes and habitat is permanently lost, the issues for management are the capacity of populations elsewhere to continue to persist once the populations are lost from some parts of their range. Of course, for very spatially localized species, no loss is tolerable and prediction of a catastrophe becomes very important.

In the latter case, there is no point in monitoring to detect an impact – the impact can be very well predicted from the nature of the planned disturbance. In the case of press and pulse disturbances, however, there is every need to decide in advance (where the disturbance is planned) exactly what sort of disturbance it will be and what sorts of effects it may have so that the sampling can be designed with appropriate time-scales to detect the effects. If a pulse disturbance is anticipated, it is essential to have sampling on sufficiently small time-scales to detect its effects, to ensure that the changes to abundances of relevant populations are not going outside the predicted ranges, or that populations are not actually heading to extinction. Otherwise, no remedial or alleviatory action could be taken and the populations will not be able to recover once the pulse has finished.

Where a press is concerned, there will be a need to sample at appropriate longer time intervals to determine the new long-run mean abundance (or the new pattern of temporal change – see below; Underwood 1991). Thus, for management of environmental impacts, a categorization of disturbances is helpful to planning, even if the disturbances do not neatly fall into the categories.

Disturbances affecting mean abundance

By and large, the literature on environmental changes that affect abundances of populations is obsessed with changed patterns of mean abundance (Fig. 6.1). For example, Green's (1979) excellent book does not consider any other types of

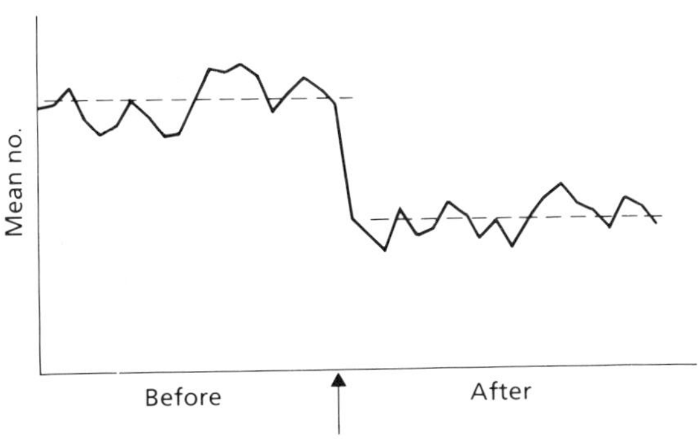

Fig. 6.1 Environmental impact (at the time indicated by the arrow) resulting in a long-term (press) decrease in mean abundance of a population.

impacts. The common notion is that a human disturbance to an environment will cause a change in average abundance. Usually, the argument becomes more irrational and the change is considered an impact only if it reduces the abundance. Increased abundances of some organisms are considered to be 'beneficial' (Underwood 1991), ignoring the obvious interactive relationships among species. For example, environmental changes that increase local habitat to favour the survival of some predatory fish (something often applauded by anglers!) must have some effect on the abundances of species of prey. These, of course, do not have political lobby-groups supporting them and therefore can be seriously depleted because of interactions with those species for which an increase in abundance is deemed desirable.

Sampling programmes to detect long-run alterations of mean abundance are widely discussed in the literature, as will be considered below. They are not the only possible environmental impacts and should not be treated as such.

Disturbances affecting temporal variances

Some potential environmental disturbances have no effect on the average numbers in a population (at least at some temporal scales) but are nevertheless extremely important for the long-term persistence or conservation of a species (Underwood 1991). Consider an impact that affects the rates of reproduction and the rates of mortality in a population. For example, construction of a sewerage outfall in a river may cause some pattern of increased flow of nutrients to an area downstream. This may well result in increased reproduction of organisms that use the nutrients, resulting in increased abundance. Every now and then, however, the discharge may be toxic (for example, there may be toxic chemicals discharged with the sewage, or the quantities of sewage may be excessive). Then the populations will crash. Thus, the impact may result in a pattern of increased and decreased abundance on a very sudden time-scale (Fig. 6.2). The average after the disturbance starts is actually the same as that prevailing before. Yet, the possibility that the population will go locally extinct is now much greater, because any new stress imposed during periods of

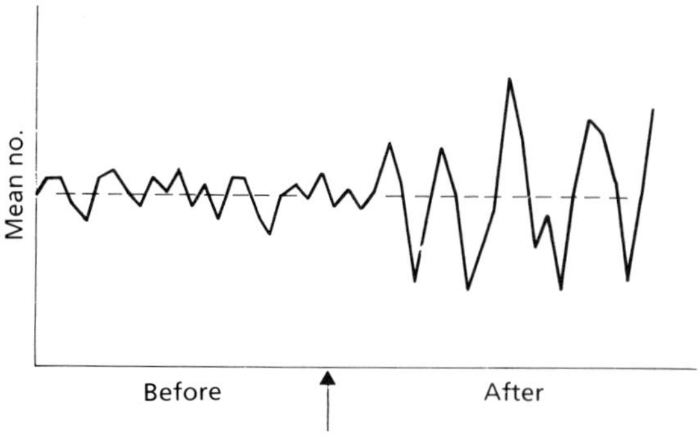

Fig. 6.2 Environmental impact (at the time indicated by the arrow) resulting in no change in mean abundance of a population, but having a long-term effect on temporal fluctuations in abundance. This would not be detected by routine environmental monitoring, unless the timing of samples happened to be appropriate.

crash will potentially reduce the already depressed abundance to zero.

Traditional methods of sampling to test hypotheses about changed mean abundance will not detect such environmental impacts, even though they may require very intense intrusive management to alleviate their effects (Underwood 1991). Instead, sampling is required that will detect changed temporal variances in numbers in the population (see later). Posing hypotheses about altered variability in time requires that the life cycle and time-courses of the populations are well-understood and that it has been realized that changed time-courses may be the effect of a disturbance.

Disturbances affecting spatial variances

Another class of disturbances does not affect overall mean abundance but instead, alter the spatial dispersion of the organisms. Again, traditional, random sampling in an area will fail to detect such impacts. Consider an organism that is scattered in clumps across the substratum, for example in those sites that provide shelter during periodic flooding of a river (Fig. 6.3). Animals do not survive outside shelters when the river floods. Some of the patches of organisms are lost because of accidents befalling some of the sheltered sites (for example, they may become scoured

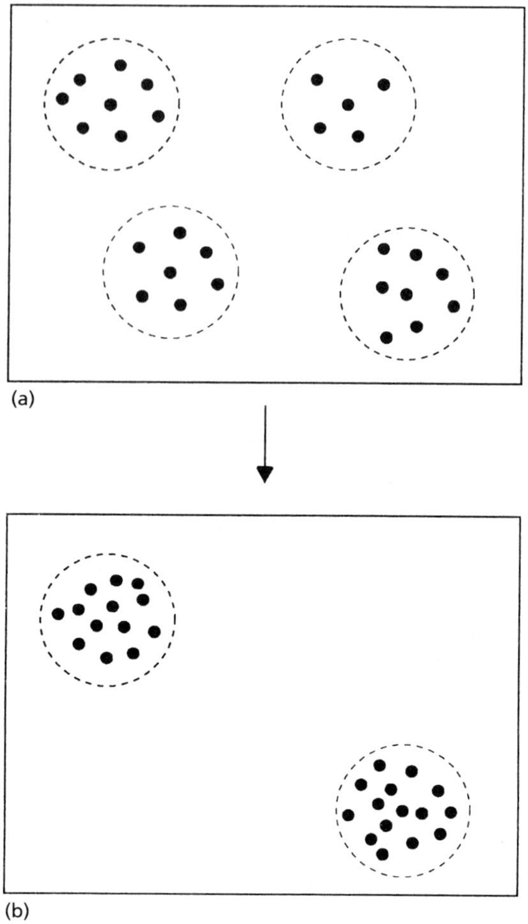

Fig. 6.3 Environmental impact (at the time indicated by the arrow) resulting in no change in mean abundance of a population, but having a long-term effect on spatial dispersion of the individuals. After the impact (b), there are less suitable patches of habitat for the organisms than was the case before (a).

by vegetation being washed down with the flood). The long-term persistence of a species is guaranteed by the existence of sufficient sheltered sites to 'spread the risk' (den Boer 1979, 1981).

Now, a human disturbance causes some alteration of some of the shelters, making them uninhabitable. For example, building flood-gates may cause sediments to fill some hollows. The animals are not actually killed or otherwise directly damaged by the disturbance – they simply redistribute themselves by moving out of the shelters and into others. This creates a greater density in the remaining shelters, but no overall change in the abundance of the population. Technically, the intensity of the pattern is altered by the disturbance (Pielou 1969). Of course, next time there is a flood, there will be greater loss of animals from the few patches that are accidentally, but naturally, demolished. If the decrease in available shelters due to the extra human disturbance is sufficient, the population may disappear at the first natural disturbance because there is no longer sufficient patchiness in its distribution. An example of this type of disturbance was investigated experimentally by Clifford (1982), who repeatedly disturbed the bed of an area of a stream. The substratum became more homogeneous (although in this case the mean abundances of many taxa also declined).

Sampling for such environmental effects requires more than simple random sampling to determine average abundance – the spatial variance of the population is the parameter affected by the disturbance.

What should be monitored?

So, the above considerations lead to the notion that environmental monitoring should only be done in response to clear advance understanding of the nature, time-courses and likely effects of some planned, or a range of possible, accidental disturbances. At this stage, the requirement is that there be much greater thought about the sorts of biological effects humans might have on the natural populations. This will then enable clear and unambiguous hypotheses to be proposed about the various possibilities (e.g. Underwood 1990). Only then can a proper plan of sampling be considered (Green 1979).

If better thought is given to the problems to which monitoring is to be addressed, there is actually little need to monitor at all, in the routine sense. Instead, the sampling is designed around specific anticipated (hypothesized) processes and results. It then fits much better into the mainstream framework of experimental ecology and can be designed much more appropriately to answer the questions.

It is interesting that to complete a piece of 'academic' experimental ecology by publishing it requires considerable demonstration (for example, by the journals) that the problems addressed and the methods used are clear and well related. Only then are interpretations accepted (although there are obviously problems with this sort of idealized approach; Hurlbert 1984; Underwood 1986). The irony is that monitoring and environmental sampling for eventual management and conservation of habitats and species can apparently often operate with a much looser framework of logic and design. This, at the very least, seems back-to-front – because, presumably, the latter type of science is potentially more important in the short term than the former!

The designs of sampling programmes discussed below are a sequence from early programmes to more recent ones, including those that are the preliminary answers to problems of environmental disturbances that do not affect long-run averages, but alter spatial or temporal variances. The solution of what to monitor is now translated to a better understanding of what sorts of effects are anticipated.

6.2 DESIGN OF SAMPLING

Unreplicated, uncontrolled before/after sampling

Some environmental assessment is based on the (incorrect) premise that populations show no natural temporal variability. A measured differ-

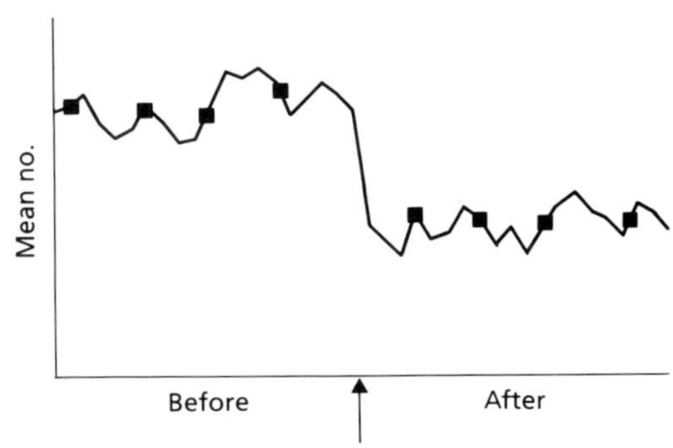

Fig. 6.4 Temporally replicated sampling to detect an impact (at the time indicated by the arrow) causing a long-term change in mean abundance of a single population.

ence in mean abundance from before to after a disturbance is then interpreted as being due to the disturbance. The situation is somewhat improved by taking replicate samples at each of several replicate times before and after the putative impact (Fig. 6.4). Then, a consistent change of mean associated with the onset of the disturbance will often be interpreted as being caused by the disturbance. This is illogical — the sampling is uncontrolled. Any environmental change that is coincidental with the start of the perturbation might have caused the sampled shift in abundance.

Unreplicated, but controlled BACI designs

To control for such natural temporal changes, Green (1979) proposed a so-called BACI design, in which there are replicate samples Before and After the planned disturbance, in each of an undisturbed control and the putatively impacted locations (Fig. 6.5). An impact due to the disturbance will cause the difference in mean abundance from before to after in the impacted location to differ from any natural change in the control. This would be detected as a statistical interaction (Table 6.1).

There are several problems with this interpretation. First, a sample taken at a single time before and again after the disturbance may not be representative of the abundance of the population. If there is much natural temporal variation, the

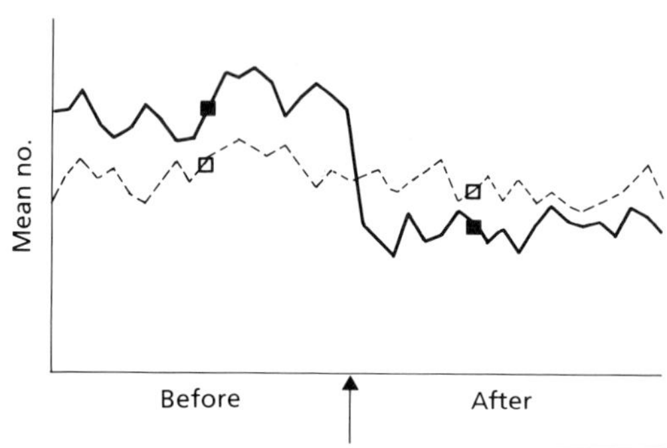

Fig. 6.5 Before–after/control–impact (BACI) sampling at a single time before and after an impact (indicated by the arrow) that affects the long-term mean abundance of a population in the impacted (solid line), but not the control (dashed line) location.

Table 6.1 Analysis of BACI design of sampling to detect environmental impact. n replicate samples are taken in each of two locations (the putatively impacted and a single control) once before and once again after the disturbance occurs (as in Fig. 6.5). Environmental impact would be detected as a significant interaction (B × L)

Source of variation	Degrees of freedom	*F*ratio versus	Degrees of freedom
Before versus after = B	1		
Control versus impact = L	1		
B × L	1	Residual	$1, 4(n-1)$
Residual	$4(n-1)$		
Total	$4n-1$		

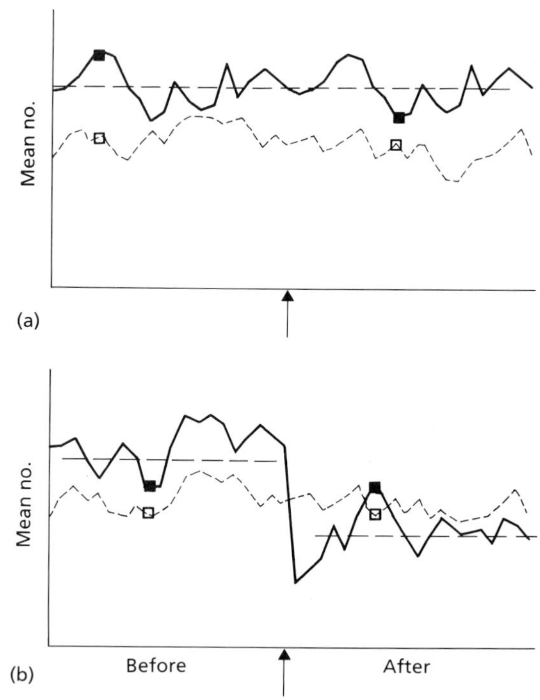

Fig. 6.6 BACI sampling to illustrate problems caused by lack of temporal replication. In (a) there is no impact and long-term mean abundance before and after a disturbance is the same. Because of temporal fluctuations, a single sample (at the time of the squares) would detect a difference in abundance in the impact (solid line), but not the control (dashed line) location. In (b) the disturbance in the impact location (solid line) causes a long-term decline in mean abundance, but this is not detected because of natural temporal fluctuations and lack of replication.

apparent or sampled difference from before to after in either location may be due to chance fluctuations (Fig. 6.6a). Alternatively, a real difference associated with the disturbance may be cancelled out by chance fluctuations (Fig. 6.6b). In either case, the wrong conclusions is reached because of lack of replication (see also Hurlbert 1984; Stewart-Oaten *et al* 1986; Underwood 1991).

A solution of the problem of lack of temporal replication was proposed by Bernstein and Zalinski (1983) and Stewart-Oaten *et al* (1986), following Green's (1979) suggestion that replicated sampling through time could be done. This involved taking samples (of several replicates each) at several times before and several times after the disturbance (Fig. 6.7). Stewart-Oaten *et al* (1986) proposed that the difference between the mean of each sample in the putatively impacted location and that in the control location be used as a variable. The average of these differences before the disturbance will be different from that after the disturbance, if it causes some impact on the population.

In the scheme of Bernstein and Zalinski (1983) and Stewart-Oaten *et al* (1986), these differences are examined by a *t*-test. Alternatively, the data can be examined by analysis of variance (Table 6.2). This analysis is identical to the *t*-test, but has the property that it can be modified for situations where it is impossible to sample both locations at the same time (Table 6.3; see Underwood 1991 for details).

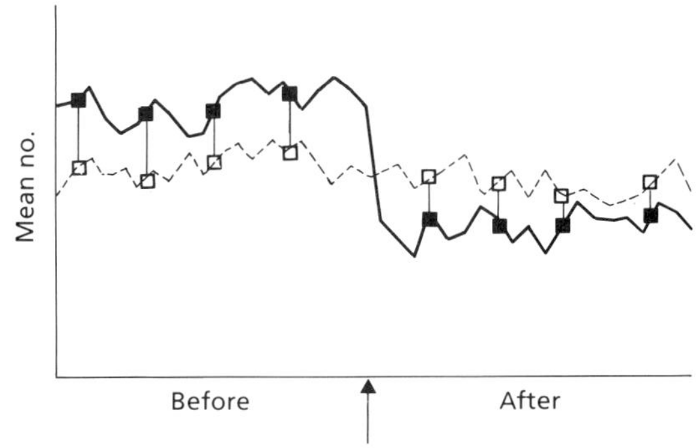

Fig. 6.7 BACI sampling with replicated temporal sampling in the impacted (solid line) and a single control (dashed line) location. Data from each time of sampling are paired (as indicated by the vertical lines) and the difference between the two locations is analysed (Stewart-Oaten *et al* 1986).

This type of sampling still suffers from *confounding*, more recently termed pseudoreplication (Hurlbert 1984). The problem is that any two places may show a difference in time that has nothing to do with the human disturbance to one location. For example, Green (1979) discussed the possibility, where an outfall pipe enters a river, of using an area upstream of the pipe as the Control for the downstream, potentially Impacted location. There are difficulties with this because of

Table 6.2 Analysis of temporally replicated BACI design of sampling to detect environmental impact. *n* replicate samples are taken in each of two locations (the putatively impacted and a single control) at *t* replicated times before and again after the disturbance occurs (as in Fig. 6.7). The two locations must be sampled at the same times. Environmental impact would be detected as a significant interaction L × T(B) (or, if there were no differences between the two locations in the time-courses of abundances before and after the disturbance, as a significant interaction B × L)

Source of variation	Degrees of freedom	*F*-ratio versus	Degrees of freedom
Before versus after = B	1		
Control versus impact = L	1		
B × L	1*	L × T(B)	$1, 2(t-1)$
Times (B) = T(B)	$2(t-1)$		
L × T(B)	$2(t-1)$*	Residual	$2(t-1), 4t(n-1)$
Residual	$4t(n-1)$		
Total	$4tn-1$		

* This is an identical test to the paired *t*-test recommended by Stewart-Oaten *et al* (1986).

Table 6.3 Analysis of temporally replicated BACI design of sampling to detect environmental impact when sampling cannot be done at the same times in each location. n replicate samples are taken in each of two locations (the putatively impacted and a single control) at t replicated times before and again after the disturbance occurs (as in Fig. 6.7). The two locations are not sampled at the same times. Environmental impact would be detected as a significant interaction B × L

Source of variation	Degrees of freedom	*F*-ratio versus	Degrees of freedom
Before versus after = B	1		
Control versus impact = L	1		
B × L	1	T(B × L)	1,4(t − 1)
Times (B × L) = T(B × L)	4(t − 1)		
Residual	4t(n − 1)		
Total	4tn − 1		

possible non-independence (see later). Nevertheless, it will illustrate the point. Suppose that the organisms being sampled are insects, with an intrinsically patchy distribution of egg-laying. At about the time the outfall pipe is built, adults happen to arrive and lay eggs in the upstream Control location, but not, by chance, in the downstream area. As a result, the subsequent sampling (after the disturbance starts and the outfall begins its discharge) will reveal an apparent impact due to the pipe. The same conclusion would be reached if adults happened to visit the downstream, but not the upstream location. The sampling is clearly confounded and any effect of the disturbance cannot be distinguished from natural variation between locations. If there is an environmental impact, it is *necessary* for it to cause a change in the differences in abundance between two locations. Such change is not, however, *sufficient* evidence that an impact has occurred — see Hurlbert's (1984) review.

One final difficulty besets this type of sampling. If there is an interaction in the differences between mean abundances in the two locations *before* the human disturbance, the data cannot be analysed for environmental impact. This is discussed in full by Stewart-Oaten *et al* (1986). Where the temporal trend of mean abundance already differs between two locations before the impact, there must be an interaction in the data. This is a major problem because many populations, particularly of short-lived or widely-dispersing species, have different temporal patterns of abundance in different patches of habitat. Thus, the Bernstein and Zalinski (1983) and Stewart-Oaten *et al* (1986) protocol will not be useful for many populations.

Beyond BACI: impact versus several controls

The solution to the spatial confounding identified above is to replicate the sampling. Obviously, it is usually impossible (and would always be undesirable!) to replicate potentially impacting disturbances solely to create a more valid statistical property of the data. There is, however, no reason that replicate control locations should not be used (Underwood 1989, 1991, 1992). What is needed is a set of locations, chosen to represent a particular habitat (i.e. with similar water-flow, depth, bottom, acidity of water, etc.). The only constraint on this set is that one of the locations must be the one in which a development or disturbance is planned (Fig. 6.8). Note that the locations are not supposed to be identical; this was not a requirement of the single control location. They must, however, come from a population of locations of which, for any variable of interest, the potentially impacted location is also a member. There will, of course, be difficulties with finding such locations, but it is not an impossible task for many habitats.

Now, the analysis is an asymmetrical one (Winer 1971; Underwood 1978, 1984, 1986, 1991, 1992) involving a contrast between the putatively impacted location and the set of controls. Of greatest importance is the analysis of interaction. An environmental impact must cause a greater

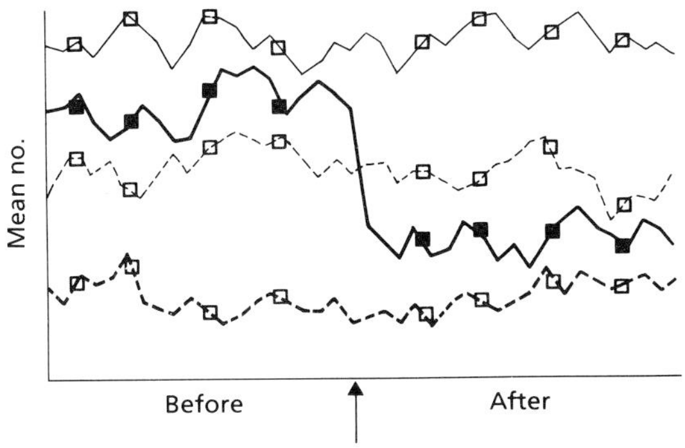

Fig. 6.8 Beyond BACI: sampling to detect effects of an environmental disturbance in the impacted (solid line) and several replicated control (dashed lines) locations. Samples are taken several times before and after the disturbance.

difference from before to after in the impacted than in the control locations. This is analysed in Table 6.4 and described in full in Underwood (1992).

Such asymmetrical analyses can also be used to solve the problem of temporal interactions occurring before the possible impact (Stewart-Oaten *et al* 1986). The details are not presented here, but an environmental impact affecting a population in one (the impacted) location will cause a different magnitude of temporal interaction after, compared with before, for that location than occurs in the control locations. These interactions can be compared in statistical tests based on the asymmetrical analysis of all the data (Underwood 1992). Thus, analyses of environmental impact can be on a more secure logical foundation than where there is only a single control. They can also use more types of data than is the case in the more usual BACI designs (Bernstein & Zalinski 1983; Stewart-Oaten *et al* 1986).

Where it is not possible to sample several locations at the same time, because of logistic constraints, these procedures can be modified to provide some partial tests for the presence of impacts (Underwood 1991).

Confusion of space and time

A major problem besetting any biological sampling is that temporal and spatial patterns are not simple to separate (e.g. Pielou 1969). For some investigations, this may not be important.

Table 6.4 Analysis of temporally and spatially replicated 'Beyond BACI' design of sampling to detect environmental impact. n replicate samples are taken in each of several locations (l) (the putatively impacted and a set of controls) at t replicated times before and again after the disturbance occurs (as in Fig. 6.8). The locations must be sampled at the same times. Environmental impact would be detected as a significant interaction B × I, indicating a different temporal pattern in the difference between the impacted and control locations after the disturbance than occurred before. If there are significant differences among the locations in the time-courses of abundances before and after the disturbance, the interaction T(B) × I can be analysed further (Underwood 1992)

Source of variation		Degrees of freedom	
Before versus after	= B	1	
Among locations	= L	$(l-1)$	
Impact versus controls	= I		1
Among controls	= C		$(l-2)$
B × L		$(l-1)$	
B × I			1
B × C			$(l-2)$
Times (B)	= T(B)	$2(t-1)$	
T(B) × L		$2(t-1)(l-1)$	
T(B) × I			$2(t-1)$
T(B) × C			$2(t-1)(l-2)$
Residual		$2lt(n-1)$	
Total		$2ltn-1$	

Understanding the natural temporal changes in populations is, however, essential for any interpretation of their ecology and, particularly, for any programme of management and conservation. It is impossible to make realistic predictions about future abundances if temporal changes are not understood.

There are two circumstances where spatial variations in abundance will become very confused with apparent temporal changes in numbers in a population. First, in a heterogeneous, patchy habitat, if only a small number of replicate samples are taken at each of several occasions, they may sample different patches. Consider the bed of a river where there are patches of habitat which support large densities of an organism and other areas where the organisms are sparse. If only a few replicated samples are used to estimate abundances, they may, by ill chance, land predominantly in one type of habitat on one occasion and in the other at some subsequent time (Fig. 6.9). The different mean abundance in the two samples will be interpreted as a change in abundance.

Where the patterns of distribution are readily visible, a 'map' can be made and sampling done in a repeated fashion by stratifying the area to be sampled and then taking the same number of replicates in each 'stratum' or type of patch during each time of sampling. Thus, if 20% of the area consists of patches of gravel (and gravel may have a different density of a sampled species than is found in other substrata), it would be prudent to put a similar number of quadrats in areas of gravel every time samples are taken. There have been many reviews of procedures for stratifying sampling (Cochran & Cox 1957; Cox 1958; Elliott 1977; Andrew & Mapstone 1987). There are different procedures when different types of information are available about the spatial variations in abundance.

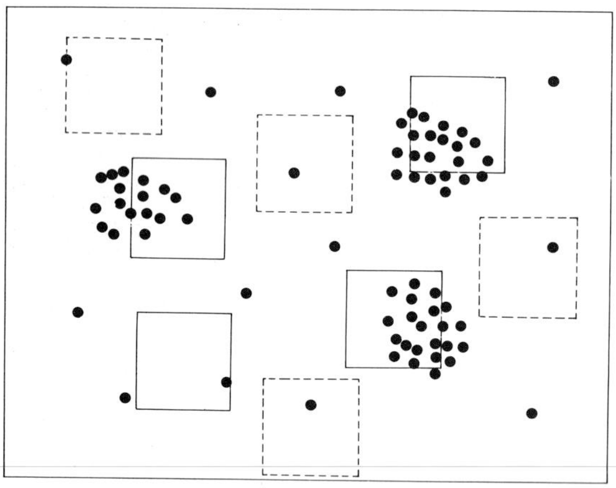

Fig. 6.9 Sampling a very heterogeneous (spatially dispersed) population with few replicates. At one time of sampling (solid squares) density is estimated to be larger than at the next time (dashed squares).

As a second situation, apparent changes of mean abundance will be seen from time to time if the organisms sampled show behavioural changes in dispersion. This is well known, but not often considered in detail. Thus, if animals are very clumped at some times of the year or at some stages of the life cycle (or under some physical conditions, such as during and immediately after floods), but are more spread out at other times (or under other weather conditions), unless very large numbers of samples are taken on each occasion, mean abundances will appear to have changed. When the organisms are clumped (unless there are very many or very large clumps), many replicate sample units will be in the sparser areas between clumps. For example, if the clumps occupy 10% of the total area being sampled, and sample units are smaller than clumps, the probability of not hitting a clump in a sample of 10 units is 0.35; more than one-third of such samples do not measure densities reliably. The mean abundance will be small (although the variance among sample units will, on average, be relatively large). When the organisms are more scattered, there will be more in most of the sample units and mean abundance will be larger (with, usually, smaller variance among sample units). The change in density is, however, not real, but simply reflective of the behavioural change in dispersion.

One method for determining whether sampling is reliably estimating abundances at appropriate spatial scales, so that temporal patterns are clearly separable from spatial differences, is pilot sampling at short time-intervals. Sampling is done at some pre-determined spatial scales of placement of sample units. Sampling is then repeated after a short time, short enough that no alteration of abundance (through recruitment, migration or mortality) could be expected. Then, if there *is* a difference between the two samples, it would identify an inadequacy of the spatial arrangement of sample units, or short-term changes in dispersion of the organisms (which could be examined by investigating the spatial variances), because it could not reflect a temporal change. This has rarely been done, but an example was described by Kennelly and Underwood (1985).

Unconfounding spatial scales

A different confusion of space and time occurs when a habitat is not sampled at appropriate spatial scales. Suppose, for example, that there is a general gradient of density across a river, such that more animals are found per unit area in its centre than at the edges. Unless some effort is made to ensure that sampling is equally representative on all occasions, there will, again, be apparent differences in abundance from time to time because of different patterns of spatial sampling. As an example in the ecology of rivers, Arthur *et al* (1982) compared the fauna in experimental channels, one at elevated temperature of the water. There was, however, only one channel at each temperature and no estimation was made of the magnitude of differences between the abundances of fauna in the two channels due to their different siting, history and any events during the study that were not identical in the two places.

This was a major problem identified by Hurlbert (1984) in his review of ecological confounding. Where geographical (or temporal) differences from plot to plot cause differences in abundance (size, growth or any other variable) from plot to plot (or time to time), comparisons between two plots, one representing one treatment (or time) and the other a different treatment (or time), cannot be used to conclude that the treatment (or passage of time) caused the observed differences. The solution, not discussed by Hurlbert (1984), is to attempt hierarchical or nested sampling (Snedecor & Cochran 1967; Green & Hobson 1970; Underwood 1981; Andrew & Mapstone 1987).

Instead of placing sampling units (quadrats, grabs, nets, trays) at random throughout the area, they are placed in sets according to a number of spatial scales (Fig. 6.10). Arbitrarily defined plots (at some spatial scale say 30 m on a side) are imposed on the area to be sampled. Some of these are picked at random to be sampled. Then, these plots are further subdivided (e.g. into sub-plots of

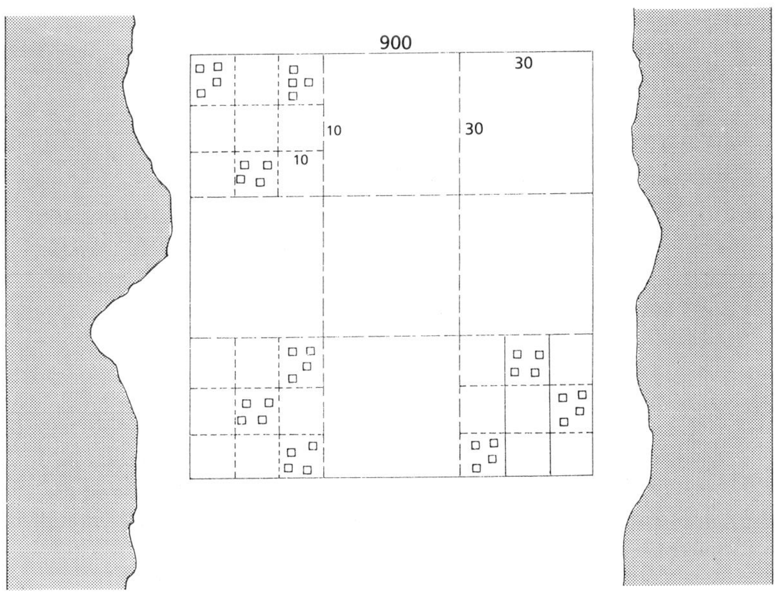

Fig. 6.10 Hierarchical (nested) sampling at several spatial scales. The area to be sampled (in a river) is divided into 30 × 30-m plots, of which three are chosen at random to sample. Each plot is divided into 10 × 10-m sub-plots, three of which are chosen at random to sample. Each sub-plot is sampled with four random, replicate sample-units. See text for further details.

side 10 m). Again, some of these are picked at random to be sampled. These could be further subdivided into areas (say, squares 5 × 5 m). In randomly-chosen areas, several replicate sampling units (nets, grabs, etc.) are used. The choice of scales is not necessarily arbitrary, but may be dictated by known biological features of the populations being sampled. For example, if the animals routinely travel 10 or 20 m up or downstream in search of food, it would be appropriate to have some sampled scale within this and one somewhat larger scale to detect any resulting spatial variances in abundance that might be related to such behaviour.

Data from such hierarchically organized sampling are amenable to nested analysis of variance (e.g. Green & Hobson 1970; Winer 1971; Sokal & Rohlf 1981; Underwood 1981). This will detect those spatial scales at which there are variations in abundance. These scales will then need to be sampled on each occasion to ensure that spatial variations do not become confused in assessment of temporal change. One great advantage of nested sampling of a hierarchy of spatial scales is that the data are readily amenable to cost-benefit procedures to determine how best to allocate effort (i.e. number of replicate samples) of each spatial scale (e.g. Snedecor & Cochran 1967; Green 1979; Underwood 1981; see Kennelly & Underwood 1984, 1985 for particular examples).

Where data are available from several hierarchical spatial scales, they can be analysed to determine whether some environmental disturbance has altered the spatial variance at some scale, regardless of whether the mean abundance has been changed (a problem raised earlier; see Fig. 6.3). The procedures are not discussed here, because of lack of space, but a similar approach was used by Green and Hobson (1970) to compare patchiness in abundance of a bivalve in different places on a beach.

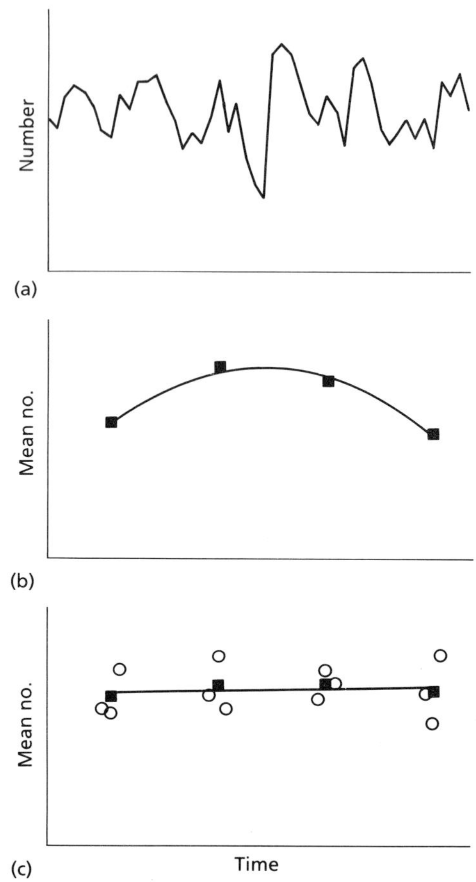

Fig. 6.11 Seasonal sampling with inadequate temporal replication. (a) True, but unknown pattern of abundance of a population; there is no seasonal pattern. (b) Apparent seasonal pattern in four widely separated samples. (c) Nested sampling (three different times in each season) gives seasonal averages that show, correctly, no pattern.

Unconfounding temporal scales

To determine temporal patterns of abundance, hierarchical (nested) time-scales are necessary. This has rarely been the case in published studies. Usually, if sampling is done, say, every 3 months, to examine seasonal trends in abundance of a population, several (spatially scattered) replicates are taken at each time of sampling. If significant differences are found among the times of sampling, these are assumed to represent seasonal differences. This conclusion is, however, not necessarily correct because the data are, in fact, confounded or pseudo-replicated (Hurlbert 1984). There is only one sample per season. The differences may not represent seasonal differences, but simply unpatterned temporal fluctuations. This is illustrated in Figs 6.11 and 6.12.

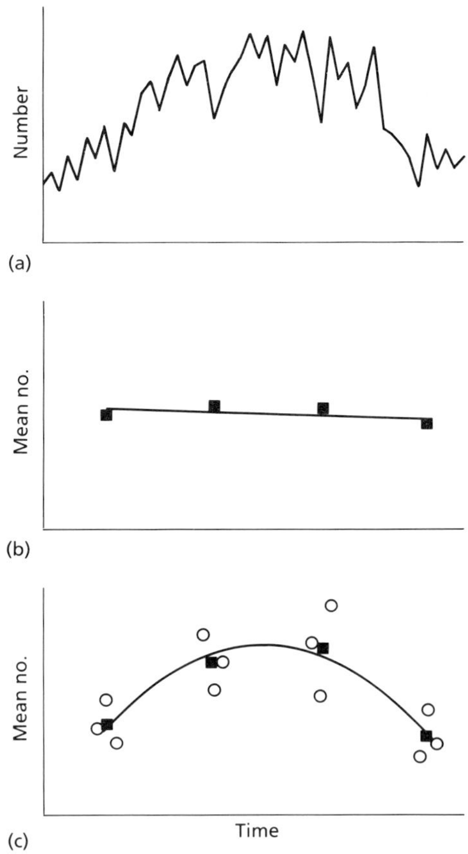

Fig. 6.12 Seasonal sampling with inadequate temporal replication. (a) True, but unknown pattern of abundance of a population; there is a marked seasonal pattern. (b) Apparent lack of seasonal pattern in four widely separated samples. (c) Nested sampling (three different times in each season) gives seasonal averages that show, correctly, the seasonal pattern.

In the first case (Fig. 6.11a), there is no temporal pattern of abundance, but the mean abundance at the four times of sampling appears to show a seasonal cycle (Fig. 6.11b). In contrast, if there is sufficient short-term temporal variation (or 'noise'), the samples may show no trend (Fig. 6.12b), even though there is a clear seasonal cycle (Fig. 6.12a). The chances of detecting statistical differences among the means of the seasonal samples is entirely dependent on the design and intensity of sampling at each time. But the replicated sampling, however the sample units are arranged, will estimate spatial variance at one time. Thus, if there is little spatial variance in abundance, there will be small within-sample variances and considerable statistical power to detect differences among seasonal samples.

What is required to be sure that there are, in fact, seasonal differences in abundance is temporal replication (see also Underwood 1991). To demonstrate seasonal variation requires that there are larger differences among mean abundances from season to season than occur at much shorter time-scales. This is illustrated in Figs 6.11 and 6.12, where samples are taken several times in each season. At each time, there are replicate (spatial) samples. The data are hierarchically arranged with two time-scales – 'Seasons' and 'Times within each season'. Only if there is significant variation among the means for each season, over and above the variation within seasons, is it possible to demonstrate that there really are seasonal differences. A nested analysis of variance is needed for these data (Table 6.5) and the previous comments about cost-benefit procedures and the efficient allocation of sampling units (how many times to sample in each season and how many replicates to take at each time) can also be applied to this problem. These nested arrangements of sampling are also necessary to detect impacts that affect temporal variation (Fig. 6.13; see Underwood 1992).

The previous discussion does not imply that sizes of samples (i.e. replication, numbers of times sampled, number of locations) need to be huge. Although considerable thought must be given to the best way to allocate time, effort and money to sampling, the power of such analyses to detect impacts is quite large, even with relatively modest sizes of replication (Underwood 1993). Insufficient examples exist to determine appropriate guidelines for designing asymmetrical 'Beyond BACI' programmes. Nevertheless, it is probable that a re-allocation of numbers of replicates from one control location to several would not dramatically increase costs. In general, it is probably better to reduce the intensity of arbitrarily-chosen temporal patterns of sampling (routine monitoring) and re-allocate effort to several spatial controls. Simulations of environmental impacts (A.J. Underwood, unpublished data)

Table 6.5 Analysis of temporally (but not spatially) replicated 'Beyond BACI' design of sampling to detect environmental impact that affects temporal fluctuations but not long-run means. Replicate samples are taken in each of two locations (the putatively impacted and a control) at two temporal scales. Samples are taken in p replicate periods before and again after the disturbance occurs (as in Fig. 6.13, where $p = 2$ for the impacted location). In each period, there are t replicate times of sampling ($t = 3$ in Fig. 6.13). n replicate samples are taken at each time. The locations must be sampled at the same times. Environmental impact would be detected as a significant interaction L × P(B) or L × T(P(B)), depending on the time-scale at which the impact altered the temporal pattern of abundance in the impacted versus the control location (see Underwood 1991 for full details)

Source of variation		Degrees of freedom	*F*-ratio versus	Degrees of freedom
Before versus after	= B	1		
Control versus impact	= L	1		
B × L		1		
Periods in before or after	= P(B)	$2(p-1)$		
L × P(B)		$2(p-1)$	L × T(P(B))	$2(p-1), 2p(t-1)$
Times within periods	= T(P(B))	$2p(t-1)$		
L × T(P(B))		$2p(t-1)$	Residual	$2p(t-1), 2lt(n-1)$
Residual		$2lt(n-1)$		
Total		$2ltn-1$		

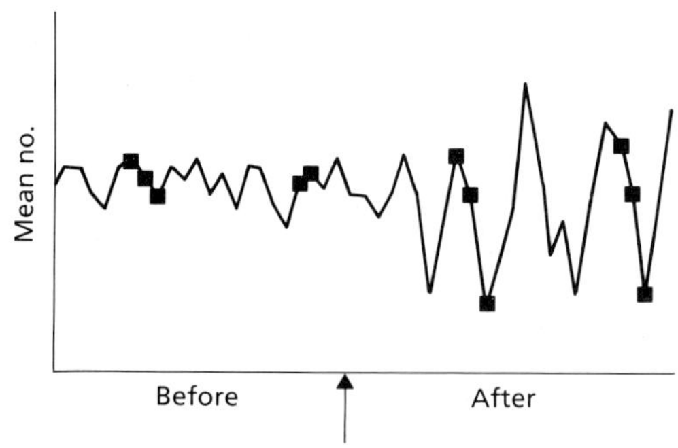

Fig. 6.13 Sampling at two different (nested) temporal scales, twice before and twice after an environmental disturbance (indicated by the arrow) to detect an impact that causes changed temporal variation, but no change in mean abundance.

suggest that pulse disturbances and large press disturbances should be detectable for many populations with only two or three control locations, sampled four to six times before and after the disturbance, with three to five replicate samples at each time in each location. What matters most is to allocate effort in the most effective and interpretable manner. Otherwise, all the effort (and budget) is expended to no purpose at all!

Independent sampling in rivers

A major problem with biological sampling in rivers is that abundances in any location are not likely to be independent of the data in samples upstream. There are many processes that can cause correlations in the abundances of a population from one place to another. In the case of environmental perturbations, for example effluents from a point source, there may be large-scale impacts, causing correlated changes in abundance at numerous locations downstream. This is well known and explains the widespread advocacy of using an area upstream of an outfall as a control location for determining any impact on populations downstream (reviewed by Green 1979). It is explicitly recognized in this procedure that no control location could be identified downstream because it is likely that any impact would probably extend a long way from the outfall (see, for example, Cuffney *et al* 1990).

What does not seem to be so widely recognized is the great potential that always exists for non-independence between two areas of a river because they are linked by the flow of water. For example, downstream drift of stream invertebrates (e.g. Cuffney *et al* 1990; review by Brittain & Eikeland 1988) will presumably link the abundances of a species in lower reaches of the stream with those in the source populations upstream. The abundances of many organisms in an upstream and a downstream location would be positively correlated.

Independence of samples, within and between groups, is a fundamental requirement of most statistical procedures (e.g. Fisher 1925; Snedecor & Cochran 1967; Winer 1971; Sokal & Rohlf 1981). It is impossible to discuss here all of the problems caused by non-independence of data.

The abundances of organisms in two locations will be positively correlated if, for example, a similar proportion of any population drifts downstream to another location. Whatever the abundance upstream, the abundance downstream will be related to it. If upstream/downstream differences are being examined in several rivers, there will be problems if the samples from each upstream and each downstream location are considered independent when they are, in fact, positively correlated.

Under these circumstances, the probability of rejecting the null hypothesis (that there is no difference in mean abundance between upstream and downstream locations) is not that given in tables for such statistics as t, F, various rank-order or binomial tests. The probability of Type II error (retaining the null hypothesis when it is, in fact, false) is enhanced (Cochran 1947).

The solution to the problem is to divide the rivers randomly into two groups. The upstream data are then taken from one half of the rivers (one group) and the downstream data from the other half. The data for the two conditions are completely independent of each other (they come from separate rivers). Alternatively, the data could be treated as pairs, so that the difference (upstream minus downstream) can be tested against the null hypothesis that the mean difference is zero (i.e. the upstream and downstream locations in each river have the same abundance, on average over several rivers).

Non-parametric, binomial and t-tests are available for either stratagem. Which of the two approaches is better depends on various factors. The second (paired) approach requires less rivers to be sampled for the same power of test. Each of n rivers furnishes upstream and downstream data. In the first approach, to obtain n replicate upstream and n replicate downstream sets of data requires $2n$ rivers to be visited (although only one or other type of data is taken from each). Apart from this, the decision about whether to use paired data rests on the degree to which the abundances in the two locations in each river are actually correlated and the variances among the replicate samples in each location. An excellent account of how to make the decision is given by Snedecor and Cochran (1967).

If the purpose of sampling is to identify in a particular river any difference between upstream and downstream that might be associated with an environmental disturbance, several control locations are needed. The task then becomes one of deciding whether the difference (upstream minus downstream) in the putatively impacted river is more or less than would be expected by chance in a population of such differences. Such a population of differences has, of course, been sampled as the control rivers.

Theoretically, upstream/downstream data could be taken in other, similar, higher stretches of the same river (they cannot be downstream in case the impact is extensive and widespread). This may, however, still result in non-independence of the data; abundances of organisms in any downstream section of a river may be correlated with those higher up.

It seems that more research is needed to determine the degree to which upstream and downstream localities do, in fact, have correlated abundances of various types of organisms. Samples from numerous upstream and downstream locations at different spatial scales apart need to be examined for positive or negative correlation. Some types of animals and plants are less likely to be correlated than others (e.g. they do not drift). At some distances apart, upstream and downstream locations are less likely to be correlated. Once it can be established that there are some species and some spatial scales along a river that will provide independent data, this information should be used in planning the sampling for assessment of potential environmental impact (or any experimental field programme requiring upstream and downstream locations to be sampled).

6.3 BASELINE DATA

Many types of disturbance are unpredictable and there are many difficulties with identifying human-induced changes to populations that have large temporal fluctuations or spatial differences. As a result, much assessment of environmental impact must be done with limited 'before' data. For accidental disturbances, there is no planned period 'before'. In the case of planned developments, there is, at best, a limited period of possible investigations before the development begins.

So, assessment of the effects of disturbances must often depend on comparisons among locations after the putative impact has occurred (see particularly Green 1979). Many useful features of the sampling designs discussed previously still apply, but the focus of the problem has, of course, shifted. There is no method for addressing the question of whether there has been a *change* in abundance in the disturbed location, different from that in control locations and coincident or following from the disturbance. The issue is now to determine whether there is some *difference* between the abundance in the disturbed location and that in the controls, or whether there is a different time-course of abundance in the disturbed location compared with that in the controls. These questions are far less indicative of an impact than can be demonstrated when data are also available from before the disturbance. There may always have been a measurable difference in the mean abundance, or in the temporal trajectory of mean abundance, between the single disturbed locations and the controls. This can never be established.

It would therefore be worth considering procedures which would allow some estimation of what might have been happening before the disturbance. By and large, there are two research programmes (Lakatos 1974) that will help. First is the general class of monitoring or baseline studies, which could be restructured to test hypotheses about the degree of spatial and temporal variability in natural, undisturbed populations (i.e. like those in control locations).

Second are intrusive or manipulative experimental programmes that seek to test hypotheses about the nature, magnitude and rates of responses to disturbance, so that better predictions can be made about the likely effects of different human activities. These will be considered later.

Estimation of background variances

It would be very informative to know the ranges and patterns of temporal and spatial variation for species with different sorts of life histories in a variety of riverine habitats. Sampling to determine these variances would, if done properly, allow general assessments of effects of disturbances in other similar habitats.

The need to assess spatial patchiness, by replicated sampling at different spatial scales, was identified earlier. If this is done in a set of undisturbed localities, it is obviously equivalent to data before disturbance in those habitats. If the spatial patterns are examined at several times (different seasons, different years), they can form a background assessment of spatial variance that should represent any similar habitat. The mean abundance of a species of mayfly in areas of hard substratum in stretches of a river of a certain range of speed of water flow, pH, oxygen concentration, etc. is, itself, a variate from a distribution (an unknown frequency distribution) of such mean abundances over all possible similar habitats. If the mean abundance is sampled in say, five independent, representative stretches of this habitat, the variation among the habitats provides a measure of the variance of the entire population of similar habitats. This is a familiar consequence of the theory of unbiased, representative sampling (Fisher 1925; Feller 1968; Johnson & Kotz 1969).

To ensure that the sample of localities examined is representative of the entire possible population of localities, they must be chosen as a 'random factor', in the terminology of analysis of variance (Scheffe 1959; Winer 1971; Underwood 1981). Once the sort of habitat being examined has been defined (i.e. the ranges of depth and speed of water, the substratum, etc., have been decided), a random sample of places that fall within the definition can be sampled.

The places must not be chosen on the basis of any particular feature that is not part of the definition of the habitat. Thus, a particular location that happens to be the nearest to a main road, or that is locally reputed to contain very large trout, etc. should not be included in the sample because the investigator thinks it would be informative to attempt to investigate the result of being near a road (or of having large trout nearby). All possible locations that fit the defined conditions should have an equal chance of being sampled.

If abundances are estimated in this sample of locations, the variation among the locations provides an estimate of the variance among all the locations in the defined distribution of localities in the defined habitat. Therefore, the variation from location to location in the sample is an unbiased estimate of the variation among locations in any other set of representative localities. Thus, once this variation has been estimated, it is known for other sets of the defined habitat—including any set of locations chosen as controls for some putatively impacted location. Thus, a properly sampled set of representative locations will provide the 'before' data for any localities of that habitat. So, it is not impossible to use properly documented estimates of variation in abundance among representative locations instead of controls for a potential impact in a particular location.

Of course, nothing is ever quite as simple as just described. The differences among locations are not constant from time to time. Consequently, the sampling just described must be repeated at several time-scales, chosen because of known time-courses of the life history of the sampled organism(s). These samples provide unbiased estimates of the temporal variation of abundances in this type of habitat. The temporal variance is, anyway, required for control locations in a study to detect environmental impact (just as the spatial differences must be known).

A temporal series of estimates of abundance in a representative set of locations of a particular habitat also provides a reliable estimate of the interactions between time and space that must be analysed for undisturbed sites in order to detect an environmental impact (Table 6.6). So, once this is sampled properly, the estimates should be reliable to use as control, before and after data for any disturbance that might affect another 'impacted' location.

The validity of using these estimates where

Table 6.6 Calculation of variances for temporal and spatial patterns of abundance of a population, using the Cornfield–Tukey calculations of mean squares estimates. The analysis is of four randomly-chosen locations (l), each sampled three times (t) in each of 2 years (y). The locations could not be sampled at the same times (hence, times are nested in the combinations of year and location). There were $n = 6$ replicate samples at each time in each location. σ^2_Y is variance between years; σ^2_L is variance among locations; σ^2_{YXL} is variation associated with the interaction between years and locations; $\sigma^2_{T(YXL)}$ is variance among times of sampling in each location and year. These data correspond to 'before', control data for any other locations that come from the habitat represented by the locations sampled here. Confidence intervals for these variance components can also be calculated (see Winer 1971 for details)

Source of variation	Df	Mean square estimates	Data from analysis
Between years	1	$\sigma^2_R + n\sigma^2_{T(YL)} + tn\sigma^2_{YL} + ltn\sigma^2_Y$	268
Among locations	3	$\sigma^2_R + n\sigma^2_{T(YL)} + tn\sigma^2_{YL} + ytn\sigma^2_L$	273
Years × locations	3	$\sigma^2_R + n\sigma^2_{T(YL)} + tn\sigma^2_{YL}$	181
Among times (years × locations)	16	$\sigma^2_R + n\sigma^2_{T(YL)}$	107
Residual	120	σ^2_R	80

Calculated estimates of variance components:

$\sigma^2_R = 80.0$

$\sigma^2_{T(YL)} = (107 - 80)/6 = 4.50$

$\sigma^2_{YL} = (181 - 107)/18 = 4.11$

$\sigma^2_L = (273 - 181)/36 = 2.55$

$\sigma^2_L = (268 - 181)/72 = 1.21$

'before' data are not available can be checked by collecting 'after' data in some control locations after the particular disturbance has occurred. All other things being correct (i.e. the current locations are from the same frequency distribution or population of localities as previously sampled), the estimates of spatial and temporal variances and their interaction in the previous, general sampling should be similar to those in the 'after' samples collected now. This proposition can be tested by analysing to compare the general data with the new 'after' data. If the proposition is reasonable, the previously gathered, general data can be used to replace the unavailable 'before' data from the not-yet-impacted and control localities.

The calculation of the appropriate variances is dependent on the number of replicate samples, the number of times sampled and the number of localities examined. For any size of sampling, the estimates of variance can be calculated from analyses of variance, using the Cornfield–Tukey formulae for estimated mean squares (Cornfield & Tukey 1956; Scheffe 1959; Winer 1971; Underwood 1981). This is illustrated in Table 6.6.

6.4 EXPERIMENTAL PROGRAMMES TO INCREASE PREDICTIVE CAPABILITY

Inertia, resilience and stability

To be able to predict the potential consequences of some disturbance to a habitat obviously requires knowledge of several attributes of the populations that might be affected. In an idealized view of populations, these attributes can be divided into three main features – the inertia, resilience and stability of the abundance of a population (although the terminology is a minefield; Underwood 1989).

Inertia is the property of a population not to respond to a disturbance (Murdoch 1970; Orians 1974). An inert population does not change in abundance when a pollutant arrives in its habitat. A less inert population declines in abundance (Fig. 6.14a). A population will be inert if, for example, the animals have behavioural adaptations to move away from excessively polluted parts of their habitat, biochemical adaptations to detoxify

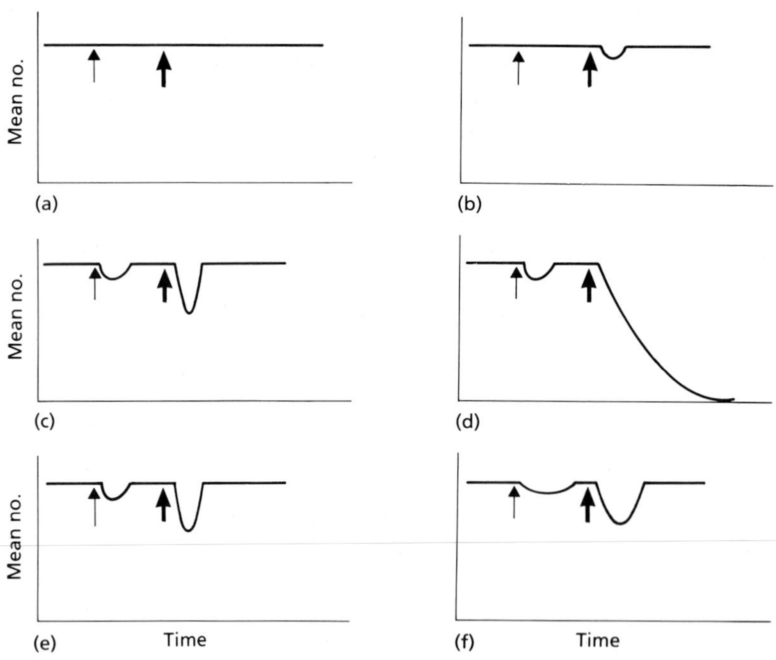

Fig. 6.14 Attributes of idealized, model, equilibrial populations. (a, b) Inertia. Population (a) is inert to a small (thin arrow) and a large (thick arrow) disturbance. Population (b) is less inert and shows a response to the larger disturbance. (c, d) Resilience. Population (c) is resilient and recovers from a small and a large disturbance. Population (d) is less resilient and cannot recover from the larger disturbance. (e, f) Stability. Population (e) is very stable and recovers quickly from two disturbances. Population (f) is less stable and takes longer to recover from either a small or a large environmental perturbation causing an impact. See text for reality.

their bodies so that pollutants do not become deleteriously concentrated, or cellular tolerances to the pollutant so that it does not interfere with any vital functions. Inertia is measurable in terms of the smallest magnitude of disturbance that elicits a response. If the inertias of populations were known, precise predictions could be made about whether a particular form and magnitude of disturbance would have any effect on abundances.

Resilience, in contrast, is the capacity to recover from a disturbance if it is large enough to cause a change (although there have been competing definitions and uses of the term: Holling 1973; Orians 1974), reviewed by Underwood (1989). A resilient population will recover from a disturbance (Fig. 6.14b). An idealized population will return to equilibrial abundance. In the real world, its abundance will converge with that of control, undisturbed populations (Underwood 1989). A less resilient population will not recover and its abundance will be permanently altered (including possibly to extinction).

Resilience is thus measured as the maximal departure from undisturbed abundances from which the populations can recover. Clearly, knowing this would allow precise prediction about the outcome of a disturbance causing a change in abundance. Alleviation and remedial action are urgently required for less resilient populations. Resilient populations will recover from the effects of disturbance (but see the consequences of instability, below).

Stability is the rate at which abundance of a population recovers from the effects of a disturbance. Other terms have been used for this (Boesch 1974; Connell & Sousa 1983). Recovery, again, can be defined in terms of idealized responses of equilibrial systems (Fig. 6.14c) or, realistically, as returning to the range of values of undisturbed controls. Knowledge of the stability of a population, which must be measured in relation to disturbances of particular magnitudes, is crucial for predicting the consequences of environmental impacts. In a stable population, recovery (if the disturbance has not moved the abundance beyond the resiliences of the population) will be rapid. Repeated disturbances will be similar. In an unstable population, subsequent disturbances will arrive before recovery from a previous one; there will then be quite different consequences.

How much perturbation can a system stand?

It follows from the previous discussion that estimates of inertia, resilience and stability of populations in relation to natural patterns of temporal and spatial variability would be extraordinarily useful in environmental assessment. The former will assist in predicting the consequences and the latter in detecting the magnitude of environmental disturbances. The determination and interpretation of these attributes is, of course, the whole point of modern ecology (e.g. Krebs 1978). So, ecology as a discipline and its application to environmental assessment can move even closer together.

Experimental perturbations will be necessary to create the stresses from which the responses of populations can be estimated and to unconfound natural variability (e.g. Underwood 1989). Intrusive, manipulative experiments have been very successful in unravelling complex biological interactions in very variable marine environments (Connell 1972, 1975; Underwood 1985, 1986). Experimental analyses of environmental disturbances in rivers and streams have an equally long and healthy history. They range from microcosmal studies of diatoms (Patrick 1975) to responses to changes in entire ecosystems (Likens *et al* 1970; Bormann *et al* 1974). Manipulations of streams (e.g. Bernard *et al* 1990) have revealed a great deal about toxins. Experimental diversions of streams (e.g. Hopkins *et al* 1989) or artificial, experimental streams (e.g. Clements *et al* 1988) have been successfully used to create replicated arrays of treatments to determine the effects of various disturbances. There should be many more such efforts.

Finally, the outcome of environmental disturbances can be examined using planned or previous cases of the disturbance. More widespread use could be made of the fact that many planned disturbances in natural systems are repeated in space and time. Thus, there are many outfalls into rivers from pulp mills or other industries. There are many industrial plants using freshwater for cooling. There are numerous weirs, dams, irrigation programmes, etc. It is appropriate to consider these as parts of large-scale experiments (Hilborn & Walters 1981). Planning appropriate experiment is not simple, but there are immense possibilities (Underwood 1989).

Choice of species

As a final comment on detection and interpretation of environmental disturbances, considerable thought must be given to appropriate species or groups of species to sample. It has been argued that abundant, widespread species are the most useful (e.g. Lewis 1976). These are, however, probably the most inert, resilient and stable – which is why they persist abundantly in numerous locations. They are therefore unlikely to be sensitive to many types of environmental disturbance.

Other authors have argued for attention to be on 'middle abundance' species that show responses to a range of disturbances (Gray & Pearson 1982; Gray 1989). There are numerous other considerations and strategies (e.g. Underwood & Peterson 1988), including the efficiency and ease of sampling, taxonomic reliability, role in structuring assemblages, or in energy transfer through food webs, etc.

To add to the list of possible criteria, the temporal variance of abundance of populations should be considered. Are relatively invariant numbers a sign of inertia (the population is not responding to the numerous local natural disturbances that beset more variable populations)? Or is it a sign of great stability (the populations are disturbed but recover so rapidly that sampling fails to reveal this)? Are populations that show great temporal variability the ones that have small inertia (they are always being perturbed), yet massive resilience (they keep bouncing back)? Experimental analyses to ascertain which is an appropriate interpretation will be very helpful. Organisms with different patterns of temporal variance may be shown by experiments to have different classes of inertia, resilience, stability or interactions among these attributes. If so, it may prove possible to discover the different features of life histories, sizes, rates of growth, position in food webs, phylogenies, etc. that are correlated with these attributes. This would be very important because it should enable predictions to be made about the effects of disturbances on populations for which there is sufficient knowledge of natural history to classify the species into appropriate groupings of inertia, stability and resilience, but for which there are insufficient quantitative data (and no time) before a disturbance begins. Such a classification would enable predictions about the effects of a disturbance even though no quantitative data are available about the population. The development of better theory relating life history to disturbance is only going to be possible through better experimental intrusions into natural populations.

6.5 CONCLUSIONS

In conclusion, the need for improved scientific input into environmental decision-making is large and urgent. Much more thought is required about the processes by which detection, interpretation and prediction of environmental threats can be improved. Considerable care is needed, but is too infrequently included, in environmental programmes of sampling (Green 1979; Hurlbert 1984; Stewart-Oaten *et al* 1986; Underwood 1981, 1986, 1991, 1992; see also Vol. 1, Chapter 13). Searches of the relevant literature in preparation of this review did not reveal many good case histories. For freshwater research, they were, perhaps, not found or not published, rather than not done. There are some interesting examples of pilot experiments on the methods to be used, their replication and sources of variability caused by their use (e.g. Furse *et al* 1981). The review by Clark (1989) considered a number of approaches to evaluations of contamination in field sites and Elliot (1977) remains the standard text.

The intellectual challenge is to design realistic field experimental perturbations to measure effects of possible impacts and to understand the requirements for management, reconstruction of habitat and rehabilitation of species. Without this, public pressure, environmentalists' emotional biases and the random outcomes of legal processes will be the only components of environmental protection. It is a collective responsibility for environmental scientists to do better science, to avoid the total demise of a rational, objective understanding of how best to conserve species in natural habitats subject to human influences. If this discussion provokes any thought about current problems of logic, design of sampling and interpretation of data and reliability of conclusions, it will have achieved some purpose.

ACKNOWLEDGEMENTS

The preparation of this chapter was supported by grants from the Australian Research Council and the Institute of Marine Ecology, University of Sydney. I am grateful for the advice from and discussions with Drs M.G. Chapman, P.G. Fairweather, K.A. McGuinness and P.A. Underwood and to K.L. Astles for considerable help in the preparation of the paper.

REFERENCES

Andrew NL, Mapstone BD. (1987) Sampling and the description of spatial pattern in marine ecology. *Annual Review of Oceanography and Marine Biology* **25**: 39–90. [6.2]

Arthur JW, Zischke JA, Erickson GL. (1982) Effect of elevated water temperature on macroinvertebrate communities in outdoor experimental channels. *Water Research* **16**: 1465–77. [6.2]

Bender EA, Case TJ, Gilpin ME. (1984) Perturbation experiments in community ecology: theory and practice. *Ecology* **65**: 1–13. [6.1]

Bernard DP, Neill WE, Rowe L. (1990) Impact of mild experimental acidification on short term invertebrate drift in a sensitive British Columbia stream. *Hydrobiologia* **203**: 63–72. [6.4]

Bernstein BB, Zalinski J. (1983) An optimum sampling design and power tests for environmental biologists. *Journal of Environmental Management* **16**: 35–43. [6.2]

Boesch DF. (1974) Diversity, stability and response to human disturbance in estuarine ecosystems. *Proceeding First International Congress of Ecology* **1**: 109–14. [6.4]

Bormann FH, Likens GE, Siccama TG, Pierce RS, Eaton JS. (1974) The export of nutrients and recovery of stable conditions following deforestation at Hubbard Brook. *Ecological Monographs* **44**: 255–77. [6.4]

Brittain JE, Eikeland TJ. (1988) Invertebrate drift – a review. *Hydrobiologia* **166**: 77–93. [6.2]

Clark JR. (1989) Field studies in estuarine ecosystems: a review of approaches for assessing contaminant effects. In: Cowgill UM, Williams LR (eds) *Aquatic Toxicology and Hazard Assessment*, vol. 12, pp. 120–33. ASTM, Philadelphia. [6.5]

Clarke KR. (1993) Non-parametric multivariate analyses of changes in community structure. *Australian Journal of Ecology* **18**: 117–43. [6.1]

Clarke KR, Green RH. (1988) Statistical design and analysis for a 'biological effects' study. *Marine Ecology Progress Series* **46**: 213–26. [6.1]

Clements WH, Cherry DS, Cairns J. (1988) Structural alterations in aquatic insect communities exposed to copper in laboratory streams. *Environmental Toxicology and Chemistry* **7**: 715–22. [6.4]

Clifford HF. (1982) Effects of periodically disturbing a small area of substratum in a brown-water stream of Alberta, Canada. *Freshwater Invertebrate Biology* **2**: 39–47. [6.1]

Cochran WG. (1947) Some consequences when the assumptions for the analysis of variance are not satisfied. *Biometrics* **3**: 22–38. [6.2]

Cochran WG, Cox G. (1957) *Experimental Designs*, 2nd edn. John Wiley, New York. [6.2]

Connell JH. (1972) Community interactions on marine rocky intertidal shores. *Annual Review of Ecology and Systematics* **3**: 169–92. [6.4]

Connell JH. (1975) Some mechanisms producing structure in natural communities: a model and evidence from field experiments. In: Cody MS, Diamond JM (eds) *Ecology and Evolution of Communities*, pp. 460–90. Harvard University Press, Cambridge, Massachusetts. [6.1, 6.4]

Connell JH, Sousa WP. (1983) On the evidence needed to judge ecological stability or persistence. *American Naturalist* **121**: 789–824. [6.1, 6.4]

Connor EF, Simberloff D. (1986) Competition, scientific method and null models in ecology. *American Scientist* **75**: 155–62. [6.1]

Cornfield J, Tukey JW. (1956) Average values of mean squares in factorials. *Annals of Mathematical Statistics* **27**: 907–49. [6.3]

Cox G. (1958) *The Planning of Experiments*. Wiley, New York. [6.2]

Cuffney TF, Wallace JB, Lugthart GJ. (1990) Experimental evidence quantifying the role of benthic invertebrates in organic matter dynamics of headwater streams. *Freshwater Biology* **23**: 281–99. [6.2]

Dayton PK. (1971) Competition, disturbance and community organization: the provision and subsequent utilization of space in a rocky intertidal community. *Ecological Monographs* **41**: 351–89. [6.1]

Den Boer PJ. (1979) The significance of dispersal power for the survival of species, with special reference to the Carabid beetles in a cultivated countryside. *Fortschritte Zoologie* **25**: 79–94. [6.1]

Den Boer PJ. (1981) On the survival of populations in a heterogeneous and variable environment. *Oecologia* (Berlin), **50**: 39–53. [6.1]

Elliott JM. (1977) *Methods for the Analyses of Samples of Benthic Invertebrates*. Freshwater Biological Association Scientific Publication No. 25, Reading. [6.2, 6.5]

Feller W. (1968) *An Introduction to Probability Theory and its Applications*. Wiley, New York. [6.3]

Fisher RA. (1925) *Statistical Methods for Research Workers*. Oliver & Boyd, London. [6.2, 6.3]

Frank PW. (1981) A condition for a sessile strategy. *American Naturalist* **118**: 288–90. [6.1]

Furse MT, Wright JF, Armitage PD, Moss D. (1981) An appraisal of pond-net samples for biological monitoring of lotic macro-invertebrates. *Water Research* **15**: 679–89. [6.5]

Gray JS. (1989) Effects of environmental stress on species rich assemblages. *Biological Journal of the Linnaean Society* **37**: 19–32. [6.4]

Gray JS, Pearson TH. (1982) Objective selection of sensitive species indicative of pollution-induced changes in benthic communities. I. Comparative methodology. *Marine Ecology Progress Series* **9**: 11–21. [6.4]

Green RH. (1979) *Sampling Design and Statistical Methods for Environmental Biologists*. Wiley, Chichester. [6.1, 6.2, 6.3, 6.5]

Green RH, Hobson KD. (1970) Spatial and temporal structure in a temperate intertidal community, with special emphasis on *Gemma gemma* (Pelecypoda: Mollusca). *Ecology* **51**: 999–1011. [6.2]

Hilborn R, Walters CJ. (1981) Pitfalls of environmental baseline and process studies. *Environmental Impact Assessment Review* **2**: 265–78. [6.4]

Holling CS. (1973) Resilience and stability of ecological systems. *Annual Review of Ecology and Systematics* **4**: 1–23. [6.4]

Hopkins PS, Kratz KW, Cooper SD. (1989) Effects of an experimental acid pulse on invertebrates in a high altitude Sierra Nevada stream. *Hydrobiologia* **171**: 45–58. [6.4]

Hurlbert SJ. (1984) Pseudoreplication and the design of ecological field experiments. *Ecological Monographs* **54**: 187–211. [6.1, 6.2, 6.5]

Johnson NL, Kotz S. (1969) *Distributions in Statistics: Discrete Distributions*. Wiley, New York. [6.3]

Kennelly SJ, Underwood AJ. (1984) Underwater microscopic sampling of a sublittoral kelp community. *Journal of Experimental Marine Biology and Ecology* **76**: 67–78. [6.2]

Kennelly SJ, Underwood AJ. (1985) Sampling of small invertebrates on natural hard substrata in a sublittoral kelp forest. *Journal of Experimental Marine Biology and Ecology* **89**: 55–67. [6.2]

Krebs CJ. (1978) *Ecology: the Experimental Analysis of Distribution and Abundance*. Harper & Row, New York. [6.4]

Lakatos I. (1974) Falsification and the methodology of scientific research programmes. In: Lakatos I, Musgrave AE (eds) *Criticism and the Growth of Knowledge*, pp. 91–6. Cambridge University Press, Cambridge. [6.3]

Lewis JR. (1976) Long-term ecological surveillance: practical realities in the rocky littoral. *Annual Review of Oceanography and Marine Biology* **14**: 371–90. [6.4]

Likens GE, Bormann FH, Johnson NM, Fisher DW, Pierce RS. (1970) Effects of forest cutting and herbicide treatment on nutrient budgets in the Hubbard Brook watershed-ecosystem. *Ecological Monographs* **40**: 23–47. [6.4]

Murdoch WW. (1970) Population regulation and population inertia. *Ecology* **51**: 497–502. [6.4]

Orians GH. (1974) Diversity, stability and maturity in natural ecosystems. In: van Dobben WH, Lowe-McConnell RH (eds) *Unifying Concepts in Ecology*, pp. 139–50. W. Junk, The Hague. [6.4]

Patrick R. (1975) Structure in stream communities. In: Cody MS, Diamond JM (eds) *Ecology and Evolution of Communities*, pp. 445–59. Harvard University Press, Cambridge, Massachusetts. [6.4]

Pielou EC. (1969) *An Introduction to Mathematical Ecology*. Wiley, New York. [6.1, 6.2]

Scheffe H. (1959) *The Analysis of Variance*. Wiley, New York. [6.3]

Snedecor GW, Cochran WG. (1967) *Statistical Methods*, 6th edn. University of Iowa Press, Ames. [6.2]

Sokal RR, Rohlf FJ. (1981) *Biometry*, 2nd edn. Freeman, New York. [6.2]

Stewart-Oaten A, Murdoch WM, Parker KR. (1986) Environmental impact assessment: 'pseudoreplication' in time? *Ecology* **67**: 929–40. [6.2, 6.5]

Underwood AJ. (1978) An experimental evaluation of competition between three species of intertidal prosobranch gastropods. *Oecologia* **33**: 185–208. [6.2]

Underwood AJ. (1981) Techniques of analysis of variance in experimental marine biology and ecology. *Annual Reviews of Oceanography and Marine Biology* **19**: 513–605. [6.1, 6.2, 6.3, 6.5]

Underwood AJ. (1984) Vertical and seasonal patterns in competition for microalgae between intertidal gastropods. *Oecologia* **64**: 211–22. [6.2]

Underwood AJ. (1985) Physical factors and biological interactions: the necessity and nature of ecological experiments. In: Moore PG, Seed R (eds) *The Ecology of Rocky Coasts*, pp. 371–90. Hodder & Stoughton, London. [6.4]

Underwood AJ. (1986) The analysis of competition by field experiments. In: Kikkawa J, Anderson DJ (eds) *Community Ecology: Pattern and Process*, pp. 240–68. Blackwells, Melbourne. [6.1, 6.2, 6.4, 6.5]

Underwood AJ. (1989) The analysis of stress in natural populations. *Biological Journal of the Linnaean Society* **37**: 51–78. [6.1, 6.2, 6.4]

Underwood AJ. (1990) Experiments in ecology and management: their logics, functions and interpretations. *Australian Journal of Ecology* **15**: 365–89. [6.1]

Underwood AJ. (1991) Beyond BACI: experimental designs for detecting human environmental impacts on temporal variations in natural populations. *Australian Journal of Marine and Freshwater Research* **42**: 569–87. [6.1, 6.2, 6.5]

Underwood AJ. (1992) Beyond BACI: the detection of environmental impact on populations in the real, but variable, world. *Journal of Experimental Marine Biology and Ecology* **161**, 145–78. [6.2, 6.5]

Underwood AJ. (1993) The mechanics of spatially replicated sampling programmes to detect environmental impacts in a variable world. *Australian Journal of Ecology* **18**, 99–116. [6.2]

Underwood AJ, Peterson CH. (1988) Towards an ecological framework for investigating pollution. *Marine Ecology Progress Series* **46**: 227–34. [6.1, 6.4]

Warwick RM. (1993) Environmental impact studies on marine communities: pragmatical considerations. *Australian Journal of Ecology* **18**: 63–80. [6.1]

Winer BJ. (1971) *Statistical Principles in Experimental Design*, 2nd edn. McGraw-Hill Kogakusha, Tokyo. [6.2, 6.3]

7: Water-Quality Monitoring

W. PARR

The examples given in this chapter are gained principally from studies and monitoring programmes undertake within the UK and, as such, deal with the problems and challenges of monitoring temperate UK-sized watercourses, i.e. headwaters to medium-sized rivers. Examples of how the results of monitoring programmes can be used to construct management plans for large rivers and catchments are given elsewhere in this book (see Part 5), but the underlying theory is the same, regardless of the size of the river.

7.1 INTRODUCTION: SCALE AND PURPOSE

Ellis and Lacey (1980) recognize nine distinct objectives of water quality sampling exercises.

1 To identify abnormal concentrations.
2 To define the peaks (and troughs, e.g. dissolved oxygen).
3 To estimate mean concentrations.
4 To detect a trend or change.
5 To monitor a well-understook process.
6 To determine fair (and unfair!) consent values for effluent.
7 To estimate a percentile.
8 To monitor for compliance.
9 To build up a picture of the process.

However, many of these aims overlap and others are a combination of two or more of the other aims; in particular, the ninth aim may be considered either as a combination of some of the previous objectives, or as a term used to cover less well-defined aims. Furthermore, all of these objectives can be summarized in three reasons for monitoring water quality, namely:

1 To assess the state of rivers and the variability of quality.
2 To determine action if necessary to maintain and enhance quality.
3 To assess the effectivenss of such action.

Whilst both flow rate and chemical load undoubtedly have enormous ecological implications, neither are considered as aspects of water quality *per se*; rather, their effects on chemical water quality are apparent primarily as a change in concentration. Thus, neither flow nor loading will be further discussed independently in this section, only their effects on concentration.

The process of designing a monitoring programme has been categorized by numerous people in recent years, including Sanders *et al* (1983), Ward and co-workers (Ward & Loftis 1986; Ward & McBride 1986; Ward *et al* 1986, 1990), Mar *et al* (1986), Hirsch *et al* (1988) and Whitfield (1988). Since all of these authors discuss the same subject, it is not surprising that a great deal of similarity exists between the views expressed, most dividing monitoring programmes up into a series of stages which include the following elements.

1 Define the objectives and the statistical methods required to achieve them.
2 Confirm that the statistical assumptions made are valid.
3 Determine where, what and when to sample/measure.
4 Draw up detailed monitoring plans.
5 Define reporting procedures.

Interrelationships between these steps are shown in Fig. 7.1.

The two primary questions to ask before undertaking any monitoring programme are 'what do

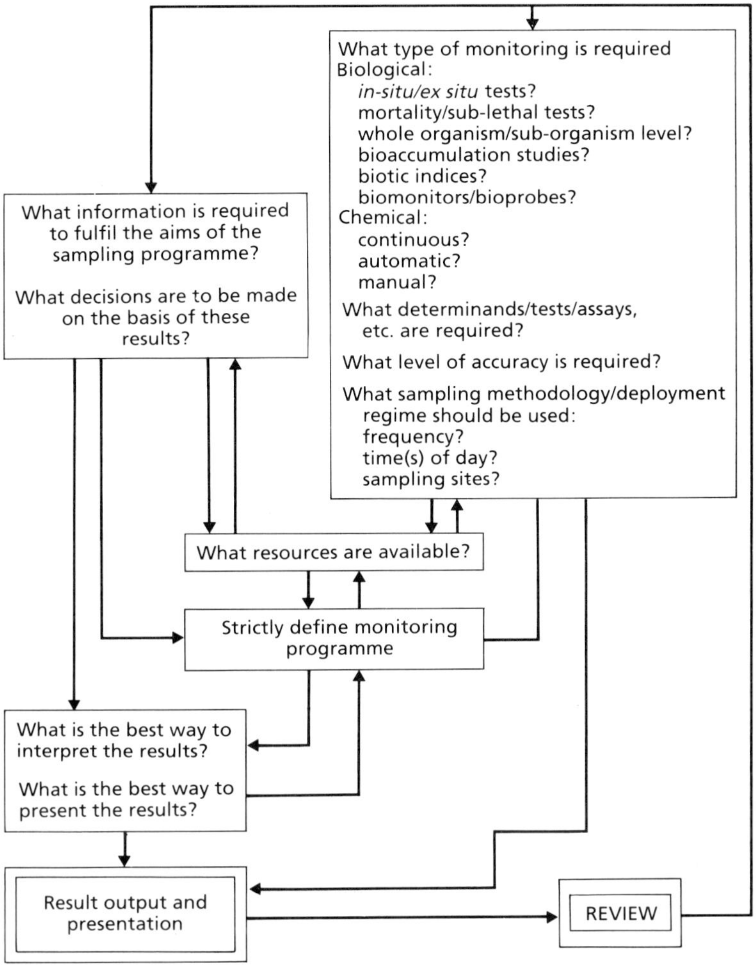

Fig. 7.1 Monitoring programme decisions.

the results need to show?' and 'what resources (monetary, staff and equipment) are available to carry out the programme?' The answer to the first question is loosely given above, but a much firmer aim is required. Ellis (1989) recommends that a stated objective should always include an explicit statistical aim and that interpretation of results should always investigate whether this aim has been achieved. The answer to the second question will be judged by the monitoring organization, be it a water company that abstracts from and/or discharges into a river, or a monitoring authority whose responsibility it is to check compliance and/or set consents to ensure that the biological integrity of receiving waters is maintained. In almost every case the answer will (and should) be: 'as little as is necessary'. To help assign a monetary value to this response, Nesbitt (1969) suggested that the monitoring programme design stage should include the question: 'would this objective still be met with 10% fewer samples?'

The design stage of constructing a water quality monitoring system must be taken into account when considering finances, since in large-scale monitoring programmes this can use up a large proportion of the budget. Ward *et al* (1990) suggest that the design stage of a long-term monitoring system (involving sample collection, chemical analysis, data handling, data analysis and reporting) will account for some 10% of the projected 20-year operating costs. This proportion may at first seem excessive, and is only applicable to large-scale, long-term chemical monitoring programmes, but may be reduced where some data already exist, i.e. where a monitoring programme already exists but requires some peripheral restructuring. However, if the whole programme requires redesigning, the effects of having operated a monitoring system that is fundamentally flawed are likely to cost a great deal more to correct in the long term if the data produced are to fulfil the original aims. For monitoring programmes with less ambitious aims and programmes to assess water quality over shorter periods of time, proportionately less effort is usually placed on the design stage.

Whilst water quality is often discussed with reference to one or more particular determinands, water quality in the broadest sense is a combination of all biological, chemical and physical parameters, of which any single determinand represents only a single facet. Therefore, while 'ideal' approaches to water quality monitoring provide a basis from which successful monitoring programmes are derived, a pragmatic (i.e. cost-effective) approach will almost certainly be chosen, since unnecessary accuracy and precision inevitably mean unnecessary cost. The detailed design of the programme will also rely heavily on experience: if equipment reliability has been a problem in the past, a new monitoring programme will try to prevent that problem by reducing the importance of that equipment in the programme as a whole, or by not using it at all, i.e. either replacing the equipment with a more reliable alternative or adopting a different approach.

There are no hard and fast rules for water quality monitoring, since every monitoring programme requires special determinands, has different aims, and needs to take account of characteristics which are unique to the particular river/catchment being studied. However, a great deal of overlap exists between programmes for monitoring quite different aspects of water quality, though these can very roughly be divided into chemical and biological categories, with sampling methodology and choice of determinands/techniques further subdividing these approaches.

7.2 MONITORING TECHNIQUES

The design of monitoring programmes involves decision-making with regard to four major factors.

1 Sampling sites.
2 Sampling frequency.
3 Sampling methodology.
4 Choice of appropriate analytical methodology (including analytical quality control (AQC), procedures, etc.).

All water-quality monitoring techniques and sampling methodologies introduce some bias. For chemical analysis, criticisms are largely concerned with the absence of any ecological impact relevance and the unrepresentative nature of discrete samples; for example, in a medium–large river such as the lower reaches of the Thames, a monthly 1-litre discrete sample may be used to represent the quality of over $0.5 \times 10^9\,m^3$ of water passing a given point! Continuous monitoring provides more representative results, but is available for very few parameters. The major advantages of chemical monitoring over biological techniques are the smaller variability between replicate determinations of a single sample and the highly specific nature of the data obtained in terms of which pollutants/toxins are present and at what concentrations. The advantages and disadvantages of biological monitoring are discussed below.

Discrete chemical monitoring: manual sampling

Some determinands require very different analytical and storage procedures to others, in terms of both technique and time-scale (taking into account the use of preservatives). Because of this a laboratory analysis procedure should be established (Fig. 7.2) to define an optimum routine which takes into account AQC procedures.

There are three broad approaches to sampling a river in a discrete chemical monitoring exercise: random sampling; random sampling with time weighting; and systematic sampling. In theory, random sampling is the best option if the aim of

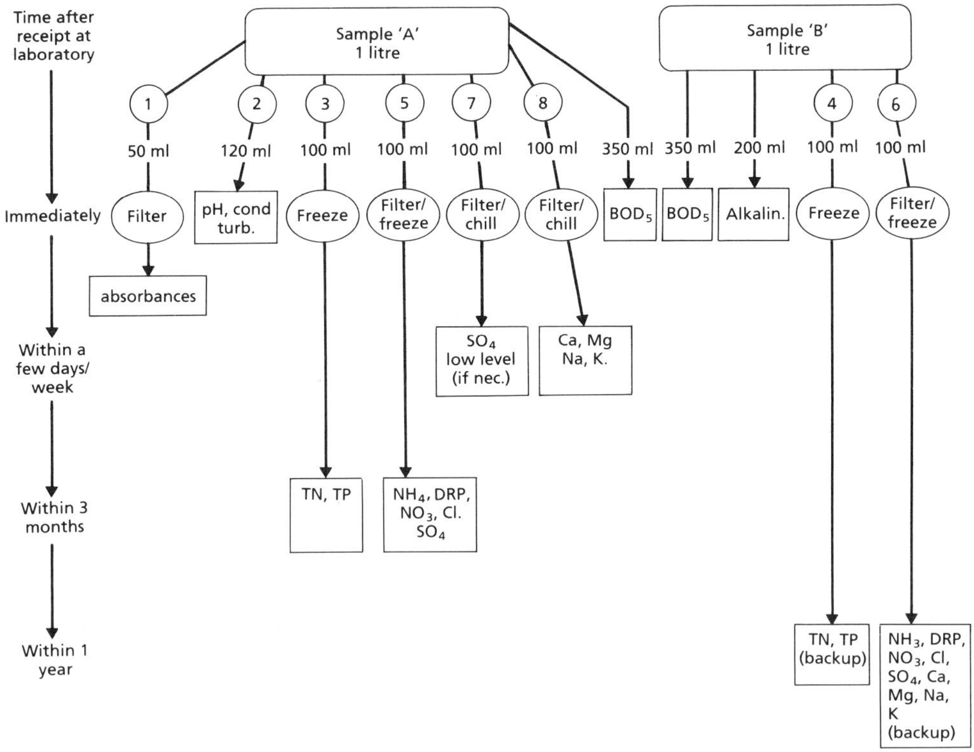

Fig. 7.2 Laboratory sample processing flow sheet for rivers (from Smith & McBride 1990).

the exercise is to provide an unbiased mean value for any water quality parameter, though in practice this choice of sampling is not suitable because of the constraints imposed upon sample collection, whether automatic or manual. Given that a random sample is one collected by a system which offers no favouritism to the collection of any discrete volume of water, sampling should be based on inter-sample volume interval. Furthermore, if the calculated mean is to be truly representative of the whole river at a particular site, a sampling site should be regarded as a transverse section through a river, not at a specified depth and width. Velocity is usually greatest below the surface above the deepest part of the channel, and slowest (due to frictional forces and eddies) immediately adjacent to the channel bottom/ sides. This has implications for physical mixing within the system and may produce lateral heterogeneity in water quality. In practice, however, random sampling may actually worsen the precision of the estimate of the mean (Ellis & Lacey 1980) when compared to flow- or time-weighted systematic sampling.

The option most often chosen for monitoring programmes is time-weighted systematic sampling, though flow-weighted sampling regimes are more likely to give representative data sets for many determinands such as nutrient concentrations. Flow-weighted sampling requires that samples be collected after a standard volume of water has passed the sampling point, and therefore necessitates flow monitoring, but offers the advantages of truly representative sampling in which storm events, droughts, snowmelt, etc. are all taken into account.

Time-weighted systematic sampling proves to be more practicable and representative than random sampling, providing that sampling is undertaken at sites where mixing is complete (both laterally and vertically), unless the aim of a study is to investigate stratification or to monitor the localized effects of a point source discharge (Montgommery & Hart 1974; see Section 7.6 below). Whilst systematic time-weighted sampling offers accuracy and practical advantages over random sampling, such sampling has to be out of step with water quality cycles, which may be natural, e.g. diurnal fluctuations in dissolved oxygen, or artificially induced, such as weekly industrial discharges from batch production processes. Thus, time-weighted sampling may be prone to bias if short-term cycles in river quality are not accounted for, unlike random or volume-weighted sampling.

Where to sample a river at any given cross-section depends as much on practicality as it does on theory. In streams and small rivers, hand-collected samples are often collected from the surface close to the bank; but as rivers widen, samples are more often collected from bridges or piers, closer to the middle reaches. In some monitoring programmes a single sampling point may be considered to adequately represent water quality at a broad cross-section, though sampling exercises with different aims may require a number of samples from each cross-section if heterogeneity is being investigated. Again, depending on the purpose of the monitoring programme, individual samples may be collected from surface water or from a range of depths.

Discrete chemical monitoring: automatic sampling

Automatic samplers may be employed for the collection of routine samples since the range of designs available offers great flexibility for programming when samples are to be collected and, once set, offer greater reliability over manual sampling, in theory, at least. Machines can be programmed for the collection of different-sized samples, for discrete or cumulative samples collected over a given time span (e.g. 100-ml subsamples collected every 5 h into a single container, with a new sampling vessel selected every 25 h). Furthermore, sample containers constructed out of a range of materials (glass, polythene, etc.) can be supplied by some manufacturers if a particular type of material is found to interfere with analytical results. However, while automatic samplers are irreplaceable in some situations where sampling frequency is high or where site access is difficult, problems with the reliability of sample collection, whether through blockage of the sampling tube, vandalism, poor site suitability so machines are prone to flooding, equipment malfunction, etc. are known to almost every worker who has used such machines.

Automatic samplers are not suitable for samples that require analysis to be started shortly after collection (e.g. for biochemical oxygen demand (BOD) or bacteriological counts); for chemicals for which no suitable preservative is available to prevent adherence on to the container walls, to inhibit complexation with other chemicals in the sample; or to stop degradation or microbial sequestration (e.g. soluble reactive phosphorus). However, those determinands for which suitable preservation techniques are available may be suitable for analysis in automatically-collected samples. For example, metals such as iron, manganese and lead are rapidly adsorbed on to the walls of glass and polythene containers, but samples to be analysed for 'total' (not 'dissolved') metals can be routinely preserved using 0.1 N nitric acid (1 ml of 10 N nitric acid placed in each 100-ml sample bottle prior to sample collection). Similarly, samples to be analysed for total phosphorus (though not soluble reactive phosphorus) can be preserved using 0.08 N sulphuric acid. These acids may, however, interfere with the analysis of other determinands, so consideration must be given to what other determinands are required before any decision on the use of preservatives is made.

The distinction between dissolved and particulate contaminants is usually an operational definition based on filter pore size (often 0.45 μm), rather than a chemical distinction between the two fractions. Thus, difficulties may arise when considering chemical–biological interactions; e.g. metals complexed with colloidal humic materials are largely unavailable for sequestration by plants, but are classed as being dissolved, along with truly dissolved metal ions which can be taken up by higher plants and algae. Furthermore, it may be necessary to treat dissolved and particulate fractions of the same parameter as being independent of each other; i.e. a different sampling design may be required for the two fractions because of the effect of water velocity on suspended solids concentrations.

Practicality is the prime factor involved when deciding where to install automatic sample collection tubes or continuous monitoring probes (see below). For example, in 'flashy' streams and rivers, it may be necessary to position sampling tubes and probes near to the bottom at the deepest part of the channel to prevent the equipment from being washed away during high flows and, conversely, to ensure that the water is sampled/monitored when the water level drops during periods of drought. In larger rivers, automatic sampling collection points and continuous monitoring probes tend to be positioned at the side of the channel, largely because it is easier to place anchoring points for sampling tubes in slower-flowing peripheral stretches of water than in more turbulent deeper waters.

Continuous monitoring

Continuous monitoring equipment is available for a limited number of determinands such as pH, ammonium, temperature, dissolved oxygen, turbidity and chlorophyll-*a* (phytoplankton standing crop), and is currently being developed for others. When working correctly, this equipment provides data which are more typical of overall quality during the monitoring period than discrete sampling, the results obtained effectively being those achieved by a very frequent discrete sampling programme. The equipment, however, requires regular calibration and maintenance (replacement of membranes, etc.) for results to be reliable.

Continuous monitoring equipment is best used for measuring those parameters such as pH and dissolved oxygen which can only satisfactorily be measured *in situ* or immediately after collection, i.e. where automatic samplers cannot be used; and for frequent monitoring, where manual sample collection and analysis is extremely expensive. Consequently, the very large cost advantages offered by measuring dissolved oxygen with an oxygen electrode often outstrip the slightly better accuracy of chemical (Winkler) titration. This does not necessarily imply that chemical titration is an inherently more accurate technique, since the improved accuracy achieved is at least partly a consequence of better quality control procedures for laboratory analysis compared to field-based monitoring using often infrequently calibrated probes.

Biological monitoring

Biological monitoring encompasses two very different aspects: the first provides a measure of ecological quality and the second uses the responses of biota to changes in water quality as a method of assessing such changes. Water quality impact may be monitored at a number of different levels, from community status (e.g. fish/invertebrate indices), through populations effects (which take into account fecundity and age-specific survivorship), to the presence/absence of indicator species and effects on the individual which range from whole-organism mortality to sub-lethal tests. This last group includes behavioural techniques, pathology, developmental tests (e.g. the oyster embryo development bioassay), biochemical assays (e.g. the induction of cholinesterase enzyme activity), physiological tests (respiratory activity, ventilation rate, etc.) and bioaccumulation studies.

Tests on individuals may be *in situ* or *ex situ*. *Ex situ* tests which rely on exposure to water samples share the drawback with spot chemical monitoring of whether the discrete water sample is truly representative, but offer advantages in terms of ecological relevance. *In situ* tests offer the following advantages.

1 The results monitor the impact of water quality directly.

2 Exposure history is summarized, thus the advantages of both cumulative and continuous chemical monitoring are applicable.

3 The effects of all toxicants are integrated.

4 Use of biotic indices as a measure of medium to long-term water quality is more cost-effective than frequent chemical monitoring.

However, biotic indices also have several disadvantages for monitoring water quality.

1 Monitoring organisms may be highly tolerant of a type of pollution (e.g. organic pollution) but be sensitive to specific toxicants, and vice versa.

2 Organisms often show a wide range of sensitivities to different types of pollutants.
3 Intra-taxon variability may be as great as inter-taxon variability with regard to pollutant sensitivity.

The reliability of biotic indices varies greatly, but large theoretical advantages are offered by those which monitor a wide range of taxa and take account of physical habitat when assessing community structure (e.g. HABSCORE for fish communities; Milner *et al* 1985), or those which are least affected by physical variables such as flow (e.g. epilithic diatom indices based on sampling of large stones; Round 1991). Alternatively, indices may be restricted to use in one type of habitat (e.g. Biological Monitoring Working Party (BMWP) score, see Chapter 8). The use of indicator species as opposed to communities offers advantages in terms of simpler data collection and less complex interpretation, though the many requirements of an ideal indicator mean that relatively few species are suitable. Hellawell (1986) states that these include:
1 organisms should be readily identifiable;
2 sampling procedures should be easy;
3 species should be widely distributed;
4 the ecology of indicators should be well understood;
5 organisms should readily accumulate pollutants;
6 laboratory culture should be easy to enable a ready supply of organisms for experimental work;
7 genotypic and phenotypic plasticity should be minimal.

Tests which measure mortality as the end-point are most widely used to assess toxicity, offering the advantages of good reproducibility, simplicity and being highly standardized and cost-effective. However, these also offer the drawbacks of being insensitive to sub-lethal effects of pollutants, and extrapolating the results from non-indigenous species to indigenous species at community-level may introduce a large source of error.

Bioaccumulation studies do not measure detrimental impacts on biota, but rather are a means of providing a cumulative sample and enabling the concentration of toxins present at sub-detection levels in the water column to levels where they can be measured. Owing to the fact that bioaccumulation does not measure biological impact directly, the levels of pollutant(s) accumulated must be calibrated against known biological/ecological effects. Similarly, accumulated levels must be calibrated against ambient concentration to enable the back-calculation of water quality, though this in itself is fraught with assumptions which may not be valid because of the often pulsed nature of pollutant dynamics and the complex exposure period/ambient concentration/uptake-excretion relationships for any given species.

Bioaccumulation studies are not an end in themselves, but they are a useful tool nevertheless. Their main value lies in the fact that they can indicate where there is a problem so that more detailed investigations can be undertaken to gauge the severity of the problem and how to control it. Whilst relatively few organisms are recognized as being ideal indicators, analysis of invertebrates, algae, higher plants, etc. can reveal contaminant 'hot spots' along a river and so be used to identify pollutant point sources.

Bioaccumulation is most useful as a monitoring tool where fluctuations in water occur relatively frequently, e.g. from several times daily to weekly, and where the collection of discrete water samples may either miss these changes or in some cases (where sampling coincides with pollution events), may exaggerate the analytical results. Bioaccumulation studies are of little use for monitoring infrequent very severe pollution events, since even if the organisms survive, accumulation tends to be greater if the same pollutant load is applied at a lower concentration over a long period than as a single very concentrated pulse. Care must also be taken over the choice of monitoring species and site of deployment within the river. For example, the accumulation rate of a filter-feeder positioned close to the benthos is likely to be highly influenced by the toxicant content of suspended sediment and bed-load, while algae take up and accumulate toxins which are overwhelmingly in a dissolved state. A large number of factors affect the reliability of bioaccumulators (and indicator species more generally) including accumulation and excretion rates, physiological condition, age/size, interference between substances, environmental variables and trophic level (Hellawell 1986).

Commercially available biomonitors and probes are generally suitable for monitoring levels of pollutants that are highly toxic to aquatic organisms, but at lower concentrations the lack of sensitivity limits their widespread use (van der Schalie 1986). The major advantage of biomonitors is the enormous range of pollutants they respond to, consequently their use as high-concentration, wide-scale (organic and inorganic) pollutant monitors cannot be matched by traditional chemical analysis. However, biomonitors can only indicate when a toxin (or toxicant) is present, not what the toxin is. Hence situations where biomonitors can be used in isolation are rare; the potential of these tools is probably best realized when used in conjunction with event-triggered automatic samplers for subsequent chemical analysis of major pollution events. The majority of commercially available biomonitors/probes rely on fish, though a smaller number rely on bacteria and algae (which offer a number of advantages for non-continuous assays), with even fewer based on mussels and cladocerans.

7.3 STATISTICAL ANALYSIS OF WATER-QUALITY DATA

Statistical analysis of water-quality data is comprehensively covered by Ellis (1989) who includes numerons 'real life' examples of questions asked of water quality data, together with advice on how to go about answering them. Ellis splits the interpretation of data into past and future tenses: how can data already collected be most sensibly analysed? and what monitoring programme design will yield the most useful data for the minimum cost? Numerous reports and items of computer software have resulted from the ongoing sampling project in which Ellis is involved.

If the limit of detection (LoD) for any determinand is not sufficiently low, then data of less than this value will inevitably arise. An account of the statistical approach used to deal with this type of problem is given by Berry and Day (1973), who give an example of how the probability distribution of concentrations can be modelled, even when a majority of the determinations are less than the LoD. It can be revealing to check how sensitive the statistical analysis is to the LoD by performing the calculations twice – once with zero substituted for determinations less than LoD, and once with the limit itself substituted. In practice, however, it is more usual (though often not best) to substitute a set value, often LoD/2, for all determinations which fall below the LoD. Less frequently, but perhaps more disturbingly, the upper detection limit may be too low for some determinations. Here, simple substitutions cannot be used since the scope for error is so much greater, and statistical modelling is the only acceptable answer if the range of detection cannot be expanded.

A number of conventional statistical packages are available for handling water quality data, such as Genstat (Rothamsted Experimental Station, UK) and Minitab (Minitab Inc.), together with spreadsheets, such as SuperCalc (Computer Associates) and Excel (Microsoft Corporation). However, such software packages tend to have been used on a restricted basis only with water quality data, since they demand a good background knowledge of the statistical analyses required. Because of this, several less versatile but more user-friendly computer programs have been written specifically to handle water quality data, amongst the earliest of which was produced by Lettenmaier *et al* (1982), though other developments include WQStat II (Loftis *et al* 1989) and AARDVARK (Ellis 1989).* This latter package has gained wide acceptance in the UK, largely because of its user-friendliness when analysing water industry-type data and its ability to interface with a range of data storage and retrieval systems. AARDVARK can be used to determine the existence of seasonal trends, to test whether water quality is changing (to what extent and how rapidly), and to show whether data is becoming more or less variable. A typical example of data analysis by AARDVARK is shown in Fig. 7.3.

More recent water quality statistical packages from the AARDVARK stable include ZEBRA and LAPWING (WRc plc).† ZEBRA is used for the nonparametric analysis of data collected for

* AARDVARK: Analyse Any Routine Data; Visually Acquire Real Knowledge.

† ZEBRA: Zippily Evaluated Binomial-Related Applications; LAPWING: Longitudinal Analysis Program With Improved NIVEAR Graphics; NIVEAR: New Improved Version Extended Along River.

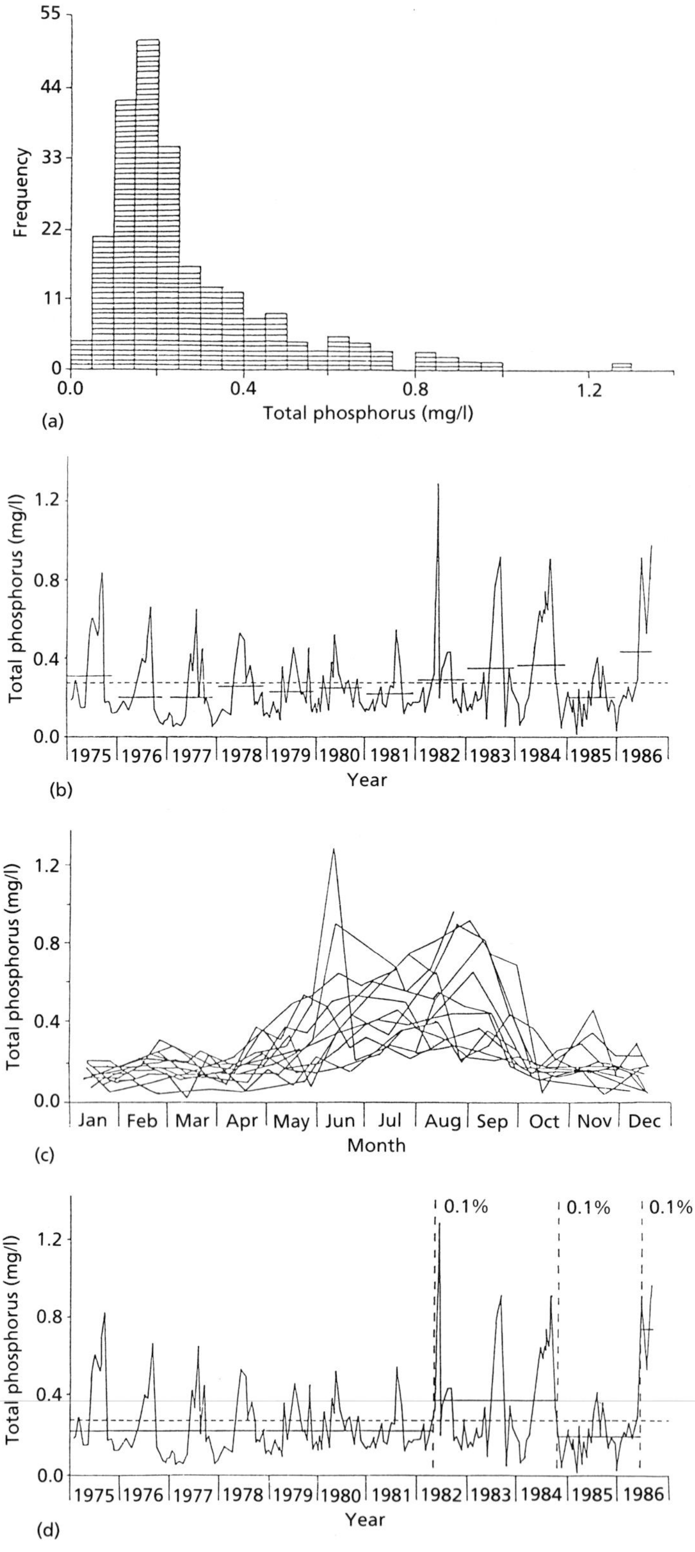

Fig. 7.3 Total P at river sampling point; (a) frequency histogram; (b) time series plot plus annual means; (c) year-on-year plot; (d) time series plot showing statistically significant changes in de-seasonalized data.

compliance with percentile-type standards, and LAPWING is used for the detection of spatial trends by determining the presence/absence of statistically significant differences between seasonally corrected means. As such, LAPWING can be used to help to determine where best to apply financial resources for optimum sampling efficiency.

7.4 ROUTINE MONITORING

Routine monitoring is required to provide a backdrop against which statistical analysis of trends can be assessed to determine whether catchment management requires changing, or to provide a baseline against which water-quality recovery can be judged following a pollution event. Monitoring also provides the means to assess recovery programmes after decades of pollution. Consequently, comprehensive routine monitoring programmes are probably the hardest to design and implement because of the large range of variables which they attempt to monitor. Despite this, data interpretation and recovery targets are eased by comparison with environmental quality standards (EQSs) specified for particular uses or laid down in legislation.

At the very least a routine monitoring programme should be able to detect changes in a set of determinands that can then be used to identify an area of chemistry or biology for more intensive monitoring. Results gained from the latter investigation(s) should then indicate what aspects of water quality are changing, and subsequently help to ascribe a cause to these changes. Thus, good routine monitoring programmes should not just be able to identify trends or abrupt changes in water quality, but should also go some way to help identify the underlying cause(s). The determinands should, therefore, be fundamental to the biological integrity of riverine communities. These include parameters which affect speciation, bioavailability and toxicity of pollutants (e.g. pH), those which directly affect biota (e.g. dissolved oxygen), together with analyses for potent chemicals (such as biocides) and chemicals that affect productivity and standing crop (nutrients). Moreover, the impacts of changing water quality on acquatic biota are often not best recorded by a suite of chemical determinands, so biological monitoring is also beneficial.

Routine monitoring programmes should also be designed to enable seasonal cycles to be monitored, even though more frequent cycles (which

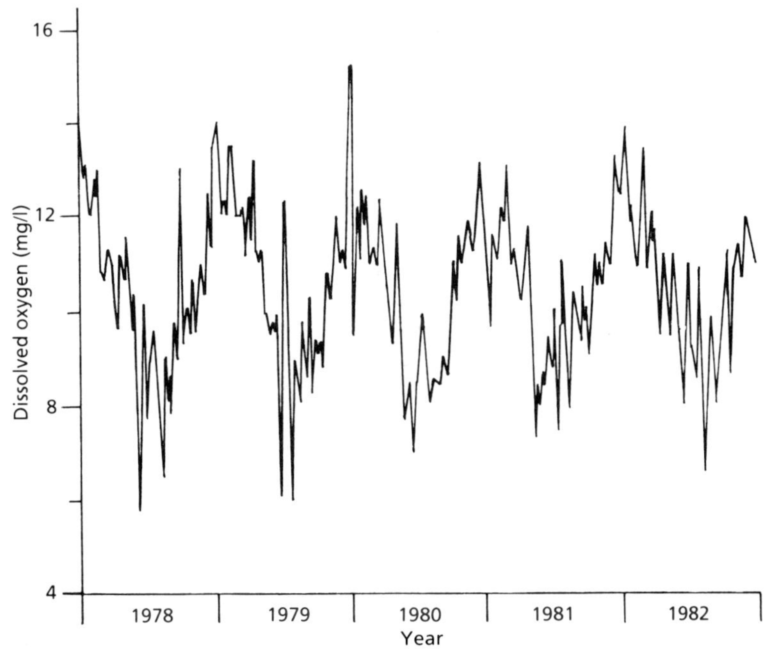

Fig. 7.4 Dissolved oxygen at river sampling point.

may be as large in magnitude) also occur. An example of how short-term and seasonal cycles may overlap is shown in Fig. 7.4, where dissolved oxygen levels in the River Ribble show large short-term fluctuations; these are due partly to sampling at different times of the light-controlled diurnal cycle of net photosynthesis/respiration, although an annual cycle is also apparent. Higher levels occur when the oxygen saturation level is maximal and microbial/photosynthetic activity is minimal during cooler months, and vice versa during summer. Whether or not the samples should be collected at different times depends on the aims of the sampling programme, since different sampling times in given 24-hour periods yield a far more accurate mean value, but also obscure the annual pattern to some extent. If all samples had been collected at the same time, the plot would show a smoother seasonal cycle, but little information would have been gained concerning short-term fluctuations, which may be more ecologically important.

Monitoring for consent compliance is included in the majority of routine monitoring programmes (see Section 7.5), so the rivers for which the best monitored (as opposed to modelled) loading data are available are those draining urban regions. Monitoring of consented effluents provides valuable data for mass-balance studies, but gives little indication of the effects on river water quality unless both effluent and river flow are measured. Alternatively, river flow can be modelled from rainfall data and catchment characteristics.

Water utilities monitor water quality at potable abstraction points, largely for chemical information, though microbiological data (possibly including information on phytoplankton numbers and community composition) are also collected. Regulators, however, are charged with monitoring river water quality and loading on a more general and ecological basis, so have a greater reliance on biological monitoring techniques.

7.5 COMPLIANCE MONITORING

When monitoring for compliance, decisions are made easier for dischargers, since the choice of determinands and sampling frequency are given as a condition of issuing numeric consents. In many cases, effluent flow monitoring may also be required to calculate load. Consequently, financial considerations are largely outside the decision-making process. If a consent requires that a determinand has to be within set limits, the organization responsible for maintaining compliance may:

1 monitor itself, reaching an agreement with the consenting authority which will have access to audit the data;

2 carry out its own completely independent monitoring to satisfy itself that it is meeting the relative standards (in which case the data will not be placed on the public register) and the consenting organization will conduct a separate monitoring programme;

3 leave all compliance monitoring to the regulators.

The National River Authority's discharge consent policy (NRA 1990) recommends that 'all environmentally sensitive discharges with numeric limits set for effluent flow and concentrations of determinands should include absolute limits not open to be exceeded at any time. Further limits in 80 or 50 percentile form can then be added to refine requirements which routine performance must stay within'. Therefore, a case may be made for an absolute maximum limit for any determinand above which ecological damage is considered likely to occur, especially in the light of current European Community concern with 'Sensitive Areas' (CEC 1991). Recommendation 11 of the NRA document goes further by ensuring that all numeric consents are load- rather than concentration-based: 'all numeric consents should include absolute limits for instantaneous effluent flow'. Some none-numeric consents may be based purely on flow; for example, where a specified volume of effluent has to be carried forward in a sewage system prior to the diversion of flow to a combined sewer overflow (CSO).

Whilst an environmental impact assessment is required for significant discharges as part of the consent setting procedure – which takes account of the range of annual flows normally experienced to ensure that relevant EQSs are met at the outer edge of the mixing zone – river flow rate is not usually required to be measured as part of the consent monitoring process. As with potable water abstraction, a suite of chemical determinands for substances with defined limits has traditionally been used to determine whether consented effluents are of acceptable quality, but several shortcomings of this approach have been identified (OECD 1987):

1 Not all organic chemical components can be readily identified or quantified using available techniques.

2 Toxicological data are non-existent or sparse for many organic components.

3 What data are known may bear little relevance to indigenous species.

4 EQSs, the maximum levels deemed acceptable for given chemicals, are derived on the basis that they are the only toxin present. Thus, synergistic, antagonistic or purely additive pollutant interactions in mixtures are not considered.

In the UK, these disadvantages have been noted. The NRA (1990) has endorsed the use of toxicity-based consents (now well established in other countries): 'for environmentally significant discharges of complex composition where not all important constituents can be individually identified and numerically limited, consents should specify a clearly-defined toxicity limit, the appropriate form of toxicity test to be used, and the linimum frequency with which it should be applied'.

7.6 USE OF WATER-QUALITY AND CATCHMENT DATA TO IDENTIFY SOURCES

In practice, the aims of a water-quality monitoring programme may be too broad, so a pragmatic approach usually yields the most useful answers. For example, if a difference in water quality is found to exist between two sampling points (20 km distant) along a river, the downstream site containing elevated levels of a pollutant, what kind of monitoring programme will determine the source(s) of this pollutant at the minimum cost? Whilst a statistician would argue that the aims are not precise enough (and who could argue with that?), it nevertheless represents an often typical starting point from which researchers have to work.

Before any further sampling is done, the integrity of the data from which the original conclusion was drawn (i.e. that water quality deteriorates at the downstream site) should be checked. Was the analytical method generic (e.g. for a group of pesticides) or does it analyse only for a specific compound? Was the sampling/storage/analytical procedure used at the two sites identical or did other factors interfere? For example, restricted access may have meant that samples were taken automatically at one site, without the use of a preservative, and not collected/analysed for several days; but at the other site the samples may have been analysed on the day of collection. Are the data affected by 'outliers'? If so, are these results trustworthy (i.e. can a suitable explanation be found for extreme values, such as changes in habitat characteristics which account for large differences in biological monitoring results), or should statistical techniques be used to remove them?

When questions regarding data reliability have been investigated and it is decided that further sampling is necessary, mapped information and local knowledge is used to look for likely sources of the pollutant – could it be derived from a single point source, from several point sources or is it diffuse in origin? Once possible sources have been identified, the reason for more careful monitoring should be addressed. This should indicate whether the pollutant was the result of a single spill, e.g. was awareness aroused by a one-off fishkill in a river which past data show to have been of good quality, or do the data suggest that spills/discharges are more common? If the latter, do the data show a regular cycle of water quality degradation/improvement (seasonal or short-term) or does the pollution appear to be more sporadic?

The list of questions and putative answers continues, but clearly any detailed investigation will rely very heavily on local knowledge (whether published or anecdotal) and on the quality of data previously collected for the river. If the pollution was caused by a single spill/discharge, finding the source will require prompt action which may not be possible – it may already be too late if the pollutant has been carried downstream and diluted. Prompt reporting is required if such events are to be traced to their source, for which regulators rely heavily on the reporting of incidents by the public, anglers, emergency services and, in some case, by the people who have accidentally caused a pollution incident. In the latter case, this is usually used as a mitigating circumstance when considering whether or not to prosecute the 'polluter'.

However, if it can be concluded that the pollution was not the result of a single event, water quality should be monitored at a number of sites between the two original sampling points. These new sampling sites should be located downstream of possible sources, including sites at the confluence with large tributaries. To identify whether the pollutant is derived from a point or diffuse source, biological monitoring is the preferred option for multi-event monitoring, preferably in conjunction with continuous or event-driven automatic chemical sampling. Source monitoring is further covered under the following two headings.

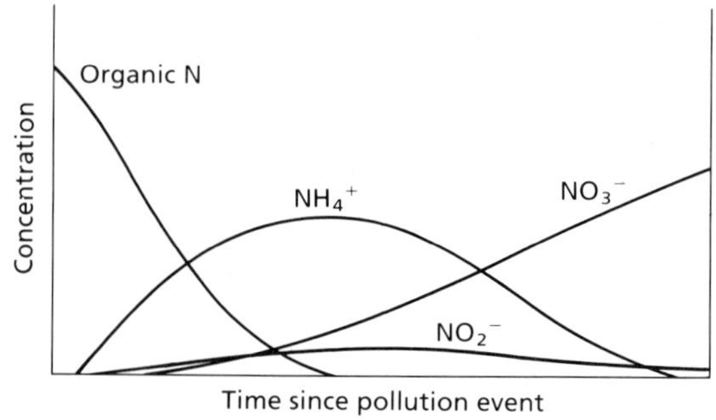

Fig. 7.5 Breakdown and oxidation of nitrogen following an organic pollution event.

Point source monitoring

At least two sites are required to monitor the effects of a point source: one upstream (a control site) and the other downstream of the source. Care must be taken over choosing the exact sampling sites, particularly the downstream site, to ensure that the samples collected are typical of river water quality at that cross-section of the river. If not, the downstream samples will either be biased towards the quality of the undiluted point source, or towards the river water upstream of the source. Thus the downstream site should be below the mixing zone, but not so far downstream that other factors (alternative sources and/or substantially increased dilution from a tributary) have an effect on water quality. In some cases, however, the proximity of the two sites may be limited by access restrictions. If this limitation does apply, a monitoring programme may be needed to ensure that the water quality at the two sites is not significantly different when the point source is not discharging.

The information gathered from a two-site sampling programme is of relevance to the sampling sites only, and even then its biological relevance is limited. The effects farther downstream are dependent to some extent on dilution, contaminant degradation rates, chemical behaviour and bioavailability, together with the toxicity of the parent contaminants and their breakdown products. For example, following an organic nitrogen pollution event, of the breakdown products produced, ammonium is likely to have the largest impact on fish populations because of its toxicity and concentration (see the idealized plot in Fig. 7.5). Nitrite is also toxic, but because it is rapidly converted to nitrate, its concentration, and its effects on fish, are likely to be lower. (The resultant decrease in dissolved oxygen levels could also have a very large impact, but this is not considered here since it is the nitrogen content of the pollutant, not the carbon content, which is under discussion.) Consequently, for a thorough investigation of the effects of a point source discharge, several monitoring sites are required downstream, as well as the upstream control site. Continuous or discrete chemical monitoring is suitable for monitoring single chemical discharges, but the use of biotic scores or deployed organisms for biological monitoring provides valuable additional information, especially for complex effluents.

Diffuse source monitoring

The detection of longitudinal trends in river water quality is the primary aim of diffuse source monitoring, rather than the determination and/or characterization of extremes. As such, diffuse source monitoring is generally long term and non-intensive. Sampling frequencies are often weekly or monthly and data collection may be considered as part of (or an adjunct to) routine monitoring. Chemicals identified in diffuse source monitoring programmes may be genuinely diffuse in origin (e.g. nitrate and phosphate derived from the weathering of nutrient-rich rocks or soil); may be derived primarily from a number of discrete sources (e.g. a series of minor consented discharges along the length of a river); or may originate from a single point source, the effects of which are diluted before entry into the river (e.g. contamination of groundwater in a chalk aquifer from a major chemical spill prior to the water reaching the surface).

The choice of determinands for a diffuse source monitoring programme is usually easy to make, since such programmes are generally undertaken to investigate the gradual build-up of specific pollutant levels in the lower reaches of rivers. The choice of sampling sites may, however, present more of a problem, since urban regions are more likely to contain point sources of the chemicals being monitored. Thus, sampling sites tend to be few and far between in the upper reaches of rivers but closer together in more heavily populated and/or more intensive agricultural regions downstream, where changes in water quality between adjacent sites are likely to be greater. Additionally, where best to obtain a 'typical' sample of water as the river increases in size may require a study of its own at each chosen site, and sampling at boreholes may be necessary to understand the contribution of groundwater.

Because diffuse source monitoring is associated primarily with trends, a number of sampling sites along a river are usually chosen and the data from these sites compared statistically to reveal which stretches of river receive the greatest pollutant loads. If necessary, more intensive monitoring of a stretch of river receiving a particularly high contaminant load may be undertaken to ensure that the source is genuinely diffuse. Thus, a diffuse source providing a large contaminant load along a small stretch of river may be considered a point source if a wider viewpoint is chosen – this all depends on the scale of the monitoring programme; but equally well, a series of point sources may be considered as part of the diffuse load if a monitoring programme is poorly designed.

To help determine whether this is the case, it may be useful to construct a pollutant budget. These are usually based on catchment physiographic factors and land use, relying on export coefficients derived from monitoring small-scale experimental plots and more reliable data from intensively monitored sub-catchments. Such budgets take into account the known load from point sources (readily controllable) and the modelled load from diffuse sources (less controllable). If large differences exist between the measured and expected results, more intensive monitoring may be called for. Budgets may be used to help predict or interpret the data gathered from monitoring programmes, with simple models often offering greater accuracy and having the additional advantage of requiring smaller datasets for calibration purposes (Leite 1990). Likewise, they may also be used to design a catchment management plan to reduce diffuse and/or point source loading, with water quality monitoring subsequently being used to determine whether the aims of the plan are being met.

7.7 EVENT MONITORING (SEASONAL)

If a seasonal event is defined as 'any change in water quality which occurs on a regular annual cycle, so allowing a period of more intensive monitoring to be planned', this includes both gradual annual cycles as discussed for dissolved oxygen levels (see Section 7.4), and episodic events which occur within a season cycle. Seasonal events are normally monitored as part of (or in extension to) routine monitoring, and are often seen to represent worst case scenarios for the water supply industry, since the peak concentrations which occur during such events may cause major treatment problems. In some cases it may be uneconomical to solve these problems by upgrading the treatment plant to cope with contaminant levels which may be troublesome for only a few days a year; instead, using an alternative source of water for such a short period may prove to be more cost-effective. Indeed, this is the principle of bankside storage for potable water supply.

An example including both short-lived and annual seasonal cycles is found in many rivers which drain organic (principally peat-based) soils. Here, colour flushes associated with elevated aluminium, iron and manganese levels occur dramatically in autumn following the first heavy rainfall after the summer period, with the most pronounced changes in water quality often lasting for less than a day. However, increased metal and colour levels may be experienced for months following this time, when compared to levels measured throughout spring and summer. This situation is frequently observed in small streams, but because even closely situated sub-catchments may have quite different rainfall patterns and soil/land use characteristics (and therefore marked differences in runoff water quality; e.g. McDonald & Naden 1987), as these tributaries join into successively larger watercourses the changes in water quality tend to be diluted by streams draining other sub-catchments. Dispersion downstream also reduces the severity of such changes in water quality. Other streams may undergo similar dramatic changes in water quality, but on different occasions, perhaps days or even weeks distant. Thus, the changes in water quality observed downstream tend not to be as dramatic as those seen in headwaters, but last for a longer period.

Where monthly samples have been collected and analysed for colour and metals during the period from winter through to summer, increasing the sampling frequency to twice weekly or daily throughout autumn will help to identify which upland rivers are most affected and therefore represent the major source(s) for a reservoir draining a number of sub-catchments. However, the data will still not be frequent enough to calculate load with any degree of accuracy, even if all the feeder streams and rivers and flow gauged, and a compromise may have to be reached by modelling the expected concentrations. Several methods are available, but the most accurate rely on calculating loads from an established statistical relationship, e.g. between flow and concentration (Ferguson 1986).

Flow and concentration correlate to some extent for many solutes (and suspended sediment), either directly or inversely, with linear or more complex relationships. Figure 7.6 shows the type of relationships encountered for four deter-

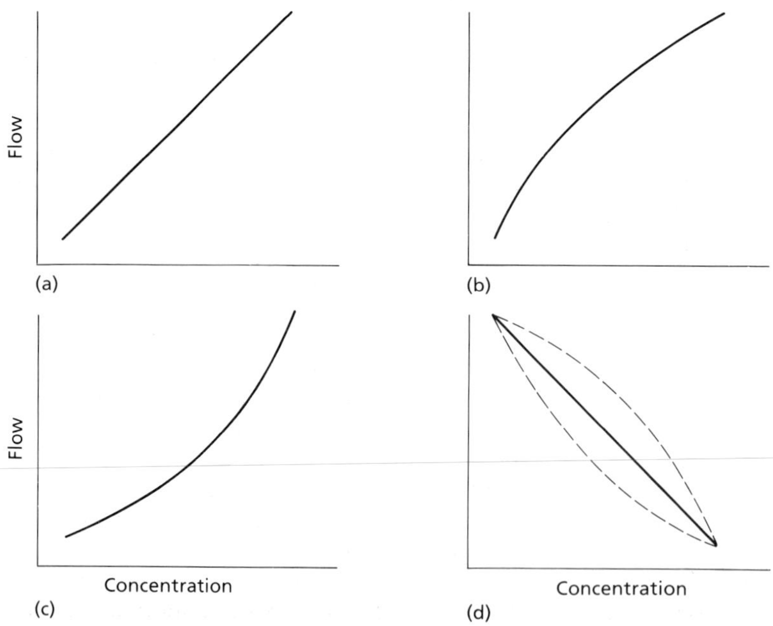

Fig. 7.6 Flow–concentration relationships for four water quality determinands: (a) total iron (peat-based catchment); (b) suspended sediment (afforested catchment); (c) total phosphorus (agricultural catchment); (d) conductivity (acid moorland catchment).

minands. These idealized plots are based on unpublished original data but no scales are given because of the large measured differences between rivers/streams draining different types of catchment. Such data are often plotted on log-log axes. Nevertheless, similar relationships hold in many watercourses, albeit individually tailored to the characteristics of specific catchments. The product of mean concentration and mean flow either over- or underestimates the load if the concentration–flow relationship is not linear and the margin of error increases for those determinands which exhibit progressively more pulsed dynamic characteristics. Clearly, the lower the frequency of data used to derive the model, the less accurate the model is likely to be.

The flow and concentration values from which loads are modelled must be frequent enough to represent accurately the conditions in the river during the period for which modelling is undertaken. For 'flashy' streams in particular, this infers continuous monitoring (or, at the least, very regular discrete monitoring) to ensure that the data collected covers periods of extreme drought and flood events.

Furthermore, because year-to-year variability in the load of some solutes exported can amount to a factor of two or more, even in quite large rivers, data from a number of years should be used to construct the model. Despite these drawbacks, an example of the potential improvement in accuracy is given by Meyer and Likens (1979), who found the amount of phosphorus exported from an afforested catchment to be three times greater when modelled using frequent flow data than when calculated directly from multiplying discrete weekly concentration values by flow.

For monitoring very intense, short-lasting seasonal events such as colour flushes, the most useful compromise is often achieved when regular time-weighted monitoring is combined with flow-triggered event monitoring in which rainfall triggers or floating tilt switches are used to start automatic sampling as the water level rises. McDonald and Kay (1981) successfully used this approach when monitoring enteric bacterial levels, the concentrations of which increased during high flow.

Relationships between different determinands established from such monitoring may also be used to determine management strategies. For example, colour correlates well with iron, manganese and aluminium in many upland streams; thus if the metals are primarily in a dissolved or colloidal state and so cause difficulties for potable water treatment, it may be possible to decide whether or not to abstract on the basis of water colour. This is less expensive to measure than several metals and, while lacking some accuracy and precision, portable equipment can be used to determine a value for colour within seconds. Almost immediate decisions can therefore be made, the advantages of which over waiting a day or longer for analytical results are obvious.

Seasonal variations in water quality in larger lowland rivers tend to occur as more gradual, smoother trends than in small upland streams. Total phosphorus concentrations in rivers draining agricultural catchments usually increase with flow, but a dilution effect occurs in rivers draining predominantly urban catchments, since the phosphorus load from sewage treatment works and industrial discharges is largely independent of rainfall (with the exception of combined sewer overflows). Thus, as flow is reduced in summer, the nutrient load remains the same and nutrient concentrations (notably total phosphorus) increase (Fig. 7.3b).

In mixed urban/agricultural catchments, phosphorus levels may show little seasonal variation, since the higher concentrations from sewage treatment effluent in summer can mask the lower levels from agricultural runoff, and vice versa in winter.

7.8 EVENT MONITORING (SPORADIC)

Episodic pollution events, as with seasonal fluctuations, are often (though by no means always) associated with heavy rainfall. These events may be concerned with a single pollutant or type of pollutant, such as pesticide runoff during a storm shortly after application to a crop, or may contain a cocktail of pollutants such as the runoff from motorway culverts following a dry period of several weeks.

Where a single pollutant or a related group of

pollutants are the sole or major group of chemicals responsible for an overall decline in water quality, monitoring of these determinands and the effects they have on other determinands (e.g. a change in pH) may be sufficient for data interpretation if the effects under investigation are purely chemical. However, where a number of different pollutants are involved, comprehensive chemical analysis is usually time-consuming and prohibitively expensive. Furthermore, data interpretation will almost certainly prove to be more difficult than when dealing with fewer determinands. In such cases it is often prudent to monitor for a limited number of the most concentrated and/or toxic pollutants, and to biologically monitor for the combined effect of all pollutants in an event (see Section 7.2). To obtain chemical data for sporadic events, either continuous monitoring equipment has to be used (which is limited because of the small range of determinands which can be monitored) or automatic samplers can be employed. Where events are linked to a heavy rainfall pattern, increases in river level or rainfall triggers can be used to initiate sampling as explained above (Section 7.7). Depending on the toxicity of the pollutant(s) and the frequency and duration of such events, biotic indices, deployment of caged invertebrates for bioaccumulation studies and/or *in situ* bioassays may provide a suitable choice for monitoring cumulative effects.

One of the major sources of this type of pollution incident in industrialized countries is combined sewer overflows (CSOs). During periods of heavy rainfall, urban runoff is combined with domestic sewage to a level where the volume of waste is too great to be handled by the receiving sewage treatment works. With unsatisfactory CSOs a backlog of untreated sewage and runoff builds up in a pre-works retention tank which eventually overflows, whereupon untreated waste is discharged into the river that usually receives treated sewage effluent. Overflows can be designed to operate at an appropriate frequency, to be screened (or equivalent) and to be located on a watercourse that is sufficiently large to assimilate the spill so that there is not a significant impact. This, however, is not always practicable.

A particular problem is caused by heavy rainfall over an urban catchment after a dry period. The overflows may operate into rivers under low flow conditions as they have not yet responded to the rainfall. The effects on aquatic ecology can be devastating, initially due to the high ammonium levels and very low dissolved oxygen levels (as a consequence of the high BOD) in both the water column and sediment. Other, less immediate effects may be caused by the increased levels of other pollutants, such as metals, and the associated high nutrient levels may have major consequences for the quality of lentic and lotic waters farther downstream.

Broadly similar monitoring programmes are necessary for events which are not triggered by rainfall, but rather are the results of poor catchment management practices. Examples include the poor maintenance of slurry stores situated near to watercourses and the disposal of spent sheepdip into rivers in agricultural catchments (though both of these may also be rainfall-driven); and the non-consented discharge of chemical waste in industrialized catchments. Where there is no warning of such events occurring or having occurred, unless the effects are reported by the public, biological monitoring (bioassays, biotic indices and/or bioaccumulation studies), together with continuous or frequent routine monitoring is needed if such events are not to be missed, especially if they do not have an immediate visual impact. However, the low cost-effectiveness of chemical monitoring is almost certain to prevent its use unless a suitable trigger can be used to initiate automatic sampling, as can be the case when using biomonitors to detect highly toxic spills.

7.9 RETROSPECTIVE MONITORING

If no water quality data have been collected for a watercourse, it may be possible to infer the historical state of water quality from chemical analysis and microscopic examination of the sediment. Such techniques are more usually applied to lentic waters, and while they do not provide a universal panacea there is no reason why such methods cannot be successfully applied to some slow-flowing, deep stretches of rivers or canals. To be able to apply a suitable scale to the results, sediment cores must be removed intact and dated, the most widely used method being ^{210}Pb radio-dating. Samples must be radio-dated from along the full length of a core, since sediment deposition rates vary greatly, both on a seasonal basis and from year to year, depending on rainfall and catchment and river management practices. Once the cores have been dated and lengths assigned to specific years, the cores can be analysed for a number of determinands, the results of which are subsequently converted into loads/unit area/year to help remove the effect of sedimentation rate.

The palaeolimnological techniques of diatom frustule and chironomid larval head capsule analyses reconstruct live biotic indices, while pollen analysis may reveal much about catchment land use. Analysis of sediment diatom frustules may be used to determine pH history, while chironomid distribution is more closely related to dissolved oxygen levels and therefore trophic status (Carter & Wood 1987). The theory behind the method is the same as that of all biotic indices, and while rarely used in lotic waters, for waterbodies with known historical data, correlations between back-calculated water quality and known chemical analytical results have generally been good.

The mammalian sterol coprostanol may be used as an indicator of sewage pollution. If high levels of this compound are found in sediments it is justifiable to assume that faecal coliform numbers have also been high in the past, though clearly historical records from sewage-treatment plants will also need to be reviewed. However, being an organic substance, coprostanol is biodegradable, so is best used as an indicator of recent water quality history.

Biodegradability is not a problem with many inorganic pollutants such as metals, however, but relating the concentration of any chemical to an exact date, rather than a range of dates such as a 5-year span, should be treated with care as bioturbation will inevitably result in sediment mixing. Also, during peak flow only in the most sheltered, slow-flowing regions of a river will the surface sediment remain undisturbed – these are the same sites where sediment is most readily deposited. Consequently, while sediment cores may be useful for gaining an overall impression of past water quality, great care should be taken in deciding where to sample, and caution should be exercised when interpreting such results.

It must be realized that retrospective monitoring of the types given above is based to some extent on conjecture, since the velocity of water flow will greatly affect the rate of sedimentation and therefore affect the deposited load of those determinands which are readily adsorbed on to sediment. Furthermore, sediment adsorption/desorption equilibria are highly susceptible to changes in water quality parameters such as dissolved oxygen and pH. The techniques have, however, been successfully used to determine the historic water quality in lakes for periods from years to thousands of years ago.

7.10 FUTURE DIRECTIONS

Because the cost of manual sample collection and chemical analysis is so high, automatic samplers are likely to be used on a wider basis than at present, especially as equipment reliability and system flexibility improves. Similarly, the development of continuous monitoring probes for a wider range of determinands is likely to continue, along with greater emphasis being placed on remote data collection. At present, downloading of data from loggers in the field requires monitoring stations to be visited on a regular basis. Whilst such visits are also occasionally needed to service the equipment, they are expensive to undertake so frequently, so remote data collection should decrease costs further in terms of staff time. Furthermore, remote data reporting offers the advantages of being able to respond to changes in water quality almost as quickly as they are recorded, provided the data is used to automatically trigger an alarm. Conversely, it should be possible to detect when monitoring equipment is malfunctioning almost as soon as such an event occurs, so preventing the loss of long data runs before the equipment is known to be broken. This approach, however, is only likely to be taken at particularly sensitive or important sites, e.g. potable abstraction points, and as with other field-based equipment, vandalism may prove to be a major factor limiting its widespread use.

Within the past 20 years, and more particularly

within the last decade, the statistical interpretation of water-quality data has been made much easier by the increasing user-friendliness of personal computer statistical software. This trend will undoubtedly continue and so make available findings from well-established programmes which have received sufficient funding for monitoring to be undertaken, but for which data interpretation has been neglected.

Chemical analysis of water quality makes the greatest contribution to almost all regular monitoring programmes and will continue to do so for the foreseeable future, but the potential advantage of biomonitors in terms of the wide range of pollutants to which they respond is great. Thus, the development of a new generation of less expensive, more sensitive biomonitors that require less regular maintenance and can be linked directly to alarm systems is currently underway. Of the methods currently being developed, immobilized microbial probes relying on changes in respiratory activity or other aspects of cellular physiology appear to offer the greatest potential in terms of portability, price and reliability.

The importance of biological monitoring will also continue in a wider sense, and the adoption of toxicity-based consents in the UK, preferably using indigenous species from a range of taxa, must now be seen simply as matter of time. The use of biotic indices, notably BMWP score and RIVPACS (River Invertebrate Prediction and Classification System) (Section 7.3), has been widespread since the methods were first published, but only comparatively recently has quality control been addressed – this is essential if the use of biotic indices is to remain credible, especially for those which rely on the identification of similar-looking species, notably diatom indices. Bioaccumulation studies have also become increasingly used, and will probably continue to be so for short-term studies because of the more representative nature of results obtained when compared to chemical monitoring. However, these studies do not differentiate between diffuse sources (including natural background levels of some pollutants) and point sources unless an upstream control is used. Also, if used for long-term monitoring it may be difficult to ensure that the external concentration/accumulation relationship is consistently maintained. Thus, bioaccumulation studies represent a method of assessing the relative concentrations of pollutants discharged from point sources, though standardized protocols for the use of a greater range of indigenous species in bioaccumulation studies are required.

The establishment of a core of bioassays to monitor all rivers and sediments with a range of other site-specific biological monitoring techniques to support these data (Dutka *et al* 1991), has much to commend it in terms of flexibility, while permitting inter-site water-quality comparisons. This approach forms a logical basis for both nationwide and international monitoring and must be seen as one of the future directions for biological monitoring, rather than the regional or local studies to which many bioassays appear to be designated.

Finally, the use of retrospective monitoring techniques as a means of assessing historical water quality is thought likely to increase, especially with the recent advances in the quantitative nature and reliability of the results obtained. Whilst the use of these techniques in rivers requires very slow-moving water and is therefore limited, the methods are especially suitable for monitoring the historical quality of water in lakes with a single inflow and outflow, and also in the upper reaches of estuaries. Inference of riverine water quality from these sites may prove to be more accurate than direct river sediment analysis.

To summarize, the strong dependence on traditional discrete sampling and chemical analysis in water-quality monitoring programmes is changing towards a multidisciplinary approach which seeks to optimally use both chemical and biological techniques. The biological aspects of water quality monitoring are likely to increase with the development of innovative bioassays and bioprobes. Such techniques will continue to be used primarily for environmental monitoring, with chemical monitoring remaining the mainstay of monitoring programmes for potable water supply.

REFERENCES

Berry G, Day NE. (1973) The statistical analysis of the results of sampling an environment for a contaminant when most samples contain an undetectable level. *American Journal of Epidemiology* **97**: 160–6. [7.3]

Carter CE, Wood RB. (1987) Chironomid distribution in lake sediments and past water quality. In: Richardson DHS (ed) *Biological Indicators of Pollution*, pp. 111–22. Royal Irish Academy, Dublin. [7.9]

CEC (Council of the European Communities). (1991) Directive concerning urban wastewater treatment. 91/271/EEC, *Official Journal* L135/40, 21 May 1991. [7.5]

Dutka BJ, Kwan KK, Rao SS, Jurkovic A, McInnes R, Palmateer GA, Hawkins B. (1991) Use of bioassays to evaluate water and sediment quality. *Environmental Toxicology and Water Quality* **6**: 309–27. [7.10]

Ellis JC. (1989) *Handbook on the Design and Interpretation of Monitoring Programmes*. Report No. NS29. WRc Medmenham, Marlow, UK. [7.1, 7.3]

Ellis JC, Lacey RF. (1980) Sampling: defining the task and planning the scheme. In: Stiff MJ (ed) *River Pollution Control*, pp. 247–74. Ellis Horwood, Chichester. [7.1, 7.2]

Ferguson RI. (1986) River loads underestimated by rating curves. *Water Resources Research* **24**: 74–6. [7.7]

Hellawell JM. (1986) *Biological Indicators of Freshwater Pollution and Environmental Management*. Elsevier Applied Science, London. [7.2]

Hirsch RM, Alley WM, Wilber WG. (1988) *Concepts for a National Water Quality Assessment Programme*. US Geological Survey Circular 1021. USGS, Denver, Colorado. [7.1]

Leite LAK. (1990) Réflexions sur l'utilité des modèles mathématiques dans la gestion de la pollution diffuse d'origine agricole. Thesis (in French). l'École Nationale des Ponts et Chaussées. [7.6]

Lettenmaier DP, Conquest LL, Hughes JP. (1982) *Routine Streams and Rivers Water Quality Trend Monitoring Review*. Technical Report 75. Charles W. Harris Laboratory, University of Washington. [7.3]

Loftis JC, Phillips RD, Ward RC, Taylor CH. (1989) WQSTAT II: a water quality statistics package. *Ground Water* **27**: 866–73. [7.3]

McDonald A, Kay D. (1981) Enteric bacterial concentrations in reservoir feeder streams: baseflow characteristics and response to hydrography events. *Water Research* **15**: 961–86. [7.7]

McDonald A, Naden P. (1987) Colour in upland sources: variations in small intake catchments. *Water Services* **91**: 121–2. [7.7]

Mar BW, Horner RR, Richey JS, Palmer RN, Lettenmaier DP. (1986) Data acquisition: cost-effective methods for obtaining data on water quality. *Environmental Science and Technology* **20**: 545–51. [7.1]

Meyer JL, Likens GE. (1979) Transport and transformation of phosphorus in a forest stream ecosystem. *Ecology* **60**: 1255–69. [7.7]

Milner NJ, Hemsworth MJ, Jones BE. (1985) Habitat evaluation as a fisheries management tool. *Journal of Fish Biology* **27** (Suppl. A): 85–108. [7.2]

Montgommery HAC, Hart IC. (1974) The design of sampling programmes for rivers and effluents. *Water Pollution Control* **73**: 77–101. [7.2]

Nesbitt EF. (1969) *The Sampling of Ores*. MSU Technical Brochure No. 3. British Steel Corporation, East Moors Works, Cardiff. [7.1]

NRA (1990) *Discharge Consent and Compliance Policy: a Blueprint for the Future*. Water Quality Series No. 1. National Rivers Authority, Bristol. [7.5]

OECD (1987) *The Use of Biological Tests for Water Pollution Assessment and Control*. Organization for Economic Cooperation and Development, Paris. [7.5]

Round FE. (1991) Diatoms in river water-monitoring studies. *Journal of Applied Phycology* **3**: 129–45. [7.2]

Sanders TG, Ward RC, Loftis JC, Steele TD, Adrian DD, Yevjevich V. (1983) *Design of Networks for Monitoring Water Quality*. Water Resources Publications, Littleton, Colorado. [7.1]

Schalie van der W. (1986) Can biological monitoring early warning systems be useful in detecting toxic materials in water? In: Poston TM, Purdy R (eds) *Aquatic Toxicology and Environmental Fate*, pp. 107–21. American Society for Testing and Materials, Philadelphia. [7.2]

Smith DG, McBride GB. (1990) Initiation of a national water quality network for New Zealand. In: Ward RC, Loftis JD, McBride GB (eds) *Design of Water Quality Monitoring Systems*, pp. 162–79. Van Nostrand Reinhold, New York. [7.2]

Ward RC, Loftis JC. (1986) Establishing statistical design criteria for water quality monitoring systems: review and synthesis. *Water Resources Bulletin* **22**: 759–67. [7.1]

Ward RC, McBride GB. (1986) *Design of Water Quality Monitoring Systems in New Zealand*. Publication No 8. Water Quality Centre, Ministry of Works and Development, Hamilton, New Zealand. [7.1]

Ward RC, Loftis JD, McBride GB. (1986) The data-rich but information-poor syndrome in water quality monitoring. *Environmental Management* **10**: 291–7. [7.1]

Ward RC, Loftis JD, McBride GB. (1990) *Design of Water Quality Monitoring Systems*. Van Nostrand Reinhold, New York. [7.1]

Whitfield PH. (1988) Goals and data collection design for water quality monitoring. *Water Resources Bulletin* **24**: 775–80. [7.1]

8: Biological Water-Quality Assessment of Rivers: Use of Macroinvertebrate Communities

J. L. METCALFE-SMITH

8.1 INTRODUCTION

At present, the monitoring, assessment and regulation of aquatic ecosystems is largely based on chemical measures of water quality. Yet, chemical parameters alone do not provide adequate information for sound management of aquatic resources because they tell us little of the effects of pollution on living organisms.

Direct biological assessments of the health of biotic communities in receiving waters offer several important advantages over chemical-based approaches. For example, organisms integrate environmental conditions over time, whereas chemical data are instantaneous in nature and require large numbers of measurements for an accurate assessment (De Pauw & Vanhooren 1983). Biological communities also integrate the effects of multiple stresses and demonstrate cumulative impact (Plafkin *et al* 1989). Biological studies can serve an early warning function by detecting intermittent pollution and subtle disruptions which would likely be missed by conventional chemical surveys (Howmiller & Scott 1977; Reynoldson 1984). Because chemical monitoring programmes are usually menu-driven, the possibility exists that the pollutants or factors responsible for environmental degradation will be excluded from consideration. Finally, it must be recognized that not all impacts are chemical in nature; biological assessments may also be able to detect the impact of flow alterations, habitat destruction, overharvesting of biological resources, etc. (Karr 1991).

At it is obviously impractical to conduct bioassessments on entire aquatic ecosystems, most workers have focused on a particular component. Hellawell (1977) and Reynoldson (1984) tabulated the advantages and disadvantages of all major groups, and a clear preference for using macroinvertebrates emerged. Benefits of using benthos include the following.

1 Macroinvertebrate communities are differentially sensitive to pollutants of various types and react to them quickly (Cook 1976), and are capable of a graded response (Pratt & Coler 1976).

2 Macroinvertebrates are present in most aquatic habitats, especially flowing water systems (Reynoldson 1984), and are abundant and relatively easy and inexpensive to collect (Plafkin *et al* 1989). Furthermore, their taxonomy is well established, although admittedly difficult at the species level for some groups (Reynoldson 1984).

3 Benthic invertebrates are relatively sedentary, and are therefore representative of local conditions (Cook 1976).

4 They have lifespans long enough to provide a record of environmental quality (Pratt & Coler 1976).

5 Finally, macroinvertebrate communities are very heterogeneous, with numerous phyla and trophic levels represented. The probability that at least some of these organisms will react to a particular change in environmental conditions is, therefore, high (France 1990).

The use of macrobenthos in bioassessment has three major disadvantages.

1 They respond to seemingly minor changes in substrate particle size, organic content and even texture. As a result, discrimination between the effects of pollution and other environmental factors is often difficult (France 1990).

2 Their life histories are complex and the results

of bioassessments can vary with season (Hellawell 1977).

3 Spatial heterogeneity is high, requiring considerable replication (Reynoldson 1984).

This chapter is largely based on an earlier review of bioassessment techniques developed and applied in Europe (Metcalfe 1989). The information has been updated and expanded to include the North American literature as well as several new and alternative approaches. Sampling design and methodology will not be specifically addressed, as these topics have been adequately covered elsewhere (e.g. Voshell *et al* 1989; Klemm *et al* 1990).

8.2 TRADITIONAL APPROACHES TO BIOASSESSMENT

The saprobic system

The term 'saprobia' refers to the dependence of an organism on decomposing substances as a food source (Persoone & De Pauw 1979). Early research efforts of two German scientists, R.W. Kolkwitz and M. Marsson, led to the classic Saprobic System, best known through the saprobic index (Pantle & Buck 1955). This is based on the presence of indicator species (mainly bacteria, algae, protozoans and rotifers, but also some benthic invertebrates and fish) which have been assigned saprobic values based on their pollution tolerance. Values range from 0 to 8; the higher the value the more tolerant the organism. Pollution tolerances of individual species are determined by observations on their relative occurrence under specifically-defined conditions of water quality. According to the saprobic system, water quality is classified into one of ten categories ranging from the purest groundwater to anaerobic sewage and industrial wastes (see Table 1 in Metcalfe 1989). No single indicator species will be representative of only one saprobic zone; rather, its distribution will follow a normal curve over a range of zones reflecting its tolerance. The shape and area of this distribution curve defines the saprobic 'valency' of the species (Zelinka & Marvan 1961), and the position of the apex is its saprobic value (Sladecek 1979). Various lists of saprobic values have been published, all for European species. Most notable is that of Sladecek (1973), which contains information for aproximately 2000 species.

Briefly, the Saprobic Index is calculated as follows:

$$S = \frac{\Sigma (s.h)}{\Sigma h}$$

where S = Saprobic Index for the site, s = saprobic value for each indicator species, h = frequency of occurrence of each species (rare, $h = 1$; frequent, $h = 3$; abundant, $h = 5$). The value of S will normally range from 1 to 4 for ambient waters.

Major criticisms of saprobic systems are:

1 The taxonomy is not far enough advanced for some groups and too controversial for others; intensive sampling is required; species lists and saprobic values will not be applicable to other geographical locations; and the system cannot be confidently applied to other types of pollution (Persoone & De Pauw 1979).

2 The pollution tolerances of species are very subjective, as they are based on observational rather than experimental data (Slooff 1983).

3 Each taxon is considered as a separate entity, therefore no information on the community as a whole is provided (Jones *et al* 1981).

Two saprobic-based systems are currently in use: the Biologically Effective Organic Loading (BEOL) method in West Germany (Persoone & De Pauw 1979) and the Quality-index, or K-index, in The Netherlands (Woodiwiss 1980). The K-index is calculated as follows. Approximately 60 indicator taxa are arranged in five groups, each of which is assigned a pollution factor (see Table 2 in Metcalfe 1989). The percentage of the total number of animals in the sample belonging to each group is then multiplied by the appropriate factor, and the group values are summed into an index value ranging from 100 (very heavily polluted) to 500 (not polluted), as follows:

$$\begin{aligned} K_{135} = {} & (\%\ \textit{Eristalis} + \textit{Chironomus}\text{-group}) \times 1 \\ & + (\%\ \text{Hirudinea-group}) \times 3 \\ & + (\%\ \textit{Gammarus} + \textit{Calopteryx}\text{-group}) \\ & \times 5 \end{aligned}$$

In 1980, the Limburg Water Pollution Control Authority began to develop a biological classification system for rivers and streams in the province of Limburg, The Netherlands. Their purposes were to determine the extent of pollution by organic wastes, evaluate the effectiveness of enforced measures to reduce pollution, and define reference communities for different types of streams which could be used as water quality objectives. To select the most appropriate bioassessment method, Tolkamp (1985a,b) applied a variety of biotic and saprobic indices to a large data set from the drainage basin of the River Geul. None performed as well as the K-index, which was more sensitive to smaller changes in the middle range of the pollution scale. Most other indices underestimated water quality because they included indicator organisms which do not normally occur in these lowland streams. Vandelannoote *et al* (1981) reported similar results for lowland streams in Belgium. The majority of biotic indices available were developed for small upland streams, and the assessment of lowland streams and rivers has generally been neglected.

Diversity indices

Diversity indices are mathematical expressions which use three components of community structure, namely *richness* (number of species present), *evenness* (uniformity in the distribution of individuals among the species) and *abundance* (total number of organisms present), to describe the response of a community to the quality of its environment. Undisturbed environments are characterized by high diversity or richness, an even distribution of individuals among the species, and moderate to high counts of individuals (Ghetti & Bonazzi 1977; Mason *et al* 1985). Organic pollution causes a decrease in diversity as sensitive organisms are lost, an increase in the abundance of tolerant organisms due to nutrient enrichment, and a decrease in evenness. In contrast, toxic or acidic pollution may cause a decrease in both diversity and abundance as sensitive organisms are eliminated and there is no additional food source for the remaining tolerant forms, and an increase in evenness (Kovalak 1981).

By far the most widely used diversity index is the Shannon–Wiener index (H′), because it is stable in any spatial distribution and insensitive to rare species (Cairns & Pratt 1986). Its formula is as follows (after Wilhm & Dorris 1968):

$$H' = -\sum \frac{Ni}{N} \log_2 \frac{Ni}{N}$$

where H′ = index value, N = total number of individuals of all species collected, and Ni = number of individuals belonging to the ith species. The higher the value of H′, the greater the diversity and, supposedly, the cleaner the environment. The reader is referred to Washington (1984) for a critical review of the many diversity indices which have been applied to aquatic ecosystems.

Diversity indices are considered to have the following advantages:

1 They are strictly quantitative, dimensionless, and lend themselves to statistical (Cook 1976).

2 Most are relatively independent of sample size (Wilhm & Dorris 1968).

3 No assumptions are made about the relative tolerances of individual species, assumptions which may be very subjective (Pinder *et al* 1987).

France (1990) notes that 'few subjects in applied ecology are as controversial as the use (and misues) of diversity indices'. Many criticisms have been levelled at diversity indices, and only the major ones will be identified here. Their most serious problem is that they reduce individual species to anonymous numbers which disregard their environmental adaptations. This could result in equating, theoretically, a pollution-tolerant oligochaete/chironomid community with a pollution-sensitive mayfly/amphipod community (France 1990). Secondly, not all undisturbed communities have inherently high diversity and, consequently, it is not always possible to correlate certain values with ecological damage (Jones *et al* 1981). For example, the oligotrophic offshore areas of large lakes (Howmiller & Scott 1977) and the headwaters of streams fed by nutrient-poor groundwater (Pinder & Farr 1987a) are naturally low in productivity. Because wide variations in diversity index values have been reported for unpolluted conditions (Cook 1976), standards set for the interpretation of values are not universally applicable.

Thirdly, diversity indices may generate 'false negatives' under certain circumstances. Moderate pollution can cause an increase in abundance without excluding species, with the result that the index value actually goes up (Cook 1976). Because H′ is more sensitive to changes in evenness than diversity, its value may be high at sites heavily contaminated by toxic chemicals (Kovalak 1981).

Finally, many studies have shown that diversity indices are insensitive and give poor site discrimination, particularly over the moderate range of various types of pollution including nutrients (Jones *et al* 1981), metals (Perkins 1983) and pesticides (Webber *et al* 1989). It is this last characteristic that limits the usefulness of diversity indices to assessing the impact of gross point source pollution of known chemical composition on relatively simple systems. They are of little use in complex systems affected by multiple and diverse stresses and/or basin-wide non-point source pollution. Some developing countries (e.g. Jhingran *et al* 1989) continue to rely on diversity indices for assessing severe sewage pollution.

Biotic indices

The biotic approach to bioassessment, as defined by Tolkamp (1985b), is one which combines diversity on the basis of certain taxonomic groups with the pollution indication of individual species or higher taxa or groups into a single index or score. Numerous biotic index and score systems have been developed, most of them in Europe and the UK. The major indices and their relationships are illustrated in Fig. 8.1. Only the most widely used systems and their recent modifications will be described. The reader is referred to Metcalfe (1989) and Washington (1984) for information on other biotic indices.

Trent Biotic Index

The Trent Biotic Index (TBI) was originally devised for use in the Trent River Authority area in England, but has since been adapted for use in many other countries and appears to form the basis for most modern biotic indices and scores (Persoone & De Pauw 1979). Organisms are collected from all available habitats by means of a kicknet, then identified to family, genus or species depending on the type of organism, but they are not enumerated. The index is based on the sensitivity of key groups to pollution and on the number of component groups in a sample. Clean streams are given an index value of 10, and this value decreases with increasing pollution. The TBI was later extended to cover a wider range of water qualities (0 to 15 instead of 0 to 10) to improve sensitivity. This version, called the Extended Biotic Index (EBI), is shown in Table 8.1. One major drawback of these indices is that

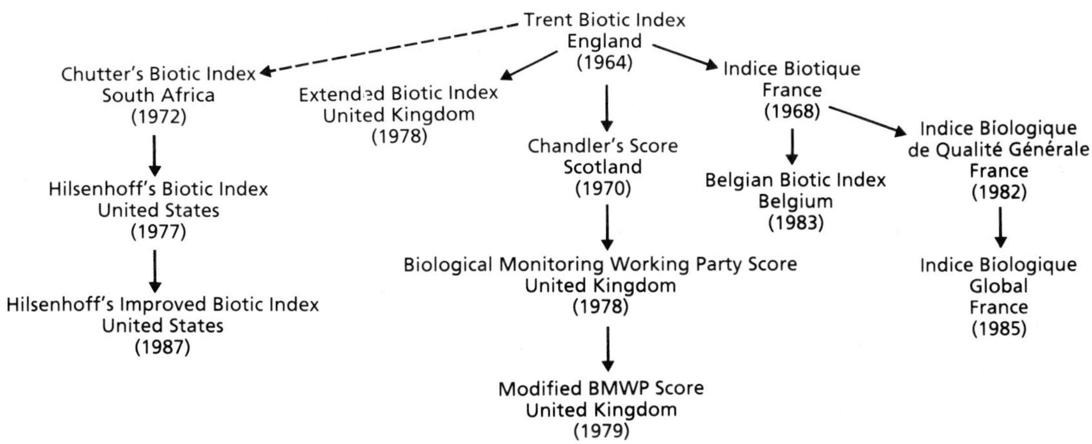

Fig. 8.1 Development of the most widely used biotic index and score systems (after Metcalfe 1989, with permission).

Table 8.1 Extended Biotic Index (from Persoone & De Pauw 1979, with permission)

Extended Biotic Index			Total number of groups present									
			0–1	2–5	6–10	11–15	16–20	21–25	26–30	31–35	36–40	41–55
Trent Biotic Index			Total number of groups present									
			0–1	2–5	6–10	11–15	16+					
Organisms in order of tendency to disappear as degree of pollution increases			Biotic indices									
Clean	Plecoptera nymphs present	More than one species	–	7	8	9	10	11	12	13	14	15
		One species only	–	6	7	8	9	10	11	12	13	14
	Ephemeroptera nymphs present (excluding *Baetis rhodani*)	More than one species	–	6	7	8	9	10	11	12	13	14
		One species only	–	5	6	7	8	9	10	11	12	13
	Trichoptera larvae or *Baetis rhodani* present	More than one species	–	5	6	7	8	9	10	11	12	13
		One species only	4	4	5	6	7	8	9	10	11	12
	Gammarus present	All above species absent	3	4	5	6	7	8	9	10	11	12
	Asellus present	All above species absent	2	3	4	5	6	7	8	9	10	11
	Tubificid worms and/or red chironomid larvae present	All above species absent	1	2	3	4	5	6	7	8	9	10
Polluted	All above species absent	Some organisms such as *Eristalis tenax* not requiring dissolved oxygen may be present	0	1	2	–	–	–	–	–	–	–

abundance is ignored. Therefore, the accidental presence of an organism in a sample (due to drift, for example), could drastically alter the value of the index (Cook 1976).

Chandler's Score system

This system was developed for upland rivers in Scotland (Cook 1976). Chandler's Score is theoretically an improvement over the TBI because it includes an abundance factor and incorporates a more detailed list of macroinvertebrates. The Score is determined by identifying the organisms present, determining the abundance classification for each group, then selecting the appropriate points for that group (see Table 5 in Metcalfe 1989). The points for all groups are added to give a site Score. Points scored increase with increasing abundance for sensitive groups and decrease with increasing abundance for tolerant groups, and the value of the site Score is unending. Criticisms of this system are that it is too complicated, the level of taxonomic identification is not uniform for all groups and also that it is geographically restricted due to the number of indicators identified to genus. However, Cook (1976) found that modifications aimed at adapting Chandler's Score to local conditions did not significantly improve the performance of the system in a New York stream. The Score was superior to the diversity index H′ in grading sections of a mildly polluted stream according to water quality. Armitage (1980) applied Chandler's Score to a zinc-contaminated river in northern England and found that Score values were lowest at sites affected by high zinc levels, suggesting that it may be adaptable to other types of pollution.

Biological Monitoring Working Party Score system

The Biological Monitoring Working Party (BMWP) was set up in 1976 to develop a standardized system for assessing the biological quality of rivers in Great Britain. Its terms of reference were 'to provide an overall view of the condition of rivers and canals and of the discharges to them and to show the effectiveness of pollution control policies' (ISO 1984). They developed a standardized score system which was a simplification of Chandler's Score, where all organisms were identified to family for uniformity, families with similar pollution tolerances were grouped together, and the abundance factor was eliminated because it was time-consuming and had only a small effect on score value. The method for scoring is as follows, using Table 8.2: list the families present in the sample, ascribe the score for each family, then add the scores together to arrive at a site score.

The 'average score per taxon' (ASPT) computation, which simply refers to dividing the total score by the number of scoring taxa, has frequently been applied to both the Chandler and BMWP Scores because it is independent of the number of taxa counted. Murphy (1978) compared the performance of Chandler's Score and ASPT, the TBI and several diversity indices over a range of polluted and unpolluted sites in Welsh rivers. The Chandler ASPT showed the least temporal variability, thus allowing the best spatial discrimination. While Chandler's Score and TBI values were depressed at headwater sites, probably reflecting physical habitat properties, ASPT values were high, reflecting good water quality. Armitage *et al* (1983) evaluated the BMWP Score vs. ASPT for classifying unpolluted sites with different physical and chemical characteristics. The results showed a steady decline in ASPT values from the upland to the lowland range of environmental features, while Score values varied. Predictive equations based on multiple regressions using physical and chemical parameters explained 65% of the variance in ASPT values as opposed to only 22% for BMWP Score values.

The performance of the Chandler and BMWP Scores and their ASPT versions were compared in a three-part study on a chalk stream in southern England (Pinder *et al* 1987; Pinder & Farr 1987a,b). They found that the BMWP Score stabilized after fewer replicates than Chandler's Score, and that the ASPT versions of both were much less affected by sample size, season and habitat. The BMWP-ASPT was most sensitive to slight changes in pollution status, ranked sites similarly to both Scores and several diversity indices, and was also best correlated with direct chemical measures of water quality. In contrast, the Chandler–

Table 8.2 The modified BMWP Score system (from Armitage *et al* 1983, with permission)

Families	Score
Siphlonuridae, Heptageniidae, Leptophlebiidae, Ephermerellidae Potamanthidae, Ephemeridae Taeniopterygidae, Leutridae, Capniidae, Perlodidae, Perlidae Chloroperlidae Aphelocheiridae Phryganeidae, Molannidae, Beraeidae, Odontoceridae Leptoceridae, Goeridae, Lepidostomatidae, Brachycentridae Sericostomatidae	10
Astacidae Lestidae, Agriidae, Gomphidae, Cordulegasteridae, Aeshnidae Corduliidae, Libellulidae Psychomyiidae, Philopotamidae	8
Caenidae Nemouridae Rhyacophilidae, Polycentropodidae, Limnephilidae	7
Neritidae, Viviparidae, Ancylidae Hydroptilidae Unionidae Corophiidae, Gammaridae Platycnemididae, Coenagriidae	6
Mesoveliidae, Hydrometridae, Gerridae, Nepidae, Naucoridae Notonectidae, Pleidae, Corixidae Haliplidae, Hygrobiidae, Dytiscidae, Gyrinidae Hydrophilidae, Clambidae, Helodidae, Dryopidae, Eliminthidae Chrysomelidae, Curculionidae Hydropsychidae Tipulidae, Simuliidae Planariidae, Dendrocoelidae	5
Baetidae Sialidae Piscicolidae	4
Valvatidae, Hydrobiidae, Lymnaeidae, Physidae, Planorbidae Sphaeriidae Glossiphoniidae, Hirudidae, Eropobdellidae Asellidae	3
Chironomidae	2
Oligochaeta (whole class)	1

ASPT consistently failed to agree with other indices. Pinder and Farr (1987b) concluded that the BMWP-ASPT was the best biotic index available, although they recommended eliminating the Chironomidae (except *Chironomus riparius*) and Oligochaeta from the scoring system. Since both of these groups include many species which are tolerant of pollution and many which are sensitive, their inclusion without further discrimination had the effect of depressing the ASPT while contributing nothing to the score. After eliminating these groups from their calculations, Pinder and Farr (1987b) achieved better site discrimination.

The Belgian Biotic Index method

The Belgian Biotic Index (BBI) method is derived from the French Indice Biotique (IB), which in turn is a modification of the Trent Biotic Index. The IB differs from the TBI in that it includes a greater number of indicator taxa (Persoone & De Pauw 1979), separates the Ecdyonuridae from other Ephemeroptera and divides the Trichoptera into species with and without cases (De Pauw & Vanhooren 1983), does not separate *Nais* from other Naididae and *Baetis* from other Ephemeroptera (Ghetti & Bonazzi 1977), and does not consider a systematic unit represented by a single individual since its occurrence may be accidental. Also, the TBI specifies that a handnet be used for sampling all habitats, while the IB calls for lentic and lotic habitats to be sampled separately and for indices from both habitats to be used in interpreting water quality (Persoone & De Pauw 1979). De Pauw and Vanhooren (1983) adapted the IB for use in Belgium, with the following modifications: sampling would be by handnet only, as this technique explores a larger array of habitats than other samplers; nematodes would be excluded from consideration, as most would not be caught in a 300–500 μm mesh handnet; the Chironomidae were divided into two systematic units, those belonging to the *thummi-plumosus* group and those not; level of taxonomic identification would generally be set at a higher level (genus or family) to avoid erroneous interpretations due to misidentification.

The BBI is calculated using Table 8.3 which has

Table 8.3 Standard table to determine the Belgian Biotic Index (from De Pauw & Vanhooren 1983, with permission)

	Total number of systematic units present				
	0–1	2–5	6–10	11–15	16 and more
Faunistic groups	Biotic Index				
1 Plecoptera or Ecdyonuridae (=Heptageniidae)					
Several SU*	–	7	8	9	10
Only one SU	5	6	7	8	9
2 Cased Trichoptera					
Several SU	–	6	7	8	9
Only one SU	5	5	6	7	8
3 Ancylidae or Ephemeroptera except Ecdyonuridae					
More than two SU	–	5	6	7	8
Two or less SU	3	4	5	6	7
4 *Aphelocheirus* or Odonata or Gammaridae or Mollusca (except Sphaeridae)					
All SU mentioned above are absent	3	4	5	6	7
5 *Asellus* or Hirudinea or Sphaeridae or Hemiptera (except *Aphelocheirus*)					
All SU mentioned above are absent	2	3	4	5	–
6 Tubificidae or Chironomidae of the *thummi-plumosus* group					
All SU mentioned above are absent	1	2	3	–	–
7 Eristalinae (=Syrphidae)					
All SU mentioned above are absent	0	1	1	–	–

* SU: systematic units observed of this faunistic group.

both rows and columns representing faunistic groups and systematic units (SUs), respectively. The seven faunistic groups are ranked in order of increasing tolerance to pollution. For groups 1–3, it is necessary to know whether there are one, two or more SUs present. The row chosen from the table is the one corresponding to the presence of the most sensitive faunistic group in the sample. The column chosen depends on the number of systematic units present in the sample. The intersection of the appropriate row and column gives the index value for the site.

The surface water quality of Belgian rivers has been routinely surveyed by the BBI since 1978, and by 1985 over 30 000 km of watercourses had been surveyed and mapped using the BBI. The programme is sponsored by the National Institute for Hygiene and Epidemiology in Brussels and is driven by the urgent need for a coordinated policy in the field of surface water sanitation and management (De Pauw & Vanhooren 1983). Goals of the programme are to obtain better insight into the self-purification of rivers and streams and to assist decision-makers in selecting sites for water purification plants and surface water reservoirs. Belgium requires a method which is equally applicable in fast-flowing shallow streams and slow-running deep lowland rivers and canals. The BBI has been declared highly successful (De Pauw & Vanhooren 1983) for this application. Identification keys and handnet specifications have been standardized, and intercalibration exercises with respect to sampling and identification gave satisfactory results. Results are reproducible over long periods of time in areas where no changes in pollution status occur, and seasonal changes are minor. A single sampling in either early summer or autumn is sufficient for a proper assessment. Artificial substrates were found to be a valid alternative for deep, lowland rivers where handnet sampling was difficult. De Pauw *et al* (1986) found that three pooled replicates of a 2250–4500 cm^3 sampler yielded BBI values equivalent to those generated by handnet collections.

De Pauw and Roels (1988) investigated the relationship between the BBI and various common chemical indicators of pollution, using data from a wide variety of polluted and unpolluted sites in Belgium, Italy, Portugal and the UK. They found that correlations between chemical variables and the BBI were consistently positive (dissolved oxygen) or negative (BOD, COD, NH_4, PO_4), but that the slopes of the regression lines varied considerably among watersheds. This indicated that the degree of stress associated with a particular chemical factor in one river was not necessarily of the same magnitude in another. They suggested, therefore, that 'biological assessments should be used as an early warning system and be the precursor of extensive chemical analyses, identifying the causes of biological stress.' The BBI has recently been shown to be applicable in other countries, including Spain, Algeria, Luxemburg, Portugal and Canada (De Pauw *et al* 1986).

Bervoets *et al* (1989) proposed several modifications to the BBI to improve reliability, save time and provide better correlation with water chemistry. BBI values were found to be higher if samples were sorted live, instead of preserved in formalin and washed through a series of sieves, and if SUs consisting of only one individual were included in the calculation of the index. With respect to sample replication, polluted sites could be accurately characterized in only two subsamples, whereas new single-individual SUs were often found in the 10th subsample from unpolluted brooks. They strongly recommended that sampling effort be standardized for area sampled rather than time, because the specified 5-minute period was insufficient for finding rarer individuals in the cleanest streams. Their modification was also more efficient, allowing more samples to be processed within a given time period.

The Indice Biologique Global is the standard method presently used in France (AFNOR 1985). It is also derived from the IB, and its application appears to be restricted to that country. Further details on the development of the French indices are presented in Metcalfe (1989).

Chutter's Biotic Index

A biotic index for South African streams and rivers was developed by Chutter (1972). European indices such as the TBI could not be readily applied, because some of the key indicators were absent (e.g. *Gammarus*, *Asellus*) or of very restricted occurrence (Plecoptera) in South African rivers, while the Baetidae fauna was much richer. Chutter assigned Quality (tolerance) values of 1(clean)–10(polluted) to all taxa collected, based on the literature. All 'pristine' species were assigned a value of 0. Each organism found in the sample was recorded at its quality value, then the values were summed for all taxa and divided by the total number of individuals. Because some taxa were extremely abundant, he included a 'sliding scale' of quality values which takes into account the abundance and diversity of these dominant taxa. Because of the instability of flows river beds in the rainy season in South Africa, the index was not reliable in recently flooded areas. Although loosely based on the TBI, this approach is unique in that every individual organism contributes to the index value. According to Washington (1984), Hilsenhoff is the only worker to consider Chutter's index and adapt it for use in another country.

Hilsenhoff's Biotic Index

In 1977, Hilsenhoff introduced a biotic index (BI) for evaluating organic pollution in Wisconsin streams based on riffle-dwelling arthropod fauna (Hilsenhoff 1987). To determine the index value for a given site, a sample of 100 or more arthropods was collected by means of a handnet and the organisms identified to genus or species. Each taxon was assigned a tolerance value ranging from 0 (most sensitive) to 5 (most tolerant) based on information from 53 streams. As per Chutter's index, the average of the tolerance values for all individuals constituted the site index value. Hilsenhoff's Biotic Index was extensively tested by the Wisconsin Department of Natural Resources and the data generated were used to improve the index. Tolerance values were revised by comparing the original tolerance value assigned to a species with the average biotic index value of streams in which it most commonly occurred. To provide greater precision, the scale was expanded from 0 to 10. To date, tolerance values have been assigned to some 400 species or genera, and these are listed in Hilsenhoff (1987). The influence of current, temperature and seasonal factors on BI values have also been evaluated (Hilsenhoff 1988b). BI values were found to be erroneously high in summer, when many sensitive species are in diapause, and in currents of less than $0.30\,m\,s^{-1}$. Table 8.4 is a guide to the water quality of streams based on this index.

Hilsenhoff (1988a) also adapted his index for a rapid assessment by providing tolerance values for families. Because family tolerance values are averages, they are lower than the tolerance values for some species in the family and higher than the values for other species. The result should be a dampening effect on the performance of the index. This was confirmed in testing on second- and third-order Wisconsin streams, where the family-level biotic index (FBI) tended to be higher than the BI at unpolluted sites and lower at polluted sites. The FBI was also more variable. However, an average of only 23 minutes was required to sample, sort and calculate an FBI in the field, as compared with at least 85 minutes to calculate a BI. Hilsenhoff (1988a) recommended using the FBI only for rapid assessment of the general status of organic pollution in streams, in essence as a screening tool to identify problem areas. He noted that if all organisms were preserved, the BI could always be calculated later if needed.

Despite all the effort which has gone into the development of biotic indices, they have serious limitations as bioassessment tools. First of all, 'they set the same target for all sites when it is clear that different physical and chemical regimes of fast-flowing mountain streams and slow-flowing lowland rivers will support totally different faunal communities' (Armitage *et al* 1990). Secondly, a primary weakness of biotic indices is the subjective approach which is often used to classify organism tolerance. Herricks and Cairns (1982) suggest that only where a long record of study is available (e.g. Sladecek 1973;

Table 8.4 Evaluation of water quality using biotic index values of samples collected in March, April, May, September and early October (from Hilsenhoff 1987, with permission)

Biotic index	Water quality	Degree of organic pollution
0.00–3.50	Excellent	No apparent organic pollution
3.51–4.50	Very good	Possible slight organic pollution
4.51–5.50	Good	Some organic pollution
5.51–6.50	Fair	Fairly significant organic pollution
6.51–7.50	Fairly poor	Significant organic pollution
7.51–8.50	Poor	Very significant organic pollution
8.51–10.00	Very poor	Severe organic pollution

Hilsenhoff 1987) is reliability high. They recommend replacing 'subjective tolerance estimates' with 'quantitative tolerance determinations', which require extensive correlations between species presence and water quality. Thirdly, biotic indices apply only to organic pollution, and their application to other types of pollution or perturbation is questionable at best and erroneous at worst. Clearly, a more diagnostic approach is needed.

Community comparison indices

Community comparison indices (CCIs), which measure the similarity of the structure of two communities, have mostly been developed for use in terrestrial ecology and have not been extensively applied to aquatic ecosystems (Washington 1984). CCIs require a clean water station for comparison, and have therefore mainly been used for upstream–downstream contrasts of the response to point source pollution. Like diversity indices, they are strictly quantitative and therefore provide no information about the actual composition of communities. Unlike biotic indices, they will respond to any perturbation that affects benthic communities, not just organic pollution. There is some evidence that CCIs may be more sensitive to subtle changes in community structure than diversity indices. For example, Perkins (1983) found that CCIs demonstrated a consistent decrease in similarity among benthic communities exposed to increasing concentrations of copper, while the diversity index H′ gave false negatives for the lower concentrations. Washington (1984) called for further evaluation of CCIs, specifically for comparisons with both diversity and biotic indices. Hellawell (1986) assessed the relative performances of many of the older 'pollution' (i.e. saprobic and biotic), diversity and community comparison indices and provided a summary of the preferred methods. He concluded that no single index would satisfy all requirements and recommended that several types of index should always be used.

A wide range of CCIs are currently available, each having somewhat different mathematical and ecological properties. To add to the confusion, some measure similarity and others dissimilarity. Because of their differing properties, various indices generate somewhat different results when applied to the same data set. Most require both validation under controlled conditions and careful consideration of their intrinsic properties to determine which index is most suited to the objectives of a particular study and the type of data available. Formulae for an exhaustive list of indices are presented in Perkins (1983) and Washington (1984). One example of the application of CCIs will be presented here.

Brock (1977) used the indices P_{sc} and B to evaluate the effects of thermal effluents on zooplankton communities in a reservoir in central Texas. The formulae are as follows:

$$P_{sc} = 100 - 0.5\sum^{k}|a - b|$$

where a and b are, for a given species, percentages of the total samples A and B which that species represents. The absolute value of their difference is summed over all species, k.

$$B = \frac{1}{k}\sum^{k}\frac{\min(Xia, Xib)}{\max(Xia, Xib)}$$

where Xia and Xib = abundances of species i at stations A and B, the smaller number being divided by the larger number for each species, and k = total number of species observed between the two stations.

P_{sc} is based on relative abundance while B is based on actual abundance, therefore the two indices respond differently to certain changes in community structure. Brock (1977) found that sites with very similar proportions of taxa but very different total abundances registered a high degree of similarity according to P_{sc} and less similarity according to B. The removal of a few rare species from a data set for two stations hardly affected the value for P_{sc}, whereas B indicated a definite change. Conversely, B was insensitive to the addition of a substantial number of individuals to a dominant taxon. Brock (1977) concluded that if B overemphasizes shifts in rare taxa and de-emphasizes changes among dominant species, then it is too sensitive to normal sampling error. It could, however, have an important application where the loss of rare or endangered species is of interest.

Camargo (1990) recently developed a new 'ecotoxicological index' for assessing the impact of a regulated and industrial area on the Duraton River in Spain. This index combines a measure of the percentage difference between the number of species occurring above and below a disturbance point,

$$\left[(A - B) \times \frac{100}{A}\right]$$

with a measure of the species substitution between the two sites,

$$\left[(A - C) \times \frac{100}{A}\right]$$

where A = number of species occurring upstream, B = number of species downstream and C = number of species common to both sites, into a single index:

$$\mathrm{EI} = \frac{(2A - B - C) \times 50}{A}$$

The values of the ecotoxicological index, or EI, range from 0 (no impact) to 100 (maximum impact). He reported that the new EI was significantly correlated (but inversely, due to their inverted scales) with H′ and Margalef's diversity index. However, the range of values was much greater for the EI than for H′, suggesting that the EI may be more sensitive to subtle changes. Interestingly, Margalef's index gave better site discrimination than H′, a finding which was also reported by Wilhm (1967).

8.3 ALTERNATIVE APPROACHES

Functional feeding groups

According to the River Continuum Concept (RCC: Vannote *et al* 1980; Minshall *et al* 1985), drainage networks form a predictable continuum of increasing channel size and associated biological characteristics. Stream morphology, current velocity, substrate composition, temperature and allochthonous vs. autochthonous food sources all interact to influence food availability to invertebrates, and these interactions vary systematically with stream order thereby regulating distribtion patterns of invertebrate functional feeding groups (Hawkins & Sedell 1981). These feeding groups are referred to as the scrapers, collector–filterers, collector–gatherers, predators and shredders (Cummins 1974). Under unperturbed conditions, headwater areas are normally dominated by shredders and collectors due to the mainly allochthonous energy source; mid-sized streams are autotrophically driven and dominated by scrapers and collector–filterers; and communities in large rivers are mainly composed of collector–gatherers due to the accumulation of fine sediment from upstream (Rabeni *et al* 1985; Cummins 1988).

Although this general pattern appears to hold worldwide, the exact nature and rate of change will vary from river to river depending on catchment characteristics and water chemistry at the origin (Ormerod & Edwards 1987) and on the efficiency of retention of sediments and organic matter (Cummins 1988). Cushing *et al* (1983) verified the changes in functional feeding groups predicted by the RCC in a study of 16 streams in Oregon, Idaho, Michigan and Pennsylvania. An important finding was that rates of change varied regionally, such that first-order streams in Pennsylvania and Michigan were more like third-order than first-order streams in Oregon. Because the RCC was developed in North America, rivers on other continents may show considerable divergence from the original model (e.g. Winterbourn *et al* 1981 and Marchant *et al* 1985, for New Zealand and Australia).

Cummins and others (e.g. Cummins 1974; Cummins & Klug 1979; Merritt & Cummins 1984; Cummins 1988) have developed a functional classification of stream invertebrates over the last 15 years in response to the inadequacies of systematic and trophic analyses; i.e. identification to the species level is still difficult and gut content analyses reveal that all invertebrates are omnivores. According to Cummins (1988), the functional view permits 'clustering of genetically and taxonomically diverse entities into groups, or guilds, which share fundamental properties — such as invertebrates having the same morphological-behavioural mechanisms of food acquisition.' Hawkins and Sedell (1981) point out that such a system reduces the variability associated with taxonomic complexity, allowing trends to be more easily recognized. This approach is also more universally applicable because local taxonomic differences do not seriously affect it. The only drawback to the functional feeding group (FFG) approach is that it is based on nutrient dynamics and can therefore only be used to assess the effects of organic enrichment.

Thus, Rabeni *et al* (1985) assessed changes in benthic community structure and function in the Penobscot River, Maine, before and after pollution abatement technology had been implemented at all of the major point sources of organic enrichment, including pulp mills and municipal sewage treatment plants. Characteristics of the study area, including the stream order, water clarity and general erosional substrate, classified the river as autotrophically-driven under the RCC scheme. At unperturbed sites, densities of scrapers exceeded those of collector–filterers and collector–gatherers combined, reflecting the normal autotrophic nature. As water quality degraded, the functional groups responded in a manner predicted by the RCC (e.g. the percentage density of scrapers decreased from a high of 45% to less than 1% at the most polluted sites) and the system became heterotrophic. This change occurred in the absence of any longitudinal gradient, suggesting that organic pollution can 'reset' the normal sequence of feeding group shifts and convert an autotrophic system into a heterotrophic state that would normally be found farther downstream in a much larger river. As water quality improved due to pollution abatement, the ratio of collector–filterers and scrapers to collector–gatherers increased, indicating that the river was returning to its normal condition. Rabeni *et al* (1985) concluded that the FFG approach is promising because it 'may reflect more ecologically significant attributes of streams and rivers than do structurally-based water-quality systems'.

Reduced assemblages

Most macroinvertebrate bioassessment techniqus are based on the response of the entire benthic community to pollution. However, focusing on a single component of the community (generally an order or family) has several attractive benefits. It can simplify the collection, sorting and identification of benthic samples, thereby reducing time and effort. Selective sampling techniques can be employed which provide more precise estimates of the diversity and abundance of the organisms under consideration. The resources saved by limiting the investigation to one group can then be redirected into more intensive or extensive studies and species-level taxonomic identification. The latter is an extremely important consideration. Indices which use species-level identification have better potential for site discrimination (Hilsenhoff 1988a; Furse *et al* 1984), since species have more precise environmental requirements than families and species belonging to a single group may have a wide range of susceptibilities to various pollutants (Slooff 1983). Working at the species level also allows the identification of indicator species for certain types of perturbations. Observational data can then be verified by laboratory toxicity tests on these species, such as those conducted by Chapman *et al* (1982a,b), in order to establish cause/effect links.

For a group to be a candidate for the reduced assemblage approach, it must be capable of representing the response of the community as a whole. Therefore, it must be a prominent group comprising a large proportion of the fauna; it must contain many ecologically different species; and individual species within the group must have a broad range of tolerances for different types of pollution. The groups which have been most successfully exploited are the oligochaetes, chironomids and caddisflies.

Oligochaetes have mainly been used in lakes because of their prominence in soft sediment communities. They have been used both for trophic classification (e.g. Saether 1979, for nearctic and palaearctic lakes; Howmiller & Scott 1977 and Lauritsen *et al* 1985, for the Laurentian Great Lakes) and for indicating different types and degrees of pollution (e.g. Lang & Lang-Dobler 1979, for eutrophication and heavy metal pollution in Lake Geneva). Saether (1979) felt that oligochaete communities could not provide as distinct a classification system as chironomid communities because their environmental requirements are less restricted. However, Chapman *et al* (1982a,b) have since demonstrated that oligochaete species have a broad range of tolerances to organic pollution and specific chemicals. Recently, there have been several studies on oligochaetes in lotic environments. Smith *et al* (1990) observed changes in species composition and dominance in oligochaete assemblages in low-order woodland streams in the Adirondack Mountains of New York, in response to the degree of acid pollution. Barton and Metcalfe-Smith (1991) found that variations in oligochaete densities in both the soft sediments and riffles in an agriculturally-polluted watershed in Quebec gave excellent site discrimination, were highly correlated with other indices and were temporally stable. It appears that oligochaetes deserve further consideration for river and stream applications. They may be particularly suitable for assessing sediment quality in large rivers. As Barton (1989) points out, a drawback to using oligochaetes is that they can only be identified if they are sexually mature, whereas chironomids, for example, can be identified at any stage of maturity after the first instar.

Chironomids are an extremely diverse group of insects, frequently accounting for 50% of the total species diversity in benthic communities (Merritt & Cummins 1984). They occur in a wide range of freshwater habitats, have representatives in all trophic groups (predators, herbivores, detritivores) and are important food items for fish (Rosenbert *et al* 1986). They have been successfully used in lake classification (e.g. Saether 1979; Johnson 1989, for Swedish Lakes), although Johnson (1989) commented that many genera and species have wide tolerance ranges and are therefore poor indicators of lake type. He recommended identifying and focusing on indicator species. In a study on the Scioto River basin in Ohio, Rae (1989) found that several groups of common chironomid genera were indicative of certain chemical conditions. *Stictochironomus* was an indicator of hard, clean, unpolluted water, while *Pentaneura*, *Cricotopus* and *Tanytarsus* were characteristic of sewage pollution (phosphates, low oxygen), *Ablabesmyia* and *Tribelos* were associated with soft water and general organic pollution, and *Procladius* and *Dicrotendipes* indicated moderate hardness and high agricultural runoff (nitrates and turbidity due to fertilizers and siltation, respectively). Thus, Rae (1989) succeeded in isolating indicator taxa which were extremely tolerant or intolerant of certain types of pollution. He makes a strong case for focusing on indicator taxa rather than studying the entire benthic community, by pointing out that little information is gained by examining the distributions of facultative organisms.

Trichoptera are also a very diverse group of insects, occupying a wide variety of habitats and trophic levels. Caddisflies are also well represented in all functional feeding groups (Cummins & Klug 1979). Basaguren and Orive (1990) investigated Trichoptera as indicators of water quality in the River Cadagua basin in Spain. They identified 33 taxa from 12 families within the basin and observed a succession of species from the headwaters to the lowland reaches in relation to selected physico-chemical features. This succession was particularly evident in the family Hydropsychidae among species of the genus *Hydropsyche*. In organically-polluted river sections, species substitutions occurred which deviated from those in unpolluted sections with the same habitat.

In an earlier study on a nearby pristine river, the River Lea, Basaguren and Orive (1989) had identified 47 taxa from 14 families and 32 genera and described, using ordination techniques, zones characterized by different communities of caddisflies. The headwaters of the main river were distinct from those of the tributaries and a downstream sequence was indicated in the main river. The main river had a more diverse fauna, and diversity increased in a downstream direction with increasing river width, water level and substrate diversity. Although this is a characteristic pattern for unpolluted rivers, it is rarely observed today due to the counter-effects of cumulative pollution.

Basaguren and Orive (1989) worked largely at the species level, and were able to describe a very detailed continuum of caddisflies in this system which should be extremely useful for assessments of similar systems in the Basque country. Higler and Tolkamp (1983) also found the

Hydropsychidae to be useful for classifying watercourses in The Netherlands. Based on historical data, the distributions of 10 native species in small streams to large rivers throughout The Netherlands were determined. A succession was again identified, and deviations from the expected pattern for a given area could be attributed to pollution sources. Although several species of Hydropsychidae were found to be tolerant of organic pollution, even these species were eliminated by severe agricultural runoff.

Ratio indices

Ratio indices express the dominance of one group of organisms over another and have been used in conjunction with both taxonomic and functional feeding group data. They are simple to determine, but can be fairly specific and therefore useful in certain situations. Thus Saether (1979) found that an increasing ratio of oligochaetes to chironomids was a good indicator of eutrophication in lakes. Winner *et al* (1980) demonstrated that the ratio of chironomids to total insects was related to the degree of metal contamination (Cu, Cr, Zn).

Several ratio indices have been incorporated into the US EPA's Rapid Bioassessment Protocols for river assessment (Plafkin *et al* 1989), which will be discussed in more detail in the next section. For example, a decrease in the ratio of scrapers to filtering collectors is used to indicate degradation from an abundance of diatoms (which are the primary food source for scrapers) to an abundance of filamentous algae and aquatic mosses (which provide attachment sites for filterers and accumulations of the fine particulate matter on which they feed). Also, communities having an even distribution of organisms among four key insect groups – the mayflies, stoneflies, caddisflies and chironomids – are considered healthy, while those skewed towards a disproportionate number of chironomids indicate environmental stress.

Whitehurst and Lindsey (1990) compared the performance of a *Gammarus*:*Asellus* ratio index with traditional biotic indices (Chandler's Score, Extended Biotic Index, BMWP Score) for assessing mild sewage pollution in the River Adur in Sussex, England. *Gammarus* is more sensitive to organic pollution than *Asellus*. Since these taxa are direct niche competitors, a reduction in the abundance of *Gammarus* due to pollution will result in a corresponding increase in the abundance of *Asellus*, thereby altering the ratio. Whitehurst and Lindsey (1990) found this ratio index to be more sensitive to mild organic pollution than the biotic indices. The reason for this may be that if an impact is subtle and affects only certain organisms in the community, then the appropriate ratio index should provide good site discrimination. However, a biotic index could mask this response because it includes information on other organisms which are not affected by the impact. In a similar study using a variety of diversity, biotic and ratio indices, Olive *et al* (1988) found that different indices did not always agree in the classification of sites, usually due to the confounding effects of habitat. They caution against using indices of community response interchangeably or relying on only one type of index for assessing the effects of pollution, and conclude that ratio indices may be most useful when combined with other indices.

8.4 RECENT DEVELOPMENTS

The purpose of biomonitoring and assessment is to distinguish degraded sites from healthy ones, identify the cause of the impact, then monitor the response of the system to remedial action. The situation is rarely simple. Rather, multiple stresses, which originate from both point and non-point sources and can be chemical or physical in nature, generally act in a cumulative manner on aquatic systems. Traditional approaches to bioassessment have been unsatisfactory for a variety of reasons. Diversity and community comparison indices respond to all types of perturbations and the normal range of environmental factors, so that except in clear-cut situations they cannot be easily interpreted. On the other hand, biotic and saprobic indices only respond to organic pollution and are geographically restricted.

Probably the major obstacle to incorporating bioassessments into water management policies has been the lack of realistic targets with which to compare index values and to serve as water quality criteria. There has been a general recognition on both sides of the Atlantic of the need for identifying and characterizing reference sites in unimpaired locations in order to define attainable water quality objectives. The UK and the USA have addressed this problem differently, but both countries have succeeded in incorporating macroinvertebrate community assessments into the water resource management process.

The multivariate approach – UK

In the early 1980s, the Freshwater Biological Association (FBA) began to explore the relationships between environmental parameters and macroinvertebrate communities using multivariate analysis techniques. The goal of this 'River Communities Project' was to prepare a biological classification of all running-waters in Great Britain. The work was conducted in three phases, the results of which have appeared in a series of papers published over the last 10 years. A good overview is presented by Wright *et al* (1989).

Armitage *et al* (1983) examined the possibility of predicting 'expected' communities from physical and chemical variables unrelated to pollution, using data collected from 268 sites on 41 unpolluted rivers. Multiple linear regressions were computed using BMWP Score or ASPT as the dependent variable and various physical and chemical parameters as the independent variables. The predictive equations for ASPT were superior; as previously noted, ASPT is less sensitive to sampling effort and seasonal change than is the Score. Approximately 70% of the variability was explained using both physical and chemical data, and 60% by physical data alone.

Wright *et al* (1984), in perhaps the 'keystone' paper of the series, then used the same data set to develop a classification of running-water sites based on all macroinvertebrate taxa (not just BMWP Scores) and to predict community type from environmental data. Using ordination (detrended correspondence analysis) and a hierarchical clustering technique called TWINSPAN (two-way indicator species analysis), the sites were classified into 16 groupings using species lists generated from three seasons of sampling. Multiple discriminant analysis was then used to correlate the groupings with 28 physical and chemical variables (Table 8.5). Using environmental data, 76.1% of the sites were predicted to the correct grouping. For a further 15.3% of the sites, the correct grouping was the second most probable.

The influence of season and level of taxonomic identification on the performance of this system was tested (Furse *et al* 1984). Qualitative species-

Table 8.5 Various environmental factors considered in the ordination and classification of running-water sites in Great Britain, and the prediction of community type (after Wright *et al* 1984, and others)

Physical variables	Chemical variables
Distance from source*	pH
Slope*	Dissolved oxygen
Altitude*	Total oxidized nitrogen*
Discharge	Chloride*
Mean channel width*	Dissolved orthophosphate
Mean channel depth*	Total alkalinity*
Surface velocity (max., min., median/mode)	
Mean substratum size*	
Dominant particle size (max., min., median/mode)	
Substratum heterogeneity	
% macrophyte cover (max., min., mean)	
Air temp. range*	
Mean air temp.*	

* Used in latest version of RIVPACS (Wright *et al* 1989; see text).

level data were found to give more reliable classifications and predictions than either quantitative or qualitative family-level data, because of the greater number of taxa and because individual species have more precise environmental requirements than families. The greatest accuracy was achieved by combining the results from all three seasons, because species which were absent from one season's data due to life cycle, drought, etc., had a good chance of being captured in another season.

Although the prediction of site groupings is useful for classification, it is only a step towards the prediction of species occurrence at sites with know environmental characteristics (Wright *et al* 1985). Thus Moss *et al* (1987) conducted field trials to test the accuracy of classification and prediction of 21 new unpolluted sites and to determine the probability that a certain species would occur at a given site. Reducing the number of environmental variables was shown to result in very little loss of predictive accuracy. For example, of all taxa predicted as having a >75% chance of occurring at a given site, using suites ranging from five physical features to 28 physical and chemical features, 87.0–89.7% actually did occur.

Moss *et al* (1987) stated that the major use of their system would be to provide a 'target' community to be used as a standard for a given site when it is unpolluted. The magnitude of the difference between the expected and observed fauna gives a measure of the loss of biological quality due to pollution or other perturbations. Subsequently, RIVPACS (River InVertebrate Prediction and Classification System) software was developed (Wright *et al* 1989) to analyse combined three-seasons data obtained by standard procedures (Furse *et al* 1981; Wright *et al* 1984). The program permits the prediction of fauna at one or all of the following taxonomic levels: species (qualitative), all families (logarithm of categories of abundance), all families (qualitative), and the more restricted listing of BMWP families (qualitative) using various combinations of environmental variables. The system also allows for determination of the BMWP score or ASPT, which is widely used for rapid site appraisal.

The predictive equations were revised after the database was expanded to 370 sites on 61 rivers, including more small streams and lowland river sites.

The latest (Phase III) version of RIVPACS is based on 438 sites from 80 rivers, and permits a classification to either 10 or 25 TWINSPAN groups. Each group contains at least six reference sites, ensuring reliability of the system. It is more flexible in that data from one, several or all three seasons can be imported, and predictions of ASPT, BMWP score and number of scoring taxa can be obtained. However, for the purpose of

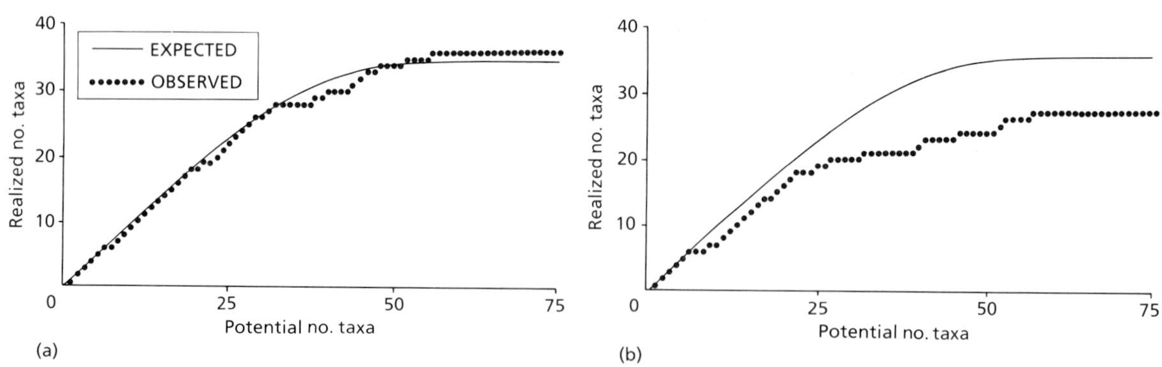

Fig. 8.2 Observed and expected BMWP family accretion curves for an upstream control site (a) and an impact site (b) which receives paper-mill effluent. 'Expected' curve accumulates the probabilities of occurrence of 75 BMWP families using the sequence given in the prediction. 'Observed' curve accumulates the number of families captured, based on the same sequence of taxa (from Wright *et al* 1988, with permission).

River : LOURO Site : 5 EIDOS

Environmental data used :

Substratum composition

Boulders+cobbles (%)	0.00
Pebbles+gravel (%)	50.00
Sand (%)	30.00
Silt+clay (%)	20.00
Mean substratum (phi)	0.57
Dist. from source (km)	18.70
Total oxidised N (ppm)	1.51
Alkalinity (ppm CaCO3)	31.42
Chloride (ppm)	28.95

Groups predicted from MDA with 5 physical and chemical variables

42 42.2%	23 28.5%	41 8.8%	44 5.1%	37 5.0%
38 3.3%	40 3.1%			

Predicted taxa, in decreasing order of probability

99.7% Oligochaeta √	48.9% Psychomyiidae
99.7% Chironomidae √	48.2% Goeridae
99.7% Elminthidae √	27.2% Planorbidae √
99.3% Baetidae	24.9% Brachycentridae
99.2% Simuliidae	23.9% Aphelocheiridae
98.7% Ephemerellidae √	22.2% Coenagriidae
96.9% Tipulidae	21.5% Odontoceridae
96.7% Leuctridae	21.1% Perlidae
95.4% Sphaeriidae √	18.0% Cordulegasteridae √
94.2% Gammaridae	17.1% Gerridae
93.9% Hydropsychidae	12.9% Beraeidae
93.9% Limnephilidae	12.3% Helodidae
93.1% Perlodidae	11.9% Piscicolidae
90.4% Dytiscidae	11.3% Dryopidae
90.1% Polycentropodidae √	8.4% Physidae √
89.8% Heptageniidae	7.3% Neritidae
89.8% Hydrobiidae √	6.2% Capniidae
87.7% Erpobdellidae √	5.9% Phryganeidae
87.6% Rhyacophilidae	5.7% Hydrometridae
86.2% Leptoceridae	4.7% Nepidae
86.2% Nemouridae	4.7% Siphlonuridae
85.6% Sericostomatidae	3.7% Dendrocoelidae
84.8% Lymnaeidae	3.6% Philopotamidae
84.6% Gyrinidae	3.4% Valvatidae
80.7% Ancylidae	2.1% Astacidae
78.8% Hydrophilidae	1.1% Molannidae
78.4% Caenidae	0.7% Unionidae
77.4% Glossiphoniidae √	0.4% Hirudidae
76.7% Ephemeridae	0.3% Viviparidae
75.3% Leptophlebiidae √	0.2% Notonectidae
73.3% Hydroptilidae	0.0% Platycnemididae
73.0% Planariidae	0.0% Aeshnidae
72.3% Chloroperlidae	0.0% Libellulidae
65.0% Sialidae	0.0% Corophiidae
62.6% Lepidostomatidae	0.0% Pleidae
58.6% Corixidae √	
56.7% Asellidae	
56.5% Haliplidae	
56.3% Taeniopterygidae	
51.9% Agriidae	

A) Predicted number of families to 50% = 33.2
B) Observed number of families to 50% = 11
C) Total observed families = 14
D) Observed BMWP score = 66
E) Observed ASPT = 4.714
F) Faunal index 11/33.2
G) BMWP score index 66/234
H) ASPT index 4.714/5.827

Expected taxa with P >=.75 = 26.9. with P >= .5 = 33.2
Predicted BMWP score = 234
Predicted average BMWP score per taxon = 5.827

Fig. 8.3 Example of a printout showing predicted families with their probability of capture for site 5-Eidos, on the Rio Louro. Taxa actually observed at the site are indicated with a √ (from Armitage *et al* 1990, with permission).

standardization, a single group of 11 physical and chemical variables is used (Table 8.5). Testing has now begun in polluted systems to demonstrate the utility of the prediction technique. Wright *et al* (1988) compared the observed vs. expected occurrence of 75 BMWP families at sites above and below the input of a paper-mill effluent on a river in southwest England. As illustrated in Fig. 8.2, the under-representation of families at the impact site compared to the control site is clearly evident.

The application of the BMWP score system and RIVPACS to the biological assessment of Spanish rivers is also currently being evaluated. Armitage *et al* (1990) applied the Phase II version of RIVPACS to the assessment of 18 sites in two Galician rivers in northwest Spain, one receiving minor organic pollution and one unpolluted. An example printout is shown in Fig. 8.3. At this particular site, Plecoptera were conspicuously absent and this was attributed to low dissolved oxygen (DO) and an unsuitable substratum. Although this study indicated that RIVPACS could be successfully applied in Spain, the most efficient application of the system will require the acquisition of a national or regional database similar to that available in Great Britain.

EPA approach — USA

The quality and quantity of water resources in the USA continue to decline despite massive regulatory efforts (Karr 1991). For example, 602 stream and river segments in the northwest USA are water-quality limited due to chemical contamination, and 56% of degraded segments nationwide have reduced fishery potential due to pollution. Hunsaker and Carpenter (1990) state that 27% of the fish fauna of the USA are endangered, threatened or of special concern, over 50% of the mollusc species of the Tennessee River system are either extinct or endangered, and 38 states reported fisheries closures, restrictions or advisories in 1985. Serious and widespread biological impairment is apparent, clearly indicating that existing monitoring and assessment programmes, which are based on chemical criteria, are inadequate.

Under the most recent amendment (1987) of the Clean Water Act, the US EPA is required to develop programmes to 'evaluate, restore and maintain the chemical, physical and biological integrity of the Nation's waters' (US EPA 1990). The inclusion of biological integrity as a goal on equal footing with chemical water quality is a major step forward. For a review of the arduous process leading up to this legislation, see Karr (1991). Biological criteria based on structural and functional attributes of resident fish and benthic invertebrate communities are currently being developed and integrated with chemical-specific criteria and whole-effluent toxicity evaluations for a more holistic approach to water management.

Under Section 303 of the Act, individual states will be required to develop narrative biological criteria by 1993, with numerical criteria to follow. Biocriteria will be used to improve water quality standards, identify impairment of beneficial uses, assist in setting programme priorities and detect problems which might otherwise be missed or underestimated. Biological criteria are especially suited for the detection of non-point source, cumulative and episodic pollution, as well as physical changes such as habitat deterioration, none of which would be detected using traditional chemical assessment methods. It has also been demonstrated that biocriteria are more sensitive than chemical criteria; in Ohio, an evaluation of instream biota indicated that 36% of biologically impaired sites had not been identified using chemical criteria (Ohio EPA 1987).

Twenty states now use some form of biological assessment and five (Florida, Arkansas, North Carolina, Maine, Ohio) are currently using biological criteria to define aquatic life use classifications and enforce water-quality standards (US EPA 1990). The application of biocriteria differs somewhat among the various states. Florida has a legal criterion based on macroinvertebrate diversity whereby impact is defined as a reduction of more than 75% below established reference values. Maine uses macroinvertebrate community data to assess attainment or non-attainment of standards for designated water uses (e.g. drinking-water supply, fish and wildlife habitat, recreational use, agriculture, industrial use). Oklahoma uses biotic information to assess long-term trends in water quality.

Ohio's programme is the furthest advanced,

and serves as an example of the successful incorporation of biological criteria in water quality regulations. Biological criteria were developed for Ohio rivers and streams using the 'ecoregion' approach, which has been adopted nationwide as a framework for defining attainable water quality objectives. An ecoregion is 'a relatively homogeneous area defined by similarity of geography, hydrology, land use, or other ecologically relevant variable' (US EPA 1990). Sites within a given ecoregion have similar ecological potentials, and 'attainable quality is then based on assessment of conditions in minimally impacted reference sites that characterize the region' (Hughes & Larsen 1988). In Ohio, parts of five ecoregions occur, these being the Eastern Corn Belt Plains, the Huron/Erie Lake Plain, the Erie/Ontario Lake Plain, the Western Allegheny Plateau and the Interior Plateau (Ohio EPA 1987).

The Ohio EPA has used biological criteria based on fish and macroinvertebrates for quantitatively determining attainment/non-attainment of designated aquatic life uses (warmwater habitat, exceptional warmwater habitat, coldwater habitat, seasonal salmonid habitat and limited resource waters) since 1980. For invertebrates, early criteria were narrative and based on simple structural measures of diversity, abundance and biomass. These have recently been replaced by the Invertebrate Community Index (ICI).

The ICI is derived from the IBI (Index of Biotic Integrity), which is based on attributes of fish communities and is described in detail in Karr (1991). The ICI consists of 10 essentially structural community metrics (Table 8.6), each with four scoring categories of 6, 4, 2 and 0 points which correspond to exceptional, good, fair and poor community condition. The scoring categories were calibrated using data from 232 reference sites in the five ecoregions. Individual metric scores are summed to generate a site ICI score which could theoretically range from 0 to 60. Quantitative sampling is conducted using multiple-plate artificial substrate samplers. However, additional qualitative sampling of the natural fauna from all available habitat types is also conducted. Metrics 1–9 are based on artificial substrate samples, while metric 10 is based on the natural fauna. Attainable values for the ICI were determined for each of the five ecoregions based on median values for the reference sites in these regions. Within each ecoregion, the lower (25%) and upper (75%) percentile values are used to determine attainment or non-attainment of criteria for warmwater habitat and exceptional warmwater habitat, respectively.

Table 8.6 Metrics for calculating the Invertebrate Community Index (ICI) (after Ohio EPA 1987)

1	Total number of taxa
2	Total number of mayfly taxa
3	Total number of caddisfly taxa
4	Total number of dipteran taxa
5	Percent mayfly composition
6	Percent caddisfly composition
7	Percent tribe Tanytarsini midge composition
8	Percent other dipteran and non-insect composition
9	Percent of tolerant organisms (oligochaetes and selected midges, limpets and pond snails)
10	Total number of qualitative EPT (Ephemeroptera, Plecoptera, Trichoptera) taxa

To determine the performance of the ICI, index values were determined for 431 sites sampled between 1981 and 1984, including 279 which had been classified as 'good', 76 'fair' and 76 'poor' quality sites based on best professional judgement. The results indicated a wide segregation of the good and excellent sites from the fair sites and the poor sites. It was concluded that the ICI provides an 'objective, quantifiable, and standardized means of evaluating biological integrity' (Ohio EPA 1987).

The US EPA has also supported the development of a hierarchy of methods for biological monitoring which is referred to as the Rapid Bioassessment Protocols (Plafkin *et al* 1989). The document presents a tiered approach to using fish and macroinvertebrate communities in biological assessment. Three protocols are present for invertebrates, each one progressively more intensive and rigorous than the previous one. Protocol III is similar to the ICI, but includes some functional metrics (Table 8.7). It was intended for riffle/run habitats in wadable streams. The original protocols were designed as inexpensive screening tools for determining if a stream is supporting or not supporting a designated aquatic life use. However, they were also found to be useful for discovering and determining the cause and severity

Table 8.7 Rapid Bioassessment Protocol III (after Plafkin *et al* 1989)

1	Taxon richness
2	Hilsenhoff's Family Biotic Index
3	Ratio of scrapers to filtering collectors
4	Ratio of EPT and chironomid abundances
5	Percentage contribution of dominant family
6	EPT index
7	Community similarity index
8	Ratio of shredders: total organisms

EPT = Ephemeroptera/Plecoptera/Trichoptera.

of impairment, evaluating the effectiveness of control actions, determining attainability of aquatic life uses and trend monitoring.

8.5 NEW DIRECTIONS

Because biological systems are complex and the causes of degradation in rivers and streams are multiple, not all measures of community structure or function will be useful under all circumstances. Instead of rejecting the more specific approaches because they cannot be generalized, Karr (1991) favours integration of the various indices and metrics to create a more robust approach to biological monitoring and assessment. What is needed most now is information on the differential responses of benthic communities to the wide range of polluted conditions and physical perturbations which occur. To date, only the responses to organic pollution (mainly domestic sewage) have been adequately documented.

It is unlikely that indices developed for assessing organic pollution will be useful for detecting toxic pollution or habitat alterations, because organisms do not respond in the same way to different stresses. For example, Chapman *et al* (1982a) exposed nine species of freshwater oligochaetes to selected chemical pollutants under bioassay conditions. They observed that eutrophic species were, as expected, the most tolerant of sewage and anoxia. However, when specific pulp-mill constituents, heavy metals and pentachlorophenol were tested, the relative tolerances of the various species were found to be pollutant-specific. This suggests that assemblages used to indicate organic enrichment are not appropriate for indicating other types of pollution. Similarly, Slooff (1983) compared the relative tolerances of 12 invertebrates from various taxonomic groups to 15 chemicals as well as surface-water concentrates from the Rhine River in The Netherlands. His most interesting finding was that organisms considered to be intolerant of organic pollution in general were sometimes very tolerant to specific toxicants, and vice versa.

Although much of the information is scattered at present, it may soon be possible to describe macroinvertebrate assemblages which are characteristic of broad categories of impacts, such as agricultural activities, heavy metals, acidic pollution and river regulation. Acidic pollution will be discussed as an example.

Several independent studies have collectively provided a profile of the characteristic responses of benthic communities to acid pollution. As previously noted (see Section 8.2), communities affected by toxic or acidic pollution are typically depauperate in comparison with those influenced by organic pollution. Reash *et al* (1988) described the benthos of a stream draining a coal strip-mining area (pH 4.5) in southern Ohio as characterized by oligochaetes, Odonata (*Pantala*, *Ishcura*), Heteroptera (*Trichocorixa*, *Notonecta*), the alderfly *Sialis*, the diving beetle *Laccophilus*, the mosquito *Anopheles* and chironomids, all in relatively low numbers. *Sialis* and the odonates were absent from the most impacted site. Tomkiewicz and Dunson (1977) investigated a tributary of the West Branch of the Susquehanna River in Pennsylvania, which was also polluted by acid strip-mine drainage. The normal pH of the system was 6.0, that of the acid feeder stream was 3.2, and that of the stream below the feeder was 4.7, recovering to 5.0. Only a chironomid, *Sialis* and the caddisfly *Ptilostomis* inhabited the acid feeder. Populations of Coleoptera, Ephemeroptera and Trichoptera showed little or no recovery as the acid pollution abated over three downstream sites. Heptageniid mayflies, and baetid mayflies, and pteronarcyid and peltoperlid stoneflies, which were present at the control site, did not re-occur. However, Diptera and leutrid and nemourid stoneflies showed a decided recovery at pH 5.0.

Peterson and van Eeckhaute (1990) determined

the distributions of 30 stonefly and 100 caddisfly species in relation to stream acidity in the Maritime provinces of Canada. Stonefly taxa most sensitive to acidic conditions were primarily the large Perlidae, but also the Chloroperlidae, Pteronarcidae and Perlodidae. Least sensitive were the Nemouridae, Leutridae and Capniidae. The Trichoptera were more difficult to categorize because several families contained both sensitive and tolerant species. The Hydropsychidae in particular had representative species in all categories (sensitive, tolerant, ubiquitous), confirming other reports of the utility of this group for pollution assessment (e.g. Higler & Tolkamp 1983; Basaguren & Orive 1990). Like Tomkiewicz and Dunson (1977), Peterson and van Eeckhaute (1990) also found the presence of *Ptilostomis* to be indicative of acid pollution.

While such descriptive studies are informative, a more precise and quantitative method for defining characteristic communities is needed. Smith *et al* (1990) employed multiple regression models to determine the response of benthic communities to acidity in low-order woodland streams in the Adirondack Mountains in New York. The findings of this study were generally in agreement with those described earlier, but the data analysis technique permitted a more precise definition of the environmental requirements of the indicator species. For example, *Cynigmula* (Heptageniidae), *Baetis* (Baetidae), Elmidae (Coleoptera) and Perlidae occurred at pH 6.2, but not at pH 5.8; mayfly densities increased steadily from pH 4.9 to pH 6.2; *Ephemerella* was excluded at pH 4.9, but not at pH 5.3; nemourid and leutrid stoneflies were found at sites ranging from pH 4.9 to 6.2. It is worth noting that the indicator value of the mayfly *Baetis* with respect to acidity is exactly the opposite of its value as an indicator of organic pollution. This genus features largely in almost all biotic indices due to its extreme tolerance of organic pollution, yet it is very sensitive to low pH. It is abundantly clear that the application of traditional biotic indices to acidic pollution would give completely erroneous assessments.

Reynoldson and Zarull (1989) demonstrated that combined analysis of physical, chemical and biological data could be used to link cause and effect between sediment-associated contaminants and benthic communities in a study on the heavily industrialized Detroit River in Canada. Using clustering techniques applied to benthic community data, five distinct benthic communities were defined, mainly based on densities of tubificids, mayflies and amphipods. These five groups were correlated individually against 18 chemical and physical characteristics of the sediment in which they live. Site groups were found to be highly correlated with Hg, Cr, Zn, Ni, hexachlorobenzene and phosphorus and also significantly correlated with five other factors, including concentrations of Pb, Cd and Cu. Multiple discriminant analysis was then used to determine the ability of these physico-chemical factors to predict benthic community structure. When all 11 environmental variables were included, every site was assigned to its correct group. Using this integrated strategy, Reynoldson and Zarull (1989) were able to associate concentrations of contaminants with specific levels of biological impact. The main variables responsible for degradation of the system were identified and could be singled out for remedial action. The method also defined a site-specific target (the cleanest group) to be used as an objective for cleanup.

This approach would be particularly suitable for large lowland rivers with significant accumulations of fine sediment, where sediment toxicity is probably more significant and limiting than water quality.

Rutt *et al* (1990) developed a similar system for assessing acid streams in upland Britain. Multiple discriminant analysis revealed that stream pH, aluminium concentration and total hardness were the major variables discriminating among groups of streams with different benthic faunas. They felt that their model could be used to forecast the effects of management decisions concerning acidifying influences such as sulphur emissions and conifer forestry.

The incorporation of pollution variables into multivariate analyses to determine the cause of community change could lead to a useful development of RIVPACS. At present, RIVPACS can predict the expected community at a given location in the absence of pollution and provide a measure of the *degree* of impact (i.e. deviation from normal). However, it cannot determine the *type* of impact.

More studies on relatively simple systems with well-known specific impacts are needed in order to define the characteristic responses of benthic communities to different types of pollution and habitat alterations. These community profiles or 'fingerprints' could then be included in models such as RIVPACS for comparison with observed communities, in order to identify the probable cause of deviation from normal conditions. The information could also be used to develop new biotic indices which respond to polluted conditions other than enrichment and for the development of new specific metrics for inclusion in the ICI or RBPs (rapid bioassessment protocols). Furthermore, characteristic community profiles could be used to identify the 'worst offender' in multiple stress situations.

It will probably not be feasible to develop biotic indices suitable for large-scale application. This task would require the generation of extensive databases on species occurrences and tolerances, such as those prepared by Sladecek (1973) and Hilsenhoff (1987), which would still be geographically restricted in their application. However, specialized biotic indices could be very useful for managing systems at the local level. For example, Rabeni *et al* (1985) developed a biotic index which was specific to the Penobscot River in Maine. Tolerance values were assigned by direct observation of the species occurring in the system, and best achievable communities were defined as those occurring at the cleanest sites. As they state, the index was 'specific for the watershed, and specific in response to pulp and paper mill effluents and municipal sewage, and therefore sensitive to improvements in water quality.'

It is perhaps too early to suggest improvements to the ICI and RBPs, since they (particularly the latter) have not been extensively tested. However, Karr (1991) maintains that the goals for the future should be to develop suites of metrics that integrate taxa (fish, invertebrates, diatoms, microorganisms) and consider biological responses to stress at all levels of organization (individual, population, community, ecosystem).

Many different approaches to bioassessment have been presented and compared in this chapter. Because the most recent systems have not yet been directly compared, it is difficult to determine whether sophisticated multivariate techniques or suites of simple, yet specific, metrics are the way of the future for large-scale monitoring and assessment applications.

REFERENCES

AFNOR (Association Française de Normalisation). (1985) Essais des eaux. Determination de l'indice biologique global (IBG). AFNOR T90-350, Paris. [8.2]

Armitage PD. (1980) The effects of mine drainage and organic enrichment on benthos in the River Nent system, Northern Pennines. *Hydrobiologia* **74**: 119–28. [8.2]

Armitage PD, Moss D, Wright JF *et al.* (1983) The performance of a new biological water quality score system based on macroinvertebrates over a wide range of polluted running-water sites. *Water Research* **17**: 333–47. [8.2, 8.4]

Armitage PD, Pardo I, Furse MT *et al.* (1990) Assessment and prediction of biological quality. A demonstration of a British macroinvertebrate-based method in two Spanish rivers. *Limnetica* **6**: 147–56. [8.2, 8.4]

Barton DR. (1989) Some problems affecting the assessment of Great Lakes water quality using benthic invertebrates. *Journal of Great Lakes Research* **15**: 611–22. [8.3]

Barton DR, Metcalfe-Smith JL. (1991) A comparison of sampling techniques and summary indices for assessment of water quality in the Yamaska River, Quebec, based on benthic macroinvertebrates. *Environmental Monitoring and Assessment* **21**: 225–44. [8.3]

Basaguren A, Orive E. (1989) Spatio-temporal changes in the caddisfly (Trichoptera) communities of the river Lea basin (Basque Country, North Spain). *Annals of Limnology* **25**: 61–8. [8.3]

Basaguren A, Orive E. (1990) The relationship between water quality and caddisfly assemblage structure in fast-running rivers. The River Cadagua basin. *Environmental Monitoring and Assessment* **15**: 35–48. [8.3, 8.5]

Bervoets L, Bruylants B, Marquet P *et al.* (1989) A proposal for modification of the Belgian biotic index method. *Hydrobiologia* **179**: 223–8. [8.2]

Brock DA. (1977) Comparison of community similarity indexes. *Journal of the Water Pollution Control Federation* **49**: 2488–94. [8.2]

Cairns J Jr, Pratt JR. (1986) Developing a sampling strategy. In: Isom BG (ed) *Rationale for Sampling and Interpretation of Ecological Data in the Assessment of Freshwater Ecosystems*, pp. 168–86. American Society for Testing and Materials, Philadelphia. [8.2]

Camargo JA. (1990) Performance of a new ecotoxicological index to assess environmental impacts on freshwater communities. *Bulletin of Environmental Contamination and Toxicology* **44**: 529–34. [8.2]

Chapman PM, Farrell MA, Brinkhurst RO. (1982a) Relative tolerances of selected aquatic oligochaetes to individual pollutants and environmental factors. *Aquatic Toxicology* **2**: 47–67. [8.3, 8.5]

Chapman PM, Farrell MA, Brinkhurst RO. (1982b) Relative tolerances of selected aquatic oligochaetes to combinations of pollutants and environmental factors. *Aquatic Toxicology* **2**: 69–78. [8.3]

Chutter FM. (1972) An empirical biotic index of the quality of water in South African streams and rivers. *Water Research* **6**: 19–30. [8.2]

Cook SEK. (1976) Quest for an index of community structure sensitive to water pollution. *Environmental Pollution* **11**: 269–88. [8.1, 8.2]

Cummins KW. (1974) Structure and function of stream ecosystems. *Bioscience* **24**: 631–41. [8.3]

Cummins KW. (1988) The study of stream ecosystems: A functional view. In: Pomeroy LR, Alberts JJ (eds) *Concepts of Ecosystem Ecology*, pp. 247–62. Springer-Verlag, New York. [8.3]

Cummins KW, Klug MJ. (1979) Feeding ecology of stream invertebrates. *Annual Review of Ecology & Systematics* **10**: 147–72. [8.3]

Cushing CE, McIntire CD, Cummins KW *et al.* (1983) Relationships among chemical, physical, and biological indices along river continua based on multivariate analyses. *Archiv für Hydrobiologie* **98**: 317–26. [8.3]

De Pauw N, Roels D. (1988) Relationship between biological and chemical indicators of surface water quality. *Verhandlungen Internationale Vereinigung für theoretische und angewandte Limnologie* **23**: 1553–8. [8.2]

De Pauw N, Vanhooren G. (1983) Method for biological quality assessment of watercourses in Belgium. *Hydrobiologia* **100**: 153–68. [8.1, 8.2]

De Pauw N, Roels D, Fontoura AP. (1986) Use of artificial substrates for standardized sampling of macroinvertebrates in the assessment of water quality by the Belgian Biotic Index. *Hydrobiologia* **133**: 237–58. [8.2]

France RL. (1990) Theoretical framework for developing and operationalizing an index of zoobenthos community integrity: Application to biomonitoring with zoobenthos communities in the Great Lakes. In: Edwards CJ, Regier HA (eds) *An Ecosystem Approach to the Integrity of the Great Lakes in Turbulent Times*, pp. 169–93. Great Lakes Fishery Commission Special Publication, Ann Arbor, MI. [8.1, 8.2]

Furse MT, Wright JF, Armitage PD *et al.* (1981) An appraisal of pond-net samples for biological monitoring of lotic macro-invertebrates. *Water Research* **15**: 679–89. [8.4]

Furse MT, Moss D, Wright JF *et al.* (1984) The influence of seasonal and taxonomic factors on the ordination and classification of running water-sites in Great Britain and on the prediction of their macroinvertebrate communities. *Freshwater Biology* **14**: 257–80. [8.3, 8.4]

Ghetti PF, Bonazzi G. (1977) A comparison between various criteria for the interpretation of biological data in the analysis of the quality of running water. *Water Research* **11**: 819–31. [8.2]

Hawkins CP, Sedell JR. (1981) Longitudinal and seasonal changes in functional organization of macroinvertebrate communities in four Oregon streams. *Ecology* **62**: 387–97. [8.3]

Hellawell J. (1977) Biological surveillance and water quality monitoring. In: Alabaster JS (ed) *Biological Monitoring of Inland Fisheries*, pp. 69–88. Applied Science Publishers, London. [8.1]

Hellawell JM. (1986) *Biological Indicators of Freshwater Pollution and Environmental Management.* Elsevier, London. [8.2]

Herricks EE, Cairns J Jr. (1982) Biological monitoring part III – receiving system methodology based on community structure. *Water Research* **16**: 141–53. [8.2]

Higler LWG, Tolkamp HH. (1983) Hydropsychidae as bioindicators. *Environmental Monitoring Assessment* **3**: 331–41. [8.3, 8.5]

Hilsenhoff WL. (1987) An improved biotic index of organic stream pollution. *The Great Lakes Entomologist* **20**: 31–9. [8.2, 8.5]

Hilsenhoff WL. (1988a) Rapid field assessment of organic pollution with a family-level biotic index. *Journal of the North American Benthological Society* **7**: 65–8. [8.2, 8.3]

Hilsenhoff WL. (1988b) Seasonal correction factors for the biotic index. *The Great Lakes Entomologist* **21**: 9–13. [8.2]

Howmiller RP, Scott MA. (1977) An environmental index based on relative abundance of oligochaete species. *Journal of the Water Pollution Control Federation* **49**: 809–15. [8.1, 8.2, 8.3]

Hughes RM, Larsen DP. (1988) Ecoregions: An approach to surface water protection. *Journal of the Water Pollution Control Federation* **60**: 486–93. [8.4]

Hunsaker CT, Carpenter DE (eds). (1990) *Ecological Indicators for the Environmental Monitoring and Assessment Program.* United States Environmental Protection Agency, EPA 600/3-90/060, Office of Research and Development, Research Triangle Park, N.C. [8.4]

ISO (International Organization for Standardization). (1984) Assessment of the water and habitat quality of rivers by a macroinvertebrate 'score'. *ISO/TC* 147/SC 5/WG 6 N 40. British Standards Institution, London. [8.2]

Jhingran VG, Ahmad SH, Singh AK. (1989) Application of Shannon–Wiener Index as a measure of pollution of river Ganga at Patna, Bihar, India. *Current Science* **58**: 717–20. [8.2]

Johnson RK. (1989) Classification of profundal chironomid communities in oligotrophic/humic lakes of Sweden using environmental data. *Acta Biologica Debrecina Oecologia Hungarica* **3**: 167–75. [8.3]

Jones JR, Tracy BH, Sebaugh JL *et al.* (1981) Biotic index tested for ability to assess water quality of Missouri Ozark streams. *Transactions of the American Fisheries Society* **110**: 627–37. [8.2]

Karr JR. (1991) Biological integrity: A long-neglected aspect of water resource management. *Ecological Applications* **1**: 66–84. [8.1, 8.4, 8.5]

Klemm DJ, Lewis PA, Fulk F *et al.* (1990) *Macroinvertebrate Field and Laboratory Methods for Evaluating the Biological Integrity of Surface Waters.* United States Environmental Protection Agency, EPA/600/4-90/030, Washington, DC. [8.1]

Kovalak WP. (1981) Assessment and prediction of impacts of effluents on communities of benthic stream macroinvertebrates. In: Bates JM, Weber CI (eds) *Ecological Assessments of Effluent Impacts on Communities of Indigenous Aquatic Organisms*, pp. 255–63. American Society for Testing and Materials, Philadelphia. [8.2]

Lang C, Lang-Dobler B. (1979) The chemical environment of tubificid and lumbriculid worms according to the pollution level of the sediment. *Hydrobiologia* **65**: 273–82. [8.3]

Lauritsen DD, Mozley SC, White DS. (1985) Distribution of oligochaetes in Lake Michigan and comments on their use as indices of pollution. *Journal of Great Lakes Research* **11**: 67–76. [8.3]

Marchant R, Metzeling L, Graesser A *et al.* (1985) The organization of macroinvertebrate communities in the major tributaries of the LaTrobe River, Victoria, Australia. *Freshwater Biology* **15**: 315–31. [8.3]

Mason WT Jr, Lewis PA, Weber CI. (1985) An evaluation of benthic macroinvertebrate biomass methodology. *Environmental Monitoring & Assessment* **5**: 399–422. [8.2]

Merritt RW, Cummins KW. (1984) *An Introduction to the Aquatic Insects of North America*, 2nd edn. Kendall/Hunt Publishing, Dubuque, Iowa. [8.3]

Metcalfe JL. (1989) Biological water quality assessment of running waters based on macroinvertebrate communities: History and present status in Europe. *Environmental Pollution* **60**: 101–39. [8.1, 8.2]

Minshall GW, Cummins KW, Petersen RC *et al.* (1985) Developments in stream ecosystem theory. *Canadian Journal of Fisheries and Aquatic Sciences* **42**: 1045–55. [8.3]

Moss D, Furse MT, Wright JF *et al.* (1987) The prediction of the macro-invertebrate fauna of unpolluted running-water sites in Great Britain using environmental data. *Freshwater Biology* **17**: 41–52. [8.4]

Murphy PM. (1978) The temporal variability in biotic indices. *Environmental Pollution* **17**: 227–36. [8.2]

Ohio Environmental Protection Agency. (1987) *Biological Criteria for the Protection of Aquatic Life: Vol. I. The Role of Biological Data in Water Quality Assessment.* Division of Water Quality Monitoring and Assessment, Surface Water Section, Columbus, Ohio. [8.4]

Olive JH, Jackson JL, Bass J *et al.* (1988) Benthic macroinvertebrates as indexes of water quality in the upper Cuyahoga River. *The Ohio Journal of Science* **88**: 91–8. [8.3]

Ormerod SJ, Edwards RW. (1987) The ordination and classification of macroinvertebrate assemblages in the catchment of *Freshwater Biology* **17**: 533–46. [8.3]

Pantle R, Buck H. (1955) Die biologische Überwachung der Gewässer und die Darstellung der Ergebnisse. *Gas Wasserfach* **96**: 604. [8.2]

Perkins JL. (1983) Bioassay evaluation of diversity and community comparison indexes. *Journal of the Water Pollution Control Federation* **55**: 522–30. [8.2]

Persoone G, De Pauw N. (1979) Systems of biological indicators for water quality assessment. In: Ravera O (ed) *Biological Aspects of Freshwater Pollution*, pp. 39–75. Pergamon Press, Oxford. [8.2]

Peterson RH, van Eeckhaute L. (1990) Distributions of stonefly (Plecoptera) and caddisfly (Trichoptera) species in three stream systems in New Brunswick and Nova Scotia, Canada, with reference to stream acidity. Canadian Technical Report of Fisheries & Aquatic Sciences No. 1720. [8.5]

Pinder LCV, Farr IS. (1987a) Biological surveillance of water quality – 2. Temporal and spatial variation in the macroinvertebrate fauna of the River Frome, a Dorset chalk stream. *Archiv für Hydrobiologie* **109**: 321–31. [8.2]

Pinder LCV, Farr IS. (1987b) Biological surveillance of water quality – 3. The influence of organic enrichment on the macroinvertebrate fauna of small chalk streams. *Archiv für Hydrobiologie* **109**: 619–37. [8.2]

Pinder LCV, Ladle M, Gledhill T *et al.* (1987) Biological surveillance of water quality – 1. A comparison of macroinvertebrate surveillance methods in relation to assessment of water quality, in a chalk stream. *Archiv für Hydrobiologie* **109**: 207–26. [8.2]

Plafkin JL, Barbour MT, Porter KD *et al.* (1989) *Rapid Bioassessment Protocols for Use in Streams and Rivers: Benthic Macroinvertebrate and Fish.* United States Environmental Protection Agency, EPA/444/4-89-001, Washington, DC. [8.1, 8.3, 8.4]

Pratt JM, Coler RA. (1976) A procedure for the routine biological evaluation of urban runoff in small rivers. *Water Research* **10**: 1019–25. [8.1]

Rabeni CF, Davies SP, Gibbs KE. (1985) Benthic invertebrate response to pollution abatement: Structural changes and functional implications. *Water Resources Bulletin* **21**: 489–97. [8.3, 8.5]

Rae JG. (1989) Chironomid midges as indicators of organic pollution in the Scioto River Basin, Ohio. *The Ohio Journal of Science* **89**: 5–9. [8.3]

Reash RJ, Van Hassel JH, Wood KV. (1988) Ecology of a southern Ohio stream receiving fly ash pond discharge: Changes from acid mine drainage conditions. *Archives of Environmental Contamination and Toxicology* **17**: 543–54. [8.5]

Reynoldson TB. (1984) The utility of benthic invertebrates in water quality monitoring. *Water Quality Bulletin* **10**: 21–8. [8.1]

Reynoldson TB, Zarull MA. (1989) The biological assessment of contaminated sediments – the Detroit River example. *Hydrobiologia* **188/189**: 463–76. [8.5]

Rosenberg DM, Danks HV, Lehmkuhl DM. (1986) Importance of insects in environmental impact assessment. *Environmental Management* **10**: 773–83. [8.3]

Rutt GP, Weatherley NS, Ormerod SJ. (1990) Relationships between the physicochemistry and macroinvertebrates of British upland streams: the development of modelling and indicator systems for predicting fauna and detecting acidity. *Freshwater Biology* **24**: 463–80. [8.5]

Saether OA. (1979) Chironomid communities as water quality indicators. *Holarctic Ecology* **2**: 65–74. [8.3]

Sladecek V. (1973) System of water Quality from the biological point of view. *Archiv für Hydrobiologie Beiheft Ergebnisse der Limnologie* **7**: 1–218. [8.2, 8.5]

Sladecek V. (1979) Continental systems for the assessment of river water quality. In: James A, Evison L (eds) *Biological Indicators of Water Quality*, pp. 3-1 to 3-32. John Wiley & Sons, Chichester. [8.2]

Slooff W. (1983) Benthic macroinvertebrates and water quality assessment: Some toxicological considerations. *Aquatic Toxicology* **4**: 73–82. [8.2, 8.3, 8.5]

Smith ME, Wyskowski BJ, Brooks CM *et al.* (1990) Relationships between acidity and benthic invertebrates of low-order woodland streams in the Adirondack Mountains, New York. *Canadian Journal of Fisheries and Aquatic Sciences* **47**: 1318–29. [8.3, 8.5]

Tolkamp HH. (1985a) Using several indices for biological assessment of water quality in running water. *Verhandlungen Internationale Vereinigung für theoretische und angewandte Limnologie* **22**: 2281–6. [8.2]

Tolkamp HH. (1985b) Biological assessment of water quality in running water using macroinvertebrates: A case study for Limburg, The Netherlands. *Water Science and Technology* **17**: 867–78. [8.2]

Tomkiewicz SM, Dunson WA. (1977) Aquatic insect diversity and biomass in a stream marginally polluted by acid strip mine drainage. *Water Research* **11**: 397–402. [8.5]

US EPA (United States Environmental Protection Agency). (1990) *Biological Criteria: National Program Guidance for Surface Waters*. United States Environmental Protection Agency, EPA-440/5-90-004. [8.4]

Vandelannoote A, De Gueldre G, Bruylants B. (1981) Ecological assessment of water quality: Comparison of biological–ecological procedures in a rain-fed lowland waterway (Kleine Nete, N. Belgium). *Hydrobiological Bulletin* **15**: 161–4. [8.2]

Vannote RL, Minshall GW, Cummins KW *et al.* (1980) The river continuum concept. *Canadian Journal of Fisheries and Aquatic Sciences* **37**: 130–7. [8.3]

Voshell JR Jr, Layton RJ, Hiner SW. (1989) Field techniques for determining the effects of toxic substances on benthic macroinvertebrates in rocky-bottomed streams. *Aquatic Toxicology and Hazard Assessment* **12**: 134–55. [8.1]

Washington HG. (1984) Diversity, biotic and similarity indices. A review with special relevance to aquatic ecosystems. *Water Research* **18**: 653–94. [8.2]

Webber CE, Bayne DR, Seesock WC. (1989) Macroinvertebrate communities in Wheeler Reservoir (Alabama) tributaries after prolonged exposure to DDT contamination. *Hydrobiologia* **183**: 141–55. [8.2]

Whitehurst IT, Lindsey BI. (1990) The impact of organic enrichment on the benthic macroinvertebrate communities of a lowland river. *Water Research* **24**: 625–30. [8.3]

Wilhm JL. (1967) Comparison of some diversity indices applied to populations of benthic macroinvertebrates in a stream receiving organic waste. *Journal of the Water Pollution Control Federation* **39**: 1673–83. [8.2]

Wilhm JL, Dorris TC. (1968) Biological parameters for water quality criteria. *Bio Science* **18**: 477–81. [8.2]

Winner RW, Boesel MW, Farrell MP. (1980) Insect community structure as an index of heavy-metal pollution in lotic ecosystems. *Canadian Journal of Fisheries and Aquatic Sciences* **37**: 647–55. [8.3]

Winterbourn MJ, Rounick JS, Cowie B. (1981) Are New Zealand stream ecosystems really different? *New Zealand Journal of Marine & Freshwater Research* **15**: 321–8. [8.3]

Woodiwiss FS. (1980) *Biological Monitoring of Surface Water Quality. Summary Report*. Commission of the European Communities, Environment and Consumer Protection Service, No. ENV/223/74-EN, Nottingham. [8.2]

Wright JF, Moss D, Armitage PD *et al.* (1984) A preliminary classification of running-water sites in Great Britain based on macro-invertebrate species and the prediction of community type using environmental data. *Freshwater Biology* **14**: 221–56. [8.4]

Wright JF, Armitage PD, Furse MT *et al.* (1985) The classification and prediction of macroinvertebrate communities in British rivers. *Annual Report of the Freshwater Biological Association* **53**: 80–93. [8.4]

Wright JF, Armitage PD, Furse MT *et al.* (1988) A new approach to the biological surveillance of river quality using macroinvertebrates. *Verhandlungen Internationale Vereinigung für Theoretische und angewandte Limnologie* **23**: 1548–52. [8.4]

Wright JF, Armitage PD, Furse MT *et al.* (1989) Prediction of invertebrate communities using stream measurements. *Regulated Rivers: Research and Management* **4**: 147–55. [8.4]

Zelinka M, Marvan P. (1961) Zur prazisierung der biologischen klassifikation der reinheit fliessender gewasser. *Archiv für Hydrobiologie* **57**: 389–407. [8.2]

Part 3: Modelling: Forecasting and Prediction

Improvement of our ability to detect, interpret and predict environmental change is a fundamental objective of scientific research. Part 2 was concerned with establishing a sound basis for monitoring. This part examines the application of mathematical models with regard to environmental impact assessment and to providing a rationale for management decisions. There are a wide range of modelling tools to investigate human impacts dealing with a range of spatial and temporal scales. Although comprehensive ecosystem models remain an objective for the future, considerable progress has been made in understanding and quantifying individual components, especially the driving forces. Chapter 9 investigates the ways in which the impacts of climatic change on catchment hydrology can be estimated and Chapter 10 examines the application of models for catchment simulation, groundwater simulation and flow routing. The modelling of water quality is reviewed in Chapter 11 and the problems of modelling sediment transport and channel stability are addressed in Chapter 12. Part 3 ends with two chapters that consider the prediction of biological responses to changes of hydrology, water quality and channel morphology. Chapter 13 presents a general overview of the problems and approaches in predicting biological responses and Chapter 14 discusses the evolution of instream flow habitat modelling. Throughout this part of the *Handbook* general principles rather than specific details are emphasized and examples are given of the application of the most widely used types of model.

9: Hydrology and Climate Change

N. W. ARNELL

9.1 INTRODUCTION

The catchment manager has always had to deal with a changing physical and social environment. The pressures on rivers and catchments have changed over time, the required standards for environmental protection have (generally) been rising and catchment management has increasingly been seen as a multi-disciplinary activity involving hydrology, geomorphology, ecology and engineering. Since the late 1980s there has also been a need to deal with possible changes in those climatic inputs previously assumed to be constant.

The threat of global warming due to an increase in the concentration of so-called greenhouse gases has attracted widespread scientific, public and political interest. One of the most significant impacts of climate change is likely to be on the hydrological system. The objectives of this chapter are therefore (i) to summarize how climate changes might affect hydrological processes and river flow characteristics, and (ii) to review how the impacts on hydrological regimes, and therefore ecological, geomorphological and water resources systems, can be estimated in a particular catchment.

Table 9.1 summarizes some of the implications of climate change for a catchment. First, however, it is useful to review briefly the processes behind the enhanced greenhouse effect and the climate changes that might result.

The greenhouse effect

The basic principles behind the greenhouse effect are well established, and are reviewed in reports published by the United Nations Intergovernmental Panel on Climate Change (IPCC 1990, 1992). A few gases within the earth's atmosphere are transparent to incoming shortwave solar radiation, but they block a proportion of the outgoing longwave radiation radiated from the earth's surface so heating up the lower atmosphere. The most important of these radiatively-active gases are water vapour, carbon dioxide (CO_2) and methane, and without them the temperature at the earth surface would be around 33°C cooler than at present. The greenhouse effect is so called because the panes of glass in a greenhouse have a similar effect to the radiatively-active gases in the atmosphere (although in practice much of the effect of a real greenhouse arises from the prevention of the movement of air).

The concentrations of carbon dioxide and methane, however, have increased since the mid-19th century as industrial and agricultural development has expanded. New greenhouse gases, such as nitrous oxide and the man-made chlorofluorocarbons (CFCs), have been added and, molecule for molecule, these are more effective at blocking outgoing longwave radiation than CO_2 (although the 1992 IPCC report suggested that the radiative effects of CFCs were offset to a large extent by the effects of the depletion of stratospheric ozone caused by CFC emissions).

There are several major difficulties in estimating the impact that increasing concentrations of greenhouse gases might have on global and regional climates.

1 There are large uncertainties in forecast rates of emission of greenhouse gases, because they depend on rates of economic development and technological change. The IPCC considered a

Table 9.1 Some implications of changes in hydrological regimes induced by climate change

Water resources
Change in reliability of supply
Change in quality of supply
Change in risk of flooding
Change in potential for power generation
Ecology
Change in instream habitats
Change in seasonal inundation of riverine areas
Change in groundwater levels
Geomorphology
Change in sediment yields
Change in erosion and sedimentation

range of emissions scenarios in 1990 and 1992, reflecting different assumptions about economic and technological change as well as a number of policies for limiting future increases.

2 The emission of some greenhouse gases depends to an unknown degree on the actual changes in climate that might take place. Methane emissions from decomposable organic matter in high-latitude tundra, for example, might rise as increases in temperatures cause permafrost to thaw: such a positive feedback would exaggerate global warming.

3 The rates at which greenhouse gases are taken up by plants and oceans, and interact in the atmosphere, are currently not well understood.

4 The global and, particularly, regional climatic consequences of increasing concentrations of greenhouse gases are very uncertain. How, for example, will changes in the radiative properties of the atmosphere affect the global temperature? How would changes in cloud cover exacerbate or ameliorate increases in surface temperature? These, and other, questions are tackled with the aid of computer models of the climate system (IPCC 1990).

A consensus has evolved amongst atmospheric scientists that the effect of increasing greenhouse gas concentrations would be to increase global temperatures and change regional climates (IPCC 1990, 1992). The IPCC calculated that, under 'business-as-usual' emissions scenarios, increasing greenhouse gas concentrations would lead to a rise in global mean temperature of approximately 0.3°C per decade, with a range of 0.2 to 0.5°C/decade (IPCC 1990, 1992). This would mean that the global average temperature by 2050 would be around 1.8°C higher than at present, which would be considerably higher than at any time in human history. The rate of change too is unprecedentedly high, and may be higher than the rate at which ecosystems (and management systems) can respond.

Table 9.2 Uncertainty in estimating the implications of climate change (from Fiering & Matalas 1990)

	Degree of belief			
Proposition	Certain	Less certain	Speculative	Uncertain
1 CO_2 is increasing	XXXX	X		
2 Global mean temperature increases	XXX	XX	X	
3 Global magnitude of change in temperature	X	XXX	XX	
4 Regional magnitude of change in temperature		XXXX	XX	
5 Mean global change in precipitation	X	XX	XXX	
6 Magnitude and sign of change in global precipitation		XX	XXX	X
7 Magnitude and sign of change in regional precipitation			XXXX	XX
8 Hydrological change given climate scenario		X	XXXX	X
9 Societal adaption		X	XXX	XX
10 Adequacy of existing and projected resource systems		X		XXXX

The predicted consequences of climate change become even more uncertain if the focus switches from global mean temperature to regional impacts and measures of climate such as precipitation. Table 9.2 (Fiering & Matalas 1990) summarizes how uncertainty increases as questions about the future become increasingly demanding.

Detecting the greenhouse signal

Analysis of observational temperature records has shown an increase in global average temperature since 1900 of around 0.5°C (IPCC 1990). This does not, however, prove that the enhanced greenhouse effect is leading to global warming, for several reasons:

1 The evidence is circumstantial: there is no proven link between increasing greenhouse gas concentrations and rising temperatures, although a physical explanation has been proposed.

2 The variations in the observed temperature record could reflect 'simple' decade-to-decade variability about an unstable mean, or might be due to long-time-scale variability associated with changes in solar radiation receipts (although the changes in solar radiation over the last few hundred years are small compared with the radiative effects of increased concentrations of greenhouse gases; IPCC 1990).

3 The observed temperature record may contain some systematic bias. Records tend to be taken in or near urban areas, but Jones *et al* (1989) estimate a maximum bias of 0.1°C per 100 years due to urban warming.

The detection of greenhouse gas-induced global warming is made very difficult by the fact that any signal will be masked by very large year-to-year variability. The IPCC (1990) predicted that the unequivocal detection of global warming will not be possible for at least a decade.

It will be even harder to find evidence for global warming in hydrological records. Not only do river flow series show the effects of year-to-year climatic variability, but they may also be affected by changes in catchment vegetation and water resource developments. Such changes may in some cases overwhelm any signal from global warming.

9.2 CLIMATE CHANGE AND HYDROLOGICAL PROCESSES

The hydrological and climate systems are interdependent. Hydrological regimes are determined by the combination of climate inputs with catchment physiography, geology and vegetation: catchment vegetation is itself dynamically dependent on both climatic and hydrological conditions. Hydrological processes control the return of energy and moisture to the atmosphere.

The dependence of hydrological systems on climatic inputs means that a change in climate (and associated change in vegetation) will affect hydrological behaviour, and the influence of catchment characteristics means that impacts will vary between catchments. This chapter concentrates on the impact of climatic changes on hydrological processes: it must be remembered that changes in hydrological processes and regimes feed back into effects on climate.

Figure 9.1 attempts to summarize the impact of a change in greenhouse gas concentrations on the hydrological system, emphasizing the interrelationships between all the elements of the system (the Figure excludes management intervention). The increase in greenhouse gas concentrations results in an increase in net radiation at the earth surface. This leads to an increase in temperature, which results in changes in atmospheric moisture contents and circulation patterns. These in turn produce (regionally-variable) changes in rainfall and evaporation regimes, and hence soil moisture regimes. Temperature, rainfall, evaporation and soil moisture all affect plant growth and therefore the ecosystem, as do changes in radiation and the atmospheric CO_2 concentration. Changes in rainfall and evaporation (compounded by changes in vegetation) result in changes in groundwater recharge and river flow which, together with changes in temperature, impact upon stream water chemistry and biology. Finally, a rise in sea level will affect hydrological characteristics in lowlying areas. The rest of this chapter examines some of these changes in more detail.

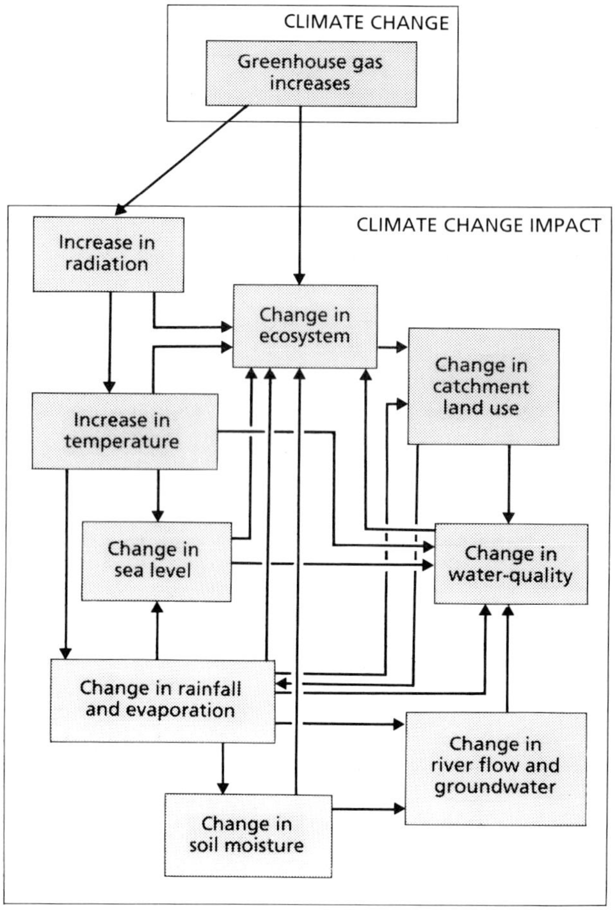

Fig. 9.1 The impact of climate change on the hydrological system.

Precipitation

Precipitation drives the hydrological system. Changes in its amount, intensity, duration and timing during the year will impact directly upon flow regimes and groundwater. The extent of impact depends, of course, on the degree of change, and different components of the hydrological regime are sensitive to different aspects of precipitation. Changes in low flows in a catchment, for example, might be most sensitive to changes in the intervals between rainfall events. In very responsive catchments (i.e. those with little storage), changes in short-duration rainfall intensities might have the greatest impact on flood characteristics, whilst in other less responsive catchments flood frequencies are more affected by changes in the rate of occurrence of prolonged wet periods. In many parts of the world summer flows are dependent on the melt of snow stored on the catchment over the winter season; flows are profoundly affected both by changes in the volume of snow stored and the rate at which water is released by melt (see Gleick 1987, for example).

The impact of a given change in rainfall depends also on the characteristics of the catchment. This is illustrated quantitatively in Fig. 9.2, which shows the effect of an increase in winter rainfall and a reduction in summer rainfall on average monthly flows in three UK catchments (Arnell *et al* 1990). The changes were simulated using a monthly water balance model, and although the assumed percentage change in

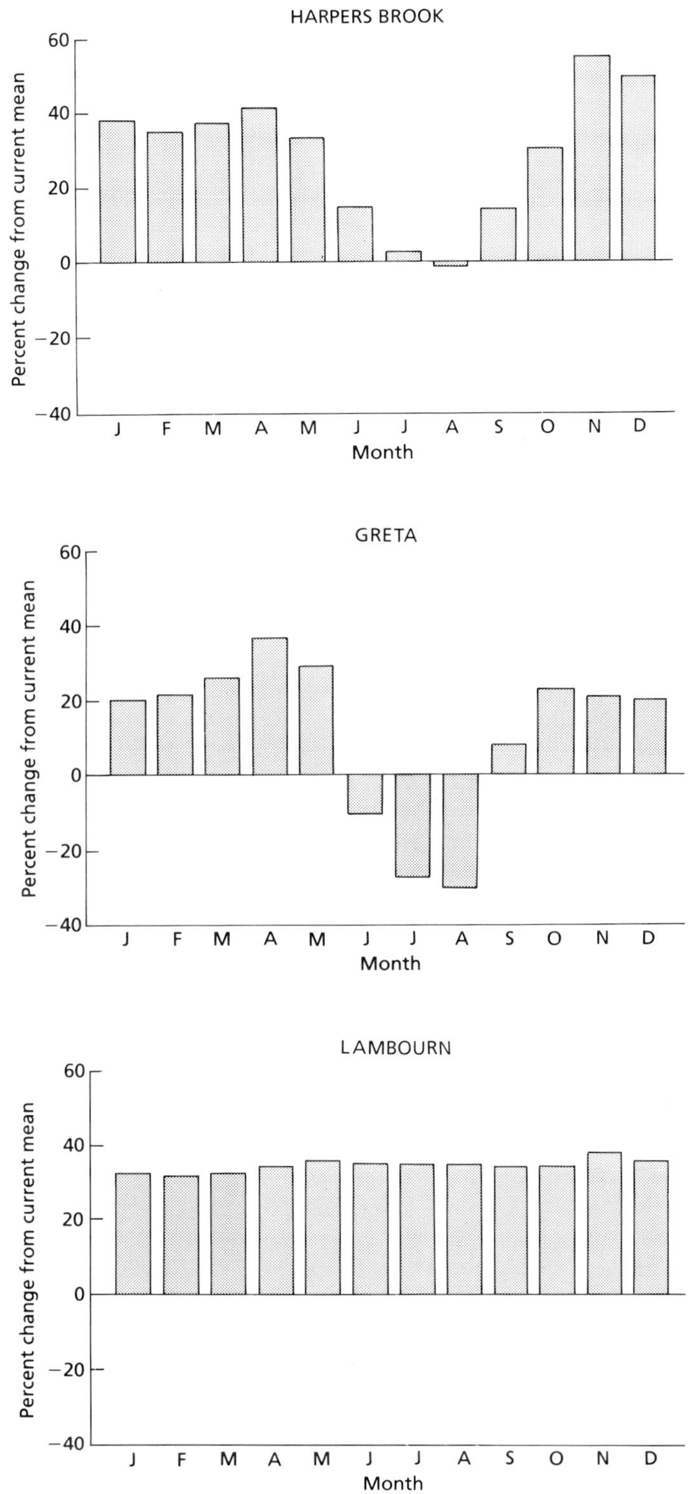

Fig. 9.2 The effect on mean monthly flow regimes in three UK catchments of increasing winter rainfall (by 20%) and reducing summer rainfall (by up to 15%). Potential evaporation is assumed unchanged (from Arnell *et al* 1990).

rainfall is the same in each catchment, differing catchment characteristics mean the effects of the change vary. The Greta is a small upland catchment (area 86 km^2) in Northumbria with a fine balance in summer between rainfall and evapotranspiration. A reduction in rainfall therefore leads to a large percentage reduction in the amount of water becoming streamflow. The Harpers Brook is an eastern lowland catchment (area 74 km^2), where rainfall in summer is insufficient to satisfy the demands of evaporation: little of the summer rainfall produces streamflow, so a reduction in rainfall in that season has little effect. Flows in the Lambourn (area 233 km^2) are dominated by a chalk aquifer, and during the drier summer are maintained by the slow drainage out of the aquifer of the extra winter groundwater recharge.

At the annual scale, the effect of a given percentage change in rainfall on runoff depends on the runoff coefficient, or the ratio of long-term average runoff to long-term average rainfall (Wigley & Jones 1985; Arnell 1992). The lower the proportion of average annual rainfall going to runoff, the more sensitive the runoff to a given change in rainfall. Studies in Britain have shown that, other things being equal, a 10% increase in average annual rainfall would result in an increase in average annual runoff of between 12 and 30% (with the highest increase in the driest regions: Arnell *et al* 1990). The greater the concentration of the additional rainfall in winter, the greater the effect on runoff.

Evapotranspiration

Potential evapotranspiration is defined as the evaporation and transpiration that would occur from an extensive short grass crop with an unlimited supply of water. If there is a shortage of water the *actual* evapotranspiration falls below the potential rate. Changes in actual evapotranspiration will therefore depend on both changes in potential evapotranspiration and changes in the availability of water through the year, which depend in turn on changes in precipitation inputs. It is possible that an increase in potential evapotranspiration at a site may be outweighed by such a large reduction in water availability that actual evapotranspiration reduces.

Potential evapotranspiration is controlled by inputs of net radiation, the ability of the air to hold moisture, the rate of renewal of air above an evaporating surface, and plant physiological properties. The plant properties of interest include aerodynamic resistance, which affects the rate of transfer of air across a leaf surface, and stomatal conductance, which controls the rate at

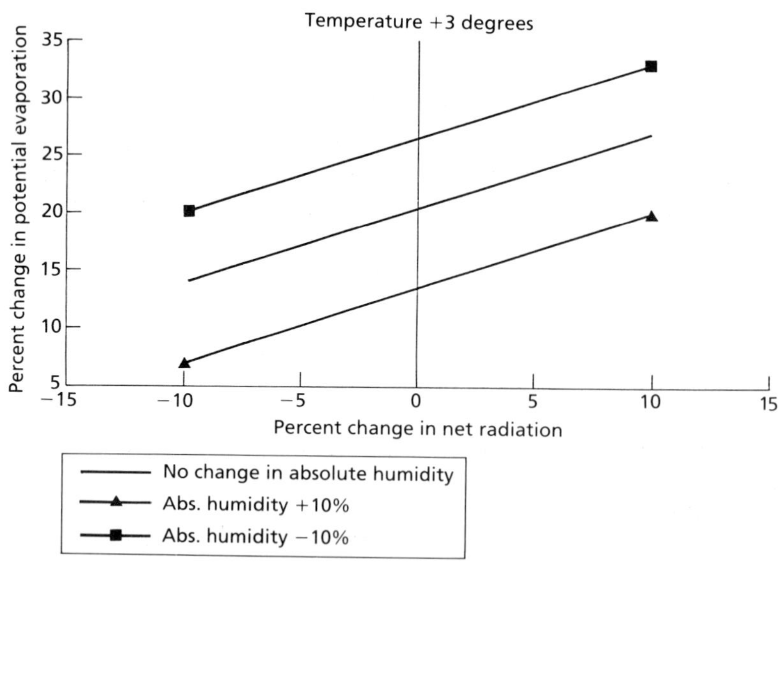

Fig. 9.3 The effect of changes in temperature, net radiation and absolute humidity on potential evapotranspiration in a wheat field in Nebraska over 10 summer days (data from Martin *et al* 1989).

which water can be transpired from the plant leaf. The leaf surface area determines the amount of rainfall that is intercepted and made available for evaporation. Climate change will affect catchment evaporation directly, through changes in hydrometeorological characteristics, and indirectly through changes in catchment vegetation.

The primary direct effect of climate change is likely to arise from the increase in the ability of air to hold moisture as air temperature rises, and Budyko (1980) estimated that potential evapotranspiration would increase by approximately 4% for every degree Celsius rise in temperature. Changes in net radiation, windspeed and plant properties, however, may serve to mitigate or exaggerate the effects of the temperature increase. Figure 9.3 shows the simulated effect of changes in net radiation and absolute humidity on the sensitivity of potential evapotranspiration from a Nebraska wheat field in summer to a 3°C increase in temperature (Martin *et al* 1989). The effect of a particular change in one of the controls on potential evapotranspiration tends to be dependent on the initial values of that and the other controls, so making generalizations difficult or impossible to express.

The characteristics and mix of catchment vegetation will change as rainfall, temperature and soil moisture regimes alter, and will also be affected by the concentration of atmospheric CO_2. Some plants grow much more vigorously in a high CO_2 atmosphere, but the effect depends on how the plant responds to the intake of CO_2. The C_3 plants, which include almost all UK plant species, are most sensitive to increased CO_2 concentrations, whilst C_4 plants (which include maize) are least sensitive (CCIRG 1991; Drake 1992). The different sensitivities among different plant types means that the mix of plants in an ecosystem and catchment may change. Increased temperatures and changed precipitation patterns can also be expected to alter the duration and timing of the plant growth cycle, and an earlier onset to the growing season could lead to significant increases in spring evapotranspiration (Allen *et al* 1991).

Experimental evidence shows that plant stomatal conductance reduces as CO_2 levels increase, and plant transpiration – and hence water use – therefore declines (Drake 1992). Kimball and Idso (1983), for example, concluded that evapotranspiration would be reduced by up to a third if the atmospheric CO_2 content was to double. However, the extra leaf area due to improved plant growth might compensate for the reduced stomatal conductance. Gifford (1988) conducted laboratory experiments that showed that plant water-use remained virtually constant as CO_2 concentrations increased, and inferred that plants tried to maintain some optimum water use. In the field, the physiological consequences of increased CO_2 concentrations might be limited by nutrient availability and changes in soil moisture regimes, and at the catchment scale may be less important than changes in the catchment vegetation mix.

Catchment land cover might also evolve in indirect response to climatic change, following the responses of farmers and policy-makers to real or perceived climate changes. Changes in the European Community's Common Agricultural Policy resulting from increased desertification in southern Europe, for example, might have implications for UK farmers and hence catchment land use. Land use change can, of course, also occur independently of climate change. Indeed in many places such change is the rule rather than the exception and its effect may be more rapid than any due to climate change; an example is catchment afforestation.

Runoff generation

Rainfall intensities in humid temperate environments rarely exceed the rate at which water can infiltrate into the soil (except in disturbed areas), and river runoff is generated by water flowing through the soil and by rainfall falling directly on saturated ground. Soil properties influence the rate at which rainfall infiltrates, the rate at which water moves through the soil, and the time over which soil becomes saturated.

Most pedological processes operate over a very long time-scale, but changes in organic matter content and soil structure may become apparent over a time-scale of less than 10 years (CCIRG 1991). Higher temperatures and increased rainfall would lead to a loss of soil organic matter and hence a decrease in the ability of the soil to hold moisture; higher temperatures would also encourage clayey soils to shrink and crack, thus assisting the passage of water into and through the soil profile (CCIRG 1991). Increased waterlogging could encourage the development of gleyed profiles, although such a change would occur over a longer time period. Changes in soil structure will probably only have a noticeable effect at the small catchment scale (unless the changes are particularly large following, for example, desertification).

Groundwater regimes

Groundwater regimes would of course be affected by changes in aquifer recharge due to alterations in precipitation characteristics, infiltration and evapotranspiration. Aquifers in most humid temperate environments are recharged seasonally. Recharge typically commences in autumn after soil moisture deficits have been removed, and ceases in spring when evapotranspiration rates reach such a level that soil moisture deficits begin to develop. Both the amount of winter rainfall and the length of the recharge season therefore influence the amount of recharge. A shorter recharge season (due to increased evapotranspiration) might offset the effect of any increase in winter rainfall.

Water in aquifers is held in fissures and pores of varying sizes. The smaller the pore, the slower it is to fill and the slower it is to drain. The recharge season might be so truncated that the smallest pores are not filled and groundwater levels therefore decline more rapidly in spring.

Finally, an increase in the proportion of winter rainfall that falls in high-intensity storm events could mean that more of the winter rainfall leaves the catchment as rapid streamflow, and therefore that less goes to recharge. The effects of climate change on groundwater recharge have so far been studied far less intensively than effects on streamflow.

It is difficult to conceive of climatically-induced changes in aquifer properties, except where saline intrusion due to sea-level rise alters local hydraulic gradients.

Water quality

Changes in climate can be expected to lead to changes in stream-water chemical and biological characteristics. Increased stream-water temperatures will affect the rate of operation of chemical and biological processes. A rise in temperature, for example, would increase the rate of biological activity (and de-oxidation) and the rate of reaeration, but because the effect on de-oxidation is greater the dissolved oxygen content in the watercourse would reduce (Jacoby 1990). Similarly, the rate of denitrification is more sensitive to temperature than the rate of nitrification so, other things being equal, higher temperatures would lead to lower nitrate concentrations (Jenkins 1991). However, in practice other things are not equal, and the effects of a change in temperature may be overwhelmed by changes in dilution, concentration and residence time associated with changes in river discharge. Lower discharges would tend to increase concentrations and lengthen residence times.

Changes in flow pathways due to alterations in

Table 9.3 A matrix of methods for estimating the hydrological impacts of climate change: most studies fall into the bottom left corner

	Change in climate inputs	Change in hydrological characteristics
Arbitrary change		
Temporal analogue		
Spatial analogue		
Use of climate model simulations•		

soil structure would change the amount of time water remains within the soil profile. Minerals could be released into soil water through sulphur and nitrate mineralization as soil structure changes (Jenkins 1991).

A change in climate might also be associated with changes in geochemical inputs to the catchment. Farmers may use more – or fewer – agricultural chemicals in response to climate change, or may change land use completely. Atmospheric deposition rates might also change as the frequency of occurrence of different air masses varies (Wilby 1989).

Sediment yields may also be susceptible to climate change, and particularly to changes in land cover and rainfall intensities and frequencies. Sediment supply in upland catchments is controlled by thresholds in slope stability and rainfall intensity, and the rate at which these thresholds are exceeded could change as climate evolves (Harvey 1991).

9.3 ESTIMATING THE IMPACTS OF CLIMATE CHANGE IN A PARTICULAR CATCHMENT

There are two main obstacles to determining quantitatively the implications of climate change for a specific catchment. Firstly, it is currently very difficult to estimate what the future climate will be in a specific catchment. Secondly, practically every aspect of the hydrological system is expected to be affected by climate change (Fig. 9.1), and the effects of change will be very dependent upon both the current characteristics of the catchment and the interactions between different parts of the system. It may therefore be very risky to transfer quantitative conclusions from another catchment.

Practical assessments of climate change impacts make use of climate change 'scenarios', which are defined to be feasible, internally consistent estimates of future climate. A climate change scenario is not to be confused with a forecast. A range of scenarios must be considered in a study, because it is valuable to investigate how different scenarios translate to differences in hydrological implications.

The four basic groups of methods of creating climate change scenarios are shown in Table 9.3. Each method can be used to estimate directly the implications of climate change for hydrological systems, or can be used to define climate inputs (such as precipitation, temperature and potential evapotranspiration) which are subsequently fed into a hydrological model. The vast majority of climate change impact studies fall into the bottom left corner in Table 9.3, although examples appear in the literature for each area.

This section first reviews briefly the four different methods of creating scenarios, and concludes by considering the types of hydrological model which are needed to convert climate inputs into hydrological outputs. For catchment management purposes, these hydrological outputs may need to be fed into ecological, geomorphological or water resource systems models: these models are not considered explicitly in this section (models to predict biological responses to change are reviewed in Chapters 13–15).

Arbitrary changes

The simplest way of determining the impact of possible climatic change is to define a number of arbitrary changes in rainfall and temperature (or potential evaporation), and estimate the implications for river flow regimes (e.g. Nemec & Schaake 1982). It is also possible to define arbitrary changes in flow regimes and determine the impacts on, for example, reservoir reliability. It is better to view such investigations as sensitivity studies rather than assessments of the implications of possible future climate change.

Temporal analogues

Climate change scenarios can be based on the transfer of data from some period in the past. Palaeoclimatic analogues use information from early warm epochs in the earth's history. Budyko (1989), for example, has proposed that the temperature and rainfall patterns pertaining during the Holocene climatic optimum around 5000 to 6000 BP provide an analogue for conditions in the year 2000. However, such palaeoclimatic anomalies were probably caused by variations in planetary orbit, and it is considered unlikely that increasing greenhouse gas concentrations will have the same climatic effect (IPCC 1990). It is also difficult to obtain hydrological information from a palaeoclimatic analogue, although geomorphological information may give insights into channel characteristics and sediment regimes.

Historical analogues based on past instrumental data (or at the least documentation) are more useful, but there are still several important problems. The period in the past with the highest temperatures locally may not coincide with the warmest period globally, and the difference between the warmest and coolest historical periods may be considerably less than the amount of warming expected due to an increase in greenhouse gas concentrations. The method is perhaps most valuable when information from individual extreme events, such as the 1975/76 drought in the UK, is used to provide insights into how hydrological, ecological, geomorphological and water resources systems might react to future climate anomalies.

Spatial analogues

Climate change scenarios can also be based on 'spatial analogues', which often appear in the popular climate change literature: it might be claimed, for example, that the climate of southern England will in the future be like that currently experienced in the Mediterranean region. However, the transfer of climate data from one region to another is only appropriate when local influences on climate are negligible. The climate of southern England will never be more than superficially similar to that of southern France because of differences in the influence of local and regional topography, the relationship between land, sea and ocean currents, and, also, the length of day. Spatial analogues might be more useful in the centre of a continent where large-scale controls on climate are dominant. Transfer of hydrological data from one area to another should not be attempted because of the importance of local catchment physical characteristics in controlling flow regime (Arnell *et al* 1990).

Use of climate model output

The final method of estimating the effects of climate change on hydrological systems uses the output from climate simulation models. Climate model output can be used in several ways. Firstly, it is possible to use directly the climate model estimates of change in river runoff and soil moisture contents (Miller & Russell 1992). Secondly, it is possible to use the climate model estimates of changes in rainfall, temperature and potential evaporation as inputs to an independent catchment hydrological model (e.g. Bultot *et al* 1988; Cohen 1991). Finally, it is possible to use climate model output to guide the development of 'generalized' estimates of possible changes in rainfall, temperature and potential evaporation (e.g. Arnell *et al* 1990; McCabe & Wolock 1992).

However, although global climate models are capable of simulating relatively accurately both global climate and general circulation patterns, they are less good at simulating regional climates (IPCC 1990). This is to a large extent due to their coarse spatial resolution: the highest resolution global climate model currently works on a grid approximately 250 × 350 km. Such a scale is far too coarse for regional hydrological investigations.

It is, therefore, unwise to take the regional output – whether of climatic characteristics such as rainfall and temperature or, especially, river runoff – from any one climate model too literally at present. Scenarios should be based on the results from several different models. As climate models improve, the potential for using climate model output directly will increase, although it is naive to expect that global climate model output will in the foreseeable future satisfy

Table 9.4 Climate change scenarios for 2050 (from CCIRG 1990)

Rainfall	Winter		
Summer	0	+8%	+16%
−16%	Driest		
0		'Best'	
+16%			Wettest

Temperature

Summer +2.1°C Winter +2.3–2.5°C

the needs of a detailed, localized water resources impact study.

Table 9.4 reproduces the set of climate change scenarios for the year 2050 used by the UK Climate Change Impacts Review Group (CCIRG 1991). The scenarios, based largely on the interpretation of climate model output, indicate that 'winter' is likely to be wetter in the UK than at present, whilst there is greater uncertainty over change in summer rainfall. The CCIRG did not estimate changes in potential evapotranspiration, but Budyko's (1980) figure of 4% per extra degree Celsius implies that annual potential evapotranspiration in the UK by the middle of the next century could be between 8 and 10% higher than at present.

Hydrological models for impact studies

Whilst a few climate change impact studies have used hydrological data from past climate anomalies to estimate directly the impacts of climate change (e.g. Palutikof 1987), the vast majority have applied catchment hydrological models with climate change scenarios derived largely from climate model output (e.g. Gleick 1987; Bultot *et al* 1988; Cohen 1991).

At one extreme, it would be possible to use a physically-based distributed hydrological model which represents all the processes involved and their interactions, and which has measurable parameters. Such a model requires very detailed input data, and its use in a catchment-scale impact assessment cannot really be justified at present given the highly generalized nature of current climate change scenarios. A physically-based model, however, can be used to determine the sensitivity of soil moisture regimes and river flows to particular changes in process or catchment structure.

At the other end of the scale are empirical models, ranging from simple relationships which estimate average annual runoff from annual climatic data to monthly flow simulation models whose parameters are estimated by multiple regression. The major disadvantage of empirical models is that model parameters are wholly dependent on current conditions, and that slight changes in either the raw data used to define the model or of model form can result in very different implied sensitivities to climate change (see Arnell *et al* 1990, for example).

Between the 'white box' physically-based models and the 'black-box' empirical models lie 'grey-box' conceptual models, which incorporate some representation of the hydrological processes operating in a catchment with a small number of parameters which must be calibrated for a specific catchment. Models used in climate change impacts studies range from two- or three-parameter monthly water balance models (Gleick 1987; Arnell *et al* 1990), to more complicated multiple-parameter daily models (Bultot *et al* 1988; Lettenmaier & Gan 1990). The key assumption is that the model calibration remains appropriate under a changed climate, but this assumption can be partially tested using data from past 'extreme' years.

The choice of a model for impact assessment should be based on the hydrological or water resources problem to be studied. If interest lies primarily in river flows, groundwater recharge or general patterns of stream-water quality, then a relatively simple conceptual model will be adequate. If, however, interest lies in a much more highly detailed, process-based problem, such as changes in soil moisture dynamics in an ecologically-important wetland, for example, then it will be necessary to adopt a physically-based model and make assumptions about possible changes in parameters and short-duration inputs.

Most climate change impact studies so far have run a hydrological model with an observed climatological time series representing current conditions, and then run the model again after perturbing the time series according to some climate change scenario (Gleick 1987; Bultot *et al* 1988; Arnell *et al* 1990). An alternative method would be to generate stochastic synthetic time series representing current and future conditions: the future time series are determined by perturbing the parameters of the stochastic model.

A stochastic approach has two main advantages. Firstly, it is possible to generate a record of any length, and the simulated time series will not necessarily be dominated by one particularly extreme event. The drought of 1975/76, for example, dominates rainfall records over the last 50 years across a large part of the UK. Secondly, it is possible to generate repeated samples and thus get some idea of the sampling variability surrounding a particular hydrological or water resources statistic as estimated under current and future conditions. It is essential to ensure, of course, that the stochastic generation model reproduces realistically the characteristics of current climatic time series, including the sum of departures from average over periods relevant to water resources assessment.

9.4 CONCLUSIONS: COPING WITH CLIMATE CHANGE

Climate drives the hydrological system, and changes in climate can be expected to have consequences for hydrology, ecology, geomorphology and water resources. The significance of these consequences, however, depends on the type and magnitude of climate change, the characteristics of the catchment and hydrological system, and the sensitivity to change of the water resources and ecological systems. The actual impact of change, expressed for example in monetary terms or square metres of lost habitat, may not be very closely related to the percentage change in climate. The more resilient the aquatic or water resources system, the greater the climate change it will be able to absorb for a given degree of impact; a fragile system – where demand for water is very close to supply, for example – might be very profoundly affected by only a very small change in climatic inputs. Meanwhile, of course, catchments, habitats and water resource systems are subject to other evolving influences which might be indirectly related to or totally independent of climate change.

So what should the catchment manager do about the possible implications of an evolving climate change? The uncertainty currently surrounding climate change means that it is not appropriate to initiate at present any changes to operational routines. It is also inappropriate to throw away past data. Scheme design should continue to be based on the analysis of all the data available, using currently accepted procedures. However, it is prudent to consider the implications of climate change when designing new schemes, and it is essential to consider climate change when engaged in long-term planning. Such an assessment would ideally contain the following stages:

1 Define a range of feasible climate change scenarios from inspection of climate model output (or, better, adopt scenarios developed by other 'responsible' organizations, such as the UK Climate Change Impacts Review Group: CCIRG 1991).

2 Run the climate change scenarios through an 'appropriate' hydrological model (bearing in mind the limitations of regression-type models and the extra data requirements of physically-based models).

3 Consider the implications of the scenarios by, for example, running the hydrological output from the previous stage through water resources or ecological models. How large are the differences between the scenarios? How do the changes compare with historical experience? Would one scenario have particularly catastrophic consequences? The scheme or strategy may also be sensitive to factors other than climate change, including for example economic assumptions.

4 Consider whether the scheme or strategy can be designed to minimize its sensitivity to climate change. What would be the implications of designing for the 'worst-case'? Is it possible to design the scheme so that it is not seriously adversely affected by possible changes in climate? Robustness to an uncertain future can be a criterion for selecting amongst alternatives.

5 Investigate the impact on the hydrological, ecological or water resource system of past extreme events.

There are, in conclusion, three main implications of climate change for the catchment manager. Firstly, it is no longer feasible to assume that climate will be constant over a planning period: this adds another dimension to the already complicated long-term planning process. Secondly, simulating the potential impacts of climate change requires realistic models of the processes operating within a catchment. These models must represent not just hydrological and water resources systems, but also the geomorphological and ecological systems. The threat of climate change is therefore yet another stimulus to multi-disciplinary model development. Thirdly, the uncertain nature of future climate change favours the adoption of flexible schemes and plans that are either robust to changing conditions or are easily adapted. In this sense too, the threat of climate change supports current developments in flexible, adaptable catchment management.

ACKNOWLEDGEMENTS

The author gratefully acknowledges the helpful comments of Max Beran, Frank Law, Alan Jenkins and Nick Reynard, all from the Institute of Hydrology.

REFERENCES

Allen RG, Gichuki FN, Rosenzweig C. (1991) CO_2-induced climatic changes in irrigation-water requirements. *Journal of Water Resources Planning and Management* **117**: 157–78. [9.2]

Arnell NW. (1992) Factors controlling catchment response to climate change in a humid temperate environment. *Journal of Hydrology* **132**: 321–42. [9.2]

Arnell NW, Brown RPC, Reynard NS. (1990) *Impacts of Climate Variability and Change on River Flow Regimes in the UK*. Institute of Hydrology Report 107. Wallingford, Oxon. [9.2, 9.3]

Budyko M. (1980) *Climate in the Past and the Future*. Gidrometeoizdat Publishing House, Leningrad. [9.2, 9.3]

Budyko M. (1989) Climatic conditions of the future. In: *Conference on Climate and Water, Helsinki, September 1989*, vol. 1, pp. 9–30. World Meteorological Society, Publications of the Academy of Finland. [9.3]

Bultot F, Coppens A, Dupriez GL, Gellens D, Meulenberghs F. (1988) Repercussions of a CO_2-doubling on the water cycle and on the water balance – a case study for Belgium. *Journal of Hydrology* **99**: 319–47. [9.3]

CCIRG (Climate Change Impacts Review Group). (1991) *The Potential Effects of Climate Change in the United Kingdom*. CCIRG for the Department of the Environment. HMSO, London. [9.2, 9.3, 9.4]

Cohen SJ. (1991) Possible impacts of climatic warming scenarios on water resources in the Saskatchewan River Sub-Basin, Canada. *Climatic Change* **19**: 291–318. [9.3]

Drake BG. (1992) The impact of rising CO_2 on ecosystem production. *Water, Air and Soil Pollution* **64**: 25–44. [9.2]

Fiering MB, Matalas NC. (1990) Decision-making under uncertainty. In: Waggoner PE (ed) *Climate Change and U.S. Water Resources*, pp. 75–84. John Wiley, New York. [9.1]

Gifford RM. (1988) Direct effect of higher carbon dioxide concentrations on vegetation. In: Pearman GL (ed) *Greenhouse: Planning for Climatic Change*, pp. 506–19. CSIRO, Melbourne. [9.2]

Gleick PH. (1987) Regional hydrologic consequences of increases in atmospheric CO_2 and other trace gases. *Climatic Change* **10**: 137–61. [9.2, 9.3]

Harvey AM. (1991) The influence of sediment supply on the channel morphology of upland streams: Howgill Fells, North West England. *Earth Surface Processes and Landforms* **16**: 675–84. [9.2]

IPCC (Intergovernmental Panel on Climate Change). (1990) *Climate Change. The IPCC Scientific Assessment*: Houghton JT, Jenkins GJ, Ephraums JJ (eds). Cambridge University Press, Cambridge. [9.1, 9.3]

IPCC (Intergovernmental Panel on Climate Change). (1992) *Climate Change 1992. The Supplementary Report to the IPCC Scientific Assessment*: Houghton JT, Callander BA, Varney SK (eds). Cambridge University Press, Cambridge. [9.1]

Jacoby HD. (1990) Water quality. In: Waggoner PE (ed) *Climate Change and U.S. Water Resources*, pp. 307–28. John Wiley, New York. [9.2]

Jenkins A. (1991) Climate change and water resources. In: Battarbee RW, Patrick ST (eds) *The Greenhouse Effect: Consequences for Britain*. Environmental Change Research Centre, University College London. [9.2]

Jones PD, Kelly PM, Goodess GB, Karl TR. (1989) The effect of urban warming on the Northern Hemisphere temperature average. *Journal of Climatology* **2**: 285–90. [9.1]

Kimball BA, Idso SB. (1983) Increasing atmospheric CO_2: effects on crop yield, water use and climate. *Agricultural Water Management* **7**: 55–72. [9.2]

Lettenmaier DP, Gan TY. (1990) Hydrologic sensitivities of the Sacramento–San Joaquin River basin, California, to global warming. *Water Resources Research* **26**: 69–86. [9.3]

McCabe GJ, Wolock DM. (1992) Effects of climatic change and climatic variability on the Thornthwaite moisture index in the Delaware River basin. *Climatic Change* **20**: 143–53. [9.3]

Martin P, Rosenberg NJ, McKenney MS. (1989) Climatic change and evapotranspiration: simulation studies of a wheat field, a forest and a tall grass prairie. *Climatic Change* **14**: 117–51. [9.2]

Miller JR, Russell GL. (1992) The impact of global warming on river runoff. *Journal of Geophysical Research* **97**: 2757–64. [9.3]

Nemec J, Schaake JC. (1982) Sensitivity of water resource systems to climate variation. *Hydrological Sciences Journal* **27**: 327–43. [9.3]

Palutikof JP. (1987) Some possible impacts of greenhouse gas-induced climatic change on water resources in England and Wales. In: *The Influence of Climate Change and Climatic Variability on the Hydrological Regime and Water Resources.* International Association of Hydrological Sciences, Publication 168, pp. 585–96. [9.3]

Wigley TML, Jones PD. (1985) Influences of precipitation changes and direct CO_2 effects on streamflow. *Nature* **314**: 149–52. [9.2]

Wilby R. (1989) Changing synoptic weather patterns, rainfall regimes and acid inputs in the East Midlands, UK. In: *Conference on Climate and Water, Helsinki, September 1989*, vol. 1, pp. 209–18. World Meteorological Society, Publications of the Academy of Finland. [9.2]

10: Modelling Hydrological Processes for River Management

N. P. FAWTHROP

10.1 INTRODUCTION

Hydrological modelling can be of use in many aspects of river management. It is an essential 'tool of the trade'. This chapter discusses the various types of models and gives examples to illustrate the sort of analyses which are possible. The chapter concludes with a section of practical advice about how to avoid some of the pitfalls and make the most of hydrological modelling at whatever level it is approached.

10.2 HYDROLOGICAL MODELLING: A CONSTITUENT OF RIVER MANAGEMENT

Water industry professionals together with their consultants and advisors are well aware of the conflicts of interest which often arise in river management (Cook *et al* 1992). In recent years that awareness has extended rapidly to the general public. There has been a dramatic increase in the prominence of environmental issues. Rivers, streams and all aspects of the aquatic environment are receiving more attention than ever before.

The science of hydrology is basic to almost every aspect of river management. It deals with the occurrence and movement of water in, on and over the land surface (details are given in Vol. 1). Without quantitative knowledge of hydrological processes most of the analysis carried out within the disciplines of Water Resources Planning, Water Quality Planning and River Engineering would not be possible.

Hydrological analysis is required for studies of all kinds. At one extreme may be the prediction of flows for the design of a channel improvement scheme, or assessment of why a small stream has dried up. At the other may be the design of a region-wide water transfer scheme involving water supply, irrigation and navigation. The two are very much connected, as the consequences of one may affect the other. Techniques appropriate to the complexity of the problem need to be applied across the whole spectrum. What techniques might be appropriate? How can all those interrelated components of the hydrological cycle be disentangled? The answer, more often than not, is by some form of modelling.

For the purposes of this chapter modelling is defined as: *Computer software which quantitatively represents the response of a water resource system (natural and/or artificial) to a sample of information*. Some of the words used here are significant. First, a model *represents* reality; it is conceptually similar to the real world, but simplifying assumptions are invariably involved. Secondly, any historical record against which a model has been calibrated is only one of many possible sequences of events; it is thus a *sample*, in time and in space. Care must be taken to ensure that this is representative of the period and location for which predictions are to be made. Thirdly, the *information* which triggers the response of the system could be virtually anything. Models are commonly used to predict the hydrological performance of a catchment consequent upon some change which could be either natural (e.g. rainfall, channel morphology) or artificial (e.g. abstraction, reservoir release). A model should never be extended too far or used to examine scenarios for which it was not designed.

These are the most important of many issues

relating to the use of models in practice, which are discussed further in Section 10.6.

Selecting the right tool for the job

The scientific literature abounds with descriptions of hydrological models to investigate every conceivable aspect of the subject, but only a small proportion are of any practical use to the applied hydrologist. The following extract is from a paper by J.C. Dooge;

> The subject of mathematical modelling in hydrology is characterised by: (1) A proliferation of approaches and techniques. (2) A further proliferation of models based on any one of these particular approaches. (3) A failure to develop adequate techniques for the evaluation of specific models and for choice between models in a given situation. (4) A widening gulf between research techniques and operational methods.

In many respects this is as true today as when it was written some 20 years ago (Dooge 1972).

It is understandable and healthy that there has been a proliferation of approaches and techniques amongst researchers and other specialists. This chapter, however, is for practising engineers and scientists, for whom the approach to modelling has to be different. Models are 'tools of the trade', and like tools in any profession they have to be robust, effective and easy to use. A large choice is unnecessary and leads only to confusion. Proof of this is found in the ready acceptance of packaged models, particularly if they form part of a 'cookbook' approach within standard procedures.

However, standard packages are not always the answer. Unless extensively developed and properly tested they often fail to deliver what is required of them. A lack of flexibility is their biggest disadvantage and there are fewer opportunities to use standard models than might be imagined. There is a preference for models which can be tailored to a particular application.

Examples of both packaged and tailored models are used in the following sections to illustrate the types of models which are available for a range of applications. They are described in broad categories of: (i) catchment simulation; (ii) groundwater simulation; and (iii) river channel models. The emphasis is upon models which are in use now, with the objective of illustrating a little of the background theory, but more importantly how they have been implemented in practice.

10.3 CATCHMENT SIMULATION

Models are needed to simulate the behaviour of catchments for a variety of purposes including:

1 water balance and water resources assessement work;

2 making forecasts of river flow;

3 quality control and infilling/extending flow records;

4 testing conceptual understanding of hydrological processes.

The objective is to simulate the land phase of the hydrological cycle, with the model output usually being a prediction of river flow. Interest in flows, and hence the emphasis on modelling, is often at one or other of the extremes of the flow range. For example, ecological studies of 'in river needs' would be mostly concerned with an analysis of low flows. Conservation studies in the floodplain would be more concerned with high flows.

Some of the terms used to describe different types of model are not always self-explanatory, so a brief summary may help. The processes which link rainfall with river flow are governed by physical principles, but solutions based upon the direct application of the laws of physics are impractical at the catchment scale. *Conceptual* models are distinguished from *empirical* models according to whether they involve consideration of physical processes, or are simply mathematical relationships based on observation and analysis. *Stochastic* models incorporate statistical chance or probability, whereas *deterministic* models do not.

Conceptual models are usually deterministic in that they seek to uniquely represent our understanding of physical processes, but their mathematical algorithms incorporate parameters which need to be calibrated. At the most simple level are 'lumped' models which employ relatively simple and spatially averaged mathematical representations of input and output to stores (or 'tanks') in order to simulate the passage of water through a catchment. At the other extreme are physically based 'distributed' models which represent processes in far more detail and with a corresponding increase in computational effort.

Lumped conceptual models

The term 'lumped' is frequently used in relation to hydrological modelling (Blackie & Eeles 1985). It indicates spatial averaging so that, for example, one store may represent soil water storage throughout the whole of a catchment covering several hundreds of square kilometres. The area covered could equally be as small as a few hectares. Flows into and out of the stores are controlled by mathematical equations. Parameters define the size of the stores and the rate of flow between them. Lumped models must be used with caution and at a level of spatial definition appropriate to the problem being investigated.

Figure 10.1 illustrates one example of this type of model in schematic form. The HYRROM model (NERC 1989) is a commercially available conceptual rainfall runoff model developed for use on personal computers (PCs). The stores fill and empty to provide flows which depend upon the volume of water they contain. Each store affects the timing and magnitude of the flow at the catchment outlet. There are nine parameters:
- one which defines the size of the interception store;
- one which controls the evaporation losses;
- one which controls the proportion of water entering the runoff store (the remainder enters the soil store);

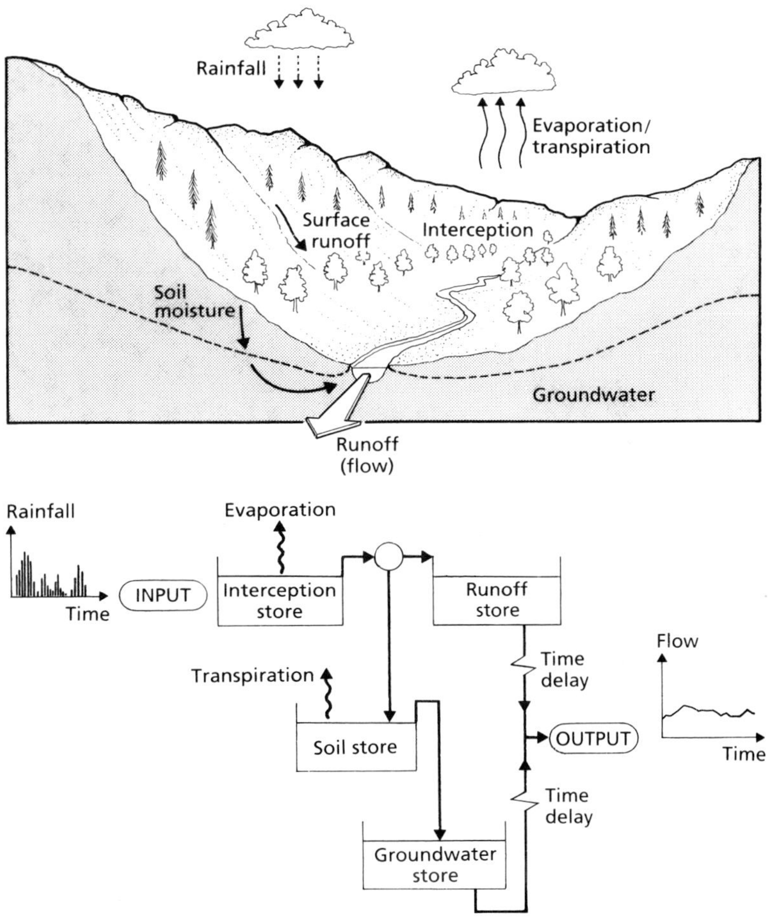

Fig. 10.1 The simulation of rainfall–streamflow processes by the HYRROM model.

- three which control the contribution of the runoff store to catchment flow;
- three which control the contribution of the groundwater store to catchment flow.

The process of model calibration involves adjusting the parameter values until acceptable agreement is obtained between the model predictions and a sample of observed flows. In this case there is an option for automatic parameter optimization. This is a useful facility, as are the user-friendly data entry screens and graphical output. These are the advantages of a packaged system, but as is often the case there are disadvantages due to the lack of flexibility. For instance, in this version of the model there is no allowance for abstractions or discharges, which limits its use to catchments with minimal artificial influences.

There are few commercially available 'packaged' lumped models available in the UK. An alternative to HYRROM which has recently been updated to run on PCs is HYSIM (Manley 1978, 1993). HYSIM has 17 parameters, but many are based on measurable catchment characteristics. Flows from separately modelled subcatchments can be passed downstream using kinematic wave routing (see Section 10.5).

The reason that there are not more general purpose models is that they nearly always have limitions of one sort or another. As many hydrologists have computer programming skills one basic model is often tailored to particular requirements. This is more likely to happen with relatively simple lumped models than it is with anything more complex and most hydrologists will have access to a catchment model of this type. Lumped models can be incorporated in to particular applications, as illustrated by the following examples.

The London Reservoirs Drought Management System

One of the main uses of hydrological models is to provide input to, or form a component of, water resources system models. Such models simulate the operation of regional schemes and may be used for yield assessment, system design, short-term operational planning or long-term strategic planning. Moore *et al* (1989) describes a decision support system which is used for the management of London's pumped storage reservoirs on the River Thames. Hydrological models form only a part of the overall system, which also includes simulation of the reservoir operations, a risk assessment procedure and a decision support user interface. A conceptual rainfall-runoff model is utilized, which had been previously developed for the River Thames (Greenfield 1984). The key stages of the decision support system are:

1 The rainfall-runoff model is run over recent time to produce a simulated flow sequence up to the current day. The results are compared with observed flow, and the conceptual stores of the model adjusted so that the modelled and observed flows agree. The conceptual model then encapsulates information on the current hydrological status of the catchment.

2 If the current day was (say) 1 July 1993 and an assessment of the next 6 months was required, the 6-monthly sequences of daily rainfall starting on 1 July would be extracted from a database for every year of record – in this case 103 going back to 1890. Each of these sequences would be run through the rainfall-runoff model to produce 103 equi-probable sequences of river flow.

3 The river flow scenarios are then used as input to the water resources system model to derive equi-probable scenarios of reservoir levels. These are then analysed statistically to provide statements about the probability of certain actions being required.

4 The method can be refined by incorporating long-range weather forecasts.

The Thames Model is now being used to simulate catchment response to climate change and land-use scenarios (Wilby *et al* 1993).

The Great Ouse Resource Model

The Great Ouse is an 8580-km^2 catchment in Eastern England. River flow records are available from some 50 gauging stations. The nature of the flow hydrographs varies considerably, reflecting the changing geology. There are three aquifers, the most important of which is the Chalk, as well as large areas of impermeable clay. This is the driest region in England and the water resources are utilized for agricultural, industrial and public water supplies. The spatial and temporal distribution of river flows are much affected by these abstractions.

Effective catchment management demanded that hydrologists were able to describe the flow regime in quantitative terms at any location. Extrapolation of data from the gauging station network was possible, but often time consuming and inconsistent. The prediction of future flow regimes consequent upon changes in abstraction and discharges was even more difficult. It was decided to develop a model of the catchment to provide the sort of information which was regularly required.

The Great Ouse Resource Model (Oakes & Keay 1990) is a semi-distributed model of the hydrological processes within the catchment. The river system is represented by 392 river reaches, linked together at node points (Fig. 10.2). The inflows and outflows for each reach are calculated, then accumulated downstream so that flows can be derived at every node in the model. This is carried out at weekly timesteps for a 26-year period.

The representation of a reach is shown schematically in Fig. 10.3. There are, in effect, 392 very simple lumped models linked together to form a distributed model. The input requirements of the program are:

- weekly values of effective rainfall, calculated using hydrometeorological data and a soil moisture accounting model (Thompson *et al* 1981);
- monthly surface water abstractions and discharges, and groundwater abstractions for each of 64 sub-catchments. These were derived from aggregated statistics based on returns from abstracters.

The abstraction and discharge data can be modified to account for anticipated future changes, and the model run for any conceivable scenario. The main outputs are flow duration statistics, which integrate the effects of all up-

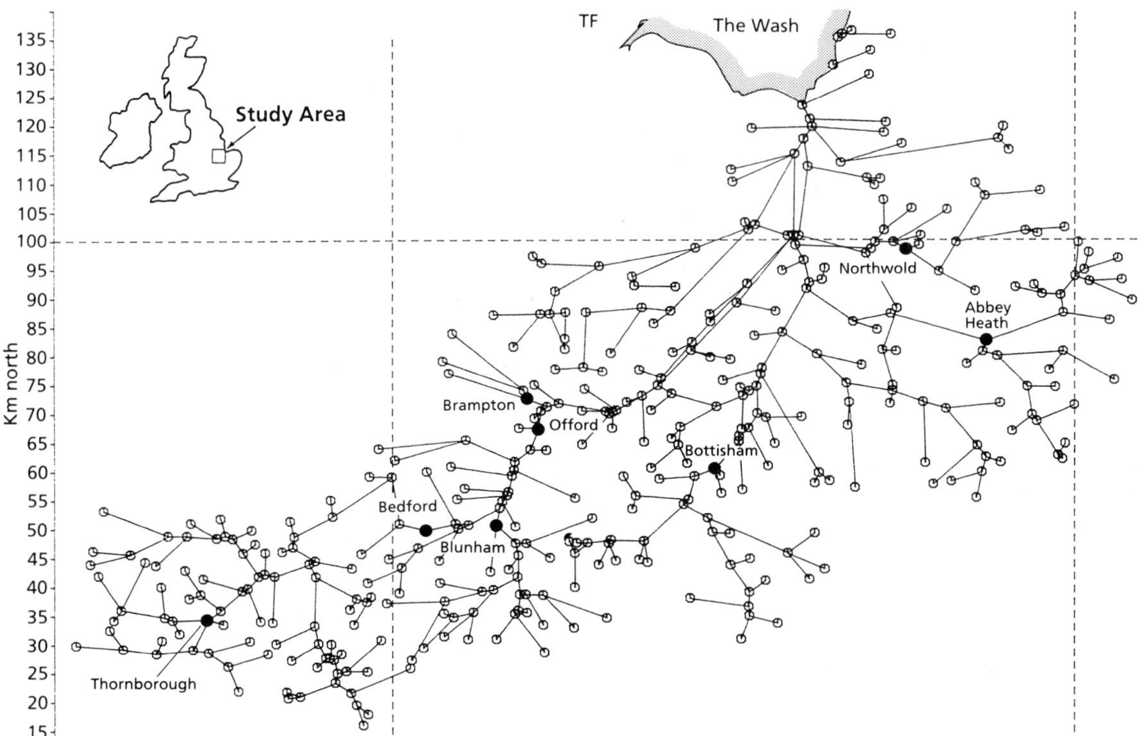

Fig. 10.2 Representation of the river network in the Great Ouse Resource Model.

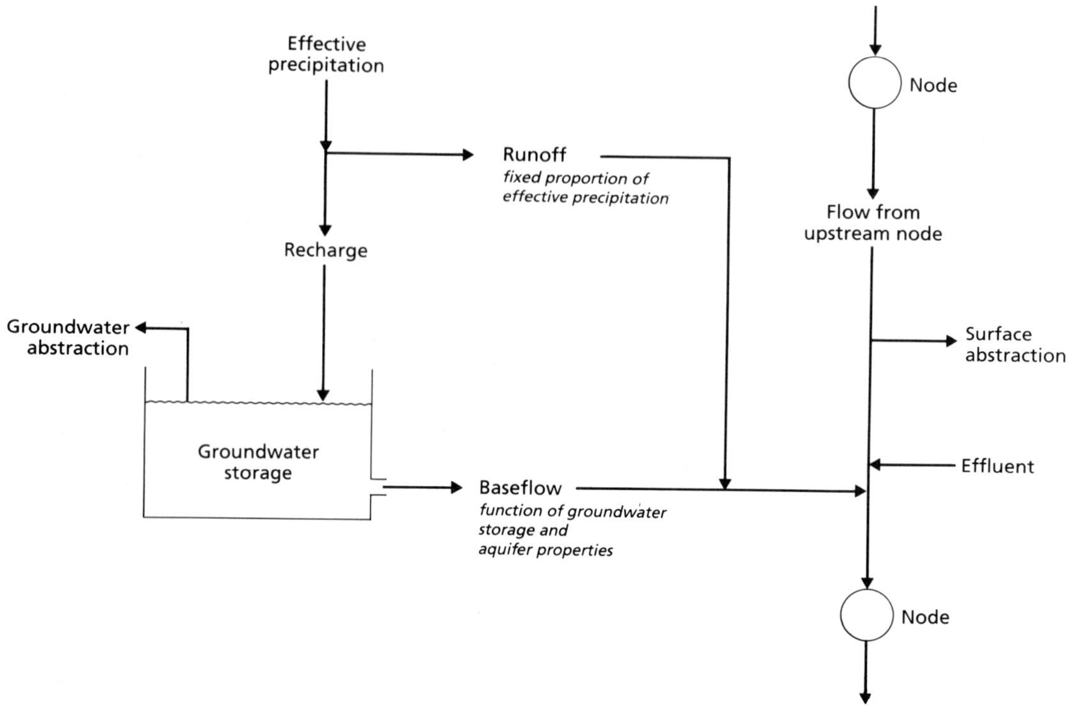

Fig. 10.3 Representation of reaches in the Great Ouse Resource Model (see also Fig. 10.2).

stream changes and summarize modifications to the flow regime. Such statistics may be used, for example, to evaluate the impact of applications to abstract water or to estimate the flows which will be available for diluting the effluent from sewage works.

The facility to calculate flow statistics so easily for so many locations is extremely useful. A user-friendly interface and graphical presentation of the results means that the model can be run by non-specialists, but the documentation draws the attention of all users to its limitations and to the degree of precision which can realistically be achieved. If greater precision is needed a more complex model will be required. An example of a model for one aquifer unit in the Great Ouse catchment is described in Section 10.4.

Physically based distributed models

A simplistic view of hydrological modelling is that there are basically two types of application. The first is to test our conceptual theories of hydrological processes and the second is to predict catchment behaviour of practical relevance. Despite significant advances in recent years the application of physically-based distributed models is still largely restricted to the former. Nevertheless, they are of interest to anyone involved in hydrological modelling, and may in the future move along the spectrum of practicality.

Physically-based distributed models usually represent surface and subsurface flows using a grid (or net) of interconnected nodes spaced only tens or hundreds of metres apart. Flows vertically and in two dimensions between the nodes are calculated using equations based on physical laws (e.g. Darcy's law). The parameters are *in theory* physically definable and measurable. The SHE (Système Hydrologique Européen) model (Abbott *et al* 1986a,b) is the most ambitious project undertaken to date to develop a physically based distributed model for commercial use. The IHDM (Institute of Hydrology Distributed Model) (Beven *et al* 1987) is the other one most often referred to in the UK.

The theoretical advantages of a distributed model are, however, often outweighed by the penalty of extremely demanding data requirements. The number of grid nodes, and hence parameters, are very high. Even on intensively monitored research catchments the number of parameters is often too high to be able to determine values in the field, and a degree of spatial aggregation is inevitable. Beven (1989) clearly spells out the problems of using these models in practice, and they are well summarized by Eeles *et al* (1990). Attention is drawn to the problems associated with using physical parameters determined at a point in a heterogeneous system and applied to a grid square. On the other hand at least one version of SHE (Danish Hydraulics Institute 1991) is commercially available and aims to provide a flexible model which can be applied at varying levels of complexity.

The manipulation of spatial data is aided by geographical information systems. These assist greatly in the storage and manipulation of the kind of data required by distributed models. Large databases of, for example, soil types or land use at grid scales less than 1 km^2, are now commercially available. Similarly, topography may be incorporated into distributed models through the use of digital terrain maps. In all fields the use of remotely-sensed data from satellites and aircraft is increasing.

A simplification of the distributed modelling approach is made in TOPMODEL (Quinn *et al* 1989). This uses a digital terrain map to determine a topographic index which is used to predict the dynamics of variable source areas. The number of parameters is kept deliberately low and is usually around three.

A yet more pragmatic approach is to utilize semi-distributed models, which are in effect lots of lumped models linked together in some way, perhaps by some form of river routeing. The London Reservoirs Drought Management System and Great Ouse Resource Model, described above, in fact fall into this category. Another good example is the RORB runoff routing model (Lowing & Mein 1981; Laurenson & Mein 1983; Rylands & Lee 1992), which uses a catchment subdivision based on isochrones.

Empirical models

This section considers how models and the processing power of computers can be used to define empirically the nature of hydrological systems.

Empirical models offer little insight to physical processes. They are sometimes referred to as 'black box' models, because inputs are converted to outputs based only on statistical correspondence. In many situations, if there is a sufficiently strong and consistent relationship between input and output an understanding of the intervening catchment processes is not necessary. Nevertheless, the empirical relationships can sometimes be interpreted in terms of implicit catchment response. This is best illustrated by means of examples.

The Unit Hydrograph Model

A milestone in the development of hydrological science was the introduction of the concept of the unit hydrograph by Sherman in 1932. This model is still in widespread use today. Figure 10.4(a) illustrates the idea of how a T hour unit hydrograph (TUH) results from a unit depth of *effective* rainfall falling in T hours over the catchment. The 'unit' is usually 1 mm and in this case the ordinates of the unit hydrograph are $m^3 s^{-1}$ per millimetre of rain. The volume underneath the curve of the hydrograph is equivalent to 1 mm depth of effective rainfall over the entire catchment area. The dimensions of the unit hydrograph will vary between catchments, depending on those characteristics which affect catchment response such as area, geology, slope, shape and land use. The ordinates of a unit hydrograph can be derived from synchronous records of river flow and rainfall. This is a purely mathematical procedure. The unit hydrograph is simply a mathematical expression of the catchment response, but relationships can be established with physical parameters of the catchment, so in that sense it does have physical meaning.

The unit hydrograph concept involves several assumptions which simplify its application to catchment scale rainfall–runoff modelling.

1 Runoff is directly proportional to effective rain-

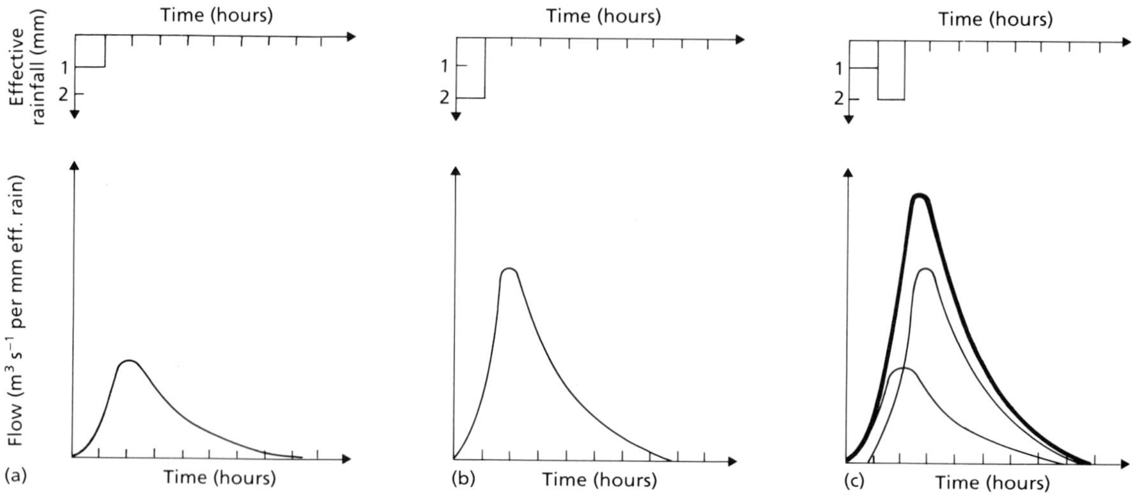

Fig. 10.4 The unit hydrograph model: (a) the 1-hour UH for 1 mm of effective rainfall; (b) superposition, the runoff hydrograph for 2 mm of effective rainfall in 1 h; (c) convolution, the runoff hydrograph for 1 mm followed by 2 mm of effective rainfall.

fall. Thus two units of effective rainfall falling in *T* hours would produce a runoff hydrograph with ordinates twice that of the unit hydrograph (Fig. 10.4b).

2 Effective rainfall in consecutive *T*-hour periods results in a runoff hydrograph which is the sum of the lagged component hydrographs for each hour (Fig. 10.4c).

3 The shape of the unit hydrograph is time invariant.

These assumptions are never entirely valid, and there are other weaknesses such as the assumption of uniform catchment–wide rainfall (which is common to most lumped models). The topic of baseflow separation and the analysis of what proportion of the total rainfall is effective rainfall would need many pages to explain fully. Nevertheless, the unit hydrograph is a sound technique which has the great advantage of simplicity, and it is central to many procedures for modelling the rainfall–runoff process. This is particularly the case in engineering hydrology where the Flood Studies Report (NERC 1975) recommends it as one of the standard approaches to flow estimation throughout the UK. As such it is likely to be used in the design of many channel improvement schemes.

An example of the use of the unit hydrograph model is given for the River Nene in the English East Midlands (Fig. 10.5). Unit hydrographs were derived from gauged flow records for each of the major catchments, and from relationships with catchment characteristics for all the ungauged sub-catchments (Cole *et al* 1981). They were then combined with hypothetical rainfalls to generate design flows for input to a mathematical hydraulic model of the main river (see Section 10.5 for a more complete description). This calculation (as simplified in Fig. 10.4c) is referred to as the 'convolution' of a time series of rainfall and unit hydrograph ordinates. Data preparation is often time-consuming, so recent developments in unit hydrograph theory as described in the next section are very welcome.

The IHACRES Model

IHACRES is an acronym for 'Identification of unit Hydrographs And Component flows from Rainfall, Evaporation and Streamflow data' (Jakeman *et al* 1990). The model used in IHACRES is a development of unit hydrograph theory, which overcomes many of the practical difficulties of that approach by utilizing an approximation based

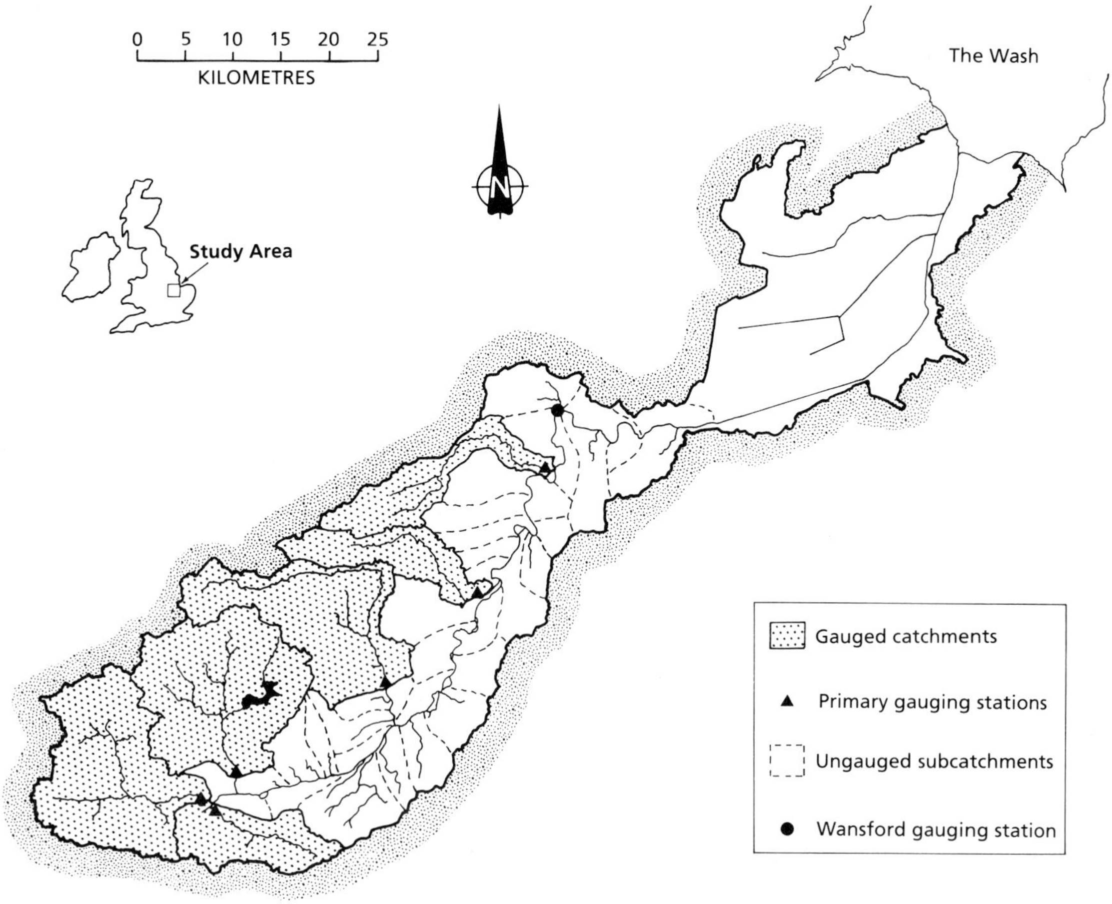

Fig. 10.5 Sub-catchments of the River Nene.

on 'transfer functions'. A PC version may soon become available commercially.

Transfer function models predict flows based on a limited number of previous rainfalls and flows. It is possible to demonstrate their mathematical equivalence to unit hydrographs, but they have numerous advantages; in IHACRES these are:

- there are fewer parameters, so it is possible to adopt a more effective optimizing approach to parameter estimation;
- predictions are made of total streamflow and the quickflow/slowflow components separately;
- data inputted are time series of *total* rainfall and river flow, not effective rainfall and rapid response runoff. Hydrograph separation is an integral part of model identification and not, as in conventional unit hydrograph derivation, a necessary prerequisite;
- continuous time series of data can be utilized rather than the discrete 'well behaved' events which are necessary to avoid instability in unit hydrographs.

Figure 10.6 illustrates one version of IHACRES in schematic form. It can be applied to catchments varying greatly in size. Jakeman *et al* (1991) give examples varying from a 743-km^2 catchment with high baseflow modelled with a daily timestep to a 0.016-km^2 flashy catchment modelled with an hourly timestep. Figure 10.7

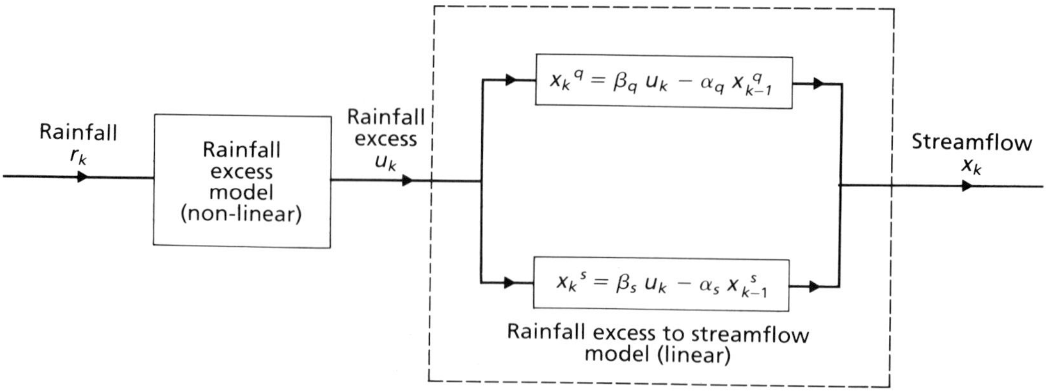

Fig. 10.6 IHACRES model.

shows the model output from the two catchments, illustrating the prediction of both total flow and the baseflow component. Further examples of use of the model, including an interesting application to baseflow separation, are given in Littlewood and Jakeman (1993).

Although the parameter values of the transfer function models have some physical interpretation it is important to realize the difference between empirical models such as this and the conceptual models described earlier. The parameter values of IHACRES summarize the catchment as it was during the calibration period, so this model would not be appropriate for predicting changes in riverflow due to changes in specific components of the water balance (such as abstractions). On the other hand IHACRES would be well suited to many other modelling applications where either characterization or comparison of catchments is required. It could be used wherever a prediction of catchment response due to rainfall is needed. For example:

- detecting the effect of land use change;
- predicting the impact of climate change;
- infilling or extending river flow records.

Models for operational flood forecasting are not considered in this chapter, but it is worth mentioning that empirical models, and particularly transfer function models, are well suited to this type of application. They need few parameters, are computationally efficient and have minimal data requirements (Cluckie & Tilford 1989).

10.4 GROUNDWATER SIMULATION

Groundwater models as described here may be thought of as catchment simulation models which focus on flows underground rather than in surface watercourses. They are useful for:

- initial evaluation and appraisal studies preceding field investigations;
- interpreting the results of a field investigation;
- prediction and forecasting to estimate future field behaviour.

For groundwater flow the physical laws which govern cause and effect are generally well understood and a deterministic approach is usually possible, albeit with numerous simplifications and assumptions. For contaminant transport there are more unknowns and stochastic techniques may form an integral part of the model in order to quantify uncertainties.

Groundwater flow models

Due to the complexity of groundwater flow mechanisms a lumped approach is usually inadequate. In order to understand aquifer behaviour it is necessary to consider aspects such as:

- properties of the aquifer such as the hydraulic conductivities and storage coefficients. These often change markedly over short distances both laterally and vertically;
- significant features such as fissures and preferential flow horizons;

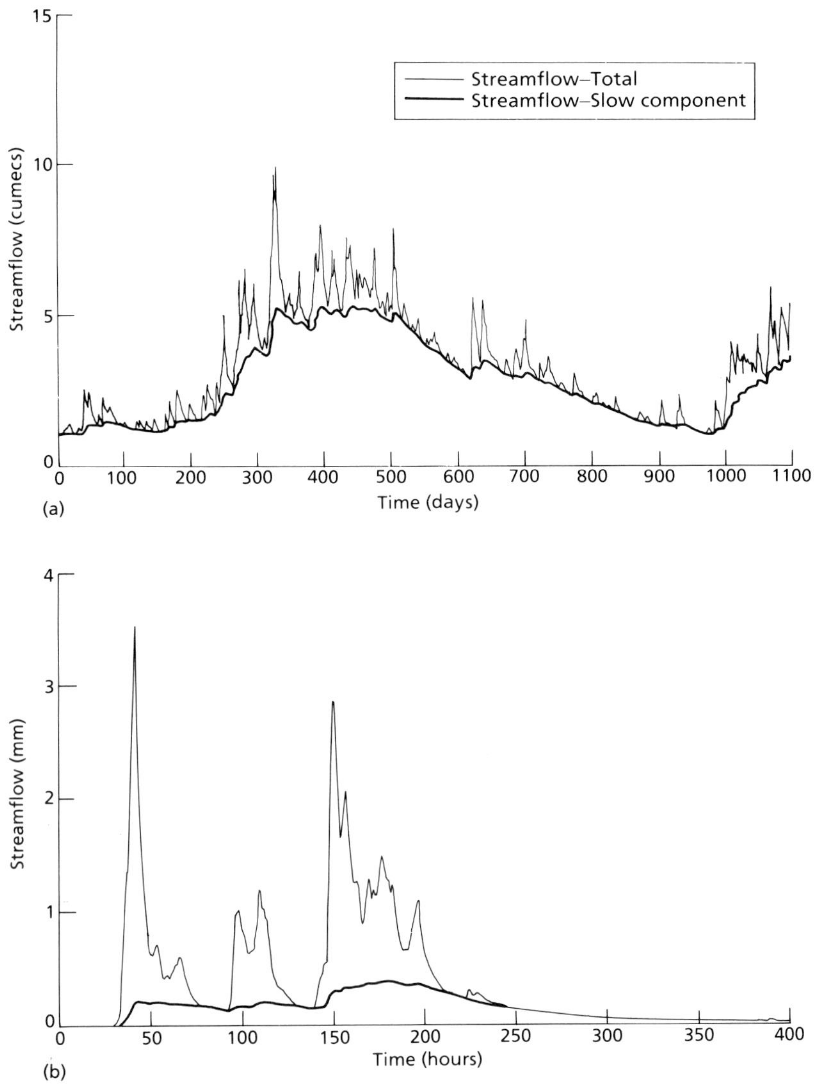

Fig. 10.7 Examples of flow simulation by IHACRES: (a) modelled flows for a catchment of 743 km^2; (b) modelled flows for a catchment of 0.016 km^2.

- recharge processes. These are many and varied, such as infiltration recharge through soils, recharge into swallow holes, recharge from leaking water supply pipes in urban areas and recharge through river beds;
- the location of features such as abstractions and springlines which can exert a controlling influence on flows.

Spatial detail must be incorporated. This is achieved by subdividing the aquifer into many 'cells' in the form of a grid or net overlaying the study area. Given estimates of the aquifer parameters, mathematical equations based on Darcy's Law define the relationships between water levels and flows between the cells. When combined with expressions for the conservation of mass and (for time-variant problems) the effect of changing storage, simultaneous equations are defined which may be solved using numerical solutions. Different numerical procedures may utilize either a rectangular grid (finite difference) or a polygonal grid (integrated finite difference or finite element). The grid need not be uniform. More detail may be required in areas around features of interest.

Modelling may be used to examine a small area rather than the whole aquifer. For example, radial flow models are used to investigate aquifer characteristics around a borehole in more detail than is possible using analytical solutions to pumping test data (Rushton & Redshaw 1979).

Models may be for steady-state conditions or be time variant. They may be one-, two- or three-dimensional and consider any number and combination of confined, unconfined and leaky aquifers. The skill required for model development reflects the complexity of the situation. The *use* of such models, however, need not be the province of specialists (Spink & Hughes 1992).

The type of model which river-management professionals are most likely to encounter is a regional aquifer model being used to predict changes in aquifer behaviour consequent upon additional or relocated abstractions. In the UK these have been in use since the early 1970s, gradually increasing in refinement as knowledge, experience and the available computing resources have all improved. Rushton (1986) reviews the approach and gives some practical considerations and examples. A review giving more emphasis to American non-proprietary models (which are increasingly being used in the UK) is provided by Anderson and Woessner (1992). The main objective is always to represent adequately the features of groundwater flow in the aquifer. The success of a model is measured by its ability to translate the results back to physical meaning and its effectiveness in answering the questions which motivated the model study. The following sections describe several such models in some detail.

Lodes–Granta Model

The Chalk northeast of Cambridge, England, is a major source of water for public supply. Abstractions have been increasing and in recent years there has been growing concern about the effect upon the watercourses in the area. The National Rivers Authority has a duty to protect the water environment whilst ensuring that water resources are utilized effectively. There is a conflict of interests in the area, with on the one hand two water companies keen to maximize their use of cheap and high-quality groundwater, and on the other local pressure groups and conservation interests very anxious to ensure that there is no further deterioration in low flows.

The maintenance of streamflows using pumped groundwater is quite common in England. It was decided to initiate a feasibility study into the possibility of this option to compensate the effect of additional groundwater abstractions. It involved detailed hydrogeological investigations and the development of a mathematical model (Rushton & Fawthrop 1991). The model is a distributed regional groundwater model based on the solution of finite difference approximations to the groundwater flow equations. It has a regular 1-km grid and uses a 7-day timestep.

It was very important to have a conceptual understanding of the hydrogeology of the aquifer before proceeding with the design of the river support scheme. The significant features in so far as they needed to be incorporated into the model are summarized below.

1 Springlines were associated with the outcrops to two bands of hard, well-fissured chalk with relatively high hydraulic conductivities. Ninety-eight springs and locations where river–aquifer interaction took place were included. Their discharge characteristics were defined by coefficients so that the modelled flows could be calibrated against periodic current meter measurements.

2 These two bands also provided important flow paths to pumped boreholes. There was no attempt to explicitly model the layering, but there was a feedback facility to restrict borehole yields if groundwater levels fell below threshold levels related to the depth of the major fissures.

3 In most Chalk aquifers, the zone of fluctuation of the water table defines a zone of high hydraulic conductivity. Consequently the transmissivity reduces rapidly as the water table falls. This was incorporated into the model by defining a relationship between hydraulic conductivity and groundwater level. It proved to be essential in order to simulate correctly the magnitude of the seasonal changes in flow towards the springs and streams.

The model was calibrated against observed groundwater levels and streamflows for the 17-year period from 1961 to 1987. The fit was sufficiently good to provide confidence that the understanding of the aquifer flow mechanisms and dynamic water balance were adequately represented.

The model was used to test whether the new boreholes for both river support and public supply could provide water reliably and without affecting existing boreholes, wells and protected wetlands. Output from the model was used to predict

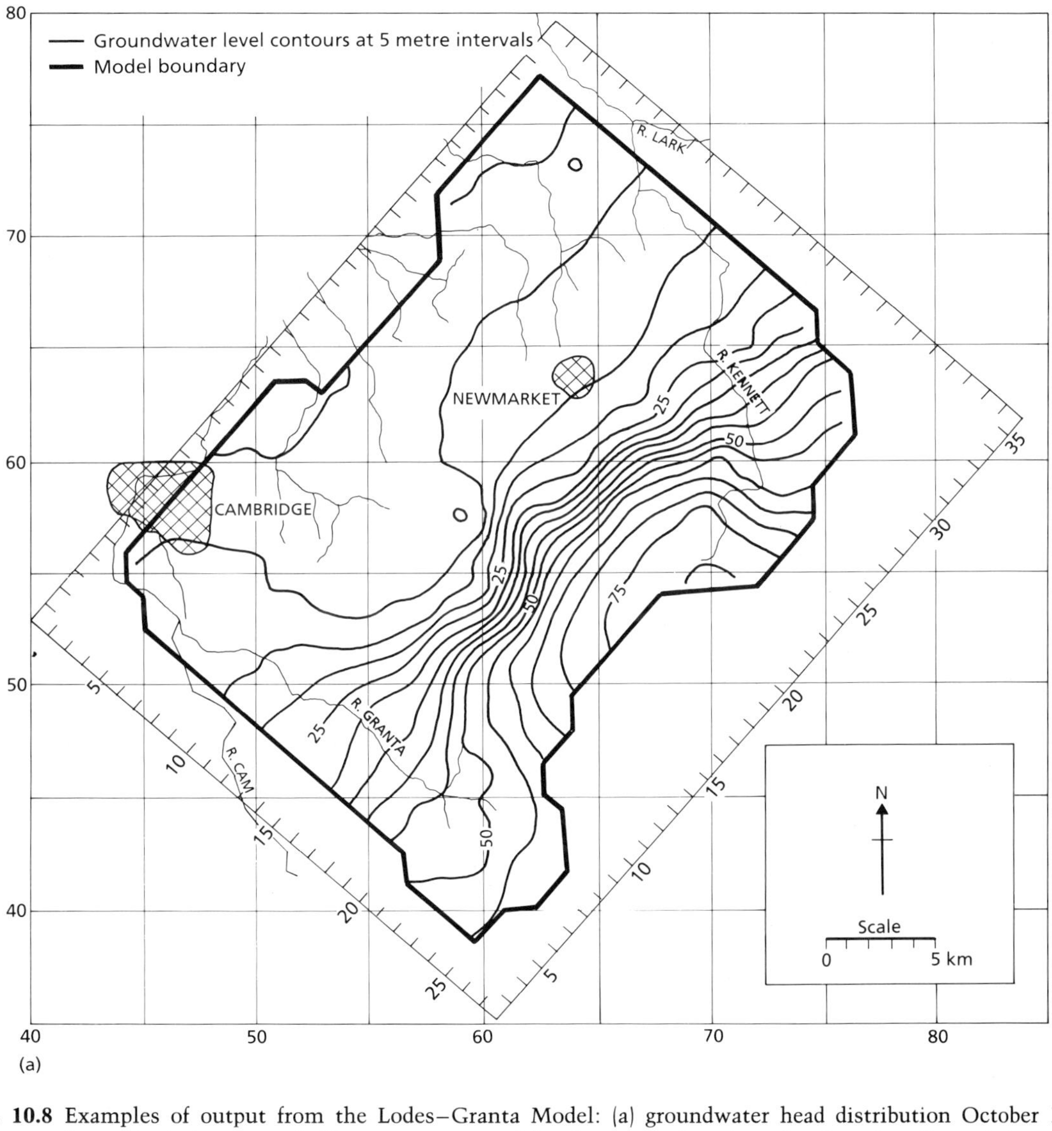

Fig. 10.8 Examples of output from the Lodes–Granta Model: (a) groundwater head distribution October 1974 (current abstraction); (*over page*) (b) a comparison of heads at one location showing the effect of increased abstraction; (c) a comparison of flows from a single spring showing the effect of increased abstraction.

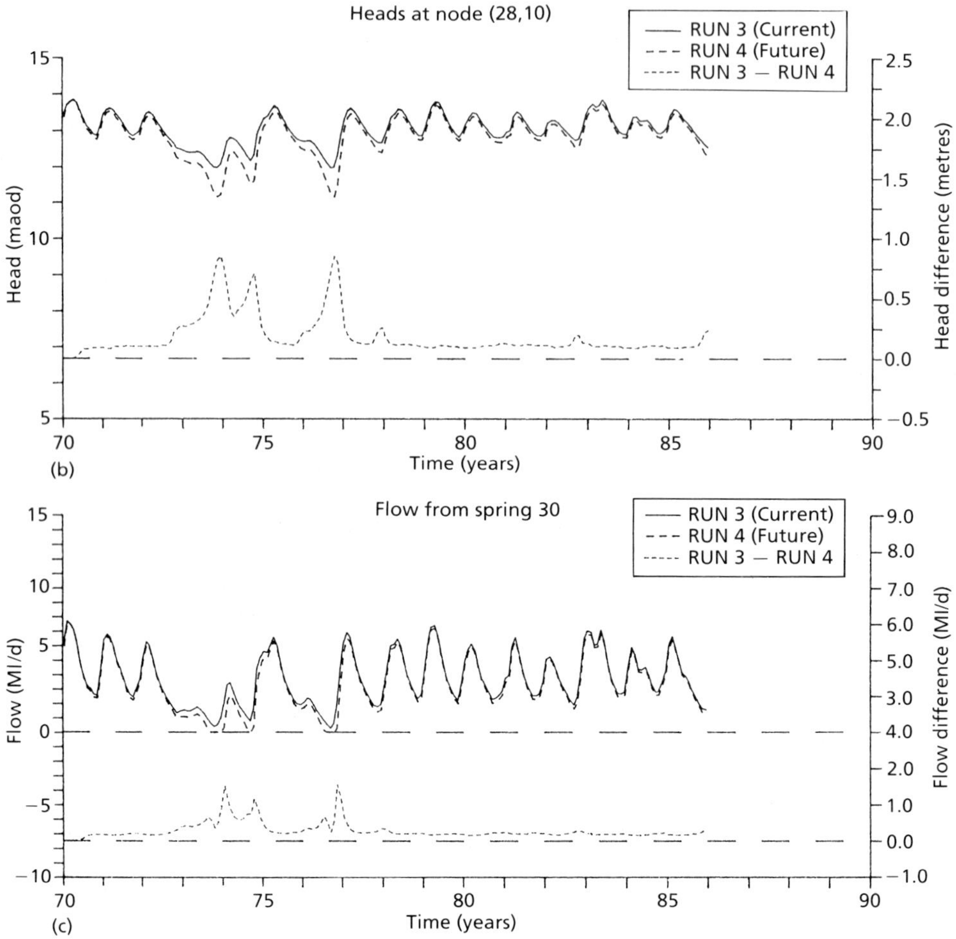

Fig. 10.8 *Continued*

the effect upon water levels and springflows at particular locations. It was invaluable in both the design of the groundwater development scheme and in illustrating its consequences during the consultation phase. Figure 10.8 shows examples of the sort of output used.

Low-flow investigations using an integrated model

Some of Britain's most pressing environmental concerns are about low flows in certain over-utilized rivers. A recent review by the National Rivers Authority identified the 40 catchments where problems are most apparent and remedial action is necessary. Although lack of rainfall causes low flows in rivers, the problem is sometimes exacerbated by over-abstraction or changing land use. It would not be possible to isolate the various components without the utilization of hydrological models.

Many of the most seriously affected rivers are in southern England. Three examples are:

1 The River Darent in Kent (Brown *et al* 1991).
2 The River Allen in Dorset (Newman & Symonds 1991).

3 The Wallop Brook and Bourne Rivulet (tributaries of the River Test) in Hampshire (Midgley & Jones 1991).

In all three cases low flows are sustained by groundwater discharges from the chalk aquifer. Hydrological investigations had to quantify the dynamic water balance, involving the assessment of recharge, runoff to the rivers, groundwater flow, springflow/seepage from groundwater to the rivers, leakage from the river back to groundwater and the impact on each of these of groundwater abstraction. The examples listed above were selected because they were all modelled using the same *Integrated Catchment Model* (Mott MacDonald 1993).

The model contains three main simulation modules, together with graphical pre- and post-processing facilities. The three modules are:

1 A lumped model to simulate the spatial and temporal distibution of recharge and runoff.

2 A model to derive the flow-level relationships for river reaches.

3 An integrated groundwater and river flow routing model.

The recharge model is a development of the Stanford Watershed Model (Crawford & Linsley 1966). This is a well known lumped model in which the water balance takes account of variations in storage and the rate of water flow both above and below the surface. Recharge estimates made in this way could be verified in the context of the total water balance for the catchment. Recharge was the primary input to the groundwater simulation model.

The groundwater model is fully distributed and can simulate complex, heterogeneous, multi layer aquifer systems. It is based on the Integrated Finite Difference Method (IFDM) which allows for the use of irregularly shaped model polygons. This permitted the more detailed investigation of selected areas (Fig. 10.9).

A distinguishing feature of the IFDM model is the representation of interaction between the river and underlying aquifers. The calculation of flow between the river and the aquifer and the routing of flow down the river network is part of the iterative computation which solves the equations governing flow between the model polygons. River–aquifer interaction depends on their relative water levels. Groundwater levels come from the groundwater flow equations. River levels come from a hydrodynamic model which relates river flow to level using open channel hydraulic computations (see Section 10.5). Flow accretion profiles (Fig. 10.10) were produced to show locations where the river gains and loses water under various water level conditions.

The benefit of using an integrated model is apparent when the relationships between the various hydrological processes need to be examined. Comparison, mapping and plotting of hydrological variables is done in the knowledge that all interdependencies have been accounted for.

There is a general trend in groundwater modelling towards a more integrated approach. The SHE model (Section 10.3) carries this concept through to its ultimate conclusion, but others have adopted a more pragmatic approach. MODFLOW (McDonald & Harbaugh 1988) is a widely used groundwater model which has several add-on modules.

The South Humberside Chalk Model

Groundwater models are usually developed at the design stage of a scheme, satisfactory calibration is achieved and predictions made. It is less common for models to be updated in order to verify performance under the new abstraction regime and even rarer for models to be sufficiently well developed to use for short-term operational planning. One exception is a model of the South Humberside Chalk (Spink & Hughes 1992), which is used routinely in the short-term local management of the aquifer.

During the 1989–92 drought in eastern England this model was run at 3-monthly intervals to establish the amounts of water which could safely be abstracted from various locations. Careful management was necessary in order to maximize drinking-water supplies from boreholes inland, whilst maintaining groundwater flow towards the sea so that saline water did not encroach upon industrial abstractions on the coast. The control of groundwater abstraction in response to hydrological conditions can prevent damage to the environment.

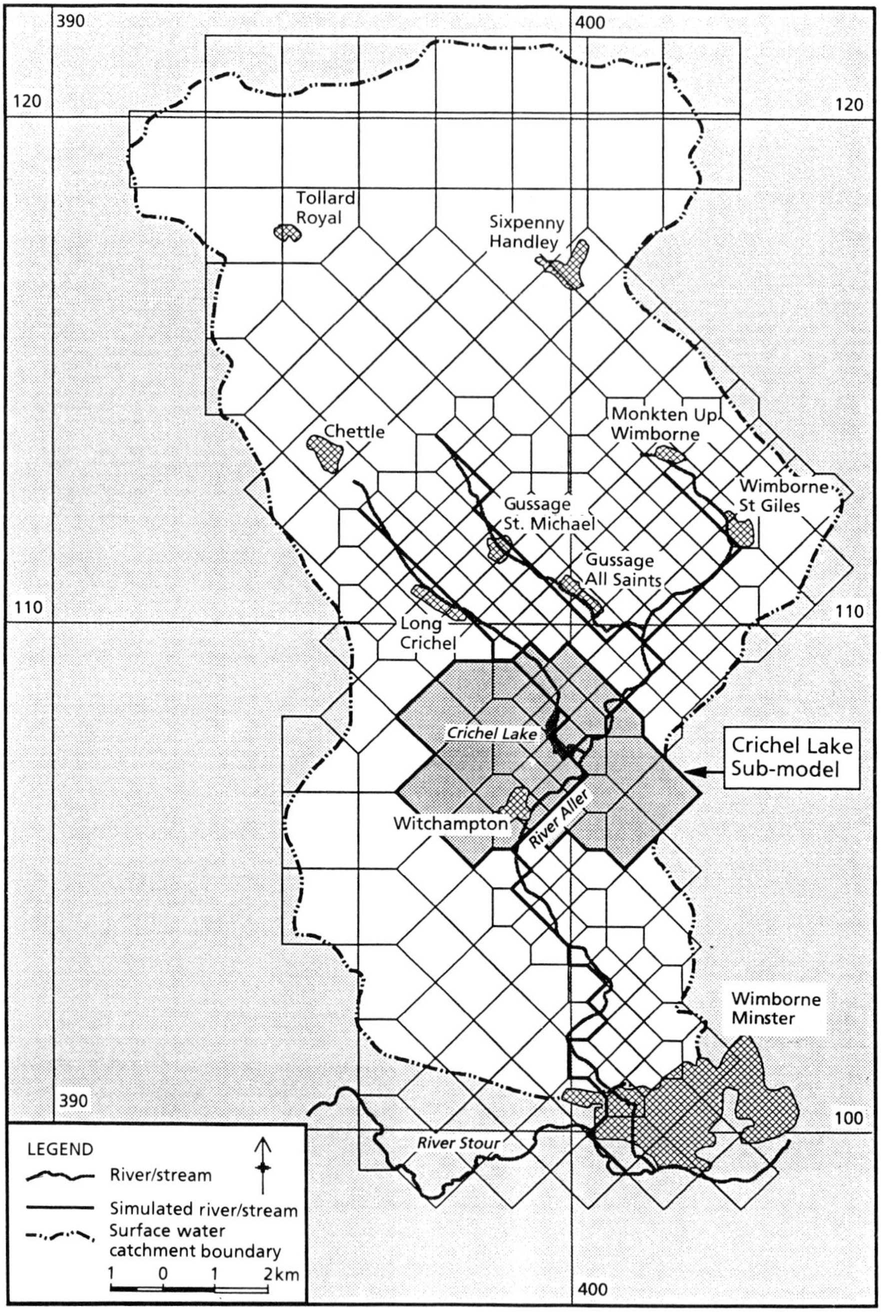

Fig. 10.9 An integrated finite difference model applied to the River Allen, UK.

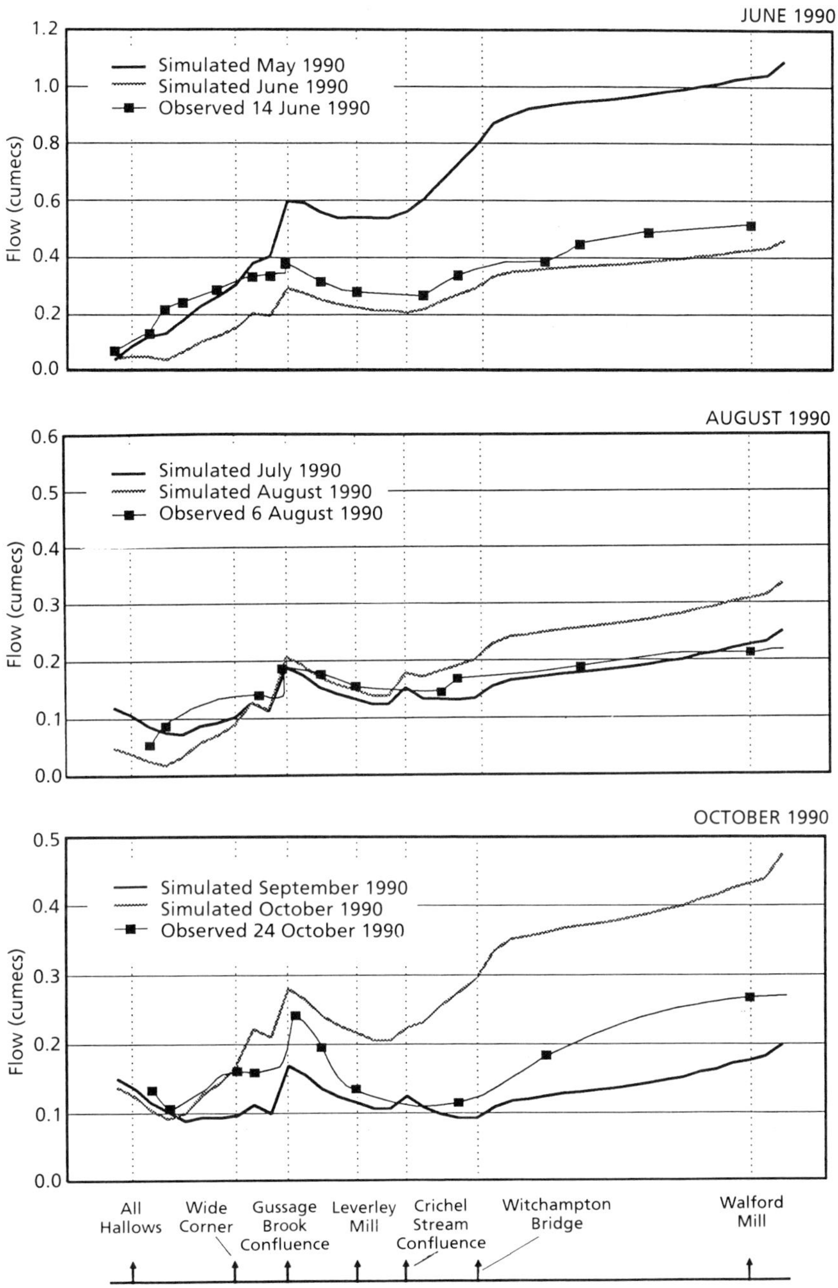

Fig. 10.10 River flow accretion profiles for 3 months in 1990 on the River Allen, UK.

Contaminant transport models

Attempts to include predictions of changing water quality adds another level of complexity. Contaminant transport models must be underpinned by reliable modelling of the regional flow mechanisms, to which are added equations representing the advection, dispersion and chemical reaction of the determinand being modelled. They have been used, for example, to estimate the rates of movement of hazardous waste from landfills, to forecast the effects of pollution incidents and to predict the intrusion of saltwater in coastal regions. Modelling the movement of solutes or contaminants by advection alone (movement with groundwater flow) is well proven through use of the 'particle tracking' technique, but groundwater quality modelling is a field where significant advances are still being made. Given the sometimes limited knowledge of chemical processes and reactions within the aquifer, limited data on the source and strengths of pollutants, and the intrinsic difficulties of modelling groundwater flow, problems concerning contaminant transport often extend mathematical models to the limit. Examples where models have been used successfully are described in Anderson and Woessner (1992) and Anderson *et al* (1992).

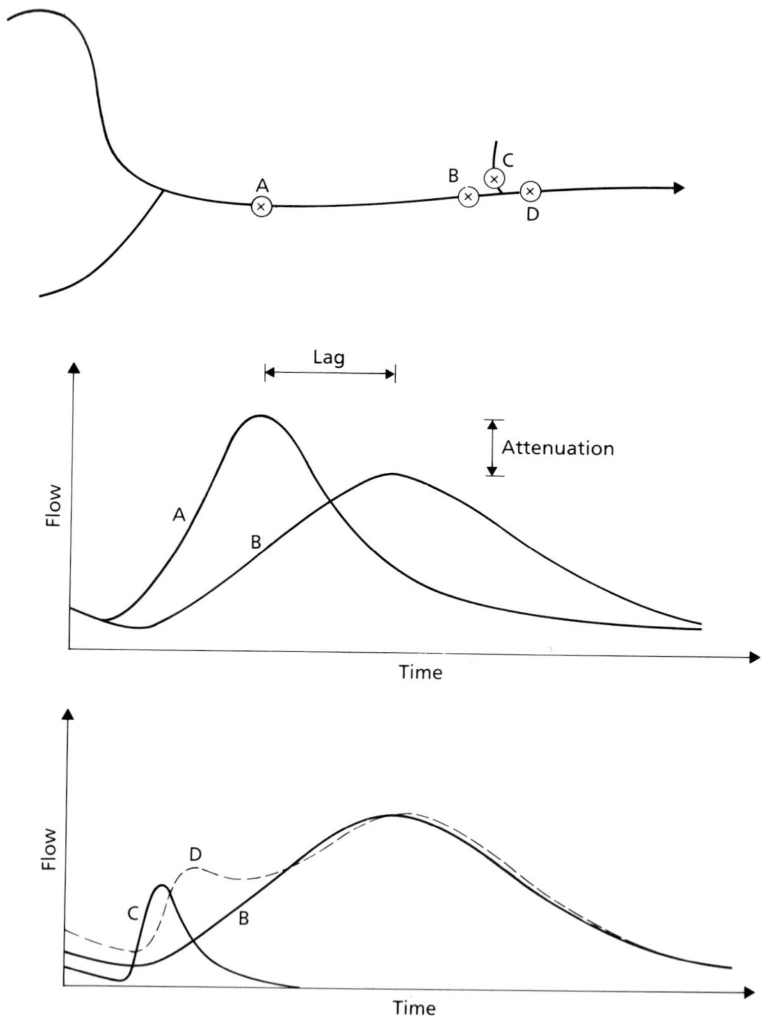

Fig. 10.11 River routing.

10.5 RIVER CHANNEL MODELS

Hydrologists are frequently asked for estimates of river flow or level along a reach of river for which no data exist. It is not always a straightforward matter to extrapolate information from a nearby site, and once again it is often necessary to employ modelling techniques of varying complexity.

Earlier chapters have referred to the characteristics of river hydrographs generated from different sizes and types of catchments. Flow 'routing' is a technique used to quantify the changing shapes of hydrographs at different points on a river system. Following a rainfall event flows increase rapidly in the smaller tributary streams. The water moves downstream at a rate controlled by the size and shape of the channel and its floodplain, the gradient of the river and the presence of any obstructions, which may be either natural or artificial. The shape of the hydrograph is different at every point on the river.

The sort of changes which occur are illustrated in Fig. 10.11. Between A and B where there are no lateral inflows the hydrograph has been modified in two ways.

1 The peak at B occurs later than it does at A. This lagging effect is known as 'translation' of the peak.

2 The peak flow is lower at B than it is at A, due to the effects of channel storage and friction. The flattening of the hydrograph is known as 'attenuation'.

As tributaries join the additional flows are superimposed upon the attenuation. The combination of B and C results in a larger hydrograph (D) which will itself be modified as it 'moves' downstream. This apparent movement of a hydrograph should not be confused with the actual movement of water. Hydrographs record the variation in flow or level through time at a point. The lag between the peaks of hydrographs at two points such as A and B is not the same thing as the time it would take a notional particle of water to travel between A and B. The translation lag is always less than the travel time.

Figure 10.11 refers only to flows. Conversion to levels is where difficulties often arise. For any point the relationship between level and flow may not be unique, for instance due to the seasonal effect of weedgrowth. The effects of weeds or any other form of obstruction are felt for some distance upstream. These are known as 'backwater' effects, and backwater analysis is commonly employed by river engineers to calculate levels under constant flow conditions.

When the additional dimension of time is introduced more complex modelling is required. Time variant modelling is needed whenever the effects of changing storage within the system need to be considered or when the 'boundary' conditions (e.g. tributary inflows or downstream levels) are changing so rapidly that constant flow assumptions are invalid. There are two broad categories of river routing models, often referred to as hydrological and hydraulic models.

Hydrological routing models

Hydrological routing models relate the inflow and outflow of a river reach using the principle of continuity and a relationship between flow and storage within the reach. Most hydrological textbooks give worked examples of this type of approach (the Muskingum method) and demonstrate how the translation and attenuation shown in Fig. 10.11 can be predicted (Shaw 1983; Wilson 1983). Such models are relatively simple and reasonably accurate, but there are limitations. They can only be used in situations where the relationship between level and flow is unique. Thus changing backwater effects from tides, confluences or moving structures cannot be incorporated, and they are not appropriate to low-gradient rivers where looped rating curves may exist.

The same limitation applies to the use of simplified hydraulic routing using the principal of kinematic waves (well described in Dingman 1984). Wave speed is derived from the flow, the dimensions of the channel and an estimate of channel roughness. More complex hydraulic models are described below.

Hydraulic routing models

In a manner very similar to developments in groundwater modelling the reduced cost of increasingly powerful computers has led to significant progress in the application of more sophisticated hydraulic models. These are based on numerical solutions to the equations describing the conservation of mass and momentum. As with groundwater modelling there are a number of mathematical approaches to the solution of the equations but most of them involve finite difference approximation and iterative solutions.

The user of results from such models will be more interested in how the river system is re-

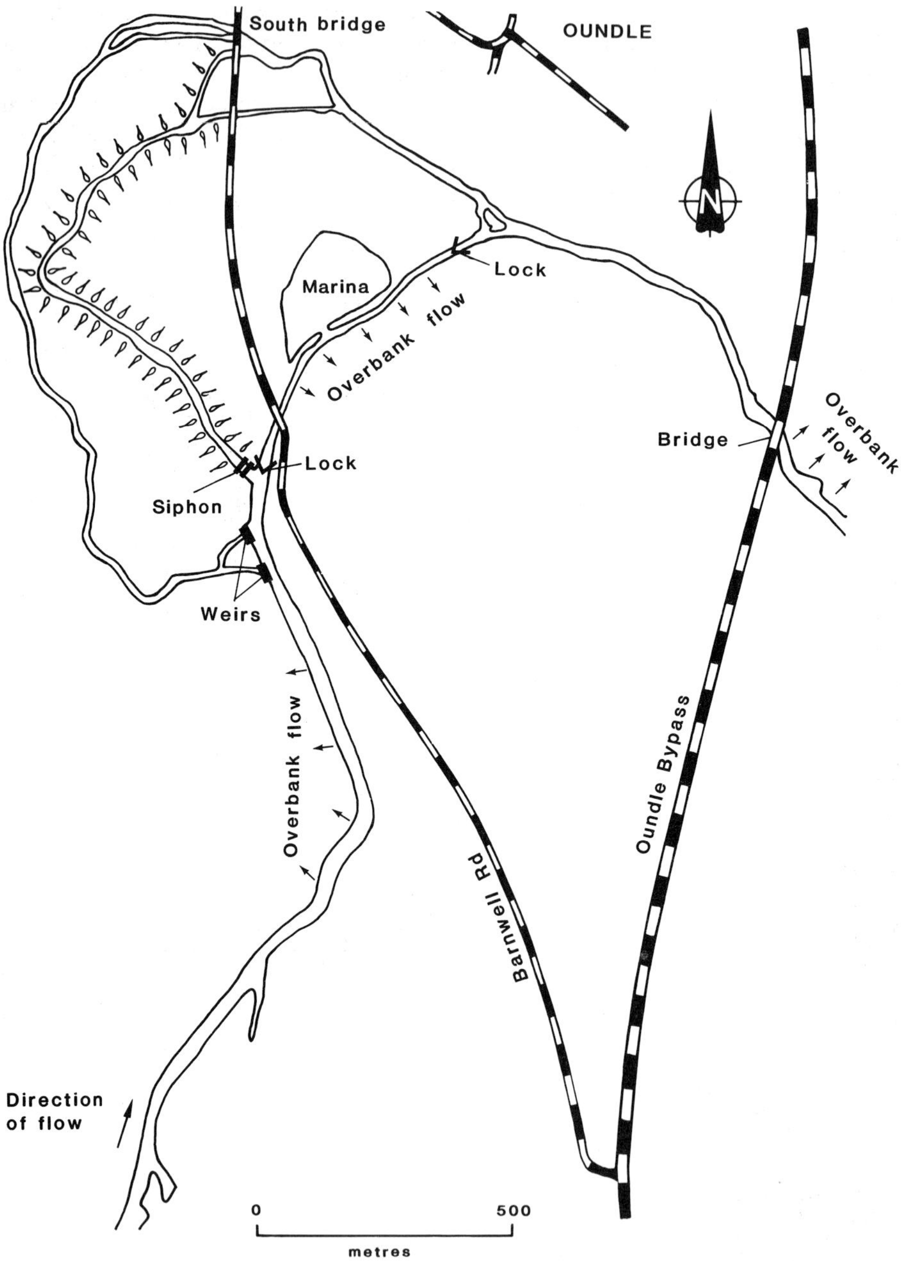

Fig. 10.12 The River Nene south of Oundle.

presented and what sort of things such models can do. The following examples illustrate some typical features.

The River Nene Model

The sort of detail that can be examined using a mathematical hydraulic model is illustrated in Fig. 10.12, which shows part of the River Nene in Northamptonshire. The river flows from south to north. During flooding the left bank of the river south of Barnwell Road becomes inundated and acts as a side weir. The water then enters the most westerly overflow channel to join with flow from over the weir control structures. The other overflow channel is fed from the main river by water passing through a set of air regulated siphons. This channel has raised banks to prevent flooding on to the surrounding land. The two overflow channels join to pass through the old arched Oundle South Bridge. The main river continues through a lock into a lower reach whose right-hand bank can also act as a side weir and then through a further lock structure before joining the now combined overflow channels. The river then continues on through a new road bridge.

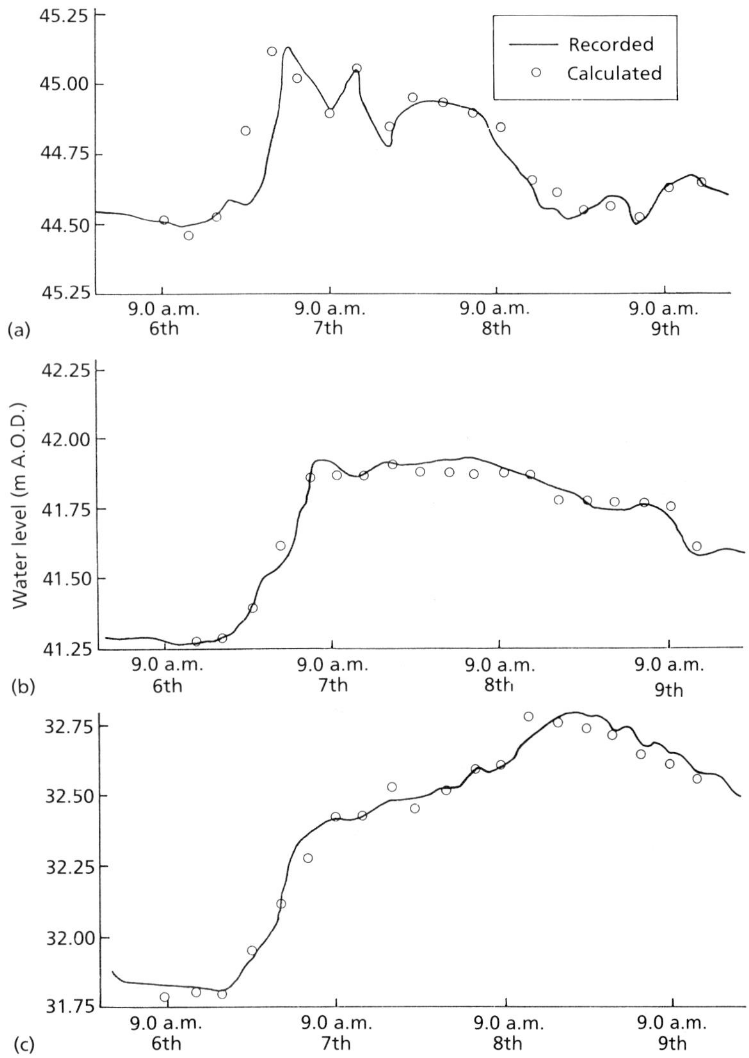

Fig. 10.13 Example output from the River Nene Model: (a) upstream of Doddington lock; (b) upstream of Wellingborough lock; (c) downstream of Upper Ringstead lock.

Time variant flows and levels can be modelled at points spaced only a few hundred metres apart, as shown in Fig. 10.13 (for elsewhere on the river). This is only 5 km out of 144 km, which includes some 120 structures and 34 bridges (Williams & Fawthrop 1988).

Such detail is necessary when analysis is required of the effects of river improvements upon other parts of the river. For example, a scheme of river widening and straightening at a particular point may reduce the degree of flooding there only to transfer the problem to a location farther downstream. Mathematical models are an effective may of examining options. This is particularly useful when it is important to look at the effectiveness and cost of alternatives, including those which minimize damage to the environment.

The Lower Colne Study

The Lower Colne Study (Gardiner *et al* 1987) is a good example of how mathematical modelling should be incorporated into clearly structured catchment management. The Colne lies to the northwest of London and forms a complex network of river channels, gravel pits and extensive floodplain (Fig. 10.14). Much of the catchment is heavily developed and major road and rail arteries cross the floodplain. It was the proposed construction of the M25 motorway on an embankment within the floodplain which initiated a comprehensive study into the river management objectives for the catchment. The impact of flood alleviation works would be of great concern to many organizations with environmental interests and it was apparent that the implementation of flood alleviation measures would be facilitated with their support. The use of a mathematical model allowed more rapid and complete evaluation of alternative options than would otherwise have been possible.

The proposals were formulated in two phases. Firstly, the model provided information on the frequency of flood risk and the extent, duration and depth of flooding in order to estimate the costs and benefits based on a preliminary scheme. At the same time an environmental database was being constructed by specialists in 12 disciplines covering all in-river and riparian interests. In the second phase the river engineers used the model to investigate options to minimize the environmental concerns raised by the specialists. The alternative engineering solutions for particular areas usually included:

- the rehabilitation or replacement of weirs and sluices;
- embankments or flood walls;
- the construction of new flood channels and associated control structures;
- improved maintenance programmes;
- localized channel enlargement.

The process of consultation was iterative until a preferred option was identified.

The model used was ONDA, a general purpose program developed by Halcrows, the consultants on this project. Other consultants have similar models. Open channels were represented by their cross-sections and roughness coefficients, and flows calculated using the St Venant equations for the conservation of mass and momentum. Structures and other constrictions such as low bridges were represented by standard hydraulic discharge equations. Although the modelling was only in sufficient detail to support the consultation phase, an assessment of the model accuracy based on the calibration runs showed that predictions were within 100 mm of the observed values in 68% of cases. The differences included errors quantifying tributary inputs, gaugeboard levelling and reading errors, and errors in representing the operation of river regulation structures. The subsequent design stage involved additional site surveys and more detailed modelling of localized areas.

Mathematical models have proved to be essential aids in catchment planning throughout the Thames Region (Gardiner 1990). Beyond the design phase of a land drainage improvement scheme they continue to play an ongoing role in adaptive river management.

Several mathematical hydraulic models are commercially available. ONDA, HYDRO, SALMON-F and MIKE 11 are used extensively in the UK. However, the magnitude of the task in setting up, calibrating and maintaining a model on the scale of either the Nene or Lower Colne examples should not be underestimated. It is a

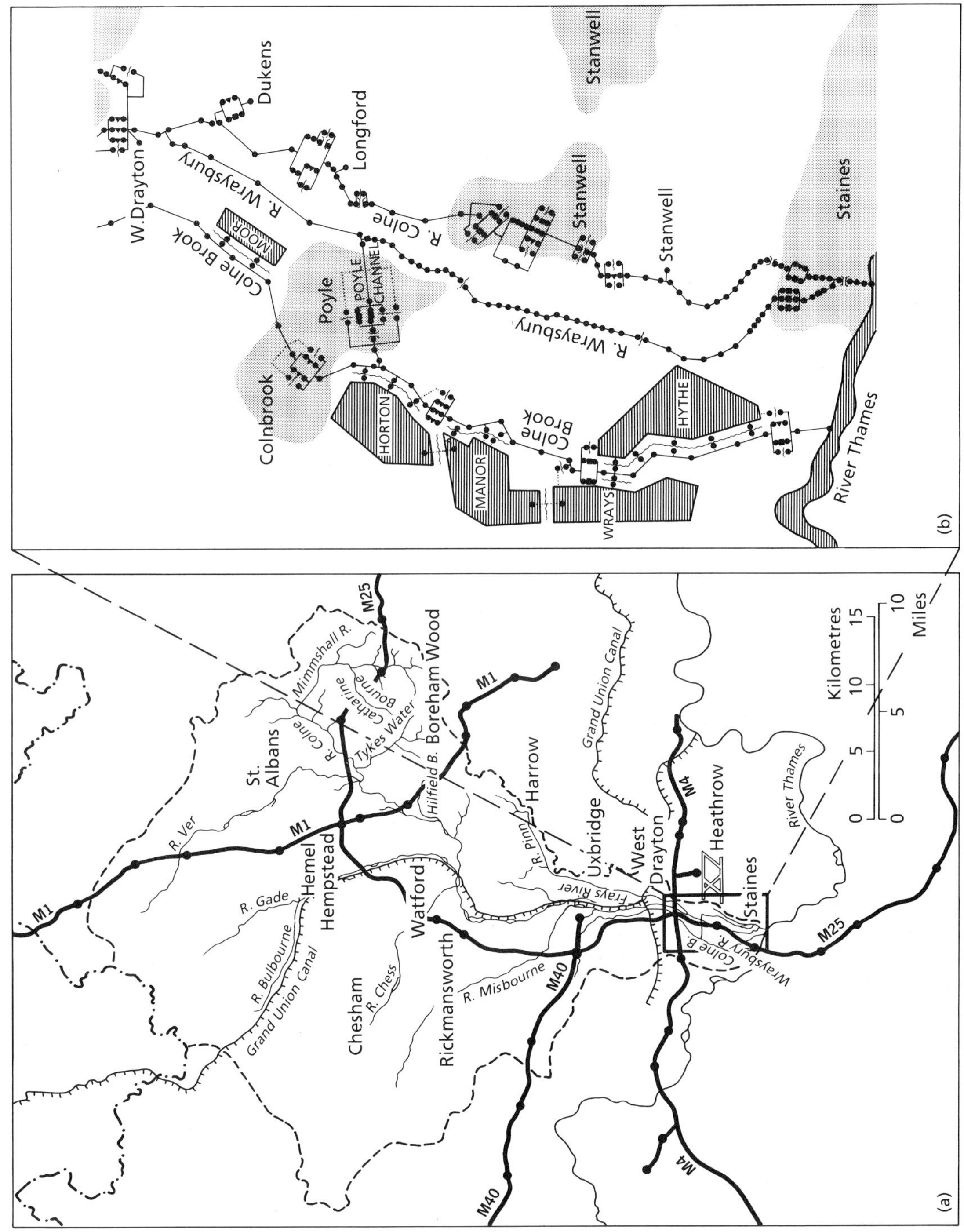

Fig. 10.14 The Lower Colne Study: (a) the Colne catchment; (b) schematic of part of the hydraulic model.

major undertaking which must be adequately resourced. Recent examples of the application of these models *by their developers* are as follow.

• The application of HYDRO to the Ouse Washes, part of the River Great Ouse in Cambridgeshire, England (Balfour *et al* 1989).

• The application of LORIS (the computational model within SALMON-F) also to the River Great Ouse in Cambridgeshire, UK (Slade & Samuels 1990).

• The application of MIKE 11 to the Kalani River catchment in Sri Lanka, in order to examine options for flood protection schemes and their impact on sediment transport (Van Kalken & Havno 1992).

10.6 MODELLING IN PRACTICE

With the growth of available databases over the past three decades, the number of applications where hydrological modelling can be useful has increased rapidly. Increasingly, hydrological models are being used to provide information for multi-disciplinary studies. Truly interdisciplinary modelling is still comparatively rare, but there are numerous examples of hydrological input to water quality or ecological models. Identification of the interface is the key issue. One common interface is the use of flow duration statistics as input to water quality planning models (see Chapter 11) or models of river ecology for quantifying in-river needs, e.g. PHABSIM (Bullock & Johnson 1991).

It would be all too easy to allow impressive examples of modelling such as those illustrated in the previous sections to create the illusion that models can be used to solve any problem. There is a danger that the jargon and technology could create a misleading and superficial gloss. The reliability of model results depends upon a great many contributing factors. These include: the soundness of the underlying theory; the skills of the modeller in translating that theory to a computer program; the quality of the input data; and the applicability of the technique to the issue in hand. If models are decision-making tools then the users of the results are, *de facto*, decision-makers. Inappropriate use of the results is often caused by ignorance of the model's limitations and its inherent assumptions and simplifications.

The important thing for users of models to remember is that you must be prepared to ask questions. As with any discipline, different hydrologists will have their own favourite techniques of analysis and modelling. These will not always be the most up to date, or the most appropriate or the best for the given problem. You should always ask why a particular modelling technique was used. There is a danger that modellers get so engrossed in the methodology and mathematics that they fail to explain the basics.

Hydrological studies are but one component in successful river management. This chapter has described the wide range of hydrological model types which are available and given examples to illustrate how they are used. An appropriate model can be an invaluable tool in helping to explain the complexities of catchment hydrology, as long as it is used correctly. It is hoped that engineers and scientists in related disciplines will now be able to better utilize the specialist hydrological advice at their disposal.

REFERENCES

Abbott MB, Bathurst JC, Cunge JA, O'Connell PE, Rasmussen J. (1986a) An introduction to the European Hydrological System – Système Hydrologique Européen 'SHE' 1. History and philosophy of a physically-based distributed modelling system. *Journal of Hydrology* **87**: 45–59. [10.3]

Abbott MB, Bathurst JC, Cunge JA, O'Connell PE, Rasmussen J. (1986b) An introduction to the European Hydrological System – Système Hydrologique Européen 'SHE' 2. Structure of a physically-based distributed modelling system. *Journal of Hydrology* **87**: 61–77. [10.3]

Anderson MP, Woessner WW. (1992) *Applied Groundwater Modeling*. Academic Press, London. [10.4]

Anderson PF, Ward DS, Lappala EG, Prickett TA. (1992) Computer models for subsurface water. In: Maidment D (ed) *Handbook of Hydrology*. McGraw Hill, New York. [10.4]

Balfour N, Guganesharajah K, Halifax P. (1989) Mathematical modelling of the Ouse Washes. In: *Proceedings of the Second National Hydrology Symposium*. British Hydrological Society/Institute of Hydrology, Wallingford, UK. [10.5]

Beven K. (1989) Changing ideas in hydrology – The case of physically based models. *Journal of Hydrology* **105**: 157–72. [10.3]

Beven KJ, Calver A, Morris EM. (1987) *The Institute of*

Hydrology Distributed Model. Institute of Hydrology Report No. 98, Wallingford, UK. [10.3]

Blackie JR, Eeles CWO. (1985) Lumped catchment models. In: Anderson MG, Burt TP (eds) *Hydrological Forecasting*, pp. 311–45. John Wiley, Chichester. [10.3]

Brown RPC, Ironside N, Johnson S. (1991) Defining an environmentally acceptable flow regime for the River Darent, Kent. In: *Proceedings of the Third National Hydrology Symposium*, pp. 2.59–2.66. British Hydrological Society/Institute of Hydrology, Wallingford, UK. [10.4]

Bullock A, Johnson I. (1991) Towards the setting of ecologically acceptable flow regimes with IFIM. In: *Proceedings of the Third National Hydrology Symposium*, pp. 2.67–2.72. British Hydrological Society/Institute of Hydrology, Wallingford, UK. [10.6]

Cluckie ID, Tilford KA. (1989) *Transfer Function Models for Flood Forecasting in the National Rivers Authority Anglian Region*. Anglian Radar Information Project Report No. 3. Department of Civil Engineering, Salford University, England. [10.3]

Cole GA, Gustard A, Beran MA. (1981) *Flood Hydrology of the River Nene*. Institute of Hydrology Report for Anglian Water Authority, Wallingford, UK. [10.3]

Cook R, Evans D, Barham P. (1992) Abstractive V's non abstractive uses for water or 'A better future for Anglian rivers'. In: *Proceedings of the Brighton Symposium*. Institution of Water and Environmental Management, London. [10.2]

Crawford NH, Linsley RK. (1966) *Digital Simulation in Hydrology: Stanford Watershed Model IV*. Technical Report 39. Department of Civil Engineering, Stanford University, USA. [10.4]

Danish Hydraulics Institute (1991) *SHE: Système Hydrologique Européen/European Hydrogical System*. Groundwater Modelling Report. DHI, Hørsholm, Denmark. [10.3]

Dingman SC. (1984) *Fluvial Hydrology*. WH Freeman & Co, New York. [10.5]

Dooge JCI. (1972) Mathematical models of hydrologic systems. In: Biswas AK (ed) *Modelling of Water Resources Systems 1*, pp. 170–80. Harvest House, Montreal. [10.2]

Eeles CWO, Robinson M, Ward RC. (1990) Experimental basins and environmental models. In: Hooghart JC *et al* (eds) *Proceedings of the International Conference on Hydrological Research Basins and the Environment, Wageningen*, pp. 3–12. Netherlands Organisation for Applied Scientific Research, The Hague. [10.3]

Gardiner JL. (1990) The River Thames Strategic Flood Defence Initiative Planning, a model influence. In: White WR (ed) *International Conference on River Flood Hydraulics*, pp. 447–58. John Wiley, Chichester. [10.5]

Gardiner JL, Dearsley AF, Woolnough JR. (1987) The appraisal of environmentally sensitive options for flood alleviation using mathematical modelling. *Journal of the Institution of Water and Environmental Management* **1**(2): 171–84. [10.5]

Greenfield BJ. (1984) *The Thames Catchment Model*. Internal Report, Thames Water Authority, Reading, UK. [10.3]

Jakeman AJ, Littlewood IJ, Whitehead PG. (1990) Computation of the instantaneous unit hydrograph and identifiable component flows with application to two small upland catchments. *Journal of Hydrology* **117**: 275–300. [10.3]

Jakeman AJ, Littlewood IJ, Symons HD. (1991) Features and applications of IHACRES: A PC program for identification of unit hydrograph and component flows from rainfall, evapotranspiration and streamflow data. In: *13th IMACS World Congress on Computation and Applied Mathematics, Dublin*. [10.3]

Laurenson EM, Mein RG. (1983) *RORB version 3. Runoff Routing Program User Manual*. Department of Civil Engineering, Monash University, Australia. [10.3]

Littlewood IG, Jakeman AJ. (1993) Characterisation of quick and slow streamflow components by unit hydrographs for single and multi-basin studies. In: Robinson M (ed) *Proceedings of a Conference of Methods of Hydrologic Basin Comparison, Oxford, 1992*, pp. 94–105. Institute of Hydrology, Wallingford, UK. [10.3]

Lowing MJ, Mein RG. (1981) Flood event modelling – a study of two methods. *Water Resources Bulletin* **17**(4): 599–606. [10.3]

McDonald MG, Harbaugh AW. (1988) *A Modular Three-Dimensional Finite-Difference Groundwater Flow Model*. Techniques of Water Resources Investigations 06-A1, United States Geological Survey. [10.4]

Manley RE. (1978) Simulation of flows in ungauged basins. *Hydrological Sciences Bulletin* **23**(1): 85–101. [10.3]

Manley RE. (1993) *HYSIM Information Pack*. RE Manley Consultancy, Cambridge, UK. [10.3]

Midgley P, Jones CRC. (1991) Rehabilitation of the Wallop Brook and Bourne Rivulet. In: *Proceedings of the Third National Hydrology Symposium*, pp. 2.37–2.42. British Hydrological Society/Institute of Hydrology, Wallingford, UK. [10.4]

Moore RJ, Jones DA, Black KB. (1989) Risk assessment and drought management in the Thames Basin. *Hydrological Sciences Journal* **34**(6): 705–17. [10.3]

Mott MacDonald. (1993) *The Integrated Catchment Model. User's Manual*. Mott MacDonald, Cambridge, UK. [10.4]

NERC (Natural Environmental Research Council). (1975) *Flood Studies Report*: 5 vols. NERC, London. [10.3]

NERC (Natural Environmental Research Council). (1989) *HYRROM. User's Manaual.* NERC, London. [10.3]

Newman AT, Symonds RD. (1991) The River Allen – A case study. In: *Proceedings of the Third National Hydrology Symposium*, pp. 2.53–2.58. British Hydrological Society/Institute of Hydrology, Wallingford, UK. [10.4]

Oakes DB, Keay D. (1990) *A Resource Model of the Great Ouse River System.* Report No. CO2504-M. Water Research Centre, Medmenham, UK. [10.3]

Quinn P, Beven K, Morris D, Moore R. (1989) The use of digital terrain data in modelling the response of hillslopes and headwaters. In: *Proceedings of the 2nd National Hydrology Symposium*, pp. 1.37–1.47. British Hydrological Society. [10.3]

Rushton KR. (1986) Groundwater models. In: *Groundwater Occurrence, Development and Protection.* Institution of Water Engineers and Scientists, London. [10.4]

Rushton KR, Fawthrop NP. (1991) Groundwater support of streamflows in the Cambridge area, UK. In: Nachtebel HP, Kovar K (eds) *Hydrological Basis of Ecologically Sound Management of Soil and Groundwater*, pp. 367–76. IAHS Publication No. 202, Institute of Hydrology, Wallingford, UK. [10.4]

Rushton KR, Redshaw SC. (1979) *Seepage and Groundwater Flow.* John Wiley, Chichester. [10.4]

Rylands WD, Lee JK. (1992) The possibility of an integrated approach to the provision of storage in the Thames Region of the National Rivers Authority. In: Saul AJ (ed) *Floods and Flood Management*, pp. 169–85. Kluwer. [10.3]

Shaw EM. (1983) *Hydrology in Practice.* Van Nostrand Reinhold (UK), Wokingham, England. [10.5]

Slade JE, Samuels PG. (1990) Modelling complex river networks. In: White WR (ed) *International Conference on River Flood Hydraulics*, pp. 351–8. John Wiley, Chichester. [10.5]

Spink AEF, Hughes AG. (1992) Development of a graphical user interface for use in groundwater management. In: Zannetti P (ed) *Proceedings of the 4th International Conference on Computer Techniques in Environmental Studies*, pp. 575–88. Elsevier Applied Science, Amsterdam. [10.3, 10.4]

Thompson N, Barrie IA, Ayles M. (1981) *The Meteorological Office Rainfall and Evaporation Calculation System: MORECS.* Meteorological Office, Bracknell, England. [10.3]

Van Kalken T, Havno K. (1992) Multipurpose mathematical model for flood management studies and real time control. In: Saul AJ (ed) *Floods and Flood Management*, pp. 169–85. Kluwer, Dordrecht. [10.5]

Wilby R, Greenfield B, Glenny C. (1993) *A Coupled Synoptic-Hydrological Model for Climate Change Impact Assessment.* NRA Research Paper, Loughborough University, UK. [10.3]

Williams JJR, Fawthrop NP. (1988) A mathematical hydraulic model of the River Nene – A canalised and heavily controlled river. *Regulated Rivers: Research and Management* **2**: 517–33. [10.5]

Wilson EM. (1983) *Engineering Hydrology.* Macmillan, London. [10.5]

11: Water-Quality Modelling

P. CROCKETT

11.1 INTRODUCTION

Since Streeter and Phelps (1925) developed the basic dissolved oxygen re-aeration equation, mathematical models have played an important role in the management of river water quality. With increasing public awareness, the introduction of pollution control legislation and improved computer techniques, the use of mathematical models to simulate river water quality has become more common. River water-quality models are now important tools that help managers and planners to evaluate water-quality issues, develop strategies for the protection and improvement of rivers and determine discharge consent conditions.

There are usually three groups of people involved with water-quality models, the model developer, the model user and the decision-maker. Each of these groups of people place varying emphases on the different components and aspects of water-quality models. In this review of river water-quality models I have concentrated on the needs of the model user and the decision-maker. Consequently, I have not included long lists of mathematical equations. Instead I have attempted to provide a framework in which the appropriate water-quality model can be selected and applied to investigate a particular water-quality problem.

This first section provides an introduction to the various terms used to describe the many diverse types of water-quality models. The next two sections detail the factors that should be considered when selecting and applying the appropriate water-quality model. I have included a section reviewing the different types of models that have been developed in response to various river water-quality problems. In this review section the emphasis has been placed on models used within Britain and to a lesser extent in North America and elsewhere. Finally, the current trends and potential future developments in water-quality models are discussed.

This chapter concentrates on mathematical models that simulate water quality in river systems upstream of the tidal limit. In addition, water-quality models developed to simulate large open water bodies such as lakes are not covered. However, the issues involved in model characterization, section and application is applicable to all water-quality models.

11.2 CHARACTERIZATION OF RIVER WATER-QUALITY MODELS

There are several ways of distinguishing between river water-quality models. These various categories are outlined below. Not every model will fit neatly into a category, and there are shades of grey between the generalized model types.

Modelling purpose

Models can be developed to help researchers gain an insight into the physical, chemical and biological processes that are active in natural rivers. River water-quality models can also be developed as tools to help managers and planners control the river system. Models developed for research are rarely suitable for management purposes. This is because the need to make the model suitable for general use is understandably given a low priority.

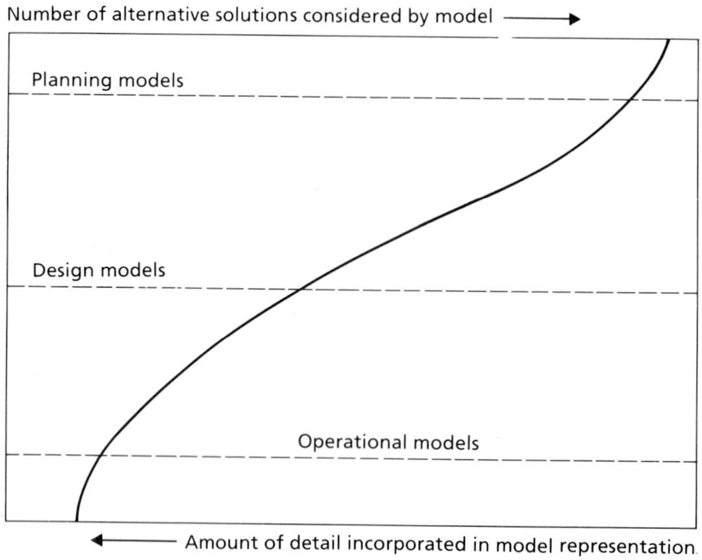

Fig. 11.1 Relationship between scale and detail for different classes of models (from Jamieson 1978).

Hierarchy of models

Jamieson (1978) suggested that there should be a hierarchy of water-quality models relating to the detail in which any scheme or proposal was being examined. This model hierarchy is shown in Fig. 11.1, and ranges from planning (or screening models), through design models to operational models. Such a sequence of modelling would allow the filtering of options so that only near-optimal solutions would be examined in more detail.

Planning models can evaluate many potential investment programmes, usually in terms of environmental impact and predicted costs. Design models enable a more detailed evaluation of various planning options. Operational models can be used in the day-to-day, or real-time, management of river systems.

Formulation

Deterministic/stochastic

Deterministic models have a fixed relationship between input and output. All the relationships are rigidly calculated and the output and input are similarly assumed not to be subject to random variations. Stochastic models allow the random fluctuations in natural systems to be reproduced. As discussed later recent river water-quality models have been developed which contain a stochastic component to cater for the uncertainty associated with the simulation of water quality.

Mechanistic/black-box

Mechanistic models attempt to represent the processes occurring within the river system. These internally descriptive models allow the effects of changing various processes and components of the system to be simulated. A black box simulates the relationships between inputs and outputs with no internal representation of the system. As a result it is only possible to simulate the system within the limits of the data upon which the model was developed.

Water-quality process

Hines *et al* (1975) diagrammatically showed the increasing complexity possible in the various biochemical interactions that can be modelled (Fig. 11.2). This ranged from the simple mass balance representation for conservative minerals, through simple dissolved oxygen (DO) models to modelling complex eutrophication processes. The data requirements will increase as the complexity of the model's representation of the system increases.

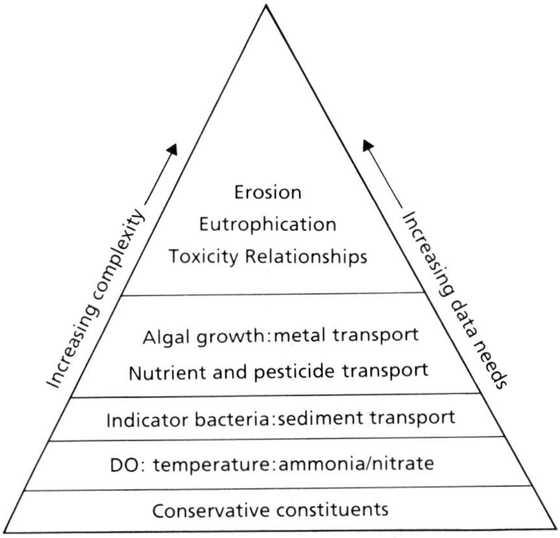

Fig. 11.2 Relative difficulty of applied modelling (from Hines *et al* 1975).

A good source for the formulation of water-quality processes is provided by Bowie *et al* (1985). This USEPA report provides an extensive review of the representation of water-quality process and provides various default values and global constants.

Physical and hydraulic representation

The prediction of water quality is dependent upon the way in which physical and hydrodynamic processes are simulated (Maskell 1991). It is important that the methods used to represent the various processes are appropriate to the application of the model.

Geometric representation

The geometric representation of a real-world water system used by a water-quality model can range from a simple zero-dimensional model, such as a constantly stirred tank reactor, to a fully three-dimensional representation of length, breadth and depth in a lake. Most river water-quality models use a one-dimensional approximation of a river by only simulating longitudinal differences in the river. This approximation assumes that vertical and horizontal variations are negligible. In larger rivers this assumption may not be valid, particularly if water quality is being simulated within the mixing zone of a discharge to the river.

Temporal variations

A water-quality determinand or process may be simulated as steady state (constant over time) or dynamic (variable over time). A water-quality model may have a mixture of steady-state and dynamic components. Where this happens, the use of the term dynamic to describe the model can lead to confusion. For example, a water-quality model may have a dynamic water-quality process (e.g. photosynthesis/respiration) while assuming a steady-state flow regime.

Steady-state models assume that all variables are independent of time. Dynamic models assume that all input parameters can vary with time. This division is not as clear as it might seem; some models are defined as pseudo-dynamic by allowing some parameters or processes to vary with time. For example, the biochemical processes can vary with time, while the flow is assumed to be steady-state.

Advection/dispersion/diffusion

Advection is the transport of particles by water currents whereas dispersion is the scattering of particles by differences in current velocities. To represent fully the impact of these processes on water particles would require a three-dimension dynamic representation of the water currents. With one-dimensional water-quality models it is common to use the diffusion process (the movement of particles from areas of high to low concentrations) to represent the process of dispersion.

Model generations

Abbott (1989) described how water-quality models have passed through various stages of development. Each stage can be described as a generation and can be related to developments in computer hardware and software. The first generation were the empirical formulae such as the Streeter and Phelps (1925) equation. The second generation occurred in the 1960s and involved the development of site-specific numerical models. During the 1970s the third generation of models were developed; these were generalized mainframe models. In the late 1980s a fourth generation of menu-driven PC- and workstation-based models have become increasingly available. A fifth generation of water-quality models will soon be developed, which will interface water-quality models with expert systems.

Comparison of model types

The complexity of all river water-quality model representations of a river system increases as the number of alternative solutions being studied decreases. The way in which this increased complexity is incorporated into the model depends on the water-quality problem being simulated. The complexity of a river water-quality model representation of a river system relates primarily to:

1 The number and type of water-quality determinands and processes being simulated. These range from the simple conservative determinands to the complex interactions involved in simulation, e.g. eutrophication.

2 How the physical and hydrodynamic processes of the river system are simulated. This ranges from zero-dimensional steady-state models to fully dynamic (flow and quality) three-dimensional models.

3 The sophistication of the model formulation, which can range from the use of simple statistical relationships to the use of complex mechanistic equations.

It is possible to combine these into a single diagram as shown in Fig. 11.3. Each axis relates to a different model component. The farther away from the origin, the more complex the model's representation and the greater the data requirements. The differences between different models can be seen in this type of diagram.

11.3 SELECTION OF RIVER WATER-QUALITY MODELS

The last 30 years have seen the development of large numbers of river water-quality models. Each of these models has been developed for a specific purpose, providing a large choice for any problem envisaged by the river quality modeller. This section aims to summarize the guidelines

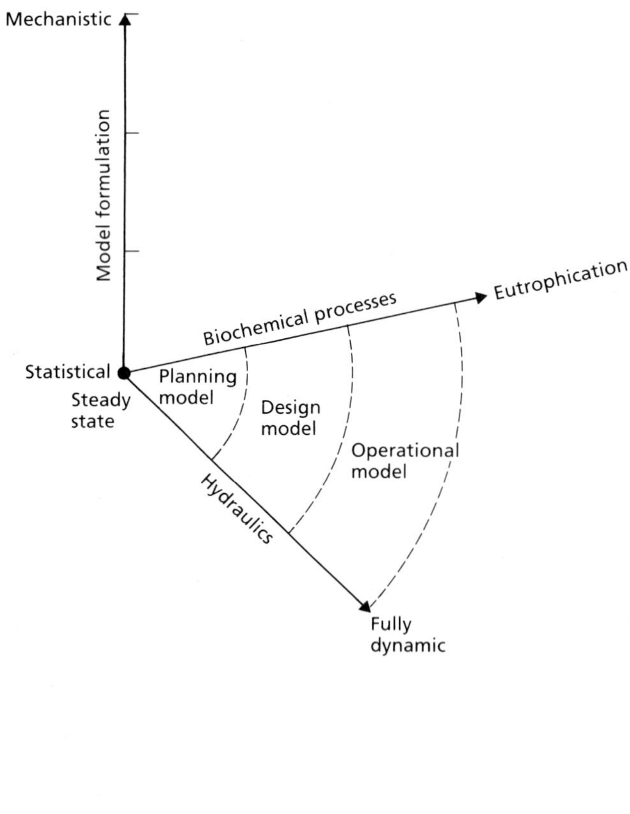

Fig. 11.3 Complexity of the various river system components in different classes of quality models (from Crabtree *et al* 1987).

that should be used in the selection of a suitable river water-quality model.

Before beginning the selection procedure there must be a clear idea of the reasons for the modelling exercise and the likely requirement for any model. This will include assessing the availability of data and deciding which is the most suitable type of model (e.g. stochastic and/or deterministic) and the relevant complexity of the model's representation of the various system components. Having decided upon the general modelling requirements, the next step is the evaluation of the river water-quality models that are available.

Grimsrud *et al* (1976) suggested a systematic procedure for the evaluation of river water-quality models. This checklist enables a comparison between various models and an assessment of the likely problems in their application. The various factors involved in this procedure are reviewed below.

Model capabilities

Is the model suitable for the river quality problem? For example, a three-dimensional model developed for an estuary is likely to be too complex for use as a river water-quality model. Conversely, a simple steady-state model may be inadequate in simulating the effect of intermittent storm overflows on a river.

Data requirements

Before attempting to apply any river water-quality model there should be a clear idea of the potential data requirements. Are the required data readily available? Are they detailed enough? If they are not available, how much will they cost to obtain? Finally, how accurate do the data need to be? Gerber (1982) cited inadequate and insufficient data as the most important factor influencing the cost and benefits of using models.

Model costs

Model costs come in two forms: the initial cost and the utilization cost. Initial costs are those involved in obtaining the model and the necessary documentation and equipment to run it. The cost of developing a model will be far greater than that of using a readily available model. The utilization costs include the labour and data collection needed to set up, calibrate and verify the model.

Model accuracy

In a modelling exercise three sources of error can affect the accuracy of a model's results.

1 The simplifications the model makes about the river system. The representation of the processes occurring within the system will be incomplete, and the model will be unable to mimic completely the changes in these processes.

2 The numerical solution technique used by the model. It should be noted that models which use analytical solutions do not have this problem. Models with finite element solution techniques can suffer from problems of stability and numerical dispersion. These numerical problems can be controlled by the sensible choice of temporal and spatial increments.

3 Inaccuracies in the input data. These can be either from inaccuracies inherent within the data used to apply a model or from the model being set up in an inappropriate way. The influence of these errors needs to be checked by carrying out a detailed sensitivity analysis.

All water-quality models will have inaccuracies. It is for the manager/model user to decide what level of error is acceptable. It is important that this associated error is quantified and remembered when the model simulation results are being analysed.

Ease of application

Much time can be wasted in a modelling exercise when a model is not, to use a much quoted phrase, 'user friendly'. Ideally a model should have readily available documentation covering in full every aspect of the model and how it should/can be set up. The program will also require a clear and complete results output. Most modelling exercises require the question 'What if?' to be asked. Consequently, the input data file should be easy to adjust to allow this sort of option-effect repetition to be carried out.

It is often better to use well-established modelling software that has been tried and tested, rather than develop a new model. This facility is increasingly provided for with 'off-the-shelf' packages such as QUAL-II (Roesner *et al* 1981) and MIKE-11 (Olesen *et al* 1989).

Summary of model selection

The relative weighting given to the various factors will vary with each application. The final choice will involve the balance of all potential costs and benefits. The guidelines for model selection and application of a river water-quality model have been summarized by Crabtree *et al* (1987) and are given in Table 11.1.

Table 11.1 Criteria for model selection and application (from Crabtree *et al* 1987)

A Define the problem and determine:
- (i) What questions need to be answered
- (ii) What information is required
- (iii) What information is readily available
- (iv) What is the required level of precision and accuracy or degree of confidence in the results
- (v) Identify what modelling and control options are available

B Apply an appropriate model (the simplest that can provide the answers):
- (i) Select or develop a model that fits the problem, not a problem that fits the model
- (ii) Use the least sophisticated model that will provide the required level of accuracy
- (iii) Do not confuse model complexity with accuracy or precision
- (iv) Question whether increased accuracy is worth the commensurate increase in effort and cost
- (v) Assess model sensitivity

C Evaluate the results and implications of the predictions produced by the model:
- (i) Consider the implications of any modelling assumptions
- (ii) Consider the implications of the degree of confidence in the results
- (iii) Assess the value of the results
- (iv) Reassess, in the light of the results, the suitability and relative significance of the available options
- (v) Identify any further possible course of action of new options

D Make recommendations or decisions

11.4 UTILIZATION OF RIVER WATER-QUALITY MODELS

The stages involved in a river water-quality modelling exercise are shown schematically in Fig. 11.4. This section covers the procedures involved in the calibration, verification and application of a river water-quality model.

Data

The importance of data in the application of river water-quality models is highlighted in Fig. 11.4. When critical data are lacking, water-quality modelling becomes more of an art than a science. Very often the model user is faced with the problem of incomplete and inadequate data sets. The more sophisticated a model's representation of a river system the greater its data requirements are likely to be. The simplifications imposed on a model's representation of a river system due to the lack of suitable data are often greater than those imposed by the model formulation.

Calibration

Several authors have described these functions in detail, e.g. Thomann (1982) and Whittemore *et al* (1982). Having developed or selected the model, the next stage is to calibrate it using field data. Calibration usually involves estimating and adjusting various model parameters followed by sensitivity analysis on the parameters. Simulated model values need to be compared with field data using graphical and statistical techniques.

Thomann (1982) suggested that in a preliminary analysis of a model an error of 10% would be acceptable. When calibrating a model, the validity of any adjustments to constants and coefficients should be checked with published values. The results of the calibration exercise may lead to a reassessment of the model structure.

Validation

Even when calibration is carried out with the greatest of care, there is still the possibility that the model will be 'rigged' to fit the observed data. Consequently, once a model is calibrated it should

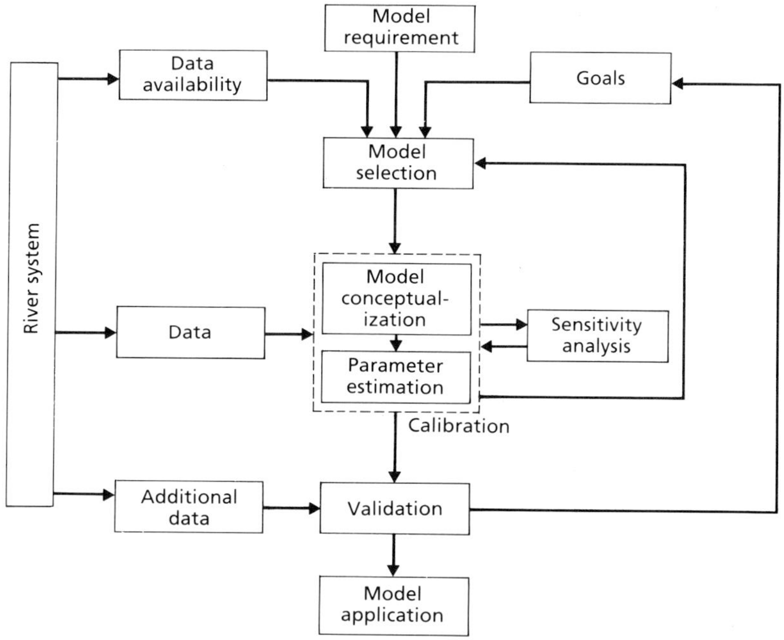

Fig. 11.4 Illustration of modelling procedure (from Linders 1980).

be verified using separate data set. Thomann (1982) reviewed several verified models and found errors of between 20 and 30%. This level of error was considered a suitable objective to achieve in river water-quality model validation. The results of this exercise may lead to further changes in the model structure.

Once a model has been calibrated and verified, it can then be used to simulate changes to the river system. However, it is often difficult to obtain a suitable data set for model verification. This will not invalidate the model but it will reduce the confidence that can be placed on the simulation results.

Communication of results

The inability of either the mathematical modeller to evaluate or the decision-maker to appreciate, the results of a model simulation is a common problem. This communication problem should decrease as the use of a model increases. The modeller must recognize the need to supply the simulation results in a clear and concise form. It is important that all significant assumptions made in the model's application are also reported. The decision-maker must take these assumptions into account when using the results of the model's simulation.

11.5 TYPES OF RIVER WATER-QUALITY MODELLING PROBLEMS

It would be theoretically feasible to develop a river water-quality model that would model every component of any river system. The resulting model would, however, be too complex for practical use. Instead there has been a tendency to develop models for specific problems or types of problems. There are examples of modelling packages that can be utilized for a range of river management tasks. These tend to be hydrodynamic models, which can be used for flood forecasting or to provide the flow input into a water-quality module.

River water-quality models tend to be developed to look at a specific type of river-quality problem. Crabtree (1986) defined six types of river-quality problem relating to the pollution source.

1 Setting effluent standards for continuous discharges.

2 Agriculture and diffuse rural runoff.
3 River impact of non-point urban discharges.
4 Real-time operational control.
5 Fate of toxic substances.
6 Pollution incidents.

The characteristics of the river water-quality model for each purpose are very different. Although there is some overlap in the processes required, their significance will vary.

Continuous discharge consent setting

This area of river water-quality modelling has seen the development of the largest number of models. This is due to the need to control continuous discharges and the relative simplicity of the required models. These types of models are typically used to simulate dissolved oxygen levels at various points along a river system. This is done by modelling the interactions of physical and biological processes affecting oxygen demand. The models are generally steady-state flow regimes and assume that longitudinal dispersion is insignificant.

Figure 11.5 shows a simplified framework for the main components of a pollution control strategy. The failure of a river to meet its required water quality will be detected by the monitoring programme. These water-quality standards will also constrain the various solution options.

The nature of the water-quality standards is central to the modelling activity. This can be illustrated by comparing the river water-quality models used in the USA and the UK for setting discharge consents. The differences in the types of river water-quality standards in these two countries have led to the development of two completely different types of river water-quality model for setting discharge consents to rivers.

Critical design period approach – US approach

Discharge standards in the USA are commonly established using fixed emission standards set for each polluting activity without reference to the receiving water. Where these standards are not strict enough for the watercourse to support fish life, more stringent quality constraints are imposed. There are several options that could be used to set these new standards (David & David 1983). Commonly, the new quality standards are set so that the river will have adequate river quality during a 1 in 10-year low flow when associated with high temperatures. This 'critical period' approach has led to the development of biochemically complex, steady-state-flow river water-quality models based mainly on modelling the interactions between biochemical oxygen demand (BOD) and oxygen. The one most widely used is QUAL-II (Roesner *et al* 1981).

QUAL-II is applicable to well-mixed dendritic stream systems. It can predict both the temporal and spatial variations of up to 13 water-quality constituents in any combination needed by the user. The model may be used in steady-state models where all the variables are solved directly,

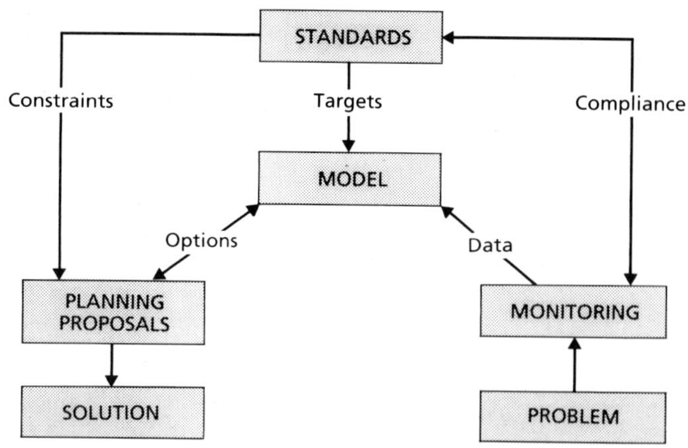

Fig. 11.5 Framework for applying river water-quality models (from Crabtree *et al* 1989).

or in a pseudo-dynamic state where climatic conditions, e.g. temperature and daylight, are allowed to vary. The model is calibrated and verified using data from two intensive field surveys taken during low-flow conditions.

Stochastic simulation – UK approach

The legislative background of pollution control in the rivers of England and Wales has produced models with quite a different structure. As consent standards are set within a probabilistic framework the model is required to predict the 95 percentile discharge quality associated with the achievement of a 95 percentile river water quality. As a result consent-setting models in England and Wales have concentrated on the correct characterization of the data inputs to a model rather than the deterministic representation of river processes.

Several consent-setting models have been developed to enable the statistically-based discharge consents to be set on a catchment scale. There is currently a choice of three: TOMCAT, developed by Thames Water (Bowen & Brown 1984); SIMCAT, developed by NRA Anglian Region (Warn 1987); and QUASAR, developed by the Institute of Hydrology (Whitehead *et al* 1991). A fundamental component of these models is the method adopted to generate representative distributions for the various model inputs and to ensure that the interaction between these inputs is considered. Each of the three models has adopted a different approach. It should be noted that QUASAR can also simulate critical design periods within a time-dependent mode.

The application of a version of TOMCAT to the River Rother in Yorkshire has been described by Lai and Vevers (1986). Incomplete data sets, particularly for effluent and tributary, caused problems in setting up the model. The accuracy of and confidence in the simulation results of these stochastically-based water-quality models are directly related to the availability of suitable data.

Impact of land use

Non-point or diffuse pollution results from changes in catchment characteristics. These may be caused by a deterioration in background or natural river quality that cannot be attributed to continuous or intermittent point-source discharges or random accidental spillage (Crabtree 1986). This definition highlights the difficulty of modelling or assessing the effect of non-point pollution. This is only possible after consideration of all known point-source discharges.

Non-point pollution is due to increased leaching of pollutants by runoff. Some commonly attributed causes are:

1 Ecological changes due to changing catchment hydrology.

2 Land use changes such as afforestation of upland catchments.

3 Acid precipitation altering the balance of geochemical processes releasing chemicals and sediments by weathering and erosion.

4 Agricultural activities releasing organic and chemical wastes and by-products.

Catchment process simulation models need to be able to model the interaction of hydrological, geochemical and ecological processes and the effects of change. To do this, a model has to be able to model rainfall, evaporation, runoff, soil systems, geology, vegetation and river processes. If they are applied to more than one catchment, the modelling framework tends to be large and requires many parameters to be assumed.

In Britain the MAGIC model has been developed by the Institute of Hydrology. The model simulates long-term changes in soil water and river water chemistry in response to changes in acidic deposition. MAGIC has investigated the impact of atmospheric pollution and afforestation in sensitive upland catchments (Whitehead 1991).

SHE is a European modelling initiative. The model is primarily directed at deciding quantitative flow predictions. Recently, the model has been developed to predict nitrate movements.

In North America there are several models available; of these HSPF (Donigan *et al* 1984) seems to have the greatest potential. Once its environmental parameters have been set, it is relatively easy to change the chemical and land use parameters, to model a variety of scenarios (Tynan *et al* 1989). HSPF does, however, require extensive field data and careful calibration and validation.

Catchment-scale land use models tend to be large, complex and require a comprehensive database. Before attempting to apply these models their data requirements should be fully comprehended and assessed.

Storm-derived urban discharges

Intermittent, rainfall-related discharges from combined and separate urban sewerage systems plus storm tank overflow discharges from sewage treatment works can have marked long- and short-term effects on the quality and associated ecology of rivers in urban areas. It is possible to recognize three different levels of sophistication in the types of models used to assess the impact of these intermittent discharges on river water quality. These three levels depend on the significance of impact of the intermittent discharge on the receiving watercourse.

Low-impact simulation

Low-impact discharges can be defined as ones where the dilution for the intermittent discharge is greater than 8:1 and there is no interaction with other discharges. In such cases it should be adequate to use the standard discharge control methods, such as Formula A (Ministry of Housing and Local Government 1970) or QUALSOC (NRA 1991).

Medium-impact simulation

Where the dilution ratio is less than 8:1, the population is equivalent to greater than 2000, and there is a cyprinid fishery it may be necessary to apply a simple modelling approach. Examples of such approaches in the UK are the Scottish Development Agency Approach (SDD 1977), QUALSOC (NRA 1991), QUALSIM (a statistical development of QUALSOC) or CARP (Crabtree *et al* 1988). Several screening models have also been put forward in the US (Heaney *et al* 1976).

High-impact simulation

Where the impact of an intermittent urban discharge is considered to have a highly significant impact on the receiving water quality, it may be appropriate to apply a more sophisticated modelling approach. Criteria for this high impact would include: a dilution of less than 2:1; interaction with other discharges; greater than 10000 population equivalent; and a cyprinid fishery.

A Danish river impact model to predict acute (short-term) and cumulative (long-term) impacts of storm sewer overflows has been developed. The total overflow volumes and loads are computed by the runoff model SAMBA (Johansen *et al* 1983). The associated river water-quality model is steady-state and assumes that the effect of the polluting load lasts considerably longer than the actual duration of the event (Johansen *et al* 1984).

To model short-term changes in the sewer and river systems requires a complex dynamic model of hydraulic performance and quality. Currently no such model exists in Britain, but in the USA, SWMM III RECEIVE attempts to simulate short-term quality changes (Huber *et al* 1981). However, SWMM only attempts to model the generation and transport of pollutant and flow, and does not have the ability to model sediments.

To accurately predict the impact of urban runoff a river water-quality model needs to be able to predict the behaviour of soluble and sediment-related determinands as well as the changes in river flow. The development of this type of river water-quality model has tended to be added to existing hydraulic design or simulation models. An example of this has been the enhancement of the Danish Hydraulic Institute model MIKE-11 (Olesen *et al* 1989).

Real-time operational control

This section is concerned with models that have been developed to analyse on-line data of past and present river water quality to predict future downstream river water-quality trends. These predictions can then be used for operational control of abstractions and discharges to the river system.

Lettenmaier and Burges (1975) suggested three levels of 'operational' model sophistication depending on the waste loading of the river. These ranged from simple monitoring for a low-use river to the use of continuous monitors. Only when the river has a high pollution risk, or when the effect of such an incident occurring with very little warning is particularly serious, should a modelling exercise be contemplated.

A real-time operational model has been developed to protect drinking water abstractions on the Bedford Ouse (Whitehead *et al* 1983). This was developed from a dynamic river water-quality model for the same river system (Whitehead *et al* 1981). This uses recursive parameter estimation to continually update or calibrate the model using data fed back from various on-line river-quality monitoring telemetry stations. The model can then predict future concentrations of BOD, dissolved oxygen and nitrate. There is insufficient information to asses whether this improved information is worth the high cost of developing and running the model.

In Britain there has recently been considerable interest in the development of river-specific pollution incident models following serious pollution incidents on major water supply rivers, for example the River Dee in North Wales. DEEMOD is one such model produced for the River Dee by Welsh Water and the University of Lancaster (Weston & Crookes 1986). The model was developed, from existing black-box statistical relationships, to predict the time of the first arrival of a pollutant at a point, the time to the pollution peak and the time of the end of the incident.

Deriving statistically-based models seems to be the best option for the development of real-time operational models. When predicting the travel times and spread of pollution slugs down a river system it is essential that the model predictions can be updated quickly as new data become available. Statistically-based models have been developed by Young, Wallis and colleagues using the ADZ model and other techniques (e.g. Young & Wallis 1986). In the future, as real-time telemetry systems become more robust the application of these statistically-based models will become more widespread.

Modelling toxic substances

The processes that decide the transport or fate of toxic substances are considerably more complex than those used in modelling the interaction of BOD and dissolved oxygen. Processes such as volatilization, absorption, desorption and sedimentation are important, and any model must be able to simulate these processes.

Most work in this area has been carried out in North America. In the USA, a general purpose model has been produced for predicting the transport and fate of toxic chemicals in the environment (Barnwell 1984). This model, called TOXIWASP (Ambrose *et al* 1983), is a dynamic model for simulating the transport and fate of toxic chemicals in water bodies. Clark *et al* (1986) described how QUAL-II (see above) was used as a screening model to identify and assess pollutants likely to affect water abstraction from a river. Chemical-specific models have also been developed, such as TOXIFATE (Halfon 1986), which was used to look at the fate of perchloroethylene in the St Clair River, Canada.

In the UK some preliminary work has been carried out at the Institute of Hydrology into modelling heavy metals (Williams *et al* 1985). This area of model development is likely to progress further in the next few years in response to the introduction of new standards.

Simulation of pollution incidents

A careful distinction must be made between the simulation of pollution incidents in real time, as discussed above, and the simulation of a potential pollution incident as part of planning procedure or to show the potential source of the pollution after the event has occurred. The latter two cases require a more sophisticated deterministic/ mechanistic model. A number of models already mentioned can be used to simulate pollution incidents; these include MIKE-11 and QUASAR.

11.6 FUTURE DEVELOPMENTS

Mathematical models provide the best method of assessing the relative impacts of proposed changes to a system. As increasing demands are placed on river systems the use of water-quality models will increase. In turn the increased utilization of water-quality models will enable them to be developed and improved. Two main areas where existing models are being improved are the user interfaces employed and the range of processes they simulate.

The development of a suitable user interface is essential to ensure the utilization of a computer model. ANNIE-IDE (Kittle *et al* 1989) has been developed by the US Environmental Protection Agency (EPA) to provide a consistent methodology for building interactive interfaces for environmental software. The package consists of a toolkit of FORTRAN subroutines and instructions on the production of the associated text file and control files. ANNIE-IDE software and manual greatly reduce the effort required by the programmer to develop an interactive user interface.

Expert systems are likely to be increasingly linked to computer models. The development of high-level computer languages and increasing computer power has made expert systems feasible. These advanced computer languages are able to make inferences using a database of knowledge. This information will eventually make it easier for a model user to apply the model to a catchment.

The settlement and resuspension of cohesive sediments often have a significant impact on several water quality processes. Increasingly water-quality models are being developed that can represent the movement of sediments and their interaction with water-quality determinands. Although data requirements are increased, more reliable results can be obtained, particularly when the flow regime in a river is being changed.

Ecological models are increasingly being added on to water-quality models. These models use the simulation water quality and flow and attempt to determine what effect they will have on the ecology of a river. A current example of this is the FGETS model developed by the US EPA (Barber *et al* 1988). This model simulates the bioaccumulation of organic pollutants in fish.

11.7 CONCLUSIONS AND SUMMARY

River water-quality models have had a long history of development. Many different forms of model have been developed to look at various pollution sources and inputs. This chapter has reviewed the types of river water-quality model available. Guidelines for model selection and application have been detailed. These highlight the many different factors that must be considered when selecting an appropriate model for a particular river or problem. The most important of these factors are a clear understanding of the modelling requirement, knowledge of the limitations of available data and the ease of model use.

Six types of river water-quality model, related to the source of river pollution, have been reviewed. This review has indicated the diversity of water-quality models that have been developed. The differences in the type of models used to define discharge consents in the USA and the UK demonstrate the importance of understanding the framework in which a model is being applied. This relates particularly to the form of the standards that are used to define what is an acceptable river quality.

The future will see the increased use of river water-quality models. In part this will be due to greater understanding of their limitations, but developments in the user interfaces and the processes simulated will also aid this process.

ACKNOWLEDGEMENT

The author is grateful to the National Rivers Authority for permission to publish this paper. The views expressed in the paper are, however, entirely those of the author.

REFERENCES

Abbott MB. (1989) Review of recent developments in coastal modelling. In: *International Conference on Hydraulic and Environmental Modelling of Coastal, Estuarine and River Waters, Bradford University*. Gower Press. [11.2]

Ambrose RB, Hill SI, Mulkey LA. (1983) *User's Manual for the Chemical and Fate Model TOXIWASP, Version 1*. EPA-600/3-83-005. USEPA Environmental Research Laboratory, Athens. [11.5]

Barber MC, Suarez LA, Lassiter RR. (1988) *FGETS – A Simulation Model for Predicting Bioaccumulation of Non-polar Organic Pollutants by Fish*. USEPA, Athens, GA 30613. [11.6]

Barnwell TO Jr. (1984) Water quality models available from the EPA. *Water Science and Technology* **17**: 1407–10. [11.5]

Bowen K, Brown SR. (1984) Relating effluent control

parameters to river quality objectives using a generalised catchement simulation model. *Water Science and Technology* **16**: 197–206. [11.5]

Bowie GL, Mill WB, Porcella DB, Cambell CL, Pagenkopf JR, Ruppen GL, Johnson KM, Chan PWH, Gherini SA, Chamberlin CE. (1985) *Rates, Constants and Kinetics Formulation in Surface Water Quality Modelling*, 2nd edn. EPA/600/3-85/040. USEPA Environmental Research Laboratory, Athens. [11.2]

Clark RM, Grayman WM, Goodrich JA. (1986) Toxic screening models for water supply. *Journal of Water Resources Planning and Management* **112**(2): 149–65. [11.5]

Crabtree RW. (1986) *River Water Quality Modelling in the UK. A Review of Past and Present Use with Recommendations for Future Requirements*. External Report ER 196E. WRc, Swindon. [11.5]

Crabtree RW, Cluckie ID, Forster CF, Crockett CP. (1987) Mathematical modelling requirements for river quality control. *Water Pollution Control* **86**(1): 51–8. [11.2, 11.3]

Crabtree RW, Crockett CP, Toft AR. (1988) *Interim River Water Quality Planning. Procedures for Controlling Intermittent Pollution from Storm Sewage Overflows*. External Report 317E. WRc, Swindon. [11.5]

Crabtree RW, Crockett CP, Ellis JC. (1989) Continuous effluent consents: Modelling and compliance testing. *Environmental Monitoring and Assessment* **12**: 149–64. [11.5]

David MH, David EL. (1983) Cost effective regulatory options for water quality limited streams. *Water Resources Bulletin* **19**: 3421–28. [11.5]

Donigan AS, Imhoff JC, Bicknell BR, Little JL. (1984) *Application Guide for Hydrological Simulation Program – FORTRAN (HSPF)*. EPA-600/3-84-065. USEPA Environmental Research Laboratory, Athens. [11.5]

Gerber A. (1982) Review and background. *Water Science and Technology* **14**: 1045–52. [11.3]

Grimsrud G, Finnemore EJ, Owen HJ. (1976) *Evaluation of Water Quality Models: a Management Guide to Planners*. EPA-600/5-76-004. US Environmental Protection Agency, Washington, DC. [11.3]

Halfon E. (1986) Modelling the pathways of toxic contaminants in the St Clair – Detroit river system using the TOXIFATE model: The fate of perchloroethylene. *Water Pollution Research Journal of Canada* **21**(3): 411–21. [11.5]

Heaney JP, Huber WC, Nix SJ. (1976) *Storm Water Management Model Level 1. Preliminary Screening Procedures*. EPA-600/2-76-275. USEPA Environmental Research Laboratory, Athens. [11.5]

Hines WG, Rickert DA, McKenzie JW, Bennet JP. (1975) Formulation and use of practical models for river quality assessment. *Journal of the Water Pollution Control Federation* **47**(10): 2357–70. [11.2]

Huber WC, Heaney JP, Nix SJ, Dickenson RE, Polmann DJ. (1981) *Stormwater Management Model User's Manual. Version III*. EPA-600/2-84-1090a. [11.5]

Jamieson DG. (1978) A hierarchical approach to water quality modelling. In: James A (ed) *Mathematical Models for Water Pollution Control*. John Wiley, New York. [11.2]

Johansen NB, Harremoes P, Jensen M. (1983) Methods for calculation of annual and extreme overflow events from combined sewer systems. In: *International Seminar on Rainfall as the Basis for Urban Runoff Design and Analysis, Copenhagen, 24–25 August*. [11.5]

Johansen NB, Hvitved-Jacobsen T, Harremoes P. (1984) Simulation of the impact of combined sewer overflow on rivers. *Nordic Hydrology* **15**: 325–32. [11.5]

Kittle JL, Hummel PR, Imhoff JC. (1989) *ANNIE-IDE, a System for Developing Interactive User Interfaces for Environmental Models (Programmer Guide)*. USEPA/600/3-89/034. USEPA Environmental Research Laboratory, Athens. [11.6]

Lai PW, Vevers AM. (1986) Ammonia modelling for the Rother catchment. The Monte-Carlo approach. *Water Pollution Control* **85**(3): 315–23. [11.5]

Lettenmaier DF, Burges SJ. (1975) Dynamic water quality management strategies. *Journal of the Water Pollution Control Federation* **47**(12): 2809–19. [11.5]

Linders JBHJ. (1980) The interaction between data collection and water quality modelling. In: Stiff MJ (ed) *Rivers Pollution Control*. Ellis Horwood, Chichester. UK. [11.4]

Maskell JM. (1991) Hydrodynamic and physical considerations for water quality modelling. In: *IWEM Water Quality Modelling Symposium, Harrogate*. The Institution of Water and Environmental Management, London. [11.2]

Ministry of Housing and Local Government. (1970) *Technical Committee on Storm Overflows*. HMSO, London. [11.5]

NRA (National Rivers Authority). (1991) *QUALSOC User Manual*. January 1991, Version 5. NRA Wales, Cardiff. [11.5]

Olesen KW, Havno K, Malmgren-Hansen A. (1989) A water quality modelling package for fourth generation modelling. In: *Proceedings of the IAHR Congress, Ottawa*. The International Association for Hydraulic Research. [11.3, 11.5]

Roesner LA, Gigure PR, Evenson DE. (1981) *Computer Program Documentation for the Stream Quality Model QUAL-II*. EPA 600/9-81-014. USEPA Environmental Research Laboratory, Athens. [11.5]

SDD (Scottish Development Department). (1977) *Storm Sewage Separation and Disposal*. HMSO, London. [11.5]

Streeter HW, Phelps EB. (1925) *A Study of the Pollution*

and Natural Purification of the Ohio River. Public Health Bulletin, US Public Health Service, Washington, No. 146. [11.1, 11.2]

Thomann RV. (1982) Verification of water quality models. *Journal of Environmental Engineering, ASCE* **108**(EE5): 923–40. [11.4]

Tynan PJ, Moore K, Watts CD. (1989) *A Review of Environmental Fate and Transport Models*. Report LR 66-M. WRc, Medmenham, UK. [11.5]

Warn AE. (1987) SIMCAT – A catchment simulation model for planning investment for river quality. In: Beck MB (ed) *Systems Analysis in Water Quality Management*. IAWPRC, Pergamon Press, Oxford. [11.5]

Weston AE, Crookes JG. (1986) Pollution incidents on the River Dee: Use of a real time model on a micro-computer. Paper presented at the Institution of Civil Engineers, Pennines Hydrological Group Meeting, January 1986. [11.5]

Whitehead PG. (1991) The role of models in environmental impact assessment. In: *IWEM Water Quality Modelling Symposium, Harrogate*. The Institution of Water and Environmental Management, London. [11.5]

Whitehead PG, Beck B, O'Connel PE. (1981) A systems model of stream flow and water quality in the Bedford Ouse River system. II Water quality modelling. *Water Research* **15**: 1157–71. [11.5]

Whitehead PG, Caddy DE, Templeman RF. (1983) An on-line monitoring, data management and forecasting system for the Bedford Ouse river basin. *Water Science and Technology* **16**: 295–314. [11.5]

Whitehead PG, Williams R, Clark J, Murdoch N, Sambrook H, Mann B. (1991) The application of QUASAR for Environmental Impact Assessment. In: *IWEM Conference on Water and the Environment, Birmingham*. The Institution of Water and Environmental Management, London. [11.5]

Whittemore RC, Hovis JS, McKeown JJ. (1982) A study of the selection, calibration and verification of mathematical water quality models. Stormwater and Water Quality Model Users Group Meeting. EPA-600/9/82/015. [11.4]

Williams RJ, Whitehead PG, Bird SC. (1985) *Strategies for Managing Nitrates and Heavy Metals in River Systems*. Institute of Hydrology Report, Wallingford, UK. [11.5]

Young PC, Wallis SG. (1986) The Aggregated Dead Zone (ADZ) model for dispersion in rivers. In: *BHRA International Conference on Water Quality Modelling in the Inland Natural Environment*, pp. 421–33. BHRA, Cranfield, UK. [11.5]

12: Sediment Transport and Channel Stability

R. BETTESS

All natural rivers transport sediment in varying quantities and, in terms of the quantity of material moved and the extensiveness of the network, they represent one of the largest transportation systems in the world (see Vol. 1, Chapter 5). One of the aspects of rivers that makes them so interesting is that changes in the sediment transport lead to changes in the river morphology (see Chapter 4; see also Vol. 1, Chapter 6) and so the shape of a river reflects the history of sediment transport within it (see Vol. 1, Chapter 7). In this chapter some of the problems of the modelling sediment transport are addressed. The most fundamental one is the prediction of sediment transport rate. This then leads to the prediction of river morphology, i.e. regime theory and changes in morphology using numerical models. An important aspect of river morphology is the planform of rivers, and methods of predicting planform are described. The types of problems and approaches to them are summarized in Table 12.1.

12.1 SEDIMENT TRANSPORT

Sediment transport has been extensively studied for over 100 years but the complexity of both the fluid flow and the interaction with the sediment is sufficiently complex that there are still aspects which are not very well understood. The prediction of sediment transport rate is based upon the assumption that for a given steady flow and a given type of sediment there is a unique sediment transport rate. Observations have suggested that the sediment transport rate of noncohesive sediments depends upon the following parameters:

1 Flow:
- velocity
- depth
- shear velocity
- kinematic viscosity
- fluid density.

2 Sediment:
- sediment size
- sediment density.

The complexity of phenomena has resulted in all sediment transport theories having an empirical content. Most theories are based upon some broad theoretical approach such as energy, efficiency or minimum rate of energy dissipation but all contain parameters or functions which have been determined empirically. In many cases this empirical content restricts the range of applicability of the theories, and the range and quantity of data on which a theory was developed are important features. Until recently most work was directed at predicting the movement of uniform sediments, i.e. sediments which consist predominantly of the same sized particles. Currently more attention is being paid to widely graded sediments which are more representative of sediments in natural channels.

For the purposes of analysis of sediment transport a division is frequently made depending upon the mode of transport.

1 *Bed load* — refers to the sediment that is moving predominantly in contact or close to the bed.

2 *Suspended load* — refers to the sediment that moves predominantly suspended in the fluid flow but interacts with the sediment on the bed; i.e. there is a continual exchange between the

Table 12.1 Problems in sediment transport and channel stability

Problem	Approach	Advantages	Disadvantages	Example
Quantify amount of sediment in motion	Use total load equation	Simple to apply	Limited accuracy	Ackers and White equation
Predict alluvial friction	Use alluvial friction equation	Simple to apply	Some theories have limited range of applicability	White, Paris, Bettess equation
Predict impact of engineering works, e.g. impact of changes in flow, sediment supply or geometry				
Predict bed level change	Numerical morphological model	Quantitative predictions, including rate of change	Large data requirements	Flumorph, HEC6
Predict armouring	Numerical morphological model	Quantitative predictions, including rate of change	Large data requirements	Flumorph, HEC6
Predict channel shape	Regime theory	Simple to apply	Predicts new equilibrium but no indication of rate of change	Lacey
Predict plan form	Regime theory	Simple to apply	Predicts new equilibrium but no indication of rate of change	Bettess and White

Table 12.2 Sediment transport theories

Author	Sediment sizes	Number of data sets used in development	Comments
Ackers and White	0.1–28 mm	925	Restricted to Froude number less than 0.8. Lightweight sediment data also used
Brownlie	0.1–2.8 mm	900	
Engelund and Hansen	0.19–0.93 mm	Approx. 140	Theory contains no threshold condition and thus predictions at low shear stresses may be subject to error
Meyer-Peter and Muller	3–28 mm	Approx. 200	Bed-load formula
Van Rijn	0.2–2 mm	130	Restricted to Froude number less than 0.9 and depths larger than 0.1 m

sediment in the flow and on the bed as sediment is removed from the bed and suspended in the flow while other sediment settles out from the flow on to the bed.

3 *Wash load* — refers to sediment that moves in suspension in the flow but is not represented in the bed of the channel. It is generally assumed that the transport of wash load is supply dependent and is independent of the local flow conditions.

4 *Total (bed material) load* — refers to the total amount of sediment in motion excluding wash load, i.e.:

$$\text{Total load} = \text{Bed load} + \text{Suspended load}$$

The division into these categories is somewhat artificial and varies both spatially and temporally as the flow varies.

There are a bewildering array of theories to predict bed load, suspended load and total load (e.g. Garde & Ranga Raju 1985; Raudkivi 1990). A small number of theories are more widely used than most and have been demonstrated to be among the most reliable. These are the theories by Ackers and White (1973), van Rijn (1984a,b) and Brownlie (1981). Due to lack of space only the Ackers and White theory is described in detail here but the major differences and relative advantages of the different theories are summarized in Table 12.2.

These theories normally provide predictions of concentration within 50% of the observations for approximately 40% of the data, while the predictions are within a factor of two for approximately 75% of the data. Laboratory experiments have indicated that for similar flow conditions the observed sediment concentration may vary by up to a factor of two, suggesting that the variability exhibited by the theories probably reflects the variability in the data.

A number of studies have been carried out concerning the accuracy of the sediment transport predictions. Though the precise details of the studies differ, the results are generally broadly similar. The best group of predictors, which includes Ackers and White, Brownlie and van Rijn, all achieve approximately the same level of accuracy, i.e. between 70 and 80% of the predictions are within a factor of two of the observed value. Other predictors can display considerably worse accuracy. Despite considerable effort over the last 20 years, there has not been a significant improvement in the accuracy of the predictions of sediment transport. This may, in part, be attributable to the limitations in the type of approach adopted and, in part, to scatter and inaccuracies in the data used to calibrate and test the theories.

It should be noted that the error quoted is for individual observations. There is evidence to suggest that, for example, where sediment transport theories are used to calculate an annual sediment load by summing daily calculated values then the accuracy of the annual load calculated is substantially better than for the individual values. This probably arises as the errors between the predicted and observed are normally distributed and by summing a large number of normal variates the standard deviation is reduced substantially. This also explains the success of numerical morphological models in making long-term morphological predictions.

Ackers and White sediment transport theory

In many methods (Task Committee 1971; Cole *et al* 1973; Raudkivi 1990), shear stress is used as the main parameter defining the stream's transporting power. However, the total shear on a deformed bed, rippled or duned, is, in part, composed of the along-stream components of the normal pressures on the irregular bed profile. These normal pressures may contribute indirectly to sediment motion through suspension, yet many methods separate the bed shear into a non-transporting form loss and the shear on the grains. As the rate of transport is very sensitive to transporting power, inaccuracy in this shear separation procedure may give large errors of prediction. In engineering practice, this factor is important because few natural streams have a plane bed.

Other methods utilize stream power or average velocity against which the transport rate is correlated. Whatever the basis, care must be exercised when applying any method because it is not always clear whether the equation relates to the total load of bed material or to some ill-defined fraction of it, for example, the contact load. It was because of deficiencies of many previous methods that Ackers and White developed a new framework for the analysis of sediment transport data, which could be used predictively.

The Ackers and White method avoided refinements that complicated the application without adding much to accuracy. The advantages of dimensional analysis were incorporated, but physical arguments were used in deriving the form of the functions. The variables used were directly related to those the engineer can readily visualize and measure, and the uncertainty of slope separation procedures was avoided. The various coefficients were expressed as algebraic functions so that the total transport of bed material may be calculated directly.

Sediment mobility

The original development of the theory is summarized in Ackers and White (1973). In essence, a coarse sediment is considered to be transported mainly as a bed process. If bed features exist, it is assumed that the effective shear stress bears a similar relationship to mean stream velocity as with a plane, grain-textured surface at rest.

A fine sediment is considered to be transported within the body of the flow, where it is suspended by the stream turbulence. As the intensity of turbulence is dependent on the total energy degradation, rather than on a net grain resistance, for fine-grained material the total bed shear stress is effective in causing sediment transport.

Sediment mobility is described by the ratio of the effective shear force on a unit area of the bed to the immersed weight of a layer of grains. This mobility number is denoted by F_{gr}, and defined by:

$$F_{gr} = \frac{V_*^n}{\sqrt{[gD(s-1)]}}\left[\frac{V}{\sqrt{32}\log_{10}\left(\frac{\alpha d}{D}\right)}\right]^{1-n} \qquad (1)$$

where V_* is the shear velocity, V is the velocity of flow, D is the sediment diameter, d is the depth of flow, g is the acceleration due to gravity and s is the sediment specific gravity. n and α are parameters, the latter usually assigned the value 10. For coarse sediments ($n = 0$) the expression reduces to the form:

$$F_{gr} = \frac{V}{\sqrt{[gD(s-1)]}}\frac{1}{\sqrt{32}\log_{10}\left(\frac{\alpha d}{D}\right)} \qquad (2)$$

and for fine sediments ($n = 1$):

$$F_{gr} = \frac{V_*}{\sqrt{[gD(s-1)]}} \qquad (3)$$

The hypothesis was that n could be evaluated by analysing flume data for a range of sediment sizes and should lie between 0 and 1, showing a continuous variation through coarse material and transitional sizes to fine non-cohesive sand or silt.

Dimensionless grain diameter

A dimensionless expression for grain diameter can be derived by eliminating shear stress from the two Shields parameters (Shields 1936): or derived from the drag coefficient and Reynolds number of a settling particle, by eliminating the settling velocity; or dimensionally, with immersed weight of an individual grain, fluid density and viscosity as the variables. The dimensionless grain diameter, D_{gr}, is therefore generally applicable to coarse, transitional and fine sediments and is the cube root of the ratio of the immersed weight to the viscous forces. The following expression characterizing sediment diameter remains constant in any one set of experiments:

$$D_{gr} = D\left[\frac{g(s-1)}{\nu^2}\right]^{1/3} \qquad (4)$$

where ν is the kinematic viscosity.

Sediment transport

The expressions for sediment transport were based on the stream power concept, in the case of coarse sediments using the product of net grain shear and stream velocity as the power per unit area of bed, and for fine sediments using the total stream power. The useful work done in sediment transport in the two cases takes account of the different modes of transport assumed, and in relation to the stream power gives an expression for the efficiency of the transport process. Efficiency was expected to be dependent on the mobility number, F_{gr}. Clearly there would be a value of F_{gr} at which sediment motion first began to take place. As F_{gr} rises above this limiting value, it was expected that the efficiency would increase.

In order to separate the primary variables, the efficiency (which is dimensionless) was combined with the mobility number, F_{gr}, to yield a general transport parameter:

$$G_{gr} = \frac{Xd}{sD}\left[\frac{V_*}{V}\right]^n \qquad (5)$$

where G_{gr} is the sediment transport rate (dimensionless), and X is the sediment concentration by weight.

General transport function

The relationship for G_{gr} was assumed to be of the form:

$$G_{gr} = C(F_{gr}/A - 1)^m \qquad (6)$$

where A, C and m are parameters. A is the value of F_{gr} at which sediment movement first begins to take place.

Numerical optimization was used to determine best fit values for the four parameters C, A, m and n. The resulting relationships were:

For transitional sizes, $1 < D_{gr} < 60$:

$$n = 1.00 - 0.56\log_{10} D_{gr} \qquad (7)$$

$$A = \frac{0.23}{\sqrt{D_{gr}}} + 0.14 \qquad (8)$$

$$m = \frac{9.66}{D_{gr}} + 1.34 \qquad (9)$$

$$\log_{10} C = 2.86\log_{10} D_{gr} - (\log_{10} D_{gr})^2 - 3.53 \qquad (10)$$

For coarse sediments, $D_{gr} > 60$:

$$n = 0.00 \qquad (11)$$

$$A = 0.17 \qquad (12)$$

$$m = 1.50 \qquad (13)$$

$$C = 0.025 \qquad (14)$$

A recent study has been carried out to up-date the values of the parameters in the light of the larger amount of data that is now available. This study recommended:

$$m = \frac{6.83}{D_{gr}} + \begin{matrix} 1.67 & \text{for } 1 \leqslant D_{gr} \leqslant 60 \\ 1.78 & \text{for } D_{gr} \geqslant 60 \end{matrix} \qquad (15)$$

$$\text{and } \log C = 2.79\log D_{gr} - 0.98(\log D_{gr})^2 - 3.46 \quad \text{for } 1 \leqslant D_{gr} \leqslant 60$$
$$\text{and } C = 0.025 \quad \text{for } D_{gr} \geqslant 60 \qquad (16)$$

A value for α of 10 was determined. This is comparable with the value of 12.3 which appears in the rough-turbulent flow equation. The transport function is plotted in Fig. 12.1.

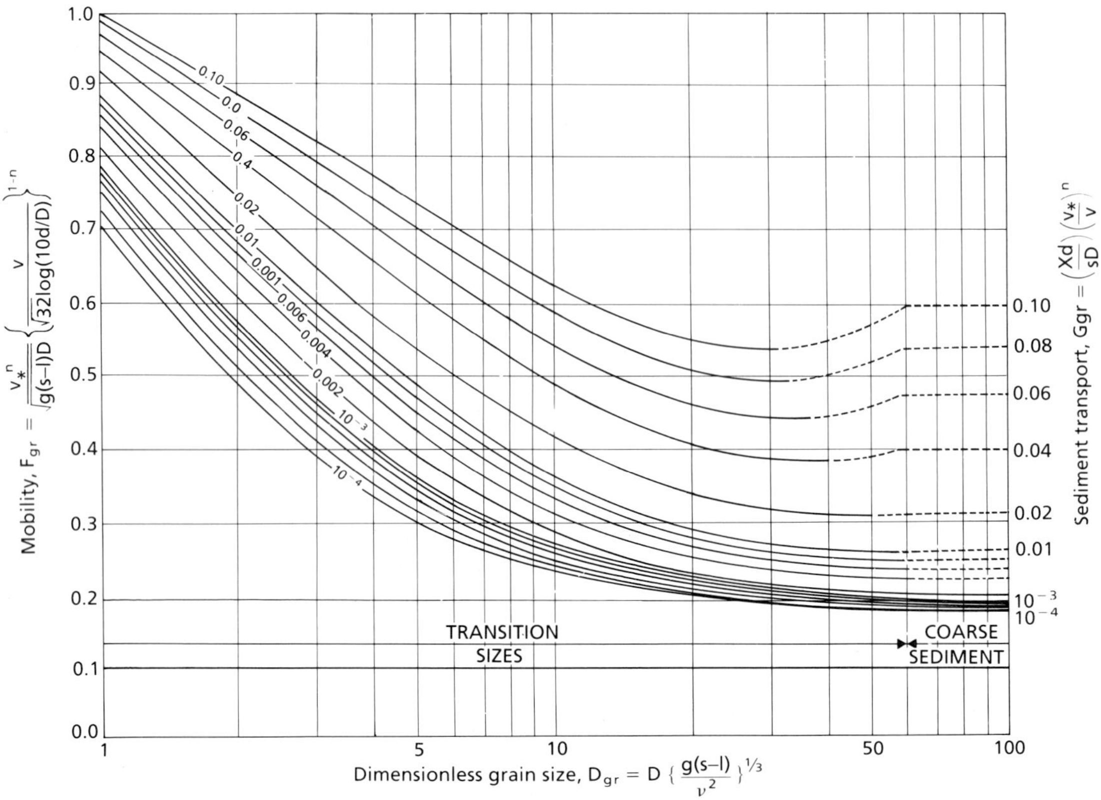

Fig. 12.1 The Ackers and White sediment transport function.

Transport of graded sediments

Many naturally occurring sediments are not uniform but contain a wide range of sediment sizes. As sediment mobility is a function of sediment size, in general the finer sediments are moved more easily than the coarser ones. This difference in sediment transport rate can lead to a number of phenomena, all described under the heading of sediment sorting. For example, bed degradation is frequently associated with the development of a surface layer of coarser sediment which can reduce or entirely prevent further degradation. Such phenomena are important in practice and to predict them one must be able to calculate the transport of widely-graded sediments.

The simplest assumption to make is that each size fraction moves independently from the other fractions. Thus, the movement of each size fraction can be determined using a uniform sediment transport theory taking due account of the proportion of material on the bed. It has been found, however, that the presence of other size fractions does influence transport. In general the transport of the coarser fractions is increased in graded sediment in comparison with a uniform sediment, while the transport of the finer fractions is reduced. To take account of this, empirical 'hiding' factors have been introduced to uniform sediment transport theories. There is limited information available for such factors and it is likely that hiding factors are specific to each sediment transport theory; i.e. a hiding function developed for Ackers and White sediment transport theory would not be applicable to van Rijn's sediment transport theory and vice versa.

Table 12.3 Numerical modelling of alluvial rivers

Type of model	Comments	Advantages	Disadvantages
1-D	Only provides information averaged over a cross-section	Easy to apply. Can be applied cheaply to long lengths of rivers and for long time periods	Only section-averaged properties
2-D, depth averaged	Provides variation across section	Easier to apply than 3-D	Difficult to represent secondary flows
3-D		Very good spatial resolution	At present, can only be applied to limited lengths of river

12.2 MORPHOLOGICAL RIVER MODELS

Attempts have been made to develop numerical models to predict river behaviour (De Vries 1976; US Army Corps of Engineers 1976; Chollet & Cunge 1980). What is required is a range of different models; the type of model to be used for a particular investigation must depend on which quantities are being considered, the details that are required and the time-scale under study. The success of such models is entirely dependent on how well they can describe the dominant physical processes in an alluvial river. The application of numerical models to such problems has not always been completely successful, which may in part be due to a failure to take account of all the significant processes involved. The roles, advantages and disadvantages of various types of model are summarized in Table 12.3.

In this section the development of a model which can predict the change in bed level of a river occurring over a number of years is described and results are presented to indicate the kind of quantitative information that may be determined. A one-dimensional, time-dependent model is discussed, i.e. a model in which all the properties of the river, such as discharge, stage and bed level, depend on only two variables: the distance along the river and the time. As the model is one-dimensional it is possible to discuss only values of variables which have been averaged over a cross-section and it is not possible to determine how values of variables change across the width of the river. The model described takes no account of bends in the river or any effects caused by bends, such as alterations in the cross-sectional shape and sediment sorting around a bend. Such a one-dimensional model cannot directly predict changes in the plan of the river. Correspondingly, alterations in channel plan do not influence such a model provided the overall length of the river under consideration does not change.

Principles of model

Mathematical formulation

The one-dimensional model is to describe the movement of the water and sediment within a river. The St Venant equations describe the slowly varying, unsteady flow of water in a fixed-bed, open channel (Chow 1959). The equations consist of a continuity equation,

$$B\frac{\partial A}{\partial t} + \frac{\partial Q}{\partial x} = 0 \qquad (17)$$

where B is channel width, A is cross-sectional area, t is time, Q is discharge rate, and x is horizontal distance; and a dynamic equation:

$$\frac{\partial Q}{\partial t} + \frac{\partial}{\partial x}\left(\frac{Q^2}{A}\right) + gA\frac{\partial d}{\partial x} = gA(S_y - S_f) \qquad (18)$$

where S_y is the bed slope, S_f is the friction slope, d is the depth of flow, and g is the acceleration due to gravity. These equations determine the discharge and depth as a function of distance and time.

If G is the volume of sediment per unit width flowing through a cross-section in unit time, the continuity equation for sediment movement may be written as:

$$B\frac{\partial z}{\partial t} + \frac{\partial G}{\partial x} = 0 \quad (19)$$

provided that the spatial variation of the suspended load is small. To determine the bed level, z, it only remains to relate the sediment movement to the hydraulic conditions in the river, i.e. to determine the sediment movement as a function of the water velocity (V), depth and slope:

$$G = G(V, d, S_f) \quad (20)$$

As discussed above there are many sediment transport theories of this form. Owing to the complexity of the processes the theories attempt to model, the predictions of the different theories in any particular case may vary considerably (White *et al* 1973).

Boundary conditions

In general the time-scale associated with changes in the fluid flow is considerably shorter than that for variations in the bed level. It follows that, if one is interested only in morphological changes, the fluid flow can be regarded as steady (De Vries 1976). Over a short period any change in bed level has little effect on the fluid flow. Thus, it may be assumed that over the period of a time-step the fluid flow may be calculated independently of the sediment transport. Hence one may calculate the fluid flows for a geometrically identical, fixed-bed channel and determine the velocity and depth. These values may then be used to calculate the sediment transport, which, together with equation 19, determines the changes in bed level. The calculation of the fluid flow thus corresponds to the familiar backwater calculation to determine velocity and depth in steady flow. To determine the fluid flow two boundary conditions must be specified; to determine the sediment transport a boundary condition on the sediment inflow must be given at the upstream end.

Sediment transport

Equation 19 implies that the only effect of any error in predicting the transport rate is to alter the rate at which the bed of the river rises or falls; thus a 10% overestimate of the transport rate leads to a 10% increase in the rate of degradation of aggregation of the bed. The final equilibrium state does not primarily depend on the transport rate although errors in the latter will affect the prediction of the time taken for this equilibrium to be achieved.

Frictional characteristics

Concomitant with sediment transport are bed features such as ripples and dunes. These features can change their size and form as the sediment transport rate varies. As such features can have a significant effect on the frictional characteristics of a channel, it follows that in an alluvial channel subject to considerable changes in sediment transport rate the roughness of the channel cannot be assumed to be fixed. To take account of these alterations a roughness relationship developed by White *et al* (1980) can be included. This is described in more detail later.

Sediment sorting

So far only the total volume of sediment that is moving has been considered and no attention has been given to how much of each sediment size is being transported. For given hydraulic conditions different sediment sizes are transported at different rates and this may lead to a restructuring of the sediment size distribution and hence to changes in the sediment sizes used in the sediment transport calculations. Consider, for example, the movement of sediment immediately downstream of a newly constructed dam and assume that there is no sediment inflow, i.e. all the sediment is trapped in the reservoir. Downstream of the dam the fine sediment will be transported more easily than the coarse sediment and so over a period of time one would expect the proportion of fine sediment to decrease, the proportion of coarse sediment to increase correspondingly, and the total transport rate to fall. Thus the sediment size distribution will change and the appropriate representative sediment sizes will alter. As the proportion of coarse sediment will increase, the representative sediment size will increase, leading to a decrease in the calculated transport rate. Thus, in this case, if one ignores the effects of sediment sorting one will overestimate the transport rate and the calculated degradation will exceed observations.

To model sediment sorting one must be able to calculate the amount of each sediment size that is moving. The Ackers and White (1973) sediment transport theory only provides an expression for the total load, but White and Day (1980) have suggested a method of adapting the theory to give transport rates of individual size fractions.

Using this method for determining transport rates a provisional model of sediment sorting has been developed. There are two main assumptions in the model: that at any given time only sediment to a particular depth in the bed, called the active depth, is affected by the water flow, and that within this active layer the sediment grading curve does not vary with depth. The active depth is related to the height of the bed features present, which in turn is assumed to be proportional to the effective roughness height. In the absence of any reliable data providing the functional relationship between active depth and effective roughness height it is assumed that the two are equal. However, if data are available another form for the relationship should be used. These assumptions together with equations of continuity for each sediment size provide enough information for determining how the composition of the bed varies. The model keeps an inventory of the sediments at different elevations within the bed and uses the appropriate sediment sizes as erosion or accretion takes place.

Numerical solutions of equations

Equations 18 and 19 may be solved separately. As steady conditions have been assumed, the determination of the fluid flow reduces to a simple backwater calculation.

The fluid flow model is steady state. However, when modelling a river over a period of time it is necessary to consider different discharges which are operative for varying lengths of time. Discharge information is usually readily available either as a discharge hydrograph or as a flow duration curve. In both cases these may be approximated by steady discharges over varying lengths of time. Where a discharge is required for a time period exceeding the maximum allowable time-step in the sediment conservation equation the time period must be divided into a number of periods, each with the same discharge.

It is normally assumed that the fluid flow can be solved independently of the sediment transport, i.e. that over the time-step considered alterations in bed level do not significantly alter the flow.

There is increasing emphasis on environmental impacts and their assessment. Frequently the flora and fauna of a channel are related both to the hydraulics of the flow and the size and nature of the channel sediments. There is thus an interest in being able to model both the flow and sediment patterns within river reaches.

One-dimensional modelling can give a generalized picture over long reaches of river. For example, modelling of degradation downstream of a dam can give predictions of the expected change in bed sediment size. This may be useful in predicting the impact of the dam on fish populations and particularly the future availability of sites for redds. Since such models can only deal in section-averaged properties, however, they frequently do not provide the detail required to look at specific habitats. Where there are significant variations on a short spatial scale, either laterally or longitudinally, one would normally have to resort to more detailed modelling using 2-D or 3-D models. Two-dimensional depth-averaged models frequently suffer from an inability to represent adequately the effect of secondary flows. In many cases, therefore, resort has to be made to 3-D modelling. The computational effort required for 3-D modelling means that at present, only restricted lengths of river can be modelled. In formulating 2-D and 3-D models a number of extra issues are raised. The impact of local bed slopes, both longitudinally and laterally, must be considered and the influence of bank effects accounted for. The formulations necessary for representing these in models are only now being developed.

Application of the model

Sand bed river

A model was applied to a reach of the Middle Loup River immediately downstream of the Milburn Dam in Nebraska, completed in May 1956. The US Bureau of Reclamation carried out a study of aggradation and degradation associated with the dam (US Bureau of Reclamation 1968) which provided cross-sections of the river at a number of sites and details of the sediment and discharge in the river. Surveys were carried out in June 1961 and June 1964 and so it was decided to simulate the degradation downstream of the dam during this period. The reach under consideration was about 5200 m long. The Bureau of Reclamation survey provided three cross-sections of the river in this length (lower range 1, 2 and 2A) and further cross-sections were linearly interpolated between these to provide details of sections at approximately every 200 m.

A time-step of 3 h was used in the model. The discharge was obtained from a flow duration curve which was entered as data into the model as a series of discharges and corresponding durations, in descending order of discharge. Measurements had been made of the quantity of sediment passing through the dam so that the sediment inflow to the upstream of the model could easily be determined. The model was arranged so that when the maximum change in bed level in any one time-step was less than 0.000 02 m it was deemed that no further change in bed level would occur in that year and the next year's series of discharges was considered. The observed and predicted bed levels are compared in Fig. 12.2; the agreement is excellent. No data were available for the change in sediment grading during the period simulated but the calculated composition of the surface layer is shown in Figs. 12.3 and 12.4. The initial grading of the bed was assumed to be independent of the depth. Figure 12.3 shows the grading curves immediately downstream of the dam after 17 days and 71 days. As a flow duration curve, arranged in descending order of discharge, was used the discharge at time *T* days is the discharge exceeded for a proportion *T*/365 of the year and the results at time *T* must be interpreted as the state of the system following all discharges in excess of that corresponding to time *T*. As *T* increases up to 365 days the discharge decreases, starting a new sequence in the next year.

Figure 12.4 shows the changes in the D_{35} and D_{50} values in time at three locations on the river. (D_{35} is the sediment diameter, in metres, for which 35% of the sample is finer; D_{50} is the sediment diameter for which 50% of the sample is finer.) The initial high flows rapidly produce an armoured surface layer immediately downstream of the dam with a less marked armouring developing farther downstream. The high flows at the beginning of the second year disturb the armoured layer, mixing it with the parent material below and causing an initial reduction in the sizes of D_{35} and D_{50}. As the flows continue

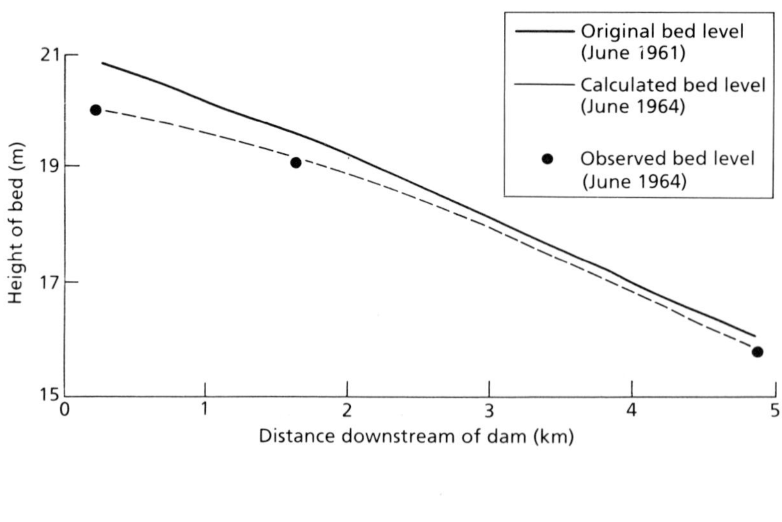

Fig. 12.2 Degradation downstream of Milburn Dam.

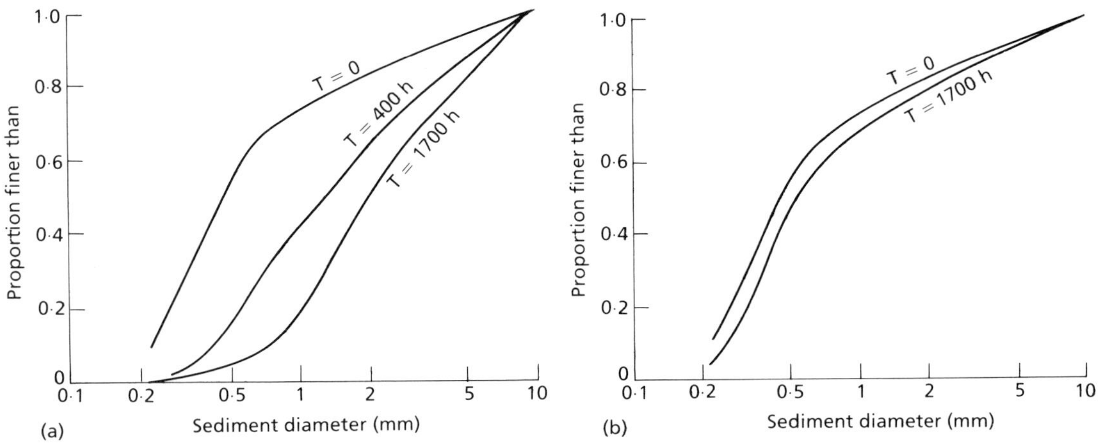

Fig. 12.3 Sediment grading courves: (a) immediately downstream of dam; (b) 2.5 km downstream of dam.

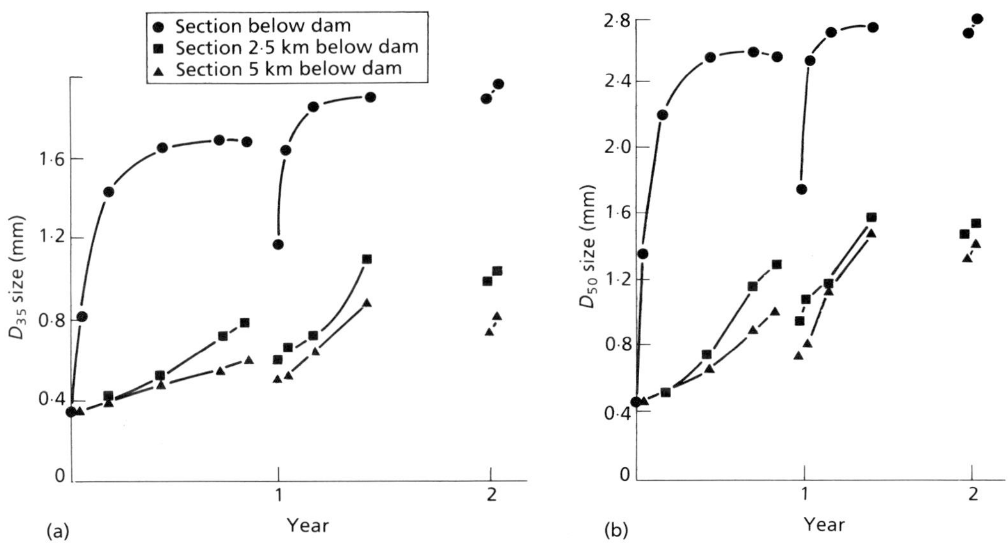

Fig. 12.4 (a) D_{35} sediment size as a function of time; (b) D_{50} sediment size as a function time.

the armoured layer is re-formed. The results cease partway through the second year as effective sediment transport ceased then. The results indicate that after a few years an armoured layer will develop which even the highest flows cannot disturb, but it should be appreciated that the results of the model may be sensitive to the order in which discharges and durations are fed in as data.

12.3 RIVER REGIME

Introduction

The problem of determining a stable cross-section geometry and slope of an alluvial channel has been the subject of considerable research over 80 years and continues to be of great practical interest (Ackers & Charlton 1970; Charlton *et al*

1978; Ackers 1980; Chang 1980). Ignoring plan geometry, an alluvial channel can adjust its width, depth and slope to achieve a stable condition in which it can transport a certain amount of water and sediment. Thus, it has three degrees of freedom and the problem is to establish relationships which determine these three quantities of width, depth and slope.

The various approaches to this problem fall into two broad categories: the regime and analytical methods. The regime method is an empirical method which relies on available data and attempts to determine appropriate relationships from the data. An early approach was made by Kennedy (1895), who collected data from stable canals and used this to derive a relationship between the mean velocity and depth of flow.

Empirical regime theory

The theory was first developed by British engineers working in the Indian subcontinent and was derived in a response to the problem of designing large irrigation canal systems. Extensive measurements were taken on existing stable channels, i.e. channels that were 'in regime'. Parameters such as the depth, width and slope were then related to the discharge. The general form of the equations was:

$$d = f_1(Q) \qquad (21)$$

$$B = f_2(Q) \qquad (22)$$

$$S = f_3(Q) \qquad (23)$$

where Q is the discharge, d is the depth of flow, B is the channel width, and S is the slope. Later, Lacey (1930) introduced a silt factor which was related to the diameter of the sediment particles. This was to take some account of the dependence of the regime of the river on sediment size and concentration.

Since then a large number of studies have been carried out utilizing data from many areas of the world and for a large range of channel sizes. The usefulness of the derived relationships depends upon the quality of the data and the validity of the assumed relationships. Also, equations should only be applied in situations similar to those for which the data were collected. The most extensive regime study carried out specifically for British rivers was undertaken by Charlton *et al* (1978).

The analytical method relies on specifying equations which describe the dominant individual processes such as sediment transport, flow resistance and bank stability. This approach can only be successful if the dominant processes are correctly identified and appropriate equations exist to describe them adequately. These approaches represent two extremes and obviously it is possible to combine aspects of both.

Analytical regime theory

In the analytical approach two sets of equations are readily available defining the sediment transport and the frictional characteristics, but it is unclear what constitutes an appropriate third equation. Several proposals have been made for a suitable equation, some concerned with bank stability (Ackers 1980), others based on some variational principle such as minimum stream power or minimum unit stream power (Chang 1980; Yang 1976). White *et al* (1981) advocated a variational principle based on the assumption that an alluvial channel adjusts its geometric characteristics and gradient in such a way that the sediment transporting capacity is maximized. No physical justification to support the principle of maximizing the sediment transporting capacity has been found but it is a useful hypothesis which leads to acceptable predictions over a large range of flow conditions. Interestingly, it can be shown that this principle is equivalent to the minimum stream power concept (White *et al* 1982). By using this variational principle together with the Ackers and White equations for sediment transport and the White, Paris and Bettess equations for flow resistance, the geometric and hydraulic parameters of a stable channel have been obtained for a wide range of practical applications.

Formulation of regime theory

Six variables are considered that describe the channel system; the average velocity, V, average depth, d, slope, S, discharge, Q, sediment concentration, X, and channel width, B. Relating these variables are equations for the continuity of water flow, sediment transport formulas, flow resistance formulas and the condition that the sediment transport should be maximized or, equivalently, stream power should be minimized. In the following treatment the discharge and slope are assumed to be imposed and the corresponding values of V, d, X and B are determined. Implicit in this treatment are the assumptions that the flow is steady and uniform and that the bed and bank material is non-cohesive.

Sediment transport. The Ackers and White (1973) equation has been used to calculate the sediment concentrations.

Frictional characteristics. By utilizing the same basic parameters as in the Ackers and White sediment transport theory, White *et al* (1980) found that a linear relationship between the sediment mobility related to the total shear stress, F_{fg}, in which:

$$F_{fg} = \frac{V_*}{\sqrt{[gD(s-1)]}} \tag{24}$$

(where V_* is the shear velocity, D is the sediment diameter, s is the sediment specific gravity, and g is the acceleration due to gravity), and the mobility, related to the effective shear stress, F_{gr}, existed with coefficients depending on D_{gr}, the dimensionless sediment diameter. An extensive correlation for a wide range of sediment sizes (0.4–10 mm) gave the equation:

$$\frac{F_{gr} - A}{F_{fg} - A} = 1.0 - 0.76\left[1 - \frac{1}{\exp(\log D_{gr})^{1.7}}\right] \tag{25}$$

where A is the threshold of motion parameter in the Ackers and White sediment transport theory. This method has been favourably compared with the traditional methods of Einstein and Barbarossa (1952), Engelund (1966, 1967) and Raudkivi (1967) and showed good agreement with data for sediment sizes in the range of 0.04–68 mm (White *et al* 1980).

Variational principle. One extra equation was needed to solve the system. Various different approaches have been used to provide the necessary relationship, some of them relying on a type of variational argument in which the maximum or minimum of some quantity is sought. Previous experience led to the consideration of whether the system might maximize the sediment-transporting capacity of the channel. More precisely, the hypothesis is that, for a particular water discharge and slope, the width of the channel adjusts itself to maximize the sediment transport rate.

If one imposes values of discharge and slope but does not impose the condition of maximum sediment transport, then there are a family of solutions each with different values of B, X, V and d, only one of which provides the maximum sediment transport rate. All the remaining solutions have sediment transport rates less than the maximum and widths both less than and greater than that provided by the maximum transport rate. These all represent possible solutions of the system if it is constrained in some way, for example by the relative erodibility of the bed and banks. Thus, a channel with banks which are less erodible than the bed will have a width smaller than that corresponding to the maximum sediment transport case, while a channel whose banks are more erodible than the bed will have a width correspondingly larger.

While regime theory assumes that the flow is uniform and does not consider the plan geometry of the river, it has been suggested that a principle of maximum sediment transport capacity is involved in determining the plan shape of a river. Onishi *et al* (1976) claim that 'a meandering channel can be more efficient than a straight one, in the sense that a given water discharge can transport a larger sediment load and, for some channel configurations and flow conditions, can require a small energy gradient'. Thus, they postulate that the plan geometry of a river represents an attempt to maximize the transport rate. This should also be considered when studying the comparison of theory and observation for natural rivers presented later. The effects of meandering may produce extremes different from those calculated on the basis of uniform flow.

Evaluation of regime theory

The method was compared with available data

and with existing empirical regime relationships derived by fitting curves to data. The field data for sand channels came from the Punjab canals. CHOP (Canals and Headworks Observation Program) canals, UP (Uttar Pradesh) and Sind Canals (ICID 1966), Pakistan canals (ACOP) (Mahmood *et al* 1979b), and the Simons and Bender data for American canals (Simons 1957). This provided a total of 213 observations. The laboratory data were taken from the work of Ackers (1964), Ackers and Charlton (1970), and Ranga-Raju *et al* (1977). The present method was also compared with data from gravel rivers in Alberta (Kellerhals *et al* 1972).

For comparison with observed data, calculations were performed in which observed values of Q and S were taken and the width, depth and sediment concentration calculated. For some of the data, a further calculation was performed in which observed values of Q and X were taken and the width, depth and slope were determined. For a fuller account of comparisons between observed and predicted data see White *et al* (1981).

For rivers, one has to select some discharge as the dominant or significant discharge. In the present treatment this was arbitrarily chosen to be the bankfull discharge. In considering the results, it must be remembered that better agreement between prediction and observation might be obtained by a different choice of dominant discharge.

In Fig. 12.5 existing regime relationships for sand channels are compared with those produced by the present method. There is reasonable agreement with the existing empirically-derived relationships for a wide range of discharges. It seems that the slope is strongly dependent on the sediment transport rate, as has been observed by other investigators (Parker 1979). The empirical relations of Lacey and ACOP do not take this into account and this could lead to errors, as shown by Ranga-Raju *et al* (1977).

Channel widths were calculated using observed values of flows and slopes, and are shown in Fig. 12.6. In the range $1 \leqslant B(\mathrm{m}) \leqslant 20$, agreement between calculated and observed values is good.

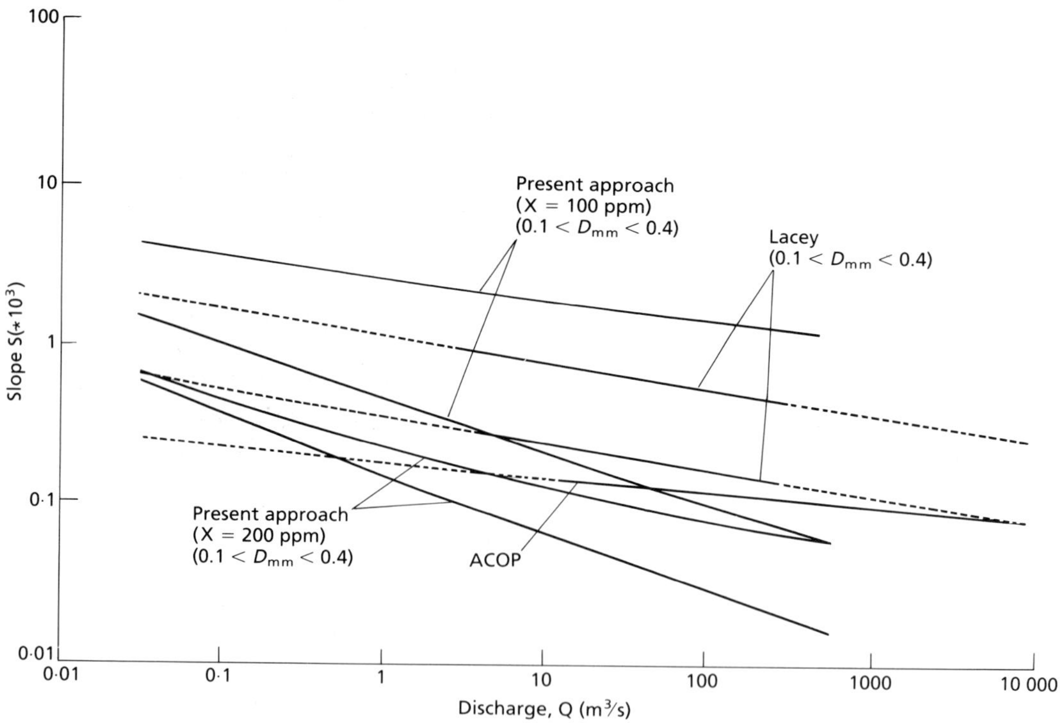

Fig. 12.5 Regime slopes for sand channels.

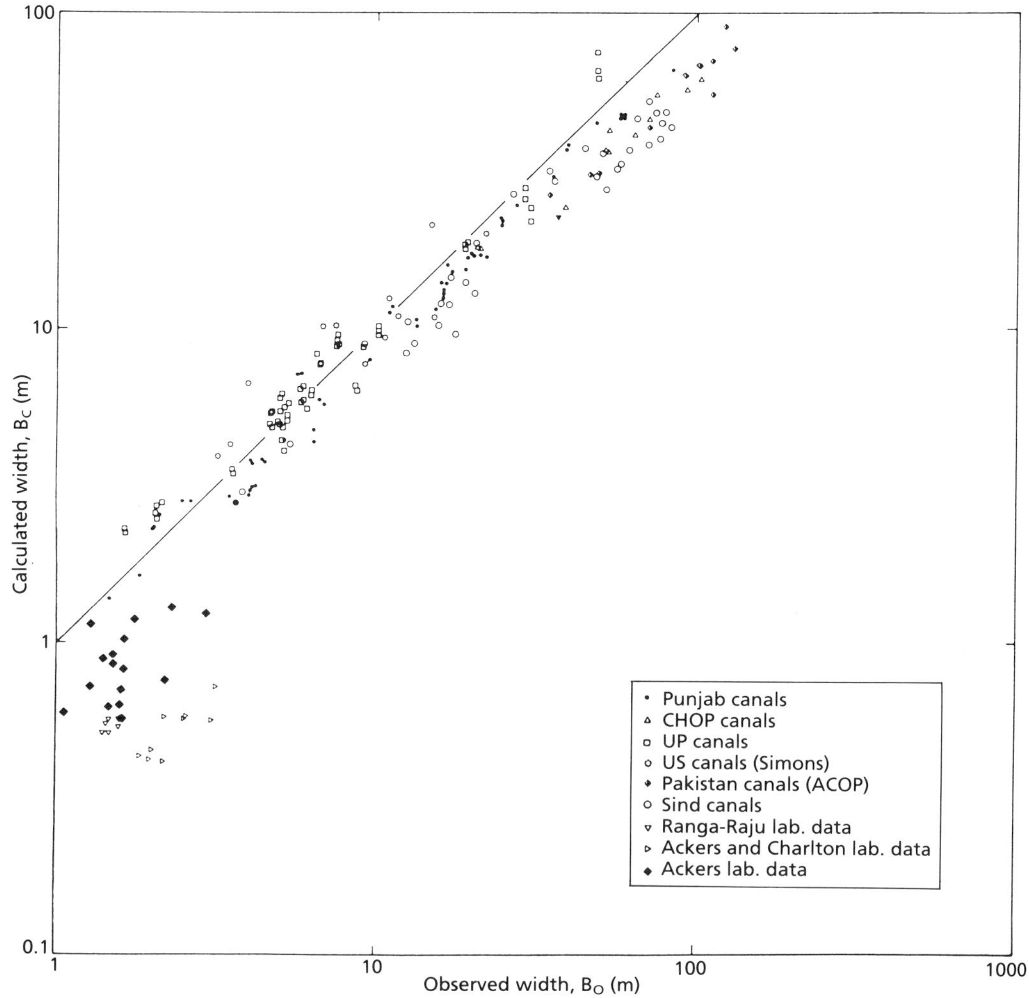

Fig. 12.6 Calculated versus observed channel width using observed slopes for sand streams.

At greater breadths, a tendency to underpredict becomes apparent. The mean discrepancy ratio is 0.86 with a standard deviation of 0.33.

Applications of regime theory

Natural rivers (dominant discharge)

The original application of regime theory was to irrigation canals. A characteristic of such canals is that the range of discharge is limited so there is little inherent difficulty in deciding the discharge to be used in the regime relations. More recently regime theory has been applied to natural rivers. By contrast, these have a wide range of discharges varying throughout the year and from year to year. It is thus more difficult to know which is the discharge that should be used in the regime theory.

It has been assumed that the dimensions of a river channel can be related to a particular discharge, referred to as the dominant discharge. Inglis (1949) suggested that

> there is a dominant discharge and its associated charge and gradient, to which a channel returns annually. At this discharge,

> equilibrium is most closely approached and the tendency to change is least. This condition may be regarded as the integrated effect of all varying conditions over a long period of time.

Unfortunately, there is no universally agreed method of determining the dominant discharge.

In a recent study a number of the different definitions proposed for dominant discharge were compared on data from large gravel rivers in Alberta. The data set was not extensive so it is difficult to be dogmatic about the results, but the best predicted expressions were those based on flow frequency. The discharge which is exceeded 0.6% of the time gave the best predictions (White *et al* 1986).

Design of physical models

As part of an investigation for a water intake for an irrigation scheme a mobile-bed model was designed of the Sabi River, a large sand river in Zimbabwe. The river channel was in regime and the physical model was designed on the basis that the model channel must also be in regime (White 1982). The resulting model successfully reproduced the behaviour of the prototype.

Assessment of morphological changes in rivers

An investigation was carried out into the impact of a proposed dam at Condo on the Sabi River. One of the effects of the dam would be to alter the distribution of flows in the river downstream. By determining how the dominant discharge would be changed the effect that this would have on the river morphology was assessed using regime theory (Hydraulics Research 1982).

12.4 PLAN FORM OF RIVERS

Although the plan shape of rivers displays a continuous variation of form it has traditionally been classified into the three broad categories of straight, meandered or braided. The division into the three categories is in part arbitrary as it is difficult to distinguish between a straight channel and a meandering channel of low sinuosity and not everybody agrees on what constitutes a meandering or braided channel. There has been much work in the past to relate the shape and meander pattern to variables such as the discharge and the slope. A successful method of predicting the meander geometry and the conditions for braiding would enable designs of alluvial channels to be made with greater confidence and allow the prediction of the effects of engineering works on a river system. At present, however, the predictions that can be made with confidence are few. What is presented here is not a definitive theory of meandering and braiding but rather a framework within which the problems of the meandering and braiding of channels in equilibrium can be considered.

It is instructive to review some of the more successful work that has been done on meandering and braiding. Leopold and Wolman (1957) plotted bankfull discharge against channel slope for a number of natural channels. This indicated that in those rivers the braided channels were separated from the meandering channels by a line given by the equation:

$$S = 0.06\ Q^{-0.44}$$

where S is the slope, and Q is the discharge in cusecs. The braided channels had steeper slopes while the meandered channels had smaller slopes. Leopold and Wolman took meandering to mean a river with a sinuosity – the ratio of the thalweg length to the straight line distance – greater than 1.5. Their conclusions, however, were tentative as they pointed out that in natural channels specific variables often occur in association. For example, steep slopes are most frequently associated with coarse bed material. They also showed that meander wavelength is related to the bankfull width of the channel and so they took it to be indirectly related to the bankfull discharge.

The approach of plotting slope against discharge has also been used by others (Lane 1957; Ackers & Charlton 1979) and lines have been drawn on such plots to distinguish areas of braided, meandering and straight channels (Schumm 1977). There is, however, considerable variation between the lines (see Fig. 12.7). While the field evidence clearly shows that for a given discharge braided channels are associated with the steepest slopes, meandering with the less steep and straight rivers with the lowest slopes, there has

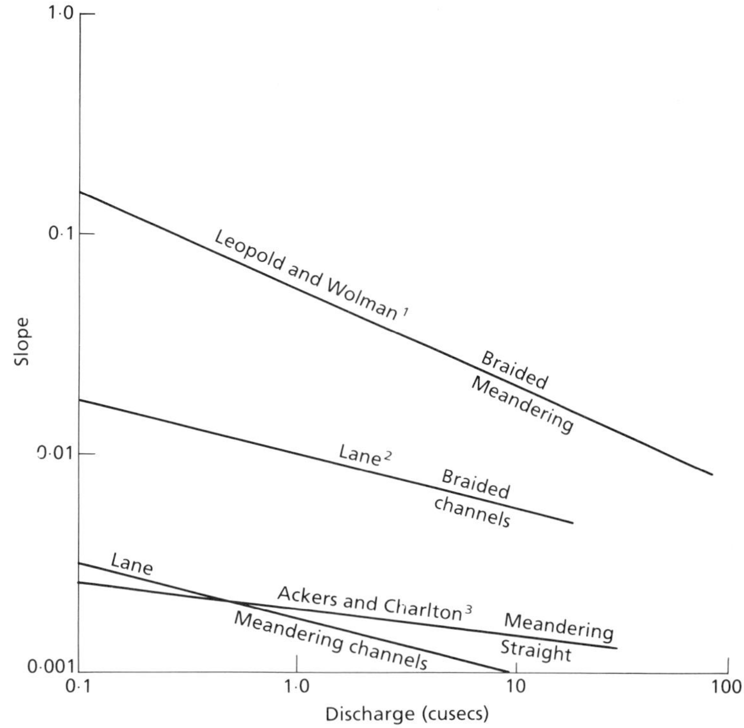

Fig. 12.7 Slope against discharge for meandering and braided channels from various works (from Schumm 1977).

always been sufficient scatter to make the distinctions seem less than satisfactory. Laboratory experiments have shown more consistent agreement but by their nature they only refer to restricted parameter ranges.

From laboratory experiments both Ackers and Charlton (1970) and Schumm and Khan (1972) reported that straight channels have no tendency to meander until the valley slope is steeper than a threshold slope above which meanders develop. Schumm and Khan further demonstrated that, as the valley slope increased, an initially meandering channel started to assume a braided pattern while the thalweg sinuosity decreased.

Chang (1979) approached the problem theoretically using a minimum stream power concept to determine a stable channel slope, which he compared with the valley slope. He derived a diagram of valley slope against bankfull discharge which is reminiscent of those of Leopold and Wolman (1957) and others but which shows two distinct regions of braiding, one being between the areas of straight and meandered channels. No observations seem to substantiate these results.

A number of authors have approached the problem using stability theory (Callander 1969; Engelund & Skovgaard 1973; Parker 1976; Ponce & Mahmood 1976; Fredsoe 1978; Ikeda *et al* 1981). All consider the equations of motions for water and sediment in an open channel. Starting with steady uniform flow they consider the effect on the system of an infinitesimal perturbation. If such a perturbation grows then this is regarded as marking the onset of meandering or braiding and the wavelength of the perturbation with the greatest growth rate is identified with, at least, the initial meander wavelength If the perturbation decays then it is considered that the channel will maintain its straight form. By using different approximations of varying complexity different authors have obtained different areas of meandering with different conditions for the threshold between meandered and braided. Almost all of these approaches show that a straight channel is unobtainable. Only Fredsoe (1978) found conditions under which a channel is stable and so retains its straight character. Ikeda *et al* (1981) extended their analysis to perturbations of channels which were initially sinuous. None of the theories adequately reproduces the sort of behaviour observed by Leopold and Wolman (1957) and others.

Theory of stable channel form

Bettess and White (1983) argued that it would seem that meandering and braiding are mechanisms by which the present conditions imposed on a river are reconciled with those conditions necessary to give a stable channel form. Thus it is necessary to base the approach on the work done on predicting stable channel forms. This has been described above in the section on regime theory. In particular, it was indicated that, for a given sediment, a river carrying a particular discharge and sediment load is stable only if the slope has a specific value. In the following treatment the analytical regime theory developed by White *et al* (1982) will be used but the essential ideas can be incorporated with any well-founded regime theory.

Application to meandering and braiding

The study of meandering and braiding of channels in equilibrium can now be based on the belief that it arises because of a discrepancy between the channel slope required for equilibrium and the valley slope. There have been suggestions that if the channel slope required for equilibrium is less than the valley slope then the river system may accommodate this discrepancy by meandering. Less attention has been paid to the separation between meandering and braiding channels. In this approach a regime theory is used to determine the equilibrium river slope and by comparing this with the available valley slope it is determined whether the river will be straight, meandered or braided.

It is assumed that the discharge and the sediment concentration in the channel are fixed and the regime theory may be used to determine the corresponding equilibrium values of S_{R1}, B, V and d, where S_{R1} denotes the equilibrium slope (for a single channel), B is the channel width, V is the velocity, and d is the depth of flow.

Given the valley slope, S_v, one can identify three different cases depending on the relative magnitudes of S_{R1} and S_v.

1 If the valley slope, S_v, is equal to the river slope then the channel will be straight and in equilibrium.

2 If the valley slope is greater than the river slope then the river can accommodate the discrepancy by meandering. Meandering reduces the slope measured along the channel in relation to the slope measured along the valley. Thus the channel slope may be reduced to S_{R1} by the river meandering with a sinuosity of S_v/S_{R1}, where sinuosity is the ratio of the length between two points measured along the channel to that measured along the valley. The difference in valley and equilibrium slope can also be accommodated in another way, by the river braiding. Since, if the sediment and sediment concentration are kept fixed, the equilibrium slope is increased by reducing the discharge by a factor of two, three or more, the discrepancy between the valley and equilibrium slopes can be removed by the river altering from a single into a multiple channel.

3 If the valley slope is less than the river slope then the river cannot achieve the required equilibrium and alterations probably involving deposition or erosion of sediment will take place in order to bring about a different equilibrium condition.

Cases (1) and (2) are illustrated in Fig. 12.8.

The possibility of the formation of multiple channels is now considered in more detail. If it is assumed that the simplest form of multiple channel is one approximating to two sub-channels each carrying half the discharge then these channels will have some equilibrium slope S_{R2}, and S_{R2} will be greater than S_{R1}. More complex braided channels can be approximated by three or more sub-channels with equilibrium slopes S_{R3}, $S_{R4}, \ldots$, forming an increasing sequence of numbers.

For valley slopes less than S_{R2} and above S_{R1}, the only stable solution is for the channel to meander with sinuosity S_v/S_{R1}. If the valley slope is equal to S_{R2} then there appears to be a choice of solutions, either a meander channel with sinuosity S_v/S_{R1}, or a straight double channel. For valley slopes in excess of S_{R2} there appear to be two possible solutions, either to meander or to

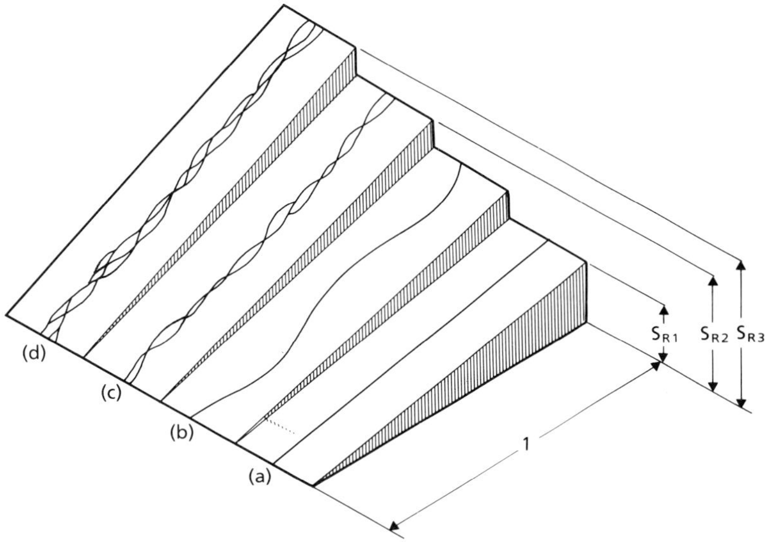

Fig. 12.8 Relationship between valley slope S_v and equilibrium water surface slope: (a) $S_v = S_{R1}$; (b) $S_{R2} > S_v > S_{R1}$; (c) $S_v = S_{R2}$; (d) $S_v = S_{R3}$.

form a braided channel. At the moment little can be said about which of the alternatives will be realized except that we would suggest that braiding is the preferred option unless there are constraints acting to maintain a meandered character.

As an example one can consider a river with a dominant discharge of $100\,m^3\,s^{-1}$ and a sediment concentration of 100 ppm running through an alluvium with a D_{35} size of 0.5 mm. The equilibrium slope is 0.0003. If the valley slope is less than or equal to 0.0003 the river will be straight. If the valley slope is greater than 0.003 then the river may meander with a sinuosity of $S_v/0.0003$. A similar channel carrying $50\,m^3\,s^{-1}$ requires an equilibrium slope of 0.000 33; for a discharge of $33\,m^3\,s^{-1}$ the slope is 0.000 35. Thus for the range of valley slopes between 0.0003 and 0.000 33 the channel will meander and for valley slopes in excess of 0.000 33 there are two possibilities, either meandered or braided.

It is of interest to consider the channel dimensions for this case. If one considers the single meandering channel carrying $100\,m^3\,s^{-1}$ with the valley slope of 0.000 33 then regime theory indicates that the width will be 44 m and the depth 2.4 m, which gives a width to depth ratio of approximately 20 to 1. If, at this valley slope, a multiple channel is formed consisting of two channels each carrying $50\,m^3\,s^{-1}$ then each channel will have a width of 31 m and a depth of 1.8 m so that considering both channels together the multiple channel will have a width to depth ratio of approximately 35 to 1. Thus the theory indicates a noticeable increase in width to depth ratio on the transition to braiding, which is frequently observed in practice.

A consideraton of stream power concepts might give an indication of why a meandering and then a braided channel are the preferred options. It has been proposed that channels adjust to minimize the stream power per unit channel length (Chang 1979). Consideration is therefore given to the stream power and stream power per unit channel length for various channel configurations. It is assumed that the valley slope S_v is larger than S_{R1} and that the valley length is L. Three cases will be studied.

1 Straight single non-equilibrium channel of slope S_v, local water-surface slope S_v.
2 Meandering single equilibrium channel, local water-surface slope S_{R1}, sinuosity S_v/S_{R1}.
3 $S_v = S_{R2}$, multiple channel, local water-surface slope $S_v = S_{R2}$, sinuosity 1.

The stream power in each case is given by $\gamma QS \times$ length of channel, and the stream power per unit channel length by γQS. The stream power and stream power per unit channel length for each case are shown in Table 12.4.

Table 12.4 Stream power and stream power per unit channel length for the three channel configurations, 1, 2 and 3, described in the text

Case	Stream power ($\gamma QS \times$ length of channel)	Stream power/unit channel lenth (γQS)
1	$\gamma QS_v L$	γQS_v
2	$\gamma Q \times S_{R1} \times LS_v/S_{R1}$ $= \gamma QS_v L$	γQS_{R1}
3	$\gamma Q/2 \times S_{R2} \times (L + L)$	$\gamma WS_{R2}/2$

For explanation of symbols see text.

Since S_{R1} is less than S_v, comparison of cases (1) and (2) indicates that a lower value of stream power/unit length is obtained by a meandering channel whose local water-surface slope corresponds to the equilibrium S_{R1} value than for a channel adopting a straight plan shape but having a less efficient cross-sectional area to obtain the water-surface slope S_v. Thus for valley slopes in the range $S_{R1} < S_v < S_{R2}$ minimum stream power/unit length would suggest that a meandering channel with dimensions and local water-surface slope given by equilibrium regime theory would be preferred.

As pointed out earlier, for valley slopes in excess of S_{R2} there are two options – the channel may meander or braid. Since, in general, $S_{R2}/2$ is less than S_{R1}, comparison of cases (2) and (3) indicates that the braided channel gives the lower value of the stream power/unit length and so would suggest that a multiple channel would be preferred to the meandered alternative.

It remains an open question, however, whether a two-branched multiple channel of the sort envisaged theoretically would be attainable practically as the system might not be stable. It is well known that two unequal branches are unstable, the larger channel tending to 'capture' the discharge of the lesser. It may be then that true braiding will only occur when the valley slope S_v attains S_{R3} or even larger values.

Since the equilibrium river slopes depend on sediment size and concentration, the foregoing suggests why there have been discrepancies between different groups of observations. Previous investigators have ignored variations in sediment size and concentration. The present approach suggests that the lines in Fig. 12.7 depend on sediment size and concentration and must therefore be replaced by a family of curves. This may explain the wide range of curves which has been presented in the past.

The counterpart of Fig. 12.7 using the present approach and assuming a sediment diameter of 0.5mm and a sediment concentration of 100 ppm is shown in Fig. 12.9. The area of braiding shown is based on the assumption that braiding occurs when the valley slope exceeds S_{R3}. If a different sediment diameter or concentration had been assumed then the location of the lines would have changed but the general configuration would have remained the same.

By its very nature any framework to analyse meandering and braiding is difficult to test. One can, however, make qualitative comparisons with observations in the laboratory and in nature.

Laboratory observations

The observations of Ackers and Charlton (1970) and Schumm and Khan (1972), already described, agree qualitatively with the proposed framework. As a further investigation into meandering and braiding a series of experiments were performed at Hydraulics Research, Wallingford. The experiments were performed in a re-circulating, tilting flume 24 m long and 2.44 m wide. Sand with a D_{35} diameter of 0.8 mm was placed in the flume. A particular discharge and sediment concentration was selected and the regime slope and channel dimensions were determined using the White *et al* (1982) regime theory. The flume was set to the appropriate slope and a channel of the appropriate dimensions was screeded down the centre of the flume. The discharge was introduced into the flume and sediment injected at the chosen rate. Various discharges were used in the range 0.001–0.0066 $m^3 s^{-1}$. The equilibrium slopes varied from 0.0056 for the smallest discharge to 0.0018 for the highest discharge. Typically, the initial channel had a width of 0.22 m and a depth of 0.04 m. Initially a straight channel established itself down the flume. As the slope of the flume was progressively increased the channel pattern went from straight to meandered, the sinuosity of the channel increasing as the

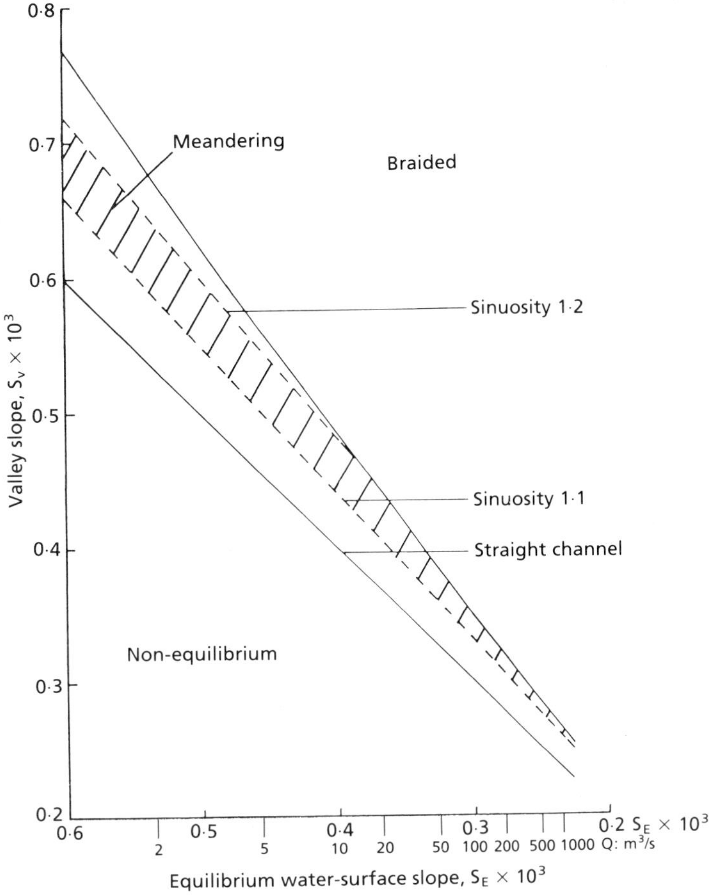

Fig. 12.9 Valley slope against discharge showing areas of meandering and braiding, $D = 0.5$ mm, X = ppm.

flume slope increased. At the steepest flume slopes the channel took on a braided character. This supports the proposed relationship between sinuosity and the ratio of valley to equilibrium slope (cf. Fig. 12.8).

Leopold and Wolman (1957) performed experiments in which straight single channels and multiple channels were obtained. The experiments were performed using a sand with a D_{35} diameter of approximately 0.5 mm. The discharges ranged from 0.0024 to 0.0026 $m^3 s^{-1}$ and the sediment concentrations from 800 to 1800 ppm. A comparison of the observed slopes with those predicted by the theory of White *et al* (1982) for the straight single channels is shown in Table 12.5. This shows reasonable agreement between the regime theory and the observed slopes. Table 12.6 shows the observed slopes for multiple channels and the corresponding predictions for S_{R1}, S_{R2}, S_{R3}, S_{R4} and S_{R6}. The corresponding width for a multiple channel with n branches was

Table 12.5 Comparison of slopes of single, straight channels observed in experiments of Leopold and Wolman (1957) and slopes predicted by regime theory of White *et al* (1982)

Q ($m^3 s^{-1}$)	X (ppm)	S observed	S predicted
0.0024	865	0.0095	0.0082
0.0024	817	0.0079	0.0079
0.0024	830	0.0081	0.0080
0.0026	1840	0.0095	0.0122

Table 12.6 Comparison of slopes and breadths for multiple channels observed by (a) Leopold and Wolman (1957) and (b) predictions for multiple channels using the present theory and the White *et al* (1982) regime theory

(a)

Q ($m^3\ s^{-1}$)	X (ppm)	S observed	S_{R1}	S_{R2}	S_{R3}	S_{R4}	S_{R6}
0.0024	865	0.0133	0.0082	0.0102	0.0118	*0.0132*	
0.0024	817	0.0104	0.0079	*0.0099*	0.0113	0.0128	
0.0024	830	0.0157	0.0080	0.0100	0.0115	0.0130	0.0154
0.0026	1840	0.0165	0.0122	*0.0154*	0.0180	0.0198	

(b)

Q ($m^3\ s^{-1}$)	X (ppm)	S observed	S_{R1}	S_{R2}	S_{R3}	S_{R4}	S_{R6}
0.0024	865	0.488	0.235	0.332	0.408	*0.472*	0.591
0.0024	817	0.421	0.234	0.332	*0.408*	0.472	0.591
0.0024	830	0.579	0.234	0.332	0.408	0.472	*0.591*
0.0026	1840	0.655	0.248	0.354	0.435	0.504	*0.651*

determined by calculating the width of a single channel carrying a discharge of Q/n and multiplying by the number of branches n. For each test the predicted slope and breadth closest to the observed values are shown in italics in Table 12.6. This small amount of experimental evidence does indeed suggest that braiding occurs for slopes in excess of some threshold in the range $S_{R2}-S_{R4}$. Similar data from experiments performed by Schumm and Khan (1972) show less satisfactory agreement. The predictions of the White *et al* (1982) regime theory for straight channels, however, are poor and suggest that shortcomings relating to the prediction of the onset of braiding may be associated with the regime theory or the basic data rather than with the framework presented here.

Field observations

Confirmation from field observations is more difficult to obtain because reliable information can only be obtained by a programme of difficult field measurements taken over a long period. Changes in flows and sediment concentrations need to be observed as well as the changes which occur in the geometry, alignment and slope of the river. Furthermore, changes in the nature of rivers are often affected by factors other than the movement of alluvial sediments. For example, both bed levels and bank alignments may be constrained by inerodible rock outcrops.

Church *et al* (1979) have reported on the morphological effects of inter-basin river diversions in Canada and their paper gives a graphic description of the major changes which can be initiated by flow diversions of this type. Church *et al* divide diversions into three types.

1 Bedrock-controlled diversion routes.

2 'Steep' diversion routes in unconsolidated materials.

3 Alluvial diversions.

The third category is the one where predictions of channel changes might be most fruitful and it is useful to compare the approach suggested in this chapter with the data given by Church *et al* (1979) for the lower Kemano river, British Columbia.

The diversion of flows from the Alcan-Nechako reservoir through the Kemano tunnels, penstocks and power house into the Lower Kemano river has not increased flood flows of that river significantly although mean flows have been trebled. After completion of this scheme in 1954 the channel has become wider. It was still a braided channel in 1975 but the sinuosity was somewhat less than in 1954. The pre-1954 mean flow in the river was $42\,m^3\,s^{-1}$ and the post-1954 mean flow has been $136\,m^3\,s^{-1}$.

If it is assumed that the diverted flow is relatively free of coarse sediment then mean flows were increased by a factor of three and mean sediment concentrations were decreased by a similar amount. Data presented by White *et al* (1982) suggest that the 'new' channel, once it has reached equilibrium, should be wider and of a lower slope. The effect of increasing the discharge alone would be to produce a new width approximately 1.8 times the old width and a new slope approximately 0.75 times the old slope. The effect of decreasing the sediment concentration alone would be to produce a new width approximately 0.9 times the old width and a new slope approximately 0.7 times the old slope. The combined effect is thus likely to increase widths by a factor of around 1.4 and decrease the equilibrium slope by a factor of around 1.4. Changes in width are likely to occur far more quickly than changes in slope because the amount of sediment which must be transported to achieve the latter is orders of magnitude higher than that required for the former.

The original mean width of the Lower Kemano river was 288 m and hence the present approach suggests a new equilibrium width of 400 m. This compares with the 1975 mean width of 370 m. The braided channel noted in 1954 suggests that, at that time, the equilibrium slope of the channel was significantly lower than the valley slope. The prediction for the post-1954 flow conditions is for a decrease in the equilibrium slope which will increase the difference between equilibrium and valley slopes. Hence the main trend in the next few decades will be for the channel to become more braided. This is already apparent in the lower reaches of this particular river. In the longer term degradation in the upper reaches and deposition in the lower reaches will tend to reduce the valley slope towards the predicted equilibrium slope. There is some evidence that this process has already started but the establishment of a new equilibrium will probably take several hundred years.

12.5 STABILITY AND INSTABILITY

Stability and instability within river systems and the related topic of thresholds have long been debated. This discussion has frequently been confused and contradictory because different notions or definitions of stability have been adopted. In this account a particular definition of stability will be adopted and the discussion will take place within this context. The reader is warned that the author's ideas on stability are personal and, some would think, idiosyncratic.

Definitions of stability

As the initial definitions of stability arose in connection with mechanical systems, these will first be reviewed and then the application of these ideas to river systems will be discussed.

Everybody has some intuitive idea of the concepts to which the words 'stable' and 'unstable' refer. In the context of mechanics, if one alters a dynamical system slightly, then, depending upon whether this alteration subsequently has a small or large effect on the system, one calls that system stable or unstable.

The state of a physical system is described by a number of quantities, such as coordinates, which are a function of time. The number of such coordinates, and consequently the dimension of the motion, depend upon the particular stability problem being considered. A motion can be seen as the trajectory of a point in a suitable space. For a finite dimensional system the space R^n is almost invariably used, and for an infinite dimensional system some function space will, in general, be needed. Suppose we have some motion $\phi(t)$ whose stability we wish to consider. Then we must study a general class of motions $u(t)$ evolving from initial conditions that are close, in some appropriate sense, to those for ϕ. The motion ϕ is stable if, by making u sufficiently close to ϕ initially, u remains close for all subsequent time.

It will be noticed that in the above we referred to two functions being 'close' so, for a discussion of stability, it is necessary to introduce the idea of nearness. Thus a distance function or metric must be introduced which represents how close or far apart two functions or points are. This distance between two functions f and g is denoted by d(f, g).

Liapunov formulated a definition of stability in the context of a system defined by a finite number of ordinary differential equations (Hahn 1963). This has since been generalized to the infinite dimensional case (Zubov 1964). A motion ϕ is said to be stable if d($\phi(t)$, $u(t)$) can be kept small be selecting the initial condition $u(0)$ to be sufficiently close to $\phi(0)$. (It may be noted that the metric selected forms an integral part of the definition of stability.)

A motion is asymptotically stable with respect to a set M if the motion is stable and d$\phi(t)$, $u(t)$ * * 0 as t** ?? for all u such that $u(0)$?? M. Thus, an equilibrium is asymptotically stable if after a perturbation the system returns to the original equilibrium state. If the system always returns to the original equilibrium, independent of the size of the perturbation then the system is globally asymptotically stable. If the system returns to the original equilibrium only for perturbations of less than a particular size then the equilibrium is stable to perturbations of that size. An example of such a form of stable equilibrium is a ball rolling in a basin. When the ball is at the lowest part of the basin it is stable. If the basin is infinite in extent then the system is asymptotically stable, since when perturbed the ball will always eventually return to the lowest point. If the basin is finite in extent then the equilibrium is asymptotically stable only for those perturbations within the bowl. For larger perturbations the ball will no longer return to the original equilibrium. The division between perturbations for which the system will return to the original equilibrium and those for which it does not could be regarded as forming some form of threshold.

Stability of channel shape

The above ideas have an obvious application to channel stability. The equilibrium channel size and shape is normally indicated by regime theory. The perturbations to this equilibrium are normally caused by changes in the discharge, although human intervention can also be a source of perturbation. The basis of regime theory is that although the channel size and shape may vary as the flow varies, they will always tend to revert to the regime conditions. This represents a system that is asymptomatically stable to all sizes of perturbation.

Stability of planform

To discuss planform stability one needs a more sophisticated approach to the definition of closeness. In the case of a ball rolling in a basin the notion of the distance of the ball from the equilibrium is easy to grasp and to quantify. If one is to consider wave motions one has to be more sophisticated. The problem arises that two waveforms, or meandering river planforms, may be identical in shape but merely translated relative to each other. Using the normal metrics these planforms would be regarded as being a long distance apart, but in morphological terms we would normally think of them as being identical. Much work has been done on the stability of waveforms (Benjamin 1972) and many of the concepts developed in that area can usefully be applied to rivers. Most waves naturally progress introducing the notion that, for the purposes of a stability analysis, two waves may be close to each other if they have the same shape even though the crests may have very different locations. Benjamin imposed a metric on the equivalence classes of waveforms, where each equivalence class consisted of translations of the same form. Thus if one wave or planform is just a translation of the other then the 'distance' between them is zero. If the profiles of two waves or planforms are very similar except for a translation then they are considered to be close. This is formally described as orbital stability.

As hinted at above, this approach has an application to meandering channels. There is a natural tendency for the planform of meandering channels to migrate downstream. If the planform remains constant subject to a translation then the system remains within the same equivalence class. If the planform is perturbed but the development of the meander pattern remains similar to that which would have occurred with the unperturbed form, subject to an arbitrary translation, then the system is stable. It can readily be seen that with this definition most meandering or straight channels are stable.

Stability of braided channels

The problem of defining a metric on a braided system presents more problems than for a meandering system. Braided channels normally display rapid changes in the size, location and alignment of individual anabranches. But, in a similar fashion to the wave case discussed above, the character of the river, in terms of the number and size of the anabranches, normally remains remarkably constant. A definition of stability should, therefore, concentrate on the essential features rather than the rapid, cosmetic changes that occur over a short time-scale but which leave the nature of the channel unchanged. Howard *et al* (1970) showed how the topology of a braided river can be described independently of the precise locations of the anabranches. Stability can then be defined in terms of a metric imposed on these topological variables. Using this approach it is found that the majority of braided systems are stable.

Nature of perturbations

In considering the stability and instability of river systems it is worth contemplating the nature of the perturbations that are imposed on the system. These can be natural, for example a large return-period flood event or a volcano suddenly increasing the sediment load substantially, or human-induced, for example the construction of a dam. Most river systems are stable to small perturbations. Thus, after a small flood or minor change, a river system normally returns towards its original state. An exception is in very flat or deltaic regions. In such areas river channels can suddenly shift from one course to another during a modest flood. To return to the analogy of a ball rolling in a basin, the situation in these flat or deltaic regions is equivalent to a ball rolling on a gently undulating table. A small perturbation, equivalent in the river system to a small flood, is sufficient to move the ball from one local minimum to an adjacent one, or to shift the channel from one course to another. It can be noted that this also explains the ease with which changes occur in a braided system.

Most rivers demonstrate a remarkable ability to recover from large perturbations. There are many cases where, despite major changes in channel size and shape occurring during a major flood, the river has returned to its pre-flood condition within a few years. In terms of the definitions above this would be described as being asymptotically stable. This asymptotic stability is frequently the cause of frustration to river engineers. There have been many cases of engineering works being carried out to change either the size or planform of rivers only for the river to revert back to its previous form in a few years.

In considering perturbations a distinction has to be made, however, between rivers which are in equilibrium and, after a perturbation, return towards the original equilibrium and rivers which are in transition due to changes in the externally imposed variables, such as discharge. In this latter case the effect of a large perturbation may be to speed up or slow down the rate of change and the river may never return towards its original equilibrium.

12.6 FUTURE TRENDS

In predicting sediment transport there seems to be two main avenues of current research. To predict sand transport, efforts are being made to combine predicted velocity and sediment concentration profiles together with bed load formulae to provide detailed sediment transport formulae. This generation of predictors are likely to be sufficiently complicated that a computer of some sort will be needed to evaluate them. The other avenue is studying the transport of graded sediments and discovering more about hiding functions and how sediment sizes interact.

Numerical morphological modelling is also progressing in two different directions. Effort is going into the development of two-dimensional sediment transport models to predict transverse as well as longitudinal variations in bed level. Meanwhile attempts are being made to include bank processes in one-dimensional models so that changes in channel width may be predicted. No doubt these same processes will eventually be included in two- and three-dimensional numerical models.

The trends in work on planform appear to be less definite. In the past stability analyses have been carried out to attempt to predict channel form but the results of many of these have been unconvincing. At present, much effort is being expended on collecting field data but with the present lack of appreciation of the mechanics involved in the various processes, it seems unlikely that much of great value will emerge.

REFERENCES

Ackers P. (1964) Experiments on small streams in alluvium. *Journal of Hydraulics Division, ASCE* **90**(HY4) 1–38. [12.3]

Ackers P. (1980) Use of sediment transport concepts in stable channel design. *International Workshop on Alluvial River Problems, Roorkee, India.* [12.3]

Ackers P, Charlton FG. (1970) The geometry of small meandering channels. *Proceedings of the Institute of Civil Engineers* Suppl. XII, Paper 73285. [12.3, 12.4]

Ackers P, Charlton FG. (1979) The slope and resistance of small meandering channels. *Proceedings of the Institute of Civil Engineers* Suppl. 15, Paper 7362 S. [12.5]

Ackers P, White WR. (1973) Sediment transport: new approach and analysis. *Journal of Hydraulics Division, ASCE* **99**(HY11): 2041–60. [12.1, 12.2, 12.3]

Benjamin TB. (1972) The stability of solitary waves. *Proceedings of the Royal Society, London, A* **328**: 153–83. [12.5]

Bettess R, White WR. (1983) Meandering and braiding of alluvial channels. *Proceedings of the Institute of Civil Engineers Part 2,* **75**: 525–38. [12.5]

Brownlie WR. (1981) *Prediction of Flow Depth and Sediment Discharge in Open Channels. California Institute of Technology* W M Keck. Lab., Report KH-R-43A. [12.1]

Callander RA. (1969) Instability and river channels. *Journal of Fluid Mechanics* **36**(3): 465–80. [12.4]

Chang HH. (1979) Minimum stream power and river channel patterns. *Journal of Hydrology* **41**: 303–27. [12.4]

Chang HH. (1980) Stable alluvial canal design. *Journal of Hydraulics Division, ASCE* **106**(HY5): 873–91. [12.3]

Charlton FG, Brown PM, Benson RW. (1978) *The Hydraulic Geometry of Some Gravel Rivers in Britain,* Report IT 180, Hydraulics Research, Wallingford, UK. [12.3]

Chollet JP, Cunge JA. (1980) Simulation of unsteady flow in an alluvial stream. *Applied Mathematical Modelling* **4**(August): 234–44. [12.2]

Chow VT. (1959) *Open Channel Hydraulics.* McGraw-Hill, New York. [12.2]

Church M, Davies LB, Hellerhals R. (1979) Morphological effects of interbasin river diversions. *Canadian Journal of Civil Engineering* **6**(1): 8–3. [12.4]

Cole CV, Tarapore ZS, Dixit JG. (1973) Applicability of sediment transport formulae to natural streams. *Proceedings of the 15th IAHR Congress, Istanbul, 1973,* vol. 1, Paper A53. [12.1]

De Vries M. (1976) *Morphological Computations.* Delft University of Technology. [12.2]

Einstein HA, Barbarossa NL. (1952) River channel roughness. *Transactions of the ASCE,* **117**: 1121–76. [12.3]

Engelund F. (1966) Hydraulic resistance of alluvial streams, JHD. *Proceedings of the ASCE* **92**(HY2): 315–26. [12.3]

Engelund F. (1967) *Proceedings of the ASCE* **93**(HY4): 287–96. [12.3]

Engelund F, Skovgaard D. (1973) On the origin of meandering and braiding in alluvial streams. *Journal of Fluid Mechanics* **57**(2): 289–302. [12.4]

Fredsoe J. (1978) Meandering and braiding of rivers *Journal of Fluid Mechanics* **84**(4): 609–24. [12.4]

Garde RJ, Ranga Raju KG. (1985) Mechanics of sediment transportation and alluvial stream problems, 2nd edn. John Wiley, Chichester. [12.1]

Hahn W. (1963) *Theory and Application of Liapunov's Direct Method.* Prentice-Hall, New York. [12.5]

Howard AD, Keerch ME, Vincent CL. (1970) Topological and geometical properties of braided streams. *Water Resources Research* **6**(6): 1674–88. [12.5]

Hydraulics Research. (1982) *Sabi River, Zimbabwe, Sedimentation Studies.* Report EX 1020, Hydraulics Research, Wallingford, UK. [12.3]

Ikeda S, Parker G, Sawai K. (1981) Bend theory of river meanders: Part 1 Linear development. *Journal of Fluid Mechanics* **112**: 363–77. [12.4]

Inglis CC. (1949) *The Behaviour and Control of Rivers and Canals.* Research Publication No. 13, Central Board Irrigation India, Simla. [12.3]

Kellerhals R, Neill CR, Bray DI. (1972) *Hydraulic and Geomorphic Characteristics of Rivers in Alberta.* Report 72-I, River Engineering and Surface Hydrology, Research Council of Alberta, Alberta, Canada. [12.3]

Kennedy RG. (1895) On the prevention of silting in irrigation channels. *Proceedings of the Institute of Civil Engineers* **119**. [12.3]

Lacey G. (1930) Stable channels in alluvium. *Proceedings of the Institute of Civil Engineers* **229**: 259–384. [12.3]

Lane EW. (1957) *A Study of the Shape of Channels Formed by Natural Streams Flowing in Erodible Material.* US Army Engineering Division, Missouri River, Corps Engineers, MRD, Sediment Series 9. [12.4]

Leopold LB, Wolman MG. (1957) River channel patterns: braided, meandering and straight. US Geological Survey, Professional Paper, pp. 282–8. [12.4]

Mahmood K *et al.* (1979a) *Selected Equilibrium State Data from ACOP Canals*. Report EWR 79.2, Civil, Mechanical and Environmental Engineering Dept., George Washington University, Washington, DC. [12.3]

Mahmood K, Tarar RN, Masood T. (1979b) *Hydraulic Geometry Relations for ACOP Canals*. Report EWR 79.3, Civil, Mechanical and Environmental Engineering Dept., George Washington University, Washington DC. [12.3]

Onishi Y, Jain SC, Kennedy JF. (1976) Effects of meandering in alluvial streams. *Journal of the Hydraulics Division, ASCE* **102**(HY7): 899–917. [12.3]

Parker G. (1976) On the cause and characteristic scales of meandering and braiding in rivers. *Journal of Fluid Mechanics* **76**(3): 457–80. [12.4]

Parker G. (1979) Hydraulics geometry of active gravel rivers. *Journal of the Hydraulics Division, ASCE* **105**(HY9): 1185–201. [12.3]

Ponce VM, Mahmood K. (1976) Meandering thalwegs in straight alluvial channels. In: *Rivers '76: Symposium on Inland Waterways for Navigation, Flood Control and Water Diversions*, pp. 1418–41. ASCE, Fort Collins, USA. [12.4]

Ponce VM, Garcia JL, Simons DB. (1979) Modelling alluvial channel bed transients. *Journal of the Hydraulics Division, ASCE* **105**(HY3): 245–56.

Ranga-Raju KG, Dhandapani KR, Kandop DM. (1977) Effect of sediment load on stable sand canal dimensions. *Journal of Waterway, Port, Coastal and Ocean Division, ASCE* **103**(WW2): 241–9. [12.3]

Raudkivi AJ. (1990) *Loose Boundary Hydraulics*, 3rd edn. Pergamon Press, Oxford. [12.1]

Schumm SA. (1977) *The Fluvial System*. John Wiley, New York. [12.4]

Schumm SA, Khan HR. (1972) Experimental study of channel patterns. *Geological Society of America Bulletin* **83**: 1755–70. [12.4]

Shields A. (1936) Anwendung der Aelinlichkeits. mechanik und der turbulenz forschung auf die Geschiebebewegung. *Mitteilungen der pruessischen versuchsanstalt für wasserbau und schiffbau, Berlin*. [12.1]

Simons DB. (1957) Theory and design of stable channels in alluvial material. PhD thesis, Colorado State University Fort Collins. [12.3]

Task Committee on Preparation of Sediment Manual. (1971) Sediment transportation mechanics: sediment discharge formulae. *Journal of the Hydraulics Division, ASCE* **97**(HY4): 523–67. [12.1]

US Army Corps of Engineers. (1976) Scour and deposition in rivers and reservoirs: computer program HEC-6. Hydrologic Engineering Centre, Davis, California. [12.2]

US Bureau of Reclamation (1968) *Behaviour of Middle Loup River Channel as Influenced by Milburn Diversion Dam*. US Department of Interior, Sedimentation Branch, Denver, Colorado. [12.2]

Van Rijn LC. (1984a) Sediment transport Part I: Bed load transport. *Journal of Hydraulic Engineering, ASCE* **110**: 1431–56. [12.4]

Van Rijn LC. (1984b) Sediment transport Part II: Suspended load transport. *Journal of Hydraulic Engineering, ASCE* **110**: 1613–41. [12.4]

White WR. (1982) A novel loose boundary model for investigating sedimentation problems at an intake. In: *International Conference on Hydraulic Modelling of Civil Engineering Structures, Coventry, England*, pp. 181–94. BHRA. [12.3]

White WR, Day TJ. (1980) Sediment transport in gravel bed rivers. In: Hey RD, Bathurst JC, Thorne CR (eds) *Gravel Bed Rivers*. Wiley, Chichester. [12.2]

White WR, Milli H, Crabbe AD. (1973) *Sediment Transport: an Appraisal of Available Methods*. Report INT 119, Hydraulics Research Station, Wallingford. [12.2]

White WR, Paris E, Bettess R. (1980) The frictional characteristics of alluvial streams: a new approach. *Proceedings of the Institute of Civil Engineers Part 2*, **69**: 737–50. [12.2, 12.3]

White WR, Paris E, Bettess R. (1981) River regime based on sediment transport concepts. Report IT 201, Hydraulics Research, Wallingford. [12.3]

White WR, Bettess R, Paris E. (1982) Analytical approach to river regime. *Journal of the Hydraulics Division, ASCE* **108**(HY10): 1179–93. [12.3, 12.4]

White WR, Bettess R, Wang Shiqiang (1986) A study on river regime. Report SR 89, Hydraulics Research, Wallingford, UK. [12.3]

Yang CT. (1976) Minimum unit stream power and fluvial hydraulics. *Journal of the Hydraulics Division, ASCE* **102**(HY7): 919–34. [12.3]

Zubov VA. (1964) *Methods of AM Lyapunov and Their Application*. Noordhoff. [12.5]

13: Prediction of Biological Responses

P. D. ARMITAGE

13.1 INTRODUCTION

The first aim of most scientific investigations is to observe and describe phenomena. Once the descriptive data have been collected attempts can be made to organize and clarify what has been observed. Then through the use of experimentation, the mechanisms which determine these observations can be investigated. These three steps of study provide the basis for developing models to predict the response of the components of the system under investigation to changed conditions. It is this latter aspect which will be addressed in this chapter.

Increasing demands on river systems for water supply and amenity have put pressure on scientists to obtain and use basic data to develop guides for the optimal management of aquatic resources and maintenance of environmental integrity. These basic data may concern different levels of organization, for example system, community, population, organism, cell or molecule, and relate to such factors as pollution, fisheries, flow regulation (supply, flood control and hydropower) and whole catchment management.

These data obviously encompass the whole gamut of lotic studies and it is beyond the remit of this chapter to cover all aspects. Instead emphasis will be placed on reviewing current methodologies and providing selected examples of their application.

Stimulus and response

In animals a stimulus, such as a variation in light levels or the introduction of a pollutant, will produce a change in part of an organism, usually a sense receptor. The animal reacts to this change through the action of muscles and glands mediated to a greater or lesser extent by the nervous and endocrine systems. The sum of these effects constitutes the behaviour of the individual. In this chapter the term 'response' will be considered to be the overall reaction of an individual or community to a stimulus or range of stimuli.

In most studies in lotic environments it is the community, population and organism levels of organization which are most studied. However, recent studies (Guyomard 1989, and references therein) have started to examine the responses of biota to environmental variables at the genetic level, and this is a developing field of research.

In a natural river system the biotic communities have evolved in response to the prevailing physico-chemical conditions (temperature, light, flow conditions, substratum, oxygen concentration, ionic composition) and biotic interactions such as competition and predation to which a biotic unit can respond. It is important to note that the stimuli are not mutually exclusive and that a biological response is more often than not a result of the various combinations of primary stimuli. Only in extreme cases, for example severe heavy metal pollution or physical disturbance, is there one main factor affecting the response.

The need for predictive models

Biological monitoring in lotic systems provides information on the quality status of the river environment and is an essential feature for the maintenance of good-quality water-courses. It does not, however, provide information on the probable impacts of changes in the environment. This type of information is essential for planning and management in the water industry. It is here that the ability to predict the biological responses of river biota or whole river ecosystems to environmental change has its main use.

Biological responses are predicted from knowledge based on field observation and experimental work. Given sufficient data on the effects of physico-chemical changes on biological responses it would be possible to develop mathematical models. These may be simple or complex, pragmatic or conceptual, or combinations of all of these. At one extreme there are explanatory or system models whose behaviour is considered to more or less duplicate the true behaviour of the modelled population; at the other extreme there are predictive models whose primary purpose is not to explain population behaviour but rather predict it in the future. The classification and description of model types is beyond the scope of this chapter and the reader is referred to Jeffers (1982) for detailed information on this subject.

Aspects of the water industry which affect the river environment include water supply, flood control, effluent purification, waste disposal and recreation. In addition, changes in river catchments such as land drainage, urban development and large-scale changes in agricultural land use have major impacts on the water courses which drain them.

The water industry has a need to predict the likely responses of river biota to all these aspects of water management. There is also a need for information on ecotoxicological aspects in order to assess the effects of toxic compounds on river biota and predict the possible consequences on the environment as a whole.

Fisheries are a major recreational use of rivers and it is important to be able to predict yields and angling success, assess stocks and predict responses of fish to changes in food and habitat availability. This latter aspect has been the subject of much research and one particular approach, the Instream Flow Incremental Methodology (IFIM: Bovee 1982) is described in Chapter 14 of this book and will not be dealt with here.

As background to these impacts and aspects of management there is an overwhelming need for careful studies which describe responses of river biota to changes in the physical and chemical environment and to other biota both in natural and disturbed systems. This basic information is the substance of all predictive and conceptual models.

It is important to recognize the central role of predictive models in biomonitoring. The concept of monitoring has no meaning unless the measurements are tested against some model of system performance (Radford 1988). In the simplest case the model may be that no change is expected with time. This is the assumption made by many monitoring techniques. In more complicated situations the results of monitoring depend on comparisons with a standard or conceptual model based on real observations. Figure 13.1 illustrates the relation between monitoring, prediction and the development of prediction into simulation procedures which can be used to forecast responses at system or lower levels of organization.

Hydrology is a crucial aspect of the lotic system and knowledge of the behaviour of hydrological variables has many applications in the prediction of biological responses. Hydrological modelling is a well-developed field of research (Reed 1987; Trapp *et al* 1990) and is the subject of Chapter 10 in this book. Similarly the concentration and distribution of chemical parameters in lotic systems will have a major influence on water quality (Tipping 1989). Neither chemical nor hydrological modelling are discussed directly in this chapter but their role in predicting biological responses is noted where appropriate.

Prediction is not only needed in monitoring activities. Predictive methodologies can offer the

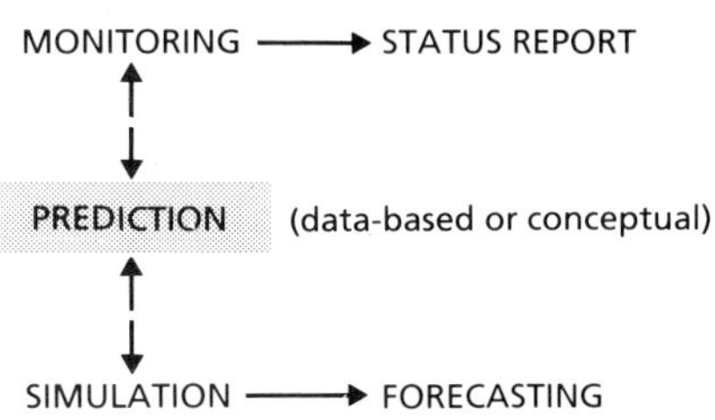

Fig. 13.1 The relationship between prediction, monitoring and simulation.

opportunity for rapid scientific development by placing individual work in context, by providing unambiguous criteria for the evaluation of theories, and by elaborating general, quantitative theories which can then be combined to yield sometimes unexpected results (Peters 1986).

13.2 PREDICTION OF BIOLOGICAL RESPONSES

General biological monitoring

A river is prone to a wide range of environmental disturbances. The river biota will respond to both natural and anthropogenic influences whether these be chemical, physical or combinations of the two. It is the role of biomonitoring to assess the extent and direction of these responses. These assessments can take place in the field, in the laboratory or in both.

Many groups of organisms have been employed to assess both environmental quality and ecotoxicity. Macroinvertebrates appear to have been used most often and their role in biological assessment is the subject of Chapter 8 in this book. For an exhaustive treatment of biological indicators of pollution and environmental management see Hellawell (1989).

Environmental assessment

In the area of environmental assessment the prime object is to identify change, the assumption generally being that no change is expected. For example, a biotic index based on macroinvertebrates could be used to type or classify a site. When this site is revisited (provided that either the period between samples is within about 2 weeks or in the same season) a similar value of the index might be expected. Divergence from this indicates change. All biotic indices, whatever group they are based on, operate largely in this way. If previous samples are available for comparison it is possible to assess the amount of change. However, such techniques are not helpful when an assessment of a new site is required; neither can they provide a measure of any perturbed site's divergence from the 'norm'.

Recently workers in the UK (Moss *et al* 1987) have used multivariate techniques to develop a software package, RIVPACS (River InVertebrate Prediction And Classification System), which can predict the probability of macroinvertebrate species occurrence at new sites from known environmental features. The database for RIVPACS consists of 438 sites on 81 unpolluted river systems covering all major types of geology and topography in Great Britain. The database comprises species lists from all sites for spring, summer and autumn samples, together with associated physical and chemical data for each location.

The various stages in the development of this package have been described elsewhere (Moss *et al* 1987; Furse *et al* 1987; Wright *et al* 1989) and will not be dealt with here. The predicted probabilities of taxon occurrence (at family or species level) can be converted into predicted biotic index values for the site. If required, RIVPACS can be adapted to provide target values for almost any macroinvertebrate-based index in common use. The ratio of observed index value over predicted value provides an index of environmental quality.

This combination of attributes makes RIVPACS a powerful tool for environmental assessment. It has been used to assess the effects of river regulation on faunal communities (Armitage 1989) and to describe the extent of pollution in rivers (Armitage *et al* 1990). A recent study explores its use in assessing physical disturbance, heavy metal pollution and organic pollution (Armitage *et al* 1992) (Fig. 13.2). The figure clearly shows the marked effects of pollution and more importantly compares this with a predicted standard.

RIVPACS can also be used to simulate the response of individual taxa or groups of taxa to single environmental variables or groups of variables (Fig. 13.3). This can provide insight into the critical values of particular variables (Armitage 1989).

Winget (1985) in the USA describes a biotic index which is sensitive to all types of environmental stress. Four variables – gradient, substrate roughness, total alkalinity and sulphate – are used to derive 'tolerance quotients' (TQs) for benthic species. The predicted community TQ is the mean of the TQs for a predicted macroinvert-

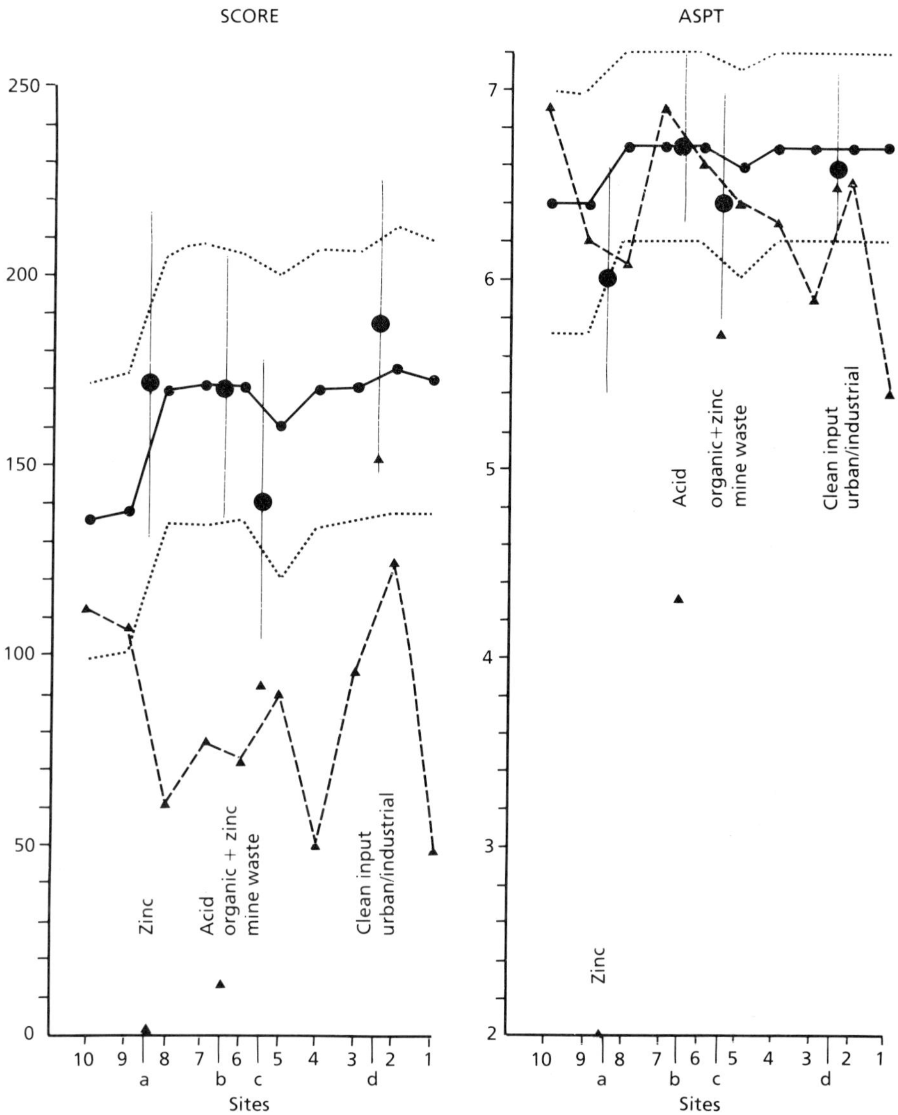

Fig. 13.2 The variation in Biological Monitoring Working Party (BMWP) Score (SCORE) and average score per taxon (ASPT) in the River Nent system. Predicted values in main stream (sites 10–1): small closed circles; in tributaries (sites a–d): large closed circles. 95% confidence limits indicated with dotted line for main stream and with fine vertical lines for tributaries. Observed values: closed triangles. Inputs into the main stream are indicated in the figure (from Armitage *et al* 1992).

ebrate community. Management success in reclamation/restoration works is judged by physical habitat and return of actual macroinvertebrate community condition to the predicted potential condition.

Both RIVPACS and the biotic condition scheme employ the same basic philosophy and use macroinvertebrates to assess environmental quality. Of course many other groups have been used for environmental assessments (Hellawell

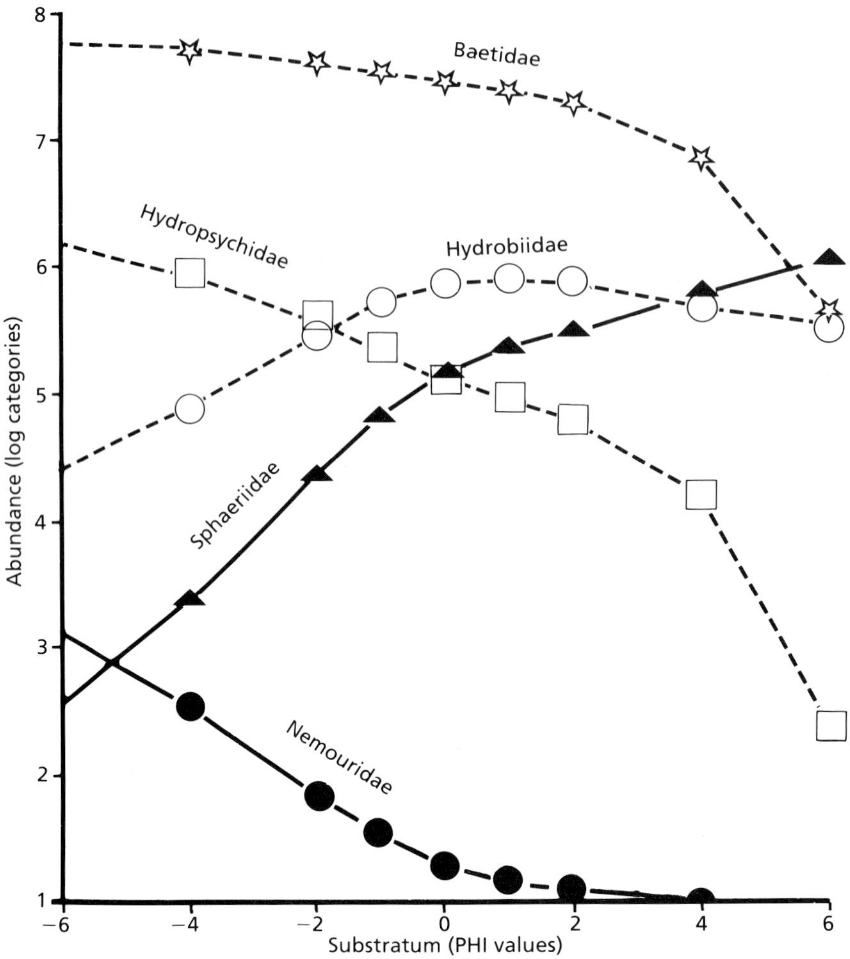

Fig. 13.3 The effect of simulated change in mean substratum particle size on predictions of abundance of five families (from Armitage 1989, by permission of CRC Press).

1989) but few of these have been able to predict target values as do the two methods described.

The only general system of assessment which features bacteria, protozoa and algae is the 'saprobien' (Kolkwitz & Marsson 1909) although this system can also use all other river biota. The use of bacteria and protozoa for general assessments (Evison 1979; Sladacek 1981) is rare but they have been widely used to examine the effects of contaminants in field and laboratory systems (Mathews *et al* 1982). Algae alone have been used to assess environmental quality in rivers (Patrick 1954; Descy 1979; Rumeau & Coste 1988). However, none of these systems offers a predictive methodology.

Macrophytes have some advantages as indicators of pollution in that they are stationary, are visible to the naked eye, and manpower demands for macrophyte surveys are low provided experienced operators are used. Harding and Whitton (1981) and Haslam (1982) have developed schemes to assess disturbance to the environment but these do not allow prediction. However, Best (1990), in reviewing 12 macrophyte models, indicates how some of them, based on oxygen flow, can be used to predict the probable effects of waterway management measures (Westlake 1966), and farming, urbanization, effluent treatment and thermal stress (Rutherford 1975).

The use of fish in routine environmental monitoring has the advantage that the public at large and other interested parties such as anglers can readily identify major disturbance to the river system by the observation of fish kills. Nevertheless fish community surveys require extensive manpower and are difficult to implement in fast-flowing deep rivers. Perhaps as a result of this fish have been used rarely in comprehensive monitoring programmes, although they are used commonly to detect contaminants (Cairns & van der Schalie 1980). There is regular monitoring of fish populations in the UK but not usually so frequently as for invertebrates. The monitoring is for fisheries management purposes rather than pollution assessment except where fish are used for bioaccumulation work. The mobility of fish away from a pollution source detracts from their value as pollution monitors. In the USA, Karr *et al* (1986) have developed an 'index of biotic integrity' based entirely on fish, which is being adopted for widespread use in the USA (Plafkin *et al* 1989). While not a predictive methodology it does provide a means of assessing the response of fish communities to disturbance. The index is based on data for species richness and composition, trophic composition, abundance and condition.

Acidification

The recent acidification of freshwaters in parts of Europe and eastern North America has had severe effects on aquatic life (Overrein *et al* 1981; Battarbee & Charles 1986; Schindler 1988). Many studies have been devoted to examining lake sediments since they contain decipherable biological and chemical information about changes in lake and watershed conditions through recent geological time (Warwick 1980; Holmes *et al* 1989). Such retrospective analyses can be used to predict or infer probable biological responses to changing environmental variables.

It is important to be able to detect and simulate change due to acidification and the advantages of model simulations are becoming apparent. Hydrochemical models such as MAGIC and ILWAS focus exclusively on deposition and acidification elements (Reuss *et al* 1986). Marmorek *et al* (1990) present a hydrochemical model that estimates the pre-acidification ('original') lake chemistry and then predicts the eventual chemistry to be expected given a specified level of acid sulphate deposition. Such models are an integral part of those models which predict biological responses. MAGIC has been used in a whole series of studies on the biological responses of stream animals to acidification in Wales (Ormerod *et al* 1988; Ormerod & Tyler 1989; Ormerod *et al* 1990a). The hydrochemical model is coupled with biological models, employing similar basic multivariate methods to those used in the construction of RIVPACS (Moss *et al* 1987), but developed for catchments in mid-Wales (Weatherley & Ormerod 1987). The technique is used to predict the response of invertebrate assemblages to alterations in stream chemistry due to acidification as a result of changing ionic deposition or land-use modifications (Rutt *et al* 1990).

In fish studies models have been used to determine egg and fry mortalities of Norwegian trout at low pH (Sadler 1983), predict changes in fish distribution under different acid loadings by simulating pH and alkalinity changes in lakes (Small & Sutton 1986), examine the loss of Atlantic salmon from acidic brown waters in Canada (Lacroix 1987), and determine fish response to acidification in lakes (Reckhow *et al* 1987). Minns *et al* (1986) have simulated changing alkalinity and total dissolved solids to assess temporal trends in the fish populations of lakes of the Canadian Shield. Kelso *et al* (1990) have reviewed the effects of atmospheric acidic deposition on fish and the fishery resource of Canada. They investigated integration of cause and effect through model development by examining 11 models which incorporate a biotic damage element. They concluded that of the models available those that attempt to predict biomass, production or yield receive limited support from available evidence, whereas those that incorporate critical thresholds of occurrence and to some extent community structure receive more support.

Ecotoxicology

The prediction of biological responses is an integral part of ecotoxicological tests and bioassays, whose aim is to assess the ecological risks of polluting chemicals. The response of organisms to specific substances together with data on the properties of those substances in field and laboratory can be used to estimate threats to the environment (see Munawar *et al* 1989 for a comprehensive presentation of these techniques).

Maltby and Calow (1989) have reviewed tests used for resolving environmental problems. They distinguished between bioassays concerned with prediction (anticipating environmental impacts) and assessment (monitoring actual impacts). An environmental impact can be considered to be a response of the biotic community as a whole to some stressor. However, over 90% of the bioassays classified as predictive by Maltby and Calow (1989) were single-species tests mainly using death as a criterion of toxicity and were carried out over short periods of time. Chronic and sublethal bioassays represented less than 20% of all single-species tests and involved a variety of response criteria including growth, reproduction, physiology/biochemistry, behaviour and morphology. Multi-species tests have been in the minority and have ranged from small mixed-flask cultures (Larsen *et al* 1986) through *in situ* trough-type mesocosms and large-scale channel macrocosms (Giesy & Allred 1985), to the manipulation of natural systems (Hall *et al* 1980; Ormerod *et al.* 1987). Criteria used in these studies have involved population density, species composition and diversity (structural properties), and primary productivity, P/R ratios, bioaccumulation and nutrient cycling (functional properties).

The relative simplicity of single-species acute tests and their replicability has meant that they have dominated ecotoxicological work. However, in these and also in multi-species tests there are difficulties in extrapolating from laboratory response to the field situation. Interpretation can be complex, sensitivity will differ between taxonomic units, and variability and replicability may cause problems (Cairns 1985; Slooff 1985). Giesy and Allred (1985) note that when copper was added to limnocorrals in an experimental heavy metal study, the population densities of phytoplankton increased. This effect was attributed to depressed grazing by zooplankton and would not have been predicted from single-species tests used to evaluate the effects of metals on phytoplankton.

Frequently, insufficient data are available to determine the long-term chronic effects of toxicants on the survival, growth or reproduction of aquatic biota. As a result the chronic effects must generally be inferred or estimated from observations recorded during short-term acute studies which may be conducted at higher levels of toxicity. These observations are then related to the chronic effects by some statistical relationship. Giesy and Graney (1989) note that if all the information about a chemical and the ecosystem with which it interacts were known it would still be impossible to predict the effect of the chemical on the ecosystem because of the vast amount of information and the ecological uncertainty principle. With these limitations the authors ask what is possible, what use can be made of short-term toxicity tests and can such tests be used to predict long-term effects. It is clearly impossible to predict the effects of many toxicants from simple single-species bioassays (Cairns 1988). Conversely more complex tests are not necessarily more predictive (Giesy 1985). In view of these limitations it is likely for the foreseeable future that reliance will have to be placed on traditional assays of acute toxicity testing.

Giesy and Graney (1989), in their detailed review, discuss the importance of different levels of organization in ecotoxicological studies. Biochemical measures of toxic effects such as protein content, RNA/DNA ratio, energy content or enzyme concentrations have been used as functional measures of effects which can be used to predict the effects of chronic exposures to pollutants on the survival and fecundity of organisms. However, Woltering (1985) suggests that before biochemical and physiological parameters can be employed to predict chronic effects it is necessary to understand the relationship between short-term biochemical effects and individual or population effects.

The situation is further complicated at community and ecosystem level (Giesy & Allred 1985) by the increase in the number of biotic interactions and habitat diversity. Some interactions which occur within and among individuals make it difficult to predict effects at one level of organization from observations at another level of organization. The example provided by Giesy and Graney (1989) concerns the partitioning behaviour and metabolism of compounds in whole animals which may be such that the dose to a particular biochemical system is much different under *in vivo* field conditions than in laboratory-based studies of tissues, cells or subcellular preparations. Similarly the prediction of survival, growth and reproduction of organisms based on results from exposure to toxicants under controlled laboratory experiments may bear little resemblance to the responses of the organisms in the dynamic conditions of their environment, population and community.

Despite the difficulties, bioassays are used routinely to assess the impact of pollution loads in water courses (Cairns 1980), and the range of studies incorporating bioassays and early-warning systems is enormous throughout the world. A review of the literature is not appropriate to this chapter and the reader is referred to Cairns (1985) and Munawar *et al* (1989) for more information.

General management

The prediction of environmental responses to management activities other than those directly concerned with toxic pollution is an important part of water resource development. Studies are particularly well advanced with respect to the effects of nutrient enrichment in lake ecosystems (see Vollenweider 1975; Aldenberg & Peters 1990). In lotic systems predictions may take the form of inferred effects from previous studies such as the prediction of the environmental impact of a major reservoir development (Edwards 1984) or the prediction of regulation effects on natural biological rhythms in south-central African freshwater fish (Jackson 1989). Alternatively historical analyses may be used to predict possible impacts.

Bravard *et al* (1986) used a historical approach to investigate the impact of civil engineering works on the successions of communities in a fluvial system. By comparing present ecosystem dynamics and ecological succession with past data from abandoned channels it was possible to predict or forecast possible ecological succession resulting from changes in river management. The value of these studies lies in their recognition of the fact that many physical and biological events occur too slowly to be perceived directly within the period available for study. The historical approach provides data on rates of change in response to agricultural development, other land-use impacts, and river use (Petts *et al* 1989), and is particularly valuable in assessing the long-term effects of climate change (Holmes *et al* 1989).

Restoration and biomanipulation

Recently there has been considerable interest in the restoration (Gore 1985a) and biomanipulation (Gulati *et al* 1990) of water bodies. Stream channel modifications can be used to enhance fish and invertebrate habitat (Wesche 1985; Gore 1985b). Case histories can be used to develop guidelines for improving techniques which will be based on past observations of the responses of fish and invertebrates to restoration measures. Gore (1985b) discusses the prediction of rates and trends in colonization processes and examines the establishment of stable communities in relation to habitat improvement. Sheldon (1984) has used data on emigration and immigration of aquatic insects to propose a model describing the colonization of stream ecosystems. The model incorporated elements of competition between species together with information on periphyton production and detritus accumulation, and was able to simulate colonization processes. However, the author concluded that environmental variability precludes the generalized application of this model.

Despite this inability to produce mathematical models with a general application Gore (1985b) notes that basic trends have been reported which will assist in the assessment and monitoring of reclamation works. A useful endpoint is the demonstration of a permanent and stable (macroinvertebrate) community; but it must be stressed that the whole question of environmental stability and hence biological stability frequently depends on the time-scale of observations (Petts 1987). For recent studies on the recovery of lotic communities and ecosystems following disturbance see Yount and Niemi (1990). Gore (1985b) suggests that the best monitoring regime for the determination of stable communities and the success of restoration works involves comparison between the restored communities and communities in unstressed source areas by means of similarity indices (Gore 1982) or biotic indices (Winget 1985).

Biomanipulation is another field of management in which predictive models can have application. Scheffer (1990), working on the restoration of eutrophied lakes, uses simple models to analyse specific ecological relationships. The models contain only four functional 'groups' – turbidity, fish (planktivorous and benthivorous), vegetation and piscivores – and the only driving variable considered is nutrient level. The models indicate that several ecological relationships in freshwater systems potentially give rise to the existence of alternative equilibria over a certain range of nutrient values. In shallow aquatic systems there appear to be two alternative stable states: a clear vegetated state with low fish stocks, and a turbid unvegetated state with high densities of planktivorous and benthivorous fish. The recognition of these stable states increases understanding and helps optimize biomanipulation/restoration measures. Biomanipulation, for example by reducing planktivorous and benthivorous fish to certain levels or introducing predatory fish, can tilt the system over a threshold to a new stable equilibrium in conditions where it is logistically impossible to reduce nutrient levels sufficiently (Hosper & Jagtman 1990).

In forested streams alterations to riparian vegetation may have considerable impact on stream ecology. Duncan *et al* (1989) modified existing techniques for studying energy flow dynamics and have attempted to evaluate these effects by the means of energy flow models which reflect negative and positive production responses with respect to estimated production values. The models predict changes in production of benthic and terrestrial invertebrates and salmonid tissue in response to alteration of riparian cover and composition, and stream nutrients.

Conservation and catchments

Managers of the environment are increasingly required to conserve areas of streams, rivers and wetlands (Boon *et al* 1992). In order to do this effectively it is necessary to know the habitat requirements of the species or communities which are to be protected. Numerous models have been developed which describe species/habitat relationships (Pearsall *et al* 1986). These models do not precisely predict biological response but instead can be used to provide an index of the carrying capacity of a particular habitat in a procedure closely akin to the habitat preference curves used in the Instream Flow Incremental Methodology (see Chapter 14). Pearsall *et al* (1986) describe a range of habitat evaluation procedures all of which use models and they conclude that the definition of carrying-capacity and the validation of species habitat models by measuring population densities is problematic. Some reasons put forward are: that stochastic and temporal effects are difficult to account for in experimental design, and site tenacity and social interactions which prevent subdominant animals using suitable habitat may result in 'habitat sinks', or areas of poor habitat and high density. This point is illustrated by a study of habitat selection in muskrats (Messier *et al* 1990) where the relative densities in less suitable habitats increased much faster than those of the other more suitable habitats.

Thermal modelling

Management activities such as clear-cut logging, removal of riparian vegetation, release of thermal effluents, hydropower generation and deep-water releases from impoundments can alter the natural thermal regime in rivers. The effects of increased temperature on aquatic dipterans has been studied by Rempel and Carter (1987) in experimental channels. The population dynamics of aquatic insects were simulated using a stage-projection matrix model. The results indicated that changes in development rate together with changes in fecundity must be considered when predicting the response of aquatic populations to altered thermal regimes.

Water management modelling

Predictive models have a role in general water quality management and the planning of water resource development (Cairns 1990). The emphasis is frequently not on predicting biological responses but rather on forecasting of river flows (Reed 1987), sediment transport (MacMurray & Jaeggi 1990), water quality impact (Crockett *et al* 1989), or pollution transport (Trapp *et al* 1990) (see also Chapter 17 of this book). Part or all of these may be incorporated into catchment models (Virtanen 1989; Bodo & Unny 1990) to aid and guide management decisions. A catchment model obviously increases in value if specific biological responses can be predicted and the inclusion of biological data in the more general models is the next step forward. The development of geographic information or expert systems is the logical evolution of model integration.

Forecasting impacts of pollution

Frequently it is necessary to be able to forecast the effects of changing environmental variables on river biota. The establishment of guidelines for limiting nutrients in running water is one such need. Welch *et al* (1989) have attempted this by modelling predictions of periphyton biomass as a function of soluble reactive phosphate (SRP). They use the SRP concentration that would produce a threshold nuisance biomass (150–200 mg chlorophyll a m^{-2}) to propose an approach to control the length of stream in which biomass exceeds nuisance level. Painter and Jackson (1989) have developed a mathematical model which simulates *Cladophora* internal phosphorus levels. The model was used to determine the necessary reduction in SRP to decrease production of *Cladophora* at a specified site. Dauta (1986) has produced a model which simulates the dynamics of a phytoplankton population in the lower course of the River Lot in France in relation to temperature, light and nutrients. The model generated phases of development and decline for blue-green algal blooms that were identical to those in the natural environment. The model, which is based on laboratory estimates of specific growth rates of algae (diatoms, Cyanophyceae, Chlorophyceae), provides a description of development conditions and succession for algae.

Reynolds and Glaister (in press) modified a model of phytoplankton growth based on lake studies and used it to reconstruct the growth of algae moving downstream according to criteria of water temperature, day length and the time of travel. They then applied the technique to investigate the problems of hydraulic retention in river channels.

Svadlenkova *et al* (1989) mathematically modelled the kinetics of radiocaesium uptake and release by the filamentous alga *Cladophora glomerata* in order to trace radioactivity in running waters. Leps *et al* (1990) have used a transition matrix model to predict changes in ephemeropteran communities, arising mainly from increased eutrophication. These models are the simplest method for predicting further development of ecological systems and have had most application in the study of plant succession (Hulst 1979; Usher 1981). The predictions are basically linear extrapolations of observed present trends.

Habitat and hydraulics

Gore (1989) presents a range of models for describing benthic macroinvertebrate habitat suitability. The basis of some of these models is the 'habitat suitability curve', which represents the preference of the species under test for a range of variables. These data are used in conjunction with predictions of habitat change in relation to flow in order to predict the impact of altered discharge regimes on river biota. This in essence is the Instream Flow Incremental Methodology (see Chapter 14). The situation is complicated by the need to define, for some organisms, habitats at different life history stages. Gore and Bryant (1990), for example, note that the crayfish *Orconectes neglectus* has some size-related flow preferences as well as dimorphic velocity preferences during the period when females migrate to riffles carrying their eggs. In addition there are few studies which quantify biotic responses to extreme discharge events. These may take the form of concentration of organisms in refuges or dead zones which may bear little resemblance to the preferred habitat estimated at 'normal' flows.

Shifts in distribution in relation to flow-related variables such as velocity depth and substratum have been the subject of much study. Much of this concerns fish habitat, particularly the Instream Flow Incremental Methodology (see Chapter 14) but there has been increased interest in the influence of hydraulic factors in determining the behaviour and hence the distribution of river biota, especially invertebrates. For a detailed presentation of these approaches see Statzner *et al* (1988).

Pest control and parasite behaviour

Control programmes for pest species have required predictive techniques to facilitate their implementation. Precise and accurate estimates of secondary production are important for the rational management of aquatic resources and the development and testing of ecological theory. Morin *et al* (1988b) described the growth rates of three blackfly species in relation to individual mass, water temperature and food quality. Allometric models of growth rates were used to estimate production. Benke and Parsons (1990) have also modelled blackfly production dynamics. Two predictive models were used together with intensive field sampling to estimate production of blackflies. One model predicts daily growth rate from temperature and hydrograph pattern, the other predicts habitat abundance (in the form of submerged wood) from river height.

Such predictions can be essential for the efficient application of control measures. In addition it is useful to have information on the ingestion rates of blackfly larvae in order to optimize application of particulate larvicides (Morin *et al* 1988a). Insecticide treatments can be further improved if predictions of blackfly susceptibility to particulate control agents are linked to a one-dimensional transport equation (Chalifour *et al* 1990). In this way it is possible to determine an optimal strategy of treatment which minimizes the quantity of insecticide applied and the logistic movement between injection sites along the river.

Another area in which modelling has contributed to 'control' is in parasitological work, and its relevance to rivers is seen in studies of schistosomiasis. A comprehensive review of helminth infections of humans (Anderson & May 1985) addresed the role of mathematical models in investigations of the population and transmission dynamics of human helminths. These authors concluded that such models can be valuable tools in the design of policies for parasite control. The usefulness of these models depends on a thorough and detailed understanding of host/parasite population dynamics and interrelations. Woolhouse and Chandiwana (1990) have developed a model which describes the population dynamics of the freshwater snail *Bulinus globosus*, which is an intermediate host of *Schistosoma haematobium*. The responses of the snail population to low and high flows are modelled and used to simulate changes in abundance over time-scales of 10 or more years.

Janovy *et al* (1990) constructed a BASIC computer simulation model to mimic the dynamic behaviour of parasite species over a range of abiotic conditions. The model data were compared to a field data set consisting of 33 samples of the fish *Fundulus zebrinus* and its parasites taken during a 7-year period from a single location on a fluctuating river. Over the long term field data did not depart significantly from that predicted by the model.

Fisheries

The close association of the human race with fish as a food source means that studies of fish population dynamics have had important applications. In order to ensure a continuing supply of this food resource it has become necessary to develop models which assess and predict the impact of fishing and fishing technologies on the stocks of fish. The historical development of fish population dynamics and models are reviewed in Gulland (1977), and the practical application of some standard methods and models for estimating production in freshwaters is described in Ricker (1958) and Bagenal (1978).

Fish stock and yield

The mathematical analysis of fish stock dynamics has been the subject of a recent American Fisheries Society Symposium (Edwards & Megrey

1989). The proceedings include numerous examples of predictive models. Most deal with marine oceanic fisheries but have some application in freshwaters, and the reader is referred to this publication for source references.

A common objective of fishery managers is the prediction of the future yield of a fish population in response to harvest effort (Gros & Prozet 1988; Vøllestad & Jonsson 1988). Stock production models (models of population growth) (Pella & Tomlinson 1969; Schnute 1977) are commonly used to estimate yield based on data on growth, mortality and reproduction of the population. Information on age structure is, however, notably lacking from these models and this has been redressed to a certain extent by the use of Leslie-matrix or Beverton–Holt (Beverton & Holt 1957) models. These procedures in turn, while employing a basic age structure, make use of the approximation that ages of animals occur in discrete groups whereas an animal can have an age represented by any positive real number. These then are the only two classes of models which are widely used for fisheries management but neither take into account the dependence of mortality, growth and reproduction on food consumption (Bledsoe & Megrey 1989).

On a broader scale it is necessary to assess the effect of general environmental impacts on fish populations and fisheries. Hilden and Kettunen (1985) have reviewed a range of models (basic, age-structured, models with a spatial resolution, multi-species, stochastic and ecosystem models) which attempt to do this. Their analysis suggests that an impact model should preserve the age structure of the population and allow studies of the spatial distribution of the impact and the fish populations. They noted that ecosystem models have been used for simulation tests which may provide information on the fish response to stress. However, a problem arises with increase in model complexity because the model itself becomes more and more difficult to analyse. Hilden and Kettunen concluded that it is almost impossible to tell which predictions simply reflect the behaviour of the model and its computer implementation and which accurately illustrate changes in natural populations and actual fisheries. They suggested that models incorporating a stochastic element are likely to be of increased value in predicting impacts because they would take into account the fact that many environmental impacts are stochastic in nature. Environmental variability tends to increase fluctuations in exploited fish populations (Horwood & Shepherd 1981).

Bioenergetics

Preall and Ringler (1989) noted that despite the fact that energy budgets have been developed for many freshwater fish species (Solomon & Brafield 1972; Niimi & Beamish 1974) little is known of the ways in which growth is governed under natural conditions where food supplies, temperature regimes and numbers of competing individuals often fluctuate unpredictably. Preall and Ringler (1989) draw on the extensive work of Elliott (1975a,b,c,d, 1976a,b,c) on feeding and growth of brown trout. They utilize his bioenergetic equations in a computer model (TROUT) which predicts the maximum growth rate of streamdwelling brown trout and compare the predicted values with actual field estimates. TROUT can be used to study interactions of brown trout with stream invertebrates or other prey and assess the impact of foraging by brown trout on invertebrate production.

Temperature is an important factor controlling growth and is an integral variable in bioenergetic studies of adult fish (Wildhaber & Crowder 1990). Temperature also influences the duration of egg development in freshwater fish (Elliott 1981) and Elliott *et al* (1987) have compared eight mathematical models for the relationship between water temperature and hatching time of salmonid eggs. Saltveit and Brabrand (1987) used a laboratory approach to predict the effects of possible temperature increases, resulting from river regulation, on the eggs of whitefish.

Closer scrutiny is now being given to the importance of activity in bioenergetics models. Boisclair and Leggett (1989) used the Kitchell *et al* (1977) and Kerr (1982) models and field data on *Perca flavescens* (yellow perch) to estimate energy allocated to activity. They concluded that this (activity) component of the bioenergetics budget of fishes can help to explain inter-population differences in perch growth and, by extension, the variance in growth of other actively foraging species.

Fish habitat conditions

Where fish habitat is likely to be changed through flow regulation or other disturbance it is useful to be able to predict the response of the fish population or individual to that change. The techniques range from those involved in the Instream Flow Incremental Methodology (Chapter 14) through simpler habitat models such as HABSCORE, a computer model which has been developed from work started by Milner *et al* (1985). The model, currently in use by the Welsh Region of the National Rivers Authority (Milner, personal communication) can predict fish densities (salmonids) from selected habitat variables.

There have also been recent studies which have attempted to predict: the impact of thermal effluent on a fish population (Shuter *et al* 1985); the effect of climatic warming on the loss of suitable habitat of brook trout (*Salvelinus fontinalis*) (Meisner 1990); the effects of draining releases from reservoirs on brown trout populations in the river below the dam (Garric *et al* 1990); the response of fish to macrophyte control in a lake (Bettoli *et al* 1990); and the growth of loach in cadmium-loaded environments (Douben 1990).

Feeding behaviour

Basic studies of fish behaviour are not necessarily directly linked with environmental impacts, as they are, for example, in pollution early-warning systems (Cairns & Schalie 1980). However, the work of Hughes and Dill (1990) on position choice of grayling (*Thymallus arcticus*) in mountain streams shows how such work can have relevance to environmental change. The position chosen by the drift-feeding fish is dependent on biotic factors, water depth and velocity. The demonstrated ability to predict position could be used to determine the relationship between flow regulation or abstraction and habitat loss. Grant and Kramer (1990) examined the role of territory size as a predictor of the upper limit to population density of juvenile salmonids in streams. Their results indicated that the territory size model has practical value for predicting maximum densities in shallow habitats. Territory size and position choice both relate to food consumption, distribution and availability. Several predictive feeding models have been used to evaluate size-dependent feeding by lentic fish (Eggers 1977; Wright & O'Brien 1984). In lotic environments there has been less effort (Dunbrack & Dill 1984). Newman (1987) has compared encounter model predictions, i.e. models based on the reactive distance of fish to prey of different sizes, with observed size selectivity by stream-dwelling trout. Such models provide insight into mechanisms of prey selection and the predictions may be useful in estimating the effects of fish predation on benthos production and population dynamics. Such studies have less to do with monitoring and more to do with the development of ecological theory.

13.3 ECOLOGICAL THEORY AND MODELLING

The main emphasis of this chapter so far has been on the application of predictive models in water resource management. However, some investigations which predict the response of an individual or population to specific variables have little direct application to management objectives. They can be thought of as providing the raw data for future predictive models and stimulating further thought and the development of ecological theories. It is beyond the scope of this chapter to present an exhaustive account of such studies and here only some recent examples are presented.

McCauley *et al* (1990) studied the physiological ecology of *Daphnia*. They found that existing models did not describe some essential features of individual growth, particularly when food availability was low. This led the investigators to examine how *Daphnia* stores energy and uses reserves. Their findings formed the basis of a new model for the growth of *Daphnia pulex*. Gurney *et al* (1990) then further developed and tested this model. Their success in matching their predictions to a target data set provides strong evidence that measurements of short-term physiological rates can be used successfully as predictors of long-term growth and fecundity. Hallam *et al* (1990), also working with *Daphnia*, developed an energetics model specifically for females. The model mimics the life history of an individual as it progresses from egg through juvenile to adult instars.

Pockl and Timischl (1990) studied the quantitative relationship between water temperature and brood development time of two species of *Gammarus* (Amphipoda), *G. fossarum*, a headwater species, and *G. roeseli*, which is usually found farther downstream. They suggested that models which can be linearized are more useful than non-linear models, and recommended the use of the 'mixed power-exponential' model for comparing intra- and interspecific variations in the relationship between brood development time and water temperature.

Kokkinn (1990) examined the hypothesis that the rate of embryonic development in *Tanytarsus barbitarsis* (Chironomidae: Diptera) is a predictor of overall development rate. The results showed that the power function that closely described embryonic development could not be fitted to the generation time data, which suggests that laboratory egg-hatching experiments could not account for the increased field development time caused by non-optimal environmental conditions.

Interest in the dynamics of ecological processes in space as well as in time (Allen 1975; Taylor 1990) has stimulated the study of competition in 'patchy' environments (Yodzis 1986; Shorrocks & Rosewell 1987). Many of these studies use models to simulate this competition and the results assist in the formulation of new theories of population dynamics. These may then be adapted for application in stream community ecology (Townsend 1989) and act as a focus for new lines of research (Pringle *et al* 1988). The inevitable disagreements in the utility of new concepts (see Frid & Townsend 1989; Downes 1990) only serve to stimulate the scientific community to more detailed examination of the concepts and reasons for disagreement.

Disturbance and stability

The effect of stress on ecosystems has been the subject of much basic scientific research (see Lugo 1978 for a review) and it is a field of considerable relevance to aquatic ecology (Rapport *et al* 1985; Resh *et al* 1988). The search for generality, which is an inherent characteristic of the scientific method and a feature which facilitates the prediction of biological responses, is made harder by the environmental and economic constraints which force researchers to examine only limited data sets and from these make inferences to larger systems (Perry *et al* 1987). Lately, however, the concern over acidification has prompted the implementation of whole ecosystem experiments involving experimental acidification (Hall *et al* 1980) and liming (Elser *et al* 1986; Ormerod *et al* 1990b). Perry *et al* (1987) have examined the effects of stress on a stream channel ecosystem and in a lake. Further disturbance studies are referred to in Perry *et al* (1987), who concluded that generalities in the ways that community and ecosystem structure respond to disturbance do follow hypothesized trends, including increased nutrient export and decreases in efficiency in system energetics, nutrient standing stock, diversity, evenness and total numbers of organisms. The consistency of response among a variety of ecosystems augurs well for the future development of predictive models, but a note of caution (Hilden & Kettunen 1985) must be added with respect to the difficulties of interpretation of complex ecosystem models.

The relationships between environmental predictability (with respect to lotic communities) and disturbance have been studied without particular reference to predictive models by Rader and Ward (1989) and Poff and Ward (1990). Weatherley and Ormerod (1990), however, address specifically the implications of the constancy (persistence) of invertebrate assemblages in softwater streams to the prediction and detection of environmental change. Disturbance is the dominant organizing factor in stream ecology (Resh *et al* 1988). Its relationship to community stability will have a powerful influence on the design and interpretation of predictive models.

On a different scale Wilson and Botkin (1990) examined the effect of ecosystem complexity on stability using models of simple plankton microcosms and addressed questions relating to the reductionist and holistic approaches to research.

They concluded from their studies that the community and ecosystem-level properties of the models were emergent (i.e. unpredictable solely from observations of the components of the system) and therefore could not in general be predicted only from a knowledge of individual species responses to different environmental conditions.

13.4 FUTURE DIRECTIONS

The future of predictive limnology (in lakes) has been discussed by Peters (1986). Six approaches are identified as being central to improvement.

1 The selection of variables which are easier to measure, less uncertain and better in prediction.

2 The use of multiple regression to focus on the effects of more than one independent variable.

3 The definition of the range over which the model is effective.

4 The replacement of simple arithmetic and logarithmic models with more sophisticated models as our mathematical and statistical knowledge increases.

5 The application of predictive methodologies to a wide range of organisms and habitats.

6 The culling of competing theories.

In addition Peters (1986) notes an increasing interest in heterogeneity at finer temporal and spatial scales. The recognition that dramatic differences in biotic responses can occur in successive years or within a single season in single lakes suggests an increasing ability to think probabilistically and the author anticipates a growth in models and analytical tools to express and use the observed levels of variation.

Peters' suggestions generally still hold true, and although there have been improvements there is a plethora of predictive models available for a variety of applications; there is probably a need for Peters' (1986) suggestion that extensive reviews of the current field should be promoted by leading journals to facilitate the evaluation of differences among similar theories.

In addition to these comments I would add a further six points for future development.

1 Despite the unglamorous nature of the work it is vital that large good-quality databases are collected. The acquisition of the database for RIVPACS depended on the infrastructure of the water industry in Great Britain, which meant that data could be collected in a standardized form throughout the country. Similar databases provide a firm foundation for model development.

2 There should be greater emphasis on linking disciplines as well exemplified by the Instream Flow Incremental Methodology (IFIM) work (Chapter 14) and the work of Statzner *et al* (1988) where hydrology is linked with animal distribution. The IFIM is strictly a management tool but nevertheless provides a new way of looking at the lotic environment and the habitat requirements of all the river biota and encourages dialogue between engineer and scientist. Divisions between basic research and management are being broken down by the need for applied research which borrows from basic research. In turn management requirements encourage new directions in both applied and basic research.

3 It is essential that managers and central government do not render research workers intellectually bankrupt by only supporting applied work. It is vital that work which aims to elucidate mechanisms of response is also funded to build a bank of knowledge on which future researchers can draw.

4 In order to predict biological responses usefully it is important to know the natural variation in response over long and short periods of time in different types of river habitat.

5 Emphasis should be placed on the development of models which predict the biological consequences of environmental disturbance. The ultimate aim of predictive models is the ability to alter variables and or groups of variables which may be affected by disturbance and estimate their consequences. This would be a pre-disturbance step and would determine the implementation and operation of any planned perturbation.

6 Integration of a wide range of models (hydrological, hydrochemical, biological) into geographic or knowledge-based systems. These systems represent an amalgamation of data from many sources which may include information on land use, water quality, and urbanization effects, and will incorporate predictive models. Their aim is to facilitate water management by making these diverse databases and models accessible and providing links between them which can be used in decision-making. However, some caution is required: as Hedgpeth (1977) points out, bureaucracies are notoriously resistant to change and once a concept or principle is accepted by bureaucrats it may persist and affect policy long after it has been passed by in the 'real world'. This puts the onus on scientists to either get it right first time (clearly unlikely) or maintain closer contacts with management in order to continually revise their predictive methods in the light of increased knowledge.

ACKNOWLEDGEMENTS

This work has been supported financially by the Natural Environment Council. I am grateful to the Council and to my colleagues at the Institute of Freshwater Ecology and Dr Jim Gore for providing information.

REFERENCES

Aldenberg T, Peters JS. (1990) On relating empirical water quality diagrams and plankton dynamical models: the SAMPLE methodology applied to a drinking water storage reservoir. *Archiv für Hydrobiologie Beiheft. Ergebnisse der Limnologie* **33**: 893–911. [13.2]

Allen JC. (1975) Mathematical models of species interactions in time and space. *American Naturalist* **109**: 319–42. [13.3]

Anderson RM, May RM. (1985) Helminth infections of humans: mathematical models, population dynamics, and control. *Advances in Parasitology* **24**: 1–101. [13.2]

Armitage PD. (1989) The application of a classification and prediction technique based on macroinvertebrates to assess the effects of river regulation. In: Gore JA, Petts GE (eds) *Alternatives in Regulated River Management*, pp. 267–93. CRC Press, Boca Raton, Florida. [13.2]

Armitage PD, Pardo I, Furse MT, Wright JF. (1990) Assessment and prediction of biological quality. A demonstration of a British macroinvertebrate-based method in two Spanish rivers. *Limnetica* **6**: 147–56. [13.2]

Armitage PD, Furse MT, Wright JF. (1992) Anexo: Calidad medioambiental y valoracion Biologica en los rios britanicos. Perspectivas de pasado y futuro. In: Departamento de Economia Planificacion y Medio Ambiente del Gobierno Vasco (ed) *Caracterizacion Hidrobiologica de la red fluvial de Alava y Gipuzcoa*, pp. 477–511. Itxarapena S.A., Zarantz. [13.2]

Bagenal T. (ed) (1978) *IBP Handbook No. 3. Methods for Assessment of Fish Production in Fresh Waters.* Blackwell Scientific Publications, Oxford. [13.2]

Battarbee RW, Charles DF. (1986) Diatom based pH reconstruction studies of acid lakes in Europe and N. America: a synthesis. *Water, Air and Soil Pollution* **30**: 347–54. [13.2]

Benke AC, Parsons KA. (1990) Modelling black fly production dynamics in blackwater streams. *Freshwater Biology* **24**: 167–80. [13.2]

Best EPH. (1990) Models on metabolism of aquatic weeds and their application potential. In: Pieterse AH, Murphy KJ (eds) *Aquatic Weeds. The Ecology and Management of Nuisance Aquatic Vegetation*, pp. 254–73. Oxford University Press, Oxford. [13.2]

Bettoli PW, Springer T, Noble RL. (1990) A deterministic model of the response of threadfin shade to aquatic macrophyte control. *Journal of Freshwater Ecology* **5**: 445–54. [13.2]

Beverton RJH, Holt SJ. (1957) On the dynamics of exploited fish populations. *UK Ministry of Agriculture Fisheries Investigations (London). Series 2*, p. 19. HMSO, London. [13.2]

Bledsoe LJ, Megrey BA. (1989) Chaos and pseudoperiodicity in the dynamics of a bioenergetic food web model. *American Fisheries Society Symposium* **6**: 121–37. [13.2]

Bodo BA, Unny TE. (1990) Linear stochastic conceptual response models for small catchments. *Revue de la Science Eau* **3**(2): 151–82. [13.2]

Boisclair D, Leggett WC. (1989) The importance of activity in bioenergetics models applied to actively foraging fishes. *Canadian Journal of Fisheries and Aquatic Sciences* **46**: 1859–67. [13.2]

Boon PJ, Calow P, Petts GE. (eds) (1992) *River Conservation and Management*. John Wiley, Chichester. [13.2]

Bovee KD. (1982) *A guide to Stream Habitat Analysis using the Instream Flow Incremental Methodology.* Instream Flow Information Paper 12, Report FWS/OBS-82/86, U.S. Fish Wildlife Service, Fort Collins, Colorado. [13.1]

Bravard J-P, Amoros C, Pautou G. (1986) Impact of civil engineering works on the successions of communities in a fluvial system. *Oikos* **47**: 92–111. [13.2]

Cairns J Jr. (1980) Estimating hazard. *Bioscience* **30**: 101–7. [13.2]

Cairns J Jr. (ed) (1985) *Multispecies Toxicity Testing.* Pergamon Press, New York. [13.2]

Cairns J Jr. (1988) Should regulatory criteria and standards be based on multispecies evidence? *Environmental Professional* **10**: 157–65. [13.2]

Cairns J Jr. (1990) Lack of theoretical basis for predicting rate and pathways of recovery. *Environmental Management* **14**: 517–26. [13.2]

Cairns J Jr, van der Schalie WH. (1980) Biological monitoring. Part I – Early warning systems. *Water Research* **14**: 1179–96. [13.2]

Chalifour A, Boisvert J, Back C. (1990) Optimization of insecticide treatments in rivers: an application of graph theory for planning a blackfly larvae control program. *Canadian Journal of Fisheries and Aquatic Sciences* **47**: 2049–56. [13.2]

Crockett CP, Crabtree RW, Markland HR. (1989) SPRAT, a simple river quality impact model for intermittent discharges. *Water Science and Technology* **21**: 1793–6. [13.2]

Dauta A. (1986) Modélisation du développement du phytoplankton dans une rivière canalisée eutrophe le Lot (France). *Annales de Limnologie* **22**: 119–32. [13.2]

Descy JP. (1979) A new approach to water quality estimation using diatoms. *Nova Hedwigia* **64**: 305–23. [13.2]

Douben PET. (1990) A mathematical model for cadmium in the stone loach, *Noemcheilus barbatulus* L. from the river Ecclesbourne, Derbyshire, England, UK. *Ecotoxicology and Environmental Safety* **19**: 160–83. [13.2]

Downes BJ. (1990) Patch dynamics and mobility of fauna in streams and other habitats. *Oikos* **59**: 411–13. [13.3]

Dunbrack RL, Dill LM. (1984) Three-dimensional prey reaction field of the juvenile coho salmon (*Oncorhynchus kisutch*). *Canadian Journal of Fisheries and Aquatic Sciences* **41**: 1176–82. [13.2]

Duncan WFA, Brusven MA, Bjornn TC. (1989) Energy flow response models for evaluation of altered riparian vegetation in three southeast Alaskan streams. *Water Research* **23**: 965–74. [13.2]

Edwards EF, Megrey BA. (1989) Mathematical analysis of fish stock dynamics. *American Fisheries Society Symposium* No. 6, pp. 1–214. [13.2]

Edwards RW. (1984) Predicting the environmental impact of a major reservoir development. In: Roberts RD, Roberts TM (eds) *Planning and Ecology*, pp. 55–79. Chapman & Hall, London. [13.2]

Eggers DM. (1977) The nature of food selection by planktivorous fish. *Ecology* **58**: 46–59. [13.2]

Elliott JM. (1975a) Weight of food and time required to satiate brown trout, *Salmo trutta* L. *Freshwater Biology* **5**: 51–64. [13.2]

Elliott JM. (1975b) Number of meals in a day, maximum weight of food consumed in a day and maximum rate of feeding for brown trout, *Salmo trutta* L. *Freshwater Biology* **5**: 287–308. [13.2]

Elliott JM. (1975c) The growth rate of brown trout (*Salmo trutta* L.) fed on maximum rations. *Journal of Animal Ecology* **44**: 805–21. [13.2]

Elliott JM. (1975d) The growth rate of brown trout (*Salmo trutta* L.) fed on reduced rations. *Journal of Animal Ecology* **44**: 823–42. [13.2]

Elliott JM. (1976a) Body composition of brown trout (*Salmo trutta* L.) in relation to temperature and ration size. *Journal of Animal Ecology* **45**: 273–89. [13.2]

Elliott JM. (1976b) Energy losses in the waste products of brown trout (*Salmo trutta* L.). *Journal of Animal Ecology* **45**: 561–80. [13.2]

Elliott JM. (1976c) The energetics of feeding metabolism and growth of brown trout (*Salmo trutta* L.) in relation to body weight, water temperature and ration size. *Journal of Animal Ecology* **45**: 923–48. [13.2]

Elliott JM. (1981) Some aspects of thermal stress on freshwater teleosts. In: Pickering AD (ed) *Stress and Fish*, pp. 209–45. Academic Press, London. [13.2]

Elliott JM, Humpesch UH, Hurley MA. (1987) A comparative study of eight mathematical models for the relationship between water temperature and the hatching time of eggs of freshwater fish. *Archiv für Hydrobiologie* **109**: 257–77. [13.2]

Elser MM, Elser JG, Carpenter SR. (1986) Peter and Paul lakes. A liming experiment revisited. *American Midland Naturalist* **116**: 282–95. [13.3]

Evison LM. (1979) Microbial parameters of raw water quality. In: *Biological Indicators of Water Quality. Proceedings of a Symposium at the University of Newcastle-upon-Tyne, UK, September 1978.* Wiley, Chichester. [13.2]

Frid CLJ, Townsend CR. (1989) An appraisal of the patch dynamics concept in stream and marine benthic communities whose members are highly mobile. *Oikos* **56**: 137–41. [13.3]

Furse MT, Moss D, Wright JF, Armitage PD. (1987) Freshwater site assessment using multi-variate techniques. In: Luff ML (ed) *The Use of Invertebrates in Site Assessment for Conservation. Proceedings of a Meeting Held at the University of Newcastle upon Tyne, 7 January 1987.* Agricultural Environment Research Group, University of Newcastle-upon-Tyne. [13.2]

Garric J, Migeon B, Vindimian E. (1990) Lethal effects of draining on brown trout. A predictive model based on field and laboratory studies. *Water Research* **24**(1): 59–65. [13.2]

Giesy JP. (1985) Multispecies tests: Research needs to assess the effects of chemicals on aquatic life. In: Bahner RC, Hansen DJ (eds) *Aquatic Toxicology and Hazard Assessment: Eighth Symposium, ASTM STP 891*, pp. 67–77. American Society for Testing and Materials, Philadelphia, PA. [13.2]

Giesy JP, Allred PM. (1985) Replicability of aquatic multispecies test systems. In: Cairns J (ed) *Multispecies Toxicity Testing*, pp. 187–247. Pergamon Press, New York. [13.2]

Giesy JP, Graney RL. (1989) Recent developments in and intercomparisons of acute and chronic bioassays and bioindicators. *Hydrobiologia* **188/189**: 21–60. [13.2]

Gore JA. (1982) Benthic invertebrate colonization: source distance effects on community composition.

Hydrobiologia **94**: 183–93. [13.2]

Gore JA. (ed) (1985a) *The Restoration of Rivers and Streams – Theories and Experience.* Butterworth Publishers, Boston. [13.2]

Gore JA. (1985b) Mechanisms of colonization and habitat enhancement for benthic macroinvertebrates in restored river channels. In: Gore JA (ed) *The Restoration of Rivers and Streams – Theories and Experience,* pp. 81–102. Butterworth Publishers, Boston. [13.2]

Gore JA. (1989) Models for predicting benthic macroinvertebrate habitat suitability under regulated flows. In: Gore JA, Petts GE (eds) *Alternatives in Regulated River Management,* pp. 253–66. CRC Press, Boca Raton, Florida. [13.2]

Gore JA, Bryant RM Jr. (1990) Temporal shifts in physical habitat of the crayfish, *Orconectes neglectus* (Faxon). *Hydrobiologia* **199**: 131–42. [13.2]

Grant JWA, Kramer DL. (1990) Territory size as a predictor of the upper limit to population density of juvenile salmonids in streams. *Canadian Journal of Fisheries and Aquatic Sciences* **47**: 1724–37. [13.2]

Gros P, Prozet P. (1988) Forecasting stochastic model of spring salmon catches (*Salmo salar*) on Alune River, Brittany, France. Elements of fishery management. *Acta Oecologica Applicata* **9**: 3–24. [13.2]

Gulati RD, Lammens EHRR, Meijer M-L, Donk van E. (eds) (1990) Biomanipulation – tools for water management. *Hydrobiologia* **200/201**: 628. [13.2]

Gulland JA. (ed) (1977) *Fish Population Dynamics.* John Wiley, London. [13.2]

Gurney WSC, McCauley E, Nisbet RM, Murdoch WW. (1990) The physiological ecology of *Daphnia,* a dynamic model of growth and reproduction. *Ecology* **71**: 716–32. [13.3]

Guyomard R. (1989) Principaux concepts et methodes relatifs à la description de la diversité génétique d'une espêce. *Bulletin Francais de la Pêche et de la Pisculture* **314**: 98–108. [13.1]

Hall RJ, Likens GE, Fiance SB, Hendrey GR. (1980) Experimental acidification of a stream in the Hubbard Brook Experimental Forest, New Hampshire. *Ecology* **61**: 976–89. [13.2, 13.3]

Hallam TG, Lassiter RR, Li J, Suarez LA. (1990) Modelling individuals employing an integrated energy response: application to *Daphnia. Ecology* **71**(3): 938–54. [13.3]

Harding JPC, Whitton BA. (1981) Accumulation of zinc cadmium and lead by field populations of *Lemanea. Water Research* **15**: 301–19. [13.2]

Haslam SM. (1982) A proposed method for monitoring river pollution using macrophytes. *Environmental Technology Letters* **3**: 19–34. [13.2]

Hedgpeth JW. (1977) Models and muddles – some philosophical observations. *Helgoländer Wissenschaftliche Meeresuntersuchungen Helgoland* **30**: 91–104. [13.4]

Hellawell JM. (1989) *Biological Indicators of Freshwater Pollution and Environmental Management.* Elsevier Applied Science, London. [13.2]

Hilden M, Kettunen J. (1985) Models for the assessment of environmental impacts on fish populations and fisheries. *Finnish Fish News* **6**: 35–42. [13.2, 13.3]

Holmes RW, Whiting MC, Stodard JL. (1989) Changes in diatom-inferred pH and acid neutralizing capacity in a dilute, high elevation, Sierra Nevada lake since A.D. 1825. *Freshwater Biology* **21**: 295–310. [13.2]

Horwood JW, Shepherd JG. (1981) The sensitivity of age structured populations to environmental variability. *Mathematical Biosciences* **57**: 59–82. [13.2]

Hosper SH, Jagtman E. (1990) Biomanipulation additional to nutrient control for restoration of shallow lakes in the Netherlands. *Hydrobiologia* **200/201**: 523–34. [13.2]

Hughes NF, Dill LM. (1990) Position choice by drift feeding salmonids: model and test for Arctic grayling (*Thymallus arcticus*) in subarctic mountain streams, interior Alaska. *Canadian Journal of Fisheries and Aquatic Sciences* **47**: 2039–48. [13.2]

Hulst R van. (1979) On the dynamics of vegetation: Markov chains as models of succession. *Vegetatio* **40**: 3–14. [13.2]

Jackson PBN. (1989) Prediction of regulation effects on natural biological rhythms in South-Central African freshwater fish. *Regulated Rivers: Research and Management* **3**: 205–20. [13.2]

Janovy J Jr, Ferdig MT, McDowell MA. (1990) A model of dynamic behaviour of a parasite species assemblage. *Journal of Theoretical Biology* **142**: 517–30. [13.2]

Jeffers JNR. (1982) *Modelling.* Chapman & Hall, London. [13.1]

Karr JR, Fausch KD, Angermeier PL, Yant PR, Schlosser IJ. (1986) *Assessing Biological Integrity in Running Waters. A Method and its Rationale.* Special Publication No. 5, Illinois Natural History Survey, Champaign, Illinois. [13.2]

Kelso JRM, Shaw MA, Minns CK, Mills KH. (1990) An evaluation of the effects of atmospheric acidic deposition on fish and the fishery resource of Canada. *Canadian Journal of Fisheries and Aquatic Sciences* **47**: 644–55. [13.2]

Kerr SR. (1982) Estimating the energy budgets of actively predatory fishes. *Canadian Journal of Fisheries and Aquatic Sciences* **39**: 371–9. [13.2]

Kitchell JF, Stewart DJ, Weininger D. (1977) Applications of a bioenergetics model to yellow perch (*Perca flavescens*) and walleye (*Stizostedion vitreum vitreum*). *Journal of the Fisheries Research Board of Canada* **34**: 1922–35. [13.2]

Kokkinn MJ. (1990) Is the rate of embryonic development a predictor of overall development rate in *Tanytarsus barbitarsis* Freeman (Diptera: Chironomidae)? *Australian Journal of Marine and Freshwater*

Research **41**(5): 575–9. [13.3]

Kolkwitz R, Marsson M. (1909) Oekologie der tierischen Saprobien. *Internationale Revue der gesamten Hydrobiologie und Hydrographie* **2**: 126–52. [13.2]

Lacroix GL. (1987) Model for loss of atlantic salmon from acidic brown waters of Canada. In: Perry R, Harrison RM, Bell JNB, Lester JN (eds) *Acid Rain: Scientific and Chemical Advances*, pp. 516–22. Selper, London. [13.2]

Larsen DP, de Noyelles F, Stay F, Shiroyama T. (1986) Comparison of single species microcosm and experimental pond responses to atrazine exposure. *Environmental Toxicology and Chemistry* **5**: 179–90. [13.2]

Leps J, Soldan T, Landa V. (1990) Prediction of changes in ephemeropteran communities – a transition matrix approach. In: Campbell IC (ed) *Mayflies & Stoneflies: Life Histories and Biology*, pp. 281–7. Kluwer, Dordrecht. [13.2]

Lugo AE. (1978) Stress and ecosystems. In: Thorp JH, Gibbons JW (eds) *Energy and Environmental Stress in Aquatic Systems*, pp. 62–101. Department of Energy Symposium Series 78, National Technical Information Service, Springfield, Virginia. [13.3]

McCauley E, Murdoch WW, Nisbet RM, Gurney WSC. (1990) The physiological ecology of *Daphnia*: development of a model of growth and reproduction. *Ecology* **71**: 703–15. [13.3]

MacMurray HL, Jaeggi MNR. (1990) Modeling erosion of sand and silt bed river. *Journal of Hydraulic Engineering* **116**(9): 1080–9. [13.2]

Maltby L, Calow P. (1989) The application of bioassays in the resolution of environmental problems: past, present and future. *Hydrobiologia* **188/189**: 65–6. [13.2]

Marmorek DR, Jones ML, Minns CK, Elder FC. (1990) Assessing the potential extent of damage to inland lakes in eastern Canada due to acidic deposition. I. Development and evaluation of a simple 'site' model. *Canadian Journal of Fisheries and Aquatic Sciences* **47**: 55–66. [13.2]

Matthews RA, Buikema LA Jr, Cairns J Jr, Rodgers JH Jr. (1982) Biological monitoring. Part IIA – Receiving system functional methods, relationships and indices. *Water Research* **16**: 129–39. [13.2]

Meisner JD. (1990) Effect of climatic warming on the southern margins of the native range of brook trout, *Salvelinus fontinalis*. *Canadian Journal of Fisheries and Aquatic Sciences* **47**: 1065–70. [13.2]

Messier F, Virgl JA, Marinelli L. (1990) Density-dependent habitat selection in muskrats: a test of the ideal free distribution model. *Oecologia* **84** (3): 380–5. [13.2]

Milner NJ, Hemsworth RJ, Jones BE. (1985) Habitat evaluation as a fisheries management tool. *Journal of Fish Biology, Suppl. A* **27**: 85–108. [13.2]

Minns CK, Kelso JRM, Johnson MG. (1986) Large-scale risk assessment of acid rain impacts on fisheries: models and lessons. *Canadian Journal of Fisheries and Aquatic Sciences* **43**: 900–21. [13.2]

Morin A, Back C, Chalifour A, Boisvert J, Peters RH. (1988a) Empirical models predicting ingestion rates of black fly larvae. *Canadian Journal of Fisheries and Aquatic Sciences* **45**: 1711–19. [13.2]

Morin A, Constantin M, Peters RH. (1988b) Allometric models of simuliid growth rates and their use for estimation of production. *Canadian Journal of Fisheries and Aquatic Sciences* **45**: 315–24. [13.2]

Moss D, Furse MT, Wright JF, Armitage PD. (1987) The prediction of the macroinvertebrate fauna of unpolluted running-water sites in Great Britain using environmental data. *Freshwater Biology* **17**(1): 41–52. [13.2]

Munawar M, Dixon G, Mayfield CI, Reynoldson T, Sadar MH. (eds) (1989) Environmental bioassay techniques and their application. *Hydrobiologia* **188/189**: 1–680. [13.2]

Newman RM. (1987) Comparison of encounter model predictions with observed size-selectivity by stream trout. *Journal of the North American Benthological Society* **6**: 56–64. [13.2]

Niimi AJ, Beamish FWH. (1974) Bioenergetics and growth of largemouth bass (*Micropterus salmoides*) in relation to body weight and temperature. *Canadian Journal of Zoology* **52**: 447–56. [13.2]

Ormerod SJ, Tyler SJ. (1989) Long-term change in the suitability of Welsh streams for Dippers *Cinclus cinclus* as a result of acidification and recovery: a modelling study. *Environmental Pollution* **62**: 171–82. [13.2]

Ormerod SJ, Boole P, McCahon CP, Weatherley NS, Pascoe D, Edwards RW. (1987) Short-term experimental acidification of a Welsh stream: comparing the biological effects of hydrogen ions and aluminium. *Freshwater Biology* **17**: 341–56. [13.2]

Ormerod SJ, Weatherley NS, Varallo PV, Whitehead PG. (1988) Preliminary empirical models of the historical and future impact of acidification on the ecology of Welsh streams. *Freshwater Biology* **20**: 127–40. [13.2]

Ormerod SJ, Weatherley NS, Gee AS. (1990a) Modelling the ecological impact of changing acidity in Welsh streams. In: Edwards RW, Gee AS, Stoner JH (eds) *Acid Waters in Wales*, pp. 279–97. Kluwer, Dordrecht. [13.2]

Ormerod SJ, Weatherley NS, Merrett WJ, Gee AS, Whitehead PG. (1990b) Restoring acidified streams in upland Wales: a modelling comparison of the chemical and biological effects of liming and reduced sulphate deposition. *Environmental Pollution* **64**: 67–85. [13.2, 13.3]

Overrein LN, Seip HM, Tollan A. (1981) *Acid*

Precipitation – Effects on Forest and Fish. Final report of the SNSF-project 1972–1980. SNSF, Oslo. [13.2]
Painter DS, Jackson MB. (1990) Cladophora internal phosphorus modeling: verification. *Journal of Great Lakes Research* **15**(4): 700–8. [13.2]
Patrick R. (1954) Diatoms as an indication of river change. In: *Proceedings of the 9th Industrial Waste Conference, Purdue University.* Engineering Extension Series No. 87, pp. 325–30. [13.2]
Pearsall SH, Durham D, Eagar DC. (1986) Evaluation methods in the United States. In: Usher MB (ed) *Wildlife Conservation Evaluation,* pp. 111–33. Chapman & Hall, London. [13.2]
Pella JJ, Tomlinson PK. (1969) A generalised stock production model. *Bulletin of the Inter-American Tropical Tuna Commission* **13**: 421–96. [13.2]
Perry JA, Troelstrup NH Jr, Newsom M, Shelley B. (1987) Whole ecosystem manipulation experiments: the search for generality. *Water Science Technology* **19**: 55–71. [13.3]
Peters RH. (1986) The role of prediction in limnology. *Limnology and Oceanography* **31**: 1143–59. [13.1, 13.4]
Petts GE. (1987) Time scales for ecological change in regulated rivers. In: Craig JF, Kemper JB (eds) *Regulated Streams – Advances in Ecology,* pp. 257–66. Plenum Press, New York. [13.2]
Petts GE, Müller H, Roux AL. (eds) (1989) *Historical Change of Large Alluvial Rivers: Western Europe.* John Wiley & Sons, Chichester. [13.2]
Plafkin JL, Barbour MT, Porter KD, Gross SK, Hughes RM. (1989) *Rapid Bioassessment Protocols for Use in Streams and Rivers: Benthic Macroinvertebrates and Fish.* Report EPA/444/4-89-001. USEPA, Washington, DC. [13.2]
Pockl M, Timischl W. (1990) Comparative study of mathematical models for the relationship between water temperature and brood development time of *Gammarus fossarum* and *G. roeseli* (Crustacea: Amphipoda). *Freshwater Biology* **23**(3): 433–40. [13.3]
Poff NL, Ward JV. (1990) Physical habitat template of lotic systems: recovery in the context of historical pattern of spatiotemporal heterogeneity. *Environmental Management* **14**: 629–46. [13.3]
Preall RJ, Ringler NH. (1989) Comparison of actual and potential growth rates of brown trout (*Salmo trutta*) in natural streams based on bioenergetic models. *Canadian Journal of Fisheries and Aquatic Sciences* **46**: 1067–76. [13.2]
Pringle CM, Naiman RJ, Bretschko G, Karr JR, Oswood MW, Webster JR, Welcomme RL, Winterbourn MJ. (1988) Patch dynamics in lotic systems: the stream as a mosaic. *Journal of the North American Benthological Society* 7: 503–24. [13.3]
Rader RB, Ward JV. (1989) The influence of environmental predictability/disturbance characteristics on the structure of a guild of mountain stream insects. *Oikos* **54**: 107–16. [13.3]
Radford PJ. (1988) Model-monitoring relationships. In: Salmons W, Bayne B, Dursma E, Foestrer V (eds) *Pollution of the North Sea: An Assessment,* pp. 666–75. Springer Publishing, New York. [13.1]
Rapport DJ, Regier HA, Hutchinson TC. (1985) Ecosystem behavior under stress. *The American Naturalist* **125**: 617–40. [13.3]
Reckhow KH, Black RW, Stockton TB, Vogt JD, Wood JG. (1987) Empirical models of fish response to lake acidification. *Canadian Journal of Fisheries and Aquatic Sciences* **44**: 1432–42. [13.2]
Reed DW. (1987) UK flood forecasting in the 1980's. In: Collinge VK, Kirby C (eds) *Weather Radar and Flood Forecasting,* pp. 129–42. John Wiley, Chichester. [13.1, 13.2]
Rempel RS, Carter JCH. (1987) Modelling the effects of accelerated development and reduced fecundity on the population dynamics of aquatic diptera. *Canadian Journal of Fisheries and Aquatic Sciences* **44**: 1737–42. [13.2]
Resh VH, Brown AV, Covich AP, Gurtz ME, Li HW, Minchal GW, Reice, SR, Sheldon AL, Wallace JB, Wissmar RC. (1988) The role of disturbance in stream ecology. *Journal of the North American Benthological Society* **7**: 433–55. [13.3]
Reuss JO, Christophersen N, Seip HM. (1986) A critique of models for freshwater and soil acidification. *Water, Air and Soil Pollution* **30**: 909–30. [13.2]
Reynolds CS, Glaister MS. (in press) Spatial and temporal changes in phytoplankton abundance in the upper and middle reaches of the River Severn. *Archiv für Hydrobiologie.* [13.2]
Ricker WE. (1958) Handbook of computations for biological statistics of fish populations. *Bulletin of the Fisheries Research Board of Canada* **119**: 1–300. [13.2]
Rumeau A, Coste M. (1988) Initiation à la systémique des diatomées d'eau douce pour l'utilisation pratique d'un indice diatomique générique. *Bulletin Francais de la Peche et de la Pisciculture* **309**: 1–69. [13.2]
Rutherford JC. (1975) *Simulation of Water Quality in the Waikato and Tarawera Rivers.* Report 119, School of Engineering, University of Auckland, Auckland, New Zealand. [13.2]
Rutt GP, Weatherley NS, Ormerod SJ. (1990) Relationships between the physicochemistry and macroinvertebrates of British upland streams: the development of modelling and indicator systems for predicting fauna and detecting acidity. *Freshwater Biology* **24**: 463–80. [13.2]
Sadler K. (1983) A model relating the results of low pH bioassay experiments to fishery status in Norwegian lakes. *Freshwater Biology* **13**: 453–63. [13.2]

Saltveit SJ, Brabrand A. (1987) Predicting the effects of a possible temperature increase due to stream regulation on the eggs of whitefish (*Coregonus lavaretus*). In: Craig JF, Kemper JB (eds) *Regulated Streams – Advances in Ecology*, pp. 219–28. Plenum Press, New York. [13.2]

Scheffer M. (1990) Multiplicity of stable states in freshwater systems. *Hydrobiologia* **200/201**: 475–86. [13.2]

Schindler DW. (1988) Effects of acid rain on freshwater ecosystem. *Science* **239**: 149–57. [13.2]

Schnute J. (1977) Improved estimates from the Schaefer production model; theoretical considerations. *Journal of the Fisheries Research Board of Canada* **34**: 583–603. [13.2]

Sheldon AL. (1984) Colonization dynamics of aquatic insects. In: Resh VH, Rosenberg DM (eds) *The Ecology of Aquatic Insects*, pp. 401–29. Praeger, New York. [13.2]

Shorrocks B, Rosewell J. (1987) Spatial pathiness and community structure: coexistence and guild size of drosophilids on ephemeral resources. In: Gee JHR, Giller PS (eds) *Organisation of Communities: Past and Present*, pp. 29–51. Blackwell Scientific Publications, Oxford. [13.3]

Shuter BJ, Wismer DA, Regier HA, Matuszek JE. (1985) An application of ecological modelling: impact of thermal effluent on a smallmouth bass population. *Transactions of the American Fisheries Society* **114**: 631–51. [13.2]

Sladecek V. (1981) Indicator value of the genus *Opercularia* (Ciliata). *Hydrobiologia* **79**: 229–32. [13.2]

Slooff W. (1985) The role of multispecies testing in aquatic toxicology. In Cairns Jr J (ed) *Multispecies Toxicity Testing*, pp. 45–60. Pergamon Press, New York. [13.2]

Small MJ, Sutton MC. (1986) A direct distribution model for regional aquatic acidification. *Water Resources Research* **22**: 1749–58. [13.2]

Solomon DJ, Brafield AE. (1972) The energetics of feeding, metabolism and growth of perch (*Perca fluviatilis* L.). *Journal of Animal Ecology* **41**: 699–718. [13.2]

Statzner B, Gore JA, Resh VH. (1988) Hydraulic stream ecology, observed patterns and potential applications. *Journal of the North American Benthological Society* **7**: 307–60. [13.2, 13.4]

Svadlenkova M, Dvorak Z, Slavik O. (1989) A mathematical model for Cs uptake and release by filamentous algae. *Internationale Revue der gesamten Hydrobiologie* **74**(5): 461–9. [13.2]

Taylor AD. (1990) Metapopulations, dispersal and predator-prey dynamics: an overview. *Ecology* **71**: 429–33. [13.3]

Tipping E. (1989) Acid-sensitive waters of the English Lake District; a steady-state model of stream water chemistry in the Upper Duddon catchment. *Environmental Pollution* **60**: 181–208. [13.1]

Townsend CR. (1989) The patch dynamics concept of stream community ecology. *Journal of the North American Benthological Society* **8**: 36–50. [13.3]

Trapp S, Brueggeman R, Muenzer B. (1990) Estimation of releases into rivers with the steady state surface water model EXWAT using dichloromethane. *Ecotoxicology and Environmental Safety* **19**: 72–80. [13.1, 13.2]

Usher MB. (1981) Modelling ecological succession, with particular reference to Markovian models. *Vegetatio* **46**: 11–18. [13.2]

Virtanen M. (1989) Mathematical models within river basin management – an overview. *Aqua Fennica* **19**(2): 145–52. [13.2]

Vollenweider RA. (1975) Input–output models with special reference to the phosphorus loading concept in limnology. *Schweizerische Zeitschrift für Hydrobiologie* **37**: 53–84. [13.2]

Vøllestad LA, Jonsson B. (1988) A 13 year study of the population dynamics and growth of the European eel *Anguilla anguilla* in a Norwegian river: evidence for density dependent mortality and development of a model for predicting yield. *Journal of Animal Ecology* **57**: 983–97. [13.2]

Warwick WF. (1980) Palaeolimnology of the Bay of Quinte, Lake Ontario: 2800 years of cultural influence. *Canadian Bulletin of Fisheries and Aquatic Sciences* **206**: 1–117. [13.2]

Weatherley NS, Ormerod SJ. (1987) The impact of acidification on macroinvertebrate assemblages in Welsh streams: towards an empirical model. *Environmental Pollution* **46**: 223–40. [13.2]

Weatherley NS, Ormerod SJ. (1990) The constancy of invertebrate assemblages in soft-water streams: implications for the prediction and detection of environmental change. *Journal of Applied Ecology* **27**: 952–64. [13.2, 13.3]

Welch EB, Horner RR, Patmont CR. (1989) Prediction of nuisance periphytic biomass: a management approach. *Water Research* **23**: 401–5. [13.2]

Wesche TA. (1985) Stream channel modifications and reclamation structures to enhance fish habitat. In: Gore JA (ed) *The Restoration of Rivers and Streams – Theories and Experience*, pp. 103–64. Butterworth Publishers, Boston. [13.2]

Westlake DF. (1966) A model for quantitative studies of photosynthesis by higher plants in streams. *Air and Water Pollution International Journal* **10**: 883–96. [13.2]

Wildhaber ML, Crowder LB. (1990) Testing a bioenergetics-based habitat choice model: bluegill (*Lepomis macrochirus*) responses to food availability and temperature. *Canadian Journal of Fisheries and Aquatic Sciences* **47**: 1664–71. [13.2]

Wilson MW, Botkin DB. (1990) Models of sample

microcosms: emergent properties and the effect of complexity on stability. *The American Naturalist* **135**: 414–34. [13.3]

Winget RN. (1985) Methods for determining successful reclamation of stream ecosystems. In: Gore JA (ed) *The Restoration of Rivers and Streams – Theories and Experience*, pp. 165–92. Butterworth Publishers, Boston. [13.2]

Woltering DM. (1985) Population responses to chemical exposure in aquatic multispecies systems. In: Cairns Jr J (ed) *Multispecies Toxicity Testing*, pp. 61–75. Pergamon Press, New York. [13.2]

Woolhouse MEJ, Chandiwana SK. (1990) Population dynamics model for *Bulinus globosus* intermediate host for *Schistosoma haematobium* in river habitats. *Acta Tropica* **47**: 151–60. [13.2]

Wright DI, O'Brien WJ. (1984) The development and field test of a tactical model of the planktivorous feeding of white crappie (*Pomoxis annularis*). *Ecological Monographs* **54**: 65–98. [13.2]

Wright JF, Armitage PD, Furse MT, Moss D. (1989) Prediction of invertebrate communities using stream measurements. *Regulated Rivers: Research and Management* **4**: 147–55. [13.2]

Yodzis P. (1986) Competition, mortality and community structure. In: Diamond J, Case TJ (eds) *Community Ecology*, pp. 480–91. Harper & Row, New York. [13.3]

Yount JD, Niemi GJ. (eds) (1990) Recovery of lotic communities and ecosystems following disturbance: theory and application. *Environmental Management* **14**: 515–762. [13.2]

14: Evolution of Instream Flow Habitat Modelling

C. B. STALNAKER

14.1 SPECTRUM OF INSTREAM FLOW PROBLEMS

Instream flow methods have predominantly been developed by biologists and hydrologists working for agencies having regulatory responsibility related to water development and management (Stalnaker & Arnette 1976). Such efforts over the last 30 years have provided the impetus for detailed ecological studies leading to a significant evolution in the understanding of the relation between streamflow and aquatic habitats. Most of the empirical evidence gathered to date has focused on fish species and benthic macroinvertebrate habitat requirements, with very recent emphasis on the relation between streamflow and woody riparian vegetation and river-based recreation (Gore 1987; Orth 1987; Brown 1992; Shelby *et al* 1992).

Minimum flow standards

Following the large reservoir and water development era of the mid-20th century in North America, resource agencies became concerned over the loss of many miles of riverine fish and wildlife resources in the arid western USA. Consequently, several of the western states began promulgating rules for protecting existing stream fishery resources from future depletions brought about by accelerated water development. Many instream flow methods appeared during the 1960s and early 1970s, primarily based on hydrological analysis of the water supply and hydraulic considerations of critical stream channel segments, coupled with empirical observations of habitat quality and historical understanding of the ecology, distribution and life history functions of riverine fish species, most notably the Pacific salmon and freshwater trouts. These efforts led to a general category of instream flow models meant to reserve a specific amount of water in the channel for the benefits of fish and general aquatic life production (Wesche & Rechard 1980; Morhardt 1986; Templin *et al* 1992). Applications usually resulted in some level less than the optimal or pristine habitat condition, and thus the terminology 'minimum flow' evolved. These methods usually result in a single threshold or 'minimum' flow value for a specified stream, below which water may not be withdrawn for consumptive water use. These 'reservations' form the basis for issuing water permits in many of the states in the USA at the present time (see MacDonald *et al* 1989; Lamb & Lord 1992, for a recent discussion of the status of State recognition and protection of flowing water for instream flow uses). More than a decade ago the threat of the 'minimum flow' concept and the potential limitations that low-flow management could place on warmwater streams in the humid eastern USA were evident (Stalnaker 1981).

Impact analyses leading to instream flow management and resource protection

Following enactment of the National Environmental Policy Act (NEPA), attention was shifted to the evaluation of various alternative designs and operations of federally-funded water projects, and a call was made for methods capable of incremental analyses for evaluating a series of possible alternative development schemes (Stalnaker

1993). This led to the development of habitat vs. discharge functions represented by life-stage-specific relations for selected species, i.e. fish passage, spawning and rearing habitat vs. flow for trout or salmon. Corroborating research took the form of analyses correlating the general well-being of fish populations (usually in terms of measured standing crop) with various physical and chemical attributes of the streamflow regime and its interaction with the stream channel structure (Binns & Eiserman 1979). Several variables contributed significantly and consistently to the variation in fish population and production. These were water velocity, minimal water depths, instream objects as cover, stream bottom substrate (particularly the amount of fines in the interstitial spaces within the coarse bed elements), water temperature, dissolved oxygen, total alkalinity, turbidity and light penetration (Gosse & Helm 1981; Shirvell & Dungey 1983). These flow functions were integrated into methodologies for analysing the consequences of proposed water withdrawal or storage-release activities and applied to numerous federal water projects operated by the Bureau of Reclamation, Corps of Engineers, and Tennessee Valley Authority (Nestler *et al* 1989).

During the late 1970s and early 1980s, an era of small hydropower development commenced. Hundreds of proposed hydro sites in the Pacific Northwest and the New England region of the USA came under intensive examination by state and federal fishery management interests. During this transition period from evaluating large federal reservoirs to evaluating licence applications for small hydro development, the Instream Flow Incremental Methodology (IFIM) was developed within the US Fish and Wildlife Service (Trihey & Stalnaker 1985). This methodology attempted to integrate the planning concepts of water supply and hydrological time series and analytical models from hydraulic and water quality engineering with empirically derived habitat vs. flow functions. The methodology produced simulations of the quantity and quality of 'potential habitat' resulting from proposed water development illustrated through a series of alternative flow regimes. Efforts involving incremental methods and time series analyses of alternatives were further enhanced during the second half of the 1980s and in the early 1990s by several hundred relicensing applications submitted to the Federal Energy Regulatory Commission. Most of these applications involved reservoirs that had been in place for 30–50 years without any instream flow considerations factored into their operating procedures. Great opportunities were seen by the natural resource agencies to restore riverine aquatic resources that had been impacted (in some cases eliminated) for many decades.

Conversely, many of the hydro companies wanted to shift their operating protocols towards peaking and pump-storage to enhance the revenues at existing installations. Many of the projects operated by private power companies or public utilities were readily accessible for recreational use and recreational interests seized on the relicensing opportunity as a means for enhancing river recreation use by canoes, kayaks and rafts.

Building on the NEPA guidelines for examining alternatives and using the hydro relicensing opportunity for taking a fresh look at the use of the flowing water resource, US decision-makers were forced to balance potential conflicts among uses of the riverine resources. In the 1980s incremental methods became the methods of choice for quantitatively describing the consequences of alternative ways of managing flowing waters. These methods set the stage for negotiation among interest groups and provided better information to decision-makers for resolving conflicts (Stalnaker 1993).

Water budgets

As the multiple-use ethic has evolved over the last two decades, it has become evident that simply allocating a portion of the water supply to various uses is no longer sufficient to resolve conflicts. Furthermore, the same water can be used many times if it is managed such that the timing of delivery can serve instream purposes while still being delivered to downstream users. This multiple-use management philosophy, illustrated by the Northwest Power Planning Act and subsequent efforts by the Bonneville Power Authority, allows federal, state and tribal fishery biologists to identify management prescriptions for restoration and enhancement of the anadromous salmon runs in the Columbia river basin. Minimum flow management did not provide sufficient protection for stream resources during drought cycles, nor did it provide the opportunity for optimal fish production during wet years. Water budgets that allocate a portion of water storage in upstream reservoirs for fishery benefits could provide water when it is most needed (Waddle 1991). When downstream water users are not calling for delivery through critical spawning or rearing reaches, leaving habitats with little water, the 'fish water' can be released to relieve these habitat-induced bottlenecks.

A recent case study by the National Research Council (1992) recognized the value of reservoir release management in alleviating conflict and enhancing multiple uses (including instream) within arid western US river basins. The Yakima River basin in Washington, focusing on Pacific salmon restoration, and the Truckee–Carson river systems in Nevada, emphasizing endangered species and terminal wetland habitat recovery, are highlighted in this study as examples of emerging multiple-use management. Assigning a reservoir water budget to the fisheries for downstream habitat management is now being seriously considered during the operations evaluation phase of many large federal water storage projects. The traditional water user groups (i.e. irrigated agriculture, municipal and industrial) may actually prefer such reservoir water budget management because the water for fish is equally weighted with other water in the system. Hence, the call for water must be determined by the fishery manager, relieving the reservoir operator and other users in the system from having to determine how and when the instream flows are to be delivered.

Shifting from a set of minimum flow constraints to a water budget set aside specifically for fisheries purposes has created a major shift in the role of the fishery manager in river systems in the USA. Henceforth, to be most effective, fishery resource managers must become water and habitat managers, and natural resource agencies must acquire a more interdisciplinary mix of employees. Resource agencies must be prepared to make decisions in real time as to the delivery of water during particular seasons and to determine what portion of the river basin fishery resource will be favoured (and what part sacrificed) during droughts. By switching the emphasis to water budgets, a mutual 'sharing' of storage in federal water projects across all user groups during droughts is facilitated. Under the traditional diversionary allocation philosophy of 'first in right first in time' practised under USA appropriation doctrine, the fishery can be given first priority during a water-short year (flows reserved from appropriation), or last (if allocated a very junior water right). By gaining a 'seat at the management table' instream fishery interests obtain a portion of the water stored during high-flow periods for release when the most critical conditions occur downstream. Sharing the storage allows for delivery to relieve these critical conditions.

Future implications of conflicting water use

Multiple-use management of water at the river basin level is widely recognized and is now being called for in the USA. The recent study by the National Research Council (1992) points out that management of reservoirs for single purposes such as irrigation or hydro-electric production is no longer socially acceptable. Efficient use of water should consider distribution of water through a river basin incorporating a multitude of instream and consumptive uses. I suggest that this will require an interdisciplinary group of professional water managers to establish procedures for evaluating the water supply, distributing the water, and sharing the consequences of low-supply years. Resolving conflicts among states and user groups sharing the same river system requires inter-jurisdictional river boards or commissions charged with managing water stored in public reservoirs for instream and out-of-stream use. Agency resource personnel will be called to apply state-of-the-art tools, tempered with a lot of judgement and experience, to real-time decision-making. We will see fishery and recreation agency managers making recommendations on seasonal and monthly bases depending on forecast water supply and available storage. Stream ecologists have a great challenge before them to provide research data to enhance the ability of resource managers to make these decisions. Research is needed to improve and validate the relationships between fish, wildlife and riparian vegetation quality and the streamflow regime, and new relationships are needed for groups of organisms and variables that are not yet included in the general decision-making process.

14.2 CONCEPTS OF INSTREAM FLOW INCREMENTAL METHODOLOGY

The enactment of the National Environmental Policy Act of 1980 mandated all federal water resource agencies to consider alternative water development and management schemes. This placed increased responsibility on natural resource agencies for methods, evaluations and recommendations related to reservoir storage and release and stream-channel depletions. The Instream Flow Incremental Methodology (IFIM) was developed by the US Fish and Wildlife Service and was founded on a basic understanding and description of the water supply and habitats in stream reaches of concern. Historical analysis of the flow regime using a monthly or weekly time-step as a means of describing the reference or baseline hydrological conditions was considered essential becasue this was the normal practice of the water resource engineering profession. Constructing a hydrological time series allows one to compare the frequency and duration of wet and dry periods, to examine the difference between snow melt and rain driven systems, and to examine the intensity of short-term events such as cloudbursts and peaking cycles. In the context of influencing operating decisions within large-scale water development settings, a tool was needed that could highlight the inherent conflicts and complementary uses among competing water uses, could consider and evaluate each user group's needs, and would be understood, accepted and used by a wide variety of agencies and decision-makers. Such decision arenas involve many disciplines, typically engineers, hydrologists, biologists, recreation planners, lawyers and political scientists.

The Service Directorate solicited direct input from other state and federal agencies for development of this methodology. Water resource professionals were detailed to work on this co-operative effort for periods of up to 4 years. Engineering, water-quality modelling and planning expertise came from the US Bureau of Reclamation, Soil Conservation Service, Corps of Engineers, Environmental Protection Agency and university scientists. Expertise in aquatic ecology, fishery biology, water law, institutional arrangements, and planning came from state agencies. The Intergovernmental Planning Act provided the vehicle for these assignments from state organizations.

This inter-agency effort led to the conclusion that an analytical methodology should handle a wide variety of instream flow problems, from simple direct diversion from the stream channel to complex storage and release schemes involving hydro-peaking schedules, pump storage, and a network of interconnected reservoirs. For such a methodology to be suitable for evaluating alternatives, it must help in identifying, evaluating and comparing potential solutions, and it must be capable of being tailored to a specific stream reach and of expanding such reach information throughout an entire river basin. With this general charter, and building on historical planning practices utilizing stream reach hydrology, the IFIM has evolved over the last 15 years into a river network analysis incorporating fish habitat, recreational space and woody vegetation responses to alternative water management schemes (Auble *et al* 1991; Bartholow 1989; Milhous *et al* 1989). Information is presented in the form of time series of flow and habitat at selected points within a river system (Milhous *et al* 1990).

Optimizing for any one use is contrary to the general philosophy of multiple use, and efficiency of use is thought of in terms of the greatest return in the number and quantity of uses with emphasis on simultaneous use. It is imprudent to simply use the intermediate output (i.e. flow vs. habitat or flow vs. recreation functions) to argue for a minimum release or flow standard chosen from the maximum value on a flow vs. habitat graph. The timing of events across seasons is critical to the reproductive success and relative strength of year classes within fish populations. The distribution of strong vs. weak year classes is indicative of the well-being of the fish population and, perhaps more importantly, determines the number of adults available to the fishery on an annual basis. To illustrate this point it is essential to combine the habitat functions and hydrological time series into a quasi-population analysis by displaying habitat quality and quantity through time and space. This methodology translates the baseline historical hydrology into a description of the available or usable habitat present during that historical period. This is often referred to as the resource benchmark from which fishery scientists identify enhancements and impacts resulting from proposed water delivery schedules.

By examining recent historical conditions (say the last 5–10 years) using the power of computer simulation, it is possible to 'calibrate' a system model by use of historical data such as annual population indices and trends, creel census data, computed year-class strength data from age and growth studies, or anecdotal information on the general well-being of fish populations in one year vs. another. If the fishery manager can demonstrate that a simulated habitat analysis over a 10–20-year historical hydrological regime generally agrees with historical information on good years vs. poor years in the fishery, much more credence can be placed in those models for comparing various alternative 'futures'. In addressing a complex river basin management issue, it is essential that such 'calibration' be performed.

The IFIM was designed for river system management. The methodology provides an organizational framework for evaluating and formulating alternative water management scenarios. It was built on a basic philosophical underpinning of hydrological analyses for understanding the limits of the water supply. Iterative analysis offers a description, evaluation and comparative display of water use throughout a river system. Emphasis is on the spatial display of usable habitat illustrated through time series across several years to capture the variability in water supply and habitat. Such comparative information enhances negotiations in the planning and management of riverine resources. Sharing of the limited water supply during drought cycles and the management of timed releases contribute to compatibility among instream and out-of-stream user groups and allow for rapid recovery of aquatic populations during wet years.

14.3 COMPONENTS OF INSTREAM FLOW INCREMENTAL METHODOLOGY

Problem identification

After a proposed change in the water system becomes known, an analysis should be conducted of the physical location and geographical extent of probable physical and chemical changes to be imposed on the river system. This analysis is often done in conjunction with a scoping meeting among the management and regulatory agencies likely to be involved with the decision. This inter-agency group conducts a legal and institutional analysis to identify:

1 the affected or interested parties;
2 information needs for constituencies;
3 the relative influence among parties;
4 the probable pathway the decision process will take (e.g. is it more likely to be a brokered or arbitrated decision?).

The primary product of problem identification is a negotiated description of the water supply, the various existing priority uses of water rights, and the existing system-operating rules.

Transforming this jointly developed product into a baseline and a proposed hydrological time series at points in the river system most likely to experience a change in the flow regime provides the basis for study planning and IFIM application. Usually, a preferred alternative is identified by the project proponent and the consequences of this alternative are translated into a hydrological time series that assumes the project is in place and operating as proposed. At this point the water user groups in the river basin can begin to identify their concerns related to the proposed project, to identify their information needs, and to recognize the value of a multi-objective planning exercise. The IFIM provides a framework for evaluating various alternatives, including the proposed project operation.

Study planning

An interdisciplinary planning team builds on the objectives and information needs of each of the parties to the decision process. The interdisciplinary team should not get trapped in trying to predict the outcome of a study, but rather should focus on the data collection and methods to be used for producing the information needs of the parties to the alternative decision. This planning process leads to the collective identification of:

1 The pertinent temporal and spatial scale of evaluations.

2 The most important variables for which information is needed (for instream fisheries and recreation it is important to identify all critical flow-related variables to be included in the ultimate habitat models used for simulation).

3 How information will be obtained (i.e. monitoring or simulation) if it does not already exist.

The hydrological information chosen to represent the baseline or reference condition is re-examined in great detail at this point in a study. All parties must understand and agree on one or more hydrological time series that will be used for comparison and serve as reference for judging potential impacts. Quite often this reference baseline condition is not the actual historical hydrology referred to in the problem identification stage, but rather a synthetic hydrological time series representing present water uses and management philosophy, superimposed on the variability found in the historical hydrological records. In some studies, the baseline hydrology includes future management schemes that are not actually imposed on the system but have been formally approved and are coming on line. It is not unusual to agree to the necessity of preparing two or more reference hydrological time series to be used for baseline conditions.

Alternative project comparisons can lead to very different responses when placed in context with other important decisions. The resource agency responsible for the fisheries must describe the biological reference or benchmark conditions. Identifying the geographic distribution and time steps within years is critical to evaluating different life-history phases of the fish species or groups. A population benchmark is constructed using historical habitat conditions (from best estimates of actual historical hydrology) and 'backcasting' to identify critical events that populations may have experienced due to physical or chemical limitations reflected as habitat bottlenecks (Burns 1971). The planning group must agree on methods of summarizing multiple reaches of the river for quantification and alternative comparisons. A written plan determines when data collection must be completed in the field; synchronizes the collection of data needed for model input, calibration and testing; and estimates the labour, equipment, travel and other costs required to produce the necessary information by the study deadline. An interdisciplinary planning effort representing all of the major interest groups can result in considerable savings of time and effort during the conflict resolution phase. A common mistake in planning is to go about describing a 'set' of data and 'standard collection procedures' appropriate for commonly-used models. This facilitates entering the field data-collection phase but often does not build a common understanding among the important players of data needed and analytical approach. Lack of group planning may lead to polarization and the stifling of a negotiated resolution process.

Implementation and quality assurance

From the field biologist and resource agency perspective implementation is often the most interesting and scientifically challenging part of an instream flow study. Proper implementation is critical to establish biological credibility in the decision process, but it cannot be done in isolation. During this phase sampling locations are determined for the purpose of collecting data for use in conjunction with computer models to describe the relation between flow regimes and stream habitat. Calibration and quality assurance are key steps when integrating large-scale habitat variables (e.g. temperature, dissolved oxygen) with small-scale habitat variables (e.g. microhabitat for fish) to obtain a quantitative description of the spatial distribution of total usable habitat for a given life stage or species throughout the system of concern.

The most important intermediate product of

an instream flow study therefore is the baseline habitat time series. This analysis determines how much habitat would be available for each life stage of the target species as a function of time over each of the previously agreed baseline hydrological time series. This baseline habitat time-series information incorporates baseline hydrology, empirically-measured channel characteristics, land-use and water-use patterns, waste loading and assimilation, and temperature conditions. It provides an information base from which rational judgements can be made about proposed alternative management schemes.

Inappropriate selection and use of models, and the failure to verify model assumptions can, at this phase, lead to major errors in application (Shirvell 1986; Scott & Shirvell 1987). Since all habitat-based instream flow models rely on empirical measurements of the stream channel structure as inputs, it is imperative that adequate understanding of the sediment transport and channel dynamics be incorporated into any habitat time series analysis. If a channel is not in some state of dynamic equilibrium the modellers may have to hedge on simulation of 'futures' and call for periodic adjustments with empirical measurements at regular intervals. Channel maintenance and sediment flushing flows may be critical components in some instream flow studies. These important physical processes are discussed by Hill *et al* (1991), Reiser *et al* (1989a), Rosgen *et al* (1986) and Stalnaker *et al* (1988).

Alternatives analysis

Usually the water project proponent will have a preferred alternative, and often a series of alternatives may be identified. Not too surprising, most parties to the decision process may identify their own alternatives. It is best to compare the baseline conditions with all offered alternatives to facilitate an understanding of potential impacts and to set the stage for negotiating and creating new alternatives more compatible with the multiple objectives of the many parties. Simulation modelling allows for relatively straightforward comparison of many alternatives, each of which is compared with the baseline conditions and examined for the following.

1 Effectiveness – Are the objectives sustainable? Is no net loss of habitat possible?
2 Physical feasibility – How often may reservoirs dry up or are priority water rights not met?
3 Risk – Both short-term risk (i.e. seasonal differences) and long-term risk (yearly differences) must be examined to determine how often an alternative or recommendation leads to failure. Examine the duration and timing of limiting events (e.g. critical high temperatures, habitat bottlenecks, low dissolved oxygen).
4 Economic considerations – Cost-benefit analysis must be done to examine not only economic gains and losses associated with proposed alternatives but also habitat costs and benefits relative to the baseline conditions.

Problem resolution

The IFIM was designed to assist in formulating and evaluating alternatives, and it relies heavily on professional judgement by the fishery resource managers in identifying the best alternative. When the present fishery is considered good to excellent the best alternative is often the one which deviates the least from the baseline habitat condition. This is often referred to as the 'no net loss of habitat' philosophy, which is followed by many resource agencies, and it is the official policy of the Canadian Department of Fisheries and Oceans. A single best minimum flow solution or optimizing for a single use are inappropriate strategies for a multiple-use river management system. Since water-use values are never truly commensurate and are difficult to compare, and since data and model outputs are never complete or perfect, different people can reach different conclusions using the same basic data set. Consequently, the IFIM has been designed for interdisciplinary teams to integrate state-of-the-art knowledge and understanding of a problem with their professional judgements on resource issues and social needs to reach a negotiated solution, implying some kind of balance among conflicting social values. The methodology is not fixed; rather it is open-ended and imaginative, and flexible solutions are encouraged. Many analytical tools and techniques have been developed over the last decade and are constantly being extended, tested and enhanced. In most applications only a few of these tools are appropriate or actually used in the analysis.

14.4 VERIFICATION AND VALIDATION

There is a large amount of uncertainty in understanding how biological systems work and relative imprecision when using small samples to represent a dynamic stream system. A sufficient amount of site-specific empirical evidence must be collected to establish validity when applying instream flow models to the decision-making process. Just as grab-sample measurements of temperature, water quality, depths and velocities are routinely used to calibrate physical and chemical models, it is equally important that samples of the aquatic organisms and their habitat use be used to 'calibrate' the habitat simulations used in the IFIM alternatives analysis. Two tests of habitat–population interactions have been developed specifically for IFIM habitat analyses focusing on: (i) short-term habitat use and individual fish behaviour; and (ii) long-term habitat availability and population response in terms of year-class strength.

Usable vs. unusable habitat

Habitat suitability criteria (HSC) have been developed for many life stages of important riverine species over the last three decades. Many of these data sets have been developed throughout the range of a species leading to considerable debate as to the transferability of habitat-use criteria (Bovee & Zuboy 1988; Heggenes *et al* 1990). Also, the concept of habitat preference and various mathematical schemes for correcting for environmental bias have been proposed. Since there is no consensus over whether analytical procedures applied to a data set collected in one stream will ensure that the criteria are transferred throughout the range of the species, it is essential that some habitat-use data be collected as a part of each application of the IFIM habitat models. Field verification is official policy of the US Fish and Wildlife Service (Mayer 1986). Unfortunately, many, if not most, applications of IFIM do not collect sufficient data on site to verify that the models are behaving as intended.

Bovee and Thomas (1993) describe an empirical test that allows the modeller to determine whether an existing set of habitat criteria are transferable to a destination stream subject to an instream flow study. This test requires a limited number of habitat use observations in the destination stream. The first element is to select criteria from a source stream that contains a variety of microhabitat conditions. Such physical diversity expands the universe of microhabitats from which the target organisms can select. The second element is the sampling design. Rather than sampling different habitat types in the proportion that they occur in the stream, equal areas of all habitat types are sampled an equal number of times. This results in a moderate adjustment of the HSC compared with traditional mathematical corrections. The equal-effort sampling approach requires less field data than the mathematical correction, and data produced by this method are easily pooled for different reaches or different streamflows.

The testing procedure involves the measurement of both occupied and unoccupied locations in the destination stream. Candidate HSC are used to calculate the relative suitabilities (e.g. optimal, marginal or unsuitable) in test stream habitat locations. A one sided chi-squared test (Conover 1971) is used to test for non-random selection of optimal over marginal locations and suitable (optimal and marginal combined) locations over those calculated to be unsuitable. To be considered transferable, a set of HSC must pass both tests. The reliability and robustness of the test procedures have been extensively evaluated in several rivers.

Long-term habitat analysis

Since one of the more important underlying assumptions of an IFIM habitat analysis is that fish populations respond to limiting habitat events, some site-specific empirical data must be gathered to determine if this assumption is valid. This is typically done through historical analyses ('backcasting') using effective habitat time series analysis (Bovee 1982). Historical records of standing crop, back-calculated year-class strength and growth histories from age and growth analysis are valuable for evaluating historical habitat analyses. The effective habitat time series analysis was specifically designed to take advantage of back-calculation techniques common to age and growth studies from fishery science and the use of stream gauge records for synthesizing hydrological time series from water resource engineering. By using recent historical hydrological time series and verified habitat models for the site under consideration, the effective habitat time series is used to identify any historical habitat limitations (habitat bottlenecks). Potential habitat bottlenecks identified through this analysis are examined for general agreement with typical indicators of population response. Poor year-class strength, low growth rates, and unusually low survival of adults as well as positive responses during excellent habitat years, can be used to 'calibrate' the habitat time series models and establish validity for the model simulations when used for comparing future alternatives.

14.5 CURRENT MODELLING EFFORTS

Contemporary research (Hagar *et al* 1988; Cheslak & Jacobson 1990; Auble *et al* 1991; EA Engineering, Science and Technology 1991) is broadening the role of the Instream Flow Incremental Methodology to provide fisheries management as well as habitat management capabilities. Reiser *et al* (1989b) surveyed IFIM users and found that the highest priority research needs are: (i) to define the relation between flow, habitat and fish production; (ii) to validate and test the relation of habitat measurements to fish production; and (iii) to develop new methods for determining flow requirements.

In riverine systems the amount and quantity of suitable habitat can be highly variable within and among years. The observed population and biomass of fish may be influenced (depressed or stimulated) by numerous preceding habitat events. Habitat-induced population limitations are related to the amount and quality of habitat available to the fish population at critical stages in its life history (Burns 1971; Mundie & Traber 1983; Morhardt & Mesick 1988). Long-term habitat reductions from reduced flows may also be important in determining the fish population and production (Bovee 1988, unpublished).

Scientists of the US Fish and Wildlife Service are testing these new concepts on anadromous salmonid populations in California and resident trout populations in Colorado (Williamson *et al* 1993). Goals of this research are:

1 the development and validation of a dynamic fishery population model including response to flow-related limiting events, specifically physical habitat and temperature;

2 testing of habitat bottleneck hypotheses;

3 the development of processes for evaluating water management strategies to achieve fish population objectives;

4 testing of strategies for long-term population support;

5 improvement to those existing components of the Instream Flow Incremental Methodology necessary to provide a smoothly working set of analytical tools for fish population analysis. Such efforts extend the state of fisheries science and the art of water management by focusing on population-limiting habitat events and water temperature effects on movement, growth and mortality rates of salmonids. This kind of direct feedback between fish populations and reservoir operations facilitates the use of water budgets for fishery management purposes.

REFERENCES

Auble GT, Friedman J, Scott ML. (1991) *Riparian Vegetation of the Black Canyon of the Gunnison River, Colorado: Composition and Response to Selected Hydrologic Regimes Based on a Direct Gradient Assessment Model.* Report for National Park Service, Water Resources Division, Water Rights Branch, Fort Collins, CO. [14.2, 14.5].

Bartholow JM. (1989) *Stream Temperature Investigations: Field and Analytic Methods.* Instream Flow Information Paper No. 13. Biol. Rept. 89(17). US Fish and Wildlife Service, Washington, DC. [14.2]

Binns NA, Eiserman FM. (1979) Quantification of fluvial trout habitat in Wyoming. *Transactions of the American Fisheries Society* **108**: 215–28. [14.1]

Bovee KD. (1982) *A Guide to Stream Habitat Analysis Using the Instream Flow Incremental Methodology.* Instream Flow Information Paper No. 12. FWS/OBS-

82/26. US Fish and Wildlife Service, Washington, DC. [14.4]

Bovee KD. (1988) Use of the Instream Flow Incremental Methodology to evaluate influences of microhabitat variability on trout populations in four Colorado streams. In: *68th Annual Conference of the Western Association of Fish and Wildlife Agencies, Albuquerque, NM*, pp. 227–57. [14.5]

Bovee KD, Thomas JA. (1993) Application and testing of procedures to evaluate transferability of habitat suitability criteria. *Regulated Rivers: Research and Management* **8**: 285–94. [14.4]

Bovee KD, Zuboy JR. (eds) (1988) *Proceedings of a Workshop on the Development and Evaluation of Habitat Suitability Criteria*. Biol. Rept. 88(11). US Fish and Wildlife Service, Washington, DC. [14.4]

Brown TC. (1992) Water for wilderness areas: Instream flow needs, protection, and economic value. *Rivers* **2**(4): 311–25. [14.1]

Burns JW. (1971) The carrying capacity of juvenile salmonids in some northern California streams. *California Fish and Game* **57**: 44–57. [14.3, 14.5]

Cheslak EF, Jacobson AS. (1990) Integrating the Instream Flow Incremental Methodology with a population response model. *Rivers* **1**(4): 264–88. [14.5]

Conover WJ. (1971) *Practical Non-parametric Statistics*, 2nd edn. John Wiley, New York. [14.4]

EA Engineering, Science and Technology. (1991) *San Joaquin River System Chinook Salmon Population Model Documentation*. Report for Turlock Irrigation District and Modesto Irrigation District. EA Engineering, Science and Technology, Lafayette, CA. [14.5]

Gore JA. (1987) Development and applications of macroinvertebrates instream flow models for regulated flow management. In: Craig JF, Kemper JB (eds) *Regulated Streams: Advances in Ecology*, pp. 99–115. Plenum Press, New York. [14.1]

Gosse JC, Helm WT. (1981) A method for measuring microhabitat components for lotic fishes and its application with regard to brown trout. In: Armantrout NB (ed) *Acquisitions and Utilization of Aquatic Habitat Inventory Information*, pp. 138–41. American Fisheries Society, Bethesda, MD. [14.1]

Hagar J, Kimmerer W, Garcia J. (1988) *Chinook Salmon Population Model for the Sacramento River Basin*. Report for National Marine Fisheries Service, Habitat Conservation Branch. BioSystems Analysis, Sausalito, CA. [14.5]

Heggenes J, Brabrand A, Saltveit SJ. (1990) Comparison of three methods for studies of stream habitat use by young brown trout and atlantic salmon. *Transactions of the American Fisheries Societv* **119**(1): 101–11. [14.4]

Hill MT, Platts WA, Beschta RL. (1991) Ecological and geomorphological concepts for instream and out-of-channel flow requirements. *Rivers* **2**(3): 198–210. [14.3]

Lamb BL, Lord E. (1992) Legal mechanisms for protecting riparian resource values. *Water Resources Bulletin* **2**(4): 965–77. [14.1]

MacDonald LJ, Rice T, Shupe SJ. (1989) *Instream Flow Protection in the West*. Natural Resources Law Center, University of Colorado School of Law, Boulder, CO. [14.1]

Mayer S. (1986) *Development of Instream Flow Incremental Methodology (IFIM) Suitability Index (SI) Curves to Meet Project Study Needs*. Habitat Resources Instructional Memorandum ES 90.01. U.S. Fish and Wildlife Service, Washington, DC. [14.4]

Milhous RT, Updike MA, Schneider DM. (1989) *Physical Habitat Simulation System Reference Manual – Version II*. Instream Flow Information Paper No. 26. Biol. Rep. 89(16). U.S. Fish and Wildlife Service, Washington, DC. [14.2]

Milhous RT, Bartholow JM, Updike MA, Moos AR. (1990) *Reference Manual for the Generation and Analysis of Habitat Time Series – Version II*. Instream Flow Information Paper No. 27. Biol. Rep. 90(16). U.S. Fish and Wildlife Service, Washington, DC. [14.2]

Morhardt JE. (1986) *Instream Flow Methodologies*. Report of Research Project 2194-2. Electric Power Research Institute, Palo Alto, CA. [14.1]

Morhardt JE, Mesick CF. (1988) Behavioral carrying capacity as a possible short term response variable. *Hydro Review* **7**: 32–40. [14.5]

Mundie JH, Traber RE. (1983) The carrying capacity of an enhanced side-channel for rearing salmonids. *Canadian Journal of Fisheries and Aquatic Sciences* **40**: 1320–2. [14.5]

National Research Council. (1992) *Water Transfers in the West: Efficiency Equity, and the Environment*. National Academy Press, Washington, DC. [14.1]

Nestler JM, Milhous RT, Layzer JB. (1989) Instream habitat modeling techniques. In: Gore JA, Petts GE (eds) *Alternatives in Regulated River Management*, pp. 295–315. CRC Press, Boca Raton, FL. [14.1]

Orth DJ. (1987) Ecological considerations in the development and application of instream flow-habitat models. *Regulated Rivers* **1**: 171–81. [14.1]

Reiser DW, Ramey MP, Wesche TA. (1989a) Flushing flows. In: Gore JA, Petts GE (eds) *Alternatives in Regulated River Management*, pp. 91–135. CRC Press, Boca Raton. [14.3]

Reiser DW, Wesche TA, Estes C. (1989b) Status of instream flow legislation and practices in North America. *Fisheries* **14**: 22–9. [14.5]

Rosgen DL, Silvey HS, Potyondy JP. (1986) The use of channel maintenance flow concepts in the Forest Service. *Hydrological Science and Technology* **2**(1): 19–26. [14.3]

Scott D, Shirvell CS. (1987) A critique of the instream flow incremental methodology and observations on flow determination in New Zealand. In: Craig JF, Kemper JB (eds) *Regulated Streams: Advances in Ecology*, pp. 27–43. Plenum Press, New York. [14.3]

Shelby B, Brown TC, Taylor JG. (1992) *Streamflow and Recreation*. General Technical Report RM-209. US Department of Agriculture Forest Service, Rocky Mountain Forest and Range Experiment Station, Fort Collins, CO. [14.1]

Shirvell CS. (1986) *Pitfalls of Physical Habitat Simulation in the Instream Flow Incremental Methodology*. Canadian Technical Report of Fisheries and Aquatic Sciences 1460, Prince Rupert, Canada. [14.3]

Shirvell CS, Dungey RJ. (1983) Microhabitat chosen by brown trout for feeding and spawning in rivers. *Transactions of the American Fisheries Society* **112**: 355–67. [14.1]

Stalnaker CB. (1981) Low flow as a limiting factor in warmwater streams. In: Krumholtz L (ed) *Warmwater Streams Symposium*, pp. 192–9. American Fisheries Society, Bethesda, MD. [14.1]

Stalnaker CB. (1993) Fish habitat models in environmental assessments. In: Hildebrand SG, Cannon JB (eds) *Environmental Analysis: The NEPA Experience*, pp. 140–62. Lewis, Boca Raton, FL. [14.1]

Stalnaker CB, Arnette JL. (eds) (1976) *Methodologies for Determination of Stream Resource Flow Requirements: An Assessment*. FWS/OBS-76/03. U.S. Fish and Wildlife Service, Washington, DC. [14.1]

Stalnaker CB, Milhous RT, Bovee KD. (1988) Hydrology and hydraulics applied to fishery management in large rivers. In: Dodge DD (ed) *Proceedings of the International Large River Symposium*, pp. 13–30. Canadian Special Publication of Fisheries and Aquatic Sciences 106, Ottawa, Canada. [14.3]

Stromberg JC, Patten DT. (1991) Instream flow requirements for cottonwoods at Bishop Creek, Ingo County, California. *Rivers* **2**(1): 1–11.

Templin WE, Herbert RA, Solley WB, Stalnaker CB. (1992) Water use. In: *National Handbook of Recommended Methods for Water-data Acquisitions*, Chap. 11. US Geological Survey, Reston, VA. [14.1]

Trihey EW, Stalnaker CB. (1985) Evolution and application of instream flow methodologies to small hydropower development: an overview of the issues. In: Olson FW, White RG, Hamre RH. (eds) *Proceedings of a Symposium on Small Hydropower and Fisheries*, pp. 176–83. American Fisheries Society, Bethesda, MD. [14.1]

Waddle TJ. (1991) A water budget approach to instream flow maintenance. In: Darling DD (ed) *Waterpower 91. Proceedings of an International Conference* on *Hydropower*, pp. 155–62. American Society of Civil Engineers, Denver, CO. [14.1]

Waddle TJ. (1992) A method for instream flow water management. PhD dissertation. Colorado State University, Fort Collins, CO.

Wesche TA, Rechard PA. (1980) *A Summary of Instream Flow Methods for Fisheries and Related Research Needs*. Eisenhower Consortium Bulletin No. 9. Water Resources Research Institute, University of Wyoming, Laramie. [14.1]

Williamson SC, Bartholow JM, Stalnaker CB. (1993) Conceptual model for quantifying presmolt production from flow dependent physical habitat and water temperature. *Regulated Rivers: Research and Management*. [14.5]

Part 4: Management Options

The earlier parts of the *Handbook* have provided the scientific basis of the evaluation of river management problems. Part 4 focuses on tools and approaches that are available for solving or at least mitigating these problems. The role of the river manager is to confront the problem of how to manage a river in an environmentally sensitive way when land management practices are continuing to alter the primary driving forces for river ecosystems. Catchment management is clearly needed if the full benefits of our advancing scientific knowledge of river ecosystems is to be realized. However, the integration of land-use policy and river management practice, whilst much needed, is only rarely an option for river managers at the present time. This part is concerned with the integration of environmentally sensitive tools and approaches into contemporary river management practices.

The problem of defining an ecologically acceptable flow regime is discussed in Chapter 15 and approaches to establishing flow regulation strategies are examined in Chapter 16. The techniques and approaches for controlling water quality are elaborated in Chapter 17. Having addressed the allocation of water and maintenance of water quality to achieve a target environmental quality for a river, or sector of river, management options turn to environmentally sensitive approaches for channel engineering (Chapter 18), the management of aquatic and riparian weeds (Chapter 19) and direct controls on fauna, the most common of which involves the use of hatcheries for fish stocking and the enforcement of fishing regulations (Chapter 20). The final two chapters consider the advancement of river management objectives through corridor, as opposed to channel, management. The roles of buffer zones are addressed in Chapter 21 and Chapter 22 illustrates the way in which river corridors may be restored.

15: Flow Allocation for In-river Needs

G. E. PETTS AND I. MADDOCK

15.1 INTRODUCTION

The structure and functioning of river systems has recently been synthezised as the 'fluvial hydrosystem' concept (Amoros & Petts 1993). A fluvial hydrosystem comprises the whole river corridor – the river channel, riparian zone, floodplain and alluvial aquifer – and is viewed as a four-dimensional system, being influenced not only by longitudinal processes but also by lateral and vertical fluxes, and strong temporal changes. Flow is important in three contexts. First, flow characteristics define the environmental domains within which biological communities develop. Secondly, the geomorphological effectiveness of flows of different magnitude – i.e. the amount of sediment transport, erosion and deposition – determines the channel and floodplain morphology, and the types and age structure of the different sediment-landform units that comprise the river corridor. Thirdly, the river and groundwater regimes determine the nature of vertical interactions within the hyporheic zone. Interactions between flow and biota are complex and highly sensitive to river regulation and abstraction.

With the increasing demands upon water resources and growing concern for environmental protection over the past decade, greater attention has been given to the allocation of water for in-river needs. Typically, this attention has been directed to maintaining habitat for a target species, usually a salmonid, during the summer, dry-weather period. However, the maintenance of minimum flow is only one of a range of considerations necessary for the protection or rehabilitation of fluvial hydrosystems. Hill *et al* (1991) argue that a holistic streamflow management framework requires the integration of flow regimes to maintain four groups of biotic and abiotic resources: (i) overbank flows that inundate riparian and floodplain areas; (ii) floodflows that form floodplain and valley features; (iii) in-channel flows that sustain the functioning of the instream system; and (iv) in-channel flows that meet critical fish requirements. To this must be added a fifth: surface-water and groundwater regimes to sustain the functioning of the hyporheic system.

15.2 THE SCIENTIFIC BASIS

Quantitative and process-based research on rivers was founded during the 1950s and 1960s. A number of early works established relationships between flow (see Fraser 1972) and current velocity (see Statzner *et al* 1988) and salmonid fish, macroinvertebrates and macrophytes. Three of the first papers to emphasize the importance of flow as an ecological factor were Phillipson (1954), Wickett (1954) and Ambuhl (1959). However, the role of flow and its associated hydraulic characteristics in structuring river ecosystems received little development until the early 1980s (Newbury 1984; Nowell & Jumars 1984). Throughout the intervening period, stream ecologists focused on energy flows, carbon fluxes and macroinvertebrate life histories (Resh 1985).

Studies of biotic responses to flows and especially to changes of flows have gained considerable momentum over the past decade. The first important catalyst for linking flows and ecology was *River Ecology and Man*, edited by R.T. Oglesby, C.A. Carlson and J.A. McCann (Oglesby *et al* 1972). At the time, however, river management was an 'art' rather than a science. This is illustrated by Fraser's statement (p. 277): 'Discharge recommendations are often based more on a biologist's or engineer's guess than on a quantified evaluation of the relationship between discharge and the ecology of the stream, its aesthetics and other in-place uses.'

During the late 1970s, a significant response to the intensifying need for an objective method to assess instream-flow allocations was the development of the Instream Flow Incremental Methodology (IFIM) by the US Fish and Wildlife Service Cooperative Instream Service Group (see Chapter 14). The methodology has become one of the most widely used methods in North America for estimating the effect of changes in flow on trout habitat (Conder & Annear 1987) and provides a useful guide for determining low-flow requirements, particularly in late summer and autumn (Hill *et al* 1991).

In the UK, advances in linking hydrology and ecology have lagged behind those in the USA. At a symposium held in 1970, on *Conservation and Productivity of Natural Waters* (Edwards & Garrod 1972), organized by the British Ecological Society and the Zoological Society of London, Morgan (p. 143) notes: 'In the long term it would be desirable to gain knowledge of the wider implications of the amounts of water let down rivers on the ecosystem as a whole.' Importantly, Morgan urged the focusing of efforts on the intricate relationships between animals and plants and their environment if better management is to be achieved. Significant advances were slow to be made although influential reviews were published by Brooker (1981), Milner *et al* (1981) and Petts (1984). Furthermore, despite detailed studies of the River Tees below Cow Green reservoir (see Armitage 1978; Crisp *et al* 1983), in a study of compensation flows in the UK Gustard *et al* (1987) concluded that the primary research need remained the development of quantitative relationships between freshwater biota and the physical and chemical variables at a scale appropriate to the river reach. A marked acceleration of research effort followed the Fourth International Symposium on Regulated Streams, held at Loughborough in 1988 (Petts & Wood 1988; Petts *et al* 1989). Subsequently, IFIM was introduced (Bullock & Gustard 1992; Petts 1992) and progress made in addressing biological responses to water abstraction (Armitage & Petts 1992; Bickerton *et al* 1993).

Research directions

Four complementary lines of scientific research are of significance for linking hydrology and ecology: (i) the flood pulse concept; (ii) the channel-forming discharge; (iii) the concept of hydraulic stream ecology; and (iv) the demonstration of the ecological significance of the hyporheic zone.

The flood pulse concept

The river margin is a water-land ecotone (see Chapter 21). The downstream transport of organic matter derived from the headwaters and nutrient spiralling, the core of the 'river continuum concept' (Vannote *et al* 1980), have little significance for floodplain systems. Large floodplain rivers derive most of their animal biomass from within the floodplain; they are dominated by lateral exchanges between floodplain and river channel, and nutrient recycling within the floodplain. The 'flood pulse concept' (Junk *et al* 1989) proposes that the biota of large floodplain rivers is determined principally by the hydrological regime. The biota is adapted to the pronounced aquatic and terrestrial phases: aquatic organisms colonize the floodplain at rising and high water levels because of feeding and spawning opportunities; terrestrial organisms occupying non-flooded habitats along the borders are adapted to exploit the floodplain at low water levels.

Floodplains and floods are now regarded as essential components of fluvial systems (Sedell *et al* 1989). Without the 'flood pulse', production within the fluvial hydrosystem is drastically reduced, and community composition and energy pathways are radically changed. The 'flood pulse advantage' (Bayley 1991) describes the degree by which annual multispecies fish yield exceeds that from a system with constant water level (Fig. 15.1). An 'advantage' is also seen in the 'subsidy' to production of higher vegetation due to seasonal

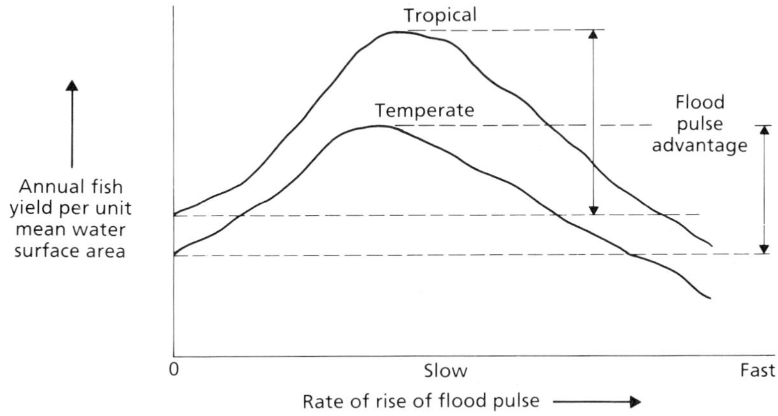

Fig. 15.1 The 'flood-pulse advantage' (after Bayley 1991). The relationship between mean water surface area of a floodplain and annual multispecies fish yield is shown to depend upon the flood pulse (no flood pulse is indicated by the abscissa = 0) and particularly the rate of hydrograph rise. The 'advantage' of the slow rise and long duration hydrograph contrasts with the disadvantage' associated with a flashy regime.

flooding of the riparian zone (e.g. Stromberg 1993). Vegetation decline consequent upon flow diversion from semi-arid streams demonstrates the limiting conditions of water availability upon riparian vegetation in these environments (e.g. Smith *et al* 1991).

A key component of the flood pulse concept is the 'moving littoral', the inshore edge of the aquatic environment that traverses the floodplain. The effectiveness of the moving littoral relates to the rate of hydrograph rise (Fig. 15.1). A slow hydrograph rise allows sufficient time for *in situ* processes, mineralized products from any preceding aquatic cycle and the current terrestrial one being released into the water. In dry periods, oxygenation of sediments promotes processing of organic material; subsequent submergence recycles plant nutrients into the water, enhancing productivity. During the flood phase, food supply in fertile floodplains can be so abundant that individual growth and population density of fish and other aquatic animals are limited by factors other than food, such as dissolved oxygen levels, availability of spawning habitat, and predation. Thus, large rivers are characterized by terrestrial and aquatic organisms with anatomical, morphological, physiological and/or ethological adaptations to utilize the opportunities created by predictable and prolonged floodplain inundation the drying out.

The occurrences, life cycles and abundances of primary and secondary producers and decomposers are determined by duration, magnitude, frequency, timing and predictability of floods. Timing and predictability may be key factors explaining differences between regions in terms of flood pulse effectiveness. Unpredictable floods include irregular events to which the biota is unable to develop profitable strategies for occupying the inundated riparian and floodplain areas, and disturbances – catastrophic events that periodically 'reset' the physical and biotic environment (see Chapters 3, 5 and 6).

Junk *et al* (1989) stated that the effect of a flood on biota is principally hydrological. From a conceptual viewpoint, flood characteristics may be the driving variable over short time-scales (1–10 years) but over longer time-scales, geomorphological processes are seen to play an increasingly significant role. Floodplain morphology, sediment characteristics and disturbance regimes reflect the geomorphological setting as well as the flow regime (see Amoros & Petts 1993; see also Vol. 1, Chapters 6 and 7).

Hydraulic stream ecology

The concept of hydraulic stream ecology is founded on the basic ecological principle that the relative difference in speed between an organism and the medium in which it lives affects the energy budget of the organism (Statzner *et al* 1988). Current velocity influences respiration and other measures of metabolism in all major groups of organisms found in running waters. Velocity also affects the feeding biology and behavioural characteristics of lotic animals, including rheotaxis, locomotory activity, schooling and territoriality. Some species undergo progressive changes in body shape, and are able to exploit different hydraulic habitats at different stages of their life history.

The morphology of a stream channel is a key component for river management, influencing the productivity and biological quality of the instream habitat. The hydraulic conditions where organisms live arise from the interaction of flow and channel form: the shape of the channel in cross-section, especially its asymmetry; bedform, such as pool-riffle development; the heterogeneity of the channel margins; and planform (see Chapter 1; see also Vol. 1, Chapters 5 and 6). Stream power – the slope-discharge product – describes the rate of energy supply at the channel bed, relates to channel stability (Brookes 1988) and is especially useful for explaining differences between river sectors (upland/lowland; meandering/braided; stable/unstable: see Chapter 1). However, at a smaller scale, velocity, depth and a range of complex hydraulic variables such as shear stress, Froude number and Reynolds number (Table 15.1) relate to the distribution of instream biota.

Gore (Chapter 3) and Brookes (Chapter 4) have demonstated the effects of changes of channel hydraulics upon biota. In channelized streams, the reduction in abundance and diversity of fish species may result from the inability of some species to maintain physiological equilibrium in conditions where the 'hydraulic fitness' of their body shape is suboptimal. In natural channels, the heterogeneous flow environment supports a biota of comparatively greater variety and abundance, and hydraulic refugia play a major role in determining the impact of hydrological changes.

Table 15.1 Common complex hydraulic variables

Froude number	(Fr)	=	$V/(gd)^{0.5}$	Dimensionless
Reynolds number	(Re)	=	VR/ν	Dimensionless
Shear stress	(r)	=	$pgSR$	dyne cm^{-2}
Manning's roughness	(n)	=	$R^{0.67}S^{0.5}/V$	Dimensionless
Bagnold power	(Ω)	=	$pgQS$	watts ($= J\ s^{-1}$)

Ω = V = velocity; d = depth; R = hydraulic radius (cross-sectional area/wetted perimeter length; for wide shallow channels this approximates to the mean depth); power; Q = discharge; S = slope; p = water density; ν = kinematic viscosity; and g = acceleration due to gravity.

Habitat preference curves.

Different species of fish and macroinvertebrates select specific habitat types in terms of preferred water velocities, depths and substrates, and these may be defined as habitat suitability criteria. The use of habitat profiles and preference curves for target species (e.g. Fig. 15.2) has become established as a means of describing habitat requirements for both fish and invertebrates. In their simplest form, habitat criteria may be derived from life-history studies in the literature or from professional experience or judgement (e.g. Chaveroche & Sabaton 1989). Alternatively, detailed field observations enable the frequency analysis of microhabitat conditions utilized by different life stages and species. In this latter case, problems may arise where preferred habitat conditions are absent or limited at the time of observation.

A variety of techniques have been used to relate presence/absence or abundance data to hydraulic parameters. The binary format provides the simplest approach, establishing a suitable range of conditions for a hydraulic parameter as it pertains to a life stage of a species; conditions are either suitable (a rating of 1.0) or unsuitable (rating of 0.0). More complex methods include an incremental approach (Gore & Judy 1981), multiple regression (Morin *et al* 1986) and bivariate depth

Fig. 15.2 (*Opposite.*) Examples of habitat suitability criteria curves. (a) Binary and simple continuous curves for Chinook salmon spawning (Raleigh *et al* 1986); (b) curves for Atlantic salmon parr showing variable preferences according to fish size (large parr and small parr or young-of-the-year) (Heggenes 1990); and (c) response surface plot for a depth and velocity relationship showing the raw data and completely smoothed data produced by application of an inverse distance squared smoothing algorithm (Lambert & Hanson 1989); various polynomial models were also used to establish bivariate response surfaces.

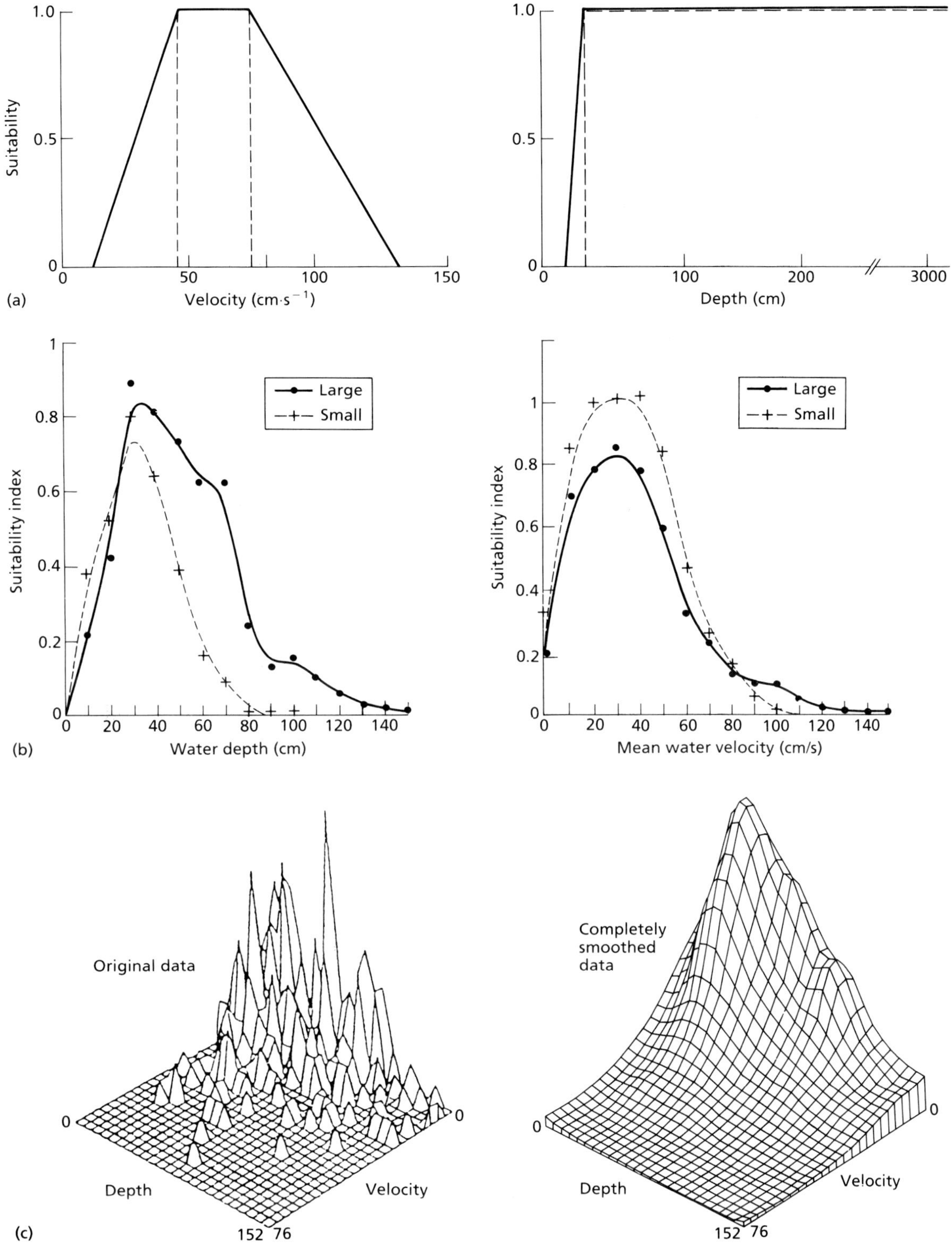
1.0
0.5
0
Suitability
0
50
100
150
Velocity (cm·s⁻¹)
(a)
1.0
0.5
0
0
100
200
3000
Depth (cm)
1
0.8
0.6
0.4
0.2
0
Suitability index
Large
Small
0
20
40
60
80
100
120
140
Water depth (cm)
(b)
1
0.8
0.6
0.4
0.2
0
Suitability index
Large
Small
0
20
40
60
80
100
120
140
Mean water velocity (cm/s)
Original data
0
0
Depth
Velocity
152
76
(c)
Completely smoothed data
0
0
Depth
Velocity
152
76

and velocity exponential polynomial models (Lambert & Hanson 1989).

The channel-forming discharges

Consideration of channel-forming flows is important if the diversity of instream, riparian and floodplain habitats is to be protected or restored. Channel and floodplain processes are intimately linked, although models to quantify these links remain in their infancy (Howard 1992). For management purposes, channel-forming flows may be viewed from three perspectives.

First, it is necessary to consider the *bankfull flow*. Although channel form is a function of the full range of discharges and sediment loads experienced, on many rivers the bankfull flow has been shown to approximate the most effective discharge for sediment transport and to re-form bar shape and orientation, re-order substrate size distribution in local areas, and maintain channel width (see Vol. 1, Chapter 6). The bankfull discharge is perhaps best viewed as a hydrological benchmark (Hill *et al* 1991), representing the range of intermediate flows that determine the size and shape of many alluvial channels. For a range of rivers, the bankfull discharge has been shown to have a return period of between 1 and 3 years, with a modal value of about 1.5 years, using the annual series. However, Chorley *et al* (1984) for 36 active floodplain rivers, showed that bankfull recurrence interval varied between 1 and 32 years, differences that may reflect variations in sediment loads and flow regimes.

Secondly, *flushing flows* must be assessed. The channel substrate provides important habitat for a wide-ranging biota and in gravel-bed rivers is a composite feature that can be divided into four components.

1 The substrate framework of coarse particles (>2 mm).

2 The substrate matrix of fines (<2 mm).

3 The surface ('armour') layer of coarse gravel which is usually incomplete and of one grain thickness.

4 Surface deposits of fine gravel and sand or superficial fines that form more or less discontinuous patches on top of the armour layer.

Flushing flows are moderately high discharges required to maintain the quality of the substrate, preventing vegetation encroachment into the channel and removing superficial and interstitial accumulations of fine organic and inorganic sediments (Reiser *et al* 1989a).

Thirdly, it is important to consider *floodplain-maintenance flows*. Floodplains are formed by a suite of accretional and erosional processes and comprise a range of sedimentary facies (see Vol. 1, Chaper 7) reflected on the surface by the floristic mosaic (e.g. Salo 1990; see also Plate 21.1). Floods disrupt the floodplain system by lateral erosion, meander cut-off and channel avulsion, bar deposition, cut-off channel siltation, and overbank sedimentation. Such disturbance sustains habitats for pioneer species and lower successional stages, and the mosaic of landform-sediment-vegetation units reflects the channel stability: some channels maintain a stable course for long periods of time (>10^2 years) whereas at the other extreme channel migration by more than 1000 m yr^{-1} has been reported.

Very high floods cause disturbance which counteracts the advantage of the flood pulse in the short term but over longer time-scales such floodplain maintenance flows are important for maintaining habitat diversity. Thus, Stromberg (1992) demonstrated that in semi-arid basins, the greatest plant richness and abundance occurred along rivers with intermediate flood magnitudes (as indexed by the magnitude of the 50-year flood). Low vegetation diversity is associated with highly stable systems, as evidenced by a decline in diversity on regulated streams where flooding has been suppressed and the geomorphological dynamics reduced (see Chapter 21).

The hyporheic zone

The channel substrate provides important habitat for a wide range of invertebrates and fish, especially salmonid-spawning habitat. Once perceived as a stable environment, the subsurface zone along the river corridor is now recognized as an important and dynamic system that is largely determined by the hydrological conditions. Research on these subsurface systems – phreatobiology – has a long history (see Obrdlik *et al* 1992 for review) and many studies have focused on the groundwater fauna of major alluvial plains along the Danube, Rhine and Rhône. However, most studies have involved taxonomic and biogeographic descriptions. Various species of the exclusively hypogeon-dwelling organisms (e.g. the amphipod group *Niphargopsis*) have been found throughout the alluvial plains of the three major European rivers. Particularly interesting groundwater faunas have been found associated with floodplain wetlands and springs.

During the 1980s a new reseach direction developed, considering the hyporheic zone as an ecotone between the surface water and groundwater (Gibert *et al* 1990). In many cases, the hyporheic zone is of limited extent bordering the channel but in some geomorphological settings, notably glacial troughs infilled with highly permeable gravels, the zone may be extensive. On the Flathead River, Montana, USA, the hyporheic zone extends 10 m vertically and about 1500 m laterally (Stanford & Ward 1988).

Gibert *et al* (1990) identify four unique features that distinguish groundwater ecotones. First, the ecotone is colonized by both epigeal and hypogeal organisms, and has a biodiversity intermediate between the rich surface-water assemblage and the low-diversity groundwater community. Secondly, the ecotone plays an important role in the connectivity and the exchange of organisms between landscape patches. Thirdly, the ecotone acts as a physical and biological filter for exchanges between surface and groundwater environments (Triska *et al* 1993). Finally, but most importantly from the perspective of determining instream flow needs, the dimensions of the ecotone vary strongly in space and time in relation to surface flow and groundwater levels.

Spatial variations occur both at the scale of an entire alluvial plain (Marmonier *et al* 1992), river reach (Creuze des Chatelliers & Reygrobellet 1990) and an individual riffle (Sterba *et al* 1992). Temporal variations relate to flow variations (Marmonier & Creuze des Chatelliers 1991); surface water infiltration during floods causes the active or passive movement of organisms within the sediments, displacing the limits of the ecotone. Two of the most important ecological factors, thermal gradients and oxygen profiles, are closely linked to the strength and direction of flow lines within the hyporheic zone (White 1990). Thus, Shepherd *et al* (1986) suggested that the influence of flow on processes and biological communities within the hyporheic zone should be of concern to all those involved in fisheries research, benthic invertebrate studies and environmental impact assessments.

15.3 METHODS FOR DEFINING FLOWS

The effect of flow on instream biota is important but the effects may be masked by environmental factors that determine regional variations and downstream patterns of biota and biological processes within catchments (Armitage & Petts 1992). The location along the river continuum, water quality and channel morphology are used most frequently for habitat evaluation in river management (e.g. Milner *et al* 1985). Altitude and a measure of distance from source, such as stream order or drainage area, are important indices of position along the river continuum, providing surrogates for temperature and organic-matter sources as well as for discharge and channel form. Important water-quality indicators include pH, alkalinity and hardness. Typically, such indicators change from source to mouth along a river, reflecting the variable contributions of runoff from different sources. Channel morphology, usually indexed by width, depth, velocity and substrate size, relates not only to the downstream continuum but also to the historical legacy of channel change in relation to climate (over the Quaternary). Today, along many rivers, the spatial pattern of water-quality and channel-form variation reflect the historical legacy of human impacts. Thus, in order to elucidate the influence of flow patterns on biota it is necessary to classify sites at the appropriate scale, i.e. to match the geomorphological setting to the hydrological processes (Hill *et al* 1991) and to minimize the influence of the higher-level attributes to biological responses.

The general principles of classifying rivers for the purpose of assessing conservation potential have been reviewed by Naiman *et al* (1992). For the assessment of instream-flow needs, a three-level classification system is proposed. First, the river may be divided into *types* (e.g. upland, intermediate and lowland), reflecting position along the river. Secondly, each type may be divided into *sectors*. Within each sector, water quality, sediment load and hydrological regime are seen as invariant between sites. Thirdly, each sector is divided into *reaches* on the basis of local variations in channel morphology or river-margin vegetation. Along natural rivers, channel morphology often has the same general form throughout each sector.

Minimum acceptable flows

Three approaches are available for determining minimum acceptable flows: hydrological methods; river assessment methods; and PHABSIM (physical habitat simulation).

Hydrological approaches

Traditionally, minimum acceptable flows have been set using discharge-based methods. These express the instream flow as a hydrological statistic: commonly either as a flow duration statistic (such as the 95th percentile) or as a fixed percentage of the average daily flow (ADF). The classic example of the later is Tennant's (1976) Montana Method (Table 15.2). Orth and Leonard (1990) demonstrated that for streams in Virginia, USA: (i) 10% ADF correctly defines degraded or poor habitat conditions; (ii) 20% ADF provides an appropriate criterion for protecting aquatic habitats; and (iii) 30% ADF defines near-optimum habitat in small streams.

Table 15.2 Examples of discharge methods based upon average daily flow (ADF)

	Percentage of ADF	
	Minimum	Maximum
Baxter (1961) flow for *Salmo salar*		
Juveniles in summer	20	25
Adult migration	30	70
Spawning	12.5	30
Incubation	10	17
Tennant (1976)		
Absolute minimum	10	
Good habitat	30	60
Optimum habitat	60	100
Flushing flow	200	

In the UK water industry, two measures of the dry weather flow (DWF), the 95th percentile flow or the mean annual minimum 7-day flow frequency statistic, have been used most frequently to set prescribed flows. An alternative approach (Drake & Sheriff 1987) employs a proportion of the dry weather flow weighted according to a range of environmental characteristics and uses. Thus the prescribed flow is set at 1.0 × DWF for the most sensitive rivers and at 0.5 × DWF for the least sensitive.

The Northeast Region of the US Fish and Wildlife Service has established an aquatic baseflow (ABF), which defines the minimum acceptable flow as the median flow in the month during which low flow conditions typically result in the most metabolic stress to aquatic organisms (Kulik 1990). Stress was recognized as being exerted by high water temperature and diminished living space, dissolved oxygen and food supply.

The justification for a hydrological approach is that over the long term, stream flora and fauna have evolved to survive periodic adversities without major population changes. In natural systems, the effects of stress on the biological communities are seen to be balanced by recovery mechanisms. However, any human impact that reduces the effectiveness of these recovery mechanisms will clearly affect the level of flow required to meet the primary objective, namely 'to sustain and perpetuate indigenous aquatic fauna'. Relevant impacts include channelization that isolates a river from floodplain backwaters or reduces instream habitat diversity, and dams that create barriers to upstream and downstream migrations. Synergistic toxicity effects may also be significant because of the water-quality changes often associated with low-flow problems.

A major criticism of discharge methods is their exclusion of any explicit consideration of actual habitat requirements. A second concern arises from the complex array of processes, influenced by flow, that can affect biota. Consideration of discharge in isolation of other water-quality and geomorphological factors can provide only a very superficial understanding of potential impacts of flow regulation. Thus, attention has been directed towards a more scientific understanding of the response of species and biological communities to hydrological change.

River assessment techniques

Recent models for river assessment are based on flow–habitat relationships and make the assumption that habitat attributes provide an index of suitability for biota. The approach implicitly assumes that the abundance data for invertebrates or fish, for example derived from field survey, are maximum values appropriate to the habitat conditions at each site. In reality this 'maximum' is a theoretical concept; population levels are usually below the maximum because of (i) variation in biotic factors, such as predation, competition and recruitment; and (ii) disturbance, such as floods, and variable rates of population recovery.

Quantitative and semi-quantitative studies have often been employed to relate a biological criterion, such as fish abundance, to habitat attributes. Quantitative approaches require a detailed field survey of the habitat attributes, including measurements of depths, widths, velocities, substrate size, composition, etc. Semi-quantitative studies combine field transect measurements of physical attributes with subjective assessments of elements of habitat quality, such as fish cover. Whilst simple reconnaissance methods providing quick assessments have met with disapproval for ignoring the specific character of the stream, more sophisticated approaches have been criticized for the length of time required to obtain meaningful results.

Simple linear regression and multiple regression have often been used in habitat evaluation studies, for example by Binns and Eiserman (1979), Milner *et al* (1985) and Scarnecchia and Bergensen (1987). Many attributes are strongly interrelated and careful selection of habitat attributes is necessary to minimize the problem of autocorrelation between variables (Table 15.3). A second problem is the non-linearity between physical and biotic variables, a problem that is usually overcome by using a rating system, such as the five-point scores of Binns and Eiserman.

Thus, Maddock (1992) developed an empirical relationship between an invertebrate-based score,

Table 15.3 Attributes commonly used in habitat-fish abundance relationships

I Catchment	*II Water quality*
Altitude	Maximum temperature
Drainage area	pH
Stream order	Alkalinity
	Hardness
III Flow regime	Conductivity
Average daily flow	Nitrate concentration
Average seasonal flow	
Baseflow index	*IV Channel structure*
	Average depth
	Maximum depth
	Width
	Velocity
	% pool
	% riffle
	% cover
	Dominant substrate

the BMWP, and habitat variables. For 28 streams in East Anglia, UK, field data were obtained during summer low-flow at approximately 100 points from five transects within a reach defined as five channel widths or one riffle-pool sequence, whichever was the greater. All sites had long-term invertebrate records showing a stable BMWP and had good-quality flow-gauging data.

$$\text{BMWP} = a + CSb + Qc + Md + He \quad (r^2 = 0.88)$$

where CS is a chemical score based upon oxygen and ammonia levels; Q is Q_{95} divided by channel width; M is an index of instream cover; H is a hydraulic index; and a–e are regression coefficients. The relationship was validated for the River Glen, to which it was then applied to assess the improvement in BMWP with flow augmentation (Petts *et al* in press).

Empirical relationships using physical habitat variables have successfully described variations of trout abundance within a uniform group of streams ($>25\ \text{mg l}^{-1}\ CaCO_3$) in Wales (Milner *et al* 1985). Fish population estimates were obtained by electro-fishing. Average daily flow was significantly inversely related to the abundance of small trout ($\leqslant 10$ cm) but in this study, ADF may simply reflect stream size. Site attributes, such as average depth, percentage of reach with depth of 46–60 cm, depth variance, were significant in explaining the distribution of larger fish, yielding explained variances of more than 80%. Milner *et al* suggested that between- and within-site variability not accounted for by habitat models may be explained by randomly changing habitat attributes and biotic factors. Further, they proposed that the difficulty in establishing fish–habitat relationships in softwater streams may reflect their more unstable discharge regime introducing a stronger stochastic element in their overall carrying capacity.

PHABSIM

Physical HABitat SIMulation (Bovee 1982) provides a methodology to integrate the changing hydraulic conditions with discharge and the habitat preferences of biota (Fig. 15.3). Detailed field-survey data are inputted to the simulation model, which relates changes in discharge or channel structure to changes in physical habitat availability for one or more chosen species. Habitat is defined as comprising two basic components: rigid structural characteristics (channel width, bed configuration and substrate composition, which are assumed to be constant over the range of flows of concern) and variable hydraulic conditions. The underlying principles of PHABSIM are:

1 the chosen species exhibits preferences within a range of habitat conditions that it can tolerate;
2 these ranges can be defined for each species;
3 the area of stream providing these conditions can be quantified as a function of discharge and channel structure.

PHABSIM can be applied in two situations: the investigation of either critical reaches which provide a particular type of habitat that is otherwise

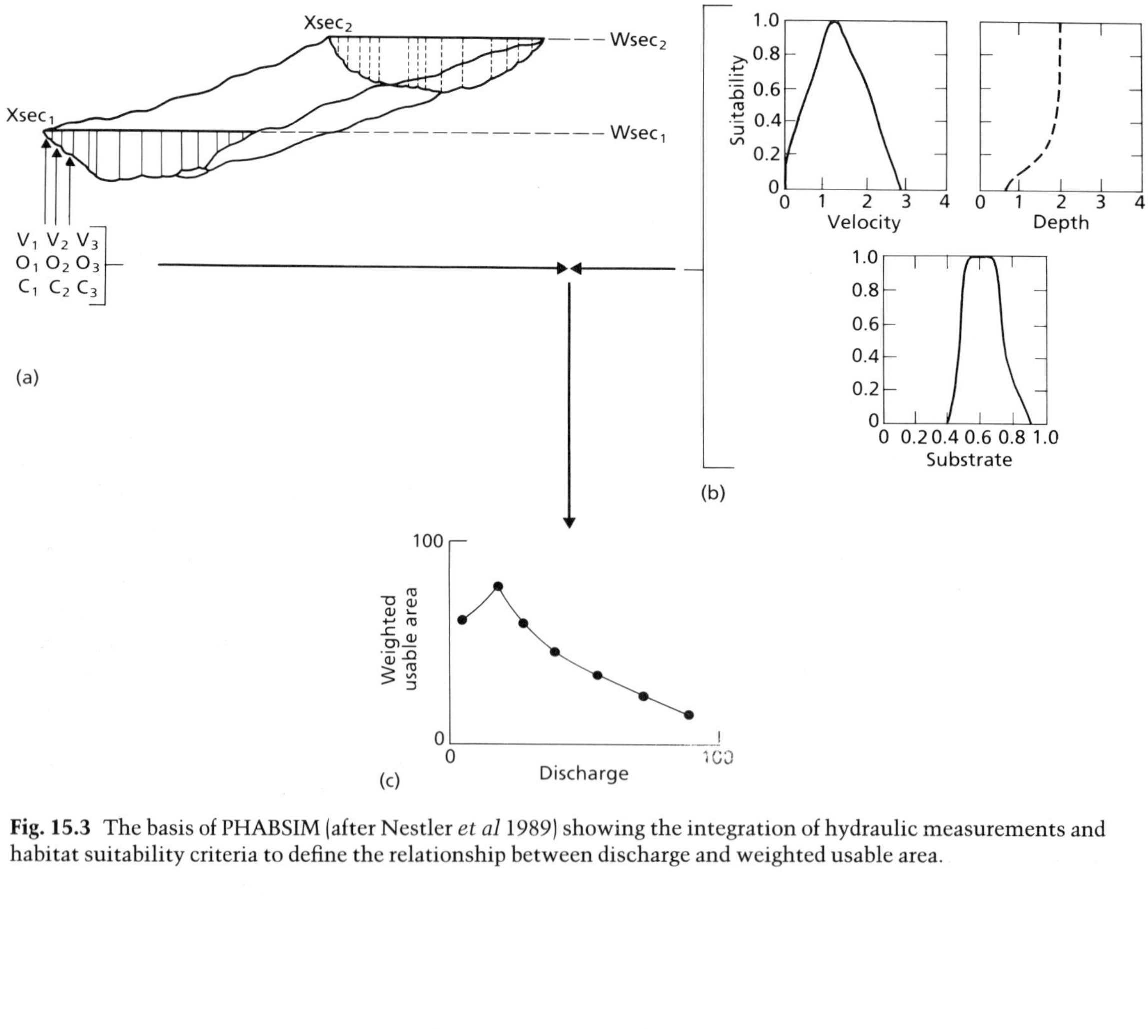

Fig. 15.3 The basis of PHABSIM (after Nestler *et al* 1989) showing the integration of hydraulic measurements and habitat suitability criteria to define the relationship between discharge and weighted usable area.

limited within the system, or representative reaches which are similar to any other reach within a sector and, ideally, contain all the habitat patches found within the entire sector. Hydraulic conditions are characterized usually at three known (calibration) streamflows from point measurements taken along transects within the reach. Transects are located so as to sample (i) all the hydraulic controls, i.e. physical aspects of the streambed that determine the height of the water surface upstream, and (ii) all habitat types that are represented along the reach. A typical reach for analysis using PHABSIM has a length of the order of 20 channel widths.

Hydraulic simulation models are used to estimate hydraulic conditions at unmeasured flows. Three types of approach are available for calculation of water surface elevation at unknown discharges: (i) the stage–discharge relationship; (ii) the use of Manning's equation; and (iii) the standard step backwater method. In many situations it may be necessary to use a mixture of models to simulate the hydraulic characteristics of the reach over the full range of flows. For instance, under low flows the stage–discharge relationship may simulate water-surface elevations and velocities most accurately whereas the standard step backwater method may be more suitable over the higher flows. The technique used for hydraulic simulation can have a significant impact on the habitat versus streamflow relationship; the correct choice of hydraulic model(s) as well as the proper calibration can be time consuming and may represent the most difficult step in the process of analysing streamflows.

Hydraulic simulation is then combined with habitat suitability curves for target organisms to define available habitat – the weighted usable area (WUA) – in the reach. PHABSIM contains a number of different programs that can be used for this purpose, each having specific conditions for their application. The most often used is the multiplicative composite suitability index:

$$WUAi = Ai \times f(v) \times f(d) \times f(s) \times f(c), \text{ etc.}$$

where the reach is divided into cells; Ai is the area of a cell (i) which is multiplied by all suitability indices used: velocity ($f(v)$), depth ($f(d)$), substrate ($f(s)$), cover ($f(c)$), etc. Effects of streamflow changes on the physical habitat available for the target organism can then be evaluated by changes in the amount of WUA. The approach may also be used to assess the habitat quality of reaches, and then sectors, where 'quality' is related to the configuration of the bed morphology. Thus, PHABSIM is able to present biological information in a format suitable for entry into the water resource planning process. The information is made explicitly in terms of changes to the physical properties of the aquatic habitat (i.e. velocity, depth and substrate) and does not predict changes in the biomass of organisms. The IFIM methodology has been applied in several countries outside the USA including Canada (Mathur *et al* 1985; Shirvell 1989), New Zealand (Scott & Shirvell 1988), Australia (Gan & McMahon 1990), Norway (Heggenes *et al* 1990), France (Chaveroche & Sabaton 1989; Courot 1989), South Africa (Gore *et al* 1992) and the UK (Bullock *et al* 1991; Bullock & Gustard 1992; Petts 1992, 1993).

Application

The results of applying the methods to define in-river needs may be illustrated by data for one reach on the River Babingley, a small but high-quality chalk stream in north Norfolk, UK, with a good stock of brown trout (*Salmo trutta*) (Petts 1993).

Long-term average monthly flows, based on the period 1976–87, compared to those experienced during subsequent years (Fig. 15.4) highlight the extent of the recent drought. Average monthly flows peak in March at a discharge of 0.770 cumecs falling to 0.339 cumecs in September. For PHABSIM, calibration flows corresponded to Q_{99}, Q_{96} and Q_{77} on the flow duration curve for the period 1963–92. Habitat suitability curves were taken from Armitage and Ladle (1991) for four life stages of brown trout (i.e. spawning, fry, juvenile and adult) and four species of aquatic invertebrates, one stonefly (*Leuctra fusca*), two caseless caddis (*Rhyacophila dorsalis* and *Polycentropus flavomaculatus*) and one pea mussel (*Sphaerium corneum*). Results of the different methods are given in Table 15.4.

An example of the discharge versus usable

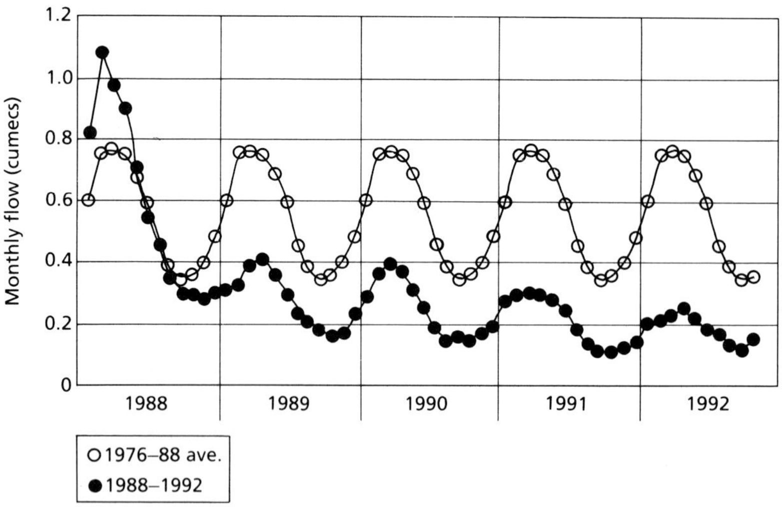

Fig. 15.4 Monthly flow data for the River Babingley contrasting the 1976–88 average data with the 1988–92 series.

Table 15.4 Recommendations for minimum flows (as gauged at Castle Rising)

Method	Record	Discharge ($m^3 s^{-1}$)
95 percentile flow	1976–85	0.300
95 percentile flow	1963–92	0.157
30% of mean flow	1976–85	0.165
30% of mean flow	1963–92	0.136
Aquatic baseflow	1976–85	0.261
PHABSIM		
Trout		
Adult		0.280
Spawning		0.280
Juveniles		0.170
Fry		0.090
Invertebrates		0.127

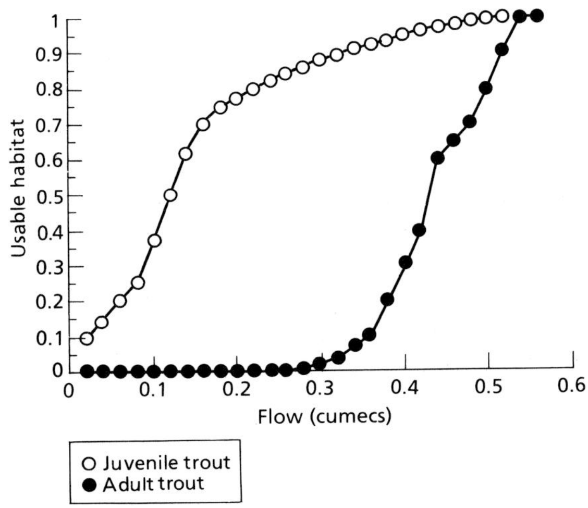

Fig. 15.5 Examples of habitat-discharge relationships for the River Babingley, for brown trout adults and juveniles.

habitat relationship is given in Fig. 15.5, and clear reductions in usable habitat can be defined for each life stage when flows fall below critical levels (Table 15.4). The results of the discharge versus percent usable area relationships were combined with the actual discharges experienced during the drought to show the effect these reduced flows had on habitat area available for life stages of brown trout (e.g. Fig. 15.6). During 1988–92 little habitat was available for adults and suitable habitat was eliminated in mid-1991. For the juvenile life stage usable habitat was consistently below average conditions but good quality habitat was sustained throughout the drought period.

One application of PHABSIM is to assess the impacts of alternative flow-management scenarios using habitat duration curves (Bovee 1982). For example, Fig. 15.7 shows two curves, one based on the flow duration statistics for the period 1963–92 and one based on a curve with similar

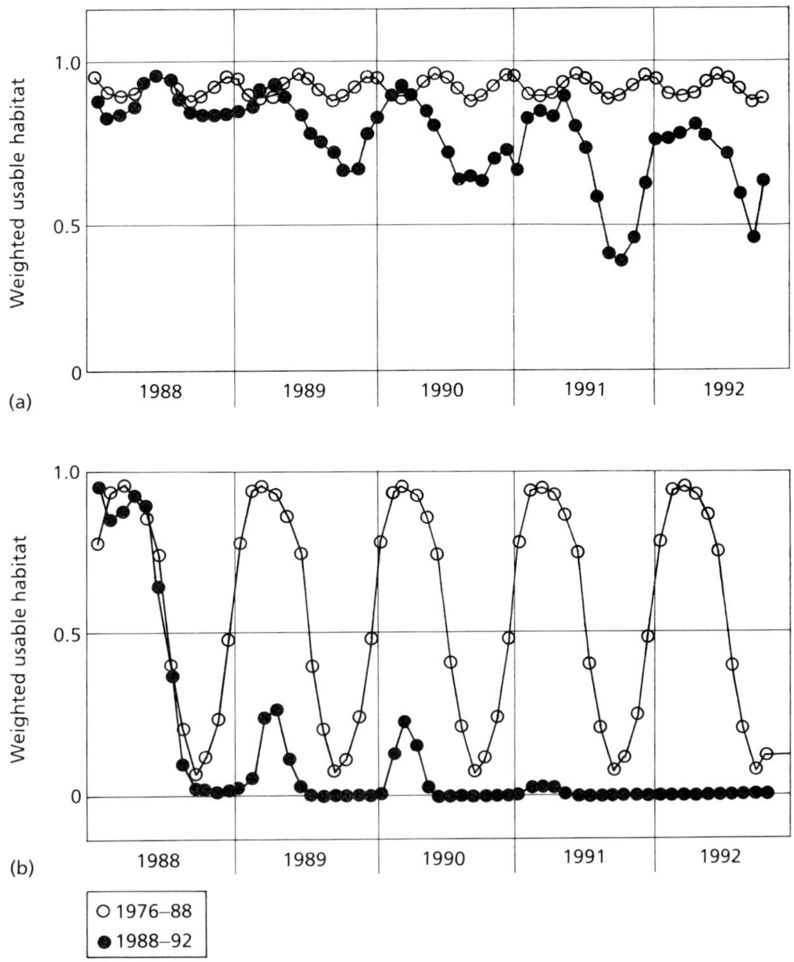

Fig. 15.6 Examples of habitat changes predicted by PHABSIM for the River Babingley, comparing the average usable habitat with that available during the 1988–92 drought for brown trout juveniles (a) and adults (b).

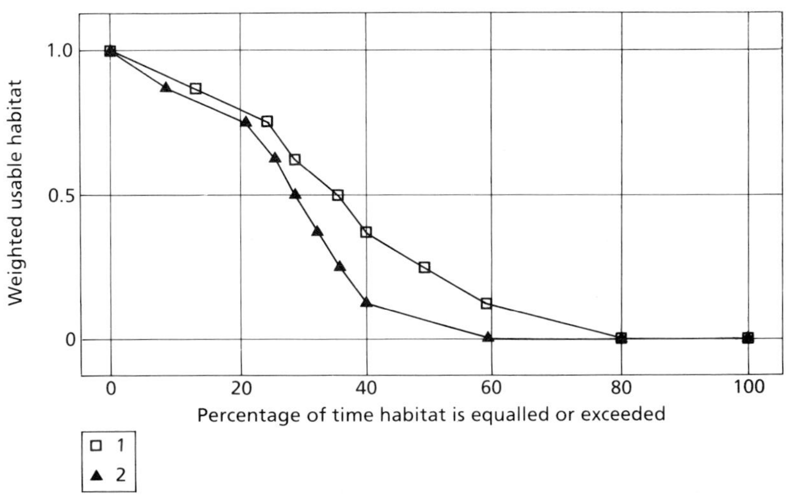

Fig. 15.7 Examples of habitat duration curves for the River Babingley, showing brown trout habitat under (1) average flow conditions and (2) a hypothetical flow management scenario (mean daily flow reduced by 30% but no change in Q_{10} or Q_{95}).

Q_{10} and Q_{95} values but with mean daily flow reduced by 30%. In this case, the most critical change in habitat would be experienced by the adult life stage. Under reduced average daily flows, the median habitat value would be reduced by 75% and zero habitat would be experienced for approximately 42% of the time.

Analyses such as those illustrated above allow definition of a range of minimum flows. For example, a *desirable* minimum ecological flow could be defined as 0.28 cumecs – the minimum flow for adult trout – which approximates to the naturalized 95th percentile flow and to the median flow in the driest month, September. The results from this one reach also show that a flow of 0.13 cumecs should sustain habitat for invertebrates and juvenile trout, even though for the latter this will be only about 50% of the potential habitat. These results suggest that exceptionally, under severe natural drought conditions, flows could be reduced to this *emergency* minimum ecological flow. Although for the duration of the drought this flow would probably eliminate habitat for adult trout, habitat for juveniles and invertebrates would be sustained and evidence (e.g. Milner, Chapter 5; Petts *et al* in press) suggests that the biota should recover rapidly once flows return to 'normal'.

High flows and the minimum streamflow hydrograph

Determination of the minimum acceptable flow is only the first step in defining a minimum streamflow hydrograph for sustaining the ecological integrity of the fluvial hydrosystem. Flood flows of up to the 50-year event can be important for rejuvenating floodplain systems whilst the riparian system and the channel morphology will respond to changes of the dominant discharge, indexed by the 1.5-year flood. Ideally, the river corridor should be delineated as the historical geomorphologically-active floodplain defined from aerial photographs or satellite imagery, cartographic evidence and field survey.

Hill *et al* (1991) developed an annual hydrograph based upon mean monthly flows for the Whitebird reach of the Salmon River, Idaho, USA (Fig. 15.8a). The average annual hydrograph was examined to determine the timing of high and low flows, the slope of the rising hydrograph and the flow recession. Minimum instream flows to maintain habitat for fish were defined as the 7-day low flow with a 1.5-year return period; a hydraulic model (HEC-2) was used to estimate the bankfull discharge and the flows required to inundate areas of valley floor of different elevations. It was recommended that daily flow recession should not exceed 10% of the previous day's flow; in general, the slower the flow recession the greater the advantage for biota by protecting fauna from stranding and providing a longer period of high flows for vegetation seeding.

For the Yampa River in Dinosaur National Monument, Colorado, O'Brien (1987) defined a minimum streamflow hydrograph (Fig. 15.8b) to maintain existing aquatic habitat, especially for the endangered Colorado squawfish (*Ptchocheilus lucius*). The hydrograph was derived using the results of a 2-year field data collection programme, a physical model, laboratory simulation of flows on a cobble substrate and a mathematical model simulation of sediment transport. It included four components: the *baseflow*, defined as the mean baseflow for the period of record, to provide food supply during the low-flow period in the fall and winter; a gradual *rising limb* that would provide backwater habitat for shelter during high-flow periods; a *recessional limb* with sufficient flow over an extended period to remove any possible sand deposition in the spawning areas; and a *peak discharge* that would inundate the active channel area to eliminate any encroaching vegetation. The peak flow was defined as the effective discharge for sediment transport, based on a sediment discharge rating curve.

Pulse releases

Where rivers are regulated by dams, artificial flow variations – pulse releases – can stimulate biological functions and advantage instream habitats. The provision of artificial flow fluctuations becomes increasingly important as the level of regulation and abstraction intensifies, particularly important for salmonids to stimulate migration and to clean spawning grounds. In a classic paper, Hayes (1953) demonstrated that large natural

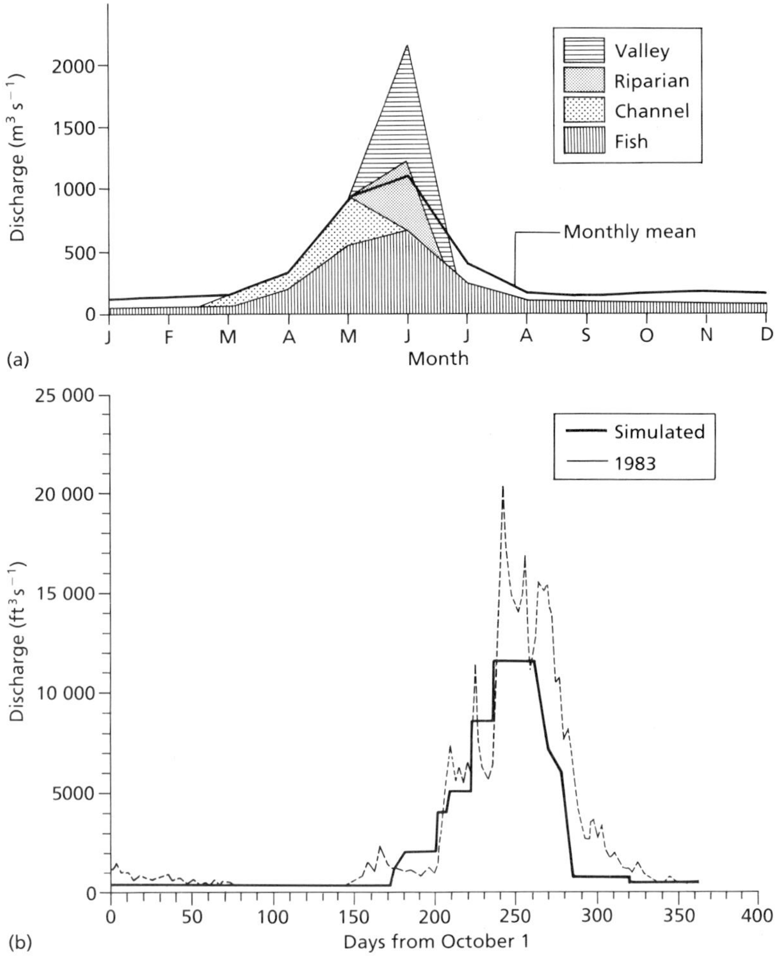

Fig. 15.8 Examples of minimum streamflow hydrographs: (a) for the Salmon River, Idaho, USA (from Hill *et al* 1991), and (b) for the Yampa River, Colorado (from O'Brien 1987).

freshets can be associated with a major run of fish into a river, as long as tides and winds are also favourable. Subsequent research has been inconclusive, but Alabaster (1970) suggested that short-term changes of flow together with accompanying changes in concentrations of dissolved substances, may stimulate entry of salmon into rivers and their upstream migration.

For channel maintenance, a general rule-of-thumb is that sediment within gravel-bed channels begins to be mobilized when flow depth is slightly greater than 80% of bankfull depth (Parker 1979, ref. 97). Flushing flows recommended by Reiser *et al* (1989a) are of two types. The first type is based on flow record and refined by field observations, and includes 200% average annual flow and the 5th $<$ 17th percentile on flow duration curve. The second type involves sediment transport methods within particular reaches involving either direct field measurements of the threshold discharge for transport, e.g. using bed-load tracers, or modelling of sediment transport to predict the discharge which causes incipient motion of particles. An application of the latter is given by Reiser *et al* (1989b). Most commonly, flows are required to flush superficial accumulations of fines and to clear the surface gravel to a depth of approximately

0.5 < 1.0 times the average diameter of surface cobbles. This has been shown to be a realistic objective (Beschta & Jackson 1979).

The timing and duration of pulse releases is also important and must avoid harm to the biotic communities, such as by dislodgement and transport of eggs and fry. For example, flushing flows may have greatest benefit if they are provided prior to salmonid spawning, cleaning the gravels and enhancing egg and alevin survival. Clearly, knowledge of life histories for target species is important for determining the optimum timing of flows, although it is often assumed that the biota is adapted to the normal hydrograph, so that coincidence of flushing flows with normal peak-flow periods should prove beneficial. In any case, test flow releases should be made to enable field observations to assess the effectiveness of the recommended flows.

15.4 PROSPECT

Linking hydrology and ecology is a complex science because of the many ways in which discharge influences biota (by determining habitat suitability for each life stage) and biotic interactions (predation, competition, etc.). In the past, river management strategies have been directed most frequently to the protection of sport and commercial fisheries. Both implicitly and explicitly, it has been assumed that this strategy would also sustain adequate habitat for other organisms.

River management requires scientifically-sound operational tools to determine the allocation of water to meet instream, riparian and floodplain needs. The rapid natural recovery of fish of following severe drought (Bayley & Osbourne 1993), of invertebrates impacted by severe low flows following flow augmentation (Petts *et al* in press), and of fish and invertebrates to flow restoration to remnant channels following channelization (Toth *et al* 1993) suggests that many natural systems can undergo self-restoration in response to rare extreme low-flow conditions. However, the assumption that enhanced minimum instream flow will result in the production of more or larger fish has not been demonstrated (Wolff *et al* 1989; Harris *et al* 1991). This arises because other habitat factors, such as cover or temperature, may be limiting or because natural variations in abundance may mask any trends in short-term data sets.

The lack of evidence that the biota responds to changes in flow and the inability to predict population responses to flow alterations remain frustrations which will persist until intensive long-term research is initiated (Orth 1987). Development of deterministic ecosystem models may be an unrealistic objective because of the stochastic nature of the driving hydrological and geomorphological processes and the complexity of biological interactions. Nevertheless, concern for sustaining the biological integrity of ecosystems requires new approaches to ensure the maintenance of ecological processes, rather than maintenance of single target species (O'Keefe & Davies 1991).

Several authors have argued that benthic macroinvertebrate communities must be protected by prescribed flows to maintain biotic integrity and ecosystem functioning (Gore 1989). The macroinvertebrates have a key role in the energy dynamics of ecosystems and provide a primary energy link to the game and forage fishes. River margins – floodplains and riparian zones – also play an important role in maintaining the productivity of rivers, and lateral and vertical interactions and exchanges need to be incorporated into approaches to assessing flow needs. There is some information on habitat change and biological responses to suggest that at the regional scale, a hydrological approach, giving due regard to floodplain, riparian, channel and minimum-flow needs, can provide a practicable option for establishing an ecologically-acceptable flow regime. In the short term, the most cost-effective approach to achieving ecologically-sensitive water development may be to use integrated biological and habitat studies to evaluate hydrological methods for instream-flow assessment based upon a regional (river type) and longitudinal, within-river (river sector) channel classification.

REFERENCES

Alabaster JS. (1970) River flow and upstream movement and catch of migratory salmonids. *Journal of Fish*

Biology **2**: 1–13. [15.3]

Ambuhl H. (1959) Die Bedeutung der Stromung als okologischer Fakton. *Schweizerische Zeitschrift für Hydrologie* **21**: 133–264. [15.2]

Amoros C, Petts GE. (1993) *Les Hydrosystèmes Fluviaux*. Mason, Paris (to be published in Engish by Chapman & Hall, London). [15.1, 15.2]

Armitage PD. (1978) Downstream changes in the composition, numbers and biomass of bottom fauna in the Tees below Cow Green reservoir and in an unregulated tributary Maize Beck, in the first five years after impoundment. *Hydrobiologia* **58**: 145–56. [15.2]

Armitage PD, Ladle M. (1991) Habitat preferences of target species for application in PHABSIM testing. In: Bullock A, Gustard A, Grainger ES (eds) *Instream Flow Requirements of Aquatic Ecology in Two British Rivers*. Institute of Hydrology, Report 115, Wallingford, UK. [15.3]

Armitage PD, Petts GE. (1992) Biotic score and prediction to assess the effects of water abstraction on river macroinvertebrates for conservation purposes. *Aquatic Conservation* **2**: 1–18. [15.2]

Baxter G. (1961) River utilization and the preservation of migratory fish life. *Proceedings of the Institute of Civil Engineers* **18**: 225–44. [15.3]

Bayley PB. (1991) The flood-pulse advantage and the restoration of river-floodplain systems. *Regulated Rivers* **6**: 75–86. [15.2]

Bayley PB, Osbourne LL. (1993) Natural rehabilitation of stream fish populations in an Illinois catchment. *Freshwater Biology* **29**: 295–300. [15.4]

Beschta RL, Jackson WL. (1979) The intrusion of fine sediments into a stable gravel bed. *Journal of the Fisheries Research Board of Canada* **36**: 204–10. [15.3]

Bickerton M, Petts GE, Armitage P, Castella E. (1993) Assessing the ecological effects of groundwater abstraction on chalk streams: three examples from eastern England. *Regulated Rivers* **8**: 121–35. [15.2]

Binns NA, Eiserman FM. (1979) Quantification of fluvial trout habitat in Wyoming. *Transactions of the American Fisheries Society* **108**: 215–28. [15.3]

Bovee KD. (1982) *A Guide to Stream Habitat Analysis Using the Instream Flow Incremental Methodology*. Instream Flow Information Paper 12, FWS/OBS-82/26. Office of Biological Services, US Fish and Wildlife Service, Fort Collins, Colorado. [15.3]

Brooker MP. (1981) The impact of impoundments on the downstream fisheries and general ecology of rivers. *Advances in Applied Biology* **6**: 91–152. [15.2, 15.3]

Brookes A. (1988) *Channelized Rivers*. John Wiley, Chichester. [15.2, 15.3]

Bullock A, Gustard A. (1992) Application of the Instream flow incremental methodology to assess ecological flow requirement in a British lowland river. In: Carling PA, Petts GE (eds) *Lowland Floodplain Rivers: Geomorphological Perspectives*. John Wiley, Chichester. [15.2, 15.3]

Bullock A, Gustard A, Grainger ES. (1991) *Instream Flow Requirements of Aquatic Ecology in Two British Rivers*. Institute of Hydrology, Wallingford, Report 115. [15.3]

Chaveroche P, Sabaton C. (1989) An analysis of brown trout (*Salmo trutta fario* L.) habitat: the role of qualitative data from expert advice in formulating probability-of-use curves. *Regulated Rivers* **3**: 305–20. [15.2, 15.3]

Chorley RJ, Schumm SA, Sugden DE. (1984) *Geomorphology*. Methuen, London. [15.2]

Conder AL, Annear TC. (1987) Test of weighted usable area estimates derived from a PHABSIM model for instream flow studies on trout streams. *North American Journal of Fisheries Management* **7**: 339–50. [15.2]

Courot A. (1989) Determination of hydraulic parameters for instream flow assessments. *Regulated Rivers* **3**: 337–44. [15.3]

Creuze des Chatelliers MC, Reygrobellet JL. (1990) Interactions between geomorphological processes, benthic and hyporheic communities: First results on a by-passed canal of the French Upper Rhône River. *Regulated Rivers* **5**: 139–58. [15.2]

Crisp DT, Mann RHK, Cubby PR. (1983) Effects of regulation of the River Tees upon fish populations below Cow Green reservoir. *Journal of Applied Ecology* **20**: 371–86. [15.2]

Drake PJ, Sherriff JDF. (1987) A method for managing river abstractions and protecting the environment. *Journal of Water and Environmental Management* **1**: 27–38. [15.3]

Edwards RW, Garrod DJ. (1972) *Conservation and Productivity of Natural Waters*. The Zoological Society of London, Academic Press, London. [15.2]

Fraser JC. (1972) Regulated discharge and the stream environment. In: Oglesby RT, Carlson CA, McCann JA (eds) *River Ecology and Man*, pp. 263–86. Academic Press, New York. [15.2]

Gan K, McMahon T. (1990) Variability of results from the use of PHABSIM in estimating habitat area. *Regulated Rivers* **5**: 233–9. [15.3]

Gibert J, Dole-Olivier MJ, Marmonier P, Vervier P. (1990) Surface water/groundwater ecotones. In: Naiman RJ, Decamps H (eds) *Ecology and Management of Aquatic Terrestrial Ecotones*, pp. 199–225. Unesco, Paris; Parthenon Publishers, Carnforth. [15.2]

Gore JA. (1989) Models for predicting benthic macroinvertebrate habitat suitability under regulated flows. In: Gore JA, Petts GE (eds) *Alternatives in Regulated River Management*, pp. 253–66. CRC Press, Boca Raton, Florida. [15.4]

Gore JA, Judy RD. (1981) Predictive models of benthic macroinvertebrate density for use in instream flow studies and regulated flow management. *Canadian Journal of Fisheries and Aquatic Sciences* **38**: 1363–70. [15.2]

Gore JA, Layzer JB, Russell IA. (1992) Non-traditional applications of instream flow techniques for conserving habitat for biota in the Sabie River of southern Africa. In: Boon P, Calow P, Petts GE (eds) *River Conservation and Management*, pp. 161–78. John Wiley, Chichester. [15.3]

Gustard A, Cole G, Marshall D, Bayliss A. (1987) *A Study of Compensation Flows in the UK*. Institute of Hydrology, Report 99. A summary is given in Gustard A. (1989) Compensation flows in the UK: a hydrological review. *Regulated Rivers* **3**: 49–60. [15.2]

Harris DD, Hubert WA, Wesche TA. (1991) Brown trout population and habitat response to enhanced minimum flow in Douglas Greek, Wyoming. *Rivers* **2**(4): 285–94. [15.4]

Hayes FR. (1953) Artificial freshets and other factors controlling the ascent and population of Atlantic salmon in the La Have River, Nova Scotia. *Bulletin of the Fisheries Research Board of Canada* **99**: 47 pp. [15.3]

Heggenes J. (1990) Habitat utilization and preferences in juvenile Atlantic Salmon (*Salmo salar*) in streams. *Regulated Rivers* **5**: 341–54. [15.2]

Heggenes J, Brabrand A, Saltveit SV. (1990) Comparison of three methods for studies of stream habitat use by young brown trout and atlantic salmon. *Transactions of the American Fisheries Society* **119**: 101–11. [15.3]

Hill MT, Platts S, Beschta RL. (1991) Ecological and geomorphological concepts for instream and out-of-channel flow requirements. *Rivers* **2**: 198–210. [15.1, 15.2, 15.3]

Howard A. (1992) Modeling channel migration and floodplain sedimentation in meandering streams. In: Carling P, Petts G (eds) *Lowland Floodplain Rivers* pp. 1–42. John Wiley, Chichester. [15.2]

Junk WJ, Bayley PB, Sparks RE. (1989) The flood-pulse concept in river-floodplain systems. *Special Publication of the Canadian Journal of Fisheries and Aquatic Sciences* **106**: 110–27. [15.2]

Kulik BH. (1990) A method to refine the New England Aquatic Base flow Policy. *Rivers* **1**: 8–22. [15.3]

Lambert TR, Hanson DF. (1989) Development of habitat suitability criteria for trout in small streams. *Regulated Rivers* **3**: 291–304. [15.2]

Maddock I. (1992) Instream habitat assessment. In: Petts GE (ed) *The River Glen: a Catchment Assessment*, Annex E. [15.3]

Marmonier P, Creuze des Chatelliers MC. (1991) Effects of spates on interstitial assemblages of the Rhône River. Importance of spatial heterogeneity. *Hydrobiologia* **210**: 243–51. [15.2]

Marmonier P, Dole-Olivier MJ, Creuze des Chatelliers MC. (1992) Spatial distribution of interstitial assemblages in the floodplain on the Rhône River. *Regulated Rivers* **7**: 75–82. [15.2]

Mathur D, Bason WH, Purdy EJ, Silver CA. (1985) A critique of the Instream Flow Incremental Methodology. *Canadian Journal of Fisheries and Aquatic Sciences* **42**: 825–31. [15.3]

Milner NJ, Scullion J, Carling PA, Crisp DT. (1981) The effects of discharge on sediment dynamics and consequent effects on invertebrates and salmonids in upland rivers. In: Coaker TD (ed) *Anvances in Applied Biology*, pp. 154–220. Academic Press, London. [15.2]

Milner NJ, Hemsworth RJ, Jones BE. (1985) Habitat evaluation as a fisheries management tool. *Journal of Fisheries Biology* **27**(Suppl. A): 85–108. [15.3]

Morin A, Harper PP, Peters RH. (1986) Micro-habitat preference curves of blackfly larvae (Diptera: Simuliidae): a comparison of three estimation methods. *Canadian Journal Fisheries and Aquatic Sciences* **43**: 1235. [15.2]

Naiman RJ, Lonzarich DG, Beechie TJ, Ralph SC. (1992) General principles of stream classification and the assessment of conservation potential in rivers. In: Boon PJ, Calow P, Petts GE (eds) *River Conservation and Management*. John Wiley, Chichester. [15.3]

Nestler JM, Milhous RT, Layzer JB. (1989) Instream habital modeling techniques. In: Gore JA, Petts GE (eds) *Alternatives in Regulated River Management*, pp. 295–315. CRC Press, Boca Raton, Florida. [15.3]

Newbury RW. (1984) Hydrologic determinants of aquatic insect habitats. In: Resh VH, Rosenberg DM (eds) *The Ecology of Aquatic Insects*, pp. 323–57. Praeger, New York. [15.2]

Nowell ARM, Jumars PA. (1984) Flow environments of aquatic benthos. *Annual Review of Ecology and Systematics* **15**: 303–28. [15.2]

Obrdlik P, Castella E, Foeckler F, Petts GE. (1992) Groundwater invertebrates of European alluvial rivers. *Regulated Rivers, Special Issue* **7**: 1. [15.2]

O'Brien JS. (1987) A case study of minimum streamflow for fishery habitat in the Yampa river. In: Thorne CR, Bathurst JC, Hey RD (eds) *Sediment Transport in Gravel-bed Rivers*, pp. 921–46. John Wiley, Chichester. [15.3]

Oglesby RT, Carlson CA, McCann JA. (1972) *River Ecology and Man*. Academic Press, New York. [15.2]

O'Keefe JH, Davies BR. (1991) Conservation and management of the rivers of Kruger National Park: suggested methods for calculating instream flow needs. *Aquatic Conservation* **1**: 55–71. [15.4]

Orth DJ. (1987) Ecological considerations in the development and application of instream flow-habitat models. *Regulated Rivers* **1**: 171–82. [15.4]

Orth DJ, Leonard PM. (1990) Comparison of discharge methods and habitat optimization for recommending instream flows to protect fish habitat. *Regulated Rivers* **5**: 129–38. [15.3]

Parker G. (1979) Hydraulic geometry of active gravel-bed rivers. *Journal of the Hydraulics Division ASCE* **105**: 1185. [15.3]

Petts GE. (1984) *Impounded Rivers*. John Wiley, Chichester. [15.2]

Petts GE. (1992) *The River Glen: a Catchment Assessment*. Report for the National Rivers Authority, Anglian Region, Peterborough. [15.2, 15.3]

Petts GE. (1993) *The River Babingley: Instream Flow Needs*. Draft Report for the National Rivers Authority, Anglian Region, Peterborough. [15.3]

Petts GE, Wood R. (eds) (1988) River regulation in the United Kingdom. *Regulated Rivers, Special Issue* **2**: 3. [15.2]

Petts GE, Moller H, Roux AL. (eds) (1989) *Historical Changes of Large Alluvial Rivers: Western Europe*. John Wiley, Chichester. [15.2]

Petts GE, Maddock I, Bickerton M, Ferguson A. (in press) Linking hydrology and ecology. In: Ferguson A, Harper D (eds) *The Scientific Basis for River Management*. John Wiley, Chichester. [15.3, 15.4]

Phillipson GN. (1954) The effect of water flow and oxygen concentration on six species of caddis fly (Trichoptera) larvae. *Proceedings of the Zoological Society of London* **124**: 547–64. [15.2]

Raleigh RF, Miller WJ, Nelson PC. (1986) *Habitat Suitability Index Models and Instream Flow Suitability Curves: Chinook Salmon*. Biological Report 82 (10.122), US Fish and Wildlife Service, Fort Collins, Colorado. [15.2]

Reiser DW, Ramey MP, Wreshe TA. (1989a) Flushing flows. In: Gore JA, Petts GE (eds) *Alternatives in Regulated River Management*, pp. 91–138. CRC Press, Boca Raton, Florida. [15.2, 15.3]

Reiser DW, Ramey MP, Beck S, Lambert TR, Geary RE. (1989b) Flushing flow recommendations for maintenance of salmonid spawning gravels in a steep, regulated stream. *Regulated Rivers* **3**: 267–76.

Resh VH. (1985) Periodical citations in aquatic entomology and freshwater benthic biology. *Freshwater Biology* **15**: 757–66. [15.2]

Salo JS. (1990) External processes influencing the origin and maintenance of inland water–land ecotones. In: Naiman RJ, Decamps H (eds) *The Ecology and Management of Aquatic–Terrestrial Ecotones*, pp. 37–64. Unesco-MAB, Parthenon, Carnforth, UK. [15.2]

Scarnecchia DL, Bergenson EP. (1987) Trout production and standing crop in Colorado's small streams, as related to environmental features. *North American Journal of Fisheries Management* **7**: 315–30. [15.3]

Scott D, Shirvell CS. (1988) A critique of the Instream Flow Incremental Methodology with observations on flow determination in New Zealand. In: Craig JF, Kemper JB (eds) *Regulated Streams: Advances in Ecology*. Plenum, New York. [15.3]

Sedell JR, Richey JE, Swanson FJ. (1989) The river continuum concept: a basis for the expected ecosystem behaviour of very large rivers? In: Dodge DP (ed) *Canadian Special Publication of Fisheries and Aquatic Sciences* **106**: 49–55. [15.2]

Shepherd BG, Hartman GF, Wilson WJ. (1986) Relationships between stream and intragravel temperatures on coastal drainages, and some implications for fisheries workers. *Canadian Journal of Fisheries and Aquatic Sciences* **43**: 1818–22. [15.2]

Shirvell CS. (1989) Ability of PHABSIM to predict Chinook salmon spawning habitat. *Regulated Rivers* **3**: 277–90. [15.3]

Smith SD, Wellington AB, Nachlinger JL, Fox CA. (1991) Functional responses of riparian vegetation to streamflow diversion in the eastern Sierra Nevada. *Ecological Applications* **1**: 89–97. [15.2]

Stanford JA, Ward JV. (1988) The hyporheic habitat of river systems. *Nature* **335**: 64–6. [15.2]

Statzner B, Gore JA, Resh VH. (1988) Hydraulic stream ecology: observed patterns and potential applications. *Journal of the North American Benthological Society* 7(4): 307–60. [15.2]

Sterba O, Uvira V, Mathur P, Rulik M. (1992) Variations of the hyporheic zone through a riffle in the R. Morava, Czechoslovakia. *Regulated Rivers* **7**: 31–44. [15.2]

Stromberg JC. (1993) Instream flow models for mixed deciduous riparian vegetation within a semi-arid region. *Regulated Rivers* **8**: 225–36. [15.2]

Tennant DL. (1976) Instream flow requirements for fish, wildlife, recreation and related environmental resources. *Fisheries* **1**(4): 6–10. [15.3]

Toth LA, Obeysekera TB, Perkins WA, Loftin MK. (1993) Flow regulation and restoration of Florida's Kissimmee River. *Regulated Rivers* **8**: 155–66. [15.4]

Triska FJ, Duff JH, Avanzino RJ. (1993) Patterns of hydrological exchange and nutrient transformation in the hyporheic zone of a gravel-bottom stream: examining terrestrial–aquatic linkages. *Freshwater Biology* **29**: 259–74. [15.2]

Vannote RL, Minshall GW, Cummins KW, Sedell JR, Cushing CE. (1980) The river continuum concept. *Canadian Journal of Fisheries and Aquatic Sciences* **37**: 130–7. [15.2]

White DS. (1990) Biological relationships to convective flow patterns within stream beds. *Hydrobiologia* **196**: 149–58. [15.2]

Wickett WP. (1954) The oxygen supply to salmon eggs in spawning beds. *Journal of the Fisheries Research Board of Canada* **11**: 933–53. [15.2]

Wolff SW, Wesche TA, Hubert WA. (1989) Stream channel and habitat changes due to flow augmentation. *Regulated Rivers* **4**: 225–33. [15.4]

16: Control Rules for Regulating Reservoirs

J. A. MAWDSLEY

16.1 INTRODUCTION

This chapter is written using terminology and examples from the UK water industry, but the principles should be applicable to most countries. The term 'flow regulation' can be used in several different ways: in this chapter it refers to reservoir releases which are made to rivers to support downstream abstractions or to maintain the flow above a defined value some distance downstream of the reservoir. Reservoirs which release water to the river to be abstracted downstream are referred to as regulating reservoirs. Most reservoirs in the UK built before 1960 were built as direct supply reservoirs: these are reservoirs where the water is abstracted from the reservoir and supplied directly to the treatment works (Fig. 16.1). Regulation of a river can also be accomplished by abstractions from aquifers, but most of the subsequent comments will refer to regulating reservoirs.

When a river is regulated by release from an upstream reservoir, there are likely to be three types of constraint imposed on the flows within the river system: compensation, minimum maintained, and prescribed flows.

For a new reservoir in the UK, a compensation flow release is specified to maintain the health of the river immediately downstream of the dam or to safeguard existing users of the downstream river. The compensation flow is the minimum acceptable flow in the river below the dam, and releases have to be made to ensure that it must at least be equalled at all times. Gustard *et al* (1987) reviewed the compensation flows for over 500 of the reservoirs in the UK, some set as early as the 1820s. During this time the main considerations in setting the compensation flow have changed from primarily flow requirements to run mills, to the requirements of flora and fauna, often migratory fish. These compensation flows are generally fixed values unchanged throughout the year, but some show seasonal variation and some allow for releases to be made to assist the migration of game fish – releases called freshets. The values that have been set for compensation flows show large variation: from a few per cent to 80% of the average daily flow, with a national mean of 16%.

For regulating reservoirs the opportunity has also been taken to help the river farther downstream by specifying a minimum maintained flow (MMF) below which the river discharge is not allowed to fall. Releases from the reservoir are used to ensure that this minimum is always exceeded at some specified point. The MMF value should be determined taking into account fisheries, amenity and ecological interests, and sometimes more specialist activities such as navigation. In many cases the MMF is also a fixed value through the year, but for migratory fish it is probably better if it varies. A fixed value has frequently been chosen because of a lack of knowledge of the needs of the fish and other flora and fauna. This lack of knowledge has resulted in rather arbitrary values having been chosen, such as some percentage of the average daily flow (ADF) (typically 15%), or of the flow which is exceeded 95% of the time (Q_{95}). It is likely that these MMF values will be reconsidered as knowledge of the needs of fish and other fauna and flora increases.

At points on a river where water is abstracted there will be abstraction conditions attached to a

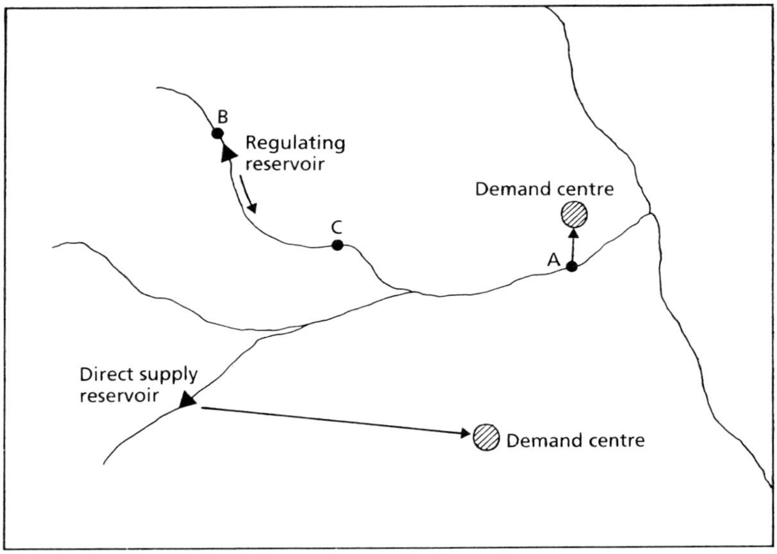

Fig. 16.1 Definition drawing of regulating direct supply reservoirs.

licence. These will generally specify a flow below which no more water can be taken from the river – this is often referred to as either a 'prescribed flow' or a 'hands-off flow'. If the river is not regulated at this point, the flow may fall naturally below this value. For many licences, if the flow is above this prescribed flow then any amount up to the licensed quantity can be taken as long as it does not force the flow below the prescribed value; but for some more complex licence conditions, only a percentage of the excess can be taken. For others, where there is regulation the abstractor can take out the water released into the river when it arrives at the abstraction point, regardless of the natural river flow. This is often referred to as a 'put and take' abstraction condition.

The release patterns produced by regulation can often be quite artificial. The river Tees in the Northeast of England is regulated mainly from Cow Green reservoir. The abstractions from the lower reaches are made only at certain times of the day. Consequently releases are made so that the water is available at these times. This produces the flow pattern shown in Fig. 16.2 at a gauging station about halfway down the river. Similar effects can be produced by releases made for hydropower generation.

The environmental effects of flow regulation are discussed in Chapter 3. Two particular problems have arisen in the UK. First, the fluctuating flow patterns can cause some difficulties for the river, particularly in the first few kilometres downstream of the dam; for example, in gravel bed rivers causing the removal of fine gravels. The reservoir can make high flows quite rare so that the larger bed material does not often get disturbed causing the interstitial speces to get clogged, causing problems for the spawning and development of salmonid eggs and for benthic fauna (Gustard *et al* 1987). Secondly, a problem observed by Haile *et al* (1989) is the change in the temperature regime of the river caused by the reservoir damping out the natural seasonal fluctuations; moreover, the level from which the water is abstracted will also strongly influence the river water temperature. The onset of feeding for fish is controlled by temperature and hence unnatural temperatures can either delay the start of feeding, or the reverse, and if the natural food is not available it can cause fish mortality.

There are clearly benefits as well as problems associated with regulating reservoirs. Jackson and Bailey (1979) perceived the advantages of regulating reservoirs to be:

1 greater yield from the same storage;
2 more re-use of water;
3 reduced capital costs as the river is used to carry the water;

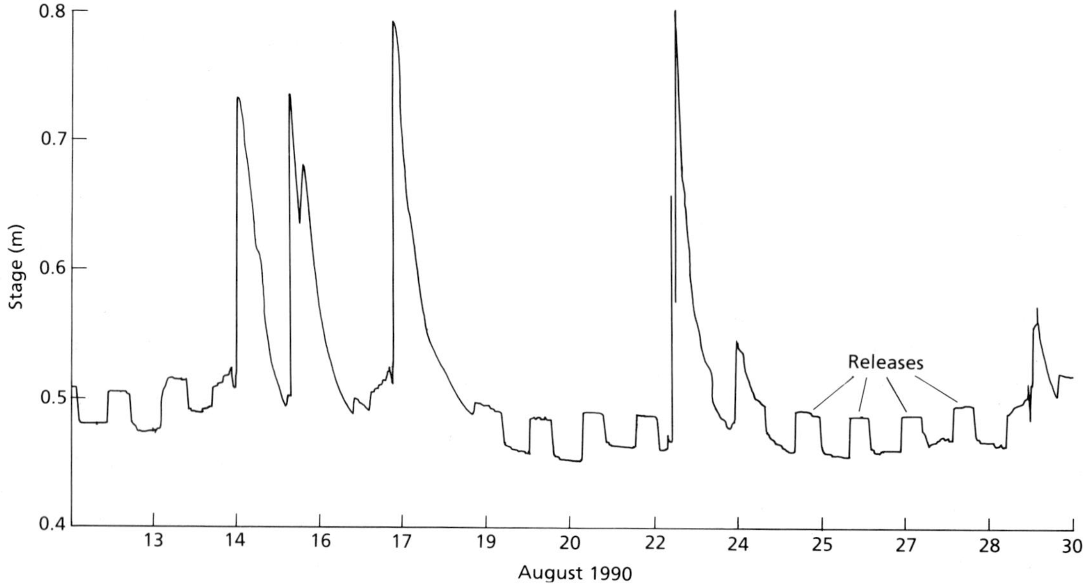

Fig. 16.2 Hydrograph showing effect of regulation releases on flow downstream of the reservoir (point C on Fig. 16.1).

4 benefits to other users as low flows are increased and high flows reduced;
5 greater operational flexibility.

The same authors cited the following disadvantages:
1 more complex management;
2 losses of water either from the river or due to unnecessary releases;
3 changed flow regime which may cause river bank instabilities, etc.;
4 increased running costs because pumping from the river is necessary;
5 water quality variations which can cause treatment problems;
6 ecological changes to the river due to changed flow regime.

The water industry in the UK has considered the advantages to outweigh the disadvantages so that most of the reservoirs built in the last 30 years in the UK have been regulating reservoirs. The increased yield has been an important factor in this balance, but it is a myth. It is based on the fact that releases from regulating reservoirs in excess of the compensation flow are only made to support downstream abstractions when there is insufficient water naturally in the river. Thus there will be periods in the year when only releases for compensation flows are made; this does not occur with direct supply reservoirs because the demand has to be met in full from the reservoir. This appears to give a higher yield from regulating reservoirs, but the extra yield occurs as a result of the conjunctive use of river and reservoir and not because of any intrinsic benefits of regulating reservoirs. If the direct supply reservoir was also used conjunctively with the river, the yield of this system would be greater than the regulating reservoir, because regulating reservoir releases will often take days to reach the abstraction point and during this time rain may fall which makes the release unnecessary. These 'losses' are unavoidables as perfect weather forecasts are, and will remain, unattainable!

But it is not the purpose of this chapter to reassess whether regulating reservoirs should continue to be built but rather to consider the often complex management of existing regulating reservoirs. For the remainder of this chapter these management issues will be considered, with some of the issues related to regulation using groundwater as the source.

16.2 MULTIPLE USES OF WATER

In the UK, regulating reservoirs have been built mainly by water supply undertakers for public water supply, but even at the time of construction other uses have often been acknowledged: primarily recreation and amenity, hydro-electricity, regulating the river for fisheries, pollution control, flood control and, mainly in countries other than the UK, irrigation and navigation. Other private abstractors on the river will also benefit from the regulation: these will include those using the water for their industrial process, cooling water for power stations, farmers for irrigation, fish farmers and gravel washings.

The water management needs for many of these uses are conflicting – under drought conditions there may be insufficient water to provide both for irrigation and cooling water and for reservoir recreation; for storage needs the reservoir would be kept full, yet for flood control the reservoir would be drawn down during the flood season. It is the resolution of these conflicting requirements which is one of the main causes of the increased complexity.

There are a number of ways to resolve these conflicts, of which the most commonly used is to give priorities to the uses. Because reservoirs in the UK have mostly been built for public water supply, this use has been given the highest priority. In addition the water industry has taken the view that the needs of the domestic consumer are paramount because of the link between clean water and health and hygiene. This link is extremely important, as consideration of health in developing countries makes abundantly clear, but it can be misleading: typical domestic consumption in the UK is about 140 litres per head per day but only a minority of this water is required for essential health and hygiene, the rest is consumed to improve the standard of living – automatic washing machines, dish washers, etc. The extra consumption has no more intrinsic merit than improved amenity or increased flows for fishing – they all contribute to an improved quality of life. Therefore, the primacy given to domestic consumption should be restricted to the proportion used for health and hygiene.

In the UK, licences are granted to those with a legitimate need for the water and the volume specified in the licence has to be available in the river, if at all possible, with generally no preference given to the different types of abstractors. Therefore, releases have to be made to the river to support these abstractions at times of low flows. Those uses which do not require abstraction licences, such as flood control and amenity or recreation, do not have the same legal protection and are put at a disadvantage in these comparisons. In some countries, e.g. the USA, recreational interests have on occasions supplied part of the money to build the reservoir and therefore have a claim on the water, with priorities presumably decided at the planning stage of the reservoir.

A second approach to resolving the conflicts between uses is to base the decision on the monetary benefits to be gained from the various uses. This idea was pioneered by Maas *et al* (1962) and has been discussed widely in the literature since (e.g. Loucks *et al* 1981). However, there are difficulties with this approach because for some of the water uses a clear market value can be assigned, e.g. water for irrigation, whilst for others the benefits cannot be easily quantified, e.g. amenity or ecological benefits to a river, or the benefits have to be inferred because no market exists, e.g. recreation or navigation. Economists have tried to overcome some of these problems, and research continues to be carried out in this area.

If benefits are quantifiable, an objective function, such as to maximize the net benefits, can be defined and an optimization procedure used to determine the sharing of the water to yield the maximum benefits. This procedure has been widely described in the research literature but rarely applied in practice. Other researchers have reservations about this process, mainly because they consider the intangible benefits in particular cannot be adequately quantified. This has led to an alternative approach – multi-objective analysis (see, e.g., Major 1972; Keeney & Raiffa 1976).

With multi-objective analysis, as the name suggests, several objective functions are defined. The first could be an economic one, similar to the one above, but a second might be to maximize the water quality in the river, and a third to maximize the amenity. The difficulty comes when one must trade an improvement in one objective function with a consequent deterioration in another. How to make this trade is the core of the multi-objective procedures. Essentially weights are assigned to the objective functions, based somehow on 'decision-makers' views, and then optimization techniques are used to find 'good' solutions. These methods are still at a fairly early research stage and it may be some time before they will be extensively applied in practice. A recent paper demonstrating their application to irrigation management was published by El-Awad *et al* (1991).

16.3 CONTROL RULE CURVES FOR PRIMARY USERS

The operation of regulating reservoirs is complex because the reservoir is often part of a system with other sources which can be operated conjunctively, because, like all reservoirs, they have to be operated in droughts with no knowledge of when the drought will end, and because of the multiple users. All of these aspects will be covered, but I will start with the second.

In the later stages of a drought, the storage in the reservoir is low, the demand will tend to increase, if measures such as hosepipe bans have not been imposed, and flexibility for the water resource manager is negligible. It is in the spring before a drought has developed that the manager has choices and, therefore, it is important that the right choices are made at that time. To help the decision-making process, reservoir control rule curves have been developed. These show the volume of storage that should be exceeded at all times if severe shortages are to be avoided towards the end of a drought (Figs 16.3 and 16.4).

As the severity of droughts appears to be a random variable, it is not possible to design these control rules for the worst drought, but rather a chosen historic one or one with a known probability of being exceeded in severity, often taken as 1% or 2% risk. If the latter course of action is taken, then a reliability of maintaining supply can be established, at least approximately. Three measures of reliability can be used (normally expressed as percentages): annual reliability (R_a), which measures how many years on average there will be no restrictions; a time based reliability (R_m), which measures how long there will be no restrictions; and volume reliability (R_v), which measures what proportion of the volume required is able to be delivered (Klemes 1981). Further, the restrictions can be classified into their types and reliabilities attached to these; for example, in the UK the levels of service guidelines (OFWATS's DG1 guidelines) state that hosepipe bans should not be imposed more than once every 10 years, on average ($R_a = 90\%$), or drought orders (requiring either rota cuts or standpipes) not more than once every 100 years ($R_a = 99\%$).

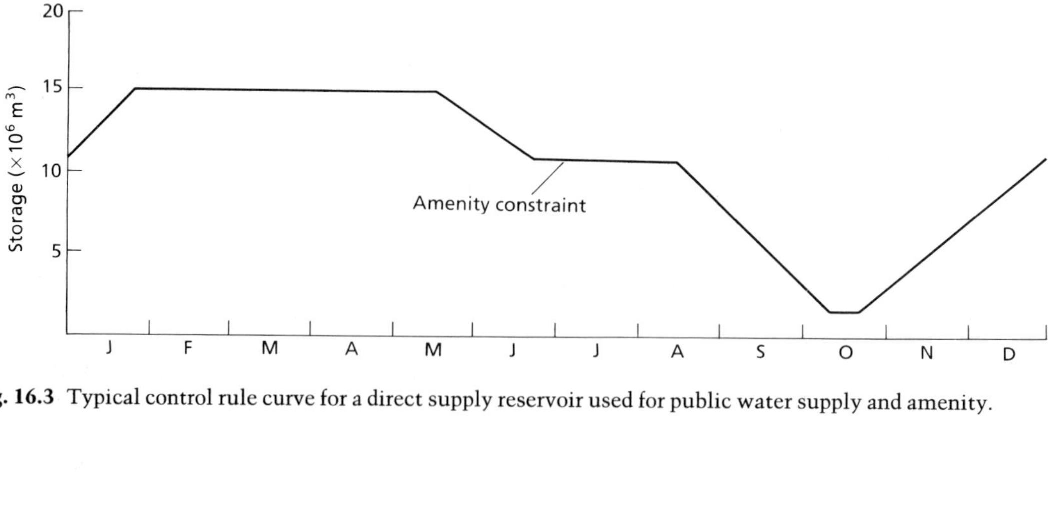

Fig. 16.3 Typical control rule curve for a direct supply reservoir used for public water supply and amenity.

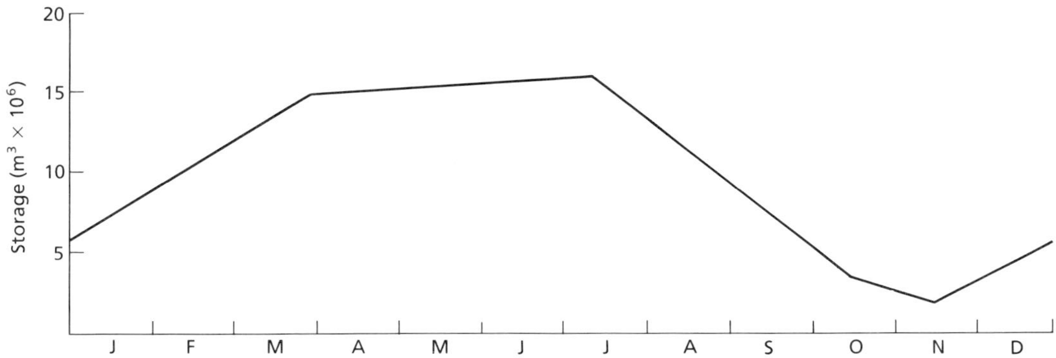

Fig. 16.4 Typical control rule curve for a regulating reservoir used for public water supply.

Control curve using historic droughts

A historic control rule curve shows the volume of storage required to just supply the full demand throughout the remainder of the worst drought on record, i.e. at the end of the drought the reservoir would be just empty, or to be precise it would have reached the dead-water storage level. For example, in Fig. 16.3 in May, the control curve shows that $15 \times 10^6\,m^3$ of water if required to just avoid emptying if the demand was maintained at its full value. In practice measures would be taken to reduce the demand before the reservoir emptied, but this can be interpreted as a safety factor on the analysis. For direct supply reservoirs the demand is assumed constant, but for regulating reservoirs it is dictated by the flows at the abstraction point downstream. This is an extra complication in control rule curve derivation for regulating reservoirs, which will be discussed more fully below.

Referring back to the June level in Fig. 16.3, the time when the reservoir would just empty is not shown, but it should be as it is important. The duration from June to the time of first emptying is referred to as the 'critical duration' for that month. It will extend to the end of the dry season, which in the UK is October or November, but not necessarily in the same year: the drought could extend through the winter to the end of the subsequent summer. At the outset of the analysis, this critical duration is not known and has to be determined.

If the reservoir has small storage relative to the mean annual flows then the critical duration is likely to be less than 12 months and hence will end in October or November of that year, but if the ratio is large it is likely to extend for over 12 months to end in the October or November one or two years hence. Those months with the critical duration in the same year are called 'within year critical'; the rest are 'over year critical'.

As reservoirs are built specifically to guarantee the supply of water in dry years, in a drought the inflow to the reservoirs will be less than that released. The storage shown in June (Fig. 16.3) is equal to the amount that the releases exceeded the inflows from June to the end of the worst drought on record. The control curve simply describes this value for all months.

A systematic way of determining the value for regulating reservoirs is to use 'net withdrawals from storage' (NWS) (sometimes the negative of this is used, i.e. the net additions to storage: Pearson & Walsh 1982). As described earlier, when operating a regulatory reservoir, releases are made to support abstraction or the MMF downstream. These are only made in excess of the compensation release if the flow downstream is insufficient to allow the full abstraction or to meet the MMF. To calculate the NWS for a historic record, the daily release values that would have been required have to be determined for every day in the flow record. Daily flows must be used to avoid a serious underestimate of the volume of releases required. The releases from the reservoir are found by comparing the measured flow at the control point to that required, paying attention to abstraction conditions and travel times; if there is insufficient flow, a quantity is released sufficient to increase the flow to the required quantity, having first made sure that losses in transmission are allowed for. Knowing the reservoir releases and the reservoir inflows, the NWS values can be calculated as the difference between these two terms.

Generally the abstraction point is well downstream of the regulating reservoir, or the point of regulation if the source of water is groundwater. Consequently the travel time for the water to reach this abstraction point can be quite long, perhaps several days. Travel times depend on the flow in the river and the channel roughness and thus they change with the seasons, which creates one of the problems of managing river regulation. On the River Wear in the Northeast of England, under low-flow conditions it takes some 3 days for the water to travel from the regulation point to the abstraction point; on the River Dee in North Wales, it takes approximately 2 days to reach the major abstractors.

A measured daily flow record is preferable both at the control point and upstream of the reservoir (see points A and B on Fig. 16.1), but if these records are not available some means of generating them can be used (see standard hydrology texts). Having calculated the daily NWS values, they should be summed to get the 5-day, or weekly or monthly totals (some of which will be negative). An example of the monthly data is shown in Table 16.1.

To calculate the control rule, the following procedure has to be repeated for each month, but as an example just June will be described. The objective is to determine the highest cumulative total of the NWS starting with June and ending some time later – probably in October or November of the same year, but possibly in the same months in subsequent years. Note that the control rule curve gives the storage required at the beginning of the month in question.

Consider as an example the year 1976 in Table 16.1: the values for June, July and August produces a cumulative total of 877 + 1645 + 1641 = 4163 Ml. In September the large negative value shows the reservoir was refilling; hence all other drought durations in 1976 starting in June yield lower totals. In 1972 there was surplus water in June (a negative NWS) but positive NWS values for the next 4 months. The cumulative total for June through October in this year is 4582, which is worse than 1976, but for all other years the total is smaller. Therefore as 1972 is the worst case for June, this value will define the storage required for the control curve, with a critical duration of 5 months.

To check durations of 17 months (i.e. June to the October of the following year), the cumulative total for the full 17 months has to be calculated. This can be time consuming but fortunately it only has to be calculated if the cumulative total at the beginning of June the following year is positive; if it is negative a worse case would obviously be the one considered above of a drought starting in the June of the following year. As it appears that in all years in this record the cumulative

Table 16.1 Net withdrawals from storage (NWS) (Ml/month)

Year	J	F	M	A	M	J	J	A	S	O	N	D
1970	−8164	−4800	−2364	−3264	−427	1259	1241	−373	673	−55	−5591	−2804
1971	−7036	−1791	−7850	−2886	318	−850	677	−7523	941	595	−2086	−282
1972	−10023	−8100	−4900	−1345	−1145	−373	923	1082	1450	1500	−1827	−3395
1973	−1564	−659	477	−2495	−1832	1141	−1386	−805	445	−977	77	−2886
1974	−9963	−5491	−3536	195	868	1436	223	1105	−91	−2982	−6350	−7027
1975	−6691	−2245	−1350	−795	−2918	972	1327	459	−759	227	191	−500
1976	−3745	−2450	−2104	−1795	−636	877	1645	1641	−6077	−15623	−1959	−3445
1977	−12618	−9482	−17595	−4623	−2218	−1623	1382	650	1241	−391	−5836	−8668
1978	−6359	−11063	−4600	−1650	−2018	1195	1036	−1627	−382	355	−1295	−13295
1979	−5945	−3968	−86	−12223	−3300	−359	1327	955	1223	−5068	−7323	−9768
1980	−4177	−7804	−6932	−427	1286	−5682	−1105	−277	409	−3436	−5827	−5009

total will be negative at the beginning of the next June, the critical duration in this case is less than 12 months.

If this procedure was repeated for all the other starting months, the control rule curve would be similar to the shape shown in Fig. 16.4. A distinctive feature of such curves is the smaller storage requirement in March than in July. This contrasts with direct supply reservoir control curves and occurs because the demands for releases in March will be much less than in July due to naturally higher flows in the river at the control point. Coupled with this are the higher inflows to the reservoir at this time, producing net additions to storage rather than net withdrawals. This tends to make regulating reservoirs much more likely to be 'one season critical'. With higher demands, the maximum storage requirement would occur earlier in the year.

These control curves tell the operator that if the reservoir is below the control line, if the worst historic drought is repeated, and if the demand is not modified, the reservoir would run out of water some months in the future. As it is possible that a worse drought than the historic one may be developing or the critical duration of a future drought may be different to the historic one, being above the line does not guarantee there will be no future problems. Lambert (1988) has suggested that in operation the manager will err on the side of pessimism and assume that the drought may go on longer than reflected in the control curve analysis: indeed a latest possible end date of the drought will be chosen. If this type of operation is adopted in practice then it could, and probably should, be reflected in the analysis. Because droughts do not generally last an exact multiple of months, if monthly data have been used a correction should be made by assuming the reservoir empties 2 weeks beyond the end of the month. This correction may also satisfy the correction for the end date referred to earlier.

If the source being used to regulate the river is groundwater, control rule curves are not normally defined. This is largely based on the assumption that the storage in the aquifer is very large so that running out of water will not be a problem. However, although there may be a lot of water stored in the aquifer, all of it may not be usable either because the borehole is not deep enough, or because with significant drawdown other undesired results are observed which constrain further abstraction. Examples of these undesired results are drying up of hydraulically connected rivers or lakes, or consequent saline intrusion. These aspects need to be considered and incorporated into control rules. If a minimum acceptable level for the water table was defined, a usable storage could be calculated and treated similar to the reservoir storage (see, e.g., Kahn & Mawdsley 1988); but much work needs to be carried out in this area before it can be applied with confidence.

Control rule with associated reliability

As the previous analysis has been carried out using the worst historic droughts, the reliability of the supply cannot be clearly established, but one can say that if a record length of 25 years was used then the annual reliability is probably in excess of 96% (i.e. one failure every 25 years or so). To improve on this estimate is difficult, as it requires a lengthy statistical analysis of the NWS values.

As an example of how this is done, consider again the months of June. Using the critical duration for June of 5 months found from the historic analysis, the cumulative NWS values for this duration and starting in June have to be calculated for each year; this will produce n values (one for each of the n years). These should be ranked in ascending order, the probability of the value not being exceeded calculated using a suitable plotting position formula (Cunane 1978), and the NWS values and probabilities plotted on probability paper to allow the determination of the NWS which would be exceeded, say, only 2% of the time. This value would presumably be lower than the historic minimum and could be used in place of it. If this was repeated for all 12 months, a new control curve could be derived which would have associated with it a probability of being exceeded, in this case 98%; it is generally assumed that this probability is equal to the annual reliability (R_a). There are one or two difficulties with this procedure in addition to the large amount of work involved: first, it assumes the critical duration for the statistical flows is the same as for the historic ones – this may not be true; second, with similar types of statistical analysis of droughts, a log-normal probability distribution has often been assumed, but some of the NWS will be negative and therefore cannot be logged. There are ways around this – fitting a line to only the positive values, or using a different distribution which does not require logged values, e.g. the extreme value for least values, or see Walsh (1971). However, most practitioners have been put off by the statistical problems associated with such analysis and have produced only the historic control curve.

An alternative procedure for assessing the reliability is to use a very long flow record (>500 years) and simulate the system for this period, deducing empirically the reliabilities from the output. In this case all three reliabilities can be calculated (R_a, R_m, R_v) but of course such very long records do not exist and have to be generated by either statistical methods or by using a rainfall-runoff model and a long rainfall series, which in turn has to be statistically generated. These techniques are well known but have serious drawbacks, particularly for daily flows as required here (Lawrence & Kottegoda 1977). As a result, they are rarely applied in practice at the present time, although much more could be done using the rainfall-runoff approach, particularly following the work of Cowpertwait *et al* (1991) on the statistical generation of rainfall series.

16.4 INCORPORATION OF RULE CURVES FOR OTHER USES

The analysis in the previous section is for a single use – water supply. The same procedure could be carried out for another use of a similar type, for example water for cooling or irrigation. The resulting storage requirements can then be added to those for public water supply. For different types of use, such as flood control which requires space to be left in the reservoir, the storage requirements must be deducted from the capacity. The resulting storage must not be exceeded at any time. If the control curves which must be exceeded do not cross those which must not be exceeded, then the reservoir can be operated to satisfy both uses; but if they do cross, priorities have to be set on the uses, as described earlier. If the total water requirement is less than the capacity, for example in Fig. 16.4 during the summer months, then the 'excess' water can be used for other purposes, such as hydropower, without jeopardizing the reliability of the water supply unduly; if retained in the reservoir, it reduces the risk of future shortages still further.

Recreational and amenity uses at reservoirs prefer the water level to be kept as close as possible to the normal maximum so that the surface area for sailing and the shoreline for fishing are at a maximum. In addition, unsightly mudflats are not revealed for the casual visitor (Hill 1988, 1991). Water supply authorities were reluctant to allow the public to use reservoirs for recreation for many years, using as explanation the pollution risk and inconvenience to them. In the UK they were forced to develop the recreational potential by the government, with the result that now many reservoirs are used intensively for recreation. However, in most cases recreation is treated as a low priority use, with no adjustment to the operation of reservoirs being carried out for their benefit. There are just a few reservoirs in the country which explicitly acknowledge recreation in their control rules (e.g. Selset Reservoir in Northeast England and Bala Lake in North Wales (Hill & Mawdsley 1987)). In a number of other countries more is done for recreational interests, and it is likely, given the pressures for recreational opportunities in the UK, that more will have to be done in the UK in the future.

Releases for helping fisheries, called freshets, fall in much the same category as recreational interests. There are a number of examples of allowances being made for a small number of releases to be made for fisheries purposes, generally to help the movement of migratory fish. Migratory fish like to move upstream when the flow is reasonably high following a small spate. The spate can be artificially created by releases, or a natural event can be supported by additional releases. A volume of water will have been identified for such purposes and it is retained in the reservoir for such uses: it is often referred to as a bank of water. Examples of reservoirs with such banks are Brenig Reservoir in North Wales and

Kielder Water in Northeast England. In the latter, water is also used for hydropower releases so that releases for fisheries may cause a loss in revenue. Therefore, fisheries are given a budget to buy water, up to a limit, for their purposes.

In North Wales a good white water canoe reach lies downstream of Llyn Celyn, where releases are made to ensure adequate flows for competitions. The water is caught in Bala Lake downstream so that at this location conflicts with other uses is generally small. Of course the fortunate presence of a second reservoir downstream is not normally available.

16.5 CONTROL RULES FOR INTERLINKED SOURCES WITH MULTIPLE USES

As mentioned earlier, most large reservoirs in the UK and abroad do not supply a demand zone on their own, but rather are part of a network of sources which together provide water for a number of demand zones. This makes water supply management easier in so far as back-up sources are available in the case of an emergency, and better in that generally the conjunctive use of sources provides a yield in total which is greater than the sum of the parts, but more difficult in that the control is more complex.

Where there are interlinked sources they will generally be of several types – direct abstraction from an unregulated river, abstractions from a direct supply reservoir, abstractions from groundwater, and of course the regulated source. It is because of the different hydrological characteristics of these different types of source that the synergism is present.

With a number of interlinked sources, there is generally a choice of sources for the required water, with constraints due to physical limitations such as treatment works capacity, or pipeline and pumping capacity. However, with this increased flexibility comes the increased risk of water being available in the system to meet all the demands but distributed in such a way that it cannot reach all the demand zones at the rates required. To avoid this, care must be taken in defining the operating policy, particularly early in the season when most choice is available. The operating policy is almost always determined by behavioural analysis (also called simulation).

In behavioural analysis a computer model is developed to describe the sources, demand zones, pipe linkages, pumping stations, treatment works, etc. for the whole system (see Fig. 16.5 for an example), and a water balance is maintained throughout the system. If the inflows to all the sources are known for a period of time and an operating policy for all the sources is defined, the performance of the system over the period of record can be assessed against predefined objectives. The output from the simulation analysis should include the water supplied to each of the demand zones, the storages in the sources, the costs of supplying the water, and of course any failures to meet the required demands.

If the primary objective is to meet the demands with a specified reliability at minimum cost, the policy can be changed until this objective is achieved. But not all of the objectives will be able to be expressed in economic terms. For example, for recreation the assessment might focus on the number of days the reservoir was drawn down below some chosen threshold level, or for ecological concerns it might be the number of days the flow at some stretch of river of particular interest was below another threshold value. The problem is then how to trade off these different objectives: for example, to enhance recreation by constraining the storage to exceed an assigned water level for the months of July and August might cost £100 000 over a 15-year period. Is this improvement in recreation worth so much money? Should the recreational users be asked to pay this sum?

In practice, the water resource manager is usually satisfied with finding what appears to be a 'good' solution using the historic analysis, but it must be recognized that it may not be as good for the future inflows. Indeed, with a more severe drought or an inflow sequence of different character, the policy may cause some serious problems. This can be explored to some extent by perturbing the inflow sequence in some way – perhaps making the drought part of the record 20% more severe. With enough trials, the manager should gain confidence in the chosen policy.

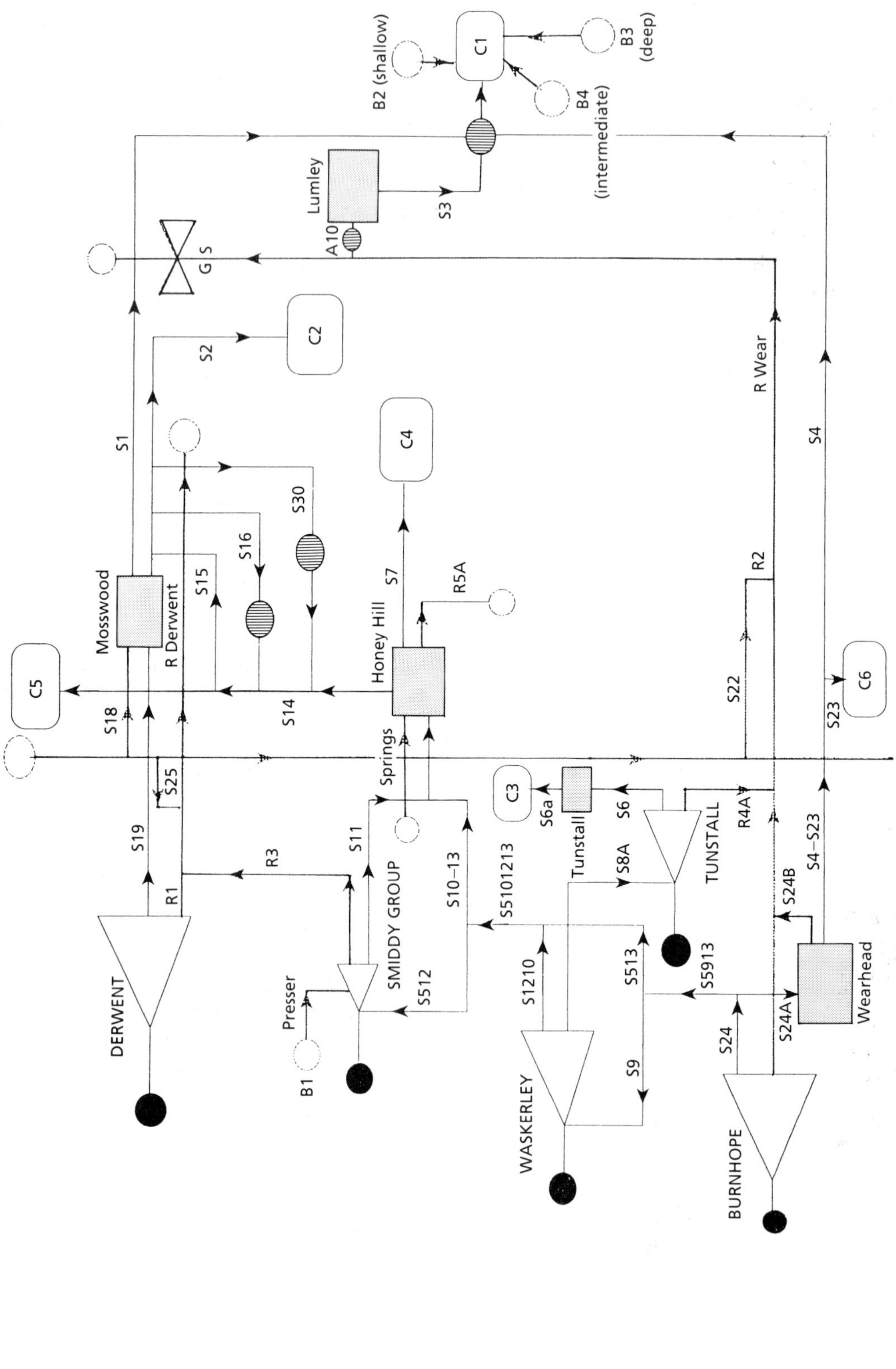

Fig. 16.5 Diagram of mid-Durham water resource system showing sources, demand zones, pipe linkages, pumping stations and treatment works.

Much research has been carried out into the use of optimization techniques with water resource systems (well described by Loucks *et al* 1981), but there have been few practical applications. All models must simplify the real system to some extent, and the techniques available also need to make further simplifications for them to be applied. Consequently, the system optimized is not quite the same as the real system, with the result that the best solution for the optimized system may not be the best solution for the real system. It is probably this aspect which has restricted the practical application of these techniques. The simpler the system the closer the match between the modelled and real system, but the easier it is to find the optimum solution by other means, such as trial and error using behavioural analysis.

An example of the pragmatic approach to evaluating trade-offs between various uses has been described recently by Glenny and Kinniburgh (1991). They describe how the River Thames is regulated for both abstraction and to maintain water quality in the tidal part of the river; in addition a constraint on the head of water is required at a point on the river for navigation. No economic trade-off has been conducted, but an agreement has been negotiated with the major abstractors and sewage works operators which produces acceptable water quality and impacts abstractors and dischargers in what they, at least, perceive to be a fair way. Pragmatic solutions of this type are the normal way to solve the complex problems.

16.6 CONCLUSIONS

Managing reservoirs is most difficult in droughts: this is when the conflicts between users are most acute and when the very real possibility exists of being unable to supply water to certain communities. Droughts develop slowly and can last many months or years. It is important to plan ahead to ensure the correct decisions are made in the spring or winter when water is more readily available: by the time the drought is clearly evident, the freedom of manoeuvre is often very restricted and the demands of the different users become most pressing. In this situation the demands of the environment have not always received as much consideration as perhaps they should or as the public now seems to require. Unfortunately the assessment or quantification of the environmental needs are not easy and will take some time to solve adequately. Much theoretical work has been done on resolving the conflicting desires of users, but little on practical solutions; this needs to be rectified.

The benefits of river releases to uses such as amenity, recreation and conservation are very real but in general the beneficiaries do not pay directly for them. To accommodate the interests of these users may require reductions in hydropower production or increased pumping for water supply, incurring direct costs to the water supply companies. Should they be compensated for these increased costs? Should the users of, say, recreation sites be asked to pay the costs? Should the costs be paid for by local government which has the responsibility for recreational provision? Or should these additional releases simply be a form of compensation for the river ecology adversely impacted by the construction of a regulating reservoir? These are issues that as yet have had little consideration in the UK. Control rules indicate when critical decisions need to be made and sometimes give guidance as to the correct decision. However, in most practical situations the final decision lies with the resource manager, who takes the control rules into account along with many other factors, including the political factors. The chapter has described how the control rules can be developed to show when these critical decisions need to be made, and what issues need to be considered when developing the rules and when making the decisions. Most systems have formally defined rules, and informal rules derived from experience incorporated in the system manager's head. Attempts are being made to capture some of the latter in expert systems.

Regulating reservoirs have been central to the issues considered in this chapter. Most of the reservoirs built in the UK in the last 30 years have been of this type, and most of them are now part of a water resource system with a number of other sources. This conjunctive operation entails complex management decisions, which control rules are designed to assist. Regulating reservoirs support downstream river abstractions, which may be vulnerable to pollution incidents between the reservoir and the abstraction point. Automatic monitoring for all possible pollutants is not practicable at present, nor is it likely to be in the near future (see Chapter 7). The best solution probably lies with biological monitors (see Chapter 8), but they require further development and testing to become sufficiently reliable.

REFERENCES

Cowpertwait PSP, O'Connell PE, Metcalfe AN, Mawdsley JA. (1991) Regionalised stochastic rainfall time series model for the UK. *Journal of Hydrology* (submitted). [16.3]

Cunane C. (1978) Unbiased plotting positions – a review. *Journal of Hydrology* **37**: 205–22. [16.3]

El-Awad OM, Gowing JW, Mawdsley JA. (1991) An approach to multi-criterion evaluation of irrigation system performance. *Irrigation and Drainage Systems* **5**: 61–75. [16.2]

Glenny CM, Kinniburgh JH. (1991) Low flow management of the lower Thames. In: *Proceedings 3rd BHS National Symposium, Southampton*, pp. 2.29–2.36. Wallingford, UK. [16.5]

Gustard A, Cole G, Marshall D, Bayliss A. (1987) *A Study of Compensation flows in the UK*. Institute of Hydrology Report 99, Wallingford, UK. [16.1]

Haile SM, James A, Sear D. (1989) *The Effects of the Kielder Reservoir on the Ecology of the River North Tyne*. Report, NRA Northumbria Region, Newcastle, UK. [16.1]

Hill AI. (1988) Reservoir operation and impact of drawdown on water-based recreation: The assessment of user reactions. In: *Water for World Development, Vol. IV of Proceedings of the 1st WRA World Congress on Water Resources, Ottawa, Canada*. [16.4]

Hill AI. (1991) Incorporation of recreation in the operation of multipurpose reservoirs. Unpublished PhD Thesis, University of Newcastle upon Tyne. [16.4]

Hill AI, Mawdsley JA. (1987) The application of amenity lines in UK reservoir operation: case studies of the Dee and Tees schemes. In: *Proceedings of the 1st BHS National Symposium, Hull*. [16.4]

Jackson HB, Bailey RA. (1979) Some practical aspects of river regulation in England and Wales. *Journal of the Institution of Water Engineers and Scientists* **33**(2): 183–99. [16.1]

Kahn LR, Mawdsley JA. (1988) Reliable yield of unconfined aquifers. *Hydrological Sciences Journal* **33**(3): 151–72. [16.3]

Keeney RL, Raiffa H. (1976) *Decision with Multiple Objectives: Preference and Value Trade-Offs*. John Wiley, New York. [16.2]

Klemes V, Strikanthan R, McMahon TA. (1981) Long memory flow models in reservoir analysis: What is their practical value? *Water Resources Research* **17**(3): 737–51, Wallingford, UK. [16.3]

Lambert AO. (1988) *An Introduction to Operational Control*. British Hydrological Society, Occasional Paper 1, Wallingford, UK. [16.3]

Lawrence AJ, Kottegoda NT. (1977) Stochastic modelling of river flow time series. *Journal of the Royal Statistical Society Series A* **140**(1): 1–47. [16.3]

Loucks DP, Stedinger JR, Haith DA. (1981) *Water Resources Systems Planning and Analysis*. Prentice-Hall, New York. [16.2, 16.5]

Maas A, Hufschmidt MM, Dorfman R, Thomas HA Jr, Marglin SA, Fair GM. (1962) *Design of Water Resources System*. Harvard University Press, Cambridge, Mass. [16.2]

Major DC. (1972) *Multi Objective Water Resources Planning*. Water Resources Monograph 4, American Geophysical Union, Washington, DC. [16.2]

Pearson D, Walsh PD. (1982) The derivation and use of control curves for the regional allocation of water resources. In: *Optimal Allocation of Water Resources*. IAHS, 135, 275–83. [16.3]

Walsh PD. (1971) Designing control rules for the conjunctive use of impounding reservoirs. *Journal of the Institution of Water Engineers and Scientists* **25**(7): 371–80. [16.3]

17: Water-Quality Control

A. J. DOBBS AND T. F. ZABEL

17.1 INTRODUCTION

Traditionally, control of the chemical and biological quality of waters in rivers to protect different uses has been considered adequate to provide full environmental protection. This can be achieved by control of point and diffuse discharges using a variety of measures. However, a profound change is occurring whereby water-quality control is considered as a necessary but not sufficient condition to protect the ecological quality of rivers and other water bodies (Newman *et al* 1992). Consideration of sediment quality and protection of riparian habitat in addition to water-quality control are now seen as requirements for full environmental protection. Doubtless this wider aspect of river-quality control will assume more importance in the future (Royal Commission on Environmental Pollution 1992), but this chapter concentrates solely on aspects of the control of river water quality itself. Such control is effected (i) by regulation of the entry or discharge of contaminated or polluted waters into the river and (ii) by control of the flow regime. The effects of abstractions and other aspects of flow regime control will not be discussed in this chapter, but it is important to emphasize that effective river management requires that control of abstractions and discharges are inextricably linked.

In the past there were considered to be two fundamentally different approaches to pollution control: the 'quality objective approach' traditionally used in the UK, which took into account the dilution capacity of the receiving waters, and the 'uniform emission standard approach' favoured elsewhere in Europe, which applied the same concentration and load conditions to discharges from the same industrial sector. Currently, these two approaches are being seen as complementary and are being applied together.

As indicated above, river water-quality control is concerned with the control of pollution which can be defined in the widest sense with reference to European Community legislation as 'the discharge by man, directly or indirectly of substances or energy into the environment, the results of which are such as to cause hazards to human health, harm to living resources and to ecosystems, damage to amenities or interference with other legitimate uses of water'. Thus any control strategy adopted must not only control those activities likely to cause pollution in the aquatic environment, but must also protect humans from exposure to pollutants. The selection of the most appropriate control strategy to minimize the effects of the discharge of pollutants will depend on the type of pollutant, its source and the nature and use of the receiving water.

The following sections summarize the different approaches which may be taken to control the entry of pollutants to surface waters. We then discuss some river engineering approaches that are used for ameliorating the effects of pollution in rivers and finally discuss some potential future developments. Before discussing the control options it is useful to consider the nature and sources of pollutants.

17.2 TYPE OF POLLUTANTS

Pollutants can be classified in a variety of different ways. Holdgate (1979) lists several of these and gives advantages and disadvantages of the alternatives. EC Directives tend to divide substances in terms of the significance of their effects into List I (black) and List II (grey) substances, based on their toxicity, persistence, bioaccumulation, carcinogenicity, teratogenicity and mutagenicity. However, little guidance is provided in the Directives on the precise means to be adopted for the selection of specific List I and List II substances. Various selection schemes have therefore been developed to identify List I substances by assigning threshold values to the appropriate parameters and estimating likely concentrations in the environment based on production and physico-chemical data. A community selection scheme is currently being developed by the EC Commission.

An alternative approach is to divide substances into those causing direct toxic effects and those resulting in indirect effects.

1 Pollutants causing direct toxic effects: most of the List I and List II substances may be classified under this heading and may include heavy metals (e.g. cadmium) organometallics (e.g. organotins), other inorganics (e.g. ammonia), persistent organic compounds (e.g. pesticides) or volatile organic compounds (e.g. industrial solvents).

2 Pollutants causing indirect adverse effects: gross pollutants causing oxygen depletion and/or sulphide formation (e.g. farm waste, sewage effluent) or those causing eutrophication (nutrients) fall into this category. Aesthetic parameters such as foaming (presence of detergents), colour, litter and oil, which although not toxic at the concentrations present, may also affect the use of the water. Bacteria and viruses may also be considered as causing indirect effects by, for instance, reducing the tolerance of animals to the effect of other pollutants present. However, they can also cause direct effects (e.g. illness in bathers).

The way in which contaminants cause effects and thereby become pollutants varies widely. For instance, the organotins are toxic at very low concentrations to mollusc species but fish are more tolerant. In contrast, DDT's main effect occurs via the food chain, because of its high bioaccumulation potential, leading to problems in higher animals, for instance the thinning of egg shells in birds of prey. In practice almost any chemical or contaminant can become a pollutant if it is present in high enough concentration, and the main value of pollutant classification is to focus pollution control activity on the most hazardous.

17.3 SOURCES OF POLLUTANTS

Whilst the distinction is not very precise it is often useful to distinguish between point and diffuse pollution sources because different pollution control strategies are required in each case. Point sources are best thought of as discharges from fixed sites, such as a factory or sewage treatment plant, which if the process involved was stopped so would the discharge. Generally speaking a modification to the nature of the process or treatment of the discharge will change the composition of a point discharge. Diffuse inputs on the other hand result from no particular process but originate from the widespread use of the substance. Using this approach, for example, agricultural run-off would be considered a diffuse source. Phosphates used in detergents would probably also be best considered as a diffuse input, at least to the sewerage system. However, the aggregate phosphate discharged from a sewage treatment plant can be considered as a point source. It is monitored in this way and is capable of reduction by tertiary treatment.

In addition to consideration of the nature of the sources, point or diffuse, which as we have just seen depends to some extent on the point in the system where the pollutant is being considered, the temporal variability of discharges needs to be considered. In general three classes of discharge can be distinguished.

1 Continuous – a discharge whose composition and flow may fluctuate but which effectively happens all the time. A sewage treatment works effluent or a cooling water discharge from a power station are good examples.

2 Intermittent – as the name implies these do not discharge all the time and they may or may not discharge regularly but their location can be identified and the discharge is to some degree predictable in terms of composition and flow. Probably the best example is a storm sewer over-

flow; during rain events of a given severity such overflows will produce a discharge.
3 Incidents – the location, nature and timing of these is unpredictable. Examples are transport accidents and fires.
Each of these have their own range of control measures. Most pollutants will arrive in rivers from a variety of sources, and control requires an understanding of all the routes if it is to be effective and efficient.

17.4 CONTROL OPTIONS

For most substances, particularly those which occur naturally, it is unrealistic to expect zero discharge to the environment. Control must be exercised and whatever control option is used, the end result must be to protect the vulnerable targets, i.e. humans and human food sources, or other living organisms at the different levels. Thus for most substances it is generally accepted (GESAMP 1986):

- that there is a minimum acceptable concentration which does not produce unacceptable effects on the target;
- the environment therefore has a certain capacity to assimilate pollutants;
- the capacity can be quantified.

This line of thought requires evidence of harm to justify controls. However, in recent years pressure has grown to move away from the system where evidence of harm is required to a precautionary approach where proof of safety is needed. In principle such a precautionary approach is sound, but in practice it is difficult to apply. In particular proof of safety is impossible; effectively you are trying to prove a negative. Also the relative safety of one substance in comparison with another is extremely difficult to quantify because it often requires the comparison of a poorly defined risk in one area with another poorly defined risk in a different area. For example, evidence to suggest a certain carcinogenicity risk may have to be compared with long environmental persistence. In the extreme the precautionary approach can lead to a requirement for zero emission of a particular substance. While this is conceivable as a *target* for particularly dangerous substances, because continued reduction can be charted, it is not a feasible *requirement* because it is simply technically impossible to show a zero concentration of anything in anything. For most substances it seems discharges will have to occur and for these a reference concentration will be needed to ensure that the selected control option is sufficient to protect humans and the environment.

Table 17.1 Limits for metals (from Miller 1991)

	Standards ($\mu g\,l^{-1}$)		
Metal	Drinking water MAC*	Freshwater EQS†	Seawater EQS†
Copper	3000	1–28	5
Zinc	5000	8–125	40
Nickel	50	50–200	30
Lead	50	4–20	25
Mercury	1	1	0.3
Cadmium	5	5	2.5
Chromium	50	5–50	15
Arsenic	50	50	25

* Maximum admissible concentration – absolute – total.
† Environmental quality standard – annual average – dissolved/total; EC/UK EQS values.

These reference values are likely to vary depending on the use of the water to be protected. For example, the standards required for metals to protect the ecosystem are often more stringent than those required for the protection of humans directly (Table 17.1), although it has to be noted that the limits are expressed in different ways (Miller 1991).

In some countries reference values are chosen on a non-environmental risk basis, e.g. expressed as a percentage of the background concentration. However, background levels can vary widely depending on the origin of the water. For instance, for metals, streams originating from a mineral-rich area will have much higher background concentrations than those from mineral-poor areas.

It is important that the reference values are based on scientific evidence and that these values are reviewed periodically in the light of new scientific evidence. Otherwise the use of the substance may be so restricted that its use is economically unattractive, leading to substitution by other substances which have different, but maybe unknown, effects on the environment. A typical example is the replacement of phosphate in detergents by alternative substances for which some of the effects are only now emerging (Dwyer *et al* 1990).

A number of different options are available to control pollutants entering the aquatic environment, which include the establishment of standards relating to different aspects of the overall production/use/disposal cycle (Holdgate 1979):

- process standards;
- product standards;
- exposure standards;
- biological standards;
- environmental quality standards;
- emission standards.

Besides these standards, two further control options need to be considered (Haigh 1985):

- preventative controls, e.g. total prohibition on the use of a substance;
- standards for total emissions for a given area, e.g. whole catchment area of a river system.

Most of these different control options have been utilized either individually or in different combinations in current EC* and national legislation. Obviously the effectiveness of the options depends critically on the enforcement.

Process standards

The process standard approach specifies the process to be used to ensure that the required quality is achieved. As such it is not really directly relevant to the process of the control of river water quality. However, the process control aspects of integrated pollution control and waste minimization are becoming increasingly important and these will have important implications for river water quality. For example, the mothproofing industry is experiencing increasing difficulties meeting the tightening consent conditions. Processes are therefore being developed to move from aqueous to solvent-based processes for the application of mothproofing chemicals to minimize their release to the aquatic environment. This could lead to the adoption of a process standard requiring solvent-based application of mothproofing agents.

Product standards

Product standards are widely used in Europe in both EC and national legislation to control the release of certain substances to the aquatic environment. One of the first examples was the EC Detergent Directive (73/404/EEC) (CEC 1973). This was introduced to prevent foaming in rivers caused by detergent residues which resisted biodegradation in sewage treatment works. The Directive requires that the detergents satisfy certain biodegradation criteria established in the form of test protocols. A more recent example is the restriction imposed on the use of tributyltin compounds in antifouling paints used on boats. This bans the use of tributyltin compounds in paints used on small boats.

Product standards are particularly useful for substances entering the aquatic environment from diffuse sources. For example restrictions on lead in petrol, Directive 85/210/EEC (CEC 1985), has resulted in reduced emissions of lead to the air which, via reductions in atmospheric deposition and run-off, has reduced the amounts of lead entering the aquatic environment.

Exposure standards

Exposure standards may be defined as primary protection standards for the target, which can be either humans or aquatic organisms. The standards laid down in the EC Directive on the Quality of Water Intended for Human Consumption (80/778/EEC) (CEC 1980a) can be considered as exposure standards. The Directive provides maximum admissible concentrations for certain parameters, taking into account exposures from other sources such as air and food.

Applying exposure standards, for instance maximum admissible concentration in drinking water, could, in the extreme, require that dis-

* European Community legislation (EC Directives) is not in itself legally enforceable in Member States. However, adoption of EC Directives by the Council of Ministers requires Member States to ensure that the national legislation includes measures no less stringent than those agreed in the Directives.

charges to receiving water used for the abstraction of drinking water may have to meet those exposure standards if insufficient dilution is available in the receiving water and the water-treatment plant is not equipped for the removal of the particular substance. Alternatively the treatment plant would need to be improved. This dilemma lies at the heart of the current problems with pesticides in Europe. A uniform non-toxicologically derived exposure standard of $0.1\,\mu g\,l^{-1}$ has been set in the Drinking Water Directive yet agricultural use frequently gives rise to run-off waters with pesticide concentrations considerably higher than this value.

The standards laid down in EC Directives on the quality of live bivalve molluscs to be placed on the market (91/492/EEC) (CEC 1991b), may be considered a further example of an exposure standard. If the standards for bacteria are exceeded, which can lead to illness in people consuming the bivalves, the bivalves must either be treated in special purification plants to reduce the levels of bacteria prior to sale, or the water quality must be improved to ensure that the standards in the bivalves are not exceeded. This particular case is more complicated though because a clear relationship between levels of bacteria in the water column and in the bivalve have so far not been established. The relationship may depend on the age of the animal, the availability of food and environmental conditions such as temperature and salinity.

Biological standards

Not many biological standards have been set so far. However, most of the daughter Directives of the EC Dangerous Substances Directive (76/464/EEC) (CEC 1976b), which lay down standards for individual List I substances, contain the statement that the concentration of the particular substances in shellfish and/or fish must not increase significantly with time, the so-called 'standstill' provision. Only the EC Directive on limit values and quality objectives for mercury discharges by the chlor-alkali electrolysis industry (82/779/EEC) (CEC 1982) contains besides the 'standstill' provision for shellfish, also a numerical standard for mercury of $0.3\,mg Hg\,kg^{-1}$ wet flesh in representative samples of fish flesh chosen as an indicator.

The EC Shellfish (79/923/EEC) and the EC Freshwater Fish (78/659/EEC) Directives also contain the general requirement that the concentration of identified substances in the flesh must be so limited that there is no effect on the quality of shellfish and fish products, respectively (CEC 1979a, 1978). However, no numerical values are laid down for individual substances.

The advantage of using biological assessments related to accumulation of the substance in tissue or blood is that it provides an indication of total exposure from all sources whether point or diffuse or from intermittent events such as storm overflows. The effect of intermittent events is particularly difficult to assess by the usually employed periodic spot-sampling methods and even diffuse inputs can vary substantially depending on the use pattern of the substances and the meteorological conditions.

The difficulties in applying biological standards are in relating tissue or blood levels to concentrations in the water or sediment, to actual effects on the ecosystem or harm to higher organisms and in obtaining representative samples for analysis from 'vulnerable' areas.

Preventative controls

Preventative controls can include a total prohibition on the use of a particular substance or partial prohibition for a particular purpose or in a designated 'protection zone'. This control can be similar to the product standard approach, as with the example of tributyltin compounds discussed above. Preventative controls can also relate to the location of a plant or the requirement that emergency plans are produced or particular storage facilities used to minimize any accidental release that could occur.

For particularly dangerous substances, total prohibition on their use may be the most appropriate control strategy. However, for other less dangerous substances partial prohibition of use may be sufficient if the permitted uses do not cause unacceptable effects and particularly if no equally effective and economical alternatives are available.

The ban on the use of certain pesticides in groundwater protection zones in Germany may be considered as an example of this approach. The ban relates to those pesticides which may reach concentrations in the groundwater above the maximum admissible concentration of $0.1\,\mu g\,l^{-1}$ for individual pesticides in the EC Drinking Water Directive (80/778/EEC) (CEC 1980a).

The various amendments of the EC Directive 76/769/EEC on the approximation of the laws, regulations and administrative provisions of the Member States relating to restrictions on the marketing and use of certain dangerous substances and preparations (CEC 1976c) prohibit the use of certain products within the EC, such as the use of mercury-containing pesticides and the use of cadmium for pigmentation in a range of products and as a stabilizer in certain plastics and surface treatments. Another example of EC legislation applying preventative control is the Directive prohibiting the placing on the market and use of plant protection products containing certain active ingredients (79/117/EEC) (CEC 1979b). For instance, the fourth amendment (90/335/EEC) (CEC 1990) bans the use of endrin, aldrin and alkyl mercury compounds.

Although preventative controls can be applied to point sources of pollution they are most appropriate for the control of substances originating from diffuse sources.

Environmental Quality Standards

The EC has adopted several use-related directives laying down standards for individual parameters which must be met in order that the water is suitable for the particular use. The use-related directives are:

1 Directive concerning the quality required of surface water intended for the abstraction of drinking water in the Member States (75/440/EEC) (Surface Water Directive) (CEC 1975).

2 Directive concerning the quality of bathing water (76/160/EEC) (Bathing Water Directive (CEC 1976a).

3 Directive on the quality of freshwaters needing protection or improvement in order to support fish life (78/659/EEC) (Freshwater Fish Directive) (CEC 1978).

4 Directive on the quality required of shellfish waters (79/923/EEC) (Shellfish Directive) (CEC 1979a).

5 Directive on the protection of groundwater against pollution caused by certain dangerous substances (80/68/EEC) (Groundwater Directive) (CEC 1980b).

These directives apply only to designated waters.

In the EC Dangerous Substances Directive (76/464/EEC) (CEC 1976b) Environmental Quality Standards (EQSs) are provided but the use-related concept has not been adopted and a distinction is only made for discharges to fresh waters, estuaries and coastal waters. The EC Dangerous Substances Directive also allows control of individual substances by Uniform Emission Standards (UES) (see below).

Confusion has arisen in the EC Member Countries relating to the definition of the term 'Environmental Quality Objective' (EQO). In the UK the EQOs laid down by the EC are used as operational standards (EQSs) for the setting of consents for discharges and the control of diffuse inputs. In most of the other EC countries the EQOs are interpreted as long-term aims to be achieved in the future. Thus, for many EC countries, particularly the Netherlands and Germany, the EQOs laid down by the EC are not sufficiently stringent because the long-term aim of these countries is to achieve 'background concentrations' in all surface waters. This has discredited the use of EQSs for the control of direct discharges, which in most other EC countries are controlled by technology-based 'uniform emission standards' (UESs).

The application of EQSs requires control over the whole catchment area of the receiving water. In the UK the National Rivers Authority in England and Wales and the River Purification Boards in Scotland, have responsibility for whole catchments. The EQS approach is less attractive in the other European countries which have to share the catchments of their surface waters with other countries and therefore do not have effective control over the incoming river quality at national borders. In these circumstances demands for treatment by 'best available technology' or for contaminants to be reduced to levels 'as low as reasonably (or technically) achievable', seems more reasonable.

Nevertheless, as discussed above, the application of technology-based effluent standards requires reference values to assess that the controls implemented are sufficient to ensure that the target species are protected. Similarly the application of EQSs requires the setting of consents for discharges to ensure that the EQS values are not exceeded.

The EQO/EQS approach

The Environmental Quality Objective (EQO)/ Environmental Quality Standard (EQS) approach is based on the premise that a minimum acceptable concentration of a pollutant can be defined which does not interfere with the use of the water. A distinction has therefore to be made between the EQO,* which defines the use for which the water is intended (e.g. abstraction for drinking water), and the EQS,† which specifies the concentration of the substance which should not be exceeded to ensure that the water is suitable for the particular designated use. For example, for the abstraction of drinking water (the EQO), the EQS for nitrate is $50\,mg\,l^{-1}$.

Thus to apply the EQO/EQS approach, the different uses of the water need to be defined, and standards, in terms of the maximum acceptable concentration of the particular pollutant to protect the different uses, must be derived. These standards must be sufficiently low to protect vulnerable targets implied or specified in the objective, including humans, human food sources or other living organisms. The standards can then be used to set consents for individual discharges, taking into account the dilution in the receiving water, but also the concentration of the pollutant already present in the receiving water from other sources. The EQO/EQS approach can therefore also deal with diffuse inputs and can be used to control multiple point sources.

In the UK the common uses of both fresh and saline surface waters have been defined for the derivation of EQS values for List II substances. These take into account the use-related directives adopted by the EC listed above in Table 17.2.

Table 17.2 Designated uses of water for derivation of EQS values

Use	Fresh water	Saline water
For direct abstraction to potable supply	√	–*
For abstraction into impoundment prior to potable supply	√	–*
As a source of food for human consumption	√	√
Protection of fish and shellfish	√	√
Protection of other aquatic life and dependent non-aquatic organisms	√	√
Irrigation of crops	√	–*
Watering of livestock	√	–*
Industrial abstraction	√	√
Bathing and water-contact sports	√	√
Aesthetic considerations	√	√

* Not applicable.

The provisions of the UK's 1989 Water Act, now incorporated into the 1991 Water Resources Act, enable a scheme of statutory water-quality objectives to be developed and used. Under the 1991 Act the Secretary of State for the Environment has the duty to set Water Quality Objectives (WQO) taking advice from the National Rivers Authority (NRA). The Secretary of State and the NRA then have a duty to ensure WQOs are met with the usual UK statutory proviso 'as far as is reasonably practical'. This scheme now seems likely to start in 1993. In response to this requirement, proposals for Statutory Water Quality Objectives have recently been published by the NRA (1991) with reference to specific water uses.‡

Derivation of EQSs

The approach adopted in the UK for the derivation of EQSs is illustrated in Fig. 17.1. Available data on the toxic effects, both from laboratory studies and field observations, are critically assessed.

* Environmental Quality Objective (EQO) is defined as the level of water quality that a body of water should achieve in order to be suitable for its agreed use (NRA 1991).

† Environmental Quality Standard (EQS) is defined as that concentration of a substance which must not be exceeded if a specified use of the environment is to be maintained.

‡ Recent developments with Statutory Water Quality Objectives, since this chapter was written are reviewed by Seager (1993).

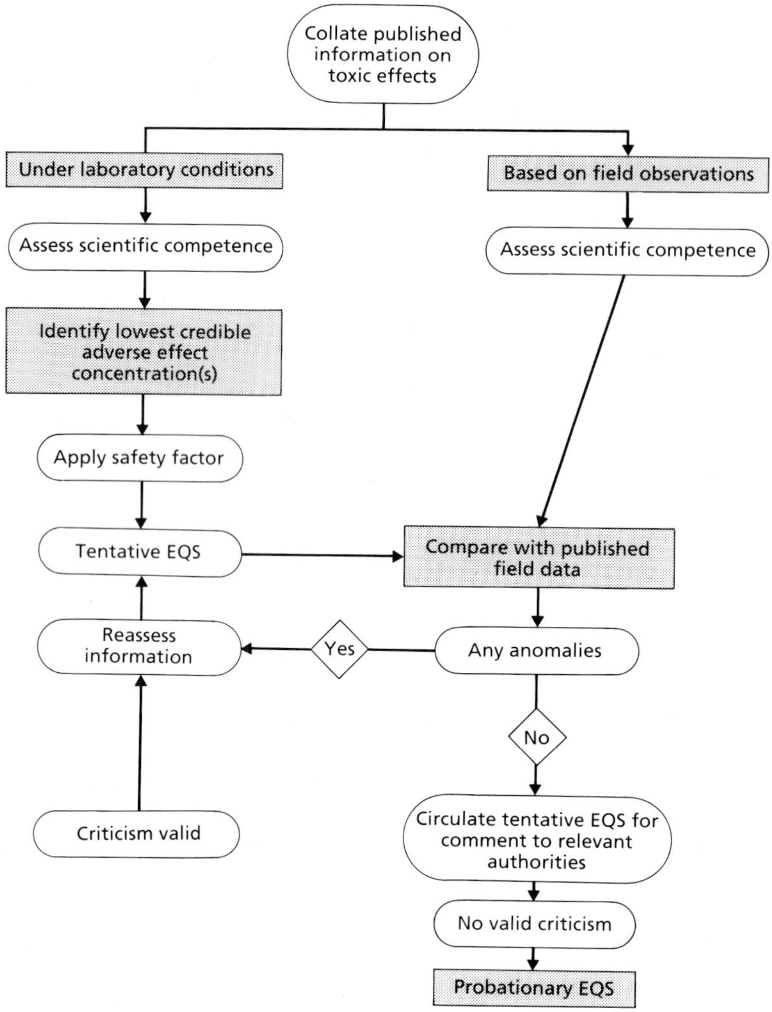

Fig. 17.1 Schematic flow chart for the derivation of environmental quality standards.

Particular attention is paid to the experimental test method used. The lowest credible adverse effect concentration is derived from the available toxicity data. Particular emphasis is placed on chronic effects and effects on sensitive life stages of the target species. A safety factor is then applied to obtain a tentative EQS. The magnitude of the safety factor depends on the quality and extent of the available toxicity data. If only acute (short-term) LC_{50} values are available, an arbitrary safety factor of 100 is usually applied. This factor tends to be reduced to 10 if chronic LC_{50} data are available for sensitive species.

For some substances 'no observable effects concentrations' (NOEC) are reported for long-term chronic studies. Despite concerns over the value of the NOEC concept these values may be used as tentative standards without applying a safety factor, provided the test design is acceptable, and provided they refer to relevant species and are supported by the other data available.

The tentative EQS derived from the laboratory toxicity data is verified by comparison with the available field toxicity data, taking care to distinguish between the effects of the particular compound and those of other toxic compounds present. It is important to compare the tentative EQS with the highest observed concentration which has no effect on the ecosystem, rather than with the lowest concentration which does have an effect. Any discrepancies between the field data and the tentative EQS require the reassessment of the available field and laboratory data and, if necessary, the application of a different safety factor.

It is important that EQS values are frequently reviewed to take account of new laboratory or field toxicity data. The response of ecosystems to pollutants is extremely complex and until we have a much better understanding of the mechanisms of ecosystem response to pollutants EQS values need to be continually reviewed.

The safety factor applied obviously has a crucial impact on the final EQS proposed. For instance, the German proposals for the derivation of Water Quality Objectives involve larger safety factors (Federal Government 1989) with the following values:

1 10 if NOEC values are available for each of the four trophic levels (bacteria, *Pseudomonas putida*, the alga *Scenedesmus subspicatus*, *Daphnia magna* and a fish species in chronic studies).

2 100 if NOEC values of at least two of the above trophic levels are available and if in addition a wide range of aquatic toxicity data is available.

3 An additional safety factor of 10 can be applied, e.g, if:

(a) data are available for organisms of ecological importance which are lower than the NOEC values obtained using the standard test methods;

(b) the substance is persistent in the aquatic environment;

(c) the substance can change into a dangerous substance in the environment.

However, the German EQS values are intended as long-term 'quality-aim' (*Qualitätsziel*) values to be achieved in the future. These values have therefore an additional built-in precautionary safety factor compared to the UK EQS values which are used for the control of discharges.

In the USA, Water Quality Criteria have been derived by the Environmental Protection Agency (EPA) for a number of substances. EPA standards are derived from acute and chronic toxicity studies with species from three trophic levels (algae, crustacea and fish), using 'uncertainty factors' to account for missing toxicity data. The values of these factors are (EPA 1984) as follow.

1 1: to data from field studies (e.g. artificial streams) as long as one/some of the most sensitive species are included.

2 10: to the lowest chronic values derived from partial or whole life-cycle studies, where the uncertainty factor allows for additional stress in the environment compared to laboratory conditions, existence of more sensitive species, other toxins in the environment and for extrapolation from partial to whole life cycles.

3 100: to the lowest acute values of the three trophic levels (based on an acute/chronic ratio of 10 and the existence of more sensitive life stages than those tested, but can be increased if acute/chronic ratio is large).

4 1000: to a single acute LC_{50} – but this is rarely used as QSAR (Quantity Structure–Activity Relation) studies are carried out to generate further data.

The EPA criteria are intended as reference values to ensure that the technology-based standards are sufficient to protect the environment.

The recently published Netherlands Environmental Water Quality Objectives 2000 are based on the NOECs from standard chronic toxicity tests with aquatic species but also include an allowance for the equilibrium partitioning of the substance between the dissolved fraction in water and the fraction adsorbed on suspended solids or sediment, the joint toxic action of related substances in aquatic species and the impact of persistent, bioaccumulating chemicals on predator species feeding on the aquatic environment (van der Gaag *et al* 1991). This approach leads to even lower concentrations as quality standards.

When setting EQSs, it is important to define explicitly the form of the chemical, e.g. as total concentration or dissolved fraction or, as in the case of ammonia, the un-ionized species. A decision has also to be made on the degree of compliance required for the standard to protect the use (100, 95, 90 or 50% compliance). This decision will have to take into account the sources of the substance and the difference between acute and chronic toxicities and the practical constraints imposed by sampling and analysis.

Although the EC has recently favoured the

technology-based fixed emission approach, e.g. with the adoption of the EC Urban Waste Water Directive (91/271/EEC) (CEC 1991a) the proposals for a Minimum Ecological Quality Directive will be based on the quality standards approach and will most likely include chemical and biological standards.

Emission standards

The EC Dangerous Substances Directive (76/464/EEC) (CEC 1976b) allows Member States to opt either for the Environmental Quality Standards (EQSs) or the Uniform Emission Standards (UESs) or limit value approach, for the control of List I or Black List substances. Most EC countries have favoured the UES approach. Standards are laid down in terms of permitted concentration of the substance in the effluent and the maximum load which may be discharged expressed in terms of load per unit of production for the different types of activities discharging the substance. Thus larger production facilities can discharge larger loads and different activities may discharge different loads.

However, the trend in Europe and the USA is to move away from controlling individual substances towards the control of effluents from industrial sectors, setting limit values for all important substances which may be present in the effluent from that particular industry. For the control of complex effluents the use of aggregate parameters such as AOX (adsorbable organic halogen) in Germany and toxicity-based consents in the UK and USA (see below) are being promoted.

Only two industrial sector EC Directives have been adopted so far. These concern the titanium dioxide industry (CEC 1989) and the treatment of urban waste water (91/271/EEC) (CEC 1991a). The Urban Waste Water Directive sets limits for biological oxygen demand (BOD), for chemical oxygen demand (COD), suspended solids (SS), total phosphorus (P) and total nitrogen (N). The values are partly related to the size of the treatment plant, but also to the sensitivity to eutrophication of the receiving water.

The most extensive industry sector-based controls are practised in Germany and the USA, whereas, for instance, France and The Netherlands have issued guidelines for only some branches of industry. Germany has adopted industry-specific standards for 52 branches of industry which include chemical standards, but also limits effluent toxicity (based on fish tests) for some branches of industry. The standards laid down are applied as minimum standards and, if the effluent contains dangerous substances, are applicable independently whether the effluent is discharged to sewer or to surface waters (Cartwright *et al* 1991).

In the UK under the Integrated Pollution Control legislation (see below) industry sector-based guidance notes are currently being issued for different industrial activities based on 'best available techniques not entailing excessive costs' (BATNEEC). These notes contain standards for emissions to water, air and in waste.

In the USA the control of discharges to surface waters is also based on 'uniform emission standards' for industrial sectors, but more stringent standards can be set if the water-quality criteria for the receiving water are not met. Under the Clean Water Act 1977 three programmes for the control of discharges have been established.

1 The National Protection Programme (NPP) which requires industry to pretreat its effluent to a technology-based standard before discharge to the sewer.

2 The National Pollutant Discharge Elimination System (NPDES) Program which provides technology and water-quality-based effluent standards that dischargers of effluents to surface waters must meet.

3 Water-quality-based effluent standards to ensure that State requirements for water quality and pollutant loading are met.

The NPDES and NPP programmes operate on the basis of 'categorical standards' (uniform emission standards) which are applied to all industries in a given category uniformly across the country. A general listing of the main industrial branches is given in Table 17.3. However, most industrial sectors are further subdivided and other industries not covered are regulated on a case-by-case basis with reference to statutory guidelines. These 'uniform emission standards' are also assessed in relation to local or State water quality criteria, which may result in more stringent discharge

Table 17.3 Industrial sectors with Uniform Emission Standards

Adhesives and sealants
Aluminium forming
Coal mining
Coil coating
Copper forming
Electroplating
Foundries
Inorganic chemicals
Iron and steel manufacturing
Leather tanning and finishing
Finishing/mechanical products
Non-ferrous metals manufacturing
Ore mining and dressing
Organic chemicals, plastics and synthetic materials
Pesticides, chemicals
Petroleum refining
Pharmaceutical manufacturing
Plastics and synthetics
Porcelain enamelling
Pulp, paper and paperboard
Steam electric power generation
Textile industry
Timber products

limits at a specific site.

The pretreatment standards are based on BAT – best available technology economically achievable – for the 126 toxic substances listed in the Clean Water Act, whereas BCT – best conventional pollutant control technology – is applied for the control of conventional parameters such as pH, BOD, COD, oil and grease, etc. Depending on the industry sector for new discharges more stringent pretreatment standards may be laid down than for existing discharges.

The emission standards approach is of course only applicable to the control of point sources. Strict implementation of emission standards may lead to the introduction of clean technology, particularly if the cost of complying with the standards is very high.

Combining the EQS and UES approaches

The protracted debate about the relative merits of the EQS and UES approaches for controlling discharges to surface waters has obscured the fact that neither method, in isolation, is fully effective for operational purposes. For receiving waters with a large dilution capacity the UES approach will lead to more stringent requirements whereas the EQS approach will lead to more stringent effluent consents for discharges to surface waters with a small dilution capacity.

As outlined above, reference values (EQS) are required to ensure that the technology-based standards are sufficiently stringent to protect the environment. Similarly, sufficiently stringent effluent standards have to be set to ensure that the EQS values in the receiving waters are met.

At the second Ministerial Conference on the Protection of the North Sea, the UK committed itself to apply a dual approach to the control of certain particularly dangerous substances, The Red List (Table 17.4). There are also indications that some EC Member States who claim to use

Table 17.4 The UK Red List substances (DoE 1989)

Substance	EC Directive adopted (List I status)
Mercury	+
Cadmium	+
Gamma-hexachlorocyclohexane (lindane)*	+
DDT*	+
Pentachlorophenol (PCP)*	+
Hexachlorobenzene (HCB)	+
Hexachlorobutadiene (HCBD)	+
Aldrin*	+
Dieldrin*	+
Endrin*	+
PCB (polychlorinated biphenyls)*	
Tributyltin compounds*	
Triphenyltin compounds*	
Dichlorvos*	
Trifluralin*	
1,2-Dichloroethane	+
Trichlorobenzene	+
Azinphos-methyl*	
Fenitrothion*	
Malathion*	
Endosulfan*	
Atrazine*	
Simazine*	

* Substances which enter the environment predominantly by indirect routes.

the UES approach are adopting a mixed system of control which takes account of the quality of the receiving water and allows the setting of stricter limit values. The system in the USA, although technology-based, also allows stricter standards to meet water-quality standards if local conditions warrant it.

Toxicity-based control

The chemical-specific approach is generally used for the control of point discharges to surface waters. However, there is increased interest in the USA and Europe in the application of toxicity-based consents (TBCs) for complex effluents, particularly those of variable nature but also for effluents containing substances for which analytical methods or toxicity data (to derive safe discharge concentrations) are not available. Using a toxicity-based consent procedure also provides some protection against possible synergistic or additive toxicological effects of the substances present and the contribution to toxicity from impurities and degradation products. The application of this approach may require that either the diluted (with receiving water) or the undiluted effluent is non-toxic to specified organisms, e.g. bacteria, *Daphnia* or fish, and that there is no evidence of chronic toxic effects. In practice the application of toxicity-based consents should be supported by chemical-specific controls wherever appropriate.

Integrated pollution control

In the UK the adoption of the Environmental Protection Act 1990 introduced the concept of integrated pollution control (IPC) for the control of the 'most polluting' (prescribed) processes. Only one process authorization for the disposal of waste to air, water and land will be required. 'Best available technology not entailing excessive costs' (BATNEEC) will have to be applied for the control of discharges and an assessment of the 'best practicable environmental option' (BPEO) will have to be carried out to avoid damaging transfer of pollution from one environmental compartment to another. In theory the need to achieve BPEO will dictate BATNEEC and what is BATNEEC for BPEO will prevail over what is BATNEEC for releases into individual media (Priest *et al* 1991).

The authorization conditions obtained by applying IPC must also be assessed to ensure that they are sufficiently stringent to meet the EQS values in the receiving water. This assessment may lead to tighter limits than those given by BATNEEC. There are also indications that the EC is moving away from controlling individual substances towards an integrated permitting system.

Effluent charges

Several EC Member States (Germany, The Netherlands, France) are operating effluent charging systems. Charges are levied at levels which depend on the pollution load discharged. Although these charges have no legal force to improve the quality of the effluent they nevertheless can provide an incentive to improve the quality of the effluent especially if the charges are sufficiently high or, as practised in Germany, if much lower charges are levied if certain effluent standards are met. It has been claimed that the introduction of effluent charges in Germany has provided an incentive to improve effluent qualities more rapidly for existing plants, which are normally allowed an extended implementation period (up to 10 years) for new legislation (Cartwright *et al* 1991).

In the UK charges for the monitoring of discharges have been introduced enabling NRA to recover the costs for processing applications and monitoring consents. The charges depend on the volume of the discharge, the substances present in the discharge and the type of receiving water of the effluent. This may incidentally also provide an incentive to improve the quality of the effluent especially as the monitoring charges for the particularly dangerous substances (UK Red List) present are quite high, reflecting the analytical costs involved. However, the Water Resources Act 1991 only allows such a cost recovery charging scheme and any move to a true pollution tax or incentive charging scheme would require new legislation.

17.5 ENGINEERING CONTROLS OF WATER QUALITY

In some cases the kinds of discharge control philosophy discussed above are insufficient. Catchments in old industrialized areas are polluted as a result of historic pollution and inadequate storm-water provisions; river water quality is often poor, in particular if the flow in the river is comprised mainly of sewage effluent. In such cases 'purification lakes' can alleviate the pollution load carried by the rivers and lead to improvements in downstream water quality. A good example is the River Tame reclamation lakes situated near Birmingham in the UK. The Upper Tame catchment above the lakes includes one of the largest urban conurbations in the UK; approximately 20% of the catchment is impermeable and run-off is rapid carrying a high polluting load into the river. The area was also highly industrialized during the 19th century, which has left a heritage of contaminated land. The first reclamation lake was completed in 1983 and a second one was brought into use in 1987. These lakes have led to significant improvements in downstream water quality as a result of reductions in the suspended solid and to a lesser extent BOD load, but at a the considerable cost of some £1 million per year, most of which results from the continuous process of removal and disposal of the contaminated sediment (Hellawell & Green 1986).

Another example of an engineering solution is the artificial introduction of oxygen into river waters. Such an approach has been successfully used to minimize the impact of pollution events such as that involving a spillage of sugar. It is also used on the River Thames in the UK to minimize the impact of oxygen depletion in the estuary resulting from storm discharges. The 'Thames Bubbler' can move up and down the estuary and inject up to 30 tonnes per day of oxygen into the water (Adams 1992).

Such engineering solutions ameliorate the effects of inadequate historic pollution control practices. They are best viewed as expedients until the origins of the pollution are identified and more effective control measures introduced.

17.6 RIVER-QUALITY CONTROL IN THE FUTURE

The technology and philosophy of point discharge control is now well advanced although at times leading to increased loads to other environmental compartments (land, air). While improvements in efficiency are to be expected in the future, improved control is largely a matter of resources and priorities. It is a sign of the state of technology and knowledge that some industrial companies can set themselves targets of zero discharge notwithstanding the technical problems posed by the term. The issues of the future are how to define river water quality; how to monitor and control it most effectively; and how to identify and control diffuse sources of pollution? With regard to the last, the use of buffer zones is attracting increasing attention (see Chapter 21).

Progress on the definition of water quality is being assisted by the recognition that both biological measures (species diversity and numbers) and chemical measures are useful and complementary. The quality objectives proposals by the NRA in the UK recognizes three main elements to the quality objectives for surface waters.

1 A 'use' category defining what the community expects to use the river for, e.g. basic amenity, cyprinoid fishery, etc. (see Section 17.4).

2 A 'regulatory' category linked to relevant EC Directives, e.g. the Dangerous Substances Directive mentioned above.

3 A classification category defined by reference to chemical water quality and the biological state of the water, based on species present.

There is possibly some value in considering extending the use classification by taking a functional view of river systems. Consideration of the basic functions that rivers have to perform may lead to new procedures to assess and monitor the overall 'health' of a river system. Such functions could include the ability to assimilate waste material at a given rate, an ability to support a population of primary producer organisms, etc. Monitoring procedures applied to such biological functions might prove to be more robust than simple species-occurrence monitoring approaches.

In monitoring chemical water quality there appears to be considerable scope for improvements both in the determinands selected and the way monitoring and classification is conceived and organized. The conventional river water quality determinands are dissolved oxygen (DO), biochemical oxygen demand (BOD) and ammonia. Dissolved oxygen is clearly of key importance for the aesthetics of water quality and aquatic species survival. Ammonia is a toxicant occurring widely so is of direct environmental relevance. However, BOD is merely a measure of oxygen depletion potential, interesting, but as long as DO is sufficiently high, of no direct biological relevance. More profound consideration of BOD use in this way seems long overdue. The requirement to measure nutrient concentrations is also likely in coming years.

Chemical monitoring in the conventional way usually involves spot sampling. This approach, while valuable, suffers from two fundamental weaknesses. The first is the low probability of detecting pollution events and the second is that the determinands selected for analysis may not include the important ones. Two developments are providing remedies for this weakness: one is the increasing use and improved basis for biological monitoring; the other is the advance in technology enabling more use of continuous chemical monitoring.

Biological monitoring, which usually involves the study of species diversity and abundance, solves these problems by integrating the effects of all pollutants and all sources whether continuous or one-off pollution events. Alternative approaches based on body burdens, physiology and biochemical changes are also starting to show promise but are not currently routinely used. Biological monitoring offers clear advantages in being capable of responding to episodic events and to unidentified pollutants and to mixtures. Its drawbacks are that causes are difficult to identify and when changes are found it can be too late.

One problem with species occurrence and diversity which has in the past limited its value is the difficulty of taking natural physical or chemical factors into account. Rivers with hard and soft water will clearly be expected to have different flora and fauna. If these natural factors could be taken into account it would be possible to attribute changes in natural population to those caused by pollution. Such an approach (RIVPACS – Wright *et al* 1989; see also Chapter 8) has been developed for predicting benthic invertebrate communities. The NRA are proposing to use this approach in the future to provide an additional biological component to the conventional data used for water quality classification (NRA 1991).

Advances in technology have enabled continuous water monitoring to develop rapidly although there is still only a small number of determinands that can be measured reliably in this way. Whereas 20 or so years ago water quality in drinking-water treatment plants was monitored continuously for process control reasons, it is now fairly commonplace for raw water quality to be monitored often using broad-spectrum sensor systems such as fish monitors to detect pollution events (Evans & Wallwork 1988). These devices are really quality-assurance tools rather than quality control, but the development of relatively small portable devices with in-built communications offers possibilities of detecting and tracing the source of pollution events. One such device uses a Cellnet phone with a microcomputer and pH, turbidity and DO probes powered by an on-board power pack which can be moored in a river. When any of the probes registers a change outside a pre-set range a sequence of samples of water can be taken and someone alerted by phone (Adams *et al* 1992). Such developments in monitoring technology will doubtless continue and will be valuable operational tools, although by the time they register a problem it is often too late.

The process of setting quality objectives, determining action plans and monitoring trends to assess achievements is much like an industrial production process. If it is viewed in this way then river quality control can be considered to be a true quality control task much like any other. Monitoring can then be viewed as quality control sampling rather than a snapshot picture of water quality, and there is the prospect of added value and better control resulting. Such an approach has been proposed for groundwater monitoring at hazardous sites (Starks & Flatman 1991) but, not as far as we are aware, for river waters. The approach would need to take account of natural variability but this could be built into a model to underlie the quality-control approach.

REFERENCES

Adams IM. (1992) Management of River Thames. In: Newman PJ, Piavaux MA, Sweeting RA (eds) *River Water Quality – Ecological Assessment and Control*. EUR 14606 EN-FR, Office of Official Publications of European Communities, Luxembourg. [17.5]

Adams JRW, Dolby JC, Williams NP. (1992) National Rivers Authority initiatives in the development of portable monitoring equipment. *Journal of the Institution of Water and Environmental Management* **6**: 64–71. [17.6]

Cartwright NG, Rees YJ, Derick M, Zabel TF. (1991) *The Application of Limit Values in Three EC Member States*. National Rivers Authority R&D Note 30, Bristol. [17.4]

CEC (Council of the European Communities). (1973) Directive on detergents (73/404/EEC). *Official Journal L347* 22 November 1973. [17.4]

CEC (Council of the European Communities). (1975) Directive concerning the quality required of surface water intended for the abstraction of drinking water in the Member States (75/440/EEC). *Official Journal L194/39* 25 July 1975. [17.4]

CEC (Council of the European Communities). (1976a) Directive concerning the quality of bathing water (76/160/EEC). *Official Journal L31/1* 5 February 1976. [17.4]

CEC (Council of the European Communities). (1976b) Directive on pollution caused by certain dangerous substances discharged into the aquatic environment of the Community (76/464/EEC). *Official Journal L129/23* 18 May 1976. [17.4]

CEC (Council of European Communities). (1976c) Directive on the approximation of the laws, regulations and administrative provisions of the Member States relating to restrictions on the marketing and use of certain dangerous substances and preparations (76/769/EEC). *Official Journal L262* 27 September 1976. [17.4]

CEC (Council of the European Communities). (1978) Directive on the quality of fresh waters needing protection or improvement in order to support fish life (78/659/EEC). *Official Journal L222/1* 14 August 1978. [17.4]

CEC (Council of the European Communities). (1979a) Directive on the quality required of shellfish waters (79/923/EEC). *Official Journal L281/47* 10 November 1979. [17.4]

CEC (Council of the European Communities). (1979b) Directive prohibiting the placing on the market and use of plant protection products containing certain active substances (79/117/EEC). *Official Journal L33* 8 February 1979. [17.4]

CEC (Council of the European Communities). (1980a) Directive relating to the quality of water intended for human consumption (80/778/EEC). *Official Journal L229/11* 30 August 1980. [17.4]

CEC (Council of the European Communities). (1980b) Directive on the protection of groundwater against pollution caused by certain dangerous substances (80/68/EEC). *Official Journal L20/43* 26 January 1980. [17.4]

CEC (Council of the European Communities). (1982) Directive on limit values and quality objectives for mercury discharges by the chlor-alkali electrolysis industry (82/176/EEC). *Official Journal L81* 27 March 1982. [17.4]

CEC (Council of the European Communities). (1985) Directive on the approximation of the laws of the Member States concerning the lead content of petrol (85/210/EEC). *Official Journal L96* 3 April 1985. [17.4]

CEC (Council of the European Communities). (1989) Directive on procedures for harmonising the programme for the reduction and eventually elimination of pollution caused by waste from the titanium dioxide industry (89/428/EEC). *Official Journal L201* 14 July 1989. [17.4]

CEC (Council of the European Communities). (1990) Commission Directive amending for the fourth time the Annex to Council Directive 79/117/EEC prohibiting the placing on the market and the use of plant protection products containing certain active substances (90/335/EEC). *Official Journal L162* 28 June 1990. [17.4]

CEC (Council of the European Communities). (1991a) Directive concerning urban waste water treatment (91/271/EEC). *Official Journal L135/40* 30 May 1991. [17.4]

CEC (Council of the European Communities). (1991b) Directive laying down the health conditions for the production and the placing on the market of live bivalve molluscs (91/492/EEC). *Official Journal L268* 24 September 1991. [17.4]

Dwyer M, Yeoman S, Lester JW, Perry R. (1990) A review of proposed non-phosphate detergents builders utilisation and environmental assessment. In: *Environmental Technology* vol. II, pp. 263–94. [17.4]

DoE (Department of the Environment). (1989) Agreed 'Red list' of dangerous substances confirmed by the Minister of State (Lord Caithness). *News Release 194*, DoE, Marsham Street, London, 10 April 1989. [17.4]

EPA (Environmental Protection Agency). (1984) *Estimating Concern Levels for Concentrations of*

Chemical Substances in the Environment. Report of the EPA Environmental Effects Branch, Washington, DC. [17.4]

Evans GP, Wallwork JF. (1988) The WRc fish monitor and other biomonitoring methods. In: Gruber D & Diamond JM (eds) *Automated Biomonitoring,* pp. 75–90. Ellis Horwood, Chichester. [17.6]

Federal Government. (1989) Federal States Working Group 'Quality Objectives' (BLAK QZ) Concept for the Derivation of Quality Objectives for the Protection of Inland Surface Waters Against Dangerous Substances, 10 October 1989. [17.4]

GESAMP (Joint Group of Experts on the Scientific Aspects of Marine Pollution). (1986) *Environmental Capacity – An Approach to Marine Pollution.* [17.4]

Haigh N. (1985) The co-ordination of standards for chemicals in the environment. Paper presented at the European Environmental Bureau Seminar on: *A European Community Policy Concerning Existing Chemicals,* Brussels, 9–11 June 1985. [17.4]

Hellawell JM, Green MB. (1986) Performance of the Tame Purification Lake. In: Solbe JG (ed) *Effects of Land Use on Water Quality.* Ellis Horwood, Chichester. [17.5]

Holdgate MW. (1979) *A Perspective of Environmental Pollution.* Cambridge University Press, Cambridge. [17.2, 17.4]

Miller DG. (1991) Environmental Issues and Concerns in Europe. Paper presented at the IWEM Scientific Section Seminar on Environmental Research In Europe, London, 24 April 1991. [17.4]

Newman PJ, Piavaux MA, Sweeting RA (eds). (1992) *River Water Quality – Ecological Assessment and Control.* EUR 14606 EN-FR, Office of Official Publications of European Communities, Luxembourg. [17.1]

NRA (National Rivers Authority). (1991) *Proposals for Statutory Water Quality Objectives.* Water Quality Series No. 5, December 1991. NRA, Bristol. [17.4, 17.6]

Priest W, Griffith R, Zabel TF. (1991) *Towards Integrated Pollution Control – BATNEEC and the Challenge for Industry.* Clifford Chance, Earnst & Young and WRc plc. [17.4]

Royal Commission on Environmental Pollution. (1992) *16th Report: Freshwater Quality.* HMSO, London. [17.1]

Seager J. (1993) Statutory Water Quality Objectives and river water quality. *Journal of the Institution of Water and Environmental Management* **7**(5): 556–66. [17.4]

Starks TH, Flatman GT. (1991) RCRA groundwater monitoring decision procedures viewed as quality control schemes. *Environmental Monitoring and Assessment* **16**: 19–37. [17.6]

van der Gaag M, Stortelder PBM, van de Kooij LA, Bruggeman WA. (1991) Setting Environmental Quality Criteria for water and sediment in the Netherlands: a pragmatic ecotoxicological approach. *European Water Pollution Control* **1**(3). [17.4]

Wright JF, Armitage PD, Furse MT, Moss D. (1989) Prediction of invertebrate communities using stream measurements. *Regulated Rivers: Research and Management* **4**: 147–55. [17.6]

18: Environmentally Sensitive River Engineering

R. D. HEY

18.1 RIVER ENGINEERING

Traditional engineering practices

From time immemorial rivers have been harnessed, managed and engineered for the benefit of humans. Within the last century river engineering works have become more intrusive, in terms of the scale of modification, as mechanization has increased our ability to modify rivers and their local environment. Early engineering practices relied on trial-and-error procedures as schemes were implemented and subsequently managed to meet and maintain design objectives. Inevitably mistakes were made as design methods evolved.

Any engineering work that modifies the river system has the potential to cause instability and adversely affect the riverine environment. Attempts to impose an unnatural condition on a river can lead to major instability problems unless the river is heavily engineered. In turn, this can cause severe environmental degradation. The types of problem that can result may be illustrated by considering a range of river engineering works.

Flood control and land drainage works have traditionally involved different combinations of channel widening, dredging and straightening (see Chapter 4). Not only does this destroy river and bankside flora and habitats, but it adversely affects the local sediment transport regime of the river and promotes channel instability (Hey 1987a). For example, long sections of the Lower Mississippi have been severely destabilized as a result of channel straightening. Headward erosion, both on the main river and its tributaries, caused severe sedimentation problems in the modified reaches. Major spur-dyke fields have been constructed to prevent bank erosion while extensive dredging is required to maintain navigation depths (Winkley 1982). On rivers which do not transport significant amounts of bed material load, for example many of the lowland rivers in Eastern England, channel resectioning may not cause serious instability problems; any siltation that does occur is routinely dredged at approximately 10-yearly intervals. The environmental consequences are, however, still very damaging as instream habitats are destroyed and any rehabilitation that does occur is short lived.

River training works are often carried out to stabilize aggrading or degrading channels. For example, braided rivers can be stabilized by increasing their ability to transport the sediment supplied from upstream. Traditionally, this involves the construction of a straight channel, to maximize the gradient, with a relatively deep and narrow cross-section. This can only be maintained if the banks are artificially lined to prevent their failure. Inevitably this totally destroys the natural channel and creates a uniform, canalized river with little or no environmental value. The formerly braided Alpine Rhine in Switzerland was extensively trained in the 18th century in this manner. Land previously destroyed as the braided river shifted its position is now intensively farmed and flood risk is considerably reduced as sediment accretion is prevented from destroying the capacity of the channel (Visher 1986).

Major water resource development schemes, either surface water reservoirs or groundwater, can also destabilize a river system by reducing flood flows. The impact depends on the degree of regulation and the nature of the tributary channels. Immediately downstream from the reservoir or abstraction zone the impact will be considerable, but this will decline as unregulated tributaries join the main river.

Reservoirs, both direct supply and regulating, can cause bed degradation for a considerable distance downstream from a dam as well as promoting a coarsening or armouring of the bed as finer material is eroded but not replenished from upstream. Eventually a state is reached when relatively little bed material transport occurs. The bed then becomes imbricated as fines fill in the voids in the gravel framework and the bed is rendered relatively impermeable and unsuited for salmon and trout spawning. In addition the width of the river is reduced by vegetation encroachment as flood flows no longer occur (Raynov *et al* 1986).

Major groundwater development schemes, by drawing down the aquifer over large areas during dry periods, encourage increased recharge and reduced run-off during wet periods. Consequently the frequency of flood flows can, potentially, be reduced. If this occurs, then the river will adjust its bankfull morphology to the new imposed condition.

Basis for environmentally sensitive approaches

The stable morphology of natural alluvial channels, defined by the bankfull longitudinal slope, cross-section and plan shape of the river, is dependent on the flow and sediment transport regime, the nature of the bed and bank sediment, bankside vegetation and the slope of the valley. Any change in the flow regime or the dimensions of the channel through engineering works can destabilize the river and promote erosion and deposition not only within the engineered reach, but upstream and downstream from it (Hey 1986). For example, a section of the Homochitto River in southwest Mississippi was straightened between 1938 and 1946 and this caused degradation up to 40 km upstream by 1974.

An understanding of the physical processes controlling channel shape and dimensions (see Chapter 12) is a prerequisite for developing and implementing river engineering schemes that maintain channel stability and the natural range of habitat features which are vital for riverine flora, fisheries and fauna. Such schemes are also likely to be more cost-effective in the longer term.

18.2 FLOOD ALLEVIATION SCHEMES

Traditional and environmental options

A post-project appraisal has been carried out on 18 flood alleviation schemes in England and Wales to establish their engineering and environmental performance (Hey *et al* 1990). Schemes were chosen to be representative of urban and rural projects, as space and design standards often require different solutions, and upland and lowland schemes; the latter essentially enabled differentiation between those rivers which transport bed material load, and therefore are more susceptible to engineering change, and those which transport sand, silt and organic material and which are likely to be less affected.

In view of the criticism of the traditional methods of increasing flood capacity, on conservation grounds, more recently advocated environmental options were also evaluated.

Traditional methods

1 Resectioning. This involves dredging and/or widening the main channel to increase its inbank discharge capacity. In addition, bed slope may be steepened to increase flow velocities and further augment flood capacity. In urban areas rectangular flume-like channels are often constructed due to limited space. In order to maintain a stable bank, vertical sheet piling, concrete or masonry must be used to line the channel.

2 Realignment. Straightening the channel increases the gradient of the river and, thereby, flow velocity and flood capacity. This is often carried out in association with resectioning.

3 Adjacent flood banks. The construction of flood banks or levees is a common engineering solution to flood control. Conventionally these are constructed close to the river and, as a consequence, they need to be higher than distant flood banks to achieve the same level of protection.

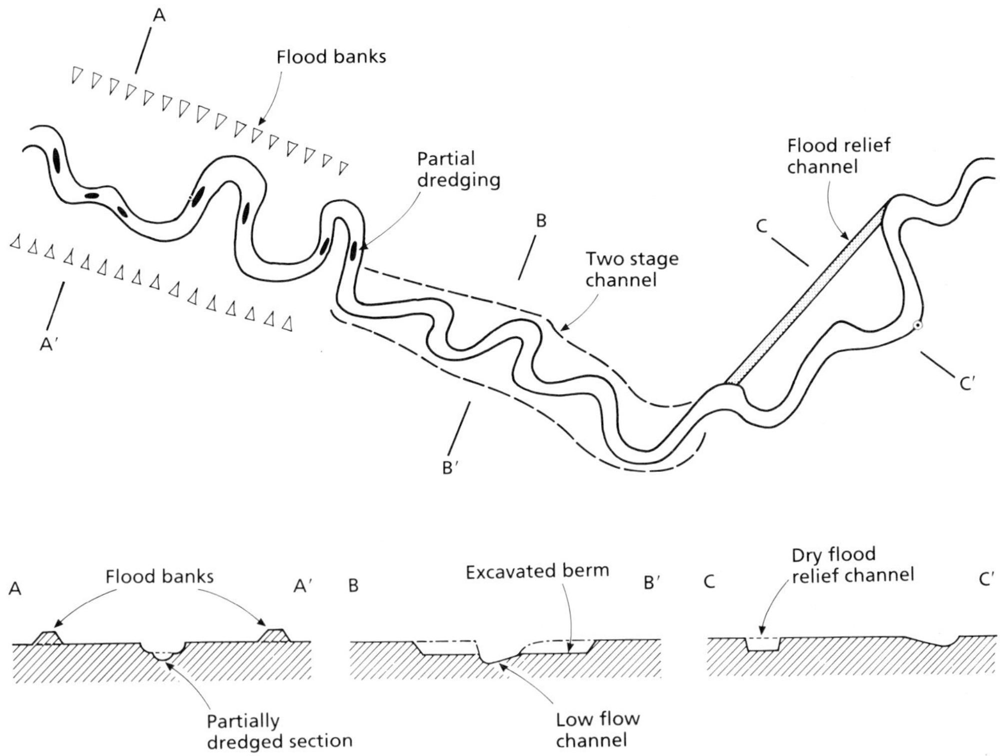

Fig. 18.1 Flood alleviation schemes: environmental options.

Environmental options

1 Relief channels. These are constructed to divert flow above a given stage away from the main channel. Such channels may be wet or dry during low flow periods. The natural channel is left intact (Fig. 18.1).
2 Partial dredging. A limited central section of the river is dredged to increase channel cross-sectional area, particularly at shallow riffle sections. The undisturbed bed acts as a biological store for recolonizing species (Fig. 18.1).
3 Distant flood banks. Where space is not limiting, the flood banks can be set back some distance from the river. Often the banks are located at the edge of the meander belt (Fig. 18.1).
4 Two-stage channels. These are created by excavating the upper section of the floodplain adjacent to the river. The natural low-flow channel is preserved while higher flows are contained within the newly created flood berm (Fig. 18.1).

Effect on channel stability

Traditional methods

Resectioning. On upland rivers this can cause severe instability in alluvial reaches as headcut erosion can promote significant deposition in the engineered reach, as the river attempts to re-establish its width, depth and slope, and erosion farther downstream (Hey 1987a). Harder engineering solutions are often employed, particularly in urban areas, to try to maintain the design channel. This often involves combinations of bank revetments and check weirs in the dredged section, and bed load traps farther upstream in order to control the transport regime of the river. Bed sills and bed load traps prevent headward erosion and limit the supply of sediment to the engineered reach, where it would be deposited, while the check weirs control local gradient at the entrance to the dredged section and prevent bed scour during floods. While this can protect the scheme, erosion can result farther downstream as sediment loads are substantially reduced.

In lowland areas instability problems are not as serious. Eventually the river would re-establish its original condition through reed encroachment from the bank and by bed aggradation. Dredging every 5–10 years can generally maintain the flood capacity of the river.

Realignment. On upland rivers, channel straightening can promote considerable instability with headcut erosion and the re-establishment of a meandering channel in the engineered reach (Hey 1987a). A bed sill at the head of the reach would prevent headcut erosion, while a series of check weirs along the straightened reach could maintain channel gradient and prevent bed scour. Bank revetments would be required to stabilize the banks.

On lowland rivers, straightening does not create the same problems unless increased gradients cause the bed to be scoured. Some form of bank protection may be required to prevent failure of the over-deepened banks.

Adjacent flood banks. The construction of flood banks adjacent to the river increases the in-bank flood capacity of the river and sediment transport capacity in comparison with neighbouring natural reaches. On upland rivers this can cause instability problems with erosion in the engineered reach and deposition farther downstream. In lowland rivers the increased velocities during flood events are unlikely to cause bed degradation and they will probably remain fairly stable.

Environmental options

Relief channels. Instability can occur at the intake into the relief channel in rivers transporting bed material load. The reduced transport capacity in the main river when the flow is divided precipitates deposition of coarse material. At the re-entry point, erosion can occur as the transport capacity of the river is significantly increased. This is especially the case if the angle of the confluence is obtuse. Secondary flows generated by the flow from the relief channel can cause significant bed scour. In lowland rivers such problems are less likely as most sediment is transported in suspension. Care still needs to be exercised to ensure that return flows from the relief channel enter the river at an acute angle to prevent bed scour.

Partial dredging. On upland rivers the partially dredged section will rapidly become filled with coarse bed material. However, on lowland rivers it can be maintained by routine maintenance dredging every 5–10 years.

Distant flood banks. Flood banks set back from the river, which allow the river to meander freely within their bounds, enable the natural channel to remain intact. During overbank flows flooding is curtailed to a limited strip of floodplain. Provided space is available such schemes represent the best solution for flood alleviation. No instability problems are generated and provided the restricted flood plain is grazed or otherwise managed, the flood capacity of the scheme will be maintained.

Two-stage channels. The construction of a two-stage channel in upland rivers is unlikely to be successful. Local change in the bedload transport capacity of the river as it enters the two-stage channel would precipitate aggradation, rapid bank erosion and destruction of the newly created berm during flood flows. On lowland rivers they have proved to be more successful but even then they have to be carefully designed and managed. If the flood berm is excavated too close to the water table, then it cannot be grazed by cattle as it will not support their weight. A major maintenance cutting programme is then required to prevent reed encroachment as this would seriously impair flood channel capacity. In urban areas the flood berm could be maintained as a linear park.

Effect on instream and bankside flora

The environmental performance of the 18 flood alleviation schemes investigated were assessed to establish their impact on instream and bankside flora. For each type of scheme the following information was provided:

1 Species diversity: this gives the number of species found per 10 m length of river. For comparison purposes it should be noted that for natural reaches the mean diversity of instream species was 4.18 species per 10 m length of river and for bank species 4.22 per 10 m.

2 Diversity ranking: rank ordering of schemes on the basis of number of species per 10 m. Natural channels were ranked third for instream species and fifth for bank species on a scale 1–7.

3 Related species: species were identified which were associated with particular engineering works irrespective of geographical location.

Traditional methods

Resectioned and realigned. These schemes considerably limit habitat diversity, particularly if the banks are lined to prevent erosion. As a result they exhibit low species diversity (Table 18.1).

Adjacent flood banks. Bank height is artificially increased and thereby bank habitat diversity. Consequently bank species diversity is relatively high (Table 18.2).

Table 18.1 Environmental parameters for resectioned/realigned schemes

Parameter	Unlined	Lined
Species diversity	Instream 3.17; bank 4.04	Instream 2.73; bank 3.18
Diversity ranking	Instream 5th; bank 6th	Instream 6th; bank 7th
Related species	None	None

Table 18.2 Environmental parameters for schemes with adjacent flood banks

Parameter	Instream	Bank
Species diversity	3.33	5.67
Diversity ranking	4th	2nd
Related species	None	Fleabane

Table 18.3 Environmental parameters for moist relief channels

Parameter	Instream	Bank
Species diversity	2.17	6.67
Diversity ranking	7th	1st
Related species	None	Teasel

Environmental methods

Relief channels. Wet relief channels are those close to or below the water table. They are difficult to maintain and create a favourable environment for such undesirable species as algae and pondweed. Dry, higher channels only contain water during periods of high flow. For moist channels the results obtained are given in Table 18.3.

Distant flood banks. These showed slightly greater species diversity than natural channels, which probably reflects some partial dredging of point bars at the sites investigated (Table 18.4).

Two-stage channels. The low-flow channel is preserved while the bank habitat is changed considerably due to the proximity of the water table.

Table 18.4 Environmental parameters for distant flood banks

Parameter	Instream	Bank
Species diversity	5.48	5.38
Diversity ranking	2nd	3rd
Related species	Jointed rush; filamentous algae	Creeping yellow cress; creeping bent

Table 18.5 Environmental parameters for two-stage channels

Parameter	Instream	Bank
Species diversity	6.51	4.63
Diversity ranking	1st	4th
Related species	Marsh woodwort; unbranched burr reed	Yellow loosestrife; lesser-tufted sedge; yellow flag

For instream species it exhibits the highest species diversity, reflecting the provision of favourable marginal habitats, but a much lower diversity for bank species (Table 18.5).

Environmental design guidelines

Distant flood banks, where the banks are located along the edge of the meander belt, are the preferred solution as they have no impact on the natural stability of the river or on the riverine environment. Stability is maintained because the natural bankfull capacity of the river is unchanged. Should space limitations preclude their use, alternative procedures are required and these increase the 'inbank' capacity of the river. Any attempt to artificially increase the natural capacity of the river will cause instability and this will adversely affect its conservation value (Table 18.6).

In lowland urban areas, two-stage channels are a viable alternative as siltation will be minimal and they can actually increase environmental value. The flood berms can be maintained as an urban park. If insufficient space is available for this approach, adjacent flood banks are acceptable provided the natural alluvial bank is preserved.

Table 18.6 Engineering and environmental performance of flood alleviation schemes

	Engineering and environmental performance		
Option	Advantages	Disadvantages	Design limitations
Distant flood banks	Natural stability and flood capacity maintained. Instream and bankside habitats and flora unaffected		*Preferred solution.* Large space requirement which may be unavailable in urban areas. Possible need to change agricultural practice in flood storage zone in rural areas
Adjacent flood banks	On lowland rivers with negligible bed material load, no significant change on instream habitats and flora. Bank diversity may be enhanced	Scouring on upland rivers which transport bed material load unless bed is armoured. May be prevented by weirs. Restricts view of river across floodplain	*Not suited to upland-type rivers without additional engineering works*
Two-stage channel	On lowland-type rivers, instream low-flow habitats and flora unaffected. In short-term floral diversity increased on berm but unlikely to be maintained should reed bed become established without heavy maintenance	On upland-type rivers deposition of bed material load at entrance to two-stage section will promote instability. On lowland-type rivers need to ensure that flood berm is sufficiently high above the water table to allow grazing/mowing and prevent reed growth. If berm surface becomes overgrown, design flood capacity is lost	*Not suited to upland-type rivers.* Larger space requirement. In urban areas could be developed as a linear park
Relief channels	Instream and bankside flora and habitats unaffected	On upland-type rivers significant deposition likely to occur at entrance to relief channel and erosion at exit	*Not suited to upland-type rivers*

continued

Table 18.6 *Continued*

Option	Engineering and environmental performance		Design limitations
	Advantages	Disadvantages	
Resectioning		On upland-type rivers severe erosion at head of dredged section, unless prevented by weir or bed sill, and deposition at downstream end. In lowland-type rivers siltation in dredged reach requiring regular maintenance dredging. Flood capacity impaired. Instream and bankside flora are totally destroyed	*Not recommended unless necessitated by lack of space.* Check sediment transport capacity for stability and ensure weirs and bank revetments have conservation/ architectural merit
Realignment		On upland-type rivers severe headward erosion, unless prevented by weir or bed sill, and deposition downstream of straightened section. Bank revetment probably required to maintain unnatural alignment and weirs to prevent bed erosion. Instream and bankside habitats and flora totally destroyed	*Not recommended unless no alternative.* Check sediment transport capacity for stability and ensure weirs and bank revetments have conservation/ architectural merit

Resectioning should only be contemplated as a last resort as it completely destroys the riverine environment and requires maintenance dredging.

At urban upland sites, adjacent flood banks are preferable to relief channels or resectioning since they have less impact on channel stability and the riverine flora. The latter can cause severe instability problems and should not be attempted. Two-stage channels are likely to cause similar instability problems.

In rural lowland regions two-stage channels can be successful provided the vegetation on the flood berm can be managed to preserve the design discharge capacity of the scheme. Instability need not be a problem, provided the flood channel is properly aligned, and the riverine environment is enhanced. Adjacent flood banks are preferable to relief channels and resectioning if space is limited. Resectioning should not be attempted, even if maintenance requirements are minimal, due to the very adverse environmental impact.

Adjacent flood banks are the only real alternative to distant flood banks in upland rural regions. Even these promote some instability depending on the degree of flood confinement. Instream flora is adversely affected although this is, in part, compensated by increased diversity of bankside species. All other procedures have been shown to cause expensive instability problems and a deterioration in riverine flora.

18.3 REGIONAL CHANNEL STABILIZATION

Nature of problem

Rivers become unstable when there is an imbalance between the sediment load supplies to the reach in question and the river's ability to transport it. Regional instability relates to large-scale and systematic degradation and aggradation, which can be natural or man-induced. It can progress downstream, due to change in sediment supply, or upstream as a result of local changes in transport capacity.

Incision along an alluvial reach of river can cause significant land loss as banks fail and the channel rapidly widens. In time the river would re-establish a regime-type condition with the river flowing within a new lower-level floodplain leaving the original floodplain as a terrace feature.

Severe aggradation, which results in the development of a braided channel form, can destroy the agricultural potential of floodplain areas due to rapid channel shift and increased flood frequencies. Eventually a new equilibrium would be achieved in which the river would either meander or form a multi-channel anastomosing system across a new, higher-level floodplain.

Erosion control options

Grade control

Incision along a reach of river is traditionally controlled by the construction of grade-control structures or check weirs at various points along its length. The requirement is to define the number, location and height of these structures to stabilize the river and prevent further headward erosion on the main river and its tributaries.

Three approaches are available: conventionally grade control is determined using tractive stress calculations or regime theory, but there is also a geomorphological alternative.

1 Tractive stress method. This determines the slope, velocity and depth of flow which will prevent the erosion of material from the bed or banks of the river (Lane 1937; Einstein 1950; Vanoni & Brooks 1957).

2 Modified regime approach. This allows sediment supplied to the reach, either upstream or from tributaries, to be transported through the system. Appropriate mobile-bed regime equations are used to predict slope, velocity and flow depth, depending on bed material size (sand, see Blench 1966; gravel, see Hey & Thorne 1986).

3 Geomorphological method. The method is based on an empirical geomorphological model of channel evolution, consequent upon the initiation of erosion, developed from field observations and measurements (Schumm *et al* 1984). Using a space for time substitution, the authors constructed a five-stage model of channel development from initial incision to final attainment of quasi-equilibrium (Fig. 18.2). For reaches that were in quasi-equilibrium, relations were obtained between channel top width, the 5-year discharge and the associated water surface slope. At unstable reaches the equation was used to calculate the allowable slope for existing top width and discharge, or the allowable width for existing top width and slope. The size, location and number of grade control structures or, alternatively, flow control reservoirs, can therefore be determined by calculating allowable slope or discharge criteria and by computing backwater profiles. Values for soil loss, from bed and banks, and the increase in top width can be ascertained which were relevant for the assessment of potential reservoir sedimentation rates if nothing was done to control the problem.

Lowered valley

For small alluvial channels (<20 m top width), where incision is of the order of the original bankfull depth of the river, it would be possible to stabilize the river by forming a new regime channel within a lowered valley in the incising section and by constructing a grade control structure to prevent continued headward erosion. Effectively this creates what nature would eventually produce in the incising section below the grade control structure without causing aggradation farther downstream.

The new valley floor would be formed at a level corresponding to the depth of incision and its width would need to accommodate the amplitude of the meanders of the new channel. Regime equations, as appropriate for the nature of the bed material, would be used to design the new channel given the bankfull discharge and sediment load for the natural reach upstream from the grade control structure. Regime equations only enable channel width, depth and slope to be

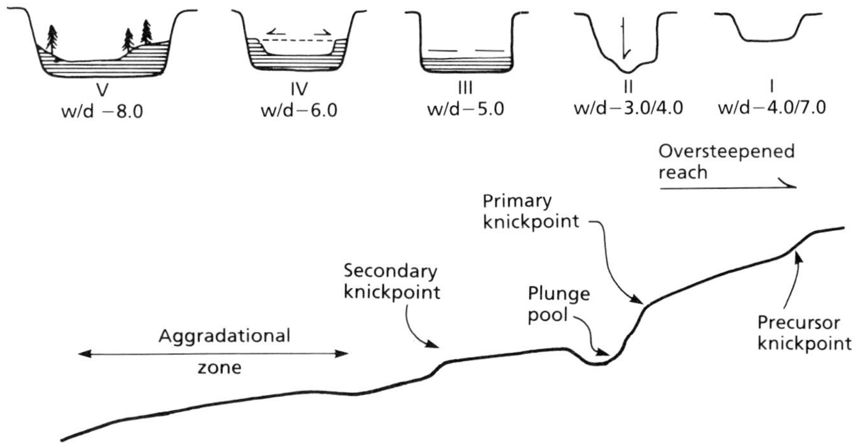

Fig. 18.2 Longitudinal and cross-sectional profiles of unstable river reach (5-stage evolution: channel width (w) to depth (d) ratios given; size of arrows indicates relative importance and direction of dominant processes) (from Schumm *et al* 1984).

determined but, as valley slope is prescribed, channel sinuosity – the ratio of channel length to valley length – can be determined from the ratio of valley slope to channel slope. This, together with meander arc length, the channel distance between inflexion points, which is given by 4–10 times channel width, uniquely prescribes the plan shape of the river (Hey 1976). Pools and riffles would need to be incorporated in the design and special attention given to the construction of meander bends to obtain a natural asymmetry and appropriate maximum flow depths.

Engineering and environmental performance

Grade control structures

The three methods outlined above were evaluated on an incising section of Oaklimiter Creek, Mississippi. Calculations showed that 15 grade control structures were required to stabilize the river according to the modified regime approach, 20 with the tractive stress method, and six with the geomorphological design procedure.

The traditional engineering solutions tend to produce an essentially constant low slope along the ponded reaches created by the weirs, while the geomorphological procedure recognizes that a natural-equilibrium longitudinal stream profile is concave.

Although the six structures were not actually constructed to enable the method to be evaluated, this study suggested that significant costs savings could be made by adopting the geomorphological approach.

The design of the actual grade control structures is also crucially important for the success of the scheme. By creating fixed points on the bed of the river, bank erosion may be precipitated at the structure due to the development of strong vortices adjacent to the bank (Fig. 18.3). Knauss and Scheuerlein (1988) advocate the use of curved submerged sills, bowed upstream with gentle side contraction. This creates convergent overspill and associated slight scouring in mid channel, while the vortices adjacent to the bank are significantly reduced as the contraction at the bank is gradual (Fig. 18.3). Where high structures are constructed as, for example, on Long and Godwin Creeks in northern Mississippi, a baffle is required in the plunge pool to dissipate energy and prevent excessive scouring.

Even though grade control structures can successfully arrest bed degradation, the heightened over-steepened banks in the ponded reaches are potentially more prone to failure following

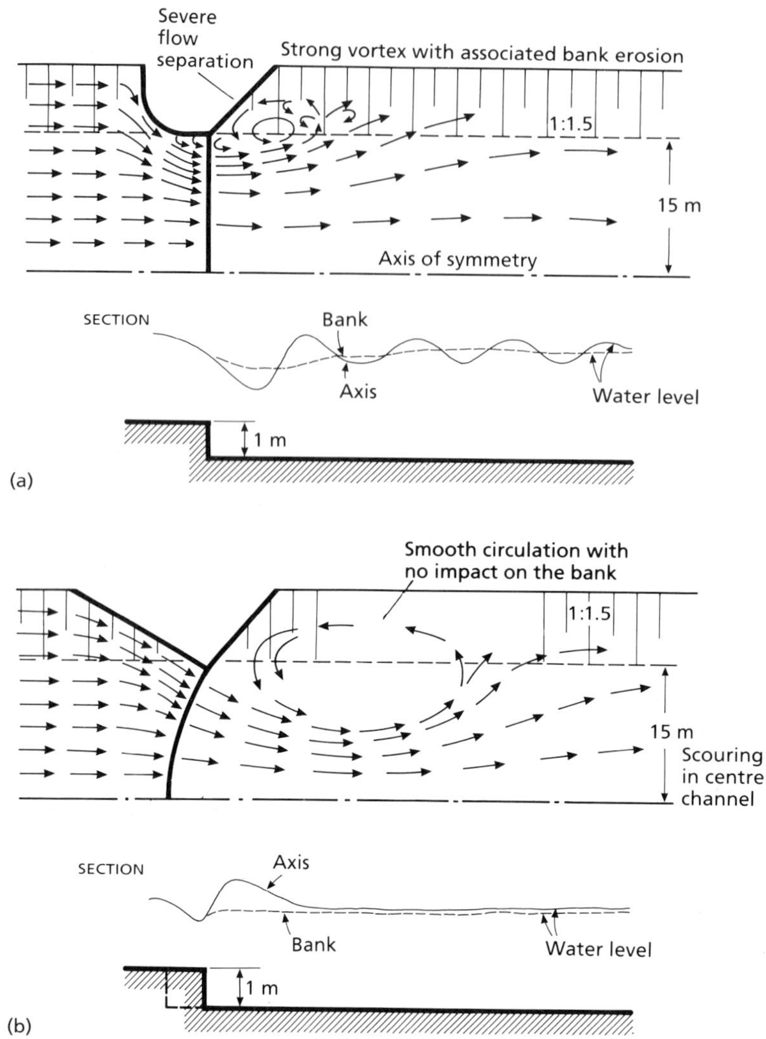

Fig. 18.3 Effect of sill geometry on bank erosion: (a) submerged sills, straight layout plan; (b) submerged sills, curved layout plan (after Knauss & Scheuerlein 1988).

incision. However, observations on Long Creek indicate that bank erosion is generally reduced since deposition in the ponded reach reduces bank heights and failed material is less readily removed (Biedenharn *et al* 1990). More recent evidence indicates that some failures are still occurring in meander bends where secondary flows are sufficiently strong to remove eroded debris and maintain local bank height.

No environmental assessments have been carried out to compare the performances of the three different approaches or to establish their absolute impact. The latter would be difficult to assess because the pre-condition is not in equilibrium and, by providing stability, there is scope for enhancement as flora, invertebrates and fisheries become established. However, as the geomorphological approach requires fewer structures and ponding behind the weirs is not total, it should create more habitat variety and have greater conservation potential.

Lowered valley

For smaller rivers where degradation would be less than natural channel depth, the lowered valley option probably represents a viable alternative to the construction of grade control structures.

A stable natural channel would be reinstated with its complementary range of instream and bankside habitat features, which would improve the conservation and fisheries value of the affected reach and enhance visual amenity.

Construction of the new channel and floodplain could constitute a problem unless the river was realigned. This would enable construction work to be carried out in relatively dry conditions and allow vegetation to become established before flood flows needed to be transmitted through the channel. Reconstructing the existing channel would involve working in difficult wet conditions and it would be prone to erosion and bank failure until vegetation became established.

Environmental design guidelines

Grade control structures represent the only sensible procedure for stabilizing large rivers and those where incision exceeds the original local bankfull flow depth. Deep excavations to create a new valley would be prohibitively expensive as large volumes of material would have to be removed (Table 18.7).

It would appear that the geomorphological approach for establishing the number of grade control structures to stabilize the river is better than alternative methods as it minimizes the number required. However, until schemes are implemented and evaluated in different river environments its potential is unconfirmed.

For smaller rivers where incision is limited, excavation of a lower-level valley would be a viable alternative, especially if it could be realigned to allow working in dry conditions (Table 18.7). Environmentally this would be the preferred solution as the reinstated river would provide a range of habitats for biological exploitation. Such an approach has been successfully employed when a stable, dredged, resectioned channel was diverted (Hey 1992) but it needs to be evaluated for stabilization schemes.

Aggradation control options

Resectioned and straightened channel

Where braiding results from sediment supplied from catchment source areas, a traditional engineering solution would be to straighten, narrow and deepen the channel in order to increase its capacity to transport sediment to match that supplied from upstream. This approach was adopted on the Alpine Rhine in Switzerland.

Natural channel

An alternative to resectioning and straightening would be to create an irregular sinuous channel. A rational design approach would be required involving appropriate flow resistance and sediment transport equations to determine bankfull depth, slope and velocity (Hey 1987b). Channel width would need to be selected, as no process-based equations are available to predict its value.

Table 18.7 Engineering and environmental performance of erosion control measures

	Engineering and environmental performance		
Option	Advantages	Disadvantages	Design limitations
Grade control	Controls incision and reduces bank erosion	Artificially ponded. Adversely affects instream and bankside habitats	*Only option for incision greater than local stable channel depth*
Lowered valley	Controls incision and reduces bank erosion. Natural channel with range of habitat features recreated	Considerable excavation to create new lower valley	*Alternative option when incision less than local stable channel depth and for rivers less than approx. 20 m top width*

For gravel-bed rivers, equations developed by Hey and Thorne (1986) could be used on the assumption that the channel would effectively be tree-lined or incised; as the river would need to be revetted to maintain its width and planform, this would be a fair assumption. Once channel width is established and the required channel slope calculated, the plan shape of the channel can be determined given the overall valley slope (see p. 345).

Bank revetments would be crucial to preserve the integrity of the scheme and it is likely that hard engineering solutions involving blockstones or gabion baskets would be necessary, particularly on the outer bank in meander bends. However, it is possible that softer, bioengineering methods could be used on straighter sections and on the inside of meander bends (see Section 18.4).

Braiding that results from incision at adjacent upstream reaches, channel-derived sediment, would be treated by stabilizing the upstream eroding reach, as outlined previously, and constructing an appropriately designed sinuous channel to accommodate the reduced sediment load.

Engineering and environmental performance

Resectioning

This represents the cheapest solution, as the new channel has the shortest possible length, and it has been successfully used on the Alpine Rhine in Switzerland. Environmentally the river lacks instream and bankside flora, as the bed has few if any exposed bars, and the hard bank revetments preclude successful colonization. In addition the straightened river has little habitat variety and amenity value.

Natural channel

No information is currently available to quantify the engineering and environmental performance of these channels. Nevertheless it is apparent that it represents a 'softer' alternative to the control of aggradation. By creating a meandering channel with a variable bed geometry, a range of instream habitats are created which are beneficial for instream flora, invertebrates and fisheries. The use of softer bank-protection measures, particularly those involving natural vegetation, provides a more sympathetic approach to bank stabilization that is aesthetically pleasing and provides cover for fish. Because of its longer length, it is more expensive to excavate than standard resectioning procedures, but savings could be made on bank protection costs and maintenance is likely to be cheaper.

Environmental design guidelines

In order to stabilize multi-channel braided systems some form of heavy engineering is required. The ecology of the riparian zone will be dramatically altered by such action but this might have to be accepted in order to satisfy economic and development interests.

The construction of a more natural sinuous channel would be preferable to resectioning as this would have greater morphological and hydraulic variability and, consequently, would have less environmental impact (Table 18.8).

18.4 LOCAL CHANNEL STABILIZATION

Nature of problem

Local instability problems are essentially concerned with bank protection and the maintenance of channel planform. Changes in bed elevation are reversible whereas width adjustments rarely are. Channels which are ostensibly in regime may also need training to prevent down-valley migration of meanders.

Bank protection options

Revetments

A range of measures are available to increase the erosional resistance of riverbanks, and it is not necessary to describe them all here. Recent publications have reviewed the available hard engineering methods and also considered more novel softer options (e.g. Hemphill & Bramley 1989).

Essentially the methods can be ranked from softer, natural vegetative treatments to harder,

Table 18.8 Engineering and environmental performance of aggradation control measures

	Engineering and environmental performance		
Option	Advantages	Disadvantages	Design limitations
Resectioning and straightening	Controls aggradation	Heavily engineered; lacks instream and bankside habitats	*Not recommended*
Natural channel	Controls aggradation. Sinuous channel created with variable bed topography	Bank revetment required	*Preferred option*

engineering methods with little potential for amelioration (Table 18.9).

Hydraulic controls

The aforementioned conventional and bioengineering approaches to bank protection provide stability by increasing the erosional resistance of the bank. Effectively this is treating the symptoms rather than the cause of the problem. An alternative is to modify the flow adjacent to the bank such that the erosional forces acting on the bank are reduced and bank failure prevented. In meander bends submerged vanes or hydrofoils have been developed which prevent bank failure by modifying the pattern of secondary flows in the meander bend (Odgaard & Mosconi 1987; Paice & Hey 1989). The structures are located (Fig. 18.4) to generate a secondary flow cell, which has opposite polarity to the main cell that results from flow curvature in the bend. Convergence of faster surface water then occurs in centre channel inside the line of the vanes and causes bed scouring. Material eroded from this zone is deposited in the original scour hole adjacent to the outer bank. As shear stresses against the outer bank and toe region and the bank height are reduced, bank retreat can be prevented and the natural bank preserved. Should deposition eventually cover the vanes the original secondary circulation pattern would re-establish itself. However, the resultant bed scour will re-expose them and further bank retreat will be prevented. The system is, therefore, self-regulatory.

Table 18.9 Bank revetment procedures ranked according to erosional resistance (after Hey *et al* 1991)

Protection type		Flowchart symbol	Costs (£/m) (May 1991)
Grass	Soft	1	<50
Grass and jute		2	
Grass and geotextile		3	
Reeds		4	
Willow/osier		5	
Ash		6	
Gravel and geotextile		7	
Mastic and geotextile		8	
Woven wooden fence		9	
Riprap	Hard	A	<50
Riprap over geotextile		B	50–100
Gabion mattresses		C	100–150
Gabion riprap		D	100–150
Grouted riprap		E	50–100
Grouted gabion mats		F	100–150
Grouted gabion baskets		G	
Jointed cellular blocks		H	
Interlocking blocks		I	
Cable-tied blocks		J	
Wooden boards		K	50–100
Sheet piling		L	>150
Masonry wall		M	
Concrete wall		N	

Engineering and environmental performance

Revetments

A range of revetment types are available and it is necessary to establish which are appropriate in different circumstances bearing in mind the need for the adoption of bioengineering solutions wherever possible. The universal application of heavy engineering methods is no longer acceptable not only on environmental grounds but also

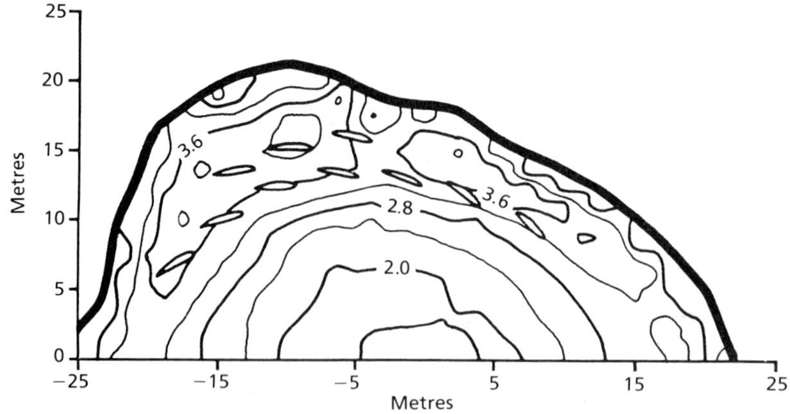

Fig. 18.4 Location of submerged vanes on River Roding, Loughton, Essex. Contours are depths below fixed zero datum (after Paice & Hey 1989).

on economic ones as many hard schemes are likely to be overdesigned.

Recent research has begun to address this issue (Hey *et al* 1991). Based on a post-project appraisal of the engineering performance of a range of schemes in different river environments in England and Wales, a series of flowcharts were developed which enable successful bank protection to be chosen for a particular site on the basis of local hydraulic, morphological, geotechnical, wave and tidal information. Effectively these enable the minimum level of protection to be specified to stabilize the bank against the factors causing failure. By entering information into the flowchart (Figs 18.5, 18.6 and 18.7; Table 18.9) a number of viable methods are identified and the softest one listed would be appropriate.

The post-project appraisal attempted to obtain precise information on the environmental impact, in terms of bankside flora, of the various revetment methods. Unfortunately pre- and post-scheme river corridor survey information was not available to allow this to be objectively carried out and recourse has had to be made to more subjective assessments (Table 18.9). Nevertheless it does enable the softest options to be chosen as appropriate for the local riverine environment.

On cost grounds the harder engineering options are often most expensive although they are generally relatively maintenance free. Table 18.9 lists the average costs of a range of protection methods, as of May 1991, but it should be recognized that considerable variation can occur around these average values due to problems of access, availability of plant, variations in bank height and economies of scale when protecting long lengths of bank.

Hydraulic controls

Submerged vanes or hydrofoils have been successfully used to reduce bank erosion on sand-and-gravel-bed rivers with widths ranging from 20 to 55 m (Odgaard & Mosconi 1987; Paice & Hey 1989). As they obviate the need to modify or cover the bank, the natural character is maintained. Over time sub-aerial weathering will cause bank angles to decline and they will eventually become vegetated. Being submerged even at the lowest flow they are not visually intrusive and actually enhance instream habitats.

As they are located towards the eroding bank and cause bed scouring in mid-channel, they can even be deployed on navigable rivers (Odgaard & Mosconi 1987).

Environmental design guidelines

Bank revetments have traditionally been used to prevent bank erosion. The requirement is to select the most environmentally acceptable type to stabilize a particular erosion problem; this can be achieved using the flowcharts (Figs 18.5–18.7). In high-energy environments, upland rivers or tidal areas, heavier protection measures may be required, whereas in lowland-type rivers

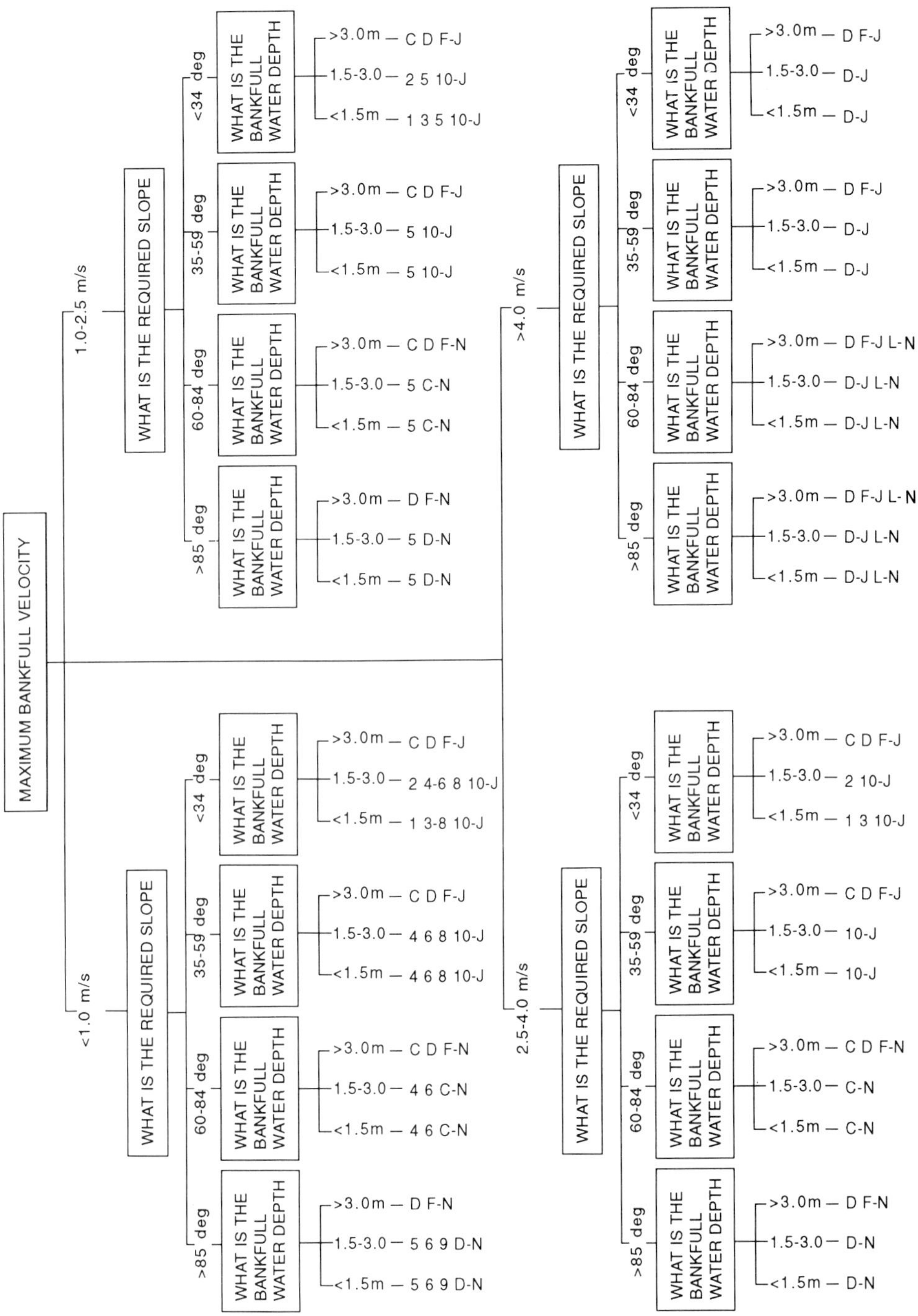

Fig. 18.5 Hydraulic and morphological controls on bank protection options (from Hey *et al* 1991).

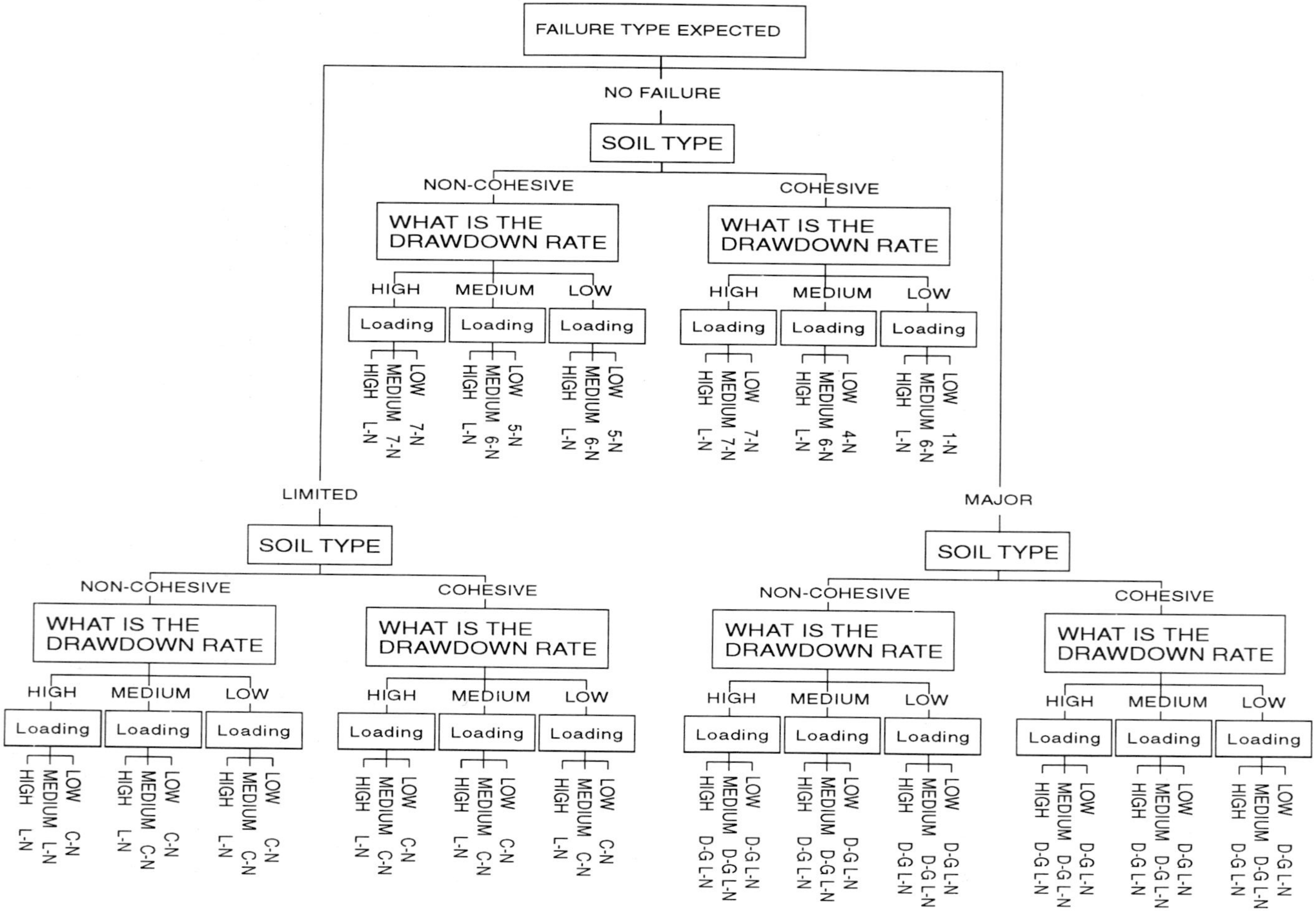

Fig. 18.6 Geotechnical controls on bank protection options (from Hey *et al* 1991).

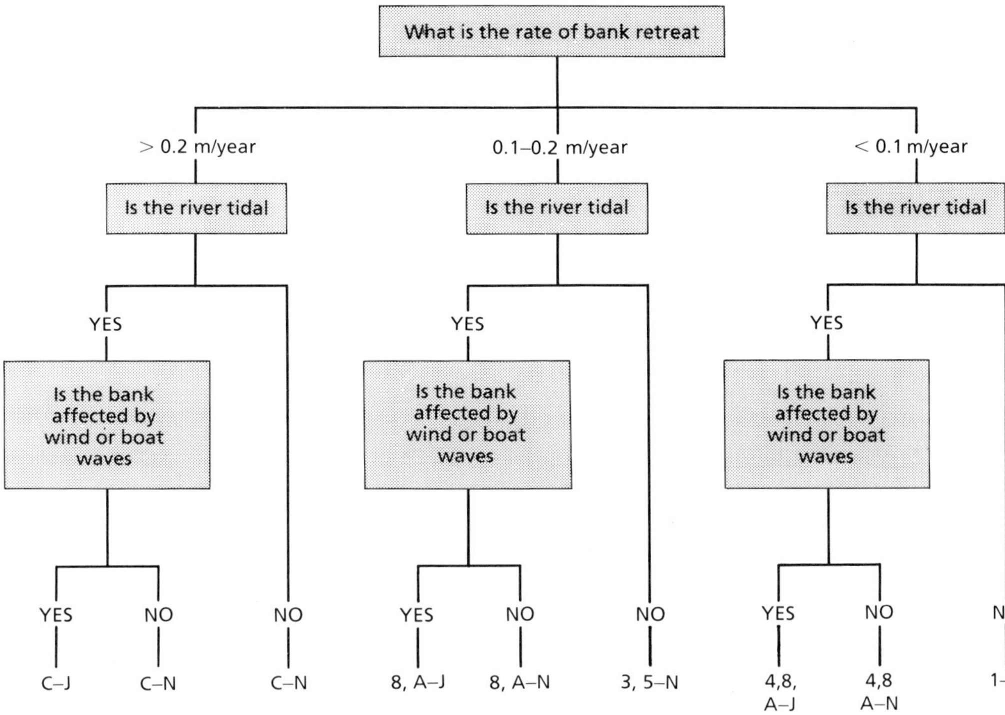

Fig. 18.7 Wave and tidal controls on bank protection options (from Hey *et al* 1991).

vegetative methods may be sufficiently robust. Where possible attempts should be made to soften harder treatments through selective planting (Table 18.10).

As the selection procedure assumes that protection works will be correctly implemented, failure may occur if this is not the case. Recommended installation practices have been identified for the various bank protection procedures, as determined from engineering experience and manufacturers' recommendations, together with methods for 'greening' harder structures (Hey *et al* 1991).

Harder bank revetment works are often an integral part of major schemes to control rivers for flood alleviation, particularly in urban schemes where space limitations preclude more natural solutions (Section 18.2), or for major stabilization works (Section 18.3). Protection is required to maintain an unnatural river condition. In such cases their impact should be ameliorated by planting or by maximizing their architectural merit.

Submerged vanes or hydrofoils are potentially cheaper than all harder revetment methods but do not have universal applications. To date they have been used to prevent bank erosion on meander bends on rivers which are ostensibly in regime, namely where the river migrates across its floodplain while maintaining a relatively stable meander pattern. They can also be used on rivers undergoing slight degradation, where the river is becoming more sinuous, provided some form of bed sill is installed to prevent further headward erosion. However, they are unlikely to be successful where major, systematic, regional erosion or deposition is occurring as they would be either undermined or submerged with sediment.

Environmentally, submerged vanes are preferable to revetment methods since the banks of the river are left undisturbed and natural recolonization is allowed to occur (Table 18.10).

Table 18.10 Engineering and environmental performance of bank protection works

	Engineering and environmental performance		
Options	Advantages	Disadvantages	Design limitations
Revetments			
Grass	Natural and unobtrusive		Lowest erosional resistance
Grass and jute			↓
Grass and geotextile			
Reeds			
Willows/osier			
Ash			
Gravel and geotextile	Natural but more obtrusive. Vegetation can become established		
Mastic and geotextile			
Woven wooden fence			
Riprap			Chosen option should be softest to prevent failure
Riprap over geotextile			
Gabion mattresses		Artificial but some vegetation can become established	
Gabion baskets			
Grouted riprap			
Grouted gabion baskets			
Jointed cellular blocks			
Cable-tied blocks			↓
Wooden boards		Artificial and cannot become vegetated	Highest erosional resistance
Sheet piling			
Masonry wall			
Concrete wall			
Hydraulic controls			
Submerged vanes	Natural bank retained. Vanes are submerged and unobtrusive. Instream habitats enhanced	Not appropriate where systematic bed degradation or aggradation	Suitable for bank erosion control in meander bends on rivers which are in regime

18.5 RIVER RESTORATON

Restoration requirements

River engineering works have, over the centuries, adversely affected hundreds of kilometres of rivers (Brookes 1988; Purseglove 1988). The ecological devastation that has been wrought by many of the traditional approaches to river engineering problems is well illustrated by the results of the environmental impact assessment of flood alleviation schemes (Hey *et al* 1990). Unsympathetic engineering treatments consistently reduced plant diversity, when compared with natural channels, while environmentally sensitive approaches actually increased diversity by creating a greater range of habitats. Limited information is currently available regarding the impact of different engineering treatments on fisheries. However, it has long been recognized that increased morphological diversity benefits the carrying capacity of a river and provides suitable habitats for the various stages in the life cycles of fishes, which is essential if the fishery is to be self-sustaining. It follows, therefore, that conventional engineering methods often have a deleterious effect on fisheries.

With the advantage of hindsight and experience it is possible to identify why problems have arisen. It is now a question of using our current level of understanding of river mechanics and its effect on channel form to restore degraded rivers to increase their environmental value in terms of flora, fisheries and amenity (Hey 1992). Essentially it is a question of removing the legacy of uniformity and recreating the diversity that is characteristic of natural channels.

Restoration options

As river shape and dimensions are essentially self-determined, then any habitat improvement or restoration scheme that imposes an unnatural condition on the river will inevitably lead to instability, loss of design performance and adverse environmental impact.

Upland rivers, because they transport considerable quantities of bed material load, are less forgiving than lowland rivers with regard to the implementation of restoration works. Badly designed schemes will quickly be destroyed as the river reacts to the imposed changes. An expensive maintenance programme would then be required to sustain the unnatural condition; this emphasizes the need for the development of sympathetic design procedures which are in harmony with local river conditions. With lowland rivers some flexibility is possible since they are less reactive to imposed change and a number of slightly different options may be feasible. Any adverse reaction by the river could subsequently be managed through a routine maintenance programme.

The basic requirement of any restoration works is to increase local morphological, hydraulic and sedimentological variability given the natural constraints of the river. Ideally any planned improvements need to be assessed with regard to their impact on channel stability, both bed and banks, and their influence on flow hydraulics (in terms of velocities, flow depths, etc.) and on the flood capacity of the channel. Only when these criteria are satisfied should the scheme be implemented. If several options are possible, then it is a question of establishing which produces the desired range of habitats. This presupposes that information is available regarding the habitat preferences of different species of fish during different stages in their life cycles. Ideally the created habitats should support every stage, from spawning to mature adult, in order that the improved reach will be self-sustaining and regular restocking will not be required. While there is considerable information available regarding habitat preferences for salmonids, in terms of velocities, flow depths, substrate, etc. (Solomon 1983) there is little comparable information for coarse fish (Smith 1989). This remains a research priority as without such data informed decisions cannot be made when choices are available.

With regard to instream flora, some species preference charts have been developed, in terms of bankfull velocity, flow depth, cross-sectional area, bed substrate and bankslope, based on data from flood alleviation schemes and natural reaches (Hey *et al* 1990). These provide some guidance on likely species response to planned changes but here again more research is required to refine the method and improve its predictive capability.

There are two basic approaches to restoration and habitat improvement works: non-structural and structural.

Non-structural restoration

This method refers to the re-establishment of natural features within a reach that had previously been dredged and/or straightened. It could involve, *inter alia*, the restoration of meanders, pools, riffles, vertical banks and dead zones. Lowland rivers are particularly amenable to this type of treatment as they are less reactive to engineering change than upland ones. With the latter, riffles will rapidly re-form after dredging if bed material is actively being transported.

Irrespective of the scale of the planned restoration works a site survey is required to ascertain the nature of the existing bed topography and substrate characteristics. The geomorphological assessment should also map bankside features: eroding, vertical and slumped banks, tree-lined sections, submerged trees, areas of bank accretion, and the extent of any backwater from mills, weirs, etc. This identifies existing habitats that need to be preserved or enhanced and provides the baseline data for designing pool-riffle sequences. Appropriate regime equations, and/or data from local natural sections upstream and/or downstream from the affected reach, should be used to define required reach average widths, depths and slopes for the river. These enable the plan geometry of the channel to be defined over the reach if that also requires modification. The location of pools and riffles needs to correspond to the meander geometry of the river, with pools at the outside of the meander bends and dredged spoil used to form the point bar or shelf on the inner bank. Channel width in the bend needs to be less than that on the riffle to prevent siltation during high flows. The riffles should be located at the inflexion point between bends and constructed at a slight angle across the river in order that flow is directed towards the outer bank in the downstream bend. This skewing of the flow helps to generate secondary currents in the bend which, in turn, encourages scouring. The riffle crest should be high enough to create ponding upstream at low flow and needs to be formed of coarse enough material to maintain its stability. Large stones, 200–300 mm diameter, would make suitable foundation material and this would need to be covered by graded gravel slightly coarser than the local bed material size. The upstream face of the riffle would be steeper than the downstream one, the latter acting as an apron for the faster-flowing water overspilling the riffle at low flow. An example geomorphological map and resultant scheme design are given in Fig. 18.8 (Hey 1992).

If pools and riffles are being re-established in a straight channel, then the riffles would need to be spaced at 4–10 × channel width. In order that the pools are located adjacent to alternate banks, the riffles need to be slightly skewed. As riffle spacing is related to channel width, care should be exercised to ensure that widths used in the calculation are natural or predicted regime values, and not local values which may be artificially large due to dredging, if a natural width is going to be allowed to re-establish itself.

Attempts to recreate a pool-riffle sequence in a reach that is located in a backwater upstream from a weir or sluice is unlikely to be very successful. The riffle will be permanently drowned out and local velocity variations will, as a consequence, be minimal. The restored riffle needs to create a step in the water surface profile, by backing up water behind it, and this would necessitate the construction of an inordinately large riffle.

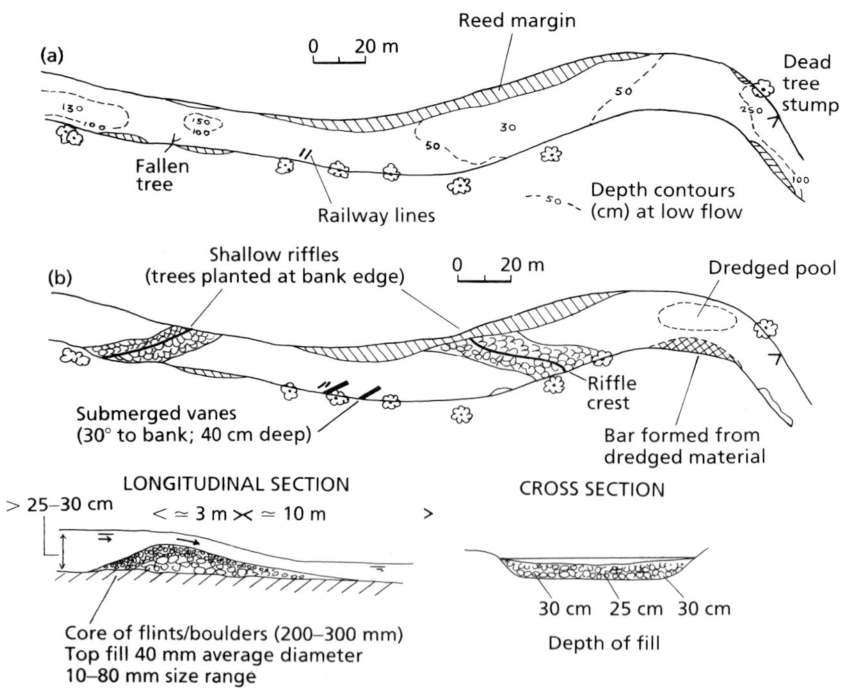

Fig. 18.8 Re-creation of pools and riffles at Lyng, River Wensum: (a) geomorphological map of original channel; (b) proposed pool-riffle sequence (from Hey 1992).

Structural restoration

These techniques refer to a range of artificial instream structures which can create ponded reaches and bar forms and prevent siltation to maintain substrate condition. Such structures are particularly useful where full restoration is not possible or, for example, where it is necessary to increase habitat diversity on urban flood alleviation schemes which have to be heavily engineered.

Weirs. A range of weirs and dams are available to create ponded reaches upstream and scouring in the overspill zone (Fig. 18.9). Full-width weirs can create different shaped plunge pools, depending on their crest form. Those V-ing downstream can precipitate bank failure as pools occur adjacent to the bank. Partial-width weirs accelerate flow througth the gap and, thereby, cause scouring. All weirs require some form of bank protection to prevent bank erosion outflanking the structure. Sedimentation will occur in the ponded reach upstream while material eroded below the structure will be deposited locally (Fig. 18.9).

Deflectors. Deflectors are structures which limit channel width and accelerate the flow through the constricted section. This causes local scouring and bar development. A variety of deflectors have been devised and these are illustrated in Fig. 18.9.

Wing deflectors can be triangular, with the leading edge angled at 45° to the flow, or straight but similarly angled, and extending up to three-quarters of the way across the channel in grossly overwidened and ponded reaches. In faster-flowing rivers a separation zone, or vortex, may develop downstream from the structure where flow expands. In extreme cases this may cause bank erosion. It can be avoided with triangular deflectors by reducing the angle of the trailing edge of the structure to allow gradual expansion to the flow. Both types have been successfully applied in a range of rivers (Hubbs *et al* 1932; White & Brynildson 1967).

At high discharges, when the structures are

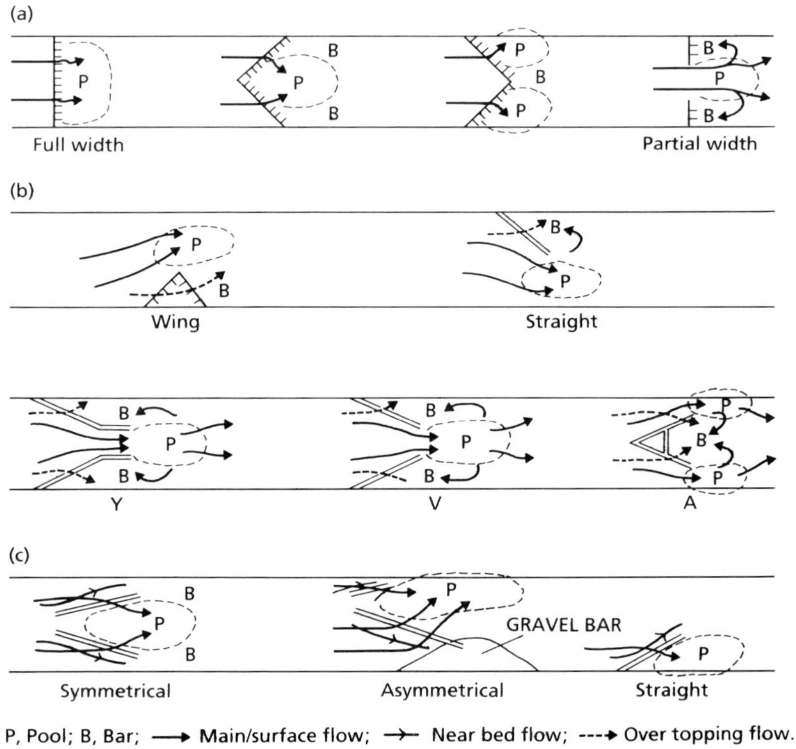

Fig. 18.9 River habitat improvement using structural measures: (a) weirs/dams; (b) deflectors; (c) vanes (from Hey 1992).

drowned out, flow is no longer fully constricted. Triangular deflectors cause overtopping flows to be directed back towards centre channel and encourage further scouring. Consequently it is not necessary for them to be constructed to bankfull level. In contrast, when the straight deflectors are overtopped, faster surface flow is directed towards the bank, which could cause bank failure if the banks are not protected. To avoid this problem, straight deflectors need to be constructed to bankfull level.

Instream deflectors take several forms: A, Y and V types, reflecting their shape. The apices of the Y and V structures lie upstream and flow accelerates through the central gap to create a scour hole. The A deflector points upstream with gaps adjacent to the bank. Pools generated adjacent to the bank may precipitate mass failure of the banks. When Y and V deflectors are overtopped, the secondary circulation that results will cause scouring at the base of the banks which may, ultimately, cause bank collapse. With the A deflector, overtopping will promote scouring in mid-channel. To avoid potential bank erosion problems with the Y and V structures, they would need to be constructed to bankfull levels. With the A deflector it is probably desirable for it to be overtopped at about half bankfull stage, otherwise bank erosion could be excessive.

Submerged vanes. Submerged vanes are instream structures which are designed to promote bed scouring by developing secondary circulation in the flow (Fig. 18.9) (Hey 1992). They are submerged even at low flow.

Pools can be created in two ways. The simplest is to install a vane, essentially a large wooden board, at 45° to the bank facing downstream and extending approximately a quarter to a third of the distance across the river. Faster surface flows overspill the vane and create a pool adjacent to the bank. Tree planting may be required to prevent bank failure and meander initiation. Alternatively two vanes can be installed facing upstream at approximately 30° to the flow. Gaps are left between the vanes and adjacent to both banks to allow for the passage of bed material load. Flow spilling over the vanes is directed towards mid-channel and this promotes bed scour.

If required twin-vane systems can be offset towards the banks, to create a series of alternate pools, or asymmetric vanes can be installed to initiate meander development.

Engineering and environmental performance

Non-structural restoration

More and more restoration schemes are being implemented but insufficient time has elapsed for any detailed assessments to be made of their engineering and environmental performance. However, provided appropriate design approaches are adopted (as outlined in Section 18.4), the restored reach should remain stable and maintain the discharge capacity of the river. By reinstating a natural range of habitat features the river will, in time, be recolonized by instream and bankside flora, invertebrates and fish. Experience indicates that this can be very rapid as freshwater crayfish, stone loaches and bullheads were observed to colonize restored riffles on the River Wensum at Lyng, Norfolk, during their construction.

Structural restoration

Weirs. Several authors (White & Brynildson 1967; Everhart *et al* 1975) argue that weirs/dams are really only beneficial in upland rivers which have steep gradients and where tranquil, ponded reaches are rare since lowland rivers are already ponded over considerable distances. However, if the original, undredged river exhibited pools and riffles, then the best restoration strategy is to let nature take its course and re-establish a natural bed topography through the reach.

On a heavily dredged reach of the River Neath at Maes Gwyn is South Wales, where riffles were removed over a 4-km reach of river, a number of blockstone weirs were installed to compensate for the loss of habitat diversity. Many of these were constructed between bends where riffles would naturally be located and, not surprisingly, they have since become submerged by gravel deposits. It should be emphasized that a natural pool-riffle sequence is unlikely to increase flow resistance during flood events, as no flow separation will occur, whereas the weirs may impair channel capacity.

The installation of weirs on rivers that are naturally relatively uniform, or devoid of pools, may be more successful provided they are appropriately located. Spacing of the weirs is critical and should be related to the width of the river as described previously. Provided they are properly spaced, there is the possibility that they will remain relatively sediment free. They are, however, likely to increase frictional resistance, even when they are drowned out at high flows, due to flow separation and, thereby, impair the discharge capacity of the river.

Many lowland rivers are ponded and it is a question of deciding whether fisheries benefits can be gained by installing more weirs. Velocities would be further reduced in the backwater zone, encouraging siltation, and scouring would be confined to the plunge pool below the weir. Anecdotal evidence suggests that the ponded section does improve coarse fish habitats and aeration is improved downstream from the dam. The structures are often constructed from wood, corrugated iron and chicken wire, and, regrettably, are often visually very intrusive. Consequently, consideration needs to be given to improvements in design techniques. The main problem with weirs on lowland rivers is their adverse effect on flood capacity, especially when they collect floating debris.

Deflectors. Wing or bankside deflectors aim to reduce the effective width of the river. As many rivers have been overwidened for land drainage purposes, this technique helps to restore a natural regime width and encourage bankside sedimentation and vegetative recolonization. Often deflectors are located on alternate sides of the river to recreate a natural pool sequence. Spacing the deflectors is crucial if this is to be successful. Where the river is grossly overwidened, there would be merit in constructing a number of deflectors, almost like a series of spur dykes, to restrict the width of the river. In ponded reaches, the channel would need to be constricted quite considerably to create any significant increase in velocity. Consequently it is anticipated that they would only maintain an existing pool in a silt-free condition. On non-ponded reaches, deflectors may be constructed to scour a new pool as the constriction could cause local ponding and increased head.

There are potential problems with regard to the application of instream deflectors on upland-type rivers. High stream energy and associated bed material transport rates will ensure that the river will rapidly respond to the imposed structures and it is vital that they are located correctly. Essentially they are required to guide and amplify the natural flow pattern along the reach in question in order to increase morphological and hydraulic diversity. On lowland rivers the flow will need to be constricted quite considerably in order to produce the required velocities to cause scouring. This will be particularly the case on ponded reaches where it may only be possible to prevent siltation in an existing pool.

All deflectors adversely affect the discharge capacity of a river, particularly if they are blocked by debris. Before installing any deflectors, geomorphological surveys should be carried out to ensure that they are sited an appropriate distance apart and that they reinforce the natural tendency of the river. Their use in active gravel-bed rivers is probably limited in view of the high bed material transport rates. Due consideration also needs to be given to the height of the deflectors bearing in mind the problems that can arise when they are overtopped.

Environmental design guidelines

River restoration is best achieved using non-structural procedures as the natural character of the river can be recreated. This would involve the reconstruction of pools and riffles and, if necessary, the recreation of meandering channels.

Where lack of space, access or finance prevents re-creation of a meandering channel, or the requirement for heavier engineering works prevents the restoration of pools and riffles, then rehabilitation could be carried out using structural procedures (Table 18.11). On upland-type rivers weirs could be deployed to provide habitat variety in heavily engineered and dredged channels, provided they are fully drowned out during flood conditions. On lowland-type rivers they are

Table 18.11 Engineering and environmental performance of river restoration works

Options	Engineering and environmental performance		Design limitations
	Advantages	Disadvantages	
Non-structural	Natural solution which recreates range of habitats and improves visual amenity. Recreates pools and riffles in dredged section, narrows widened channels, and restores meanders in straightened rivers		*Preferred solution.* For straightened channels needs sufficient space to restore meanders
Structural			
Weirs	In heavily managed and dredged upland rivers where riffles removed, fish weirs may provide some local habitat variability. In heavily engineered urban upland reaches, weirs can provide some habitat variety and visual improvement	Can impair discharge capacity of river and trap sediment and floating debris	*Blockstone weirs beneficial on heavily engineered and managed upland rivers. On lowland rivers little benefit as long reaches already ponded*
Deflectors	Wing deflectors can create and sustain pools in overwide lowland channels. Can create sinuous course in straight reaches. Increases habitat diversity	A, Y and V deflectors very intrusive and unnatural. In ponded, backwater reaches significant constriction required. Likely to be destroyed on upland rivers. Can impair discharge capacity of channel	*Wing deflectors beneficial on non-ponded lowland rivers*
Submerged vanes	Unobtrusive. Create pools by guiding and amplifying natural scour processes. Increases habitat diversity.		*Simple wooden structures suitable for non-ponded lowland rivers. On upland-type rivers larger blockstone structures required*

unlikely to provide significant habitat improvement as ponded conditions are quite prevalent. Deflectors and submerged vanes are best suited to non-ponded lowland-type rivers as they create a local increase in the energy of the flow by narrowing the river (deflectors) or generate secondary flows (submerged vanes) to cause scouring of the bed.

18.6 FUTURE RESEARCH REQUIREMENTS

Research needs to be carried out to develop and define environmentally sensitive options for flood alleviation and river training. While preliminary results indicate that more natural approaches to river management are preferable to harder engineering solutions on conservation and economic grounds, few if any schemes have been implemented using such procedures to enable comparisons to be made with more traditional options and establish their range of application.

The basic requirement is for field trials to be carried out to evaluate the natural options across a range of river environments. For flood alleviation schemes and bank protection works this would supplement existing information (Hey *et al* 1990, 1991) but for channel stabilization the more natural methods have yet to be implemented and evaluated. Pre- and post-scheme surveys would be required to enable the engineering and environmental performance of each scheme to be assessed for comparison with traditionally designed schemes. In addition the cost-effectiveness of the various options needs to be determined based on both capital and maintenance costs.

Further research is also required to develop river restoration procedures and evaluate the impact and range of application of structural rehabilitation measures. Although more restoration and rehabilitation schemes are now being implemented many of these are based on visual and aesthetic approaches and lack any rigorous scientific base. In addition the lack of any baseline surveys precludes any objective evaluation of their merit.

A number of trials need to be carried out to evaluate natural restoration procedures in both upland and lowland rivers. As the former transport bed material load, their shape and dimensions are closely prescribed and design guidelines need to be strictly adhered to (Hey & Heritage 1993). With the latter there is more flexibility with design options.

Structural approaches to rehabilitate and soften heavily engineered reaches also merit further research, particularly the deployment of submerged vanes as their potential has not been fully evaluated. Attention should be focused on the design and size of each structure in order to create the required habitat, and how this may vary between different types of river, their location in order to be sustainable and their effect on the discharge capacity and stability of the river.

Finally, as all the recommended design procedures are based on knowledge of natural channel processes, it follows that further improvements are dependent on continued basic research.

REFERENCES

Biedenharn DS, Little CD, Thorne CR. (1990) Effects of low drop grade control structure on bed and bank stability. In: Chang HH, Hills JC (eds) *Hydraulic Engineering*. Proceedings of National Conference on Hydraulic Engineering, ASCE, San Diego, August 1990, pp. 826–31. [18.3]

Blench T. (1966) *Mobile Bed Fluviology*. University of Alberta Press, Edmonton. [18.3]

Brookes A. (1988) *Channelized Rivers*. John Wiley, Chichester. [18.5]

Einstein HA. (1950) *The Bedload Function for Sediment Transportation in Open-Channel Flow*. US Dept. of Agriculture, Technical Bulletin No. 1026. [18.3]

Everhart WH, Eipper AW, Youngs WD. (1975) *Principles of Fishery Science*. Cornell University Press, Ithaca. [18.5]

Hemphill RW, Bramley ME. (1989) *Protection of River and Canal Banks*. Butterworths, London. [18.4]

Hey RD. (1976) Geometry of river meanders. *Nature* **262**: 482–4. [18.3]

Hey RD. (1986) River mechanics. *Journal of the Institution of Water Engineers and Scientists* **40**(2): 139–58. [18.1, 18.3]

Hey RD. (1987a) River dynamics, flow regime and sediment transport. In: Thorne CR, Bathurst JC, Hey RD (eds) *Sediment Transport in Gravel-Bed Rivers*, pp. 17–37. John Wiley, Chichester. [18.1, 18.2]

Hey RD. (1987b) Regime stability. In: Brandon TW (ed) *River Engineering, Part 1, Design Principles*, pp. 139–47. The Institution of Water and Environmental Management, London. [18.3]

Hey RD. (1992) River mechanics and habitat creation. In: O'Grady KT, Butterworth AJB, Spillett RP, Domaniewski JCJ (eds) *Fisheries in the Year 2000*, pp. 271–85. Institute of Fisheries Management, Nottingham, UK. [18.3, 18.5]

Hey RD, Heritage GL. (1993) *River Engineering Works in Gravel-Bed Rivers*. R & D Project Record 387/1/W, 96, National Rivers Authority Bristol. [18.6]

Hey RD, Thorne CR. (1986) Stable channels with mobile gravel-beds. *American Society of Civil Engineers, Journal of Hydraulic Engineering* **109**(6): 842–51. [18.3]

Hey RD, Heritage GL, Patteson M. (1990) *Design of Flood Alleviation Schemes: Engineering and the Environment*. Ministry of Agriculture, Fisheries and Food, London. [18.2, 18.5, 18.6]

Hey RD, Heritage GL, Tovey NK, Boar RR, Grant A, Turner RK. (1991) *Streambank Protection in England and Wales*, R & D Note 22, National Rivers Authority, London. [18.4, 18.6]

Hubbs HD, Greeley JR, Tarzwell CM. (1932) Methods for the improvement of Michigan trout streams. *Bul-*

letin of the Institute for Fisheries Research **1**: 1–54. [18.5]

Knauss J, Scheuerlein H. (1988) Erosion control and environmental protection of low slope river ranges with flood plains. *Proceedings of the International Conference on Fluvial Hydraulics '88*, pp. 185–90. Research Centre for Water Resource Development, Budapest. [18.3]

Lane EW. (1937) Stable channels in erodible material. *Transactions of the American Society of Civil Engineers* **102**: 123–94. [18.3]

Odgaard AJ, Mosconi CE. (1987) Streambank protection by submerged vanes. *American Society of Civil Engineers, Journal of Hydraulic Engineering* **113**(3): 520–36. [18.4]

Paice C, Hey RD. (1989) Hydraulic control of secondary circulation in meander bends to reduce outer bank erosion. In: Albertson ML, Kia RA (eds) *Design of Hydraulic Structures 89*, pp. 249–54. A.A. Balkema, Rotterdam. [18.4]

Purseglove J. (1988) *Taming the Flood*. Oxford University Press, Oxford. [18.5]

Raynov S, Pechinov D, Kopaliany Z, Hey RD. (1986) *River Response to Hydraulic Structures*. UNESCO, Paris. [18.1]

Schumm SA, Harvey MD, Watson CC. (1984) *Incised Rivers*. Water Resources Publications, Colorado. [18.3]

Smith RP. (1989) Habitat requirements and distribution of Chub in some rivers in Eastern England. Unpublished PhD thesis, University of East Anglia. [18.5]

Solomon DJ. (1983) *Salmonid Enhancement in North America*. The Atlantic Salmon Trust, Pitlochry, Scotland. [18.5]

Vanoni VA, Brooks NH. (1957) *Laboratory Studies of the Roughness and Suspended Load of Alluvial Streams*. US Army Corps of Engineers, Missouri River Division, Sediment Series No. 11. [18.3]

Visher D. (1986) *Swiss River Training in the 18th and 19th Centuries*. Publication No. 14 of the Hydrology Centre, Christchurch, New Zealand. [18.1]

White RJ, Brynildson OM. (1967) *Guidelines for Management of Trout Stream Habitat in Wisconsin*. Technical Bulletin 39, Department of Natural Resources, Division of Conservation, Wisconsin. [18.5]

Winkley BR. (1982) Response of the Lower Mississippi to river training and realignment. In: Hey RD, Thorne CR, Bathurst JC (eds) *Gravel-Bed Rivers*, pp. 659–81. John Wiley, Chichester. [18.1]

19: Management of Macrophytic Vegetation

P. M. WADE

19.1 INTRODUCTION

Consensus as to the acceptable level of vegetation occurring within and along a river channel is rarely achieved. At one extreme the flood defence engineer and navigation manager wish to minimize plant growth, at the other the nature conservationist argues for minimum interference. First and foremost, the vegetation is a key component of the riverine environment, being the primary producer and a significant factor in maintaining channel stability. Vegetation fixes solar radiation, making this energy available for a wide range of herbivores including invertebrates, fish, birds and mammals. The aquatic vegetation can be important in aerating the water, providing shelter and a spawning or egg-laying medium for fish and freshwater invertebrates. Emergent and marginal plants provide shelter and nesting habitat for a variety of fauna including birds and invertebrates. The vegetation is also important in the consolidation of the river bed and banks.

A number of negative effects brought about by the vegetation of the river channel and its margins can be identified. Direct effects include the impedance of the flow of drainage and irrigation water; the hindrance of navigation; an increase in sedimentation by the trapping of silt particles; a decrease in human food production, especially fish; harbouring host organisms for various diseases; interference with recreation such as water skiing and angling; restriction of access to the river including a decrease in the possibilities for washing and bathing; and deoxygenation resulting from the decomposition of large amounts of vegetation within the water.

In the temperate zone, flow resistance brought about by vegetation is regarded as the main problem associated with riverine vegetation; this is reviewed by Pitlo and Dawson (1990). The presence of such vegetation may affect flow in a channel by:

1 reducing water velocities and consequently raising water levels in the channel and in the water table in adjacent land causing it to become waterlogged;
2 increasing the incidence of flooding;
3 encouraging the deposition of suspended sediment;
4 changing the magnitude and direction of currents thus causing local erosion or reducing bank erosion.

Interference with the movement of boats is mainly a problem of tropical areas but can also be serious in temperate navigable waterways used for seasonal recreation (Murphy & Eaton 1983). Floating mats of weeds in particular may cause serious problems as they can cut off remote settlements from the outside world.

This chapter recognizes the different uses to which a stretch of river is put and outlines the various forms of vegetation management which can be employed to maintain a particular use, paying attention to the effects of such management on the ecology of the river. Plants and plant communities growing in a river can be inimical to human activity, but in other situations they might need to be fostered and encouraged. The chapter does not seek to prescribe an ideal balance between these various demands; such a decision should be based on the integrated knowledge of any given river and its catchment.

Aquatic plant management or weed control?

Vegetation management for rivers is often referred to as weed control (e.g. Seagrave 1988; Pieterse & Murphy 1990) with an emphasis on weed cutting and the use of herbicides. In many instances, however, the need to manage the vegetation arises from inappropriate uses or management of the river as a system. Eutrophication by effluents from sewage treatment works, river regulation and the siting of human settlements in river floodplains, all result in changes to the river system which can induce the need to manage vegetation either because biomass has increased to a nuisance level or because hydrological criteria for the river have changed. The management of river systems increasingly acknowledges the desirability to manage the river as nature intended, for example the introduction of tertiary treatment at sewage reclamation works associated with ecological water quality objectives for a particular river or section of river. The benefits of such an approach are not expected to be immediate due to such factors as the accumulation of nutrients in the substrate of the river; nevertheless over time a more sustainable approach to the river resource can be achieved.

The avoidance of the need for weed control through these means is not yet possible for many rivers and given certain hydrological criteria related to, say, flood defence in a heavily engineered stretch, direct manipulation of the vegetation is and will remain necessary for rivers.

The first two stages in this management process are the identification of the species and community or communities of plants which need management, and an assessment of the extent and distribution of the beds of macrophytes.

Plant recognition

Successful vegetation management needs to be based on the correct taxonomic identification of the plant species. This information enables the selection of appropriate management strategies such as a control agent for use against an unwanted stand of a particular species (e.g. MAFF 1986; Westerdahl & Gestinger 1988). Details of the species occurring within a system will give an indication of the habitat conditions (Melzer 1976; Standing Committee of Analysis 1987) and consequently the likely pattern of recolonization. Cook *et al* (1974) and Cook (1990) provide comprehensive catalogues of the genera of the aquatic macrophytes of the world, though none of the continents enjoys a full description of aquatic plant species within the confines of a single publication. Hotchkiss (1972) comes close to this for the USA and Canada, as does Aston (1973) for Australia, and Casper and Krausch (1980, 1981) for Europe. Wade (1987) reviewed the provision made for the identification of aquatic and wetland plants around the world and listed over 100 manuals and other publications to aid the recognition of aquatic flora.

Vegetation assessment

The assessment of marginal and emergent vegetation is relatively straightforward, being achieved through methods similar to those used for terrestrial vegetation. Submerged and floating vegetation is more difficult to assess as it is often inaccessible and in the case of the former out of sight. A basic methodology is provided in such texts as Wood (1975), Haslam (1978), Hellawell (1978) and Murphy (1990). More recently river corridor surveys have been pioneered by the National Rivers Authority (1992), a survey technique which includes a description of the riverine vegetation. Haslam and Wolseley (1981) describe a methodology for British rivers, and techniques for working in chalk streams have been developed through such studies as Wright *et al* (1982). Larger, deeper rivers pose more of a challenge especially as substantial beds of macrophytes can be present but unseen by the assessor, for example the submerged forms of *Nuphar lutea* and *Sagittaria sagittifolia*, both commonly overlooked in turbid eutrophic rivers.

Increasing use is being made of aerial photography for extensive surveys of marginal, emergent and floating vegetation (e.g. Haslam 1979) including the use of aerial video film of the river corridor (V. Holt, personal communication). Remote sensing by aerial photography and satellite survey methods using colour, false-colour and infrared photography offer the possibility of rapid, large-scale monitoring anywhere in the world over whatever time-scale is required (Murphy 1990). An important aspect of these assessment techniques is that they enable an accurate record to be established of the vegetation at a given point in time by which to measure the success of vegetation management techniques.

19.2 MAINTENANCE OF SUSTAINABILITY

The aim of the river manager should be the development of a strategy for given sections of river corridor or river catchment with agreed levels of usage including drainage function, potential for water abstraction and reception of discharges, and recognition of such resources as nature conservation and fisheries. Such approaches to sustainable resource utilization are being adopted throughout Europe, Australia and North America and are rapidly becoming part of river and vegetation management practice. Although an acknowledged improvement in management strategy, such practices struggle to recognize the fundamental factors controlling plant growth, namely nutrient loading and hydrological characteristics. Both the standing crop and the species composition of aquatic plant communities are controlled by these factors and regulation of nutrient inputs and flow manipulation could be used to manage the nature of the aquatic plant element of any given river. Likewise strategies are being developed for riparian and floodplain vegetation to further the sustainable approach to river management.

Nutrient availability

The first stage in developing a vegetation management strategy is to agree upon the target nutrient loading and flow characteristics for a given river in relation to the desired plant communities. Data already collected for rivers can provide basic water quality characteristics and a crude indication of past plant species composition. Only very rarely, however, do data exist for the size of standing crops and other such quantitative estimates. Nutrient concentration data are, of themselves, of limited value in that they need to be converted into mass flow and give no indication as to nutrient availability in sediments. This is of potential significance as roots have been identified as an important site of nutrient uptake for certain aquatic macrophyte species (Chambers *et al* 1989), a fact also established for rooted aquatic plants growing in lakes (Barko & Smart 1981; Smith & Adams 1986). Studies on European streams and rivers have shown that aquatic macrophyte biomass and tissue nutrient concentration are not correlated with open-water nutrient concentrations (Owens & Edwards 1961; Ladle & Casey 1971), again pointing to the importance of sediment nutrients for plant growth in flowing waters. However, Kern-Hansen and Dawson (1981) observed that aquatic plant biomass and tissue concentration in 19 Danish streams were not correlated with either openwater or sediment concentrations, suggesting that some other factor or factors limited growth. If root uptake is important for riverine macrophyte nutrition, the impact of decreased nutrient loading on growth will depend upon nutrient availability and particle size of the river-bed sediments.

In one attempt to implement aquatic plant control through nutrient limitation, results were unclear (Charlton & Bayne 1986). Water quality and nuisance aquatic plant growth were monitored in the Bow River, Canada, after the installation in 1982–83 of phosphorus removal at Calgary's two sewage treatment works. Charlton and Bayne (1986), in an analysis of the results of the first 2 years, 1984–85, did not show a consistent reduction in plant biomass in spite of an 80% reduction in total phosphorus loading. An evaluation of nutrients and plant growth in the river from 1986 to 1988, however, revealed that a significant decrease in macrophyte biomass had occurred since 1983 at six of the 22 regularly sampled sites, with biomass also appearing to decline at some other sites (Sosiak 1990). Most of the sites with a significant decline in biomass were in the mixing zone for Calgary's sewage treatment effluent.

Aquatic vegetation and flow characteristics

As early as the 1920s, Butcher (1933) recognized that changes in water flow or velocity can alter the biomass and species composition of the submerged plant community, though it is only recently that this relationship has been explored in any detail. Physiological studies have shown that photosynthetic rates are positively correlated with current speeds over the range likely to occur in weed beds (i.e. 0–0.2 m s^{-1}) (Westlake 1967; Madsen & Sondergaard 1983). Ecological studies, on the other hand, have failed to show any consistent relationship between water movement and the distribution and abundance of submerged plant communities. In the River Lambourn, UK, the 1971–80 expansion and recession of *Ranunculus* beds was positively correlated with discharge rate (Ham *et al* 1981), whereas in the River Wye, Wales, macrophyte cover and peak biomass were negatively correlated with flow (Brooker *et al* 1978). Similarly, in a Swedish stream, Nilsson (1987) observed that macrophyte cover increased at low current velocities (<0.3 m s^{-1}), yet decreased at higher speeds. Biomass and shoot density in the Bow River, Alberta, Canada, decreased with increasing current velocity within a weed bed over the range 0.001–1 m s^{-1}, and at current speeds in excess of 1 m s^{-1} aquatic macrophytes were rare (Chambers *et al* 1991). Although the impact of flow characteristics on plant growth in rivers is not sufficiently well understood to use manipulation of such factors for the control of plant growth in all rivers, the potential does exist. The effect of the current can be direct-acting upon plant metabolism, but current velocity may also exert an indirect effect on aquatic macrophyte growth by altering the composition of the river bed and hence nutrient availability (Chambers *et al* 1991). Spillet (1981) used wooden groynes to direct water flow to influence downstream weed beds and silting to improve the fishery in the River Thames.

Identifying appropriate plant communities

Deriving targets for the species content, extent and biomass for beds of vegetation will be a particular challenge. Earlier descriptions of the flora of the river will be valuable in reconstructing past community composition and extent. An investigation to determine the previous flora of the Rivers Test and Itchen, UK, involved consulting previous reports on the river vegetation, the records of botanists, the use of local floras and plant records acquired from herbaria (E. Cranston, personal communication). However, information on the biomass for any given sections of the river were almost non-existent. Where management necessitates influencing community composition, reintroductions may be necessary and it will be important to refer to those agencies regulating or co-ordinating introductions. Recognition is needed of the dynamic nature of riverine plant communities and the need for management to either direct successional processes or to slow them down in line with the strategy for that particular river or catchment.

Bank protection

Although relatively little attempt has been made to regulate plant growth through nutrient control, the need to maintain ecological stability has been identified for certain components of the river habitat, especially the riverbank.

The use of vegetation for bank stabilization and protection is described in detail by Hemphill and Bramley (1989). Four zones of vegetation are identified in relation to water level, the dominant factor affecting plant growth along the river margin. The four zones are based on the vegetation horizons along a river bank (Seibert 1968) (Fig. 19.1).

1 Aquatic plant zone in which the plants can provide protection near the water surface, an effect which decreases with increasing depth.

2 Marginal and emergent zone, comprising plants which either grow up out of the water with emergent leaves and stems or grow along water margins with roots extending below the water table. Emergent plants grow in depths up to 2 m, but most species are confined to depths less than 0.5 m.

3 Damp, seasonally flooded zone, the principal zone for natural bank protection, supporting grass and fast-growing shrub and tree species.

4 Dry, occasionally flooded zone in which plant life is relatively unaffected by the water level regime.

This chapter, being concerned with aquatic and

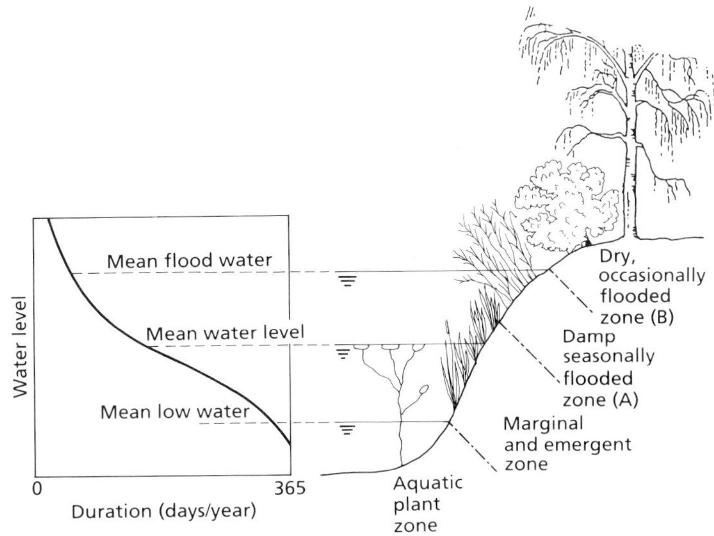

Fig. 19.1 Vegetation horizons along a riverbank (after Seibert 1968).

marginal plant management, focuses on zones 1, 2 and 3.

The aquatic plant zone is rarely encouraged as a deliberate river stabilization policy even though it has a protective influence on the emergent vegetation, for example through the damping effect of species such as *Sagittaria sagittifolia* and water-lilies such as *Nuphar lutea*. Such plants also contribute positively to the visual impact of the river. The disinterest in these plants is mainly because they have a significant potential to reduce channel capacity. Elimination of in-channel vegetation, especially in small unshaded channels, using a weed-cutting boat or herbicides can, however, put the bank at greater risk, placing emphasis on the need for active management of the marginal and emergent zone and the damp, seasonally flooded zone. On the other hand, a fringe of submerged and floating aquatic plants is often maintained in a large waterway such as the River Thames to prevent bank erosion due to boat wash.

Emergent riparian aquatic plants can be important in protecting riverbanks and the floodplain from damage due to high flows and, in navigable rivers, boat traffic (Siebert 1968; Keown *et al* 1977). Murphy *et al* (1980) assessed the protective potential of reedswamp species on the lower slopes and stoloniferous grasses on upper slopes of canal banks in Britain: *Glyceria maxima* and *Sparganium erectum* grow in this marginal zone effecting good protection. *Agrostis stolonifera* is useful on the bankward part of this zone due to its binding root system and pliable upper parts.

Studies on the River Thames, UK, have investigated the dissipation of energy waves produced by boats for various species. Sixty per cent of boat-wave energy is dissipated by a 2-m band of *Phragmites australis*, 70% by a 2–5-m band of *Scirpus lacustris*, 75% by a 2–4-m band of *Acorus calamus* and 60% by a 2–3-m band of *Typha angustifolia*. Such vegetation also encourages siltation by absorbing current-flow energy, and thus reducing the sediment-carrying capacity of the flow (Hemphill & Bramley 1989). Figure 19.2 illustrates a range of different planting situations for the marginal and emergent zone.

The damp, seasonally flooded zone is traditionally regarded as the principal zone for natural bank protection, supporting grass and fast-growing shrub and tree species. The favoured vegetation type is that of grass-based communities which can be mown to encourage rooting and to minimize potential resistance to flow in periods of high precipitation. More recently it has become acceptable on certain lowland rivers to alternate bank cutting on a year-to-year basis, i.e. leaving one bank uncut for a given year. The stimuli for

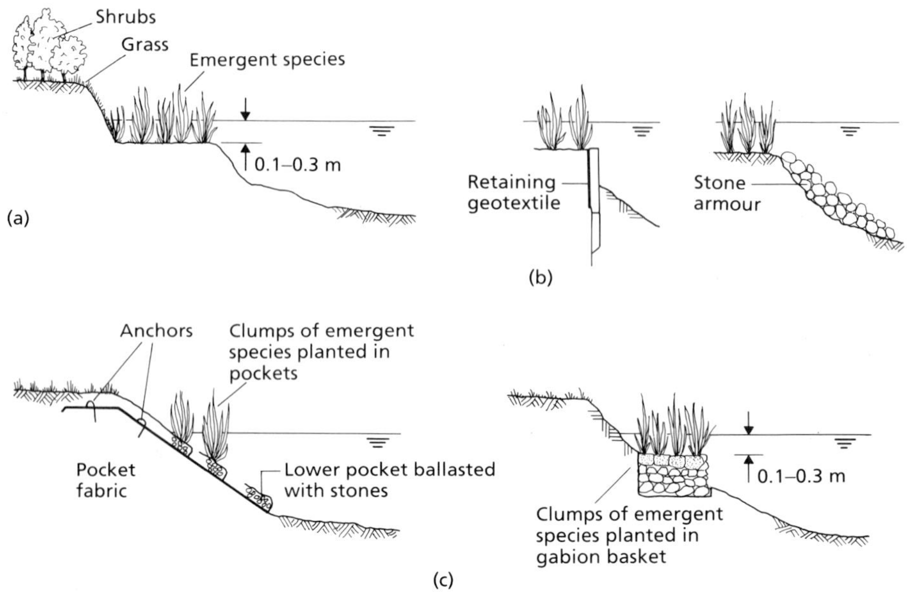

Fig. 19.2 Different planting situations for the marginal and emergent zone (after Hemphill & Bramley 1989).

this change in management have been economics, nature conservation objectives and possibly lower flows in the channel. For some rivers it is acceptable to plant trees along the bankward edge of this zone. These can be useful in shading the river and reducing plant growth as well as stabilizing the bank, creating wildlife habitat and areas of river out of the sun which are preferred by certain species of fish.

Planting can be achieved manually or by using machinery such as a dredger. The dredged plant material can be used either to return to sections of the river being managed, or for transplantation to another stretch of the same or other river. Some species colonize more aggressively than others, and care must be taken in choosing the most appropriate species to prevent problems arising through excessive growth and channel obstruction. It is best to use fresh, native, and preferably abundant plants, taken where possible from the same river or one of its tributaries. This approach is both practical and conforms to good practice from the nature conservation viewpoint. Some countries have legislation restricting the removal of plants. In the UK, for example, Section 13 of the Wildlife and Countryside Acts 1981 prohibits the uprooting of any plants without the permission of the landowner or occupier. The first prosecution under this Act was for the removal of *Stratiotes aloides* from a watercourse by a landscape gardener.

The planting of mixtures of species rather than single stands leads more rapidly to a more balanced community with a reduction in the likelihood of excessive growths of one species. Planting in early spring as the shoots first appear above ground level is recommended (Lewis & Williams 1984), and it is preferable to transfer sediment with the plant material to aid establishment and protect the plants from any late frosts. Cuttings or clumps of plant material should be placed in a hole in shallow water and firmed with the heel of a boot (Lewis & Williams 1984). Hemphill and Bramley (1989) and Kite (1980) provide specific advice on the water depth, soil type, water quality, flow velocity, planting method and growth for a range of emergent species commonly used in bank protection (Table 19.1). Most aquatic plants are easy to propagate from cuttings, whole plants or rhizome fragments

Table 19.1 Range of species used for planting in the reed-bank zone, with their habitat requirements and planting method (data from Hemphill & Bramley 1989; Kite 1980)

Species	Water depth	Soil type	Water quality	Flow velocity	Remarks	Planting method	Growth
Common or Norfolk reed (*Phragmites australis*)	Very shallow, 0.1 m max.	Sand	Clean	Very low	Least useful emergent. Does not like water motion	Rhizome and shoot clumps Shoot cuttings	Poor* Poor
Common club-rush (*Schoenoplectus lacustris*)	Deep to medium 1.0 m max.	Clay silt	Clean to moderate	Low to fast	Very useful emergent. Good resistance to flowing water	Rhizome and shoot clumps	Moderate*
Reed mace or bulrush (*Typha latifolia*)	Shallow 0.5 m max.	Clay silt	Clean	Low	Good bank protector	Rhizome and shoot clumps	Excellent
Greater pond-sedge (*Carex riparia*)	Medium 0.5 m max.	Clay silt Peat	Clean to poor	Low to moderate	Can be invasive. Useful bank protector	Rhizome and shoot clumps Shoot cuttings	Excellent Excellent
Lesser pond-sedge (*Carex acutiformis*)	Medium 0.5 m max.	Clay silt	Clean to poor	Low to moderate	Can be invasive. Useful bank protector in moderate flows	Rhizome and shoot clumps Shoot cuttings	Very good Very good
Slender tufted-sedge (*Carex acuta*)	Medium 0.5 m max.	Clay silt	Clean	Low to moderate	Needs constantly high water level	Rhizome and shoot clumps Shoot cuttings	Good Poor†
Reed sweet-grass (*Glyceria maxima*)	Deep to medium 1.0 m max.	Clay silt	Clean to poor	Low to moderate	Good bank protector	Rhizome and shoot clumps	Excellent*
Yellow iris or yellow flag (*Iris pseudacorus*)	Medium 0.5 m max.	Clay silt	Clean	Low to moderate	Good bank protector	Rhizome and shoot clumps	Good*
Galingale (*Cyperus longus*)	Shallow 0.2 m max.	Clay silt	Clean	Low	Typically a species of southern England and Wales	Shoot cuttings	Very good*
Sweet-flag (*Acorus calamus*)	Shallow 0.5 m max.	Clay silt	Clean to poor	Low	Useful in slower flowing stretches	Rhizome and shoot clumps	Good
Hard rush (*Juncus inflexus*)	Very shallow–dry, 0.1 m max.	Clay silt	Clean to poor	Low to moderate	Good for protecting the drier margins	2-year-old plants	Moderate
Branched bur-reed (*Sparganium erectum*)	Shallow to medium, 0.6 m max.	Clay silt	Clean to poor	Low	Can be invasive in shallow waters. Good bank protector	Rhizome and shoot clumps.	Moderate

* Grazed by birds or water voles but not killed. † Smothered by overhanging vegetation.

(Lewis & Williams 1984). *Phragmites australis* is an exception, being noted for its unreliability as a species for planting. Propagation can be achieved from planting clumps, cuttings, by layering and using seed (Lewis & Williams 1984). Planting clumps and cuttings was found to be usually successful. It was important to plant clumps above the mean water level (rhizomes planted below water level will usually rot), to avoid damage by trampling to the shoots in the donor stand and in the planted material, and in the case of cuttings, to plant at water level and set them so that at least three leaf nodes are buried. Advice on the best time of year for planting varies: March to May (Hemphill & Bramley 1989) or October to March (Lewis & Williams 1984). Lewis and Williams (1984) prescribe the conditions necessary for germinating *P. australis* seeds.

Protection of the planted material is often necessary to ensure establishment of the emergent vegetation. Fencing might be necessary to prevent grazing or excessive trampling, and might aid against vandalism. Protection from wave action can be achieved in different ways, the nature of the protection depending upon the severity of scour (Hemphill & Bramley 1989). Care should also be taken to avoid the invasion of the planting by alien species such as *Fallopia japonica* (*Reynoutria japonica*), *Impatiens glandulifera* and other such alien species which readily establish in such conditions (Child *et al* 1992).

Nature conservation, wildfowl and fisheries resources

Given the damage and changes to which rivers have been subject, there is a growing need in certain circumstances to repair the river habitat, whilst in others it requires protection from what some users would regard as damaging activities. Managing for nature conservation can involve a number of objectives: the maintenance of plant and animal communities as a nature conservation resource; the encouragement of certain plant species to act as a food source or for nesting or spawning; and establishing roosting or refuge for wildfowl and fish. These objectives are not always compatible. Rehabilitation and conservation range from the management of short stretches of river, including the designation of restrictive practices, to the reclamation of significant lengths of river. Most such management has focused on certain sections or components of rivers, the marginal zone being one of the more common foci (Lewis & Williams 1984; Hemphill & Bramley 1989). The submerged and floating communities have received relatively little attention and rarely feature in practical guidance. Very few examples of the restoration of plant communities have been reported in the literature. Communities have been manipulated in order to try to improve water quality and to achieve bank consolidation (Oksiyuk *et al* 1978).

Landscape

Vegetation is a key component of the river landscape, good examples being the floating, emergent and marginal zones of lowland rivers incorporating such species as *Nuphar lutea*, *Phalaris arundinacea* and *Lythrum salicaria*. These species contribute both form and colour to the landscape, elements recognized by workers such as Burkle (1978) and Kijlstra (1978).

19.3 CONTROL OF VEGETATION

River management is typically based on a traditional approach related to the main function of a river, e.g. navigable sections of river and canals for transportation. The drainage function of rivers has been dominant for many years, supporting agricultural development and serving to limit flooding of property, a consequence of which is that all aquatic and riparian plants have been seen as weeds in need of control. In recent decades rivers have increasingly been recognized as multi-use resources and not simply transportation or drainage structures. The National Rivers Authority in the UK, for example, manages certain rivers for drainage, angling, nature conservation, leisure boating and informal countryside recreation leading to management strategies based on an understanding of the landscape ecology of the resource.

Many rivers have developed plant communities which require direct control of marginal and/or in-channel vegetation. Different river conditions will dictate the type of vegetation control to be applied, and a number of organizations have drawn up guidelines and strategies for those rivers and waterways under their management (e.g. National Rivers Authority, undated; British Waterways Board 1981).

Mechanical control

Mechanical control of unwanted plants in rivers dates back to when people first used their hands and sticks to clear away weeds blocking the passage of boats or denying access to fishing. Manual methods provide a highly selective approach to weed control and can be applied to the management of fisheries, e.g. in chalk streams (e.g. Ham *et al* 1981), wildfowl and stretches of river with high nature conservation value (e.g. Lewis & Williams 1984). Seagrave (1988) provides a useful description of such techniques as cutting and hoeing, raking, pulling and the use of booms. The increasing cost of labour in many countries has made this method expensive compared with other techniques. In other countries with cheap labour, especially those in the developing world, manual control remains the most economic method (Wade 1990).

The mechanization of aquatic plant control is achieved by adapting agricultural machinery for the cutting and harvesting of bankside and aquatic vegetation. With the application of hydraulics, considerable control and sophistication can be achieved from both boat- and bank-based machinery.

Specialized weed-cutting boats are available for purchase or hire from specialist engineers and contractors (Fig. 19.3a). Such boats are craft with shallow draft driven by a paddle-wheel or screw. Most models have reciprocating cutting blades on a U- or ⊥-shaped beam. Using hydraulics this beam can be adjustd to cut at any depth down to a maximum of approximately 2 m. The boat should be light and easily handled by one person, being transportable by trailer between rivers or from one part of a river to another. The swath widths of the cutter bar on these boats range from 1–2 m to 4–5 m, the larger sizes being fitted on to barges to improve their capacity for cutting weed. Wade (1990) reviews a range of these machines, considering their advantages and disadvantages.

A bewildering range of cutting mechanisms has been developed which can be operated from the riverbank (Fig. 19.3b,c). The variety of machinery is associated specifically with the cutting and harvesting of the vegetation. The technique of dredging, which removes both vegetation and sediment, is typically of a similar design and nature.

The reach of such cutting and harvesting machinery varies: that of a tractor-mounted flail mower is approximately 7 m; a weed-cutting bucket mounted on a hydraulic excavator, 11 m; and a weed-cutting bucket on a dragline, 18 m (Cave 1981). The most widely used device is the weed-cutting bucket attached to the hydraulic jib of a tractor or excavator. The lower edge of the bucket has a cutter bar, which may range from 2 to 4 m in length, and it is lowered to the substrate surface and pulled towards the excavator cutting the weeds in its path. The bucket is able to cut weeds on the banks and the bed of the river and, given a wide enough reach, both banks can be managed in one sweep. The cut weeds can be dumped on the bank or in a truck.

Rotary, reciprocating and flail cutters provide an important range of machinery for cutting emergent and bankside vegetation (Robson 1975) (Fig. 19.3d,e). Small self-propelled pedestrian and ride-on cutters can be used on slopes with a gradient less than 2:1, although under certain conditions Allen motor scythes can work on steeper slopes (Price 1981). A review of bank-based machinery undertaken by Murfitt and Haslam (1981) indicated that reciprocating drum and disc mowers could be used in any section of the channel. The rotary devices, although needing higher power, were very much more robust than the reciprocating cutters. Flail cutters and mowers were limited in use to the area above the water.

The criteria for an ideal machine able to deal with both bank and channel are that it should be robust and reliable, be able to control rooted and non-rooted weeds at one pass without affecting the stability of the banks, have variable geometry to cater for a channel bed of 0.6–1.2 m and bank slopes of 30–45 degrees, and remove weeds to a

(a)

(b)

(c)

(d)

(e)

Fig. 19.3 A range of machines for managing aquatic vegetation. (a) A weed-cutting boat. (b) Weed basket with cutter bar (InteDrain (England) UK); (c) channel clearance (Bicycle tractor, InteDrain (England) UK); (d) flail mower Bradshaw H-6710 tracked mower with three different types of flail cutters; (e) bank clearance (AEBI TT90 InteDrain (England) UK).

stable position above 1.2 m up the bank (Wade 1990). Price (1981) presented data on the characteristics of eight weed-control machines related to these criteria giving advantages and disadvantages. The two best machines had an estimated output of approximately 2.4 km day^{-1} and 4.2 km day^{-1} allowing for normal down-time and obstructions.

A significant disadvantage of cutting and harvesting vegetation, particularly submerged and floating species, is that the underground material remains. More thorough control is achieved by dredging, which removes both plant material and accumulated sediments. Draglines have the advantage of a considerable reach whereas hydraulic excavators have greater control over the form of the bank created. The intervals between dredging are much longer than intervals between weed cutting or herbicide treatment. Where the depth of water is increased through the removal of sediment, the amount of light reaching the bottom may be reduced. Disadvantages associated with dredging are the cost and time involved and the problem of disposing of the spoil.

A number of shortfalls can be identified in the efficiency of currently available mechanical techniques. These include the following:

1 Timing. The optimal time for managing vegetation, especially the submerged and floating plants, is poorly understood; indeed, certain cutting regimes actually promote plant growth. A good example is the management of *Ranunculus* beds in chalk streams in the UK (Westlake & Dawson 1982, 1986).

2 Removal of cut material. Machinery can be ineffective in harvesting the cut plants. This is particularly so with boat-based techniques, though recent developments have overcome some of the inherent difficulties in handling large amounts of wet material (H. Miller, personal communication) (Fig. 19.3a).

3 Training. The efficient use of machines can be improved through training. This should include recognition of plant species, understanding of the role of vegetation and recovery processes, reduction of down-time, and recognition of different management objectives including nature conservation (Newbold *et al* 1989).

4 Development of new machines. Current machinery is based on the design of agricultural equipment and is seen as a means of eliminating plant species. A more sound approach would be to consider such techniques as a part of environmental management, for example shifting a river from one developmental stage with associated plant community to another with a more acceptable community structure given the use of the river. This method requires a different approach to the machinery and is the basis for emerging new designs (Wade 1990).

The ecological effects of mechanical control

Evidence of the ability of macrophytes to recover from aquatic plant management is limited, with most studies of river management being of the impact of dredging. Such fundamental management alters the habitat, and recovery of the aquatic plant community can take a long time. The recolonization of the Boro River, Botswana, after dredging was still only in its early stages 8 years after dredging, with the re-establishment of small colonies of *Rotala myriophylloides*, *Ottelia kunenensis*, *Najas pectinatus*, *Potamogeton thunbergii*, *P. trichoides* and *Nymphaea caerulea* (Lubke *et al* 1981, 1984). The dredging had resulted in the creation of a deep channel in the centre with steep sides, which, coupled with turbidity of the water, limited regrowth. The removal of natural meanders increased water flow considerably, further exacerbating recovery (see also Brookes 1988). At the other end of the spectrum, the recovery from dredging of a millstream in southern England dominated by *Elodea canadensis* with some *Nuphar lutea* and *Potamogeton pectinatus* was a faster process (Crisp & Gledhill 1970). The predredging community was succeeded by sparse growth of *E. canadensis* in the spring following management, extending through the summer to cover most of the bed despite a minor incident of pollution. The *E. canadensis* was removed by spate flow and immigrant *Ranunculus penicillatus* var. *calcareus* developed extensive growth in the second year after dredging, being replaced by *E. canadensis* and floating mats of *Apium* during the summer of that year.

Research into the recovery of chalk streams subjected to annual cutting regimes found that previously managed streams which were left unmanaged had a wider, more extreme range of both dissolved oxygen and water temperature than regularly managed streams (Dawson *et al* 1981; Dawson & Henville 1985). The morphological changes associated with the management of streams tended to create stress within the system because the system is modified to cope with winter flooding rather than allowing the channel to become naturally adjusted by flood erosion. The biomass levels at which management occurred coincided with the maximum biomasses found in unmanaged streams. Thus lack of management did not cause flooding and the fauna did not suffer the destabilizing effects of plant removal (Dawson 1989).

Concern has been expressed that cutting, harvesting and dredging remove most of the invertebrates attached to the vegetation and living in or on the mud of a watercourse; the evidence here is more extensive. The impact of dredging in certain cases can be long lasting, in others recovery is rapid. The macroinvertebrate community usually recovers rapidly after weed cutting. Free-swimming species, such as Coleoptera and Hemiptera, are left behind although in reduced numbers, though recolonization by flying and swimming species does not usually take place until suitable habitat conditions have developed in the channel. For example, oviposition by Odonata will not occur unless the required type of vegetation is available (Corbet *et al* 1960). Recolonization by species with limited powers of movement, e.g. Gastropoda and Lamellibranchia, is an even slower process, and if dredging is carried out on an extensive scale, the status of such animals will decline and, in intensively managed areas, become eradicated altogether (George 1976). Toner *et al* (1965), investigating a tributary of the River Moy, Ireland, observed that the recovery of the invertebrates after dredging was slow although large numbers of Chironomidae and Ephemeroptera were found 2 years after management. Trichoptera larvae were more seriously affected than other taxa.

In other situations macroinvertebrate re-establishment has been rapid and recovery has occurred within 12 months despite the severity of the management. Crisp and Gledhill (1970) considered that the recovery of the macrofauna of a millstream in southern England was complete about 1 year after draining and dredging. Chironomidae, Oligochaeta and Mollusca together formed 85% of the benthos and all three groups were present from the autumn of the year of management, but whereas the Chironomidae were numerically important from the outset, the Oligochaeta and Mollusca took almost a year to build up their numbers. In a similar investigation on a canalized section of the River Hull, UK, no dramatic changes in community composition were shown. Most species were recovering their numbers 6 months after dredging and both variety and numbers were expected to have returned to normal by 8–9 months after dredging (Pearson & Jones 1975). As well as removal by the dredger, it was suggested that a large number of animals were lost in downstream drift.

The effects of weed cutting on the macroinvertebrate fauna of the River Hull (Pearson & Jones 1978) were similar to those made in the earlier dredging study (Pearson & Jones 1975). As the substrate was not disturbed the recovery was more rapid although large numbers of invertebrates were removed in the cut weed. The timing of weed cutting proved to be important. In this case, Chironomidae and *Caenis horaria* had emerged prior to weed cutting with no impact on the breeding population. The effect of weed cutting on invertebrate drift was investigated by Kern-Hansen (1978) in four small Danish lowland streams. Total macrophyte clearance increased the drift of *Gammarus pulex* while partial cutting programmes resulted in only small increases in drift density. Similar increases in invertebrate drift density have been recorded in a small stream in Germany (Statzner & Stechmann 1977). In both cases the biomass of invertebrates removed from the river was significant.

Fish are not able to respond as rapidly to changes resulting from weed control as are invertebrates and macrophytes. Hence fishermen often regard mechanical weed control as deleterious to their fishery. On the Hampshire Avon, UK, anglers felt that the timing, extent and severity of weed cutting had caused destruction of cyprinid spawn, removal of habitat for fry development, a decrease of invertebrate diet and loss of cover for adult fish (Frake 1979). This view can certainly be substantiated for rivers. Swales (1982) has shown how potentially serious mechanical weed control is, through disruption of feeding, reproduction and other behaviours. Cutting, harvesting and dredging can result in the removal of large numbers of invertebrates with the weed, and increases in the level of invertebrate drift (Kern-Hansen 1978; Pearson & Jones 1975, 1978; Statzner & Stechmann 1977). A 30% decrease was observed in the fry of salmonid fish after dredging, though part of this was due to natural mortality, and within 2 years population estimates compared favourably with pre-dredging estimates (Toner *et al* 1965). A cut of *Callitriche*, *Berula* and *Ranunculus* in a tributary of the River Itchen, UK, to improve fishing did not encourage increased use of the stretch for the spawning of *Salmo trutta*, although it increased the area of exposed gravel. A count of redds, however, showed no increase over the previous years (Soulsby 1974).

Some fish, such as *Leuciscus leuciscus* and *Tilapia* spp., utilize aquatic macrophytes as a direct food source (Swales 1982; Lubke *et al* 1981). Removal of large amounts of vegetation can, therefore, have serious effects on fish feeding and growth. Macrophytes are also an important spawning substrate for phytophilous fish (Mills 1981) and weed cutting and dredging may remove potential spawning areas and incubating eggs, and eliminate the slack water needed for fry development. Mortensen (1977) found that weed cutting and the clearing out of stream weed beds increased mortality of *Salmo trutta*. Reductions in water level through weed cutting may adversely affect spawning by stranding incubating eggs (Mills 1981).

A crucial factor determining the impact of weed cutting on fisheries and spawning in particular is the timing of the weed control (Pearson & Jones 1978). Swales (1982) attributed low population levels of *Rutilus rutilus* in the River Perry, UK, to the effects of annual weed cutting on the reproductive success of the species. He also showed fish distribution in uncut and partially cut sections of river to be strongly associated with weed cover, particularly cover provided by overhanging bankside vegetation. Increased fish movements were caused by the operation of the weed-cutting launch and manual cutting was suggested as a preferable form of control because it is more selective. Indeed, selective weed cutting can be used to improve species diversity and fishing. This is difficult to achieve and is not normally attempted in the UK except in some of the more valuable trout fisheries (Barrett 1978). Frake (1979) described a reduction of the weed cut on the Hampshire Avon, UK, based on a flexible programme fitted to the pattern of weed growth and land drainage requirement with marginal strips of weed left on one or both banks where possible and in some cases a strip down the centre.

Chemical control

The use of herbicides in or near the aquatic environment has always been a sensitive issue, never more so than today when a number of agencies in Europe and North America are severely restricting herbicide application in or adjacent to waters, if not banning such use altogether. Different herbicides have been cleared for use in or near water in different countries (e.g. MAFF 1986); their most intensive use is in lowland alluvial plains where their application is compatible with the use of the water and where the plants can create a serious flood hazard (Haslam 1987). Murphy and Barrett (1990) conclude a review of the chemical control of aquatic weeds: '[herbicides] are not a panacea solution to all problems, nor are they to be rejected out of hand because of their potential ecological side-effects. As with any weed control measure, aquatic herbicides have their limitations and dangers but the record of their use to date in terms of both efficacy and environmental safety, is generally good'.

The treatment of vegetation in or near water is well described in such texts as Seagrave (1988) and Riemer (1984). The control of submerged vegetation with a herbicide in rivers limits the applicability of most herbicides, effective treatment being achieved with special formulations such as alginate and granules. It is emphasized that there is considerable legislation dictating which herbicides can be used in or near water, requirements prior to application, and the method of application itself. This legislation varies from one country to another, or from one administrative area of a country to another, as in the various USA states.

Bankside and emergent vegetation is treated using standard techniques although the range of herbicides for use near water is typically limited. In the UK herbicides such as dalapon, glyphosate and 2,4-D amine can be used for emergent and bankside vegetation. Maleic hydrazide can be used for the suppression of grass growth, fosamine ammonium for control of deciduous trees and shrubs, and asulam for bracken and dock control. These chemicals and their application are considered in detail by Seagrave (1988).

Both the bipyridinium herbicides, diquat and paraquat, are readily absorbed by submerged aquatic plants but despite this they are not generally effective in water flowing at more than about $100\,\mathrm{m\,h^{-1}}$, probably because the availance is insufficient (Murphy & Barrett 1990). This problem has been overcome by adding sodium alginate to the formulation, converting it into a sticky viscous solution (Barrett 1978). The formulation is applied from a strengthened or special applicator as a pencil jet on to the water surface where it breaks up to form discrete strings and droplets in the water. The jet can be projected up to 10 m. The strings and droplets are dense and sink rapidly into the weedbeds where they stick to the plants and release the active ingredient. Diquat-alginate effectively controls weeds in fast-flowing rivers (Barrett & Murphy 1982; Fox *et al* 1986). Diquat is strongly adsorbed and inactivated by clays and other organic particles (Narine & Guy 1982) and adsorption on to suspended material is very rapid. Hence the performance of diquat is greatly reduced in turbid waters or where plants are covered by deposits of silt and salts. High water hardness also reduces the uptake of this herbicide (Parker 1966).

Acrolein acts as a contact herbicide and has the ability to give useful weed control in fast-flowing waters because it needs only a very short contact time (Bowmer & Sainty 1977; Gaddis & Kissel 1982). Acrolein can be injected at a high concentration such that a pulse of herbicide with a peak concentration of about $15\,\mathrm{mg\,a.i.\,l^{-1}}$ moves downstream killing all aquatic plant growth as it passes (Bowmer & Smith 1984). The disadvantage of acrolein is that it is toxic to both plants and animals and the above management practice would also kill much of the animal life in the treated section of the river (Murphy & Barrett 1990). It is also hazardous to work with, being a lacrimatory agent and causing skin and lung damage; it is also explosive. Concentrations of less than $15\,\mathrm{mg\,a.i.\,l^{-1}}$ in irrigation water are generally safe for crops (USDA 1963) and it is rapidly lost from water by degradation to non-toxic residues and by volatilization (Bowmer & Higgins 1976). It is banned in many countries but is still used in certain situations in Australia, Canada, Egypt, Argentina and the USA.

Terbutryne, one of the methylthio-triazines, has a primary mode of action through photosynthetic inhibition with little translocation or activity away from the leaves (Murphy 1982). The principal target weeds are submerged and free-floating species. Terbutryne is also a potent algicide (Robson *et al* 1976). The herbicide is applied in a granular form spread evenly from the bank or a boat in order to achieve an availance of $0.05-1.0\,\mathrm{mg\,a.i.\,l^{-1}}$ for 7–16 days. It can be used in situations with flowing water so long as this concentration can be maintained. The efficacy of terbutryne can be reduced by its adsorption to clays, uptake by phytoplankton and chemical and microbial decomposition. It is also possible that the chemical is photodegraded. These processes tend to reduce its persistence considerably in the field (Murphy & Barrett 1990).

Dichlobenil is preferentially taken up by plant roots and subsequently transported around the plant. The formulation recommended for aquatic use is a clay-based granule containing 20% active ingredient. The granules have a slow-release property enabling them to carry the chemical down into the root region and release it slowly into the hydrosoil/water interface region (Tooby & Spencer-Jones 1978; Murphy & Barrett 1990). The properties associated with this particular formulation are such that it can be useful in slowly flowing rivers. Treatment is most effective early in the growing season when the chemical is translocated rapidly to growing points. Dichlobenil must not be used in potable water supplies nor in water to be used for irrigation nor for livestock watering. Fish from treated waters should not be used for human consumption or for animal feed for a period of 90 days following application (Riemer 1984).

Murphy and Barrett (1990) identify a number of areas of current research into herbicides for use in or near water. Those which are applicable to the management of rivers include the following:

1 Improvement of application methodology such as advances in controlled-release technology with the aims of maximizing impacts on target weeds and minimizing effects on non-target organisms.

2 Development of more rigorous assessment procedures to determine the environmental safety of aquatic herbicides.

3 Finding and developing new selective compounds to control major nuisance species or groups, one outstanding need being an algicide.

4 Determining the compatibility of herbicides with other aquatic weed control methods for use in integrated management programmes.

Effects of the use of herbicides

The impacts of herbicides upon aquatic ecosystems have been considered in detail by Brooker and Edwards (1975), Newbold (1976) and Robson and Barrett (1977). The effects are either direct toxicity to both target and non-target species, or indirect effects brought about by either the death/damage or removal of the target species.

Herbicide treatments can affect non-target species in two ways. First the herbicide might have an impact on non-target plant species, including algae, occurring in the section of river treated. The susceptibilities of the more common and abundant macrophyte species are given in manuals and guides for the use of herbicides (e.g. MAFF 1986). Likewise the toxic effects of most aquatic herbicides on non-target phytoplankton or other algal populations are well documented (e.g. Robson *et al* 1976). It is the less common and rare macrophyte species for which little information is available. Secondly, herbicide moving outside the treated area by drift or movement in water currents can have ecological significance, and there has been concern over long-term implications for non-target plant communities of the accumulation of inactivated residues of the bipyridinium herbicides where regular treatment has occurred (Corwin & Farmer 1984).

The direct toxic effects on herbicides on micro-organisms, invertebrates, fish and other vertebrates are comprehensively reviewed by Murphy and Barrett (1990).

The short-term indirect effects of herbicidal control are concerned with the following factors:

1 Alterations in dissolved gas and pH brought about by decaying vegetation (see Brooker & Edwards 1975; Marshall 1981; Murphy *et al* 1981).

2 Toxin and nutrient release from decaying vegetation (see Ahmed *et al* 1981, 1982).

3 Habitat loss or alteration for non-target organisms (see Brooker & Edwards 1973a,b; Hanbury *et al* 1981; Clare & Edwards 1983).

4 Increases in detritus (see Carter & Hestand 1977; Newbold 1975; Strange 1976).

5 Development of replacement growth of opportunist plant species (see Carter & Hestand 1977; Hestand & Carter 1977).

Other forms of control

Outside of the more traditional physical and chemical control methods, there are other techniques which can be used to limit vegetation growth in rivers. These are shading, hydrological manipulation and biological control.

Shading

Shading is an effective control measure which, in some rivers, has been used for many years to limit weed growth and to draw current to the weedless areas, e.g. the chalk streams of southern England (National Rivers Authority, undated). Krause (1977) and Lohmeyer and Krause (1974, 1975) advocated full shade to remove vegetation, rather than partial shade, with the retention of some vegetation for a better ecological balance. On the sandy unstable banks of streams in the Luneberger Heide, Germany, planting was achieved with *Alnus glutinosa*, a species chosen because of its tolerance to damp conditions and its deep-rooting stabilizing roots. A 10–15-year rotation of tree cutting prevents farmland being shaded excessively. The costs were less than half those of weed cutting and dredging (excluding the loss of a small amount of agricultrual land), though even with shading some macrophytes occurred beneath occasional breaks in the canopy. The autumn leaf fall was not found to hinder flow and added organic matter to the stream compensating for the loss of that from river plants.

Dawson and Kern-Hansen (1979a,b) developed these initial ideas, leading to a recommendation of light reduction of approximately 50%. This achieved a satisfactory ecological balance with planting on the north and south sides of stream sections of various widths flowing in an east–west direction. The light available at the water surface is related to the effective tree and bank height, and the orientation of the bank and its vegetation in relation to the width of the river, emphasizing the importance of channel morphology. Narrow streams with high banks can be deeply shaded by the bank and low vegetation, e.g. the sunken streams of Norwegian farmland (Haslam 1987). On lowland streams approximately 2 m in width with banks 1–2 m wide, little other control is needed where there is no grazing and tall herbs can grow. Streams 3–8 m wide will need bushes or small trees for effective shading, and those 15+ m wide may require mature trees (Dawson 1978). With full shading from the south, aquatic growth can be considerable in east–west streams because of the light availability from the north. In a chalk stream with macrophytes present in winter and an annual average of 15% of light reaching the water, the amount of light in summer was as low as 6%, and in winter it reached 30–40%¨ Westlake & Dawson 1976).

A summary of the influence of channel orientation on effectiveness of weed control is given in Fig. 19.4a. Dawson (1976, 1980) shows that leaf fall provides adequate organic matter to compensate for the loss of that from macrophytes within the stream and better conditions for breakdown of organic matter.

The provision of gaps in a shade belt to allow patches of aquatic plants to grow maintains a

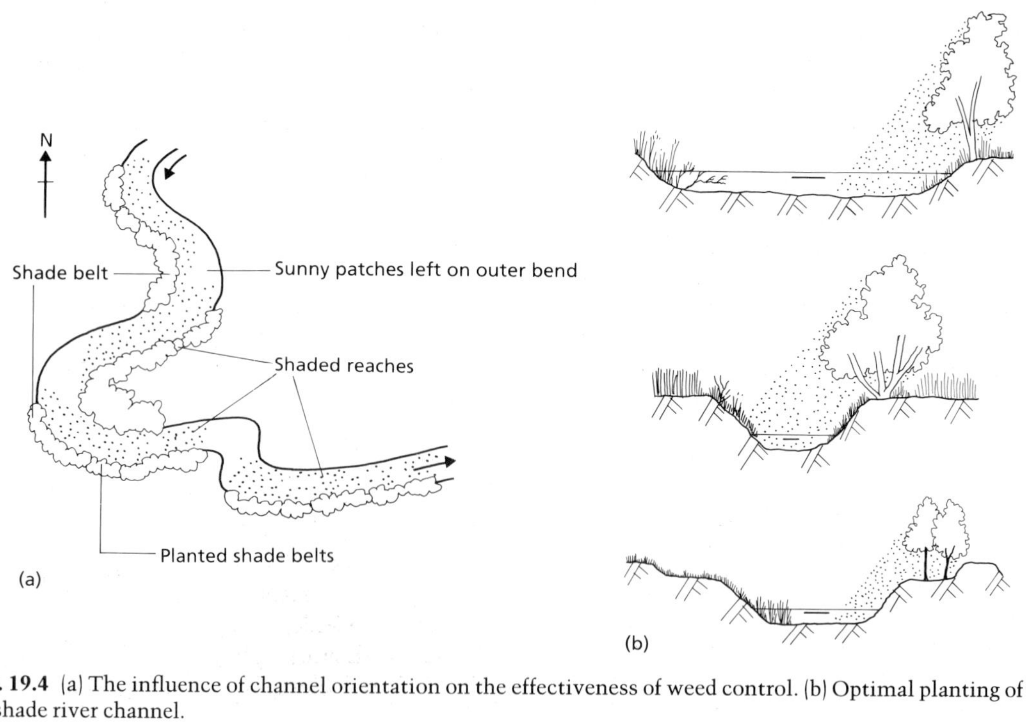

Fig. 19.4 (a) The influence of channel orientation on the effectiveness of weed control. (b) Optimal planting of trees to shade river channel.

better balanced habitat in which leaf litter from the trees decomposes more easily; otherwise it tends to form a dense, slimy deoxygenated mass on the channel bed (Fig. 19.4b) (Lewis & Williams 1984). A survey of the River Tone, England, in which tree-shaded and open sections of river were alternated at 40–50-m intervals, demonstrated that a greater variety of habitats for nature conservation is created compared with continuously shaded sections and that this approach maintains an attractive landscape for people (Lewis 1981).

Floating vegetation, e.g. water-lilies, can also provide shading to suppress submerged plant growth (Pitlo 1982). This approach is unlikely to be useful in rivers in general as floating species are not adapted to growing in flowing water. However, there is the possibility of using floating leaved plants in slow-flowing lowland rivers.

Investigation has also been undertaken of the application of artificial covering to shade out weeds (Dawson 1981; Dethioux 1982; Dawson & Hallows 1983). Dawson and Hallows (1983) describe the use of lengths of gas-permeable opaque material to control vegetation in narrow watercourses. The material is moved from one section of channel to another on a rotational basis, remaining in one place for 4–12 weeks, the duration varying with species (e.g. *Ranunculus* species are killed sooner than *Elodea canadensis*) and whether or not temporary die-back or elimination is required.

Dawson and Haslam (1983) comment that although some data are available to guide the choice of vegetation and the amount of shade provided by such vegetation (e.g. Krause 1977), detailed information dealing with all aspects of half-shade conditions is not currently available. Lewis and Williams (1984) provide a useful consideration of a number of tree species including *Alnus glutinosa*, *Salix* tree and shrub species, *Populus nigra*, *Acer pseudoplatanus* and *Fraxinus excelsior*.

Shading can present problems to river managers, especially with respect to other forms of river maintenance, exacerbating the risk of flooding and complicating landscape management. As they age, dead and diseased limbs may fall off into the river; with time the channel may be undercut or erode behind a mature tree until it falls into the channel. Likewise trees below flood level will catch trash and this may contribute to their falling in severe storms. Such dead and fallen wood is nevertheless an important natural component of rivers. Along lowland rivers in particular, access is required for bankbased machinery such as dredgers for the maintenance of sediments and banks. Trees and shrubs can prevent such access and can contribute to the hydraulic roughness of the channel or its floodplain, increasing the likelihood of flooding. Close liaison between those managing the aquatic plants and the engineer is essential to achieve a positive approach to shading as a form of plant control.

Hoogerkamp and Rozenboon (1978) drew attention to the inappropriate nature of trees in certain landscapes, in this case the flat Dutch landscape. Dawson and Haslam (1983) illustrate the alternative view that in other situations such planting can enhance the landscape. Hoogerkamp and Rozenboon (1978) also expressed concern that trees might interfere with crops through shading and reduction of drying period, increase insect nuisance to cattle, and increase the incidence of lightning damage. Hermans (1975) and Pitlo (1978), on the other hand, advocated control by shading in the Netherlands, the former finding that the cost is less than half that of other methods of weed control.

The basic principles of shading control can be summarized as follows:

1 Half shade by deciduous vegetation is the optimum for aquatic plant control in streams and rivers.

2 Vegetation, if on the south bank, should at maximum be as high as the distance to the north bank; otherwise more vegetation is required and this should aim to intercept up to two-thirds of the light near the centre of the channel around mid-summer; it is more efficient to intercept the sun path near midday.

3 The vegetation should be in regularly intermittent stands occupying two-thirds of one bank.

4 Vegetation should be considered as part of a planned cycle of planting and cropping; replanting of trees should begin in and around the gaps in those stands halfway to maturity; alternative weed control is particularly important during the first few years following establishment.

Channel deepening

The deepening of the river channel by dredging or raising water levels can effectively put plant growth out of reach of incident light, an effect observed to occur naturally in deeper rivers. This effect has been observed in a number of situations (e.g. Lubke *et al* 1981; Pitlo 1982), occasionally being enhanced by an increase in turbidity due to alterations in flow characteristics. In order to compensate for loss of the submerged plant communities as a result of channel deepening, it is preferable to introduce a shallow-water berm in which replacement plant communities can develop (Lewis & Williams 1984). This can be important for nature conservation, wildfowl and fisheries uses.

Biological control

The most widely used biological control agent for aquatic vegetation in Europe and North America is the grass carp (*Ctenopharyngodon idella* Val.), a fish which under the right conditions can effect good control of water weeds (Seagrave 1988). Limiting the movement of fish stocked in a given piece of water is essential for two reasons. First, the fish need to concentrate at an appropriate stocking density to effectively control the weeds. If the fish stock is dissipated over a too wide an area it will not achieve the level of control required. Despite the unlikely event of this fish species reproducing – it requires very specific conditions – there is concern that the grass carp, if unrestricted in its movement, could cause damage to sensitive plant communities, and at worst, despite predictions to the contrary, could establish a breeding population. It is primarily for these reasons that grass carp are most unlikely to be used to any extent in river systems. There is some potential for their use in smaller lowland rivers, canals and drains. Successful use has been demonstrated in small drains in the Netherlands and England.

Integrated management

There is an increasing recognition of the value of integrated control in river management. In its simplest form this can be a combination of two management techniques, for example weed mowing for the banks and emergent vegetation and the application of diquat alginate to control submerged vegetation. Integrated management should recognize the uses agreed for a river and the ecological processes occurring within and associated with the river, and can range from control of the nutrient budget to agreeing acceptable levels of aquatic plant biomass within the channel. The achievement of an acceptable river vegetation could take a number of years, and the management techniques would need to change over this period; for example, the growth of trees and shrubs would alter the intensity of management needed for aquatic vegetation. This period of time should also be used to understand better the dynamics of the vegetation in relation to management. This has been well illustrated in the case of chalk stream vegetation typically dominated by *Ranunculus* spp. and the advantages and disadvantages of cutting at different times of the year (Dawson 1989).

19.4 PROSPECT

The management of the aquatic and riparian vegetation of rivers has been regarded as simply weed control for too long, and there is a pressing need to put this management into a broader context in which the vegetation is seen as a key component of the river, and one which should be incorporated into an agreed strategy for that river or even the catchment as whole. River management is undertaken by a number of different agencies, ranging from the national agencies such as the National Rivers Authority in England and Wales, through to those responsible for drainage districts such as the Internal Drainage Boards and the individual landowners managing streams and smaller watercourses. Dawson (1989) reports an estimate of the total cost of aquatic plant control in flowing waters in England and Wales, made by the Water Research Centre; this is between 45 and 75 million pounds per annum and could be as high as £100 million. Main rivers were costing £8 million and streams managed by groups or individuals also £8 million. Given the relatively large resources spent on this aspect of river management by independent and often uncoordinated agencies, significant environmental and economic benefits could be derived through the development of integrated and ecologically-based river vegetation management.

REFERENCES

Ahmed SA, Ito M, Ueki K. (1981) Water quality as affected by water hyacinth decomposition after 2,4-D and ametryne application. *Weed Research* **25**: 386–93. [19.3]

Ahmed SA, Ito M, Ueki K. (1982) Water quality as affected by water hyacinth decomposition after cutting or 2,4-D application. *Weed Research* **27**: 34–9. [19.3]

Aston HI. (1973) *Aquatic plants of Australia*. Melbourne University Press, p. 368. [19.1]

Barko JW, Smart RM. (1981) Sediment based nutrition of submersed macrophytes. *Aquatic Botany* **10**: 339–52. [19.2]

Barrett PRF. (1978) Aquatic weed control: necessity and methods. *Fisheries Management* **9**: 93–101. [19.3]

Barrett PRF, Murphy KJ. (1982) The use of diquat-alginate for weed control in flowing waters. In: *Proceedings EWRS 6th Symposium on Aquatic Weeds 1982*, pp. 200–8. Novi Sad. [19.3]

Bowmer KH, Higgins ML. (1976) Some aspects of persistence and fate of acrolein herbicide in water. *Archives of Environmental Contamination and Toxicology* **5**: 87–96. [19.3]

Bowmer KH, Sainty GH. (1977) Management of aquatic plants with acrolein. *Journal of Aquatic Plant Management* **15**: 40–6. [19.3]

Bowmer KH, Smith GH. (1984) Herbicides for injection into flowing water: acrolein and endothal-amine. *Weed Research* **24**: 201–11. [19.3]

British Waterways Board. (1981) *Vegetation Control and Management*. British Waterways Board, London. [19.3]

Brooker MP, Edwards RW. (1973a) Effects of the herbicide paraquat on the ecology of a reservoir I. Botanical and chemical aspects. *Freshwater Biology* **3**: 157–76. [19.3]

Brooker MP, Edwards RW. (1973b) Effects of the herbicide paraquat on the ecology of a reservoir. II. Community metabolism. *Freshwater Biology* **3**: 383–92. [19.3]

Brooker MP, Edwards RW. (1975) Review paper: aquatic herbicides and the control of water weeds. *Water Research* **9**: 1–15. [19.4]

Brooker MP, Morris DL, Wilson CJ. (1978) Plant flow relationships in the R. Wye catchment. In: *Proceedings EWRS 5th Symposium on Aquatic Weeds*, pp. 63–70. Amsterdam. [19.2]

Brookes A. (1988) *Channelized Rivers. Perspectives for Environmental Management*. Wiley, Chichester. [19.3]

Burkle F. (1978) Natural construction on watercourses. *Garten und Landschaft* **1**: 18–24. [19.2]

Butcher RW. (1933) Studies on the ecology of rivers. I. On the distribution of macrophyte vegetation in the rivers of Britain. *Journal of Ecology* **21**: 58–91. [19.2]

Carter CC, Hestand RS. (1977) Relationship of regrowth of aquatic macrophytes after treatment with herbicides to water quality and phytoplankton populations. *Journal of Aquatic Plant Management* **15**: 65–9. [19.3]

Casper SJ, Krausch H-D. (1980, 1981) *Susswasserflora von Mitteleuropa*. 23, 1 (Lycopodiaceae bis Orchidaceae), pp. 1–403 (1980); 23, 2 (Saururaceae bis Asteraceae), pp. 404–943 (1981). Fischer-Verlag, Jena. [19.1]

Cave TG. (1981) Current weed control problems in land drainage channels. In: *Proceedings of the Conference on Aquatic Weeds and their Control, 1981*, pp. 5–14. Association of Applied Biologists, Wellesbourne, UK. [19.3]

Chambers PA, Prepas EE, Bothwell ML, Hamilton HR. (1989) Roots versus shoots in nutrient uptake by aquatic macrophytes in flowing waters. *Canadian Journal of Fisheries and Aquatic Sciences* **46**: 435–9. [19.2]

Chambers PA, Prepas EE, Hamilton HR, Bothwell ML. (1991) Current velocity and its effect on aquatic macrophytes in flowing waters. *Ecological Application* **1**: 249–57. [19.2]

Charlton SED, Bayne D. (1986) *Phosphorus Removal: the Impact Upon Water Quality in the Bow River Downstream of Calgary, Alberta. Bow River Data Base 1980–1985*. Pollution Control Division, Alberta Environment, Edmonton. [19.2]

Child L, de Waal LC, Wade PM. (1992) Control and management of *Reynoutria* species (knotweed). *Aspects of Applied Biology* **29**: 295–307. [19.2]

Clare P, Edwards RW. (1983). The macro-invertebrate fauna of the drainage channels of the Gwent Levels, S. Wales. *Freshwater Biology* **13**: 205–27. [19.3]

Cook CDK. (1990) *Aquatic Plant Book*. p. 228. SPB Publishing, The Hague. [19.1]

Cook CDK, Gut BJ, Rix EM, Schneller J, Seitz M. (1974) *Water Plants of the World: a Manual for the Identification of the Genera Freshwater Macrophytes*, p. 561. Junk, The Hague. [19.1]

Corbet PS, Longfield C, Moore NW. (1960) *Dragonflies*. Collins, London. [19.3]

Corwin DL, Farmer WJ. (1984) Nonsingle-valued adsorption-desorption of bromacil and diquat by freshwater sediments. *Environmental Science and Technology* **18**: 507–14. [19.3]

Crisp DT, Gledhill T. (1970) A quantitative description

of the recovery of the bottom fauna in a muddy reach of a mill stream in Southern England after draining and dredging. *Archiv für Hydrobiologie* **64**: 502–41. [19.3]

Dawson FH. (1976) Organic contribution of stream edge forest litter fall to the chalk stream ecosystem. *Oikos* **27**: 13–18. [19.3]

Dawson FH. (1978) Aquatic plant management in semi-natural streams: the role of marginal vegetation. *Journal of Environmental Management* **6**: 213–21. [19.3]

Dawson FH. (1980) The origin and composition of downstream transport of plant material in a small chalk stream. *Freshwater Biology* **10**: 419–35. [19.3]

Dawson FH. (1981) The reduction of light as a technique for the control of aquatic plants – an assessment. In: *Proceedings of the Conference on Aquatic Weeds and their Control, 1981*, pp. 157–64. Association of Applied Biologists, Wellsbourne, UK. [19.3]

Dawson FH. (1989) Ecology and management of water plants in lowland streams. *Annual Report Freshwater Biological Association* **57**: 43–60. [19.3, 19.4]

Dawson FH, Hallows JB. (1983) Practical applications of a shading material for macrophyte control in watercourses. *Aquatic Botany* **17**: 301–8. [19.3]

Dawson FH, Haslam SM. (1983) The management of river vegetation with particular reference to shading effects of marginal vegetation. *Landscape Planning* **10**: 147–69. [19.3]

Dawson FH, Henville P. (1985) Characteristics, calibration and output optimisation of Mackereth-type dissolved oxygen sensor. *Journal of Physics E: Scientific Instruments* **18**: 526–31. [19.3]

Dawson FH, Kern-Hansen U. (1979a) The effect of natural and artificial shade on the macrophytes of lowland streams and the use of shade as a management technique. *Internationale Revue gesamten Hydrobiologie und Hydrographie* **69**: 437–55. [19.3]

Dawson FH, Kern-Hansen U. (1979b) The effect of natural and artificial shade on the macrophytes of lowland streams. In: Symoens JJ, Hooper SS, Compère P (eds) *Studies on Aquatic Plants*, pp. 214–21. Royal Botanic Society of Belgium, Brussels. [19.3]

Dawson FH, Kern-Hansen, Westlake DF. (1981) Water plants under temperature and oxygen regimes of lowland streams. In: Symoens JJ, Hooper SS, Compère P. (eds) *Studies on Aquatic Vascular Plants. Proceedings of International Colloquium on Aquatic Vascular Plants*, Brussels, 214–221. [19.3]

Dethioux M. (1982) Les saules dans nos payages. *Parca Nationaux* **37**: 111–18. (*Notes techniques du centre d'écologie forêtière et rurale (ISRA), 43.*) [19.3]

Fox ÅM, Murphy KJ, Westlake DF. (1986) Effects of diquat alginate and cutting on the submerged macrophyte community of a *Ranunculus* stream in Northern England. In: *Proceedings EWRS/AAB 7th Symposium on Aquatic Weeds, 1986*. pp. 105–12. Loughborough. [19.3]

Frake A. (1979) The Hampshire Avon Survey. In: *Proceedings of the 1st British Freshwater Fisheries Conference, 1979*, pp. 100–5. University of Liverpool, Liverpool. [19.3]

Gaddis CW, Kissel CL. (1982) Acrolein in irrigation waterways. In: *Proceedings EWRS 6th Symposium on Aquatic Weeds, 1982*, pp. 164–70. Novi Sad. [19.3].

George M. (1976) Mechanical methods of weed control in watercourses – an ecologist's view. In: *Proceedings of Symposium on Aquatic Herbicides*. Monograph 16, pp. 91–9. British Crop Protection Council, Oxford. [19.3]

Ham SF, Wright JF, Berrie AD. (1981) Growth and recession of aquatic macrophytes on an unshaded section of the River Lambourn, England, from 1971 to 1976. *Freshwater Biology* **11**: 381–90. [19.2]

Hanbury RG, Murphy KJ, Eaton JW. (1981). The ecological effects of 2-methylthio triazine herbicides used for aquatic weed control in navigable canals. II. Effects on macro-invertebrate fauna, and general discussion. *Archiv für Hydrobiologie* **91**: 408–26. [19.3]

Haslam SM. (1978) *River Plants*. Cambridge University Press, Cambridge. [19.1]

Haslam SM. (1979) Infra-red colour photography and *Phragmites communis* Trin. *Polskie Archiwum Hydrobiologie* **26**: 65–72. [19.1]

Haslam SM. (1987) *River Plants of Western Europe*. Cambridge University Press, Cambridge. [19.3]

Haslam SM, Wolseley PA. (1981) *River Vegetation: its Identification Assessment and Management*. Cambridge University Press, Cambridge. [19.1]

Hellawell JM. (1978) *Biological Surveillance of Rivers*. Natural Environment Research Council/Water Research Centre, Stevenage. [19.1]

Hemphill RW, Bramley ME. (1989) *Protection of River and Canal Banks: A Guide to Selection and Design*. Butterworths, London. [19.2]

Hermans LCM. (1975) Levendgroen, II. Groene beken in Limburg. *Tijdschrift der Nederlandsche Leidemaatschappij* **86**: 476–81. [19.3]

Hestand RS, Carter CC. (1977). Succession of various aquatic plants after treatment with four herbicides. *Journal of Aquatic Plant Management* **15**: 60–4. [19.3]

Hoogerkamp M, Rozenboon G. (1978) The management of vegetation on the slopes of waterways in the Netherlands. In: *Proceedings EWRS 5th Symposium on Aquatic Weeds, 1978*, pp. 203–11. Amsterdam. [19.3]

Hotchkiss N. (1972) *Common Marsh, Underwater and Floating-leaved Plants of the United States and Canada*, p. 124. Dover Publications, New York. [19.1]

Keown MP, Oswald NR, Perry EB, Dordeau JDA. (1977)

Literature Survey and Preliminary Evaluation of Stream-bank Protection Methods. Technical Report H-77-9, May 1977. [19.2]

Kern-Hansen U. (1978) Drift of *Gammarus pulex* L. in relation to macrophyte cutting in four small Danish lowland streams. *Verhandlungen der Internationalen Vereinigung für theoretische und angewandte Limnologie* **20**: 1440–5. [19.3]

Kern-Hansen U, Dawson FH. (1981) The standing crop of aquatic plants of lowland streams in Denmark and the inter-relationships of nutrients in plant, sediments and water. In: *Proceedings EWRS 5th Symposium on Aquatic Weeds, 1978*, pp. 143–50. Amsterdam. [19.2]

Kijlstra BC. (1978) Nature conservation and aspects of landscape planning. In: *Proceedings 5th Symposium on Aquatic Weeds, 1981*. Amsterdam. [19.2]

Kite D. (1980) Getting to the root of river ecology. *Surveyor* 10 July: 24–25. [19.2]

Krause A. (1977) On the effect of marginal tree rows with respect to the management of small lowland streams. *Aquatic Botany* **3**: 185–92. [19.3]

Ladle M, Casey H. (1971) Growth and nutrient relationships of *Ranunculus penicillatus* var. *calcareous* in a small chalk stream. In: *Proceedings EWRC 3rd Symposium on Aquatic Weeds, 1971*, pp. 53–64. Oxford. [19.2]

Lewis G. (1981) *Somerset Rivers Survey: A Survey of the Rivers Tone, Yeo, Cary and Northmoor Main Drain, to Map and Describe their Flora and Fauna*. Somerset Trust for Nature Conservation/Wessex Water Authority (unpublished report). [19.3]

Lewis G, Williams G. (1984) *Rivers and Wildlife Handbook: A Guide to Practices Which Further the Conservation of Wildlife on Rivers*. Royal Society for the Protection of Birds/Royal Society for Nature Conservation, Bedford. [19.2, 19.3]

Lohmeyer W, Krause A. (1974) Uber den Geholzbewuchs und Kleinen Fleissgewassern Nordwestdeutschlands und seine Bedeutung für den Uperschutz. *Naturschutz und Landschaftsplanung* **49**: 323–30. [19.3]

Lohmeyer W, Krause A. (1975) Uber die Auswirkungen des Geholzbewuchses an kleinen Wassertaufen des Munsterlandesauf die Vegetation im Wasser und anden Boschungen in Hinblick auf die Unterhaltung der Gewasser. *Schriftenreike fuer Vegetationskunde* **9**: 1–105. [19.3]

Lubke RA, Raynham PE, Reavell PE. (1981) Reassessment of plant succession on spoil heaps along the Boro River, Okavango Delta, Botswana. *South African Journal of Science* **77**: 21–3. [19.3]

Lubke RA, Reavell PE, Dye PJ. (1984) The effects of dredging on the macrophytic vegetation of the Boro River, Okavango Delta, Botswana. *Biological Conservation* **33**: 211–36. [19.3]

Madsen TV, Sondergaard M. (1983) The effects of current velocity on the photosynthesis of *Callitriche stagnalis* Scop. *Aquatic Botany* **15**: 187–93. [19.2]

MAFF (Ministry of Agriculture, Fisheries & Food). (1986) *Guidelines for the Use of Herbicides on Weeds in or Near Water Courses and Lakes*. MAFF Booklet 2078. [19.1, 19.3]

Marshall EJP. (1981) The ecology of a land drainage channel. 1. Oxygen balance. *Water Research* **15**: 1075–85. [19.3]

Melzer A. (1976) Makrophytische Wasserpflanzen als Indikatoren des Dewasserzustandes oberbayerischer Seen. *Dissertationes Botanicae* **34**: 1–195, Cramer, Vaduz. [19.1]

Mills CA. (1981) The spawning of roach, *Rutilus rutilus* L. in a chalk stream. *Fisheries Management* **12**: 49–54. [19.3]

Mortensen E. (1977) Density-dependent mortality of trout fry (*Salmo trutta* L.) and its relationship to the management of small streams. *Journal of Fish Biology* **11**: 613–17. [19.3]

Murfitt RFA, Haslam SM. (1981) Some unanswered questions relating to the mechanical control of weeds in water channels. In: *Proceedings of the Conference on Aquatic Weeds and their Control, 1981*, pp. 87–91. Association of Applied Biologists, Wellesbourne, UK. [19.3]

Murphy KJ. (1982) The use of methylthio-triazine herbicides in freshwater systems: a review. In: *Proceedings EWRS 6th Symposium on Aquatic Weeds, 1982*, pp. 263–77. Novi Sad. [19.3]

Murphy KJ. (1990) Survey and monitoring of aquatic weed problems and control operations. In: Pieterse A, Murphy KJ. (eds) *Aquatic Weeds – the Ecology and Management of Nuisance Aquatic Vegetation*, pp. 228–37. [19.1]

Murphy KJ, Barrett PRF. (1990) Chemical control of aquatic weeds. In: Pieterse AH, Murphy KJ (eds) *Aquatic Weeds*. Oxford University Press, Oxford. [19.3]

Murphy KJ, Eaton J. (1983) Effects of pleasure-boat traffic on macrophyte growth in canals. *Journal of Applied Ecology* **20**: 713–29. [19.1]

Murphy KJ, Bradshaw AD, Eaton JW. (1980) *Plants for the Protection of Canal Banks*. Report, Department of Botany, University of Liverpool. [19.2]

Murphy KJ, Hanbury RG, Eaton JW. (1981) The ecological effects of 2-methylthio triazine herbicides used for aquatic weed control in navigable canals. I. Effects on aquatic flora and water chemistry. *Archiv für Hydrobiologie* **91**: 294–331. [19.3]

Narine DR, Guy RD. (1982) Binding of diquat and paraquat to humic acid in aquatic environments. *Soil Science* **33**: 356–63. [19.3]

National Rivers Authority. *Managing Aquatic Plants*. Information Sheet. Education Series 14. National

Rivers Authority (Southern Region). [19.3]
National Rivers Authority. (1992) *River Corridor Surveys. Methods and Procedures.* Conservation Technical Handbook No. 1. NRA, Bristol. [19.1]
Newbold C. (1975) Herbicides in aquatic systems. *Biological Conservation* 7: 97–118. [19.3]
Newbold C. (1976) Environmental effects of aquatic herbicides. In: *Proceedings of Symposium on Aquatic Herbicides*, pp. 78–90. Monograph 16, British Crop Protection Council, Oxford. [19.3]
Newbold C, Honnor J, Buckley K. (1989) *Nature Conservation and the Management of Drainage Channels.* Nature Conservancy Council and Association of Drainage Authorities, Peterborough. [19.3]
Nilsson C. (1987) Distribution of stream-edge vegetation along a gradient of current velocity. *Journal of Ecology* **75**: 513–22. [19.2]
Oksiyuk OP, Merezhko AA, Volkova TF. (1978) Use of higher aquatic plants in canal bank consolidation. *Water Resources* **5**: 556–62. [19.2]
Owens M, Edwards RW. (1961) The effects of plants on river conditions II. Further crop studies and estimates of net productivity of macrophytes in a chalk stream. *Journal of Ecology* **49**: 119–26. [19.2]
Parker C. (1966) Water hardness and phytotoxicity of paraquat. *Nature (London)* **212**: 1465. [19.3]
Pearson RG, Jones NV. (1975) The effects of dredging operations on the benthic community of a chalk stream. *Biological Conservation* **8**: 273–8. [19.3]
Pearson RG, Jones NV. (1978) Effects of weedcutting on the macro-invertebrate fauna of a canalised section of the River Hull, a North England chalk stream. *Journal of Environmental Management* **7**: 91–9. [19.3]
Pieterse AH, Murphy KJ (eds). (1990) *Aquatic Weeds.* Oxford University Press, Oxford. [19.1]
Pitlo RH. (1978) Regulation of aquatic vegetation by interception of daylight. In: *Proceedings EWRS 5th Symposium on Aquatic Weeds, 1978*, pp. 91–9. Amsterdam. [19.3]
Pitlo RH. (1982) Flow resistance of aquatic vegetation. In: *Proceedings EWRS 6th Symposium on Aquatic Weeds, 1982*, pp. 225–34. Novi Sad. [19.3]
Pitlo RH, Dawson FH. (1990) Flow resistance of aquatic weeds. In: Pieterse AH, Murphy KJ (eds) *Aquatic Weeds*, pp. 74–84. Oxford University Press, Oxford. [19.1]
Price H. (1981) A review of current mechanical methods. In: *Proceedings of the Conference on Aquatic Weeds and their Control, 1981*, pp. 77–86. Association of Applied Biologists, Wellesbourne, UK. [19.3]
Riemer DN. (1984) *Introduction to Freshwater Vegetation.* AVI Publishing, Westport, USA. [19.3]
Robson TO. (1975) *A Survey of Aquatic Weed Control Methods used in Internal Drainage Boards, 1973.* Technical Report No. 35, Agricultural Research Council Weed Research Organisation, Oxford. [19.3]
Robson TO, Barrett PRF. (1977) Review of effects of aquatic herbicides. In: Perring PM, Mellanby K (eds), *Ecological Effects of Pesticides*, pp. 111–18. Academic Press, London. [19.4]
Robson TO, Fowler MC, Barrett PRF. (1976) Effect of some herbicides on freshwater algae. *Pesticide Science* 7: 391–402. [19.3]
Seagrave C. (1988) *Aquatic Weed Control.* Fishing News Books, Farnham. [19.1, 19.3]
Siebert P. (1968) *Importance of Natural Vegetation for the Protection of the Banks of Streams, Rivers and Canals. Freshwater.* Council of Europe Nature and Environment Series No. 2, pp. 33–67. [19.2]
Smith CS, Adams MS. (1986) Phosphorus transfer from sediments by *Myriophyllum spicatum. Limnology and Oceanography* **31**: 1312–21. [19.2]
Sosiak AJ. (1990) *An Evaluation of Nutrients and Biological Conditions in the Bow River, 1986 to 1988.* Internal report of Environmental Quality Monitoring Branch, Environmental Assessment Division, Alberta Environment, Edmonton. [19.2]
Soulsby PG. (1974) The effect of a heavy cut on the subsequent growth of aquatic plants in a Hampshire chalk stream. *Journal of the Institute of Fisheries Management* **5**: 49–53. [19.3]
Spillet PB. (1981) The use of groynes and deflectors in river management. In: *Proceedings of the Conference on Aquatic Weeds and their Control, 1981*, pp. 189–98. Association of Applied Biologists, Wellesbourne, UK. [19.2]
Standing Committee of Analysts (1987) Methods for the examination of waters and associated materials. In: *Methods for the Use of Aquatic Macrophytes for Assessing Water Quality 1985–6*, p. 176. HMSO, London. [19.1]
Statzner B, Stechmann DH. (1977) Der Einfluss einer mechanischen Entkrautenseebach. *Faunistisch-Okologische Mittelungen* **5**: 93–109. [19.3]
Strange RJ. (1976) Nutrient release and community metabolism following the application of herbicide to macrophytes in microcosms. *Journal of Applied Ecology* **13**: 889–97. [19.3]
Swales S. (1982) Impacts of weed cutting on fisheries: an experimental study in a small lowland river. *Fisheries Management* **13**: 125–38. [19.3]
Toner ED, O'Riordan A, Twomey E. (1965) The effects of arterial drainage works on the salmon stock of a tributary of the River Moy. *Irish Fisheries Investigations Series A (Freshwater)* **1**: 36–55. [19.3]
Tooby TE, Spencer-Jones DH. (1978) The fate of the aquatic herbicide dichlobenil in hydrosoil, water and roach (*Rutilus rutilus* L.) following treatment of areas of a lake. In: *Proceedings EWRS 5th Symposium on Aquatic Weeds, 1978*, pp. 323–31. Amsterdam. [19.3]
USDA. (1963) *Chemical Control of Submerged Water-*

weeds in Western Irrigation and Drainage Canals. US Department of Agriculture Agricultural Research Service and US Department of the Interior: Reclamation Joint Report ARS 34–57. [19.3]

Wade PM. (1987) A review of the provision made for the identification of wetland macrophytes. *Archive fur Hydrobiologie Ergebnisse Limnologie* **27**: 105–113. [19.1]

Wade PM. (1990) Physical control of aquatic weeds. In: Pieterse AH, Murphy KJ (eds) *Aquatic Weeds*, pp. 93–135. Oxford University Press, Oxford. [19.3]

Westerdahl HE, Gestinger KD. (1988) *Aquatic Plant Identification and Herbicide Use Guide. 2. Aquatic Plants and Susceptibility to Herbicides.* Technical Report A-88-9, US Army Engineer Waterways Experiment Station, Vicksburg, Mississippi. [19.1]

Westlake DF. (1967) Some effects of low-velocity currents on the metabolism of aquatic macrophytes. *Journal of Experimental Botany* **18**: 187–205. [19.2]

Westlake DF, Dawson FH. (1976) The construction and long term field use of inexpensive aerial and aquatic integrating photometers. In: Evans GG, Bainbridge R, Rackham O (eds) *Light as an Ecological Factor II*, pp. 27–42. British Ecological Society Symposium, 16. Blackwell Scientific Publications, Oxford. [19.3]

Westlake DF, Dawson FH. (1982) Thirty years of weed cutting on a chalk stream. In: *Proceedings EWRS 6th Symposium on Aquatic Weeds 1982*, pp. 132–40. Novi Sad. [19.3]

Westlake DF, Dawson FH. (1986). The management of *Ranunculus calcareus* by pre-emptive cutting in Southern England. *Proceedings EWRS/AAB 7th Symposium on Aquatic Weeds 1986*, pp. 395–400. Loughborough. [19.3]

Wood RD. (1975) *Hydrobotanical Methods.* University Park Press, Baltimore. [19.1]

Wright JF, Cameron AC, Hiley PD, Berrie AD. (1982) Seasonal changes in biomass of macrophytes on shaded and unshaded sections of the River Lambourn, England. *Freshwater Biology* **12**: 271–83. [19.1]

20: Direct Control of Fauna: Role of Hatcheries, Fish Stocking and Fishing Regulations

D. P. DODGE AND C. C. MACK

20.1 INTRODUCTION

This chapter describes the benefits and deficiencies inherent in directly controlling biological populations, as opposed to habitat manipulation, as part of a scheme to manage riverine resources. Here, the use of hatchery-reared fish is discussed. The key to successful management of fish in any catchment is cut from three principles that support all animal culturing whether intensive feed-lot operations for domestic stock or open-water harvest of pelagic fishes.

First, a manager must have extensive understanding of the habitat and general environmental conditions into which it is intended to introduce a reared species. Will the existing habitats limit the success of a stocking programme? Are the limiting factors irreversible, or can selected undertakings alter conditions to increase the probability of success? Might extremes in environmental conditions occur that could prevent success even when average conditions appear suitable? Can these extremes be mitigated? Is there sufficient knowledge about the habitat requirements of the selected species in order to match them with location and quantity of riverine habitat?

Secondly, the manager needs knowledge about the species to be cultured. How adaptable is the chosen species to episodes of stress resulting from extreme environmental fluctuations? Are there developed and tested gene pools or stocks of the species that are adapted to the kinds and quantities of habitat available in the river system? Does the manager want self-sustaining stocks or will the programme depend on reared stocks or on some combination of the two? Is the selected species compatible with native species in the system, or should one expect such success that diversity of native species will decline? Will the reared individuals interbreed with wild individuals of the same species or with similar species to produce hybrids? Can any of these negative effects be mitigated?

Finally, the public must be consulted and the manager must ensure that the public understands the risks and benefits of stocking.

This chapter will propose answers to some of these questions. As will also become apparent, there may not be any answers to some of the questions, until assessments produce data about stocking activities sufficient to make adequate analyses of these kind of projects.

Scope and context

One cannot overestimate the value of planning, not only as a means of putting ideas on paper, but also as a way of testing those ideas. To realize the potential for fish production in any river system, managers should prepare plans for catchments which also have regard for creating and protecting stable and diverse fish assemblages as part of the bigger picture of water and land management. Also, a manager needs to appreciate that the productivity of a river depends on nutrients dissolved in the terrestrial systems of its catchment, and how these nutrients reach the aquatic environment depends on the interface between land and water (Zalewski *et al* 1991). Total reliance on fish culture for recruitment is fraught with as many pitfalls and subsequent disasters as for any other artificially sustained system. Wherever possible managers should select goals that advocate managing fish on the basis of self-sustaining, naturally reproducing native fish populations. Hatchery fish are a means to achieve this kind of goal, but are not by themselves the *raison d'être* for fish management. Obviously we are advocating rehabilitation over mitigation as the first choice for managers of riverine fisheries. However, there will be many situations where mitigation using hatchery fish is the only means available for river management, e.g. where spawning habitat is permanently lost or water quality changes prevent self-sustained populations. The demand for high-quality, inexpensive and readily available protein may eliminate most if not all options except the use of hatchery fish for a put-grow-and-take fisheries management system. In the Orient, and in many European and African river systems, fish communities necessary for human survival would fail without the support of stocking.

Our examples focus on multi-agency experiences derived from nearly 150 years of managing fish communities and their fisheries in the Laurentian Great Lakes Region and other parts of North America. On the one hand, this chapter may not serve managers from Europe and Asia who are working with rivers with longer management experiences and already altered fish communities. However, we feel the North American experience is relevant for river managers in developing countries facing major demands for increased fish extraction from severely-stressed native fish communities, and who are trying to develop a place and purpose for fish culture to maintain some form of sustainability. North Americans made horrendous decisions in the mid-1800s that continue to this day to thwart sustainability of fish communities in the Great lakes (e.g. introduction of the smelt, *Osmerus mordax*). These lessons, and the 1990's approach to managing the Great Lakes ecosystem with hatcheries as a major component, can help river managers throughout the world.

For the purposes of this chapter, we have confined our discussion to the intentional release of chosen species. There is no doubt that similar and often more disastrous effects can be traced to introductions of new species that have escaped confinement or that were accidentally introduced by an unsuspected vector. One hundred years of rainbow trout (*Oncorynchus mykiss*) culture and stocking worldwide have created naturalized, often self-sustaining, populations which have already caused most of their potential effects. However, rainbow trout escapes from intensive pond culture (e.g. in southern England) threaten native brown trout (*Salmo trutta*). Similarly, the introduction of ruffe (*Gymnocephalus cernus*) into Lake Superior from ballast water taken at a European port carries the potential for an ecological disaster for native Great Lakes percid communities (e.g. *Stizostedian vitreum*, *Perca flavescens*).

All this points up the importance of planning, of understanding the needs of the fish being cultured (or being considered for culture and introduction) and of assessing the potential effects on endemic fish communities from the stocked fish before one puts great reliance on a fish culture system to sustain a fishery.

This chapter is divided into sections that examine first the basic principles of fish culture and the product desired; secondly, the factors that influence the survival and growth of the fish; and finally the role of regulations.

20.2 HATCHERIES AND FISH STOCKING

History of stocking

The art and science of fish culture are ancient pursuits. The Chinese cultured carp (*Cyprinus carpio*) and goldfish (*C. crassus*) for food and ornamentation as early as 4000 B.P. (MacCrimmon 1968; Balon 1974). Although little evidence is available showing that these and other species were purposely introduced into 'wild' environments, there is no doubt that ponds and raceways failed as frequently then as they have in modern times so that cultured fish were released to new environments to start new populations.

Despite a common misconception, trout and salmon culture is a relatively young industry. In the mid-1700s, Stephen Jacobi successfully stripped and fertilized trout eggs and noted the effects of light and temperature on the developing embryos. Although his successes in Germany were well publicized, a Frenchman, Joseph Remy, is credited with the establishment of the first trout farm. Remy, a professional fisherman, noticed declining fish stocks and in 1840 decided to try to breed trout artificially. Assisted by friends and interested parties with influence, the success of his operation induced the French government to fund the construction of a large fish farm.

Following the success and publicity of Remy's operations, trout farming started developing in Britain, Europe and Russia. Researchers began detailing operations, describing facilities and recognizing the need for proper management of wild stocks, so intensive fish culture grew and flourished worldwide.

Hatching success was low (10–20%) in the early operations, until in 1850 a Russian (Wrasski) discovered the technique of dry spawning. This method has been used ever since in salmonid culture with good results.

Except for a greater understanding of that physiological entity called a 'fish', modern fish culture continues to work with the same basic factors of growth, survival and feed conversion as did the earliest culturists.

Fish stocking of natural waters to provide better commercial fishing or angling became an important tool for fisheries in the mid-1800s with the development of fish hatchery systems in Europe and North America. Of special significance was the evolution of cold-water culture techniques that allowed the raising of very large numbers of salmon and trout which survived long distances and time from hatchery to planting sites. Seth Green and Samuel Wilmot led these advances in North America. Tremendous increases in ranges through successful introductions of salmon and trout (MacCrimmon & Marshall 1968; MacCrimmon 1971) resulted from the application of their techniques.

Stocking objectives

Stocking is a method used in the management of fish stocks as integral parts of aquatic ecosystems. Stocking programmes must share the objectives of effective aquatic ecosystem management in order to provide benefits on a sustainable basis.

Protecting and rehabilitating fish stocks

Protection of aquatic ecosystems should be a fundamental management objective of any fisheries programme. Fish culture can play an important role in protecting endangered stocks by acting as a reservoir of genetic material. The establishment of broodstocks of selected species and strains is an effective way of protecting endangered fish stocks. Although there is always a risk that artificial systems like hatcheries may select towards a lower level of fitness, broodstock husbandry ensures that some of the original genetic material is always available. For example, the Aurora trout, a unique strain of brook trout (*Salvelinus fontinalis*), was approaching extinction due to acidification of the Northern Ontario lakes it inhabited. Hatcheries played a major role in preserving this strain until rehabilitation of its habitat allowed successful natural reproduction.

Rehabilitation is an important objective in managing aquatic ecosystems where they have been degraded. Stocking can play an important role in rehabilitating existing native fish stocks, providing it is used in conjunction with efforts to restore habitat and control exploitation. Examples include re-establishing fish stocks after major pollution events or when destroyed spawning habitat has been restored.

Rehabilitating fisheries with introductions

In many areas, degradation of aquatic ecosystems has been so severe that the restoration of native fish stocks has been deemed unfeasible. Fisheries managers faced with this situation have often chosen to 'rehabilitate a fishery' by using species other than those originally present. An example of this management technique was the introduction of chinook and coho salmon (*Oncorhynchus tshawytscha* and *O. kisutch*) to the Great Lakes and its tributaries in North America. Unfortunately, the introduction of these exotic species has drastically altered the fish community structure of these water bodies, and may be a major impediment to restoring native fish stocks when human commitment and technological advances enable the rehabilitation of these ecosystems.

In Africa, the primary purpose of introductions

of fish species was to maintain or increase yields and harvests. Some introductions met this purpose but others may have compromised long-term sustainable harvests, in part by altering the aquatic environment and competing with native species. The short-term benefits have increased the fish protein supply in these countries but more research and assessment are necessary to understand how introductions serve the reaching of long-term objectives (Ogutu-Ohwayo & Hecky 1991). Holcik (1991) identified at least 134 fish species that had been introduced or relocated within 29 European countries, especially Central and Eastern Europe. He noted that poor success was recorded for most cases, as well as measurable ill-effects on native fish and their habitat.

For further reading, especially for analyses about the implications to fish genetics and aquaculture from introductions, the publication by Billington and Hebert (1991) is an excellent reference.

It is imperative that fisheries managers use some disciplined decision-making when considering introductions as a means to re-establish a fishery. One approach applied by the Province of Ontario (Canada) as a procedure to evaluate proposals to stock species beyond their current range in Ontario waters is a sequential decision model adapted from Kohler and Stanley (1984). Using this model, the decision process continues until the first 'REJECT' decision is reached. Successful completion of the decision model would lead to a decision to approve the stocking (see also Fig. 20.1).

1 (a) Are reasons for introduction valid (including demonstrated need which cannot be met by other management actions)? [No – REJECT]
(b) Would adequate precautions be taken to safeguard against introduction of disease and parasites? [No – REJECT]

2 (a) Would the species be able to survive in the range of habitats that would be available? [No – REJECT]

3 (a) Would the species have adverse ecological impacts? [Yes – REJECT]
(b) Would the species potentially be hazardous to humans? [Yes – REJECT]
(c) Is the species endangered, threatened or rare in its native range? [Yes – REJECT]
Unless:
(i) purpose is to create a refuge, to prevent species extinction, and/or
(ii) source of fish is a hatchery stock or naturalized population(s) which is (are) not threatened in any fashion.

4 (a) Was database adequate to develop a complete species synopsis? [No – FURTHER RESEARCH]

5 Based on all available information, do the benefits of the fish introduction outweigh the risks? [No – REJECT; Yes – APPROVE; Unclear – FURTHER RESEARCH]

Hatchery-dependent fisheries

Stocking is also used to provide a fishery in water bodies where natural reproduction cannot occur. These hatchery-dependent fisheries often provide important social and economic benefits, and the lack of natural reproduction may actually be beneficial since it may limit the damage caused by bad stocking decisions.

However, hatchery-dependent fisheries have often been used without adequate consideration of biological, social or economic implications. For example, the adaptability of introduced species to reproduce naturally in water bodies where they 'couldn't' has tested the credibility of many respected fisheries biologists. An example of casual approaches to a potential disaster is the treatment of the spread of the wide-ranging cyprinid, the rudd (*Scardinius erythrophthalmus*), throughout North America. Although the species arrived in the USA in 1916 (Cahn 1927), Courtenay *et al* (1986) considered the species as part of a group of non-native fish whose populations were in decline, as if there were unlikely now to be any effects. However, the rudd was being cultured in ponds in central USA and distributed extensively to anglers, who preferred it as bait to catch bass. Now Burkhead and Williams (1991) report that the rudd has interbred with the native cyprinid golden shiner (*Notemigonus crysoleucas*) to produce the first non-salmonid intergenic hybrid in North America. The potential effects on any fish community where golden shiner is the major forage species may be horrific; the loss of the genetic pool called golden shiner is immoral.

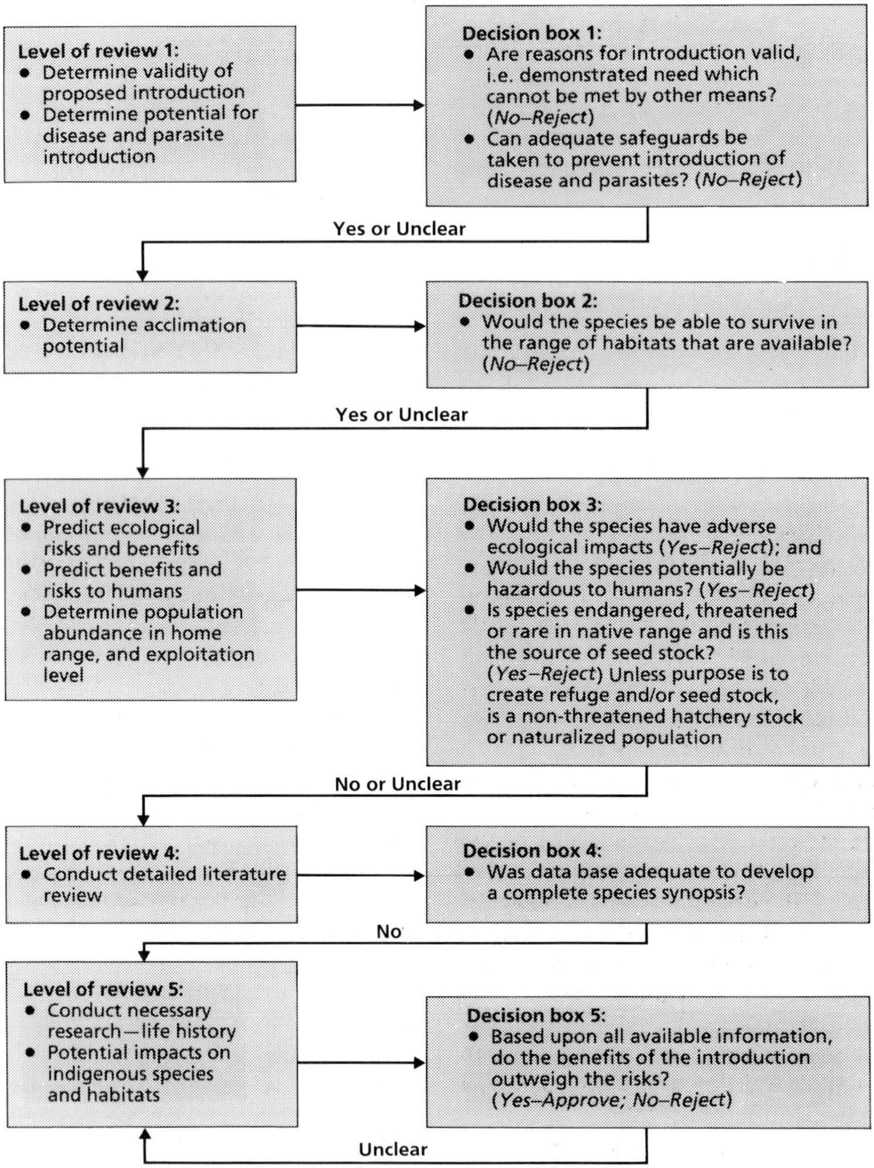

Fig. 20.1 Sequential decision model for evaluating fish species introdutions (after Kohler & Stanley 1984).

Supplemental stocking over native populations

Perhaps the most damaging stocking practice still in common use is supplemental stocking over existing self-sustaining native populations. This practice often occurs in water bodies where demand for fishing exceeds supply, particularly when exploitation cannot be effectively regulated. Environmental and genetic effects on fish reared in hatcheries often result in a loss of fitness compared to their wild counterparts. Stocking hatchery fish over native populations could ultimately result in degradation of the health of the

Table 20.1 Impacts of hatchery-reared sportfish on wild fish populations (from Marnell 1986)

Introduction of pathogens and parasites
Hybridization, introgression, intergradation
Trophic alterations:
Predation, cannibalism
Food competition
Spatial alterations
Territorial shifts, altered dominance hierarchies
Stress and aggression
Altered movements, temporal changes in activities
Altered growth and survival characteristics
Altered growth rates
Changes in natural mortality
Changes in vulnerability to angling
Displacement, replacement, extinction

native stock by adversely affecting the genetic diversity of the native gene pools. The potential disadvantages of this practice range from minor interruptions in resident fish communities to drastic changes in the aquatic systems including major genetic shifts and even extinction (Table 20.1)

The debate about this issue is causing concern amongst fisheries professionals in the USA. Some scientists see the debate as counterproductive and consuming so much energy that political and economic decisions are being made without adequate biological input, because scientists are spending so much time debating hatcheries and their effects on wild stocks (Martin *et al* 1992).

Again, the manager must balance many aspects of fish production and riverine management before deciding if supplemental stocking is worth the risk. In some cases environmental changes have precluded natural reproduction of the native fish so that the remnant population must be sustained by plantings of hatchery fish. Moreover, in an economic and social climate where fish are essential for human survival, planting hatchery fish over native stocks must happen when no viable alternatives exist to provide the needed protein.

Another risk, potentially lethal, is also associated with intensive aquaculture especially in subtropical climates. Scholtissek (1992) has postulated that the mixing of birds, swine and fish farming sets in motion a process whereby swine act as mixing vessels in which lethal new forms of human Type A influenza viruses are produced.

In summary, stocking can be an effective tool in protecting and rehabilitating existing native fish communities, and can provide increased social and economic benefits. However, stocking fish species which are more adaptable to surviving in degraded aquatic ecosystems, without parallel programmes to restore the degraded systems, and supplemental stocking that degrades existing native populations are two practices which are incompatible with the objective of providing sustainable benefits from healthy aquatic ecosystems.

Stocking strategies

Several strategies are available to fisheries managers using stocking as a management tool. These include selection of species, strains and life stages which are best suited to meet management objectives. An ecosystem approach to fisheries management requires that fish stocking only be used as a tool to achieve balanced fish communities. Fish community objectives should, therefore, form the basis for identifying stocking requirements.

Fisheries managers considering a stocking programme should first evaluate the possibility of using other management options. Increased enforcement, habitat rehabilitation, and harvest control techniques are just three examples of other management tools which are often cheaper, less damaging to native stocks, and more effective than stocking.

Species selection

Fish culture has played a major role in the introduction or range extension of fish species into waters in which the species did not occur naturally. While some of these introductions have created naturally reproducing populations which are generally deemed as beneficial to society, others have resulted in irreversible damage to native fish communities. For example, rainbow trout (*Oncorynchus mykiss*), introduced to tributaries to the Great Lakes, have provided very large, self-sustaining fisheries, but in the process have significantly reduced the range and numbers of the native charr (*Salvelinus fontinalis*). Similar effects are predicted to happen to brown trout (*Salmo trutta*) in the UK as rainbow trout become more established.

In several jurisdictions, a more conservative approach is being used to select species for fisheries management programmes. Many governments have enacted legislation and policies which restrict the movement of fish within their jurisdictions. In cases where waters are shared between nations, such as the Great Lakes in North America, co-operative policies and protocols have been developed to ensure that introductions or range extensions of species are not made without adequate consultation with stakeholders, and that proper assessment of the environmental risks involved is undertaken. Another example of multinational protocols is the Code of Practice adopted by the European Inland Fishery Advisory Committee (EIFAC) in 1987 for the regulation and control of fish introductions.

For most stocking applications, only native fish species should be considered. Native species will provide the most predictable results, with the least risk of environmental consequences. However, most species of fish are not homogeneous in terms of their characteristics; fish stocking programmes which use the wrong genetic strain for a specific purpose can be just as damaging to local, native fish communities as if a non-native species were introduced.

Genetic strain

A fish species may consist of many unique genetic strains which have evolved as a result of distinct geographical, behaviourial or temporal factors. Ignorance of the importance of protecting fish stocks has led to the loss of genetic diversity in many fish populations, causing loss of population fitness. Recent developments in fish culture, such as rotational line crossing of broodstock lines to reduce inbreeding, and the use of large numbers of randomly selected wild brood fish to maximize genetic diversity, can reduce the loss of genetic diversity in fish populations. Selection cannot be eliminated entirely. Nature does not always act randomly, and fish collection techniques are often selective. Even if we were able to maintain genetic diversity, environmental factors occur in fish culture which are fundamentally different to those in a wild environment.

Broodstock management

The management of broodstocks in a fish culture programme should be aimed not only to ensure an adequate and consistent supply of eggs for rearing and eventual stocking, but also to establish careful breeding programmes, and to maintain the genetic diversity of those stocks over several generations. Genetic variability is important for maintaining maximum hatchery production levels, for disease resistance, and for the ability of progeny to adapt to new environments after stocking.

Artificial propagation and rearing procedures used in contemporary fish culture tend to reduce the genetic diversity of captive broodstocks, resulting ultimately in inbreeding and associated loss of vigour. The protection afforded by the hatchery environment enables less fit individuals to survive, and because fewer parents produce relatively higher numbers of offspring than in natural conditions, the potential for siblings to be crossed one with another is greater.

To avoid the potential inbreeding effects of fish culture practices a number of principles should be considered in establishing and propagating broodstocks held in fish culture stations.

In collections of eggs from the wild to establish a new broodstock, and in propagating captive broodstocks in hatchery:

1 parents should be a random sample from the spawning population with no selection on the basis of size, appearance or other characteristics;

2 select from the entire spawning season or run, preferably with 25% of parents spawned when 15% of females are ripe, 50% spawned when 40% are ripe, and the remaining 25% when 70–80% of females are ripe;

3 a minimum of 50 spawning pairs should be used;

4 matings should be on a single pair basis, i.e. single males should be used to fertilize the eggs of single females with no pooling of eggs or milt;

5 individual families should be kept separate from fertilization through to incubation;

6 at eye-up, or preferably swim-up, equal numbers of eggs/fry from each family should be randomly taken to create the broodstock pool of individuals so that no family is represented more than any other – the number of eggs or animals to be taken per family will depend upon the number of families and the target number for a broodstock lot.

Thus, in a single year, a culturing station can establish one line of breeding for a broodstock. A minimum of two lines are required (three to five are preferred) in order to minimize losses of genetic variability.

The sources of establishment of these other lines of breeding are many.

1 In succeeding years follow the same process as described above so that in 3 years you will have established three lines; and/or:

2 Fish culture facilities exchange eyed-eggs at that time in the year when lines-of-breeding are being established. Disease control and certification systems are necessary to make this exchange safe for recipient fish culture facilities. It is advisable, also, to have some isolation facilities for rearing exchanged lines. It would be possible by this method to establish the minimum lines-of-breeding to maintain broodstocks in 1 year; or:

3 Use a co-operative, centralized facility to maintain lines-of-breeding.

Rotational line crossing

Once a captive broodstock is established in a fish culture station, in addition to the above procedures for the establishment of a stock, a breeding scheme is required which will prevent the mating of siblings or half-sibs. Thus inbreeding is minimized through successive generations. The technique is known as rotational line crossing (Kincaid 1977).

At the time a broodstock is established, three distinct lines are created from a single year class, or three lines are created in succession with three separate year classes. The three lines, which may be called A, B and C, must be held in separate units or marked by fin clips, branding or some other means by which they can be distinguished one from another.

All three lines may be used for producing eggs for production purposes in any given year, but when the time comes to produce the next generation of broodstock, a specific breeding scheme must be followed. For example, the females of line A are crossed with males from line B; the females of B are fertilized with males from C; and the females of C are fertilized with line A males. This creates three new lines, AB, BC and CA. The next generation is then produced from those three lines, again by crossing females and males from different lines. The creation of three

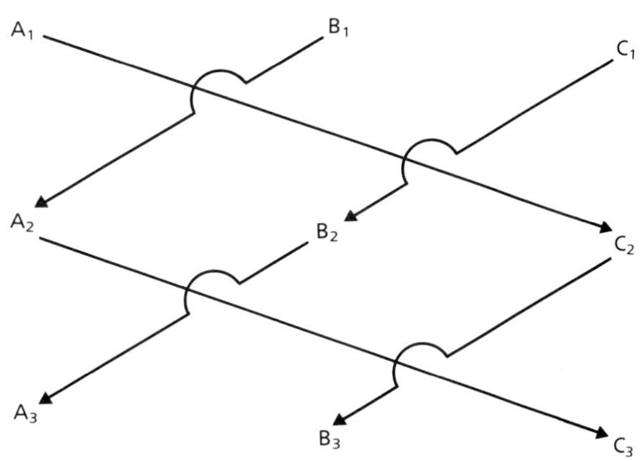

A_1 male is bred with C_2 female	*then*	A_2 male is bred with C_3 female
B_1 male is bred with A_2 female		B_2 male is bred with A_3 female
C_1 male is bred with B_2 female		C_2 male is bred with B_3 female

Fig. 20.2 Rotational line crossing (after Kincaid 1977).

lines for each generation may be carried out in a single year, or as is preferred for longer-living species, in three successive years (Fig. 20.2).

Evaluation

Evaluation of success of programmes that rehabilitate or introduce species using hatchery fish is fraught with many problems in river management. Sampling is very difficult, especially for small fish, and population estimates are usually very unreliable except by direct count methods of migratory fish at fishways (McMenemy & Kynard 1988; Beamesderfer & Rieman 1988). Sampling people who catch or eat fish may provide useful information about riverine fish, especially with artisanal fisheries (Bayley & Petrere 1989) but the quantitative methods to estimate fish production, survival and yield are poorly developed relative to approaches for the same parameters for lakes, small streams and reservoirs. Casselmann *et al* (1990) surveyed 73 scientists from 12 countries for their experience in sampling large river systems. This survey reviewed and evaluated many types of sampling gear but concluded that none was suitable for sampling large river systems in all types of habitat and situations.

20.3 FACTORS INFLUENCING SUCCESS

As stated earlier in this chapter, there must be a commitment by river catchment managers to include fisheries management as a component of their plans. Therefore, the catchment planning process becomes the most important factor that could influence the success of any fish stocking scheme. From the very beginning of this process, it should be evident from the catchment plan as to what quality and quantity of fish habitat will be delivered by the plan. Further, the plan must indicate where the various fish habitats will be in the catchment, what variability can be expected and when, in any one year, those changing conditions are likely to occur. Nelson *et al* (1992) used a hierarchical system to classify locations and sites in northeastern Nevada, USA. The system is based on the integration of geology, geomorphology and local habitat features. The authors were able to develop preliminary criteria for assessing a fishery on a large scale (km^2) to save time and money, by analysing detailed habitat attributes on those sites most likely to be important for the chosen fish community.

The application of the ideas and concerns outlined in the remainder of this section is predicated singularly on the forewarning any catchment plan should provide. It is at this level of preparation that the public and their legislative representatives become involved. Consultation over and acceptance of catchment plans and their fish management components provide greater assurance that fiscal resources and commitment to sustainability will be forthcoming.

The concept of sustainability is an important part of planning for fisheries in a river catchment. Gardiner (1991) argues for ecological sustainability that includes social, political and legal components where economic and environmental sustainability are very much interrelated. His paper is an excellent treatise on the value of planning at the catchment level to achieve holistic management of riverine systems.

Water quality and quantity

Water-quality management for fish production and survival of hatchery fish includes the optimizing of not only discharges for current and volume, but also of the water levels and wetted margins of streams. The complicated interrelationships between instream flow and fish production have caused many problems in fish and river management. Often fisheries managers have had neither expertise nor data readily available to help plan for the effects of changes in hydrological regimes. However, Gustard (1992) reported significant advances in the understanding of river regimes, including international databases that could help in the understanding of global influences on floods and their frequency.

Water abstraction schemes, e.g. cooling, drinking-water supplies, especially those that withdraw large volumes ($>500\ m^3\ min^{-1}$), present major hazards to fish, and could 'harvest' major portions of planted species intended to rehabilitate a river system. Various fish protection devices have been designed for water extraction systems, but few, if any, prevent entrapment and impingement of all species of fish (Fletcher 1985).

Even with the most efficient and effective fish culture programme, the success of planted fish depends equally on the water quality of the river system into which they are introduced. Many factors interact with the fish as they strive to establish themselves within the existing fish communities and their environment.

Nutrient concentrations and the resulting primary productivity of a river limits the growth and development of the hatchery fish. Correlations between fish production and nutrients in streams have been well documented (Welcomme 1985).

Many water quality and physical characteristics affect the survival and growth of fish introduced into river systems. In fact, the concentrations of certain parameters will dictate whether or not a hatchery-supported fisheries can ever be successful. For example, the penetration of light and its association with primary and secondary production is a key component to the survival of salmonids (Lloyd 1987). Besides affecting the quality and quantity of light penetration, suspended particles in the water column influence the proficiency with which fish see and catch prey. Also, particles, especially colloidal clay materials, can accumulate on gill filaments and other body parts to such an extent that fish die or are easy prey for larger predators. Readers should refer to Walling and Webb (1992) and Webb and Walling (1992) for more detailed descriptions of water quality parameters and their ranges expected in riverine systems.

Habitat suitability

Essential to this matching of fish with habitat is the need to understand the effects on fish production of physical processes in rivers (Heede & Rinne 1990). Similarly, the generation, transport and deposition of organic and inorganic particles, so much related to hydrological processes, can affect the success of fish stocking into rivers (Newcombe & MacDonald 1991).

River and basin management schemes also influence the quantity and size of sediment that travels suspended and/or settles in the river bed. These factors also have great influence on the survival, growth and reproductive potential of hatchery-reared and wild fish (Reiser & White 1988). Large rivers move large particles, and especially during flood events, can redefine very large portions of river bottoms. These major alterations frequently remove and destroy essential spawning, nursery and staging areas for riverine fish. Losses of gravel, for example, preferred for spawning salmonids, could mean the demise of any annual recruitment based on natural reproduction. Deforestation of headwaters of tributaries to Lake Ontario in North America created severe erosion that eventually covered spawning habitat for landlocked Atlantic salmon with several metres of sand, silt and clay. Similar changes to the quality and quantity of essential habitat are well-documented worldwide (e.g. Ward & Stanford 1989).

The essence of successful fisheries management in rivers is the matching of fish stock with suitable habitat. In many cases, however, habitat requirements for many fish species, especially needs for early-life survival and growth, are not usually understood. The interrelationships of riverine fishes with their environments, and the spatial and temporal scales one must use to understand these relationships and how fish species have adapted, are reviewed in detail by Bayley and Li (1992).

Allocation/overharvest

Management of riverine fish stocks involves more than just planting hatchery-reared fish and harvesting these stocks for food, recreation and revenue. Certain basic decisions are necessary before committing resources to hatcheries and fish production. First, are artificially-reared hatchery fish the best means to optimize fish production or should managers be rehabilitating habitat and changing fishing scenarios to allow existing stocks to stabilize and thus optimize the potential for riverine production based on those existing stocks? Inherent in this decision is the question of how much of the annual production of fish should be allocated to the long-term maintenance of the resource. Of the remainder of fish production, what quantity of fish would be ecologically and economically sound to allot to fishing, and when and how could this resource extraction take place that would maintain these two principal allocations?

Welch and Noakes (1991) designed an optimal

economic and resource protection policy for harvesting Adams River sockeye salmon (*Oncorhynchus nerka*). The policy proposed that a subdominant run be fished at 50% harvest rate in the first year and a moratorium be imposed on fishing off-year runs for a period of 8–12 years. This proposal would put all the gains from a reduced rate of harvest into increasing the size of the spawning population. When an optimal escapement level was achieved, harvesting would be reinitiated. Unfortunately, senior management appeared to be more sensitive to the reduction in fishing times and local employment, so the policy was considered impractical even though the proposal could contribute up to $600 million (Can) to the net value of the fish stock.

Major river diversions on the scale of those in Siberia and northern Quebec (Kierans 1988; Berkes 1989) present different hazards to fish management dependent on fish stocking. As well as significant shifts in habitat, in local climates and in stage discharge patterns, diversions eventually change the character of the native fish assemblages such that new or existing species increased by fortuitous habitat shifts may prevent the successful management of the intended fish species. Reliance on cultured fish to offset these changes is inevitably misplaced.

However, once major ecologically profound changes have been accomplished within a river catchment, fisheries managers have to get on with the job of managing this altered environment to maximize fish production on a sustainable basis. In many cases, fish are an essential food item so that restoration of fish production as soon as possible is often close to an emergency. At the same time, it would be a risky investment to take just any approach. Planning will lead one through a logical sequence asking questions and seeking answers until the best solution can be identified for the management of the fish community in the new environment.

Conflicting land/water use

The best fish culture strategies and practices in the world rarely offset the detrimental results of major land and water management practices which destroy certain components of a river system essential to the production and maintenance of fish stocks. For example, Raymond (1988) showed that major alterations in dam and impoundment management combined with large releases of smolt from hatcheries offset the decline of steelhead (*Oncorhynchus mykiss*) in the Columbia River, but failed to compensate for major losses of returning chinook (*O. tschawytscha*). Also, most returning adults were of hatchery origin, suggesting native gene pools had also been lost. Here may be a case for arguing for the establishment of hatchery stocks from broodstocks of local origin.

Management of established reservoirs can influence the survival and growth of hatchery-reared fish. Fisheries previously dependent in the main on hatchery productions can be boosted and even made self-sufficient by changes and innovations in reservoir management. In mid-western America, populations of centrachids (Centrachidae) dominate impoundment fisheries and are enhanced by increased time of storage of flood waters (Mitzner 1981). At the same time, cool and coldwater fish populations usually decline so that any maintenance of these species becomes almost exclusively hatchery dependent. Even coldwater tailrace fisheries are usually based on stocking.

Genetic integrity of wild and hatchery stocks

The overall fitness of any species and of any of its many stocks should be a major concern of a manager using hatchery fish to rehabilitate and maintain a riverine fish community. Undoubtedly, there are major differences among various species and stocks in their predilection to adapt to rigorous environmental conditions in any river over any annual cycle. Testing of stocks for these abilities would be an ideal screening mechanism to increase probability of returns of planted fish and also to reduce costs for fish culture programmes. Challenge-testing of stocks and their responses to success-limiting stresses showed stocks of coho salmon (*O. kisutch*) had different capabilities (McGeer *et al* 1991). Such differences should influence which stock is chosen to culture.

The danger to native populations from large-scale escapes of farmed fish reflects the same hazard that comes with intentional use of hatchery fish from inappropriate gene pools. In Norwegian rivers, farmed Atlantic salmon (*Salmo salar*) are threatening native fish through cross-breeding of wild with farmed fish, with the subsequent risk of losing genetic material essential for the survival of local populations (Gausen & Moen 1991). The river catchment manager must always be aware of this kind of threat.

The dangers inherent in introductions are further enhanced by the blind faith put in recent genetic 'breakthroughs' where managers were made to believe large populations of desexed fish established by stocking would never produce off-spring. An excellent example of this fallacious assumption was the release of (reportedly sterile) grass carp (*Ctenopharyngodon idella*) into the Mississippi River and the subsequent colonization of the system as naturally-reproduced populations established rapidly (Brown & Coon 1991), all of this despite caution advised by scientists well-placed to predict ecological disasters.

Effect of environmental contaminants

Residues of persistent, toxic substances that bio-accumulate have been linked to the reproductive failures of wildlife (e.g. bald eagle *Haliaeetus leucocephalus*) and fish (e.g. lake trout *Salvelinus naymaycush*). Obviously the body burden carried by any hatchery fish and the potential to accumulate more in the environment into which it has been introduced must be considered as factors affecting the overall success of fish survival and growth. It is possible that mega fish stocking programmes will fail to re-establish self-sustaining fish populations in rivers because of uncontrolled toxic effluents and high concentrations of contaminated sediment (Gilbertson 1992).

One pertinent criterion often ignored by managers is whether or not the species chosen for a river fish culture programme has the propensity to accumulate persistent toxic substances that may threaten human health. Notwithstanding any effects that biomagnified toxic body burdens may have on the general health of the carriers and on the specific reproductability of mature fish, there is the fundamental issue of effects from toxic substances in fish on humans, especially women and children. Swain (1988) identified significantly damaging effects on maternal health and showed the risks associated with trans-placental transmission to infants *in utero* and for nursing infants.

20.4 ROLE OF REGULATIONS

Regulations are an important mechanism for controlling risks to native fish communities which may result from stocking-related activities. Regulations can be used directly to control the choice of species or strains used for stocking, including the use of exotics, or to prevent the introduction or spread of serious fish disease agents.

Indirectly, regulations play a major role in contributing to the success of a stocking programme. Fishing seasons and access can be controlled to protect stocked fish when they are most vulnerable, such as at the stocking site, or in the case where anadromous species are stocked in a river as pre-smolts, prior to their movement downstream to a lacustrine environment. More fundamental regulations controlling exploitation can be used to eliminate the need for starting a stocking programme.

Regulations are needed to provide protection at continental, regional or local levels. Man-made boundaries (i.e. national, territorial or local government) are the main mechanism used to develop, implement and enforce regulations. The fact that fish do not recognize man-made boundaries poses a major challenge to the effectiveness of regulations. For this reason, international commissions such as the North Atlantic Salmon Conservation Organization, EIFAC and the Great Lakes Fishery Commission play an important role in developing international agreements and co-ordinated strategies which assist in managing large watersheds.

National or territorial controls provide the major framework for implementing legislation, regulations and policy. These controls are often easier to enforce, because they are based on clearly defined boundaries. Territorial legislation can be used as the basis for regulations which control at the local level, such as specifying the location of fish sanctuaries.

Fish disease regulations

The risk of transferring serious fish disease agents with live or dead fish has been recognized as a serious problem by many agencies. Until recently, fish diseases have largely been regarded as a hatchery-related problem. Recognition of the importance of fish diseases in the wild environment is a recent phenomenon which resulted from the occurrence of large epizootics such as ulcerative dermal necrosis (UDN) in the UK, and the huge mortalities of chinook salmon in Lake Michigan which were, in part, attributable to bacterial kidney disease. These and other events have shown that fish diseases can have serious economic and environmental consequences.

Harvest control/allocation

Fishing regulations that restrict the harvest of both hatchery and wild fish are developed to protect fish stocks from depletion. Most regulations use controls on fishing methods, seasons and bag limits as the major means of controlling harvest. In many cases, these regulations never achieve their intended purposes because total harvest (annual) is never controlled. Recent changes in regulations in North America, especially those related to catch-and-release (Anderson & Nehring 1984), suggest fishing quality and angler satisfaction can be maintained along with maintenance of appropriate fish stocks.

20.5 CONCLUSIONS

For a chapter dedicated to fish stocking, we are likely to be accused of being very much anti-stocking. But we have seen many native stocks lost to stocking and have witnessed many wasted efforts on ill-planned and poorly monitored schemes for fish stocking. We wish managers to view fish stocking as only one of many tools available for fish management and not the only and best tool.

When managers are required to use stocking to sustain use of riverine fish populations, they must ensure that they match the appropriate scheme with the management plan for the river catchment. Then, managers need to match the selected fish species and strain with the habitat quality and quantity available in the river. Here, we emphasize that money and resources should always go as a priority to promote the restoration and conservation of native fish species before putting any reliance on the culture and introductions of non-native species.

ACKNOWLEDGEMENTS

M. Succi typed many drafts and made order and legibility from scribbles and notes. We also acknowledge the comments by an anonymous reviewer of an earlier draft. Any bias we project in this chapter results from our collective experiences being centred in the Laurentian Great Lakes of North America.

REFERENCES

Anderson RM, Nehring RB. (1984) Effects of a catch-and-release regulation on a wild trout population in Colorado and its acceptance by anglers. *North American Journal of Fisheries Management* **4**: 257–65. [20.4]

Balon EK. (1974) *Domestication of the Carp* Cyprinus carpio *L.* Life Science Miscellaneous Publication, Royal Ontario Museum, Toronto. [20.2]

Bayley PB, Li HW. (1992) Riverine fishes. In: Calow P, Petts GE (eds) *The Rivers Handbook, Vol. 1*, pp. 251–81. Blackwell Scientific Publications, Oxford. [20.3]

Bayley PB, Petrere M Jr. (1989) Amazon fisheries: assessment methods: current status and management options. In: Dodge DP (ed) *Proceedings of the Large River Symposium*, pp. 385–98. Canadian Special Publication Fisheries and Aquatic Sciences 106, Ottawa. [20.2]

Beamesderfer RC, Rieman BE. (1988) Size selectivity and bias in estimates of population statistics of smallmouth bass, walleye and northern squawfish in a Columbia River reservoir. *North American Journal of Fisheries Management* **8**: 505–10. [20.2]

Berkes F. (1989) Editorial: impacts of James Bay development. *Journal of Great Lakes Research* **15**: 375. [20.3]

Billington N, Hebert PDN (eds). (1991) International symposium on the ecological and genetic implications of fish introductions (FIN). *Canadian Journal of Fisheries and Aquatic Sciences* **48** (Suppl. 1): 181. [20.2]

Brown DJ, Coon TG. (1991) Grass carp larvae in the lower Missouri River and its tributaries. *North American Journal of Fisheries Management* **11**: 62–6. [20.3]

Burkhead NM, Williams JD. (1991) An intergenic hybrid of a native minnow, the golden shiner, and an exotic minnow, the rudd. *Transactions of the American Fisheries Society* **120**: 781–95. [20.2]

Cahn AR. (1927) The European rudd (*Scardinius*) in Wisconsin. *Copeia* **162**: 5. [20.2]

Casselmann JM, Penczak T, Carl L, Mann HK, Holcik J, Woitowich WA. (1990) An evaluation of fish sampling methodologies for large river systems. *Polskie Archiwum Hydrobiologii* **37**(4): 521–51. [20.2]

Courtenay WR Jr, Hensley DA, Taylor JN, McCann JA. (1986) Distribution of exotic fishes in North America. In: Hocutt CH, Wiley EO (eds) *The Zoogeography of North American Freshwater Fishes*, pp. 675–98. John Wiley, New York. [20.2]

Fletcher RI. (1985) Risk analysis of fish diversity experiments: pumped intake systems. *Transactions of the American Fisheries Society* **114**: 652–94. [20.3]

Gardiner JL. (1991) Towards sustainable development in river basins: new professional perspectives? Paper presented to the International Symposium on Effects of Watercourse Improvements: Assessment, Methodology, Management Assistance, September 1991, Namur, Belgium. [20.3]

Gausen D, Moen V. (1991) Large-scale escapes of farmed Atlantic salmon (*Salmo salar*) into Norwegian rivers threaten natural populations. *Canadian Journal of Fisheries and Aquatic Sciences* **48**: 426–8. [20.3]

Gilbertson M. (1992) PCB and dioxin research and implications for fisheries research and resource management. *Canadian Journal of Fisheries and Aquatic Sciences* **49**: 1078–9. [20.3]

Gustard A. (1992) Analysis of river regimes. In: Calow P, Petts GE (eds) *The Rivers Handbook, Vol. 1*, pp. 29–47. Blackwell Scientific Publications, Oxford. [20.3]

Heede BH, Rinne JN. (1990) Hydrodynamic and fluvial morphologic processes; implications for fisheries management and research. *North American Journal of Fisheries Management* **10**: 249–68. [20.3]

Holcik J. (1991) Fish introductions in Europe with particular reference to its central and eastern part. *Canadian Journal of Fisheries and Aquatic Sciences* **48** (Suppl. 1): 13–23. [20.2]

Kierans T. (1988) Recycled run-off from the North. *Journal of Great Lakes Research* **14**: 255–6. [20.3]

Kincaid HL. (1977) Rotational line crossing: an approach to the reduction of inbreeding accumulation in trout brood stocks. *Progressive Fish Culturist* **39**: 179–81. [20.2]

Kohler CC, Stanley JG. (1984) A suggested protocol for evaluating proposed exotic fish introductions in the United States. In: Courtney WR Jr, Stauffer JR Jr (eds) *Distribution, Biology and Management of Exotic Fishes*. Johns Hopkins University Press, Baltimore. [20.2]

Lloyd DS. (1987) Turbidity as a water quality standard for salmonid habitats in Alaska. *North American Journal of Fisheries Management* **7**: 34–45. [20.3]

MacCrimmon HR. (1968) *Carp in Canada*. Bulletin 168, Fisheries Research Board of Canada, Ottawa. [20.2]

MacCrimmon HR. (1971) World distribution of rainbow trout (*Salmo gairdneri*). *Journal of the Fisheries Research Board of Canada* **28**: 663–704. [20.2]

MacCrimmon HR, Marshall TL. (1968) World distribution of brown trout, *Salmo trutta*. *Journal of the Fisheries Research Board of Canada* **25**: 2527–48. [20.2]

McGeer JC, Baranyi L, Iwama GK. (1991) Physiological responses to challenge tests in six stocks of coho salmon (*Oncorhynchus kisutch*). *Canadian Journal of Fisheries and Aquatic Sciences* **48**: 1761–71. [20.3]

McMenemy JR, Kynard B. (1988) Use of inclined-plan traps to study movement and survival of Atlantic salmon smolts in the Connecticut River. *North American Journal of Fisheries Management* **8**: 481–8. [20.2]

Marnell LF. (1986) Impact of hatchery stocks on wild fish populations. In: Stroud RH (ed) *Fish Culture in Fisheries Management*, pp. 339–47. American Fisheries Society, Bethesda, Maryland. [20.2]

Martin J, Webster J, Edwards G. (1992) Hatcheries and wild stocks: are they compatible. *Fisheries* **17**: 4. [20.2]

Mitzner L. (1981) Influence of floodwater storage on abundance of juvenile crappie and subsequent harvest at Lake Rathbun, Iowa. *North American Journal of Fisheries Management* **1**: 46–50. [20.3]

Nelson RL, Platts WS, Larsen DP, Jensen SE. (1992) Trout distribution and habitat in relation to geology and geomorphology in the North Fork Humboldt River drainage. *Transactions of the American Fisheries Society* **121**: 405–26. [20.3]

Newcombe CP, MacDonald DD. (1991) Effects of suspended sediments on aquatic ecosystems. *North American Journal of Fisheries Management* **11**: 72–82. [20.3]

Ogutu-Ohwayo R, Hecky RE. (1991) Fish introductions in Africa and some of their implications. *Canadian Journal of Fisheries and Aquatic Sciences* **48** (Suppl. 1): 8–12. [20.2]

Raymond HL. (1988) Effects of hydroelectric development and fisheries enhancement on spring and summer chinook salmon and steelhead in the Columbia River Basin. *North American Journal of Fisheries Management* **8**: 1–12. [20.3]

Reiser D.W., White R.G. (1988) Effects of two sediment size-classes on survival of steelhead and chinook salmon eggs. *North American Journal of Fisheries Management* **8**: 432–7. [20.3]

Scholtissek C. (1992) Cultivating a killer virus. *Natural*

History January 1992; pp. 2–6. [20.2]

Swain WR. (1988) Human health consequences of consumption of fish contaminated with organochlorine compounds. *Aquatic Toxicology* **11**: 357–77. [20.3]

Walling DE, Webb BW. (1992) Water quality I. Physical characteristics. In: Calow P, Petts GE (eds) *The Rivers Handbook, Vol. 1*, pp. 48–72. Blackwell Scientific Publications, Oxford. [20.3]

Ward JV, Stanford JA. (1989) Riverine ecosystems: The influence of man on catchment dynamics and fish ecology. In: Dodge DP (ed) *1989 Proceedings of the International Large River Symposium (LARS)*, pp. 56–64, Canadian Special Publication Fisheries and Aquatic Sciences 106, Ottawa. [20.3]

Webb BW, Walling DE. (1992) Water quality II. Chemical characteristics. In: Calow P, Petts GE (eds) *The Rivers Handbook, Vol. 1*, pp. 73–100. Blackwell Scientific Publications, Oxford. [20.3]

Welch DW, Noakes DJ. (1991) Optimal harvest rate policies for rebuilding the Adamo River sockeye salmon (*Oncorhynchus nerka*). *Canadian Journal of Fisheries and Aquatic Sciences* **48**: 526–35. [20.3]

Welcomme RL. (1985) *River Fisheries*. FAO Fisheries Technical Paper 262, Rome. [20.3]

Zalewski M, Thorpe JE, Gaudin P. (eds) (1991) *Fish and Land/inland-water Ecotones*. UNESCO MAB, tódź, Poland. [20.1]

21: Rehabilitation of River Margins

A. R. G. LARGE AND G. E. PETTS

21.1 INTRODUCTION

Since Welcomme's (1979) influential work on the importance of annually-inundated floodplains for sustaining fish production, attention has increasingly focused on the role of river margins in sustaining the biological integrity of both riverine and terrestrial systems (Naiman & Décamps 1990; Lachavanne 1993). Rivers, together with their marginal, ecotonal, systems, provide corridors through the landscape; their margins provide buffers between the watercourse and the variety of land uses within the catchment. However, throughout Europe and North America, this intimate relationship between land and water has been interrupted, degraded, and in extreme cases destroyed, by human activity (e.g. Bravard *et al* 1986; Petts *et al* 1989; Dister *et al* 1990; Petts 1990a). Swift (1984) has catalogued the nationwide loss of riparian habitat in the USA, estimating that of 75–100 million ha only 25–35 million ha remain. Rehabilitation of river margins has two important roles in river management: (i) to enhance values of nature conservation, and (ii) to improve water quality. This section assesses the benefits for river management of preserving, rehabilitating and restoring river margin ecosystems.

The spatial context

A river corridor comprises active and abandoned channels, the aquatic margins of these channels, the riparian zones along the channel banks and any floodplains. Although complicated by the legacy of climatic change over the Quaternary, and longer time-scales in some cases, and by the effects of neotectonic activity and human interference, a river may be viewed as a longitudinal continuum changing in a more or less predictable way from source to mouth (Vannote *et al* 1980; and see Vol. 1, Chapter 11). Along this continuum, the corridor can be seen to comprise a changing set of soil/vegetation patches – patches of different type and of different age that reflect the underlying geomorphological structure (Salo 1990; Petts 1990a; Cummins 1992; and see also Vol. 1, Chapter 7).

For simplicity, the river continuum may be divided into three primary zones (see Fig. 1.2a). First, headwater channels – first- to third-order streams – may be viewed as the production zone: the major source area tapping water, solute and sediment supplies from the hillslopes. Secondly, the middle river, or transfer zone, is characterized by high environmental heterogeneity and high species richness. Thirdly, the lowland floodplain river represents the depositional or storage zone. Rivers of fourth order and above are usually isolated from adjacent hillslopes by the river margin systems, which become especially important along reaches with major floodplains, although throughout the temperate zone, and increasingly elsewhere, these are the reaches that have been most severely impacted by human activity. The potential for managing river margins will differ in each of these three main zones of the river network.

21.2 THE CASE FOR RIVER MARGINS IN RIVER MANAGEMENT

The protection, rehabilitation and restoration of river margins should have a high priority in

Table 21.1 The role of riparian vegetation in natural river ecosystem dynamics (after Large & Petts 1992)

Primary roles	Functions addressed
Provision of important habitats for flora and fauna	High biological diversity High biological productivity Source for species dispersal
Regulation of ecosystem dynamics	Influence movement and migration of fauna Control surface runoff Regulate sub-surface flows Provide organic matter to the river Store water
Enhancement of visual quality and amenity value	
Enhancement of ecological patch diversity of floodplain rivers	Incorporate a range of habitat patches, especially small woods, cut-off channels and wetlands of different successional stages

catchment-based river management. River margins have a range of important characteristics (Table 21.1) typical of ecotonal systems (Naiman & Décamps 1990). The case for including river margins in river management is based on the four key functions: (i) water-quality control; (ii) nature conservation; (iii) instream habitat enhancement; and (iv) recreation and amenity. Functions (i) and (iii) are most effective within Zone 1 whereas functions (ii) and (iv) offer opportunities for environmental enhancement throughout river networks.

River margins for water-quality control

In temperate situations, it has been clearly demonstrated that riparian buffer zones, particularly those with organic-rich riparian soils, can significantly reduce the concentration of nutrients in surface water and groundwater entering streams (Howard-Williams *et al* 1986; Cooke & Cooper 1988; Pinay & Décamps 1988; Fustec *et al* 1991), and more specifically influence the quality of water passing from agricultural systems into the aquatic system (Haycock & Burt 1990, 1991).

Grassed buffer zones can reduce sediment-bound phosphorus and nitrogen inputs to streams by 80–87% and groundwater nitrate inputs by more than 90% (Smith *et al* 1989; Cooper 1990). Alluvial forests are known to play a particularly important role in reducing diffuse-source pollution (Labroue & Pinay 1986; Welsch 1991). Rhodes *et al* (1985) found nitrate removal from Zone 1 streams in the Sierra Nevadas to be 99% through uptake by riparian forests and wetlands. In one case study of the effectiveness of forested wetlands, along the Rhode River, Maryland, USA, Peterjohn and Correll (1984) found annual removal of nitrate-N was of the order of $45\,kg\,ha^{-1}\,yr^{-1}$ in subsurface runoff and 11 kg particulate organic-N in surface runoff. Thus, Peterjohn and Correll concluded that coupling natural systems and managed habitats within river basins may effectively reduce diffuse-source pollutions.

Three primary processes are active in controlling water quality in areas of semi-natural vegetation along river margins.

1 Retention through interception of sediment-bound nutrients, pesticides and other contaminants transported by surface runoff, which is particularly important in the case of phosphorus.

2 Uptake by vegetation or microbes of soluble nutrients – the primary process for nitrate removal.

3 Absorption by organic and inorganic soil particles.

However, the key parameter is the flow path (Hill 1990; Haycock 1991). Groundwater flow is rarely perpendicular to the channel flow, and the effect of variable permeability of floodplain sediments influencing oblique pathways means that actual retention times of groundwater are often greater than expected. Furthermore, channelled surface runoff may pass rapidly through buffer zones during storm runoff and care needs to be taken to ensure that buffers are not by-passed by pollutant transport pathways (Osborne & Wiley 1988; Risser 1990).

River margins for wildlife conservation

River margins provide both important habitat and linear connecting corridors through the landscape. The importance of the interaction between geomorphology and vegetation in floodplain environments for maintaining biotic diversity has been displayed (Bravard *et al* 1986; Salo *et al* 1986; Amoros *et al* 1987b; Pautou & Décamps 1989; Castella *et al* 1991; Greenwood *et al* 1991; Petts *et al* 1992). Furthermore, recent research has clearly demonstrated the need to consider river corridors as three-dimensional systems, especially where deep valley fill deposits create major alluvial aquifers (see *Regulated Rivers* Special Issues 6.4 and 7.1). In such systems, Stanford and Ward (1988), for example, have pointed to the importance of the hyporheic zone (the interstitial habitat influenced by surface water–groundwater interactions) of floodplains for maintenance of invertebrate diversity.

Animals

River margins are especially important for mammals, reptiles and amphibians (e.g. Brinson *et al* 1981; Cross 1985; Dickson & Huntley 1985) and river margins support some of the richest terrestrial vertebrate faunas (Risser 1990). With respect to large mammals, particular attention has been given in North America to the beaver (Naiman *et al* 1988) and in Europe to the otter (MacDonald & Mason 1983).

Birds. Much attention has focused on breeding-bird densities (Anderson & Ohmart 1977; Johnson *et al* 1977; Knopf 1985; Knopf *et al* 1988; Triquet *et al* 1990) because the loss of river margin systems can severely impact bird populations. Thus, Hehnke and Stone (1978) have described how the removal of riparian vegetation along the Sacramento River resulted in 95% fewer birds and 32% fewer species.

Fish. Many authors describe the value of hydraulically-diverse river margins for fish (e.g. Copp 1989). Along large Zone 3 rivers, especially those in the tropics with a predictable flood season, fish productivity is highly correlated with flood magnitude and the area inundated (e.g. Welcomme 1979; Copp & Penaz 1988). Thus, Bayley (1991) defined the 'flood-pulse advantage' as the amount by which annual multi-species fish yield exceeds that from a system lacking annual floodplain inundation.

Riparian zones to protect instream habitats

Riparian vegetation is especially important in determining instream habitats in Zone 1 streams, providing shade, cover and organic debris. The input of particulate organic matter (mainly twigs and leaves) has been established as being particularly important for aquatic ecosystems, especially invertebrate populations (e.g. Bilby & Likens 1980; Gregory 1992), since organic carbon is the most important fuel for running-water food webs (see Vol. 1, Chapter 14). The provision of shade, regulating instream temperatures and limiting growth of macrophytes and algae, is also an important function of riparian vegetation (e.g. Burton & Likens 1973; Karr & Schlosser 1977). Thus, in the UK, the Salmon Advisory Committee (1991) have described the introduction of riparian buffer zones as an attainable benefit of an integrated approach to catchment management. They conclude that the benefits of such an approach would include (i) allowing bankside vegetation to develop to provide cover; (ii) provision of food in the form of invertebrates; (iii) reduction of fine solid material; (iv) reduction of the risk of chemical pollution; and (v) reduction of the input of fertilizers.

Recreation and amenity

In river corridors two features particularly influence landscape quality: the existence of the land–water interface and its degree of naturalness (see Petts 1990b; Cosgrove & Petts 1990). Diversity (i.e. 'patchiness') and nature conservation interest are also important.

Researchers in both the UK and the USA (Sweep *et al* 1985; Simcox & Zube 1989; Cook 1991; Green & Tunstall 1992) highlight positive public attitudes towards preservation of multiple-use river corridor amenties including wildlife habitat, riparian vegetation and open space. Thus, multipurpose management functions have been advocated for riparian areas in lowland, urban and upland situations (Budd *et al* 1987; Cohen *et al* 1987; Rinne 1990), stressing the aesthetic and recreational potential of such areas, as well as their role in water quality and erosion control.

21.3 PROCESSES INFLUENCING NATURAL RIVER MARGINS

The degree of heterogeneity is an important feature of the riparian ecotone. In the natural situation, such heterogeneity is determined mainly by flow variability, topography (i.e. elevation in relation to river level), sediment permeability, and the frequency of destructive events (floodplain erosion and channel cut-off). It is the geomorphological structure of the river margin that provides the template for the operation of other physical, chemical and biological processes.

Geomorphological processes

Natural river margins consist of a complex mosaic of patches of different type, size and age (Plate 21.1, facing p. 408 and Table 21.2; see also

Table 21.2 Vegetation types mapped using remote sensing and large-scale mapping of the alluvial plain of the lower River Ain (from Roux 1986)

Patch number	Dominant vegetation type present
A Patches at the aquatic–terrestrial interface	
26	*Phalaris arundinacea*
9	*Phragmites australis, Urtica dioica, Calystegia sepium*
12	*Phragmites australis, Scirpus lacustris*
17	*Urtica dioica*
24	*Melilotus alba*
26	*Phalaris arundinacea*
B Patches with shallow ground water-tables, inundated for lengthy periods	
13	*Salix viminalis, S. triandra*
14	*Alnus incana, A. glutinosa, Ulmus minor*
16	*Salix alba, Carex acutiformis*
20	*Salix alba, Salix triandra*
21	*Salix alba, Impatiens glandulifera*
28	*Salix alba, Phalaris arundinacea*
36	Diverse *Salix* spp.
45	*Salix elanganos*
55	*Bromus erectus*
69	*Salix alba, Alnus glutinosa*
C Patches with deeper water-table and less persistent inundation than group B	
7	*Quercus robur, Populus alba, Carex pendula*
21	*Salix alba, Impatiens glandulifera*
29	*Ulmus minor*
31	*Fraxinus excelsior*
32	*Populus nigra*
43	*Bromus erectus*
48	*Fumana procumbens*
57	*Crataegus monogyna*
62	*Poa trivialis, Dactylis glomerata* (pasture)

Fig. 4.6) that reflect the geomorphological setting. The development of riparian systems through time is part of a directional sequence known as the reversible process concept (Amoros *et al* 1987a,b) within which the directional sequences are rejuvenated by erosion, deposition and flood disturbance. Of sediment reaching streams, 29–93% is stored in alluvial wetlands (Phillips 1989a). Although progressive sedimentation leads to terrestrialization and a loss of wetland habitat, new 'wet' habitat is created by channel cut-off and lateral migration elsewhere along the river sector so that, over time, optimal areas for particular fauna shift to different parts of the hydrosystem. Periodic replacement of older soil-vegetation complexes by pioneer patches (e.g. gravel or sand deposits) leads to an increase in both the biodiversity and the primary biological productivity of these systems.

Hydrological processes

Flooding is important in providing seasonal connectivity both between the different components of the drainage network and locally between patches, and in stimulating biological responses. Morphological and physiological adaptations directly related to flooding and waterlogging occur in riverine plants and collectively these adaptations determine a species' success in different moisture conditions (Ernst 1990). These adaptations are mainly concerned with the change of several of the nutrients present from an oxygenated to a reduced state. Life cycles of many fish species are also highly adapted to changes of flow: in temperate rivers salmonids require freshets to stimulate upstream migration (Fraser 1972) and many tropical fish are highly adapted to the predictable annual flood (Welcomme 1979; Bayley 1991).

Biogeochemical processes

Plant growth in waterlogged soils may be greatly affected by reduced minerals such as ammonium, manganous and ferrous ions, and sulphide (Ernst 1990; Laanbroek 1990). The length of the waterlogged period will influence the necessary physiological adaptations to these chemically reduced states. The principal effect of flooding is to reduce primary production in terrestrial species in the riparian and floodplain zone as these can only use CO_2 as a carbon source (whereas aquatic plants can utilize HCO_3^-). A feature of actively growing aquatic plants is that they release generally between 1 and 10% of their photosynthetically-fixed carbon to the water in the form of dissolved organic compounds (Carpenter & Lodge 1986).

In terms of nutrient dynamics, riparian wetlands can be described as valves (Kibby 1978), taking in nutrients in spring and early summer and releasing them in late summer and autumn. With regard to phosphorus for example, the rate of uptake correlates with measures of metabolic activity (see Vol. 1, Chapter 18), with the phosphorus content of leaf detritus then increasing during decomposition. The significant removal of nitrate in wetland soils is long evident in the literature (Lee *et al* 1975; Patrick & Reddy 1976; Richardson *et al* 1978). Flooding provides an input of organic carbon into the system, fuelling denitrification processes within anoxic floodplain soils (Cooper *et al* 1987). The ability of semi-aquatic communities characterized by high transpiration to absorb large quantities of water, assisting in absorption of nutrients, is also important here (Pautou & Décamps 1989), with accumulation of nitrogen in nitrate form during winter in soils with less waterlogging (Pinay *et al* 1989). Vegetation uptake will retain nitrate in the summer period, as denitrification is often lowest at this time of year (Ambus & Lowrance 1991) due to aeration of the soil.

21.4 RIVER MARGINS AS BUFFER ZONES FOR RIVER MANAGEMENT

It is the ecotonal qualities of natural river margins that have particular value for river and landscape management (Petts 1990b). Protection, rehabilitation and restoration of river margins can be partially achieved by using 'structural' measures but will be fully realized only if the appropriate flow variability and geomorphological instability are sustained.

Table 21.3 Some recommended widths for naturally-vegetated zones to enable ecosystem function to be carried out in river margins

Function	Study	Relevant details	Recommended width (m)
Wildlife protection	Brown *et al* (1990)	Provision of food, water, cover	99–169
	Rabeni (1991)	Fish, amphibians, birds	7–60
	Newbold *et al* (1980)	Invertebrate populations	>30
	Brinson *et al* (1981)	Mammals, reptiles, amphibians	200
	Stauffer & Best (1980)	Breeding bird communities	11–200
	Budd *et al* (1987)	Organic debris provision	15
	Cross (1985)	Small mammals	9–20
Fisheries protection	Williamson *et al* (1990)		10 < 20
	Karr & Schlosser (1977)		
Streambank stability	Erman *et al* (1977)	Low order streams	30
	Rabeni (1991)		<5
Water-quality control	Ahola (1989, 1990)	General improvement	50
	Hoek (1987)	As above	150
	Rabeni (1991)	As above	Various
	Pinay & Décamps (1988)	As above	1–2
	Keskitalo (1990)	As above	30
	Pinay & Décamps (1988)	As above	Narrow
	Peterjohn & Correll (1984)	Nitrate control	20–30
	Correll & Weller (1989)	As above	30
	Rhodes *et al* (1985)	As above	1–2
	Peterjohn & Correll (1984)	Phosphorus control	
	Smith *et al* (1989)	As above	
Sediment control	Rabeni (1991)	US State Legislation	23–183.5
	Brown *et al* (1990)	As above	213
	Peterjohn & Correll (1984)	Nutrient control	19
	Erman *et al* (1977)	Control stream turbidity	30
	Budd *et al* (1987)		15

Structural characteristics of river margins

Minimal area and shape of wildlife habitat are important determinants of species productivity, and this is especially true in relation to the narrow, linear habitat of stream corridors (Diamond 1975; Simberloff 1976). In the urban situation, management strategies need to protect food resources and include a corridor concept connecting fragmented habitat to larger rural populations (Greer 1982). Décamps *et al* (1987) found that avian species richness increased following fragmentation of the riparian corridor, but studies reviewed by Brinson *et al* (1981) suggest that 5–6-ha riparian habitat 'islands' are needed to support near maximum bird diversity, with larger areas required to support raptors. However, in terms of cost, restoration of a large contiguous river-floodplain area may be more cost-effective than restoring a similar area comprising smaller, disjointed areas along the river (Bayley 1991). Recommended dimensions of buffer zones are given in Table 21.3.

Continuity of the buffer zone is an important variable, not least if optimum management of riparian and instream biota is to be achieved. In one example, Dawson (1978) has outlined a plan where trees are planted along the south side of a stream to provide shade, but gaps of approximately 20 m are left at 70-m intervals to permit instream primary production, especially macrophyte growth, which provides shelter for invertebrates and fish and is an important source of autochthonous detritus. Certainly along floodplain rivers, the buffer zone should be a mosaic of patches rather than a continuous strip, thus maximizing heterogeneity.

Buffer width for water-quality control

In some studies, the recommended buffer width to protect wildlife habitat has depended on the extent of groundwater draw-down and the slope of the groundwater table. Where groundwater pollution is a major problem, wider buffer zones are required. Work in the Netherlands (van der Hoek 1987) has discussed the disappearance of vulnerable and rare plant species due to eutrophication of groundwater, recommending a buffer zone of 150 m to improve groundwater flow, and to increase floristic diversity in the riparian fringe.

With reference to specific width recommendations to achieve the function of water-quality control, a fairly narrow riparian strip can be highly effective (Table 21.3). In a study in France, a 30-m width of alluvial forest was shown to be sufficient to remove all nitrate with denitrification of up to 50 mg of $N_2\,m^{-2}\,day^{-1}$ being observed (Pinay & Décamps 1988). However, not all studies point to definite trends towards an increase in water quality. Studies in the eastern USA have concluded that the 23-m buffer zone width provided for in State legislation is inadequate for non-point pollution removal and recommend instead an 80-m width, particularly in estuarine situations (Phillips 1989b).

US Department of Agriculture recommendations for minimum widths to minimize sedimentation problems in streams range from 23 m for coarse sand to 138.5 m for silt, concluding that control over clay sedimentation cannot be met by buffers alone but requires wider-scale land management strategies. Buffers have been shown to fail in specific cases due to channelization influences (Dillaha *et al* 1986), and some sediment will almost certainly be remobilized during high flows and flood events; most, however, may be retained for a prolonged period of time (Lowrance *et al* 1986).

Overall there is wide variability in the effectiveness of buffer zones, with a range of 15–80 m being appropriate for various land/soil/vegetation complexes in riparian situations (Phillips 1989c). The pollutant-holding characteristics of emergent aquatic and wetland vegetation has been exploited in another way through the construction of artificial wetlands, and these may be particularly useful as barriers to protect streams from well-defined sources, such as drains and ditches, or any first/second-order stream (e.g. Hammer 1989).

Wildlife criteria

The size, arrangement and connectivity of patches depend on the geographical setting and should give due regard both to the nature of the natural riparian zone and to the existing features of the proposed buffer. The relationship between patches is especially important as a particular target species could be sustainable in a relatively small area if it is found in association with, or connected to, certain other patches. The factors which need, therefore, to be addressed in any scheme for ecological enhancement or river restoration include both the minimum area and configuration necessary to retain both plant and wildlife values in different riparian habitats, as well as the maximum distances which can separate islands of given habitat type before adverse effects to, and even loss of wildlife species or populations occur (Johnson *et al* 1977; Knopf *et al* 1988). In addition, the optimal as well as minimal requirements for enhancing wildlife values for a given habitat type (e.g. plant species type and cover) need to be determined in addition to information relating to their resilience under unfavourable conditions. Quite small semi-natural and natural patches on the floodplain have the potential to act as refugia (Sedell *et al* 1990; Large *et al* 1994). These patches can be defined as conveying spatial or temporal resistance or resilience to biotic communities impacted by biophysical disturbances.

Rabeni (1991), in a review primarily focusing on water quality protection, but also including habitat values for wildlife, quotes examples of mandatory buffer widths in the USA. Widths were shown to vary from 1.5 m (for forested streams in Idaho) to 92 m (in Wisconsin). Between these bounds there is a wide variation in buffer width, even in those recommended for similar land uses in different States. In Iowa, bird species diversity has been found to increase with increasing width of wooded riparian habitats (Stauffer & Best 1980) but Brinson *et al* (1981) showed most mammal, reptile and amphibian species to concentrate within 60 m of the edge of the stream. In the USA, the Washington Department of Ecology (1981) found wildlife habitat to extend beyond the stream bank a minimum of 27.4 m into the upland areas of the river corridor. Relatively wide buffers are required to provide sufficient habitat for riparian wildlife and plants and to function as wildlife corridors linking larger areas of riparian habitat (Rabeni 1991), a feature also seen in discussions of buffer zones for wildlife protection in east central Florida (Brown *et al* 1990).

River margins and fluvial processes

Vegetation has been shown to exert significant control over fluvial processes in five main ways: flow resistance, flow interruption by log jams, interception and storage of sediment, bank strengthening, and concave bank bench deposition (Hicken 1984; Gregory & Gurnell 1988). Vegetation exerts a particularly important control on bank erosion rates (Hemphill & Bramley 1989) but in general there is little information available to show how wide buffers need to be specifically for bank protection. Whipple *et al* (1981) found a correlation between buffer width and bank stability, while for low-order north California streams, Erman *et al* (1977) found 30-m vegetated buffers adequate for controlling bank erosion.

Corridor restoration

In order to effectively implement restoration, attention needs to be paid to the already-existing 'natural' and semi-natural communities present along river margins. In Germany, Arnold *et al* (1989) have argued for at least the creation of 'pocket-sized' wet biotopes and for the broadening of floodplain to at least five times the width of the main channel for any ecological improvement to be possible through re-creation of meanders, riparian forests and meadows. In the UK this aspect relating to riverine plant communities has become more topical with the introduction in the 1980s of detailed river corridor surveys (Gardiner 1991), which have complemented other works discussing distribution of herbaceous and woody riparian vegetation along a series of studied reaches in both upland and lowland situations (Mason *et al* 1984; Curry & Slater 1986; Mason & MacDonald 1990).

Individual conflicts arise. In discussing the benefits of riparian trees for ecological management of watercourses in the UK, Mason *et al* (1984) urge replanting or encouraging natural regeneration of *Quercus* species, *Acer pseudoplatanus* and *Fraxinus excelsior*. The mature root systems of these species provide secure holts for otter, but these trees are rejected as unsuitable by the engineer, as their horizontal root systems permit undercutting of the channel bank. A balance has therefore to be struck by planting a strategic mix of both vertical and horizontal rooting species. Elsewhere some workers advocate setting aside strips of land uncultivated (Newbold & Rush 1989; Nature Conservancy Council 1991), while others, for example Risser and Harris (1989) in the USA, discuss use of specific riparian species to restrict recreation access to sensitive areas as well as improving degraded wildlife habits and improving bank stability.

21.5 PRACTICAL IMPLEMENTATION OF RESTORATION

River margins have considerable ecological importance throughout the drainage network providing (i) buffer zones between the terrestrial and aquatic systems, and (ii) corridors through the landscape. However, research on river margins highlights four key points. First, river margins have an important vertical dimension extending from the subsurface groundwater zone, to the above-ground vegetation structure. Thus, Anderson *et al* (1978) discuss the importance of revegetating riparian floodplain areas for wildlife, with vertical foliage density being positively correlated with the number of species in an area. Secondly, the replacement of naturally-occurring riparian woody species with commercial crop alternatives can lead to loss of both species abundance and diversity in riparian areas (Geier &

Plate 21.1 Vegetation patch mosaic within the corridor of the River Ain at the Rhone confluence, France. Vegetation patches (defined in Table 21.3) reflect the underlying geomorphological structure. Patches are of different type and age. (From Girel *et al.* 1986.)

Table 21.4 Design options for integrated catchment management incorporating river margins

Zone	Ecological significance
1 Production zone	Buffer *zones* significant in maintaining the quantity and quality of the river flows downstream, and for conserving running-water habitats, as well as providing conservation areas in their own right. Bank stability is also improved by increasing vegetation cover. Buffers act as barriers to eutrophication of small streams
2 Transfer zone	*Strips* provide important linear habitats, not only within the riparian fringe but also along the channel margin, including shading and cover for fish. Floodplains can occur here and will be geomorphologically active. Buffers can aid channel stability in these cases
3 Storage zone	Riparian *strip* has important ecological value but considerable opportunities exist for environmental enhancement in former floodplain areas by managing *patches* for nature conservation

Best 1980). Thirdly, a significant feature of river margin systems is their inherently dynamic nature (Naiman & Décamps 1990), a property that has direct implications for river restoration and management. Fourthly, the primary reasons for river margin restoration differ in relation to location within the drainage network: water-quality control is the primary role in headwater systems and nature conservation becomes most important along larger rivers. Table 21.4 summarizes design options for integrated catchment management using restored river margins.

Headwater streams

All vegetated river margins act as buffers between land and water ecosystems and have benefits for water-pollution control, but several water-quality problems, such as pollution by nutrients such as nitrates, phosphates or pesticides, relate to diffuse sources and require management within the headwater production zone.

Land management practices on hillslopes can affect river water-quality in four ways: via overland flow, shallow subsurface flow, pipe flow and deep subsurface flow. Buffer zones can be very effective where problems are caused by diffuse overland flow or shallow subsurface flow. However, they may be by-passed by channelized surface runoff, pipe flow or deep subsurface flow. In many lowland areas the vegetated buffer could be by-passed by flow in under-drains. New drain designs and the incorporation of reconstructed wetlands and catch-water systems to intercept pipe flow, are being investigated to overcome this problem.

To maximize their effectiveness, all streams and ditches within the production zone need to be bounded by buffer zones; this represents a considerable length of channel. One alternative option is to combine buffer zones along second- and third-order streams with the creation of wetlands and/or storage ponds to intercept flow from first-order streams and ditches. Constructed wetlands and reed beds have been advocated as land-treatment alternatives in sewage systems for removal of biochemical oxygen demand (BOD) and solids leading to downstream improvement of water quality (e.g. Gersberg *et al* 1984a,b; Burka & Lawrence 1990; Upton & Griffin 1990). Although some studies show the potential of these schemes for nitrogen removal is uncertain (Bayes *et al* 1989), others (Boyt *et al* 1977; Tilton & Kadlec 1979; McIntyre & Riha 1991) demonstrate that constructed vegetated riparian areas have distinct potential for water-quality control. In relation to seasonal factors, leakage, especially of phosphorus, may occur (Richardson 1985; Mann 1990; Vanek 1991) following senescence and decomposition of the plant material. However, this can be controlled by maintenance through harvesting of biomass and, in extreme cases, removal of saturated sediment.

Middle and lower river

In the mid-channel situation, vegetated buffers will have value for water quality but their main

Fig. 21.1 Components of a rehabilitated river margin (after Petts 1990b): (1) channel with restored pattern and bedform; (2) restored riparian zone (see Chapter 19); (3) flood berm excavated and planted to create new wetlands; (4) flood embankments set well back from the river's edge; (5) additional embankment set away from river to create (6) emergency flood control channel and/or floodplain lake with link to main channel controlled by sluices, weirs or gated ducts.

role would be for environmental enhancement. Even narrow riparian strips along riverbanks having little management are important for many species, and would markedly improve the conservation value along river reaches that are currently cultivated up to the channel banks. The value of buffer zones for environmental enhancement increases directly with habitat diversity. Along the channel margin of the floodplain river, primary habitats of gravel bars, sandbanks, eroding banks in alluvium, cut-off channels and backwaters linked to the main channel are of particular value. While these habitats have been severely affected by recent and historic river management, studies of remnant patches have demonstrated their considerable importance for nature conservation (e.g. Petts *et al* 1992). Wetland and woodland patches also have particularly high conservation value.

Management requires three distinct levels of

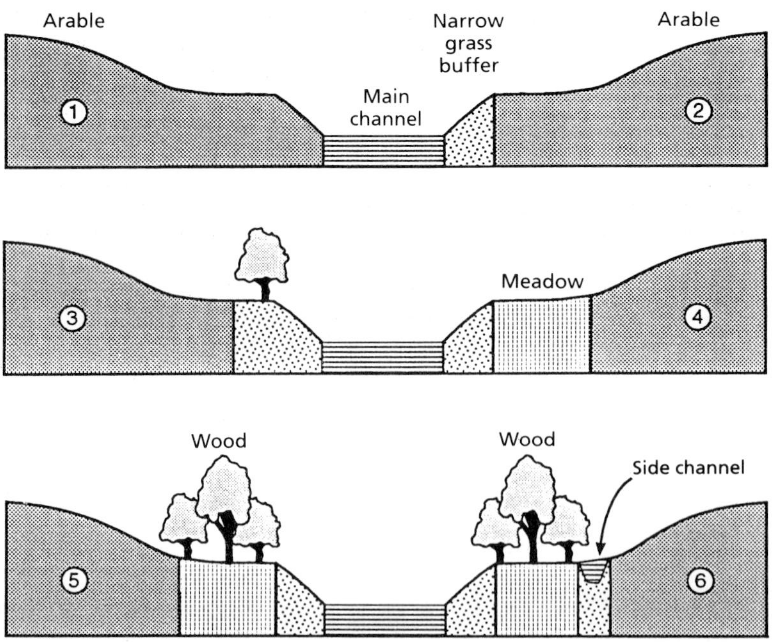

Fig. 21.2 Scenarios for river margin rehabilitation. The functions provided by each scenario are summarized in Table 21.5 (after Large & Petts 1992).

involvement. First, options should be considered to sustain the hydrological and geomorphological processes that determine the ecology of the river margin, riparian zone and floodplain. Secondly, management should seek to maintain the diversity of patches within the buffer zone. Finally, it is necessary to develop an ecologically sustainable maintenance programme to maximize the structural and sustainable functional characteristics. The structural components of ecologically sensitive, artificial river margins are illustrated in Fig. 21.1.

The landscape scale

Buffer zones fulfil a number of functions in the riparian landscape. These include increasing biotic diversity and enhancing the nature conservation value of the riparian zone and reducing water-quality problems both above ground in surface flows and groundwater quality. The benefits to bank stability and increase in water storage capability increase as the complexity of the buffer design increases. Attendant improvements in recreation potential and aesthetic quality can also be achieved.

Figure 21.2 illustrates different types of rehabilitated river margin which are suitable for use in a range of situations. Table 21.5 shows the functions achieved by the different buffer types. These range from the situation where no buffer is present and the adjacent land uses impact directly on the aquatic system (Scenario 1), to the situation where an ecologically diverse buffer incorporating a range of habitat types and including wetland patches has been established in the river corridor (Scenario 6).

A situation where no buffer is present (Scenario 1) scores zero for the seven functions outlined. Many floodplains have old channels cutting across them, often at the base of the hillslope. Such channels act as natural interceptors for surface and subsurface runoff from cultivated hillslopes and should be exploited, perhaps by dredging to create wetland areas for diffuse-source pollution control and conservation. With full restoration of floodplain function, including use of old channel courses, a range of vegetation patches to enhance biotic diversity and to aid control of both surface and subsurface runoff, scores for each of the functions can be maximized (Scenario 6).

Minimizing the flood problem

The major perceived disadvantage of river margin restoration is reduction in flood control. In several instances, the increasing hydraulic roughness of the channel margin and adjacent floodlands has been raised as an argument for *not* enhancing the ecological quality of river corridors. Any ecological advantages provided by a

Table 21.5 Functions carried out by vegetated buffer strips and patches in the riparian landscape. Scenarios range from no buffer present (1) to ecologically-complex protection mechanism (6). Maximum scores allocated in each scenarios represent the optimum management attainable using a structure of that type. This will not necessarily correspond to 100% efficiency in practice

		Scenario: increasing buffer complexity (score)					
Function	Maximum score	1	2	3	4	5	6
Biotic diversity/environmental enhancement	3	0	0/1	1/2	1/2	2	3
Overland flow control	3	0	1	1	2	3	3
Subsurface flow control	3	0	0	0/1	0	1 < 3	3
Bank stability	3	0	1	2	1	3	3
Water storage capability	3	0	0	1	1/2	3	3
Recreation provision	3	0	1	2	2	3	3
Aesthetic value	3	0	1	2	2	3	3

continuous belt of riparian forest could be outweighed by enhanced flood problems. However, the feasibility of using land adjacent to a river for flood storage and the use of the patch concept for environmental enhancement could minimize these problems. The ecological benefits of a series of small patches, connected by ditches, floodways, wetlands or grassland, designed to minimize hydraulic resistance at the landscape scale, would be considerable. In such situations the buffer zone may be visualized as a series of patches that should be connected but may appear as a 'string of beads' (Large & Petts 1992) in a landscape sense.

Longer-term maintenance

In any plan for managing river margins, commitment to habitat maintenance is obviously important and this is especially so if the natural hydrological and geomorphological dynamics that determine the natural disturbance regime cannot be fully restored. Without maintenance the conservation value may actually become quite depressed through internal competition reducing floristic diversity. Planning is required to ensure that any such maintenance is minimal, but this planning needs to be set in a long-term context. In reality, the maintenance of river margins may simply involve the redistribution of costs from current maintenance programmes.

Particular attention needs to be paid to the potential problem of invasion by undesirable species (Brock *et al* 1993). For example, *Impatiens glandulifera*, *Reynoutria japonica* and *Heracleum mantegazzianum* are examples of vascular species which have proved to be a problem along the waterways of England and Wales. Leaving areas fallow along the margins of waterways will provide potential sites for invasion, and thus a programme of continued management to prevent colonization will need to be implemented in areas where spread of such species is a potential threat.

Year-round management of water regime and the vegetation structure therefore is essential to maintain the effectiveness of any buffer for water quality. Decomposition processes and nutrient release in freshwater and wetland systems have a strong seasonality (e.g. Brinson 1977; Polunin 1982, 1984). Furthermore, in unmanaged riparian buffer zones, nutrient release may occur due to senescence and decomposition of the vegetation *in situ*.

With regard to the spiralling of phosphorus, for example, wetland systems tend to be quite leaky with a significant release from the dead plant community throughout the year (Magette *et al* 1989; Dillaha & Hayes 1991). There is also evidence to suggest that some floodplain wetlands used as a sink for phosphorus in time will become ineffective, releasing phosphorus back into circulation in the system. Examples of other problems are: (i) that some vegetated zones may not be efficient enough through nitrate uptake to reduce significantly the nitrate load during the growing season; (ii) that soil waterlogging may not be long enough to provide the anaerobic conditions for denitrification of sufficient duration to reduce loads; and (iii) that organic carbon provided by flood events and through litter fall may not be sufficient to sustain microbial respiration necessary for denitrification. Long-term effects of near-stream vegetation in reducing stream nutrients may be enhanced if buffers are used in conjunction with other practices, such as timing nutrient applications on a wider catchment basis to maximize uptake by crops and minimize loss to the aquatic system.

21.6 PROSPECT

To be effective, restoration of river margins requires legislation at governmental level to encourage landowners to set land aside for enhancement of the conservation value of the riparian zone and the river corridor in general. It has been shown that putting arable land over to grass as part of a set-aside programme is of value for nature conservation. Along larger rivers, such a policy might require arable land to be converted to grassland on all floodplain and former floodplain areas within a statutory distance of the channel bank. This distance could be fixed, a function of channel width or stream order, or based on historic flooding or channel migration (as determined from old maps and field observation).

In Europe, Dister *et al* (1990) have described the reduction of habitat on the Rhine floodplain by transformation of floodplain forests into monoculture plantations, and call for 'renaturalization' quoting the changing nature of EEC policy as a spur for implementation. Management perspectives based on isolated components of the catchment are ecologically incomplete. Phillips (1989d) recommends management in the context of the catchment rather than discrete units, and Lowrance *et al* (1985) conclude that management must occur on a catchment-by-catchment basis Nilsson *et al* (1988) highlight the fact that each river has largely unique features. However, thus far in the area of resource management, more emphasis has been placed on gathering information than on effective and efficient use of available information for decision-making and problem-solving. This is especially true with respect to whole-catchment management.

Whilst attempting to maintain the integrity of land–water interactions in otherwise altered catchments (Karr & Schlosser 1978), management must remain within the constraints posed by multiple-use objectives for the catchment (Ward & Stanford 1989). River margins have important benefits for society and progress in catchment management could be achieved by developing a programme of river margin rehabilitation. The starting point is to conceptualize river margins as the primary ecotone between the aquatic and terrestrial systems. Along many rivers of the temperate zone only a thin strip of natural vegetation remains due to a long history of fluvial landscape management. Nevertheless, these fringing zones along with small, presently isolated, wetland and woodland patches can act as refugia for flora and fauna already at the limits of their ecological tolerance limits. In many cases these refugia will provide logical starting points for rehabilitation programmes.

REFERENCES

Ahola H. (1989) Vegetated buffer zone project of the Vantaa River river basin. *Geografisk Tidsskrift* **89**: 22–5. [21.4]

Ahola H. (1990) Vegetated buffer zone examinations of the Vantaa river basin. *Aqua Fennica* **20**: 65–9. [21.4]

Ambus P, Lowrance R. (1991) Comparison of denitrification in two riparian soils. *Soil Science Society of America Journal* **55**: 994–7. [21.3]

Amoros C, Rostan J-C, Pautou G, Bravard J-P. (1987a) The Reversible Process Concept applied to the environmental management of large river systems. *Environmental Management* **11**: 607–17. [21.2, 21.3]

Amoros C, Roux AL, Reygrobellet JL, Bravard J-P, Pautou G. (1987b) A method for applied ecological studies of fluvial hydrosystems. *Regulated Rivers: Research and Management* **1**: 17–36. [21.2, 21.3]

Anderson BW, Ohmart RD. (1977) Vegetation structure and bird use in the Lower Colorado River Valley. In: Johnson RR, Jones Jr DA (eds) *Importance, Preservation and Management of Riparian Habitat: a Symposium*, pp. 23–4. USDA Forest Service General Technical Report RM-43. [21.2]

Anderson BW, Ohmart RD, Disano J. (1978) Revegetating the riparian floodplain for wildlife. In: Johnson RR, McCormick JF (eds) *Strategies for Protection and Management of Floodplain Wetlands and Other Riparian Ecosystems*, pp. 318–31. USDA Forest Service, General Technical Report WO-12. [21.5]

Arnold U, Höttges J, Rouvé G. (1989) Removing the strait-jackets from rivers and streams. *German Research* **1**: 22–4. [21.4]

Bayes CD, Bache DH, Dickson RA. (1989) Land-treatment systems: design and performance with special reference to reed-beds. *Journal of the Institution of Water and Environmental Management* **3**: 588–97. [21.5]

Bayley PB. (1991) The flood-pulse advantage and the restoration of river-floodplain systems. *Regulated Rivers: Research and Management* **6**: 75–86. [21.2, 21.3, 21.4]

Bilby RE, Likens GE. (1980) Importance of organic debris dams in the structure and function of stream ecosystems. *Ecology* **61**: 1107–13. [21.2]

Boyt FL, Bayley SE, Zoltek J. (1977) Removal of nutrients from treated municipal wastewater by wetland vegetation. *Journal of the Water Pollution Control Federation* **49**: 789–99. [21.5]

Bravard J-P, Amoros C, Pautou G. (1986) Impact of civil engineering works on the successions of communities in a fluvial system: a methodological and predictive approach applied to a section of the Upper Rhône River. *Oikos* **27**: 92–111. [21.1, 21.2]

Brinson MM. (1977) Decomposition and nutrient exchange of litter in an alluvial swamp forest. *Ecology* **58**: 601–9. [21.5]

Brinson MM, Swift BL, Plantico RC, Barclay JS. (1981) *Riparian Ecosystems: Their Ecology and Status*. US Fish & Wildlife Service FWS/OBS 82/17. [21.2, 21.4]

Brock JH, Childs LE, de Waal LC, Wade PM. (eds)

(1993) *Ecology and Management of Invasive Aquatic Plants*. John Wiley & Sons, Chichester. [21.5]

Brown MT, Schaefer J, Brandt K. (1990) *Buffer Zones for Water, Wetlands and Wildlife in East Central Florida*. Center for Wetlands Publication 89-07, University of Florida, Gainsville. [21.4]

Budd WW, Cohen PL, Saunders PR, Steiner FR. (1987) Stream corridor management in the Pacific Northwest, I. Determination of stream corridor widths. *Environmental Management* **11**: 587–97. [21.2, 21.4]

Burka U, Lawrence PC. (1990) A new community approach to waste treatment with higher water plants. In: Cooper PF, Findlater C (eds) *Constructed Wetlands in Water Pollution Control*, pp. 359–71. Pergamon Press, Oxford. [21.5]

Burton TM, Likens GE. (1973) The effect of strip-cutting on stream temperatures in the Hubbard Brook experimental forest, New Hampshire. *BioScience* **23**: 433–5. [21.2]

Carpenter SR, Lodge DM. (1986) Effects of submersed macrophytes on ecosystem processes. *Aquatic Botany* **26**: 341–70. [21.3]

Castella E, Richardot-Coulet M, Roux C, Richoux P. (1991) Aquatic macroinvertebrate assemblage of two contrasting floodplains: The Rhône and Ain Rivers. *Regulated Rivers: Research and Management* **6**: 289–300. [21.2]

Cohen PL, Saunders PR, Budd WW, Steiner FR. (1987) Stream corridor management in the Pacific Northwest. II. Management strategies. *Environmental Management* **11**: 599–605. [21.2]

Cook EA. (1991) Ecosystem modeling as a method for designing synthetic fluvial landscapes: a case study of the Salt River in Arizona. *Landscape and Urban Planning* **20**: 291–308. [21.2]

Cooke JG, Cooper AB. (1988) Sources and sinks of nutrients in a New Zealand hill pasture catchment. III. Nitrogen. *Hydrological Processes* **2**: 135–49. [21.2]

Cooper AB. (1990) Nitrate depletion in the riparian zone and stream channel of a small headwater catchment. *Hydrobiologia* **202**: 13–26. [21.2]

Cooper JR, Gilliam JW, Daniels RB, Robarge WP. (1987) Riparian areas as filters for agricultural sediment. *Soil Science Society of America Journal* **51**: 416–20. [21.3]

Copp GH. (1989) The habitat diversity and fish reproductive function of floodplain ecosystems. *Environmental Biology of Fishes* **26**: 1–27. [21.2]

Copp GH, Penaz M. (1988) Ecology of fish spawning and nursery zones on the flood plain, using new sampling approach. *Hydrobiologia* **169**: 209–24. [21.2]

Correll DL, Weller DE. (1989) Factors limiting processes in freshwater wetlands: an agricultural primary stream riparian forest. In: Sharitz RR, Gibbons JW (eds) *Freshwater Wetlands and Wildlife*, pp. 9–23. USDoE Symposium Series 61. [21.4]

Cosgrove D, Petts GE (eds). (1990) *Water Engineering and Landscape*. Belhaven. London. [21.2]

Cross SP. (1985) Responses of small mammals to forest riparian perturbations. In: Johnson CD *et al* (eds) *Riparian Ecosystems and Their Management: Reconciling Conflicting Land Uses*. USDA Forest Service General Technical Report RM-120, Fort Collins, Colorado. [21.2, 21.4]

Cummins K. (1992) Catchment characteristics and river ecosystems. In: Boon PJ, Calow P, Petts GE (eds) *River Conservation and Management*, pp. 125–36. John Wiley & Sons, Chichester. [21.1]

Curry P, Slater FM. (1986) A classification of river corridor vegetation from four catchments in Wales. *Journal of Biogeography* **13**: 119–32. [21.4]

Dawson FH. (1978) Aquatic plant management in semi-natural streams: the role of marginal vegetation. *Journal of Environmental Management* **6**: 213–21. [21.4]

Décamps H, Joachim J, Lauga J. (1987) The importance for birds of the riparian woodlands within the alluvial corridor of the River Garonne, SW France. *Regulated Rivers: Research and Management* **1**: 301–16. [21.4]

Diamond JM. (1975) The island dilemma: lessons of modern biogeographic studies for design of natural reserves. *Biological Conservation* **7**: 129–46. [21.4]

Dickson JG, Huntley JC. (1985) Streamside management zones and wildlife in the South Coastal Plain. In: Johnson CD *et al* (eds) *Riparian Ecosystems and Their Management: Reconciling Conflicting Uses*. USDA Forest Service General Technical Report RM-120, Fort Collins, Colorado. [21.2]

Dillaha TA, Hayes PE. (1991) *A Procedure for the Design of Vegetative Filter Strips*. US Soil Conservation Service. [21.5]

Dillaha TA, Sherrard JH, Lee D. (1986) *Long-Term Effectiveness and Maintenance of Vegetative Filter Strips*. Virginia Water Resources Research Centre Bulletin No. 153. Blaksburg, Virginia. [21.4]

Dister E, Gomer D, Obrdlik P, Petermann P, Schneider E. (1990) Water management and ecological perspectives of the Upper Rhine's floodplains. *Regulated Rivers: Research and Management* **5**: 1–15. [21.1, 21.6]

Erman DC, Newbold JC, Roby KB. (1977) *Evaluation of Streamside Buffer Strips for Protecting Aquatic Organisms*. California Water Resources Center, University of California, Davis, California. [21.4]

Ernst WHO. (1990) Ecophysiology of plants in waterlogged and flooded environments. *Aquatic Botany* **38**: 73–90. [21.3]

Fraser JC. (1972) Regulated discharge and the stream environment. In: Oglesby RT, Carlson CA, McCann JA (eds) *River Ecology and Man*, pp. 263–86. Academic Press, New York. [21.3]

Fustec E, Mariott A, Grillo X, Sajus J. (1991) Nitrate

removal by denitrification in alluvial groundwater: role of a former channel. *Journal of Hydrology* **123**: 337–54. [21.2]
Gardiner JL. (1991) *River Projects and Conservation: A Manual for Holistic Appraisal.* John Wiley & Sons, Chichester. [21.4]
Geier AR, Best LB. (1980) Habitat selection by small mammals of riparian communities: evaluating effects of habitat alterations. *Journal of Wildlife Management* **44**: 1–16. [21.5]
Gersberg RM, Elkins BV, Goldman CR. (1984a) Wastewater treatment by artificial wetlands. *Environmental Science and Technology* **17**: 443–50. [21.5]
Gersberg RM, Elkins BV, Goldman CR. (1984b) Use of artificial wetlands to remove nitrogen from wastewater. *Journal of the Water Pollution Control Federation* **56**: 152–6. [21.5]
Green CH, Tunstall SM. (1992) The amenity and environmental value of river corridors in Britain. In: Boon PJ, Calow P, Petts GE (eds) *River Conservation and Management*, pp. 425–42. John Wiley & Sons, Chichester. [21.2]
Greenwood MT, Bickerton MA, Castella E, Large ARG, Petts GE. (1991) The use of Coleoptera (Arthropoda: Insecta) for patch characterisation on the floodplain of the River Trent, UK. *Regulated Rivers: Research and Management.* **6**: 321–32. [21.2]
Greer DM. (1982) Urban waterfowl population: ecological evaluation on management and planning. *Environmental Management* **6**: 217–29. [21.4]
Gregory KJ. (1992) Vegetation and river channel process interactions. In: Boon P, Calow P, Petts GE (eds) *River Conservation and Management*, pp. 225–70. John Wiley & Sons, Chichester. [21.2]
Gregory KJ, Gurnell AM. (1988) Vegetation and river channel form and process. In: Viles HA (ed) *Biogeomorphology*, pp. 11–42. Blackwell Scientific Publications, Oxford. [21.4]
Hammer DA (ed). (1989) *Constructed Wastelands for Wastewater Treatment.* Lewis, Chelsea, Michigan. [21.4]
Haycock NE. (1991) Riparian land as buffer zones in agricultural catchments. Unpublished DPhil Thesis, University of Oxford. [21.2]
Haycock NE, Burt TP. (1990) Handling excess nitrates. *Nature* **348**: 29. [21.2]
Haycock NE, Burt TP. (1991) The sensitivity of rivers to nitrate leaching; the effectiveness of near-stream land as a nutrient retention zone. In: Allison RJ, Thomas DSG (eds) *Landscape Sensitivity*, pp. 261–72. John Wiley & Sons, Chichester. [21.2]
Hehnke M, Stone CP. (1978) Value of riparian vegetation to avian populations along the Sacromento River system. In: Johnson RR, McCormick JF (eds) *Strategies for protection and Management of Floodplain Wetlands and Other Riparian Ecosystems*, pp. 228–35. USDA Forest Service, General Technical Report WO-12, Fort Collins, Colorado. [21.2]
Hemphill RW, Bramley ME. (1989) *Protection of River and Canal Banks.* CIRIA and Butterworths, London. [21.4]
Hicken EJ. (1984) Vegetation and river channel dynamics. *Canadian Geographer* **28**: 111–26. [21.4]
Hill AR. (1990) Groundwater cation concentrations in the riparian zone of a forested headwater stream. *Hydrological Processes* **4**: 121–30. [21.2]
Hoek D van der. (1987) The input of nutrients from arable lands on nutrient poor grassland and their impact on the hydrological aspects of nature management. *Ekologia* **6**: 313–23. [21.4]
Howard-Williams C, Pickmere S, Davies J. (1986) Nutrient retention and processing in New Zealand streams: the influence of riparian vegetation. *New Zealand Agricultural Science* **20**: 110–14. [21.2]
Johnson RR, Haight LT, Simpson JM. (1977) Endangered species vs. endangered habitats: a concept. In: Johnson RR, Jones Jr DA (eds) *Importance, Preservation and Management of Riparian Habitat: a Symposium*, pp. 68–79. USDA Forest Service General Technical Report RM-43, Fort Collins, Colorado. [21.2, 21.4]
Karr JR, Schlosser IJ. (1977) *Impacts of Nearstream Vegetation and Stream Morphology on Water Quality and Stream Biota.* EPA-600/3-77-097. Environmental Research Laboratory, Environmental Protection Agency, Athens, Georgia. [21.2, 21.4]
Karr JR, Schlosser IJ. (1978) Water resources and the land–water interface. *Science* **201**: 229–34. [21.4, 21.6]
Keskitalo J. (1990) Occurrence of vegetated buffer zones along brooks in the catchment area of Lake Tuuslanjärvi, South Finland. *Aqua Fennica* **20**: 55–64. [21.4]
Kibby HV. (1978) Effects of wetlands on water quality. In: Johnson RR, McCormick JF (eds) *Strategies for Protection and Management of Floodplain Wetlands and Other Riparian Ecosystems*, pp. 289–98. USDA Forest Service General Technical Report WO-12, Fort Collins, Colorado. [21.3]
Knopf FL. (1985) Significance of riparian vegetation to breeding birds across an altitudinal cline. In: Johnson RR *et al* (eds) *Riparian Ecosystems and Their Management: Reconciling Conflicting Uses*, pp. 105–11. USDA Forest Service General Technical Report RM-120, Fort Collins, Colorado. [21.2]
Knopf FL, Johnson RR, Rich T, Samson FB, Szaro RC. (1988) Conservation of riparian ecosystems in the United States. *The Wilson Bulletin* **100**: 272–84. [21.2, 21.4]
Laanbroek HJ. (1990) Bacterial cycling of minerals that affect plant growth in waterlogged soils: a review. *Aquatic Botany* **38**: 109–25. [21.3]
Labroue L, Pinay G. (1986) Epuration naturelle des

nitrates des eaux souterraines: possibilités d'application au réménagement. *Annales Limnologie* **22**: 83–8. [21.2]

Lachavanne J-B (ed). (1993) *Biodiversity in Land/Inland Water Ecotones.* UNESCO-MAB, Paris. [21.1]

Large ARG, Petts GE. (1992) *Buffer Zones for Conservation of Rivers and Bankside Habitat.* R&I Project Record 340/5/Y, National Rivers Authority, Loughborough. [21.2, 21.5]

Large ARG, Prach K, Bickerton MA, Wade PM. (1994) Alteration of patch structure and boundary types on the floodplain of the River Trent, England. *Regulated Rivers: Research and Management* **9**(1). [21.4]

Lee GF, Bentley E, Amundson R. (1975) Effects of marshes on water quality. In: Hasler AD (ed) *Coupling of Land and Water Systems*, pp. 105–27. Springer-Verlag, New York. [21.3]

Lowrance RR, Leonard R, Sheridan J. (1985) Managing riparian ecosystems to control non-point pollution. *Journal of Soil and Water Conservation* **40**: 87–91. [21.6]

Lowrance RR, Sharpe JK, Sheridan JM. (1986) Long-term sediment deposition in the riparian zone of a coastal plain watershed. *Journal of Soil and Water Conservation* **41**: 266–71. [21.4]

MacDonald SM, Mason CF. (1983) Some factors affecting the distribution of otters (*Lutra lutra*). *Mammal Review* **13**: 1–10. [21.2]

McIntyre BD, Riha SJ. (1991) Hydraulic conductivity and nitrogen removal in an artificial wetland system. *Journal of Environmental Quality* **20**: 259–63. [21.5]

Magette WL, Brinsfield RB, Palmer RE, Wood JD. (1989) Nutrient and sediment removal by vegetated filter strips. *Transactions of the American Society of Agricultural Engineering* **32**: 663–7. [21.5]

Mann RA. (1990) Phosphorus removal by constructed wetlands: substratum adsorption. In: Cooper PF, Findlater C (eds) *Constructed Wetlands in Water Pollution Control*, pp. 97–105. Pergamon Press, Oxford. [21.5]

Mason CF, MacDonald SM. (1990) The riparian woody plant community of regulated rivers in eastern England. *Regulated Rivers: Research and Management* **5**: 159–66. [21.4]

Mason CF, MacDonald SM, Hussey A. (1984) Structure, management and conservation value of the riparian woody plant community. *Biological Conservation* **29**: 201–16. [21.4]

Naiman RJ, Décamps H. (eds) (1990) *The Ecology and Management of Aquatic–Terrestrial Ecotones.* Parthenon Publishing, Carnforth. [21.1, 21.2, 21.5]

Naiman RJ, Johnston CA, Kelley JC. (1988) Alteration of North American streams by beaver. *Bioscience* **38**: 753–62. [21.2]

Nature Conservancy Council. (1991) *Nature Conservation and Pollution Fron Farm Wastes.* Nature Conservancy Council, Peterborough. [21.4]

Newbold C, Rush A. (1989) Set aside and extensification of agricultural production: implications and opportunities for nature conservation and the river engineer. In: *Proceedings of Institute of Water and Environmental Management Annual Symposium, York 1989.* IWEM, London. [21.4]

Newbold JD, Erman DC, Roby KB. (1980) Effects of logging on macroinvertebrates in streams with and without buffer strips. *Canadian Journal of Fisheries and Aquatic Sciences* **37**: 1076-85. [21.4]

Nilsson C, Grelsson G, Johnansson M, Sperens U. (1988) Can rarity and diversity be predicted in vegetation along riverbanks? *Biological Conservation* **44**: 201–12. [21.6]

Osborne LL, Wiley MJ. (1988) Empirical relationship between land use/cover and stream water quality in an agricultural watershed. *Journal of Environmental Management* **26**: 9–27. [21.2]

Patrick WH, Reddy KR. (1976) Nitrification–denitrification reactions in flooded soils and water bottoms: dependence on oxygen supply and ammonium diffusion. *Journal of Environmental Quality* **5**: 469–72. [21.3]

Pautou G, Décamps H. (1989) Ecological interactions between the alluvial forests and hydrology of the Upper Rhône. *Archiv für Hydrobiologie* **104**: 13–37. [21.2, 21.3]

Peterjohn WT, Correll DL. (1984) Nutrient dynamics in an agricultural watershed: observations on the role of a riparian forest. *Ecology* **65**: 1466–75. [21.2, 21.4]

Petts GE. (1990a) Forested river corridors – a lost resource? In: Cosgrove D, Petts GE (eds) *Water Engineering and Landscape*, pp. 12–34. Belhaven, London. [21.1]

Petts GE. (1990b) The role of ecotones in aquatic landscape management. In: Naiman RJ, Décamps H (eds) *The Ecology and Management of Aquatic–Terrestrial Ecotones*, pp. 227–62. Parthenon Publishing, Carnforth. [21.2, 21.4, 21.5]

Petts GE, Moller H, Roux AL. (eds) (1989) *Historical Changes of Large Alluvial Rivers: Western Europe.* John Wiley & Sons, Chichester. [21.1]

Petts GE, Large ARG, Greenwood MT, Bickerton MA. (1992) Floodplain assessment for restoration and conservation: linking hydrogeomorphology and ecology. In: Carling P, Petts GE (eds) *Lowland Floodplain Rivers: Geomorphological Perspectives*, pp. 217–34. John Wiley & Sons, Chichester. [21.2, 21.5]

Phillips J. (1989a) Evaluation of North Carolina's estuarine shoreline area of environmental concern from a water quality perspective. *Coastal Management* **17**: 103–17. [21.3]

Phillips J. (1989b) Nonpoint source pollution risk assessment in a watershed context. *Environmental Management* **13**: 493–502. [21.4]

Phillips J. (1989c) Non-point source pollution control effectiveness of riparian forest along a coastal plain river. *Journal of Hydrology* **110**: 221–37. [21.4]

Phillips J. (1989d) Fluvial sediment storage in a wetland. *Water Resources Bulletin* **25**: 867–73. [21.6]

Pinay G, Décamps H. (1988) The role of riparian woods in regulating nitrogen fluxes between the alluvial aquifer and surface water: a conceptual model. *Regulated Rivers: Research and Management* **2**: 507–16. [21.2, 21.4]

Pinay G, Décamps H, Arles C, Lacassin-Seres M. (1989) Topographic influence on carbon and nitrogen dynamics in riverine woods. *Archiv für Hydrobiologie* **114**: 401–14. [21.3]

Polunin N. (1982) Processes contributing to the decay of reed (*Phragmites australis*) litter in fresh water. *Archiv für Hydrobiologie* **94**: 182–209. [21.5]

Polunin N. (1984) The decomposition of emergent macrophytes in fresh water. *Advances in Ecological Research* **14**: 115–66. [21.5]

Rabeni CF. (1991) *Buffer Zones for Riparian Zone Management: A Literature Review*. Department of the Army, New England Division, Corps of Engineers, Waltham MA. [21.4]

Rhodes J, Skau CM, Greenlee D, Brown D. (1985) Quantification of nitrate uptake by riparian forests and wetlands in an undisturbed headwaters watershed. In: Johnson RR *et al* (eds) *Riparian Ecosystems and Their Management: Reconciling Conficting Uses*, pp. 175–9. USDA Forest Service General Technical Report RM-120. [21.2, 21.4]

Richardson CJ. (1985) Mechanisms controlling phosphorus retention capacity in freshwater wetlands. *Science* **228**: 1424–7. [21.5]

Richardson CJ, Tilton DL, Kuclee JA, Chamie JDM, Wentz AW. (1978) Nutrient dynamics of northern wetland ecosystems. In: Good RE, Whigham DF, Simpson RL (eds) *Freshwater Wetlands: Ecological Processes and Management Potential*, pp. 217–42. Academic Press, New York. [21.3]

Rinne JN. (1990) The utility of stream habitat and biota for identifying potential conflicting forest land uses: Montane riparian areas. *Forest Ecology and Management* **33**: 363–83. [21.2]

Risser PG. (1990) The importance of land/land-water ecotones. In: Naiman RJ, Décamps H (eds) *The Ecology and Management of Aquatic–Terrestrial Ecotones*, pp. 7–22. Parthenon Publishing, Carnforth. [21.2]

Risser RJ, Harris RR. (1989) Mitigation for impacts to riparian vegetation on western montane streams. In: Gore J, Petts GE (eds) *Alternatives in Regulated River Management*, pp. 235–65. Boca Raton Press. [21.4]

Roux AL. (1986) *Recherches Interdisciplinaires sur les Ecosystemes de la Basse-plaine de l'Ain, France. Potentialities Evolutives et Gestion*. Laboratoire de Biologie Végétale, Saint-Martin d'Hères. [21.3]

Salmon Advisory Committee. (1991) *Factors Affecting Natural Smolt Production*. Ministry of Agriculture, Fisheries & Food, Lowestoft. [21.2]

Salo J. (1990) External processes influencing origin and maintenance of inland water–land ecotone. In: Naiman RJ, Décamps H (eds) *The Ecology and Management of Aquatic–Terrestrial Ecotones*, pp. 37–64. Parthenon Publishing, Carnforth. [21.1]

Salo J, Kalliola R, Häkkinen I, Mäkinen Y, Niemelä P, Puhakka M, Coley PD. (1986) River dynamics and the diversity of Amazonian lowland forest. *Nature* **322**: 254–8. [21.2]

Sedell JR, Reeves GH, Hauer FR, Stanford JA, Hawkins CP. (1990) Role of refugia in recovery from disturbances: modern fragmented and disconnected river systems. *Environmental Management* **14**: 711–24. [21.4]

Simberloff DS. (1976) Island biogeography theory and conservation practice. *Science* **191**: 285–6. [21.4]

Simcox DE, Zube EH. (1989) Public value orientations toward urban riparian landscapes. *Society and Natural Resources* **2**: 229–39. [21.2]

Smith CM, Williamson RB, Cooper AB. (1989) Riparian retirement: the effects on streambank stability and water quality. In: *Changing Times. Proceedings of the New Zealand Soil and Water Conservation Annual Conference*, pp. 27–35. Nelson, New Zealand. [21.2, 21.4]

Stanford JA, Ward JV. (1988) The hyporheic habitat of river ecosystems. *Nature* **335**: 64–6. [21.2]

Stauffer DF, Best LB. (1980) Habitat selection by birds of riparian communities: evaluating effects of habitat alterations. *Journal of Wildlife Management* **44**: 1. [21.4]

Sweep D, Zilincar JM, Smith B, Hardy R. (1985) Integration of riparian management strategies within the context of multiple land management programs in southwestern Wyoming. In: Johnson RR *et al* (eds) *Riparian Ecosystems and Their Management: Reconciling Conflicting Uses*, pp. 371–3. USDA Forest Service General Technical Report RM-120, Fort Collins, Colorado. [21.2]

Swift BL. (1984) Status of riparian ecosystems in the United States. *Water Resources Bulletin* **20**: 223–8. [21.1]

Tilton DL, Kadlec RH. (1979) The utilization of a freshwater wetland for nutrient removal from secondarily treated wastewater effluent. *Journal of Environmental Quality* **8**: 328–33. [21.5]

Triquet AM, McPeek GA, McComb WC. (1990) Songbird diversity in clearcuts with and without a riparian buffer strip. *Journal of Soil and Water Conservation* **45**: 500–3. [21.2]

Upton J, Griffin P. (1990) Reed bed treatment for sewer dykes. In: Cooper PF, Findlater C (eds) *Constructed*

Wetlands in Water Pollution Control, pp. 391–8. Pergamon Press, Oxford. [21.5]

Vanek V. (1991) Riparian zone as a source of phosphorus for a groundwater-dominated lake. *Water Research* **25**: 409–18. [21.5]

Vannote RL, Minshall GW, Cummins GW, Sedell JR, Cushing CE. (1980) The river continuum concept. *Canadian Journal of Fisheries and Aquatic Science* **37**: 130–7. [21.1]

Ward JV, Stanford JA. (1989) Riverine ecosystems: the influence of man on catchment dynamics and fish ecology. *Canadian Special Publication of Fisheries and Aquatic Sciences* **106**: 55–64. [21.6]

Washington Department of Ecology. (1981) Western Washington urban stream assessment. Washington Department of Ecology, Office of Water Programs. Water Quality Planning. Olympia, WA. [21.4]

Welcomme RL. (1979) *Fisheries Ecology of Floodplain Rivers*. Longmans, New York. [21.1, 21.2, 21.3]

Welsch DJ. (1991) *Riparian Forest Buffers: Function and Design for Protection and Enhancement of Water Resources*. USDA Forest Service. Northeast Area. Report NA-PR-07-91, Radnor, Pennsylvania. [21.2]

Whipple W, DiLouie JM, Pytalr T. (1981) Erosion potential of streams in urbanizing areas. *Water Resources Bulletin* **17**: 36–45. [21.4]

Williamson RB, Smith RK, Quinn JM. (1990) *The Effects of Riparian Protection on Channel Form and Stability of 6 Grazed Streams, Southland, New Zealand*. Water Quality Centre Publication 19, DSIR, Hamilton, New Zealand. [21.4]

22: Restoration of River Corridors: German Experiences

P. LARSEN

22.1 INTRODUCTION

Until the end of the 1970s river training in Germany was based on technical criteria, sometimes supplemented with aesthetic considerations. Emerging public concern for the environment, awakened by works such as Rachel Carson's book *Silent Spring*, first published in 1962, gradually had its impact on the profession and on the water authorities. Around 1980 the first guidelines, instructions and laws regarding 'nature-related river training' and 'maintenance' were issued. The emphasis on 'renaturalization' has now (1992) been strengthened by laws in most of the German states.

The earliest attempts to restore stretches of smaller watercourses were undertaken by enthusiasts, who were looked upon with benevolent amusement by the authorities and by hydraulic engineers. Such projects date back to the 1960s and 70s. Today, ecological criteria are considered on an equal footing with technical criteria.

Much of the information in this chapter is based on experience of river restoration in the south-western part of Germany. The approach has been developed over the last decade and has proved to be effective in ways that may well make it relevant to other parts of the temperate zone. Restoration of a river corridor requires both the knowledge and the co-operation of specialists representing several disciplines. The approach is necessarily comprehensive because in highly developed and densely populated countries, such as Germany, very little can be left to chance. Its application to other countries would require some modification because of the inevitable variety of regulations, authorities involved, type of watercourse, aim of restoration, economic constraints, etc.

The term 'renaturalization' should be properly interpreted. Even if a 'natural' condition could be defined, the goal of 'back to nature' is usually impossible to reach. People are now and have been for centuries modifying their ecosystem. Not many stretches of the world's rivers could be called natural in their present states. Much effort has already been spent in acquiring knowledge on various aspects of river corridor restoration, but the state of the art is still in its infancy. In Germany, hundreds of kilometres of river channel have been engineered to states that are far from natural. The experience gained so far with their restoration – the reversal of this process – may be useful to planners elsewhere. Planners in countries in which river systems may be in a more natural state should also take note: an ounce of prevention is worth a pound of cure, i.e. protection against degradation will always prove to be more cost-effective and ecologically effective than restoration.

22.2 BACKGROUND TO RIVER RESTORATION IN GERMANY

History of degradation

In many areas, particularly in the industrialized part of the world, rivers and watercourses have undergone progressive changes away from their natural state. Mankind has been attracted to natural waterways to enjoy their scenic value and to use them for fishing, power, transport, waste disposal, etc. Settlements began near rivers. As they grew, many areas that previously had been flooded during high water became more and more densely populated. The resultant protection of these areas reduced the river valley's natural retention of floodwater. Flood peaks downstream were thereby increased, and still further protection was needed for people living in other flood-prone areas. A second major activity, hydropower utilization, often created dams with large storage areas upstream and sections of the river downstream were deprived of normal water flows. Also the discharge of wastes produced different but seriously detrimental effects. Rivers flowing through farmland have been altered to improve drainage and to improve the layout of fields used for crops. Competition in food production led to increases in the usage of fertilizers and of pesticides, causing diffuse loads of pollutants added to those of community wastes. Such activities as these have had far reaching ecological consequences. Figure 22.1 shows, as an example, endangered plant species contained in the German Red Data Book and the causes for their reduction.

The foregoing changes were originally perceived as improvements, but the decisions that led to them were based solely on technical and economic concepts; their planning, some years ago, failed to include proper attention to possible ecological consequences. The result, in many areas, has been serious ecological deterioration, such as reductions in the variety of plant and animal life, and even the extinction of many species. Also, aesthetically pleasing features have been lost: straight channel alignment and geometric homogeneity of cross-section have replaced the winding river with its charming—and ecologically important—contrasts. Rows or groves of trees, which once marked the river path, are gone; natural habitats for birds and animals have been destroyed.

As these consequences became more and more serious, public awareness led to requests, and later to demands, for a more comprehensive consideration of ecological aspects. Public pressures in this direction were probably stronger in West Germany than in any other country; the main reasons were the extensiveness of the rectification

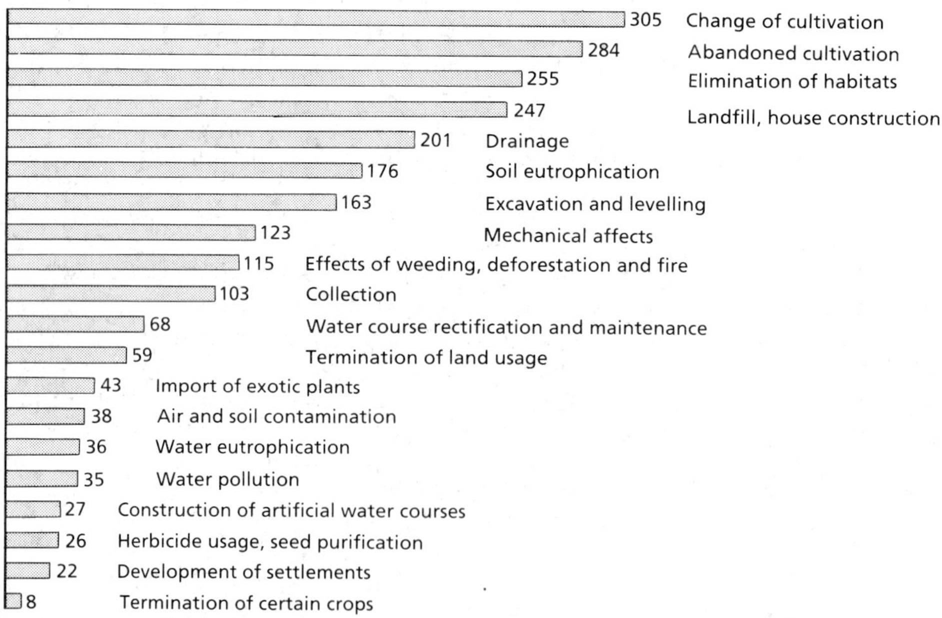

Fig. 22.1 The causes that endanger 711 species (from the *Red Data Book*, Germany) arranged in descending order of the endangering cause. The sum of the figures shown exceeds the number of species because some are endangered by more than one cause. (Source: Bundesforschungsanstalt für Naturschutz und Landschaftsökologie, 1989.)

of running waters and of the reclamation of wetlands that had occurred, the high population density and the generally high standard of living.

History of restoration

In the state of Baden-Württemberg, river restoration projects, primarily those on the Murr river near Stuttgart (Bürkle *et al* 1985), were begun in the 1970s. Guidelines and regulations were issued by the German Association for Water Resources and Land Improvement (DVWK) and by water authorities in several states. In the course of time, the emphasis on naturalization was strengthened by the enactment of legal regulations.

The proposed (1991) modification of the water law of Baden-Württemberg requires that a nature-like condition of the watercourses must be strived for; the conservation and restoration of the ecological function of a watercourse has to be given priority. The natural retention of floodwater must be preserved or increased, and increased discharge capacity is allowed only if important reasons, like the protection of settlements from flooding, can be documented. Those responsible for the maintenance of rivers shall, provided no overriding reasons in terms of such public disadvantages stand against it, convert rectified watercourses into nature-like ones within a reasonable period of time. Their efforts are to be monitored by those water authorities that have the responsibility to enforce compliance.

In the mid-1980s, the Ministry of Environment (then the Ministry of Nourishment, Agriculture, Environment and Forestry) of Baden-Württemberg inaugurated a research project proposed by the Institute of Hydraulic Structures and Agricultural Engineering at the University of Karlsruhe. Its purpose was to obtain information and experience with the renaturalization of the watercourses within the State. Fifteen sections of streams distributed throughout the State were selected as pilot projects, and these are to be executed and monitored during a 10-year period. A few of these projects have reached the monitoring stage, i.e. ecological and morphological consequences of the alterations are being surveyed.

A 2-km section of the River Enz was restored in 1990 after a period of comprehensive planning and detailed hydraulic model studies. Figure 22.2 shows a part of this section as it looked before, in 1989, and after modification in the summer of 1991. The difference will be even more marked once the vegetation becomes fully developed.

The Danube within Baden-Württemberg, a 160-km stretch of the river between its source and the point where it crosses the houndary with Bavaria, is in various stages of investigation and pre-planning; restoration is the ultimate goal. A 2-km subsection was selected as a pilot project, and detailed planning, including hydraulic model studies, has been carried out. Land adjacent to the river has been purchased and implementation began in 1991.

Also for the biggest river, the Rhine, that flows along the southern and western border of the State, restoration programmes have been established. Thus an Integrated Rhine River Programme covering a reach of about 220 km between Basle and Mannheim and an international programme Salmon 2000, covering the entire river downstream of Lake Constance have been formulated.

In present-day planning of coventional hydraulic projects ecological requirements play a role comparable to those of techniques and costs. Permission for project implementation requires, by law, thorough studies of the environment and the expected impact of the project on it. Regulations often call for proposals of alternative measures to compensate for ecological features that may be lost. These are detailed in the environmental impact assessment law. Binder (1984) presents some examples of projects on the River Lech in Bavaria, in which environmental considerations have played an essential role.

Other German states have undertaken developments similar to those in Baden-Württemberg. Bavaria has been a forerunner in the renaturalization and care of rivers. As part of the Baden-Württemberg research project, a survey of projects in progress or being executed, the state of the art in some of the German states and existing regulations was conducted (Kern & Nadolny 1986). The information thus gained was used in defining goals and methods, and in selecting the 15 pilot sections mentioned above.

All available experience with river restoration

(a) (b)

Fig. 22.2 A section of the River Enz, Baden-Württemberg, before (a) and after (b) renaturalization.

gained from projects in Baden-Württemberg has been documented, and from the information thus compiled, a guide to river restoration, contained in the same volume, has been prepared (Ministerium für Umwelt 1992).

Components of river restoration planning

The classification of running waters has proved to be an important prerequisite to the process of corridor restoration. The state of Baden-Württemberg contains a large variety of landscapes and types of watercourses. A research project is in progress on two natural units that were selected because they have landscapes of distinctly different characteristics. These are being comprehensively surveyed with the aim of establishing a methodology for stream classification adapted to the requirements of restoration planning. One goal is to typify about 20 varieties of streams in the State, and to do so in such a way that the process does not require undue effort. The results of the research project are compiled in Forschungsgruppe Fliessgewässer 1993 (Bostelmann *et al.* 1993). Following a review of stream classification methodologies the main issues are interdisciplinary characterization of stream types and leitbild development.

An important part of river restoration is to provide for the migration of both fish and invertebrates. Numerous river-training projects have effected longitudinal slope corrections in a stepwise manner by the construction of a series of drop structures, and these often obstruct fish migration. They must be replaced by structures that are passable by fish – gently sloping ramps that are covered with rip-rap made of varying stone size plus a finer substrate. At weirs and dams, by-pass structures with the same function and effectiveness must be provided. A comprehensive research project on this topic has been completed (Gebler 1991). Figure 22.3 shows an appropriate drop structure constructed to replace a vertical, impassable structure in the River Elz. This was one of several full-scale structures designed and tested in that study.

Channelization has caused a reduction in the variety of habitats and the loss of a number of species of both flora and fauna. Figure 22.4 shows the threat to fishes in German waters. Uniform channels that have been constructed cause the substrates to become impoverished and uniform. Deposition of fine material clogs the substrate surface and seals off the paths to deeper layers. Straight channels that eliminate meanders increase the hydraulic gradient and thereby aug-

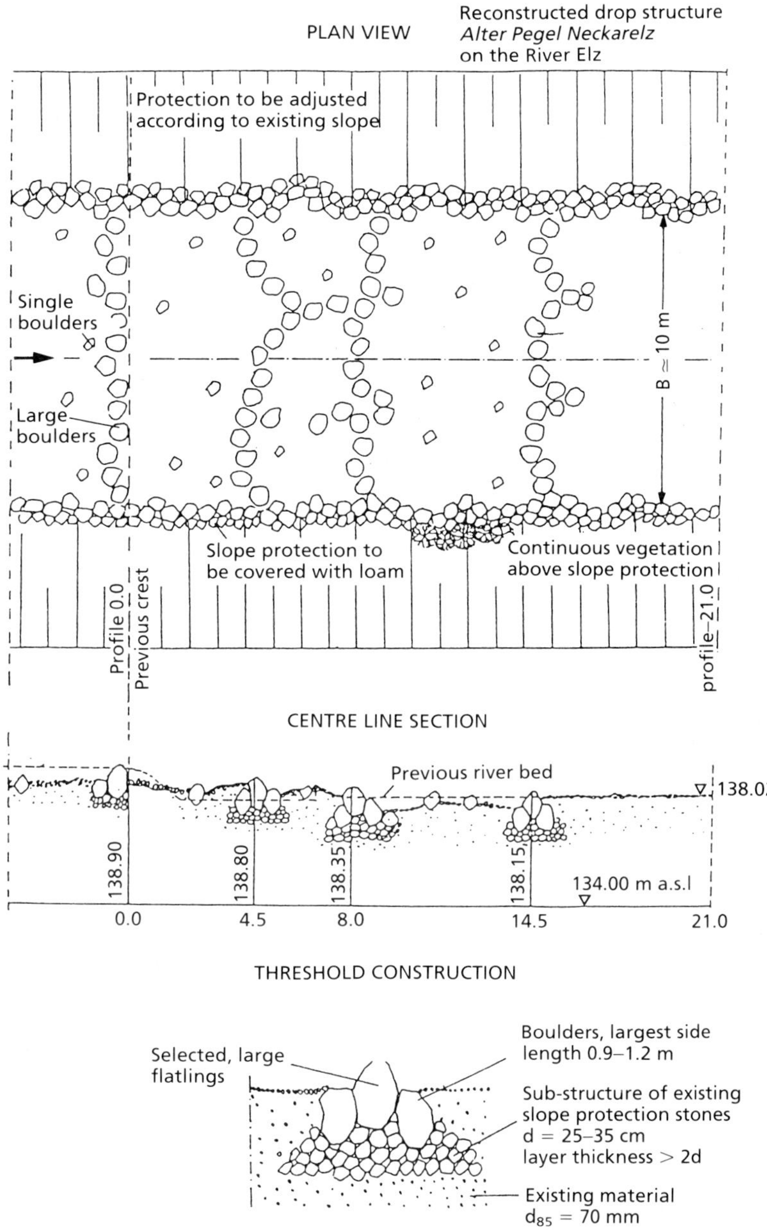

Fig. 22.3 Drop structure designed to allow passage of the aquatic fauna.

ment the transport of sediment. Lower bed levels and the accompanying lower water levels cause the groundwater table to be lower also. The capacity of aquifers is diminished and the floodplain wetlands, which contribute to ecological quality with their special habitats, are lost. The intermittent flooding of the floodplains adjacent to a natural river is extremely important to its ecosystem. The lateral exchanges between river and floodplain of inorganic and organic matter, living and dead, are key features of ecosystem dynamics. Dister *et al* (1990) give a number of reasons why the inundation of floodplains is so vitally important for balancing out anthropogenic effects.

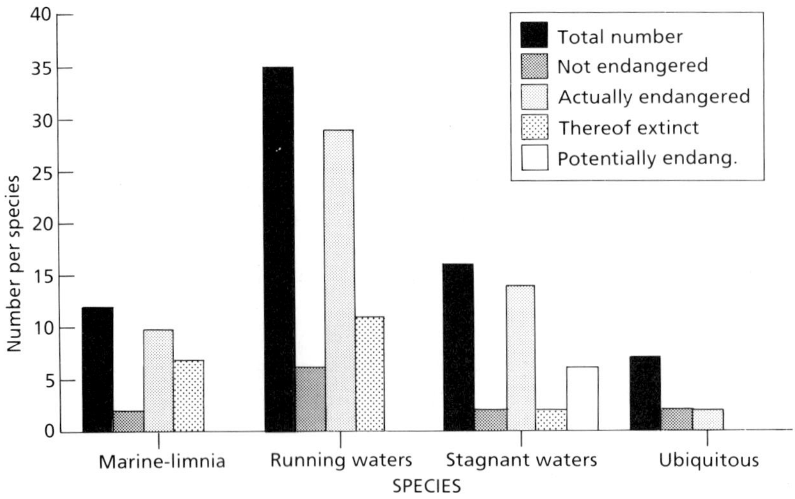

Fig. 22.4 Threat to fishes according to environment (Germany). (Source: Bundesforschungsanstalt für Naturschutz und Landschaftsökologie, 1988–89.)

Where the connection between the main channel and abandoned channels has been closed off spawning possibilities for fish are reduced and exchanges of aquatic life are eliminated. A variety of wetland plants depend for their continued existence on annual inundations, thus an altered succession takes place on previously flooded areas, marked by dying off of existing natural species. Many fishes, not only the salmonids, depend for their existence on migration and movement to refugia. Re-establishment of aquatic life after accidental losses, for example caused by chemical spill, is dependent on the existence of refugia such as marginal backwaters and floodplain channels.

River regulation that prevents the flooding of the areas outside the main channel precludes this exchange. The two-stage channel, which is favoured by some specialists (e.g. Hey *et al* 1991; see also Chapter 18), has limited value in river corridor restoration. This conclusion is supported by experience with such channels in Germany. They have been constructed in large numbers for decades, and now considerable effort is required to alter them. The limited discharge capacity, accurately determined during engineering hydraulic design, severely constrains the possibilities of renaturalization.

22.3 RESTORATION GOALS

The primary goal of river restoration is to re-establish the variety of habitats required for the natural diversity of biota in and along the river.

Knowledge of the consequences of river construction projects on the environment suggests some important goals for restoration: the primary one is the conservation of nature; the others are water resources management and landscape planning.

Conservation of nature

The re-establishment and preservation of the natural environment is the dominant objective of restoration. However, it is important to define the target end-point (see Chapter 5). A claim that maximum possible species variety should be a central goal must be stated in relative terms: each river is special and the ecology along a given river in its natural state varies with location. Therefore, it is not the species variety *per se*, but the site-specific flora and fauna that should be supported.

The features of dynamic river morphology must be re-established. Artificial bank protection should be eliminated so that streams, where possible, are free to develop their natural courses. The land required for natural channel changes to occur may need to be purchased from the present owners. Exchanges with intermittently inundated floodplains should also occur naturally. Hence, some dikes will have to be removed and, where water levels or grades require adjustment, ecologically compatible structures should be used. In addition the channel capacity has to be reduced to natural conditions to secure frequent flooding of the adjacent wetlands. The natural development of site-specific flora and fauna is based on free migration in and along the river channel. Migration paths, suitable for both fish and invertebrates, must be continuous. The complex ecosystem of a natural river depends not only on the properties of the river channel and its floodplains but also on exchanges with its tributaries. Thus restoration planning requires consideration of some ecological features of the whole river valley.

Water resources management

The primary purpose of flow management is usually the reduction of flooding. Normally a discharge of a certain probability of occurrence, depending on the values of the areas to be protected, determines a maximum discharge up to which flood protection should be guaranteed. Reduction of this design discharge, often a politically determined quantity, can be achieved by increased retention of water, i.e. physically reducing the discharge of a certain probability of occurrence. Also, the broadening of the basis for the balance between costs of flood damage and costs of prevention affects the magnitude of the design discharge. For example, the use of lower values for farmland, a common tendency, works in favour of more retention storage. Also, where the cost of flooding of farmland is the determining factor, a lower design discharge may be selected since reimbursement for flooded crops and other damages will be less.

Determination of the discharge capacity (the flow rate for a given water level) of a renaturalized watercourse to ensure that the design discharge can pass, presents the hydraulic engineer with a difficult problem. Its technical aspects are discussed further in Section 22.5, below. If the groundwater table has been lowered by the degradation of the river bed, it may require corrective actions to raise it again. Any structures installed to raise the water level need also to satisfy the requirement that the aquatic fauna can migrate along the river.

Water quality is, of course, a paramount factor in restoration planning. Although self-purification is enhanced by naturalization, it is not a primary goal of renaturalization. If the water quality is poor because of discharges of wastes from human activities, renaturalization is doomed to fail and any money to be spent on it would be better invested in waste-water treatment.

Improved quality of landscape

Restoration targets can extend from local improvements in appearance near settlements to the renaturalization of a whole stream. Whatever the goal or its extent, the changes made will affect the landscape. Renaturalization should include the creation of areas suitable for human recreation of various kinds. In this task, the conflict between the ecological requirement of undisturbed habitats and that of free human activity must be considered and a suitable compromise be found.

Model image

Even though 'back to nature' is not a realistic objective, a survey of a river's historical development is an important part of planning, as will be explained below. Such studies must include a summary of the recent geological history, natural morphological changes with time, man-made alterations, previous land use and historical changes of vegetation. Information of this kind forms the basis on which to establish a 'model image', i.e. a description of the river as it would have appeared, had it not been subjected to human activities. [The German word *Leitbild* (Kern *et al* 1986), analogous to the word *Leitmotif*, used also in English, describes this concept well.]

The model image serves as a benchmark. It is a description of an idealistic goal that helps to guide the planner. It will normally not be achievable because of a number of constraints caused by irreversible changes of abiotic and biotic factors, and by consequences of cultural ecology. Examples of the former are changes of the runoff regime and the sediment load, and rigid boundaries like bridge crossings and roads, that prohibit lateral movement of the stream bed and thus limit the dynamics of natural morphology. An example of the latter is the extinction of species due to effects of modern agricultural techniques.

22.4 A PROGRAMME FOR RESTORATION PLANNING

The planning of corridor restoration is distinct from engineering river training in that it cannot be limited to the space between the river banks; it must include the entire floodplain. Technical data, such as significant discharges, river stages and sediment properties, make up only part of the information on which such panning should be based. The necessarily more diversified approach requires additional information about geology, morphology, biology, pedology and landscape. The hydraulic engineer, who would traditionally be in charge of such projects, now finds himself to be only one member of a team that includes botanists, limnologists, geologists, geographers and landscape architects.

The planning and execution of a river corridor restoration project includes the following steps:

1 establishment of model image;
2 preliminary assessment;
3 survey and evaluation;
4 pre-design with possible options;
5 final design and acquisition of permits;
6 planning of project execution;
7 construction and supervision.

Obviously, the extent of the various activities will vary in accordance with the actual conditions. The boundaries between the steps are not rigid: some activities may run in parallel. Discussions of each of these steps follow.

Historical maps

A succession of historical maps may help to establish the model image. They may provide not only the present planform of the river, but also its past forms and thus illustrate the mobility of the river bed. Another important source of information is aerial photography, including infrared images; these may reveal previous oxbows and other features that could not be detected by direct observation.

In some cases, where a section of river, either of the same river or of a neighbouring one that flows through a similar natural region, has been left in its natural state or nearly so, its study may provide some of the information required to establish the model image.

Preliminary assessment

In a preliminary assessment, the general situation is assessed to obtain a first indication as to the feasibility of a restoration project. Items to be studied include water quality, availability of land for purchase, boundary constraints like roads and bridges, buried conduits, power cables and land borders. The results of the preliminary assessment should be discussed among representatives of the various authorities and professional disciplines that will be involved, owner associations and, of course, the project sponsor. Based on the assessed need for the restoration and expectation of success the essential goals of the project should be stated and the areas to be included in the project should be agreed upon.

Survey and evaluation

The goals and extent of the project will determine the type and comprehensiveness of field investigations. As distinct from traditional engineering projects, the major tasks are the collection of ecological data and a comprehensive pre-design, comprising various alternatives. An extensive investigation and evaluation of the present environmental conditions is a necessary prerequisite in any restoration of a river corridor. Subdivision of this task into collection of data on the catchment, on the riverine area and on the river, is convenient.

Catchment

The purpose of a catchment inventory is to collect information on *all* factors that influence the ecology and the flow regime in the river corridor to be restored.

Relevant data describe the size of the catchment; the extent of its lakes, climate, geology and relief; and existing natural units, and land utilization (forest, crop fields and pastures, villages and housing developments, sports fields, etc.). Natural and artificial drainage systems with possible waste-water influx should be noted as well as water utilization and hydraulic structures. Also important are existing wetlands, nature reserves, occasionally inundated areas and habitats of special concern.

A substantial part of this information can be obtained from maps – topographical, geological, pedological, and vegetation maps – and from aerial photographs. Inventories made by the water authorities may provide information on water quality, existing water rights, water utilization and water management structures. Field visits and interviews with local residents may help to round off the inventory.

Riverine area

The riverine area is the central object of the restoration; it includes the land that in its natural state was occasionally or regularly flooded, and the stretch of river within it. The purpose of its inventory is to assess the possibility of establishing habitats and their potential for development. Information on natural biotic communities and their development is essential to this process.

The inventory should comprise information on the natural situation: valley morphology, geology, soil types and textures, groundwater table, existing natural vegetation and the potential for other desirable vegetation. It should also include the historical development of the cultivated landscape down to its present state. The latter would include former land use, infrastructure development, dumps and landfills, including those that have been abandoned and possibly covered. Present land use and official plans for future developments are important as regards their possible incorporation into the restoration activities. Accordingly, land ownership, private and public, must be ascertained.

Landscape elements worthy of protection, such as small lakes, ponds, oxbows and wetlands with their specific vegetation, are important components of the inventory. A faunistic survey comprising typical insects, birds and mammals, and of course those species belonging in the *Red Data Book*, is normally part of the survey.

Data should also be included on drainage systems, natural and artificial, on the groundwater table, including its annual variations, and on water protection areas.

The sources of information are much the same as those mentioned above for 'Catchment'. Land ownership can be obtained from land registers and data on groundwater from well observations.

River

Information on the river is needed to perceive its past development, to determine the need for restoration, and to examine limiting conditions that may form constraints on proposals for conservation measures.

The category to which the river belongs needs to be determined in accordance with a standard classification system. Information is collected on previous rectification works, structures for regulation purposes, weirs and dams for water utilization, etc. If possible, morphological changes before human actions changed the state of the river should be documented.

Details of present utilization obviously belong to this inventory. These include waterpower, abstractions, sources of waste water, fishing and other recreational activities. Also necessary is information on hydraulic structures, such as flood control structures, plans for future developments, and ownership of water and fishing rights.

Regarding hydraulic data, discharge values and corresponding water levels at selected sites for significant probabilities of occurrence must be obtained. Water-quality observations include degree of pollution, usually in terms of saprobic level, and types of pollutants and their origins.

The morphological condition of the river channel is characterized in terms of sediment load, aggradation or degradation, boundary stability, substrate types and their structures. The extent of erosion protection and the materials used must be recorded; likewise the tendencies

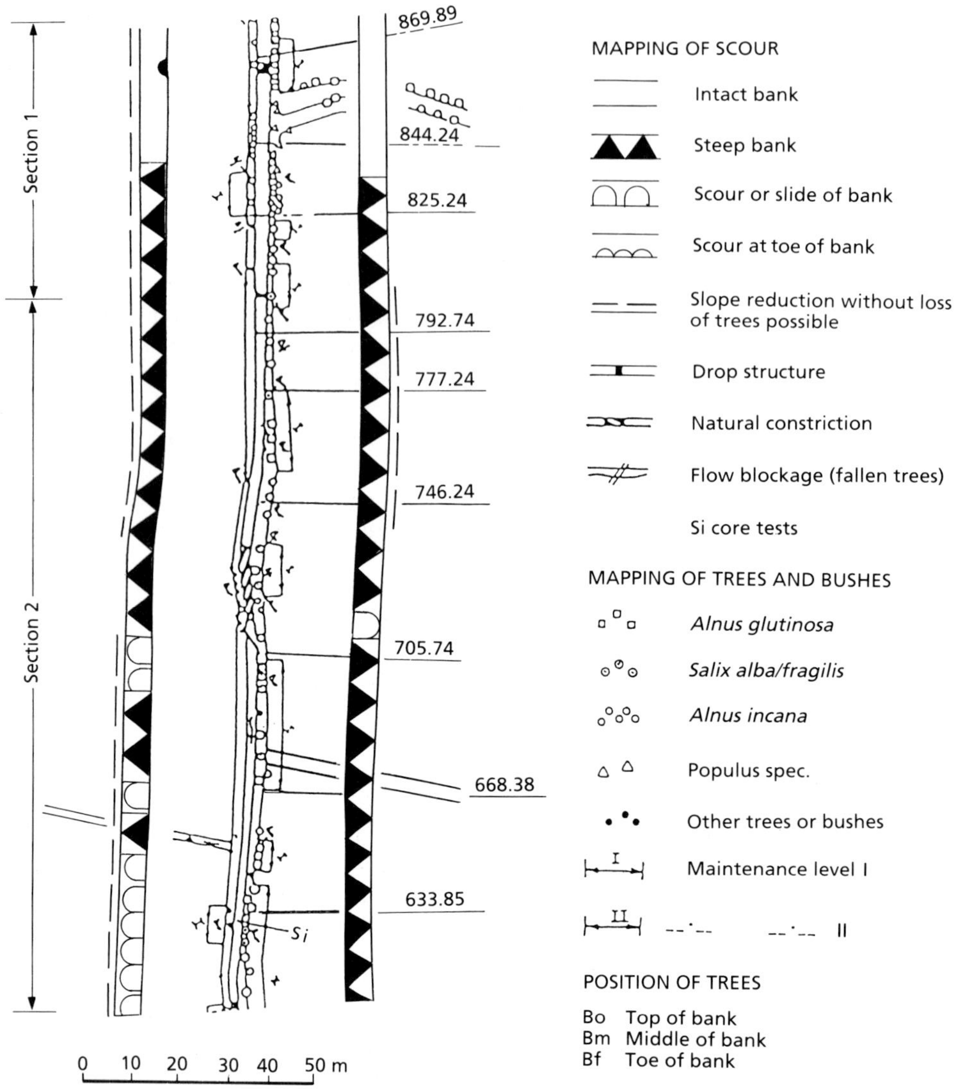

Fig. 22.5 Cutout of a map of scour and vegetation along a brook.

for channel alignment to change on unprotected strectches of the riverbanks should be noted.

The report should describe the system for the maintenance of the river, detailing the nature of the works and costs of upkeep; it should also include information on who is responsible for maintenance.

A biological report concludes the survey. It should contain information on fish and invertebrates and on plant communities – in the river, on its banks and on adjacent flood-prone areas.

The huge amount of data collected in the various surveys requires suitable reduction and thorough editing to produce a clear account of the important facts. The explanatory text should be accompanied by a variety of maps. Figure 22.5 shows one such specialized map.

Pre-design with possible options

The facts derived from the evaluation of the survey form the basis for the elaboration of measures that will lead to more nature-like conditions. These are presented in a preliminary design, possibly including alternatives, that is the basis for presentations and discussions of the project with those responsible for it. This should lead to a decision on whether and – if it is a positive decision – how to carry out the project.

The pre-design shall explain the achievable goals and the means of reaching them under the constraints imposed by water management, ecology, engineering and economy. The goals have various time perspectives: immediate, intermediate and long term. The immediate goals are achieved by the engineering and planting operations to be carried out, while the further goals will be achieved through suitable maintenance. Long-term goals, to be reached through immediate operations, and maintenance works, are conceived in terms of the model image described in Section 22.3, above.

An important need is for open discussions of the plans. All parties involved or affected by the restoration measures should be given the opportunity to participate in the deliberations. The reasons for selecting the project layout, its goals and any restrictions, should be clearly presented. Such openness may help to defuse opposition and save considerable time later, particularly in the course of obtaining the implementation permit.

Final design and acquisition of permits

The project option that is chosen in the pre-design phase is further elaborated and described in the text and supplemented by detailed drawings. Since these documents form the basis for the project initiator's final decision and for that of the various authorities regarding the execution permit, they need to be both complete and final. In the case of governmental subsidy, usually the case in Germany, they also constitute the basis for such allocations. If land is to be purchased the information needed for legal procedures comes partly from these documents.

The final design is determined while maintaining close communication among the various specialists and the project initiator. The final design documents comprise: (i) a written report; (ii) a set of drawings; (iii) hydraulic computations; (iv) design and stability analysis of hydraulic structures; (v) cost analysis; and (vi) reports and comments from experts.

Written report. This explains the reasons for, and recounts the deliberations that have led to, the selection of the actual project layout and includes a full description of the project. It also includes lists of measures to be executed, their location on a survey, cross-sections, scour protection measures, planting schemes, expected developments of morphology and vegetation as well as directions regarding future maintenance.

Drawings. The drawings that are included comprise an overall plan view and maps showing design, land requirements, ground acquisition, stock and vegetation, areas under protection, and locations of adjacent agricultural areas on which farming measures are to be modified.

Hydraulic computations. These substantiate the required flood discharge capacity (to be maintained after the vegetation has reached maturity!), expected sediment transport and any measures to resist scour.

Design and stability analyses. These are required for structures used to regulate flow or water levels, such as gently sloping drop structures, fish paths, etc.

Cost analysis. A detailed cost analysis must comprise costs of acquiring land, legal permits, planning, execution (including supervision), materials, plants and expected costs of maintenance.

Expert reports. Reports and comments on the various fields represented by the multidisciplinary team members engaged in the project evolution should be appended to the final design report.

Based on this comprehensive set of information, tender documents can be prepared. These include a time schedule for the various activities, specification of materials to be used and the quality the finished work is to achieve.

Construction

Experience shows that the construction phase also differs from that for conventional engineering projects. The success of restoration depends greatly on the supervisory participation of experts during the execution. For example, excavator operators are accustomed to be guided by straight lines and constant slopes or levels, which are often not suitable for this type of project. If such guidance is not provided, and often it is not, greater understanding and flexibility on the part of the operator is required. He has to work in close co-operation with the designer, who in turn needs to comprehend field practices in some detail. Planting operations must also be supervised by people with horticultural or silvicultural skills, and these people must also know how to work effectively with the contractor. Multidisciplinary co-operation thus prevails throughout all phases of the river corridor restoration project.

Contrary to normal river construction projects, execution of, the restoration proceeds from upstream to downstream. In this way completed areas are essentially spared sedimentation of silt and sludge, brought into suspension during the various operations.

The time schedule of the execution must take account of planting seasons, of expected periods of low discharges and of the possibility of utilizing the winter period.

22.5 HYDRAULIC COMPUTATIONS AND MODEL TESTS

The conveyance capacity of the watercourse is normally required to comply with the stated design discharge, for example Q_{50} or Q_{100}, including a certain allowance for a freeboard. The hydraulic computations are straightforward as far as regular, engineered channels are concerned: experience with such computations is usually available in the form of Manning coefficients for both single-stage and two-stage channels. Backwater computations render reliable relations between stage and discharge for a range of flows.

With natural or renaturalized watercourses the situation is different: little is known about flow resistance caused by vegetation in its varied forms and stages of growth. Considerable effort was invested in laboratory studies in the early and mid-1980s, financed by the German Research Foundation. However, the results have only limited practical value because the tests were performed using simulated bushes or tree trunks in rows, and in channels that were sufficiently long to produce uniform flow. The results are valuable in so far as semi-theoretical models could be developed and checked against them. These rather complicated models have been used to produce computer programs capable of predicting flow resistance under these particular conditions (Pasche 1984; Arnold *et al* 1987). However, their utility is quite limited in studies of actual rivers with groups of trees, irregularly located bushes (Fig. 22.6), patches of rushes, etc. in a channel that is anything but straight and uniform.

The importance of effects due to these types of 'roughness' is illustrated by results of model tests performed in connection with the reconditioning of the Enz, mentioned in Section 22.2 and shown in Figs 22.2 and 22.7. The two-stage channel with a width of about 90 m had a conveyance capacity of $800\,m^3\,s^{-1}$, far in excess of the design discharge of $450\,m^3\,s^{-1}$. The freeboard with this discharge was about 1 m in excess of that normally required. Hence, a large increase in flow resistance could be tolerated, even though the reconditioning was to take place only between the flood levees. However, just the process of reshaping the low-stage channel to create 'meanders' and a few islands raised the floodwater level by 0.4 m at the upstream end of the 1.3-km river section. Another 0.6 m of depth was shown to be required to allow for flow resistance caused by short rows of willows, essential for erosion protection, to be planted on alternate sides of the channel. The limited extent of mature willows to be tolerated was determined in the model tests, shown in Fig. 22.7.

The simulation of natural vegetation in model tests is difficult. It requires that a sufficient num-

Fig. 22.6 Group of willows resisting flow (the River Enz, $230\,m^3\,s^{-1}$).

ber of field measurements be made to provide calibrations at various discharges. The process could take decades. At present, bushes are simulated by models that essentially block the flow. This is a condition that occurs in nature if debris, brought by the flood flow, catches on the bushes. The advantage of physical model tests is that the model reproduces the combined effects of channel curvature, branching channels, patches of vegetation, etc. At the present state of mathematical modelling, such complexity cannot be accounted for.

The present approach to computing water levels is like that used in model testing: it is assumed that groups of trees or bushes block the flow, and the corresponding cross-sectional areas are subtracted from the flow section. This rather coarse and somewhat subjective method tends to render results that are 'on the safe side', i.e. they overestimate the height of the water level.

An extensive series of model tests with two-stage channels is in progress at the Theodor Rehbock Laboratory, University of Karlsruhe. Their aim is to improve the ability to account for natural flow resistance by providing data on which a Manning roughness coefficient can be estimated by the superposition of the various effects that contribute to the overall resistance. A similar approach is contained in Chow's book (1959) on open channel flow; however, the data on flow resistance due to vegetation are rather limited.

With two-stage channels, where the flood levees cannot be removed, or even displaced, as was the case for the Enz, restoration is essentially limited to measures between the levees. Normally these measures are severely constrained because the carrying capacity was accurately determined in the original design to meet the prescribed flood discharge. In this respect, the Enz, with its excess capacity, was an exception. If the flood levees are not raised, a measure that conflicts with renaturalization, the only way to achieve additional carrying capacity is to lower the floodplains. Increased cross-sectional area, and thus decreased average water velocities, can provide for reshaping the low-stage channel and for limited planting of vegetation whilst still maintaining the necessary capacity to carry the design discharge.

Restoration and renaturalization, for areas not restricted by flood levees, must provide ample areas to carry the flow. The discharge computations, required to show that the watercourse can convey the design flood, must take account of a

Fig. 22.7 Model (scale 1 : 40) of a section of the River Enz.

Fig. 22.8 Map of Baden-Württemberg.

future state with fully grown vegetation. Subsequently, maintenance may require thinning of the growth to maintain carrying capacity.

22.6 CASE-STUDY: PLANNING THE RESTORATION OF THE DANUBE IN BADEN-WÜRTTEMBERG

The Baden-Württemberg section of the Danube is an illuminating example of the planning of a river corridor restoration. The Danube originates in the Black Forest (Fig. 22.8). Two tributaries, the Brig and the Bregach, join at the town of Donaueschingen; their junction is the beginning of the Danube. It flows through the Swabian Alb in a deep and narrow valley, geologically a karst region. In this section seepage causes a substantial loss of discharge, a flow that returns to the surface within the catchment of the River Rhine. The Danube then flows along the edge of the Swabian Alb and passes the Bavarian border at the city of Ulm. The distance along the river between Donaueschingen and Ulm is 160 km.

The geomorphology is rather complicated because successive glacial periods have left their marks on the landscape: ancient fluvial systems have eroded the higher ground and transported the sediments down to the valleys. The catchment of the Danube was once much larger and had a much higher sediment yield.

After leaving the Jurassic formation of the Swabian Alb, the river enters rather soft layers of molasses, a fossiliferous sediment found in and near Switzerland. At the edge of the Swabian Alb it again hits resistant Jurassic formations. These almost entirely prevent erosion and lead to gradients of about 0.001. In this section the morphology of the river is similar to that of lowland river; also similar are the widths of the valley and extent of the floodplains.

Figure 22.9 is a map of a section of the Danube, 7 km long, near the town of Blochingen. The path of the river in 1840 and the channel of today are shown, as are numerous traces of oxbows in the floodplain area. Regulation works began in the middle of the 19th century, and they included the cutting off of all meanders along the edge of the Swabian Alb. These changes shortened the river channel by about 20% and increased the gradient accordingly. As a result of the higher gradient, erosion has lowered the bed by as much as 2.5 m in some places during the past 90 years. The water levels dropped by the same amount, and this in turn lowered the groundwater table. The few wetland habitats that had escaped cultivation were thereby further endangered.

Confronted with this undesirable outcome, in 1988 the water authorities embarked on a restoration programme that has gradually been extended to include the entire stretch of the Danube within Baden-Württemberg. The programme is in various stages of completion and includes the implementation of a pilot section. Planning was initiated by selecting a 40-km

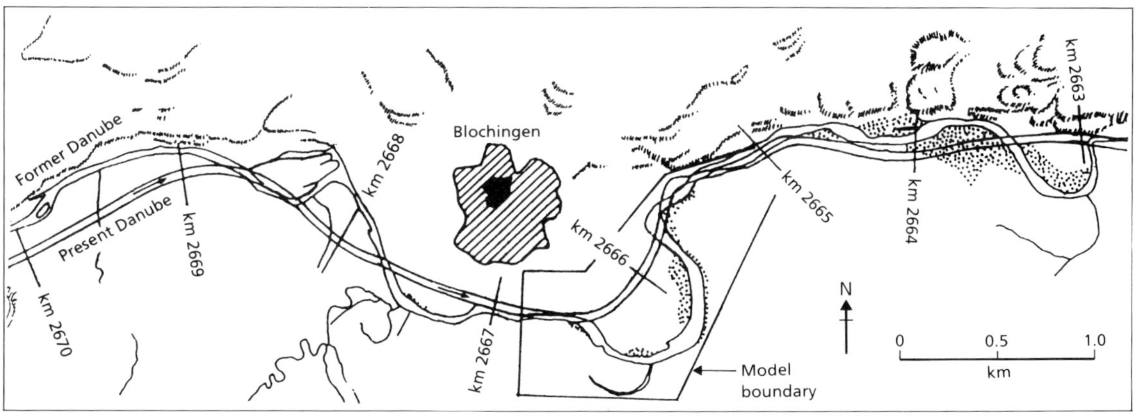

Fig. 22.9 Old map of a stretch of the Danube near the town of Blochingen.

Table 22.1 Project planning schedule for the restoration project of a 40-km section of the Danube (after Kern & Schramm 1988)

			Investigation phase			Planning phase	Implementation phase
Investigations, surveys for planning basis	pre-investigation (p-i)	p-i 1	*Survey and evaluation*	(s-e)	*Model image* (m-i)		
			River morphology	s-e 1	'Old Danube'		
		p-i 2	Landscape ecology	s-e 2	m-i 1		
		p-i 3	Water resources management	s-e 3	Deficiencies and restrictions		
			Limnology	s-e 4	m-i 2		
Planning		p-i 4 p-i 5	*pilot project* p-p 1 p-p 2 p-p 3 p-p 4	special investigations s-e 5	Concretization of planning goals m-i 3	*Master planning* m-p m-p 1	*Planning of specific measures*
Implementation	Purchase of land, preferably at the river, former river beds, oxbows, wetlands. Steps toward extensive agriculture, particularly in flood-prone areas						
				Pilot project p-p 5			*Implementation of specific measures, supervision*
Monitoring of development			*Pilot project* monitoring of development p-p 6		*Monitoring of development*. Observation of changes in river morphology, flow regime, groundwater levels, flora and fauna. Survey and data collection of initial conditions after implementation. Possible maintenance measures		

Explanations to abbreviations included in Tables 22.2 to 22.6.

section between the towns of Sigmaringen and Zwiefaltendorf and assignments were given to various bureaus and institutes. The Institute of Hydraulic Structures and Agricultural Engineering, University of Karlsruhe, was commissioned to collect the available data and propose a detailed planning procedure for the restoration of the Danube corridor. This proposal resulted in specialist studies listed below. The survey and its partial evaluation are included in a report of 130 pages, of which 50 pages are a record of source material used and information on where it is available. The report essentially follows the outline presented in Section 22.4. The proposed planning schedule included may serve as a condensed summary of the report. It lists the various project phases and the activities to be executed in chronological order (Tables 22.1 to 22.6).

Table 22.2 Pre-investigation (p-i)

p-i 1
Compilation of data on water resources management and landscape ecology

p-i 2
Pre-evaluation focusing on river morphology

p-i 3
Task identification; statement of planning goals

p-i 4
Consideration of possible solutions for proper river morphology

p-i 5
Structuring of planning activities

Table 22.3 Pilot project (p-p)

p-p 1
Water resources management and ecological surveys

p-p 2
Project planning; consideration of pilot-project goals

p-p 3
Hydraulic model tests to verify and optimize the design

p-p 4
Presentation of the pilot project to the public, including visit to the hydraulic model

p-p 5
Statutory permits and execution

p-p 6
Development monitoring

Table 22.4 Survey and evaluation (s-e)

s-e 1: River morphology
Survey of the river bed: size distribution of sediment, strata, erodibility, natural armouring
Investigation of sediment budget: sediment transport, sediment balance
Morphological development; information from old maps, etc.
Adoption of sediment transport model: calibration, range of applicability

s-e 2: Landscape ecology
Mapping of present and previous usage of the corridor
Mapping and evaluation of biotopes: wetlands, drylands, other biotopes
Mapping and evaluation of tributaries: morphological features, sediment delivery, water discharge
Mapping of stand characteristics: soil properties, humidity, climate, potential for natural vegetation
Changes in stand properties due to human activities: water management, artificial fertility, reversibility

s-e 3: Water resources management
Investigation of groundwater management; limits on water level
Investigations of flooding and flood protection
Establishment of a register of sewage and waste-water discharges

s-e 4: Limnology
Survey of the Danube, its tributaries and still waters
Chemical and physical parameters
Fish fauna, invertebrates, macrophytes

s-e 5: Special investigations
Potential for re-establishing plant societies
Biotope requirements of (especially) endangered species: grassland birds, fishes, amphibians

The following investigations were included in the survey and evaluation phase concerning the 40-km section:
- analysis of the historical river morphology;
- mapping of riparian vegetation and of ecology of riverbanks;
- physical and chemical parameters;
- insect larvae and fish species;
- weirs and drop structures in respect of migration;
- geohydrological survey;

Table 22.5 Model image: 'old Danube' (m-i)

m-i 1: Description of the old Danube landscape as it appeared at the beginning of the 19th century: river morphology and ecological aspects of landscape
- Watercourse system
- Uses of land and water
- Biotope conditions
- Riverine dynamics
- Assumed species diversity
- Landscape structure

m-i 2: Deficiencies and restrictions
- Deficiencies in comparison with present conditions
- Restrictions due to present usage
- Changes that cannot be reversed

m-i 3: Confirmation of planning goals
- Landscape ecological and landscape planning goals
- Goals for water resources management and river morphology

Table 22.6 Master planning (m-p)

m-p 1: Short, intermediate and long-term measures
- Restoration of river morphology along the entire Danube section (specification of measures to be carried out)
- Restoration of former river courses
- Regeneration of riverine woodlands, typical of the old Danube
- Biotope development and interconnection for biotopic continuity
- Landscape improvements

- vegetation and land use of the former floodplain;
- selected groups of insects and birds;
- hydrological and hydraulic studies for flood protection.

The last point was added after a devastating flood ravaged the upper Danube in February 1990.

A pilot project, to be implemented in advance, is incorporated in the schedule, and is described in more detail in the following section.

Tables 22.1 to 22.6 constitute a list of activities and measures to be taken throughout the project development, and they are ordered chronologically. This compilation may well serve as a checklist for any river corridor restoration project, with omissions and additions as appropriate in specific cases, of course.

The pilot project

The Institute was asked to prepare a design proposal for a 1.4-km subsection of the Danube and to validate its hydraulic performance through model tests in the Theodor Rehbock Laboratory. The planning of this project, situated downstream of the town of Blochingen (Fig. 22.9), is presented in more detail. Figure 22.9 shows that the short-cutting of a large bend in 1874 shortened the Danube substantially. An aerial view of that section revealed remains of the old channel, partly lined by a row of trees. Only extreme floods, like the one in 1990, show the depressions in the floodplain that mark the former channels. The land between the old channel and the present one has been cultivated, but with little success because of the poor soil – coarse, very permeable fluvial sediments. The high permeability and the low level of the groundwater, lowered as the river eroded, rendered this land quite dry.

What would the 'model image' of this section look like? Morphologically, the soil would pose little resistance to bank erosion; thus the channel could readily shift laterally. Old maps substantiated this trend: once, the channel had moved 100 m in only 30 years. Extreme floods caused short-cuts to occur, leaving oxbows behind, and these eventually were filled in either by siltation or by human activity. In the model, therefore, the river bed would be rather broad with gravel bars and possibly split by islands. The channel would probably carry little more than the annual flood. The water surface and the groundwater level would vary within a limited range just below the level of the floodplain. Nevertheless, the habitats adjacent to the river would be rather dry, except in former oxbows, because of very permeable sediments. The floodplain and the riverbanks would be forested.

What irreversible damage has the river system suffered? What restrictions must be considered in developing a restoration scheme for this section of the Danube? Not much is known about sediment transport in the 'old Danube'. The banks and many of the tributaries have been protected against erosion, and weirs for water-power plants located upstream retain sediments artificially. Consequently, it is reasonable to assume that

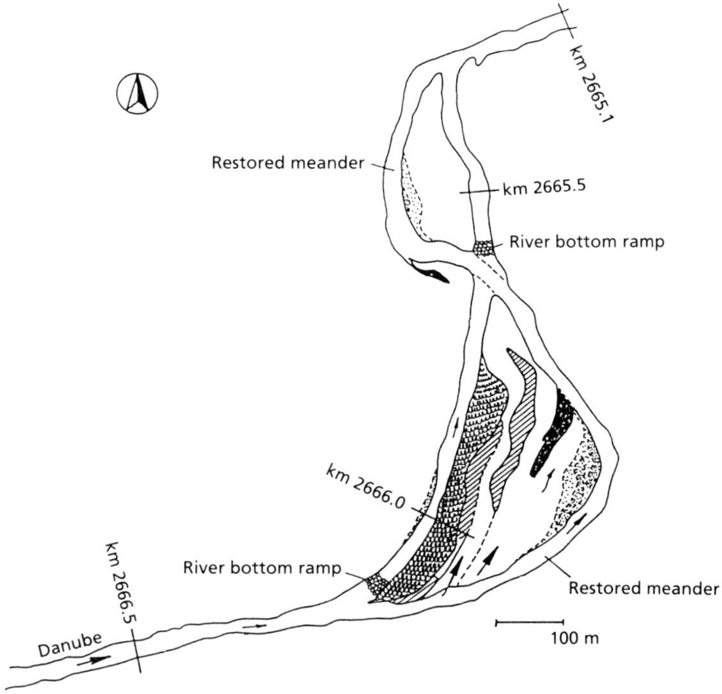

Fig. 22.10 Pilot project layout, adopted solution.

sediment transport has been substantially reduced.

Some formerly endemic fish species are now extinct; other species have been introduced. Water quality is satisfactory, classified as 'less polluted'. Flood protection for the nearby village of Blochingen has to be considered: present flood levels must not be raised.

The solution that was adopted is a river diversion that reactivates the old channel (Fig. 22.10). This included two migratable drop structures in the existing river channel, one at each diversion. The two islands thus formed and the adjacent floodplains will be landscaped with ridges and depressions. No bank protection is to be provided along the new channel and part of that along the existing channel will be removed and used for constructing the gently sloping drop structures.

The hydraulics of this concept were optimized in a model study performed at a scale ratio of 1 to 40 in the Theodor Rehbock Laboratory. The modification raised the water level at mean flow ($24\,m^3\,s^{-1}$) by almost 1 m. The bed had been lowered through erosion by between 1.1 and 1.3 m in this section of the river and by up to 2.5 m farther downstream. Although the water levels were raised up to a flow of $160\,m^3\,s^{-1}$, the design permits high flows to pass with lower levels than before. The stage for the 100-year flood ($400\,m^3\,s^{-1}$) will be 0.3–0.4 m lower at the village of Blochingen. The improved flood conditions helped convince the public to view the project favourably.

The existing (artificial) channel will be used mainly for flood relief; about 75% of the 100-year flood will flow in it. In contrast, the mean flow will pass almost entirely through the restored meanders.

What will be achieved?

- Restoration of the former average water level and unrestricted morphological dynamics in the new channel.
- Raising of water levels for flows of up to $160\,m^3\,s^{-1}$, which corresponds to the average annual flood.
- Relief for floods higher than the annual flood.
- Channel banks will be able to erode freely; a

strip of land 100 m wide along the outer bank of the meander has been purchased to allow for later movement.

- Control of bed erosion (as deduced from velocities derived from the model study).
- Raising of groundwater levels in riparian areas.
- Improvement of floodplain habitats.
- Improvement of aquatic habitats.

What concessions must be made?

- Drop structures are not typical for the Danube in this section.
- The habitat of the present channel will correspond neither to that in the still water at the oxbows nor to that of free flow of a channel.
- Recolonization of vegetation and fauna will necessarily take place under present-day conditions, rather than under the original ones.
- Costs for restoration are high: the estimated cost for this 1.3-km section is 1.3 million DM.

The project, implemented in 1991, will also be used as a full-scale test model. It includes intensive surveys of the morphological and ecological developments. The experience from this project will serve as a guide to further restoration projects, on the Danube and elsewhere. The government has allocated a budget of 100 million DM for the entire 160 km of the Danube within Baden-Württemberg, and the planning of restoration measures is now in progress.

REFERENCES

Arnold U, Gabrbrecht G, Pasche E *et al.* (1987) *Hydraulische Probleme beim naturnahen Gewässerausbau.* Forschungsbericht DFG, VCH Verlagsgesellschaft mbH, Weinheim Germany. [22.5]

Binder W. (1984) Der Lech zwischen Landsberg und Augsburg – Umgestaltung einer süddeutchen Flusslandschaft in den letzten 100 Jahren. In: *Fluss und Lebensraum.* Verlag Paul Parey. [22.2]

Bostelmann R, Braukmann U, Briem E, Humborg G, Nadolny I, Steib K, Weibel U. (1993) *Forschungsgruppe Fliessgewässer 1993. Fliessgewässertypologie – interdisziplinäre Studien an naturnahen Gewässern und Auen in Baden-Württemberg mit Schwerpunkt Buntsandstein-Odenwald und Oberrheinehene.* Reihe Umweltforschung Baden-Württemberg, Ecomed Verlag, Landsberg am Lech, Germany, p. 226. [22.2]

Bürkle F, Arnold O, Buck H *et al.* (1985) *Ökologische Untersuchungen an der unteren Murr, Landkreis Ludwigsburg 1977–1982.* Landesanstalt für Umweltsschutz Baden-Württemberg, Karlsruhe. [22.2]

Chow VT. (1959) *Open-Channel Hydraulics.* McGraw-Hill, New York. [22.5]

Dister E, Gomer D, Obrdlik P, Petermann P, Schneider E. (1990) Water management and ecological perspectives of the Upper Rhine's floodplains. *Regulated Rivers, Research and Management* **5**(1): 1–15. [22.2]

Gebler R. (1991) *Naturnahe Bauweisen zur Vernetzung von Fliessgewässern.* Mitteilungen Heft 181, Institut für Wasserbau und Kulturtechnik, University of Karlsruhe. [22.2]

Hey RD, Heritage GL, Patteson M. (1991) *Design of Flood Alleviation Schemes: Engineering and The Environment.* Ministry of Agriculture, Fisheries and Food, London. [22.2]

Kern K, Nadolny I. (1986) *Naturnahe Umgestaltung ausgebauter Fliessgewässer – Projectstudie.* Mitteilungen Heft 175, Institut für Wasserbau und Kulturtechnik, University of Karlsruhe. [22.2]

Kern K, Schramm M. (1988) Sanierung der Donau zwischen Sigmaringen und Zwiefaltendorf. Institut für Wasserbau und Kulturtechnik, University of Karlsruhe (unpublished report to the Ministy of Environment, Stuttgart). [22.6]

Kern K, Anselm R, Bürkle F *et al.* (1986) *Naturnahe Umgestaltung ausgebauter Fliessgewässer – Beiträge zum wasserbaulichen Kolloquium am 14. Febr., 1986, Karlsruhe.* Mitteilungen Heft 174, Institut für Wasserbau und Kulturtechnik, University of Karlsruhe. [22.3]

Ministerium für Umwelt. (1992) *Handbuch Wasserbau, Naturnahe Umgestaltung von Fliessgewässern.* Heft 2: Teil I: Leitfaden, Teil II: Dokumentation ausgeführter Projekte. Ministerium für Umwelt, Baden-Württemberg. [22.2]

Pasche E. (1984) *Turbulenzmechanismen in naturnahen Fliessgewässern und die Möglichkeit ihrer mathematischen Erfassung.* Mitteilungen Nr. 52, Institut für Wasserbau und Wasserwirtschaft, TH Aachen. [22.5]

Part 5: Case Studies

The *Rivers Handbook* has presented a scientific basis for river management with particular reference to temperate environments although many of the underlying principles may be applicable to other geographical regions. This final part of the *Handbook* presents an example of the progress that has been made in the development of an ecological approach to river management: the Upper Mississippi (Chapter 23). In order to contrast the problems and needs of temperate rivers with those of other regions, Chapters 24 and 25 consider rivers in cold environments and drylands, respectively. There is no doubt that whilst the immediate task for river management in the temperate zone involves rehabilitation and restoration, in other environments protection and mitigation are the priority concerns. But in all cases, further scientific advances are needed to improve our ability to evaluate the short- and long-term consequences of our actions in developing the earth's land and water resources.

23: Management of the Upper Mississippi: A Case History

J. L. RASMUSSEN

23.1 INTRODUCTION

Human exploitation of the Upper Mississippi River as a waterway for transportation, trade and settlement predates recorded history. Europeans arrived on the continent in the late 1400s, and by the mid-1600s England, France and Spain were engaged in a struggle to control the New World, and to find a profitable water route to the Orient. The Joliet-Marquette journals of 1673 provide one of the best early documentations of Mississippi River exploration (Petersen 1968a). This French expedition entered the river from the Great Lakes by way of Green Bay, the Fox River and the Wisconsin River, returning to Lake Michigan by way of the Illinois River (Fig. 23.1).

Fur was the first resource to attract European explorers and exploiters. The French and English played the greatest roles in these early ventures, and by the mid-1700s the French were attempting to draw fur trade away from English settlements on the Great Lakes by establishing trading houses (Fig. 23.1) at Lake Pepin, Prairie du Chien, St Louis and St Genevieve (Rahn 1983). France ceded its colonial possessions and control of the fur trade to Britain (east of the Mississippi) and Spain (west of the Mississippi) after the French and Indian War.

Although the French ploughed no land and cleared no forests, they had seduced the Indians into depending on European trade goods (including whiskey), and taught them to strip the river of its furbearers rather than harvest only enough for their immediate needs (Davis 1982). The British followed the French lead, and in the words of one British trader: 'When a Nation becomes addicted to drinking, it affords a strong presumption that they will become excellent hunters – in order to satisfy their craving' (Davis 1982).

The American colonies began expanding westwards after winning independence from England in 1776. The first, and perhaps most far-reaching, document to impact the Upper Mississippi River development was the Northwest Ordinance (1787). This document laid down the principles of the American colonial system and dealt with settlement of United States' territory northwest of the Ohio River. These 'Northwest Territories', bounded on the west by the Upper Mississippi (Fig. 23.1), included all or portions of the present States of Minnesota, Wisconsin, Illinois, Indiana, Michigan and Ohio. The Ordinance states that:

> The navigable waters leading into the Mississippi and St. Lawrence, and the carrying places between the same, shall be common highways and forever free, as well to the inhabitants of the said territory as to the citizens of the United States, and those of any other States that may be admitted into the confederacy, without any tax, impost, or duty therefore.

The Ordinance further states that it was intended '. . . to fix and establish those principles as the basis of all laws, constitutions, and governments, which forever hereafter shall be formed in the said territory.' The Ordinance thus formed the basis for the strong foothold and long-lasting influence commercial navigation interests have had over Mississippi River management and politics. In fact, even to this day, the Ordinance plays a major role in legal matters regarding River navigation.

Spanish policies towards the River were fickle, designed principally to stem the tide of American

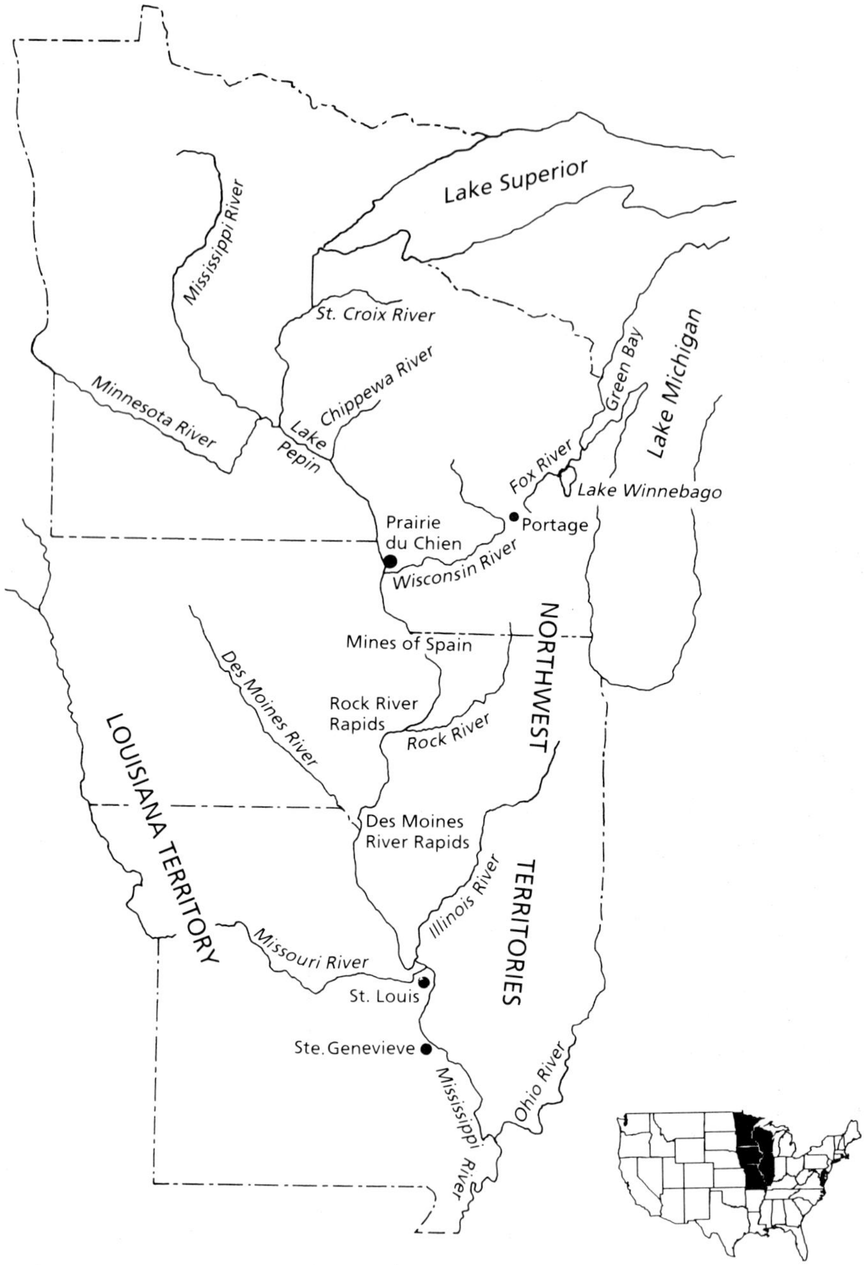

Fig. 23.1 Upper Mississippi River, major tributaries, and French fur-trading houses and routes of the 18th and 19th centuries shown within current State boundaries.

settlement (Rudd 1981), and to locate Eldorado. However, with their sale of the Louisiana Territory to America in 1803, Spanish influence on the Upper River waned. America thus gained control of both Mississippi River banks, and 'river improvement' was about to begin.

This chapter provides a brief overview of major cultural and developmental events (Table 23.1)

Table 23.1 Major cultural and developmental events in the history of Upper Mississippi River resource management

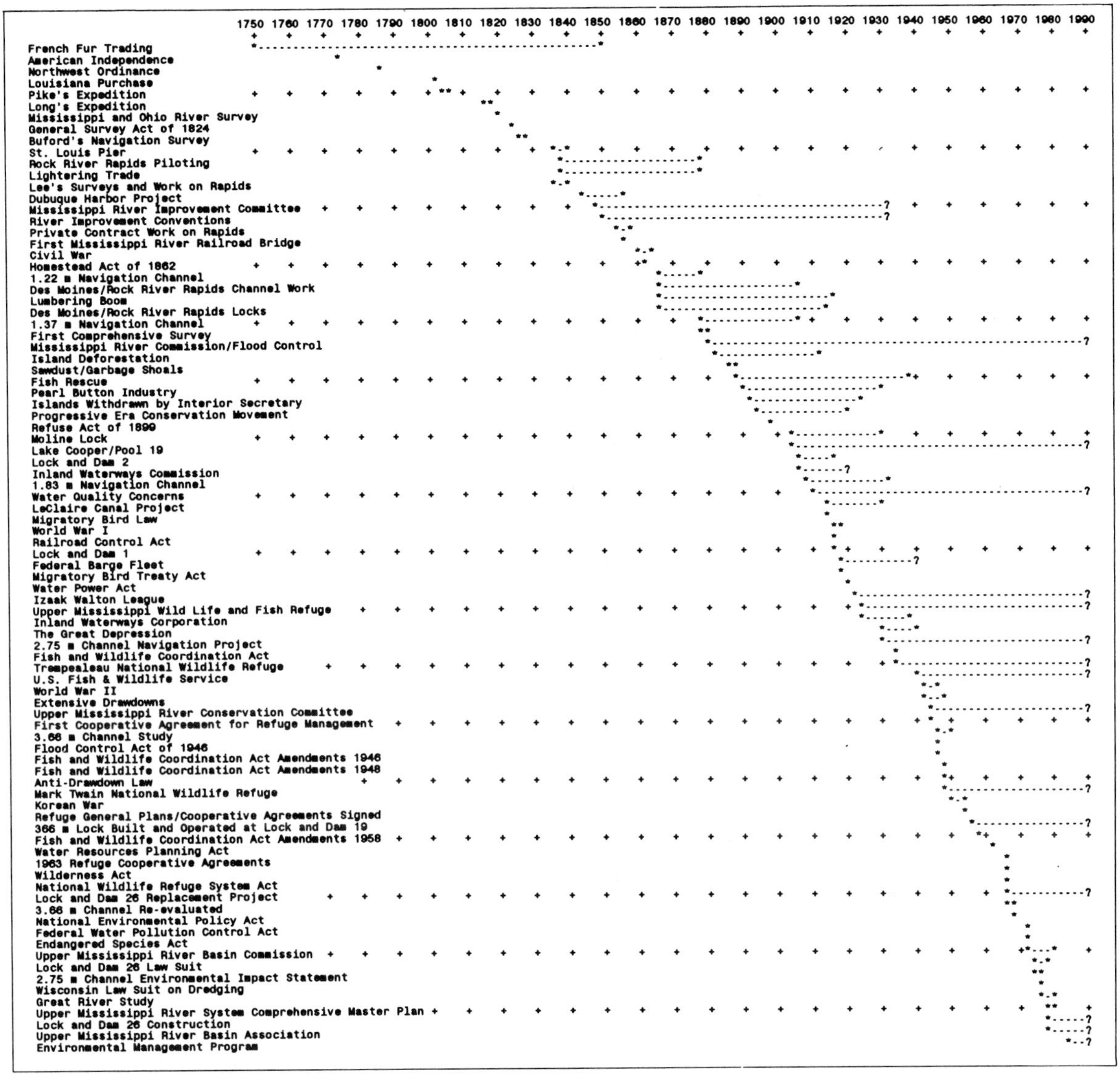

that followed, and the evolution of Upper Mississippi River politics and management for navigation, commerce, conservation and environmental awareness.

23.2 THE NATURAL RIVER

In 1805, while Lewis and Clark were exploring the Missouri River (Fig. 23.1) and the Louisiana Territory, Lieutenant Zebulon Pike led an expedition up the Mississippi to map its course, look for military sites, procure mineral and plant specimens, establish Indian contacts, and establish American authority over the fur trade.

Pike found the Des Moines River rapids (Fig. 23.1) most difficult to navigate: 'The river all the way through is 3–4 miles wide. The rapids are 11 miles long, with successive ridges and shoals extending from shore to shore' (Petersen 1968b).

He found the Rock River rapids (Fig. 23.1) less difficult: 'These shoals are a continued chain of rocks, extending in some places from shore to shore, about 18 miles in length. They afford more water than those of De Moyen, but are much more rapid.'

Pike met with Julien Dubuque at the 'Mines of Spain' (Fig. 23.1) near the future Wisconsin–Illinois border (and city of Dubuque) to inquire about the legality of his grant to these important lead mines. At the time the mines were producing 9–18 000 kg of lead annually (Petersen 1968b). A decade later Major Stephen H. Long sketched the river's course and examined it for defence sites, and by 1820 the Corps of Engineers (Corps) were examining both the Ohio and Mississippi rivers for navigation improvements (Tweet 1984).

In the late 1820s Lieutenant Napoleon B. Buford found the Upper Mississippi navigable for 8 months, except at the Des Moines and Rock River rapids, where he recommended alternatives for improvements. 'This early emphasis on transportation helped formulate the long-lasting Corps' policy of single-purpose improvement projects on the Mississippi' (Tweet 1984).

23.3 CATCHMENT CHANGE: SETTLEMENT, AGRICULTURE, LOGGING AND INDUSTRIALIZATION

The fur trade faded by the mid-1800s, not for lack of fur, but for a much greater interest in land development and harvest of the vast Wisconsin and Minnesota pine forests (Davis 1982). Settlement of the five Upper Mississippi River States proceeded rapidly, and by the time of the Homestead Act (1862), little Illinois land and less than 2.5 million ha of Iowa land were acquired under its provisions (Rahn 1983). Native Americans (representing at least 14 different tribes) were all but gone, and the plough and 'modern' agriculture had begun to reshape the prairie landscape and its rivers through subsequent erosion and sedimentation.

Public sentiment was pro-railroad in 1856 when the first Mississippi River bridge opened at Rock Island, Illinois (Fig. 23.2), but rivermen complained bitterly about its location as a navigation obstruction. Within months the first of many steamboats hit the bridge, and a lawsuit followed involving Abe Lincoln, but no conclusion as to fault was ever reached (Tweet 1984). To this day, however, bridges are considered obstructions to navigation and are maintained by their owners at no expense to the navigation industry for any towboat/bridge collisions.

After the Civil War, the Mississippi's obvious importance as a tool in reunifying North and South, brought a flood of immigration, industry and manufacturing. Further impetus for channel improvement came with low water in 1864 when navigation was almost completely suspended. That water level became the mark against which all future improvements have been measured (Tweet 1984).

The Des Moines River Rapids, the most serious obstruction, had weathered several inundation attempts by private individuals and groups to build dams for power and navigation. None was successful, and the lightering trade was booming, with costs to farmers, merchants and steamboatmen skyrocketing. By 1866 the Upper Mississippi River States were producing 30–50% of the nation's produce; and the lumber industry was coming into its own with more than 400 million board feet rafted downriver annually (Tweet 1984). Between 1875 and 1915 millions of board feet of logs and lumber were floated down river to markets and sawmills as far south as St Louis. Logging impacted the economy, growth of river towns, shape and size of river bridges, and Corps improvements.

In determining the curve of a river bend or shape of a channel crossing, Corps' engineers paid attention to both the manoeuvrability of a 92-m packet, and to the larger, far less manoeuvrable log raft (Tweet 1984). The largest log raft ever on the Mississippi was 82 m wide by 473 m long (Davis 1982). Loose logs floated down tributaries and into Mississippi River sloughs or chutes also caused problems for navigation, especially near the mouth of the Chippewa River (Fig. 23.2).

Mediating between packet and logging interests was a touchy business for the Corps (Tweet 1984). The 1900 River and Harbor Bill stated 'that it shall not be lawful to float loose timber and logs, or to float what is known as sack rafts of timber

Fig. 23.2 Upper Mississippi River showing towns and features of significance to the 1.22, 1.37 and 1.83-m navigation projects and creation of the Upper Mississippi River National Wild Life and Fish Refuge.

and logs in streams or channels actually navigated by steamboats in such a manner as to obstruct, impede, or endanger navigation' (Tweet 1984). The Corps had sympathy for lumbering interests, but not necessarily friendly relations. When the Corps refused to dredge a blocked opening in Beef Slough, near the Chippewa River, the Weyerhauser Logging Company responded by blowing up a Government dam across the entrance to West Newton Slough in order to use that area to tie up logs. The Company was taken to court, but the case was dropped after 2 years (Tweet 1984).

Most of western Minnesota's oak, maple, walnut and elm was simply put to the torch to clear the way for growing wheat (Davis 1982). This was encouraged by lumber companies, because they wanted farmers to quickly settle southern and western Minnesota in order to reduce the costs of shipping logging crew provisions in from far down river. Little reforestation occurred in the logged-over areas, and much of the land was simply abandoned by the lumber companies and allowed to revert to State ownership. The public supported such plunder because of their widely held belief that the nation's forests were inexhaustible (Davis 1982).

By the mid-1800s growth of interest-group-oriented politics reflected a national trend towards fragmentation of society (Scarpino 1985). Grain producers, dissatisfied with Congressional action to improve river commerce, formed the Mississippi River Improvement Committee in 1845 (Rahn 1983). By the 1860s cities along the river were routinely holding river improvement conventions to push for Federal money (Tweet 1984). Conventions held in 1865 (St Louis) and 1866 (Dubuque) called for immediate Congressional action to remove the Des Moines and Rock River obstructions. Concern was also expressed over the number and design of proposed bridges, limited river access to Eastern markets and the need for increased competition to lower rail rates (Tweet 1984).

In 1901 commercial clubs and chambers of commerce from St Louis to Minneapolis (Fig. 23.2) joined to form the Upper Mississippi River Improvement Association to pressure Congress to deepen the Mississippi River channel (Scarpino 1985). The National Association of Button manufacturers joined the US Bureau of Fisheries in 1908 to secure funds for artificial propagation of freshwater mussels, and donated land near Fairport, Iowa (Fig. 23.2) for the Bureau's Fairport Biological Station constructed in 1912. Mississippi River mussels had become the mainstay of America's pearl button industry. Such politics continue to the present day.

23.4 RIVER ENGINEERING

1824–65

A Board of Engineers was created by the General Survey Act of 1824 to determine the most suitable methods of improving the Ohio and Mississippi rivers. The Board advertised in newspapers throughout the country offering a $1000 reward for 'the plan, machines, or instrument best calculated to remove sawyers, planters, and snags in the Mississippi River, so as to render navigation free and safe' (Tweet 1984). Numerous ideas and plans were submitted, and one even suggested that the money for removing obstructions be raised by placing a toll on steamboats using the waterway. The idea of a 'user fee' to defray maintenance costs (outlawed by the Northwest Ordinance) was long before its time, but would resurface in the form of a fuel tax more than 150 years later.

Shoaling in the St Louis area in 1835 threatened to cut off river traffic, so Congress authorized construction of a 'pier' to deflect the current and scour the old channel (Tweet 1984). Plans for the pier also recommended improving Des Moines and Rock River Rapids navigation by stationing government-salaried pilots at each to guide boats through the narrow and shifting channels. This technique was employed at the Rock River Rapids, but through brisk private enterprise rather than government service (Tweet 1984).

Des Moines River Rapids navigation was more difficult, and there a channel marked with buoys to guide pilots was also recommended. Instead, though, the 'lighter trade' developed, employing small (6×31 m) horse-drawn flatboats large enough to relieve packetboats of cargo (Tweet 1984). Lightering was slow and expensive, costing about $1.10 per 1000 kg for upbound cargo and about $1.38 per 1000 kg for downbound.

Lieutenant Robert E. Lee came to the Upper Mississippi in 1837 to finish the St Louis pier and resurvey the Des Moines and Rock River Rapids. Funding for Lee's work was piecemeal and unsuccessful, and his report to Congress stressed the need for abandoning the small appropriations process in favour of a larger project (Tweet 1984). Lee also pointed out the importance of improving the rapids to open the upper midwest to settlement (Scarpino 1985).

Work began on both rapids in 1854, under private contract, but within a year westward railroad expansion began competing for both contractors and labour. Also, public interest in river improvement waned when shippers realized the railroads would guarantee direct access to Eastern markets, releasing their expensive ties to St Louis and New Orleans (Tweet 1984). The first Upper Mississippi River harbour dredging project (1844) was at Dubuque, Iowa (Fig. 23.2), near the Mines of Spain. It was designed to open a channel for large steamers from the city of Dubuque to the main River channel near the Illinois shore (Tweet 1984). By 1853 the project was expanded to include access to the contemplated terminus of the westerly branch of the Illinois Central Railroad.

The 1.22 and 1.37-m navigation channel

The beginnings of permanent navigation improvement came almost immediately in 1866, when Congress made appropriations for surveys of the Mississippi and its tributaries, with the understanding that a 1.22-m channel was to be an eventual goal (Tweet 1984).

Work also began in 1866 on the Des Moines and Rock River Rapids (Fig. 23.2) to enlarge the natural channel to 61 m × 1.22 m, and use excavated material in the river bed to check cross-currents and confine water to the main channel (Tweet 1984). Work progressed slowly, with deepening and widening of the worst reaches undertaken first, and continued until 1907, when the 1.83-m channel was authorized (Tweet 1984).

The Des Moines River Rapids project (constructed between 1870 and 1877) included a canal with two lift locks and one guard lock. It was built largely using hand labour and railroads for transportation. The locks measured 95 m × 24 m and were used until 1913, when they were flooded by Lake Cooper (Fig. 23.2), an impoundment created for hydro-electric power (Tweet 1984).

River improvement conventions, again pressuring Congress, won authorization of a 1.37-m low-water channel from St Paul to St Louis in 1878. This project changed Corps operations from many scattered duties to one unified project (Tweet 1984). It also generated the first comprehensive survey of the Mississippi from St Paul, Minnesota, to Grafton, Illinois (1878–79).

The Upper Mississippi's natural width in the late 1870s varied from 107 m in northern reaches to 427 m at the mouth of the Missouri River. Wing dykes had been used seasonally by lumbermen to improve the channel for log rafting. During low water these crude brush dykes, held in place by stakes, deflected water into the channel and helped wash out sand bars. The dykes themselves were, in turn, washed out by high seasonal water (Tweet 1984).

Wing dykes, of a similar but more permanent design, were used by the Corps to create a channel 366 m wide or less, constricting the channel as necessary to produce 1.37-m depths. Wing dykes, submerged during high water, became exposed as water levels dropped, confining the current and causing substrate in the channel to be loosened and carried away.

Development of the 1.37-m channel proceeded slowly, allowing the dykes time to scour the channel. This proved exceptionally economical since the dykes were made of alternate layers of willow mats cut from river islands and crushed or broken stone taken from the bluffs.

Willows used for wing dykes had to be recently cut and trimmed so that poles could be gathered into tight bundles or fascines 6 m long and 30–38 cm in diameter. Tied with lath yarn or wire at 1.2-m intervals, these bundles were lashed together to form mats 3.7 m long. They were then floated into position and sunk with rock hauled in by barge. In northern reaches they were also built in winter on the ice and allowed to sink in place with the spring thaw. The rock had to be of 15.24 cm^3 or its equivalent, but not larger than 25.4 cm^3 (Tweet 1984).

Although not their original purpose, today these wing dykes provide important fishery habitat when submerged. Revetments or bank protection was used to prevent shoreline erosion caused when wing dykes deflect sharp currents at the stream bank (Tweet 1984). When submerged, and used sparingly, these revetments can also provide important aquatic habitat.

Closing dykes, similar to wing dykes, were much more dangerous to build because of the swift currents produced when a channel is completely shut off.

Willow mats were used for dyke construction until 1911, when supply could no longer meet demand, and the islands were deforested. As early as 1881 the Corps had experimented with developing willow nurseries and reforesting sandbars with willow cuttings, but to no avail (Tweet 1984). Successful experiments were made using lumber mats instead of willow, and any handy wood (maple, elm, oak, ash, cottonwood and pine) was used. Economically, lumber mats compared favourably to willow, but did not conform as well to the substrate (Tweet 1984).

Some work was successful using sand wing dykes covered by a layer of brush and weighted with rock. Both willow and sand wing dykes proved amazingly permanent, even though often damaged by ice and passing boats. The willow submerged in the dykes did not rot, and even today, more than 100 years later, pieces cut from some of the willow will still float (Tweet 1984).

By the time the 1.83-m project was authorized, a total of 9 422 217 m^3 of brush had been put into 541.3 km of wing and closing dykes and 317.5 km of shoreline protection (Tweet 1984).

The Rock River Rapids problem resurfaced in the late 1800s when the rest of the channel was deepened for the 1.37-m channel. Consequently, Congress authorized construction of the Moline, Illinois Lock in 1905 (Tweet 1984). Completed in 1907, the project consisted of a lock and dam and dredged channel 76 m × 1.22 m.

Lock and Dam 2, between Minneapolis and St Paul, constructed to breach St Anthony Falls (a natural barrier to upstream fish movement), was also finished in 1907, but was inundated in 1917 by Lock and Dam 1, 'a high dam' or hydropower project (Tweet 1984).

The Corps and flood control

Average river width between the Ohio River confluence and St Louis increased from 1098 m in 1821 to 1716 m in 1888 (Simons *et al* 1975). There is some disagreement among hydrologists, but it is believed that much of this widening was caused by a combination of large floods which occurred between 1844 and 1888, floodplain development, and floodplain deforestation to provide fuel for river boats.

Congress established the Mississippi River Commission in 1879, marking the beginning of Corps involvement in flood control and channelization (Tweet 1984). The Commission was composed of three Army engineers, one member of the Coast and Geodetic Survey, two civil engineers, and one political appointee; the Chief of the Army Corps had veto power over Commission decisions (Davis 1982). The Commission's only job was to prescribe standard grades (i.e. specifications for levee height and width, based on previous flood heights). Commissioners recognized, but ignored, the fact that levees increase flood stage, and the river responds by setting new high-water records (Davis 1982). After each inundation the Commission reacted by simply raising the levee grade. The 'River and engineers were engaged in an escalating cold war' (Davis 1982). River width at St Louis was reduced from 1281 m in 1849 to 641 m by 1888 (Simons *et al* 1975).

A cataclysmic flood in 1927 caused the Corps to rethink their 'levees only' policy and they began to include other measures such as cut-offs and floodways in their projects (Davis 1982). By the 1930s the Mississippi River between St Louis and the Gulf had been shortened by 370 km through elimination of cut-offs and levee construction. Since that time the Corps has constructed numerous flood control reservoirs on Mississippi River tributaries, and in the 1970s even moved some small towns to the uplands, reclaiming some of the River's floodplain.

Today, the Corps works extensively with local municipalities and individual levee and drainage districts throughout the Upper Mississippi, protecting cities, factories and farms against 50- and 100-year floods. This process has effectively isolated the Upper Mississippi downstream of Dubuque, Iowa (Fig. 23.2) from most of its natural floodplain; a project feature which came back to haunt the Corps during the 1993 floods. Many of these levees broke under the force of the flood, causing extensive damage to 'protected' homes and land. A federally sponsored effort is currently underway to rethink national floodplain management policy, including the purchase of levee districts in order to partially reconnect the river with its floodplain and provide enhanced flood storage capacity.

The Mississippi River Commission still exists, but acts in an advisory capacity for flood control and navigation projects only on the

Lower Mississippi (downstream from St Louis). Each Corps District on the upper river works independently with individual flood control applicants.

The Inland Waterways Commission and the 1.83-m channel

High water experienced in the 1880s, followed by a decade of low water created navigation problems. As the 1.37-m channel neared completion, navigation interests and Congress felt the shift in river traffic to deeper draft barges, and increased competition from railroads, dictated the need for an even deeper channel. Additional impetus came when the railroads were overburdened with the record 1906 crop (Tweet 1984).

The 1906 St Louis River Convention recommended the need for a commission to draw up a comprehensive, basin-wide improvement plan for inland waterways (Tweet 1984). President Theodore Roosevelt formed such a Commission in 1907, composed of representatives from the Corps, conservation groups, and others interested in river planning.

In establishing the Commission he wrote: 'It is not possible to properly frame so large a plan as this for the control of our rivers without taking account of the orderly development of other natural resources' (Wengert 1957). This multipurpose plan, the first of its kind, was part of Roosevelt's Progressive Era Conservation Movement. But as he spoke, the River and Harbor Act of 1907 had already authorized a Mississippi River 1.83-m navigation project, thus 'end running' the planning process. The 1.83-m channel was to be constructed using 2000 wing dykes (31–92 m long), 209 km of bank revetment, dredging, and locks at Keokuk and Rock Island to breach both rapids (Tweet 1984).

The Keokuk (Iowa) and Hamilton (Illinois) Water Power Company intervened in these plans by obtaining a perpetual franchise (1905) to construct a power dam at the foot of the Des Moines River Rapids (Scarpino 1985). This project, completed in 1913, solved a major part of the Corps' 1.83-m channel development problem. Called Lake Cooper (Fig. 23.2), the project provided a high dam (12.2 m) with a single navigation lock. Thus the Des Moines River Rapids as a navigation obstruction and smallmouth bass fishery was gone forever, and a serious shift in direction for navigation improvement had begun (Tweet 1984). Until this time the entire length of the Mississippi River, below St Anthony Falls at Minneapolis, had been free of dams.

The new high dam at Keokuk provided the 'mystery of electricity' to the region, creating a host of greed-related social problems and a new problem for navigation (Scarpino 1985). In 1908 the Corps promulgated a ruling that allowed the power company to store water at night to meet early-morning power demands. This caused numerous downstream towboat groundings. Navigation soon regained supremacy, and the grounding problem was eliminated when the Corps disallowed such overnight water storage. In the process, however, a new navigation benefit was found – if a commercial boat became grounded downstream, the dam's gates could be opened for an hour to give the boat a flash flood of 0.8 m of water, flushing her off the bar (Tweet 1984; Scarpino 1985).

Once the Des Moines Rapids was breached, the Rock River Rapids became the Corps' most difficult problem in creating the 1.83-m channel. The portion of Rock River Rapids improved by the Moline Lock was the only reliable reach for navigation (Tweet 1984). The LeClaire Canal project, at the upper end of the Rock River Rapids, authorized in 1914, provided a 1.83-m channel from LeClaire, Iowa, to the Hampton (Illinois) Pool (Fig. 23.2). Interrupted by World War I, however, the LeClaire Canal and its 24 × 37 m lock chamber were not finished until 1922 (Tweet 1984).

The Federal Barge Fleet and the 2.75-m channel

Commercial use of the Mississippi did not increase as much as hoped after World War I. This was largely due to a lag in adapting boat design to the 1.83-m channel and lack of rail-to-river terminals (Tweet 1984). The federal government led the way in revitalizing traffic when Congress, as part of the Railroad Control Act of 1917, authorized a Federal Barge Fleet. The fleet, organized and operated by the Government, started service in 1918 on the Lower Mississippi (Robinson 1980).

Federal involvement expanded in 1924 when Congress authorized formation of the Inland Waterways Corporation to encourage increased use of commercial shipping on the Mississippi from New Orleans to St Louis. At that time, not one common carrier barge line had begun operations on the River. The Inland Waterways Corporation was to help stimulate such operations by demonstrating the economic feasibility of inland waterway transportation (Tweet 1984).

Upper Mississippi interests were not long in requesting similar federal aid. By 1926 Minneapolis business groups had achieved agreement with the Inland Waterways Corporation to operate a towing fleet on the Upper River. This traffic stimulated development of terminals (Fig. 23.2) at Burlington, Dubuque and Rock Island by 1928 (Tweet 1984).

Strong pressure for a 2.75-m channel, similar to that already authorized and nearing completion on the Ohio River, came from the waterway industry (Fugina 1945). Congress authorized a study of the channel's economic feasibility in 1927. The study did not find the project (requiring massive construction of locks and dams) economically warranted, and Chief of Engineers General Edgar Jadwin concurred (Tweet 1984). Jadwin was supported by conservation and recreation groups such as the Izaak Walton League of America ('Ikes') who were concerned about the welfare of the recently authorized Upper Mississippi River National Wild Life and Fish Refuge.

Conservationists opposing the navigation project claimed it would 'completely destroy bass fishing' (Fairchild 1982). The Bureau of Biological Survey, ambivalent about the project, recognized that most refuge land would be flooded, destroying existing wildlife habitat and timber, but also saw the opportunity to greatly increase refuge wildlife value if the Corps was co-operative (Fairchild 1982).

The Corps recognized wildlife damage done by locks and dams installed at Hastings, Minnesota (Fig. 23.2), and Keokuk, Iowa. The pools formed behind these dams slowed flowage, caused silting and accumulation of pollutants, and decreased dissolved oxygen levels (Fairchild 1982). In addition, because those locks and dams were not equipped to facilitate fish movement, as many as 12 species of migratory fish were being threatened (Carlander 1954).

Project supporters criticized the Corps report, and in 1929 when Herbert Hoover, a native Iowan, became President, General Jadwin was quickly replaced with a new Chief of Engineers more favourable to the project. The 2.75-m channel was authorized by Congress in 1930 without final examination of Corps' surveys, and despite feelings of concern within the Corps itself (Tweet 1984). Victory in winning project authorization was shared by President Hoover, his Secretary of War, the Mississippi Valley Association, the Minneapolis Real Estate Board, the Mississippi and St Croix River Improvement Commission and congressional representatives from Minnesota (Merrit 1980).

The 1.83-m project was only 82% complete when Congress authorized the new 2.75-m channel (Tweet 1984). The authorized project specified a lock size not less than 34 × 183 m, to standardize with the Ohio River navigation project. The Mississippi River 2.75-m project was completed in a single decade during the Great Depression (1930–40). Much of the improvement works for the 1.37-m and 1.83-m channels (wing dykes and excavations) were inundated by the 2.75-m project, and the new channel altered the river's shape along nearly every mile, inundating bottomlands and creating countless islands and backwater marshes.

Both natural conditions and population centres determined the design, location and number of dams in the project (Fig. 23.3). Wherever possible, dams were located just upstream from towns so as to minimize waterfront changes. Low natural stream banks in front of a heavily cultivated floodplain, and encroachment of railroad tracks and towns precluded construction of high dams that could have also served power or flood-control needs (Tweet 1984).

Frequent occurrence of flood discharges dictated the need for movable dams that could be raised entirely out of the water. Spring ice breakage also dictated that dams be both strong, and have wide openings to avoid causing ice jams (Tweet 1984).

The Corps worked closely with the bureaus of Biological Survey and Fisheries, making design

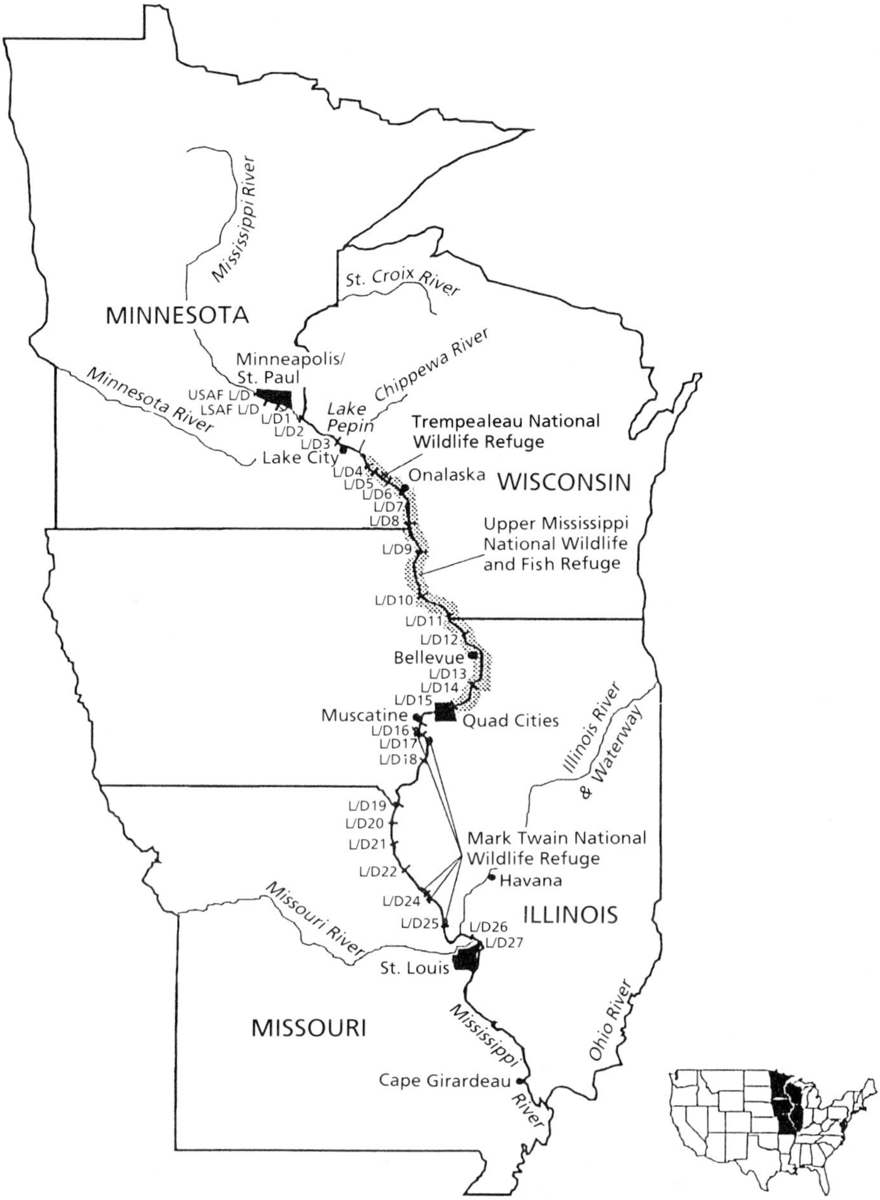

Fig. 23.3 Upper Mississippi River showing cities, locks and dams (L/D –), and features of significance to the 2.75-m navigation projects and the system of national wildlife and fish refuges.

modifications to aid area ecology (USACE 1951). For instance, roller gate design was selected to permit fish migration, help stabilize water levels, pass silt and sewage, and aerate water. National Park Service co-operation was obtained to ensure that areas needed for navigation, but not continuously overflowed, would be put to maximum recreational use consistent with the project.

With the exception of Dam 15, all dams were designed with a combination of tainter and roller gates. The newer roller gate, structurally sounder than tainter gates, could be made longer, withstand hard usage, and provide a wider space between piers to pass ice and drift. Dam 15, located in the narrowest part of the channel (at the old Rock River Rapids) and subject to ice jams, was built entirely with roller gates (Elliot 1941). Tainter gates (an old French design) were used wherever possible because they were less expensive, and did not require royalty payments (Tweet 1984).

Roller sections were originally intended to pass normal flows, reserving tainter gates to pass flood or high-water stages. However, such an uneven flow distribution caused extensive scour, endangered structures and played havoc with the channel. Consequently, all gates in each dam have been kept at about the same level (Tweet 1984).

Completion of the 2.75-m channel (1940) created a bottleneck at Lock 19 (Fig. 23.3), the site of Lake Cooper and the old Des Moines River Rapids, because of its higher lift. District studies (1945) showed a clear trend towards heavier and longer Mississippi River tows, so a new 366-m lock (twice as long as the old lock) was completed at Lock 19 in 1957 (Tweet 1984). The 12.2-m drop at Keokuk was by far the largest one in the project and enlarged lock capacity was especially time-saving, eliminating the need to double-lock large tows.

23.5 GROWTH OF ENVIRONMENTAL AWARENESS

Channel improvements made above Lake Pepin in the 1880s revealed shoals composed largely of sawdust and sawmill refuse. In 1888 the Corps, called to remove a shoal forming near St Paul, discovered that it was formed entirely of garbage. The smell was so objectionable that private citizens had obtained an injunction against both St Paul and Minneapolis for dumping more than 454 000 kg of garbage/day into the river (Tweet 1984).

The Rivers and Harbors Act of 1890 gave the Corps authority to establish and enforce rules and regulations for all navigable streams. From this came their present-day responsibility for the entire inland waterway system (Smith 1966). The Rivers and Harbors Act of 1899, 'The Refuse Act', prohibited discharge (except from public streets and sewers in liquid form) into any navigable water of the United States without a Corps permit (Tweet 1984). This regulation, revived by conservationists in the early 1970s, has seen little use.

Conservation forces were concerned about Upper Mississippi River wildlife as early as 1900. Today one can only imagine the river's appearance in 1911 when all the islands were nearly denuded of willow, and huge log rafts were being floated downstream. In some places even the bluffs were denuded of hardwood vegetation cut for steamboat fuel, use by the railroads, or for wing dyke material.

Commercial fish and mussel harvesting, regulated by the Department of Commerce, were major Mississippi River industries at the turn of the century. The Commerce Department's Bureau of Fisheries also managed a series of Mississippi River fish rescue and propagation stations between 1889 and 1939 (Carlander 1954). Freshwater mussel propagation was underway in the Commerce Department's lab at the Fairport (Iowa) Biological Station (Carlander 1954), and by 1910 concerns were raised over deterioration of the river's water quality (Scarpino 1985).

Interest in protecting Mississippi River mussels and fish was a commercial issue as well as one that grew out of citizen complaints about fish kills caused by water level fluctuations. Concern for waterfowl came from over-hunting and exploitation for commercial markets (Fairchild 1982).

The visual pollution of wing dykes, exposed at low river stage, was also of concern as noted in 1914 by Mr W.F. Daubenberger of McGregor, Iowa:

> With the automobile in such common use, people by the hundreds, within a radius of fifty miles, come in to spend a day or week on the river. Then they look across and see a pile of rocks sticking out of the water that makes them think the armies of Europe were using it for a cemetery and were burying their dead there. (Tweet 1984.)

The Izaak Walton League of America ('Ikes') was founded in Chicago in 1922 to lobby at local,

State and national levels for preservation of outdoor recreation opportunities.

The Migratory Bird Law, passed in 1913, initiated federal control over waterfowl hunting and prohibited commercial market hunting. This law's principles were carried over into the Migratory Bird Treaty Act in 1918 for protection of brids migrating between Canada and the United States (Rahn 1983). Also, the nation was in the middle of Theodore Roosevelt's 'Progressive Era Conservation Movement' (Scarpino 1985). This movement was characterized as having a deep faith in applied science, and advocating rationally planned, efficient (multiple-purpose) resource development to provide the 'greatest good for the greatest number'. The greatest good in this case was equated with economic growth and a rising material standard of living. However, passage of the Water Power Act of 1920 marked the end of the era because the Roosevelt conservationists were unable to forge a constituency among the nation's disparate resource-user groups. The concept of multiple-purpose development offered little to attract single-purpose organizations such as the National Rivers and Harbors Congress or the Upper Mississippi River Improvement Association (Scarpino 1985).

With the defeat of Roosevelt's conservation movement, single-purpose development was inevitable, and the river's fish and wildlife habitat was being threatened not only by over-hunting but also by pollution and wetland drainage for agriculture and industrialization. Inpetus for protecting the river's smallmouth bass fishery, already significantly impacted by Lake Cooper, came from the Ikes (Scarpino 1985). The Upper Mississippi was ideal for this gamefish because it was a swiftly-flowing, relatively shallow river, and creation of a refuge for this species became the Izaak Walton League's first major project.

The Ikes found eager converts among businessmen and professionals in cities and towns along the river. A decade earlier, these same individuals had joined commercial clubs and routinely boosted river-development projects such as wetland drainage, navigation improvements and the Keokuk dam (Lake Cooper). Their mass conversion to active conservationism reflected an underlying fear that the qualities of the river they had long prized, but taken for granted, might soon cease to exist (Scarpino 1985).

The Ikes' campaign began with a lawsuit filed in the Wisconsin Supreme Court by their McGregor, Iowa, Chapter to stop drainage of Winneshiek Bottoms, a 30-mile reach below Lansing, Iowa (Fig. 23.2). The Ike's proposed refuge stretched from Rock Island, Illinois, to Wabasha, Minnesota. Will Dilg, founder of the Ikes and an avid fisherman, carried the issue to Congress and gained passage of the Upper Mississippi River Wild Life and Fish Refuge Act in 1924 (Scarpino 1985).

The Act divided Refuge administration between the departments of Commerce (Bureau of Fisheries) and Agriculture (Bureau of Biological Survey), giving both police powers to enforce fish and game laws. Commerce, concerned only with fish and mussel propagation, and without authority to purchase land, received as much funding in 1924 for Refuge purposes as did Agriculture. Thus, legislative history and subsequent appropriations indicate that, as the Ikes envisioned, a major purpose for the Refuge was fish propagation (Fairchild 1982).

The Act required State consent for land acquisition and inclusion in the Refuge, and all lands had to be acquired for less than $5.00 per 0.4 ha. Consent of the four States (Minnesota, Iowa, Illinois and Wisconsin) was received by 1925, but Iowa and Wisconsin were the most restrictive, with Wisconsin explicitly reserving title to the fish (Fairchild 1982).

Some land was acquired by gift, and other land, most typically islands formed in the river after statehood, were granted to the United States by Iowa and Minnesota (Fairchild 1982). Unpatented Mississippi River islands above Cairo, Illinois (Fig. 23.2), withdrawn by the Secretary of the Interior in 1891 to serve the interest of navigation, had been kept out of private ownership and maintained in a relatively undisturbed state. These were excellent for refuge purposes. Other federally owned islands were withdrawn by Presidential Executive Order in 1925.

The maximum $5.00/0.4 ha purchase price, impossible to achieve, was amended in 1925. Today, the Refuge includes 79 000 ha of wooded islands, water and marshes extending over 457 km

(Fig. 23.3), making it the longest inland federal refuge in the country, and the only federal refuge bisected by a major commercial navigation project (EWT 1981).

Between 1924 and 1934, changing river conditions (i.e. impoundment by the new dams) and reorganization of federal executive agencies reduced the Department of Commerce's involvement in Refuge management. Concern for Refuge welfare, however, was heightened because land acquisition was not complete when the Corps began purchasing lands for the 2.75-m project. This led to potential competition between the Agriculture and War Departments.

Agriculture wanted well-defined boundaries; to eliminate non-wildlife interests in Corps' land; and for the Corps to eliminate slack water areas by either dredging channels through all lakes, sloughs and pools to the main channel or by stabilizing water levels (Fairchild 1982).

The Corps, however, wanted all project lands to be readily available for development purposes, while avoiding the administrative difficulties of land policing and maintenance (Fairchild 1982). These conflicts became increasingly formal and appeared to spell doom for the Refuge.

Early in 1931 the War Department agreed to purchase lands in fee and allow these to be used as part of the Refuge, provided that purchase price for fee interest was not materially greater than for flowage easements (Fairchild 1982). In return, Agriculture agreed to police the lands, to allow the Corps to flood some Refuge lands, and to place structures on other land. This 'amicable understanding', as it was later called (Fairchild 1982), was the first in a series of agreements for Refuge management.

The Fish and Wildlife Coordination Act, passed in 1934, directed federal agencies to make use of impounded waters resulting from federal activity'... for fish-culture stations and migratory-bird resting and nesting areas as are not inconsistent with the primary use of the waters and/or the constitutional rights of the states' (Fairchild 1982). The Act applied to Upper Mississippi River projects, but it did not specifically require Corps' concessions or affirmative actions for the benefit of wildlife preservation.

Three executive orders issued in 1935 and 1936, document the first formal agreement between Agriculture and the Corps. The orders reserved Corps' lands for the Refuge as breeding places for birds, animals, fish and other aquatic life, and for conservation of wild flowers and aquatic plants. They also noted, however, that these lands 'are primarily under the jurisdiction of the War Department' and reserved the Corps' right to pursue its activities without interference (Fairchild 1982).

In 1938 Congress included in the Corps' general grant of authorization for navigation improvement activities the requirement to give 'due regard to wildlife conservation' in pursuit of its activities (Fairchild 1982). However, no standard or enforcement authority was provided, and no documentation is available to indicate that it had any effect on Corps activities in the Refuge.

Impoundment for the 2.75-m channel reduced the need for fish rescue and propagation because fish were no longer stranded during summer low-water periods. So in 1939 all Commerce Department Refuge management activity was transferred to the Agriculture Department (Fairchild 1982). A year later the Bureaus of Fisheries (Commerce) and Biological Survey (Agriculture) were combined to form the Fish and Wildlife Service (Service) within the Interior Department (Fairchild 1982). Executive Order 8331 noted the administrative change and reserved additional Corps lands for Refuge use, but still allowed the War Department to retain use for navigation.

The Corps leased some lands, and owned only flowage easements on others. Also many Corps' fee interests contained renewable reservations for hunting, fishing and farming in favour of former owners (Fairchild 1982). Such reservations reduced the purchase price of overflow lands, and the sale of hunting rights was a revenue source the Corps used to justify maintenance of their land acquisition staff. The Service protested, and the Corps' commissioned officers agreed that such rights should be terminated where possible, but the practice continued well into the 1940s (Fairchild 1982).

Interspersion of Refuge lands with private lands or lands remaining entirely under Corps control made wildlife protection difficult. Of further Service concern was Corps' use and management of the river and surrounding lands for navigation (Fairchild 1982).

Towards the end of the 1930s, the States (most

notably Wisconsin and Iowa) expressed interest in retaining or recovering control over establishment of harvest seasons. The Service denied such rights, but Refuge hunting seasons were set to match those of the States, and local sportsmen continued to pressure the Refuge into reducing areas closed to hunting (Fairchild 1982).

The States also asserted management claims to Refuge furbearer trapping, but in this instance the Service maintained more control. By 1929 a Refuge permit system opened trapping to the public. Reducing Refuge furbearer populations also reduced populations on surrounding lands and made adjacent landowners more willing to sell (Fairchild 1982).

Service and Corps negotiations, without State participation, were held from 1941 to 1945 to develop a more structured arrangement for transfer and administration of federally-owned land. These negotiations concluded in 1945 with the signing of the first co-operative agreement (Fairchild 1982). That agreement categorized and zoned lands within the Upper Mississippi refuges for administration, hunting and trapping. The Corps retained the right to administer timbering programmes on all lands it had originally purchased and to transfer lands under 5-year revocable permits (Fairchild 1982).

Leasing authority retained by the Corps interfered with Refuge administration, and concern was raised as to whether the Corps could transfer lands directly to the States for administration, or if the transfer had to be made through the Service. The 5-year revocable permits on some lands also interfered with long-term Refuge planning. Refuge administration was also impeded by numerous administrative land categories resulting from Executive Orders, co-operative agreements, licences, permits and purchases (Fairchild 1982).

The Trempealeau National Wildlife Refuge (Fig. 23.3) was established in 1934 under authority of Executive Order with the purchase of 284 ha of land. An additional 2024 ha were purchased in 1979. Immediately adjacent to the Upper Mississippi Wild Life and Fish Refuge, Trempealeau's primary objective is waterfowl management.

The 2.75-m navigation project had brought changes in both the amount and method of river transportation. The steam engine and paddle wheel were replaced with larger and larger diesel towboats (the modern towboat measures 11 × 50 m, has a draft of almost 2.5 m and is powered by as much as 5000 horsepower, turning propellers up to 2.7 m in diameter). Additionally, the number of terminals in the Rock Island Corps District alone increased over 17-fold between 1940 and 1960 (Tweet 1984). The navigation project altered Refuge habitat and purpose, relationships between Federal agencies, relationships between agencies and States, and relationships between the Refuge and surrounding landowners (Fairchild 1982).

State biologists became increasingly concerned about the welfare of fish, wildlife and recreation in these multi-use conflicts and formed the Upper Mississippi River Conservation Committee (UMRCC) in 1943. With support of both the Corps and the Service, heads of State conservation agencies from Illinois, Iowa, Minnesota, Missouri and Wisconsin met in Dubuque, Iowa, with the two-fold objective: 'To promote the preservation and wise utilization of the natural and recreational resources of the upper Mississippi River' and 'To formulate policies, plans and programs for conducting cooperative studies; *for the* benefit and appreciation of people interested in the natural amenities and values of the River, including fishermen, hunters, boaters, naturalists and others.'

The UMRCC initiated the river's first long-term resource monitoring effort in the late 1940s and early 1950s, and to this day maintains one of the best and most complete sets of large river commercial fisheries data in the world (Welcomme, personal communication).

The 1946 Fish and Wildlife Coordination Act Amendments required provisions on federal lands for 'conservation, maintenance, and management of wildlife, resources thereof, and its habitat', and defined 'wildlife' to 'include birds, fishes, mammals, and all other classes of wild animals and all types of aquatic and land vegetation upon which wildlife is dependent.' The 1946 and 1948 Amendments also imposed affirmative duty on agencies or departments that impound, divert or control waters to make adequate provision for wildlife (Fairchild 1982). The 1948 Amendments included the Anti-Drawdown Law, which required the Corps to maintain Mississippi River navigation pools 'as though navigation was carried on throughout the year'. In earlier years, winter drawdowns to increase capacity for spring floods had caused devastating fish and wildlife losses (Greenbank 1946).

Negotiations between the Corps and the Service for Refuge land transfers, general plans and new co-operative agreements began in 1950. The Corps took the position that the 1946 Coordination Act Amendments were inapplicable to the 2.75-m project. They maintained that the Flood Control Act of 1946, approved 3 weeks before the 1946 Amendments, directed them to ensure that water areas of all such impoundments be open to general public use (Fairchild 1982). The Corps was interested in maximizing control over land that might be useful for a 3.66-m navigation project then under study.

However, after continued negotiation, the Interior and Army Secretaries agreed to dispose of wildlife lands in accordance with the 1946 Coordination Act Amendments (Fairchild 1982). General Plans were signed in 1954 and Public Land Order 936 placed all transferred lands under authority of the 1954 Cooperative Agreement, revoking all other executive order and public land order transfers (Fairchild 1982). Thus a unified system of administration over Corps land was created.

Shortcomings of the new agreements became readily apparent because the Corps specifically reserved the right to change water surface elevations, dredge and develop lands for commercial and industrial sites, issue leases for public use, and maintain that Refuge use could not interfere with navigation (Fairchild 1982).

Timber cutting and harvesting was a Refuge concern because the Corps ran the programme as a revenue source, making sole determinations of what was 'merchantable' (Fairchild 1982). Another major issue was the Refuge sharecropping programme. The Service leased certain Corps' lands to local farmers or State agencies for farming in exchange for a percentage of the crop, or allowing fields to stand after harvest. The Corps insisted that all proceeds from the sale of such crops be turned over to the Division Engineer rather than be used to defray Refuge expenses (Fairchild 1982).

The 1958 Coordination Act Amendments directed that consultation and modification requirements be applied retroactively to major federal projects not yet 60% complete, and that government agencies give 'full consideration' to the Interior Secretary's report regarding modification of water projects for wildlife protection. This Amendment clearly applied to the 2.75-m project as well as to future Upper Mississippi River Corps' activities. The Corps was thus required to act on Interior recommendations to the extent necessary to comply with the full consideration requirement. Merely consulting with Interior was insufficient.

The 1958 Amendments also settled the dispute over the relationship between the Coordination Act and the 1946 Flood Control Act, stating that: 'Federal lands acquired or withdrawn for Federal water-resource purposes and made available to the states or to the Secretary of the Interior for wildlife management purposes, shall be made available for such purposes in accordance with this Act, notwithstanding other provisions of law.' This effectively prohibited the Corps from unilaterally issuing cottage-siting or other public use leases or licences on land turned over to the Refuge for wildlife management. Additionally, the 1958 Amendments clearly authorized direct land transfers for State administration.

The 1958 Amendments also allowed the Service to begin providing a coordinator and office facilities for the UMRCC. Through the focus provided by a permanent coordination office, this State-managed group has since been an effective advocate for ecosystem protection and conservation, and has provided important leadership on major river issues where individual States or agencies could not (UMRCC 1944–90).

The Mark Twain National Wildlife Refuge was created in 1958 from 27 135 ha of river bottomland purchased by the Corps, but not needed for navigation. This land is contained in several large units lying adjacent to the River between Muscatine, Iowa and St Louis, Missouri (Fig. 23.3). The transfer was made under authority of the 1946 Flood Control Act and 1946 Fish and Wildlife Coordination Act amendments. Corps/Service management problems on this refuge have paralleled those of the Upper Mississippi

Wild Life and Fish Refuge. The Service also used Federal Duck Stamp funds to acquire an additional 2349 ha. The Refuge is managed primarily for waterfowl, with 20 250 ha transferred to the States for operation and management.

The Water Resources Planning Act of 1962 established a National Water Resources Council. This group, similar to Teddy Roosevelt's Inland Waterways Commission of 1907 and composed of the heads of all major federal water-resource agencies, was to oversee and establish national policies, standards and procedures for preparation of comprehensive regional or river basin plans. Along with the basin States, this group was involved in preparation of the Upper Mississippi River Comprehensive Basin Study (UMRCBS Coordinating Committee 1972), marking the beginning of modern-day co-operative Mississippi River Basin planning.

The Service and Corps developed a new public use co-ordination system in the 1963 Cooperative Agreement. The Corps agreed to stop issuing cottage site leases and to phase out existing leases and agricultural leases (Fairchild 1982). In their stead, some cottage sites would be converted into public access, camping, picnicking or boat launching areas. It was also agreed that both agencies must consent to any rights of way for roads, telephone lines, power lines or other uses over either Corps or Service lands.

Explicit Service authority was also provided to prevent and eliminate trespass or unauthorized use of property made available through the Cooperative Agreement (Fairchild 1982). But the Corps retained authority to determine sales of land interest or land disposal. In addition, the Service gained use of the proceeds from agricultural surpluses to cover the costs of wildlife programmes in general. The 1963 Cooperative Agreement thus led to an uneasy peace between the Corps, the Service and the States (Fairchild 1982).

The US Department of the Interior was directed in 1964 to evaluate all Federal lands for reservation as wilderness areas. This included the Upper Mississippi Wild Life and Fish Refuge, but the Refuge was found to be an unlikely wilderness candidate (Fairchild 1982).

Congress passed the National Wildlife Refuge System Act in 1966 to consolidate authorities relating to fish and wildlife conservation (Fairchild 1982). The Act also provided Interior with authority to acquire land or interests in land in exchange for acquired land. The Act was amended in 1976 by what became known as the Game Range Amendments. These amendments provided that suitable lands acquired through cooperative agreement were also part of the National Wildlife Refuge System, but could be disposed of in accordance with terms of the cooperative agreement (Fairchild 1982).

Increased river use for commercial transportation through the 1950s and 1960s led to calls for expanded barge-fleeting and renewed interest in the 3.66-m project, first proposed by Congress in the 1940s, but not pursued in the late 1940s or early 1950s because of its high costs in the face of the Korean War (Fairchild 1982).

Public frustration with growing federal bureaucracies and conflicting Congressional policies affecting Upper Mississippi River use for navigation, recreation and wildlife preservation increased in the 1960s. Dredging and spoil disposal in wetlands and backwaters had become an increasing problem and the navigation pools were ageing rapidly as agriculture and non-point source pollution in the watershed intensified. These concerns boiled over when the proposal for the 3.66-m channel was revived and a project to replace Locks and Dam 26 (Fig. 23.3) was proposed. The new locks were designed to meet the needs of a 3.66-m channel, and lock capacity was to be at least doubled by constructing two 366-m locks to replace the existing 92-m and 182-m locks.

Massive dredging required for a 3.66-m project prompted UMRCC biologists to initiate studies to analyse the problem (Robinson 1970; Sternberg 1971). It was perceived that the new locks would open the entire upper river to vastly increased commercial traffic (two to four-fold), because delays created at smaller upstream locks would dictate the need to enlarge all the locks, one after another, systemwide. UMRCC studies, subsequent public opposition, and economics stopped Corps' plans for the 3.66-m channel upstream from St Louis, but planning for the new lock continued. Bill Dieffenbach (Missouri), Jack Robinson (Missouri) and Willis Fernholz

(Wisconsin), all State UMRCC biologists, were strong opponents of the project. The environmental concerns they raised were key to stimulating public sentiment against the 3.66-m project.

Even without a planned 3.66-m project, dredged spoil disposal practices were major issues during preparation of the 2.75-m project Environmental Impact Statement (EIS), prepared in response to the National Environmental Policy Act of 1969. Controversy was such that the State of Wisconsin filed a lawsuit against the Corps in 1974 prohibiting further spoil disposal in Wisconsin's wetlands. Willis Fernholz again provided key testimony in this action. Minnesota followed Wisconsin's lead and left the Corps with the dilemma of either not meeting their federal mandate to provide a 2.75-m channel, or violating State law.

The matter was resolved later in 1974 when the Corps and Service joined hands with the States and other federal agencies in the St Paul Corps District to form the Great River Environmental Action Team (GREAT). GREAT, an attempt to resolve conflicts and produce a 'total river management plan', was authorized by Congress for the entire Upper Mississippi in 1976 as the Great River Study. It was placed under the auspices of the recently formed Upper Mississippi River Basin Commission (UMRBC). The UMRBC, reporting to the National Water Resources Council, was one of several such Commissions formed nationwide to implement basinwide planning on the nation's major river basins. The UMRBC was composed of voting members from all federal agencies and basin States with the goal of building a consensus on river management planning and decision-making.

GREAT, split into three parts according to Corps District boundaries, became a precedent-setting planning and management tool, opening up communications between public, transportation and environmental interests, and effectively solving the dredged spoil disposal issues. The Study, completed in 1982, formed the basis for on-going channel maintenance practices. Under the GREAT programme, beneficial uses have been found for dredged spoil, dredging volumes have been reduced, and consensus for spoil site selection is achieved through on-going interagency on-site inspection (GREAT-1980a,b, 1982). The study did, however, fall short of producing the desired 'total river resource management plan'.

While the dredging issue was being resolved by GREAT, the Corps continued to press for replacement of Locks and Dam 26 (Fig. 23.3). Charles Kulp, Kent Keenlyne and Jerry Rasmussen, successive UMRCC Coordinators and Service biologists in Rock Island, Illinois, played major roles in bringing the Lock and Dam 26 issue to the public eye. The threat of increased navigation traffic systemwide was of concern to both railroad economic interests and to environmental interests. Consequently, a joint lawsuit to stop construction of the new locks was filed against the Corps by a coalition of Western Railroads, the Ikes and the Sierra Club.

Environmental interests were concerned that the river would become an interstate highway for barges. The increased number and size of boats (with 2.7-m props) using the river would drastically increase turbidity and kill massive numbers of fish eggs and larvae. This, coupled with the concern that the huge towboats would keep sediments stirred up so enabling these to move laterally into fragile backwater habitats, would disrupt river ecology beyond tolerable limits.

As the lawsuit continued Congress, under intense lobbying pressure for a compromise, authorized a new lock and dam in 1978, but with only one 366-m lock. Included in the package, as a concession to those who opposed subsidizing growth of private industry at taxpayers' expense, was a 'user fee' in the form of an excise tax on the fuel used by commercial vessels (Davis 1982). Imposition of lockage fees or other more direct user fees were prevented by the language of the Northwest Ordinance passed in 1787, and referred to earlier. This fuel tax, to reach \$0.10 per 3.8 litres over several years, accumulates in a Waterway Trust Fund and is used on a cost share basis to fund new navigation improvements (including any additional lock at Lock and Dam 26). So the 'user pay' idea brought to the Board of Engineers 150 years before had finally become a reality.

Congress covered environmental concerns by also requiring preparation of a Comprehensive Master Plan. This Master Plan, to be prepared under the auspices of the UMRBC, would determine the ecological carrying capacity of the Upper Mississippi River System (UMRS) and make a recommendation on whether the second lock should be 183 or 366 m in length. The UMRS included all of the navigable reaches of the river and its tributaries (exclusive of the Missouri River), upstream from its confluence with the Ohio. This included the entire Illinois River and Waterway to Lake Michigan (Fig. 23.3).

Despite this legislation, the lawsuit continued until 1979, with Dr Richard Sparks (Illinois) and Dr Tom Claflin (University of Wisconsin-La Crosse), researchers who were working on the Master Plan, providing major testimony on the environmental impacts of increased navigation traffic. 'To the surprise of those who followed the trial' (Davis 1982), the environmentalists lost the lawsuit and on 25 April 1980, the Corps drove the first piling for the new lock.

Construction began in earnest and the Master Plan's environmental studies, under intense political pressure, were cut short to meet lock construction schedules. Consequently, the Master Plan's Environmental Work Team (EWT) was forced to forge an environmental compromise based on expert testimony.

The EWT, originally chaired by Jerry Rasmussen, representing Interior and serving as UMRCC Coordinator, was restructured. Rasmussen stepped down from the chairmanship under orders from the Department of the Interior (Reagan Administration), and Dr Howard Markus (Missouri) assumed the chairmanship. Markus was a strong environmental advocate and within 60 days was removed from the Project by his agency under pressure from Lock and Dam 26 advocates in St Louis (Congressman Robert Young). Gene Hollenstein (Minnesota) then assumed the chairmanship and saw the project to its conclusion. Other key members included Dr David Kennedy (Wisconsin), Norm Stucky (Missouri), Bill Boyd (Illinois), Kevin Szcodronski (Iowa) and John Marlin (Public Member).

At the EWT level, the Corps representative voted against nearly every environmental issue, concern and recommendation. Despite Corps opposition, this group, unique in its comradeship and dedication to its mission, came forward with a bold request for a 10-year, $300 million environmental management programme (EWT 1981).

The UMRBC studied the proposal and after lengthy deliberation reached a political compromise between environmental and transportation interests. The Commission recommended construction of a second 183-m lock, implementation of a 10-year $190 million Environmental Management Program (EMP), a $900 million watershed erosion control programme, and a waiver of the EIS requirement for the second lock (UMRBC 1982). The EIS waiver was recommended because it was felt that there were inadequate data available to write an appropriate report. The $300 million environmental programme recommended by the EWT was reduced to $190 million to equal the level of the 1982 estimated cost of constructing the recommended 183-m lock.

The new lock and the EMP were not linked, so political forces were rallied independently by both transportation and environmental interests for authorization of their respective components. Transportation interests were largely centred in the labour and trade unions of St Louis and Minneapolis/St Paul, while environmental interests were centred in the UMRCC, the Ikes, and conservation groups in western Wisconsin.

Dr David Kennedy (Wisconsin) and Jerry Rasmussen (UMRCC Coordinator and Service biologist), together played major roles in organizing conservation agency and group support for the Program. Commissioners of the Minnesota/Wisconsin Boundary Area Commission (Fred Funk, Bill Howe and Mary Swanger) and Dr Kennedy were largely responsible for gaining the support of Congressman Steve Gunderson (Wisconsin) as the lead Congressional advocate of environmental interests. Congressman Robert Young (Missouri) led the navigation interests.

Extensive lobbying by the Upper Mississippi River Basin Association (UMRBA), Minnesota/Wisconsin Boundary Area Commission (MN/WI BAC), Ikes, UMRCC, American Waterways Operators, Association for the Improvement of the Mississippi River and others occurred in the early 1980s. James Harrison, Executive Director of the MN/WI BAC, played a key role in this effort. The UMRBA is the replacement for the UMRBC, which was deauthorized upon completion of the Master Plan in 1982. The UMRBA is State funded and federal agencies participate as non-voting members.

A final compromise was reached in 1986 when Congress passed the Upper Mississippi River Management Act, authorizing both the 183-m lock and the EMP. Congress did not include a waiver for the second lock EIS nor the recommended watershed erosion control measures in the compromise. Law makers reportedly felt that the EIS waiver was too precedent-setting, and other watershed legislation had been passed (i.e. Farm Bill, Sodbuster Bill, Swampbuster Bill, etc.) to set aside farmland, stop wetland destruction, and reduce watershed erosion and sedimentation.

23.6 MODERN APPROACHES

Since the 1840s, and the first River Improvement Convention, the history of Upper Mississippi River resource management has thus been a classic example of legislative interaction caused by competition, conflict and confrontation between special interest groups, State and federal agencies, and the public. River management has evolved into a federal agency/States/public checks-and-balances system. The Corps regulates the waterway system, the EPA administers water quality programmes, and the Service and States manage fish and wildlife programmes. All, including the public, provide input and comment to the other through the federal and State permit processes.

Significant federal mandates include the following:

1 The National Environmental Policy Act (NEPA) of 1969 requires the filing of EISs for every future major federal action and federal project which may produce significant impacts on the human environment. This included the Mississippi River 2.75-m project.

2 The Federal Water Pollution Control Act (FWPCA), enacted in 1972, prohibits discharge of any pollutant (including dredged material and fill) by anyone (including the Corps) without a permit. This nationwide programme is administered by the US Environmental Protection Agency (EPA) and the Corps under the National Pollution Discharge Elimination System (NPDES). The FWPCA has been amended several times to clarify and strengthen its authority.

3 The Endangered Species Act of 1973 provides a means to preserve ecosystems upon which endangered and threatened species depend.

Implementation of the Upper Mississippi River Environmental Management Program (EMP) began in 1987 through the Corps (USACE 1990), with the Service and States co-operating. The entire EMP is overseen by the UMRBA. After a slow start, the Program reached full funding in fiscal year 1992, and received a 5-year extension to cover funding losses from 1987–91. This extends the Program to the year 2002.

As envisaged by the EWT, the EMP was to:

1 Collect long-term data sets to document change over time, 'taking the pulse' of the river ecosystem in order to identify problems before they become catastrophic.

2 Evaluate and provide solutions for short-term problems (i.e. sedimentation, navigation effects, water level fluctuations, lack of aquatic vegetation, and reduced fisheries populations).

3 Construct and evaluate habitat projects to develop tools to offset or mitigate future environmental impacts.

4 Develop a computerized data management system which could be readily accessed by interagency users.

It was felt that these components would give resource managers the information needed to better evaluate the impact of future navigation expansion projects, the tools necessary to make legitimate mitigation recommendations, and the space-age technology to make appropriate recommendations without unnecessary delays.

About two-thirds of the EMP budget goes for habitat projects and one-third for data collection and monitoring. The Corps manages construction of habitat projects proposed by inter-agency committees, while the Service manages the data collection components (Long Term Resource Monitoring Program, LTRMP) under co-operative agreement with the Corps and the States. Both portions of the Program have had their problems.

The first problem encountered by the habitat management programme was the requirement of 'cost sharing' on non-federal lands imposed by the Reagan Administration. In the recessionary economy it has often been difficult for the States to come up with their share of project costs, so most projects are being built on federal lands (which in most cases are refuge lands). Federal refuge managers have taken the position that projects constructed on refuge lands must address the goals and objectives of refuge management (i.e. shallow water and moist soils for migratory waterfowl). Fisheries interests have thus often taken second place or have even been excluded from projects (which in some cases were originally proposed for fisheries enhancement). This has not gone down well with some State biologists, nor with some of the original EWT members, because the need for the EMP in the first place was based largely on fishery habitat loss to sedimentation and navigation development. Also, the Ikes' originally proposed need for the refuge (in the 1920s) as a sanctuary for smallmouth bass spawning would seem to have been lost completely with the passage of time.

A second major problem has been with implementation of the LTRMP's eight resource monitoring components (i.e. water quality, vegetation, invertebrates, fish, birds, mammals, public use and land use). This portion of the LTRMP is called Resource Trend Analysis (RTA). The RTA, requiring a large infrastructure of both equipment and personnel, consumed much of the early LTRMP budget. Five State-operated field stations were established at strategic locations along the Mississippi River (i.e. Pools 4, 7, 13 and 26 and the unpooled river), and one on the Illinois River. Components monitored to date include water quality, vegetation, invertebrates, fish and recreation. Land use is being monitored by a major Geographic Information System (GIS), headquartered in the Service's Environmental Management Technical Center (EMTC) in Onalaska, Wisconsin (Fig. 23.3). State-operated field stations are located at Lake City, Minnesota; Onalaska, Wisconsin; Bellevue, Iowa; St Louis, Missouri; Cape Girardeau, Missouri; and Havanna, Illinois (Fig. 23.3). Advisory teams have been established to critically monitor both data being collected and methods used to collect those data in an effort to improve scientific efficiency and reduce costs.

While extremely valuable, collection of resource monitoring data is a slow, tedious process which does not produce immediate tangible results. This, together with its high start-up costs, have made many federal bureaucrats nervous, and have kept the RTA in a continued state of controversy and confrontation between federal interests (who have focused on short-term budgetary problems and development of products) and the States (who have favoured collection of RTA date for its long-term value to resource conservation). The LTRMP has also been hampered by unforeseen administrative (overhead) costs, and the parochial attitudes of some bureaucrats who would rather use LTRMP funds for current, perhaps more popular, agency-focused projects, than for the broader, systematic purposes envisaged by its creators.

So the EMP, just as the entire history of the Upper Mississippi, has been embroiled in political controversy and conflict between special interests. A few habitat projects have been built and are being monitored; trend analysis continues; some problem analysis studies have been funded; and only recently, a complete scientific staff has been assembled at the EMTC to process and evaluate data being collected.

The jury is still out on the success and magnitude of contribution the EMP will eventually make to preservation of the river's habitat diversity over the short as well as the long term. Upper Mississippi River history suggests that the Program will survive, but likely not in its originally envisaged form. The critical period for the EMP will occur 2 or 3 years from now when the first real products start to come in, and Congress and the public evaluate the Program's mid-term success.

In the meantime, the Corps continues to evaluate methods to maintain and improve the 2.75-m project and the States, agencies and public continue to review Corps' proposals. These include replacement of existing locks with new 366-m units, construction of mooring cells, use of helper boats, channel improvements and dredging, computerized traffic management and control systems, and construction of auxiliary locks to reduce delays caused by congestion of recreational craft.

Preparation of the second lock EIS included recommendations for a $29 million navigation effects study proposed for funding by the St Louis Corps District. This study, however, has not been funded and the Corps has brought some pressure on the LTRMP to focus its problem analysis efforts on evaluation of navigation impacts.

Despite its continued interest in improving the navigation system, the Corps seems to be intent on developing a new environmental image, and currently operates in a much more co-operative mode than in the distant past. GREAT and the EMP have vastly improved communications between transportation and environmental interests, and both sides seem to be working together for a more harmonious future.

Activity of special interest groups, however, is not likely to diminish. But perhaps through the enhanced avenues of communication, co-ordination, consensus and co-operation developed in recent decades, future conflicts, at least between navigation and environmental interests, can be minimized. Of course, all of this will depend on changing public attitudes and the state of national as well as global economics and politics.

Our world has become much smaller since European explorers first set out to exploit the New World and the Mississippi River Valley, and the impact we have on one another has become far greater than in the past. What happens today economically and politically in one corner of the earth, can have far-reaching, worldwide ripple effects, even affecting large river ecosystems such as the Mississippi. We should not expect this situation to change in the 21st century, but enhanced communications should go a long way towards improving worldwide co-operation, and perhaps help stave off environmental catastrophes here and elsewhere!

REFERENCES

Carlander HB. (1954) *History of Fish and Fishing in the Upper Mississippi River.* Spec. Publ. Upper Mississippi River Conservation Committee, Rock Island, IL. [23.4, 23.5]

Davis ND. (1982) *The Father of Waters, a Mississippi River Chronicle.* Sierra Club Books, San Francisco, CA. [23.1, 23.3, 23.4, 23.5]

Elliot M. (1941) *Upper Mississippi River Navigation Improvement, Providing a 9-foot Channel Depth between Minneapolis and the Missouri River.* US Army, Corps of Engineers, St Louis: Office of the Division of Engineering, Vicksburg, MS. [23.4]

EWT (Environmental Work Team). (1981) *Environmental Work Team Technical Report. Vol. 1.* Prepared for the Upper Mississippi River Basin Commission for the Comprehensive Master Plan for the Management of the Upper Mississippi River System. Upper Mississippi River Basin Commission, Bloomington, MN. [23.5]

Fairchild M. (1982) The legal and administrative history of the Upper Mississippi River Wild Life and Fish Refuge. Unpublished. Available through the library of the Upper Mississippi River Conservation Committee, Rock Island, IL. [23.4, 23.5]

Fugina F. (1945) *Love and Lure of the Upper Mississippi River.* Frank Fugina, Winona, MN. [23.4]

GREAT (Great River Environmental Action Team). (1980a) *A Study of the Upper Mississippi River,* 9 vols. USACE, St Paul Dist., St Paul, MN. [23.5]

GREAT (Great River Environmental Action Team). (1980b) *Great II Main Report and Appendices.* USACE, Rock Island Dist., Rock Island, IL. [23.5]

GREAT (Great River Environmental Action Team). (1982) *Great River Resource Management Study–14028.* USACE, St Louis Dist., St Louis, MO. [23.5]

Greenbank J. (1946) Effects of midwinter drawdowns of the Upper Mississippi River on aquatic wildlife. In: Keenlyne KD (ed) (1974) *Investigational Reports,* Upper Mississippi River Conservation Committee, Rock Island, IL. [23.5]

Merrit R. (1980) *Creativity, Conflict & Controversy, a History of the St. Paul District U.S. Army Corps of Engineers.* Government Printing Office, Washington, DC. [23.4]

Petersen WJ. (1968a) The Joliet-Marquette expedition. *The Palimpsest.* **XLIX**(10): 416–46. Iowa State Historical Society, Iowa City, IA. [23.1]

Petersen WJ. (1968b) The Zebulon M Pike expedition. *The Palimpsest.* **XLIX**(2): 41–80. Iowa State Historical Society, Iowa City, IA. [23.2]

Rahn M. (1983) *A History of Wildlife and Hunting on the Upper Mississippi River.* Spec. Publ. Upper Mississippi River Conservation Committee, Rock Island, IL. [23.1, 23.3, 23.5]

Robinson JW. (1970) *The 1969 Upper Mississippi River Dredge Spoil Survey (from Hastings, Minnesota to Cairo, Illinois).* Spec. Publ. Upper Mississippi River Conservation Committee, Rock Island, IL. [23.5]

Robinson MC. (1980) The Federal barge fleet: an analysis of the Inland Waterways Corporation, 1924–1939. In: *National Waterways Roundtable Papers.* Institute

for Water Resources, Norfolk, VA. [23.4]
Rudd CG. (1981) Upper Mississippi. *Naturalist* **32**(1): Spring 1981 Upper Mississippi Special Issue. [23.1]
Scarpino PV. (1985) *GREAT RIVER – An Environmental History of the Upper Mississippi, 1980–1950.* University of Missouri Press, Columbia, MO. [23.3, 23.4, 23.5]
Simons DB, Lagasse PF, Chen YH, Schumm SA. (1975) *The River Environment – A Reference Document.* US Dept. of the Interior, Fish and Wildlife Service. Contr. No. CER 75–76 DBS-PFL-YHC-SAS-14. Twin Cities, MN. [23.4]
Smith F. (1966) *The Politics of Conservation.* Pantheon Books, New York [23.5]
Sternberg RB. (1971) *Upper Mississippi River Habitat Classification Survey.* Spec. Rept. Upper Mississippi River Conservation Committee, Rock Island, IL. [23.5]
Tweet R. (1984) *A History of the Rock Island District U.S. Army Corps of Engineers 1866–1983.* Government Printing Office, Washington, DC. [23.2, 23.3, 23.4, 23.5]
Wengert N. (1957) The politics of river basin development. *Law and Contemporary Problems.* **XXII** (Spring) 267. [23.4]
UMRBC (Upper Mississippi River Basin Commission). (1982) *Comprehensive Master Plan for the Management of the Upper Mississippi River System.* UMRBC, Bloomington, MN. [23.5]
UMRCBS Coordinating Committee. (1972) *Upper Mississippi River Comprehensive Basin Study* (several volumes). US Army Corps of Engineers, North Central Division, Chicago, IL. [23.5]
UMRCC (Upper Mississippi River Conservation Committee). (1944–1990) *Proceedings of Annual Meeting(s) of the Upper Mississippi River Conservation Committee.* UMRCC, Rock Island, IL. [23.5]
USACE (US Army Corps of Engineers). (1951) *Report of the Federal Civil Works Program as Administered by the Corps of Engineers, U.S. Army, Part 1, Vol. III.* Government Printing Office, Washington, DC. [23.4]
USACE (US Army Corps of Engineers). (1990) *Fifth Annual Addendum – Upper Mississippi River System, Environmental Management Program.* US Army Corps of Engineers, North Central Division, Chicago, IL. [23.6]

24: River Management in Cold Regions: A Case Study of the River Laxá, North Iceland

G. M. GÍSLASON

24.1 INTRODUCTION

In recent decades traditional means of making a living have given way to industrialization and urbanization throughout the world. Humans are exploiting more parts of the globe than ever before. The Arctic is no exception. The rivers are sources of fish and freshwater, and hydro-electric development of arctic rivers began during this century. Mining companies are currently exploiting the Canadian Arctic, Greenland and Arctic Russia. Oil drilling in Alaska requires a freshwater supply (Mellor 1987), and roads and pipelines may also change the nature of a river's catchment.

Large-scale plans have been considered that would divert 120 $km^3 yr^{-1}$ of river water from the Arctic Ocean to the agricultural areas in Eurasia (Ryan 1986). The impact of these river diversions is uncertain. Calculations by researchers at the Arctic and Antarctic Institute in St Petersburg indicate that a 1% reduction of inflow to the Kara Sea could result in 1.5% increase in the summer extent of ice cover. The plan, when fully implemented, would reduce the river flow to the north in European Russia and Siberia by as much as 69% and 16% respectively. Such river diversions could have great effect on the climate of the northern hemisphere.

A primary aim of management of natural resources is to reconcile human activity with the conservation of the biota. In carrying out this objective, a major task is to devise means of identifying changes brought about by modern man, as opposed to changes occurring naturally (see Orians *et al* 1986).

Studies in the ecology of arctic rivers are relatively few. Some studies are available for Alaskan streams (Slack *et al* 1977; Stewart *et al* 1990) and the McKenzie River (Rosenberg & Barton 1986; Reist & Bond 1988). Subarctic rivers have been studied more extensively (e.g. Ulfstrand 1968). Arctic rivers are usually frozen over from October to June. Water discharge is usually severely reduced during winter and the main activity of the biota occurs over the short summer. In the temperate zone, rivers are usually covered with ice for a short period in mid-winter, and the summer temperatures become much greater than in the Arctic. The subarctic falls in between these two types of running waters.

The Icelandic situation

Iceland lies in the North Atlantic, between 63°23′ and 66°32′N. The position indicates that Iceland is within the arctic region, but due to the Gulf stream, which comes to the southeastern shore and flows clockwise around the island, the climate is milder than the position indicates. The climate is maritime with cool summers and mild winters. The island is mountainous and therefore considerable differences in the climatological elements are found (Eythórsson & Sigtryggsson 1971). The climate of the interior of northeast Iceland has continental characteristics (Einarsson 1979). The climate has been classified as subarctic in the lowlands, but in the central highland plateau, the climate is arctic (Sjörs 1963). However, since the island is virtually without a forest (cover less than 1%) (Steindórsson 1964) and most of the highlands have little vegetation cover (Table 24.1), it bears all resemblance to an arctic environment, even though the lack of forest

Table 24.1 Division of catchment area of the River Laxá, number of inhabitants, division of employment and number of domestic animals. (Vegetation cover based on Landsat satellite images: Landsat 1 MSS, 19 August 1973 and Landsat 5 TM 217/14 11 August 1987; information on area of cultivated land and on number of domestic animals from Jónasson *et al* (1986); and information on number of inhabitants and division of employment from the Institute of Regional Development, Reykjavík)

	Mývatn	Kráká	Laxá	Mýrarkvísl	Reykjadalsá	Sandvatn	Total
Total area (km^2)	1359	179	313	231	228	75	2385
Vegetated area (km^2)	234	65	278	139	201	19	936
Cultivated area (km^2)	5	1	14	6	11	0	37
Green houses (m^2)	250	0	0	4700	0	0	4950
Inhabitants 1990	458	8	393	106	306	0	1271
Employment (%)							
Farming	21	100	42	57	36		35
Industry	45		17	14	12		26
Services	32		37	15	38		33
Other	2		4	14	14		6
Domestic animals							
Cattle	59	57	1188	359	860		2523
Sheep	5090	664	5858	2734	4163		18509
Horses	119	10	214	48	89		480

(*Betula pubescens*) in the lowlands and the lack of vegetation in the highlands is mainly due to overgrazing by sheep. The southern boundary of the mild-Arctic is difficult to fix because forest gives way to tundra through a broad ecotone. Controversy exists in defining the climatic region of some northern tundra, e.g. in Iceland, northern Scandinavia and around the southwestern shores of Hudson's Bay (Longton 1988). Tundra regions beyond polar forest lines with mean monthly air temperatures at sea level below 12°C are regarded as polar. The 12°C isotherm line for July lies south of Iceland (Eythórsson & Sigtryggsson 1971).

The physical characteristics and biota of Icelandic rivers resemble rivers of the arctic and alpine regions of northern Norway and the mountain chain that forms the border between Sweden and Norway, and are grouped with them (Petersen *et al* 1994). However, rivers from spring-fed lakes in the volcanic zone of Iceland have higher production than other streams in the group.

Rivers in Iceland are divided into non-glacier-fed rivers and glacier-fed rivers (Pálsson 1792). The glacier-fed rivers originate as melt water from glaciers. Fine silt remains suspended in the water, and sand and silt are deposited in large quantities on the river bed. Along their courses the river changes repeatedly from a braided river into a swift current in a gorge (Rist 1970). Their flow is uneven, greatest in summer due to melt water and least through the winter (Rist 1956).

The non-glacier-fed rivers are generally clear all year round but may become brown in colour during floods. These rivers can be divided further into (i) spring-fed rivers and (ii) direct runoff rivers (Kjartansson 1945). Rist (1956) worked out the hydrological classification based on differences in flow and other attributes, and Gardarsson (1979b) and Gudjónsson (1990a,b) established their biological classification.

Direct runoff rivers have diffuse origins, made up of a network of small tributaries. These rivers are mainly confined to the older rock formations of Iceland outside the volcanic active zone. Their discharge fluctuates with daily precipitation and is uneven, with greatest peaks in spring and autumn (Rist 1956, 1979a). The productivity of these rivers is related to their length, steepness and the topography of the land through which they flow (Gardarsson 1979b). Usually their low nutrient status and fluctuating discharge make them rather unproductive (Gardarsson 1979b). Some of the more productive ones drain highland wetlands or lakes. Direct run-off rivers are ice-covered in the highlands during winter, but in the lowlands, ice cover forms during severe frost periods and floods occur during spring.

Spring-fed rivers originate in the volcanically active areas, where water percolates through recent lavas and palagonite rock formations. The water flows subterraneously until it emerges in artesian springs. Usually, the flow reaches full capacity not far from the head springs. The discharge is even all through the year, and floods rarely occur. These rivers are unaffected by weather and seasonal variations are smoothed out (Rist 1956). Spring-fed rivers remain open all year round, except during long and severe frosts, when frazil ice and anchor ice are formed. The ice disappears as soon as the temperature rises again and approaches 0°C. If ice cover is formed on spring-fed rivers, the water 'eats' the cover when the frost becomes milder and the rivers become susceptible to problems with ice, e.g. slush ice. There are two main types of spring-fed rivers, those which flow directly from the springs to the sea and those which flow through lakes (Gardarsson 1979b). The first group of spring-fed rivers does not differ much from direct runoff rivers in productivity, but the latter group includes many of the more productive rivers in Iceland (Gardarsson 1979b). This is mainly due to the biological production of the lakes, which contributes drifting organic matter to the rivers. The River Laxá is a spring-fed river affected by Lake Mývatn.

Sporadic permafrost is common in the central highlands, but rarely occurs below 500–600 m above sea level. In the Lake Mývatn district (ca. 300 m a.s.l.), frost mounds* are found in a few places, but along the River Laxá, permafrost has not been encountered. The annual mean temperature of the Lake Mývatn district is around 2°C, compared with 4–5°C in southern Iceland (Einarsson 1979). Lake Mývatn is within the region of Iceland showing the highest annual range of temperature (13–15°C), which is the difference between the average temperatures of the warmest month (July, average temperature 10.2°C) and coldest month (February, average temperature −4.2°C). The highest temperature ever recorded in the period 1937–76 was 25.9°C and the lowest −30.9°C. The average date of the last frost in spring for the period 1951–60 was 12 June and of first frost in autumn, 5 September (Einarsson 1979).

Large-scale river management for hydro-electricity began in the 1930s. Initially, lake-fed rivers were taken for his purpose, but since 1967 glacial rivers have also been utilized. The River Laxá flows out of Lake Mývatn on the border of the arctic and the subarctic, but flows in a subarctic environment (Fig. 24.1). One of its tributaries, the River Kráká, a spring-fed stream, originates at 430 m a.s.l. and the catchment of Lake Mývatn is mainly between 400 and 600 m a.s.l., with mountain peaks reaching up to 1500 m a.s.l.

The River Laxá has been under threat from hydro-electric development, soil erosion and the introduction of Atlantic salmon (*Salmo salar* L.) to areas that are not accessible to migratory fish species, but support a local brown trout stock (*S. trutta* L.). Lake Mývatn is threatened by sediment dredging, increased urbanization and tourism.

Knowledge of the ecosystem is necessary because management can only be as good as the knowledge it is based upon. Physical parameters of the River Laxá have been monitored for 20–40 years. The oldest discharge station was set up in 1947 (Rist 1979a), and temperature and chemistry of the river has been closely followed since 1971 (Ólafsson 1979b). Animal populations of Lake Mývatn and the River Laxá, have been studied from 1971 (Jónasson 1979) and population sizes have been monitored since 1975–77 (Gardarsson *et al* 1988; Gíslason & Gardarsson 1988). This chapter presents a review of the knowledge of the River Laxá ecosystem and provides a model which may be helpful for the management of subarctic and arctic rivers.

* A frost mound is a heap, often about 1 m high and a few metres in diameter, formed by an ice lens in wetlands, often under a pond that does not thaw in summer. The soil rises above the surrounding area and stays there for a year, while the surface soil and vegetation insulates the ice from the summer heat.

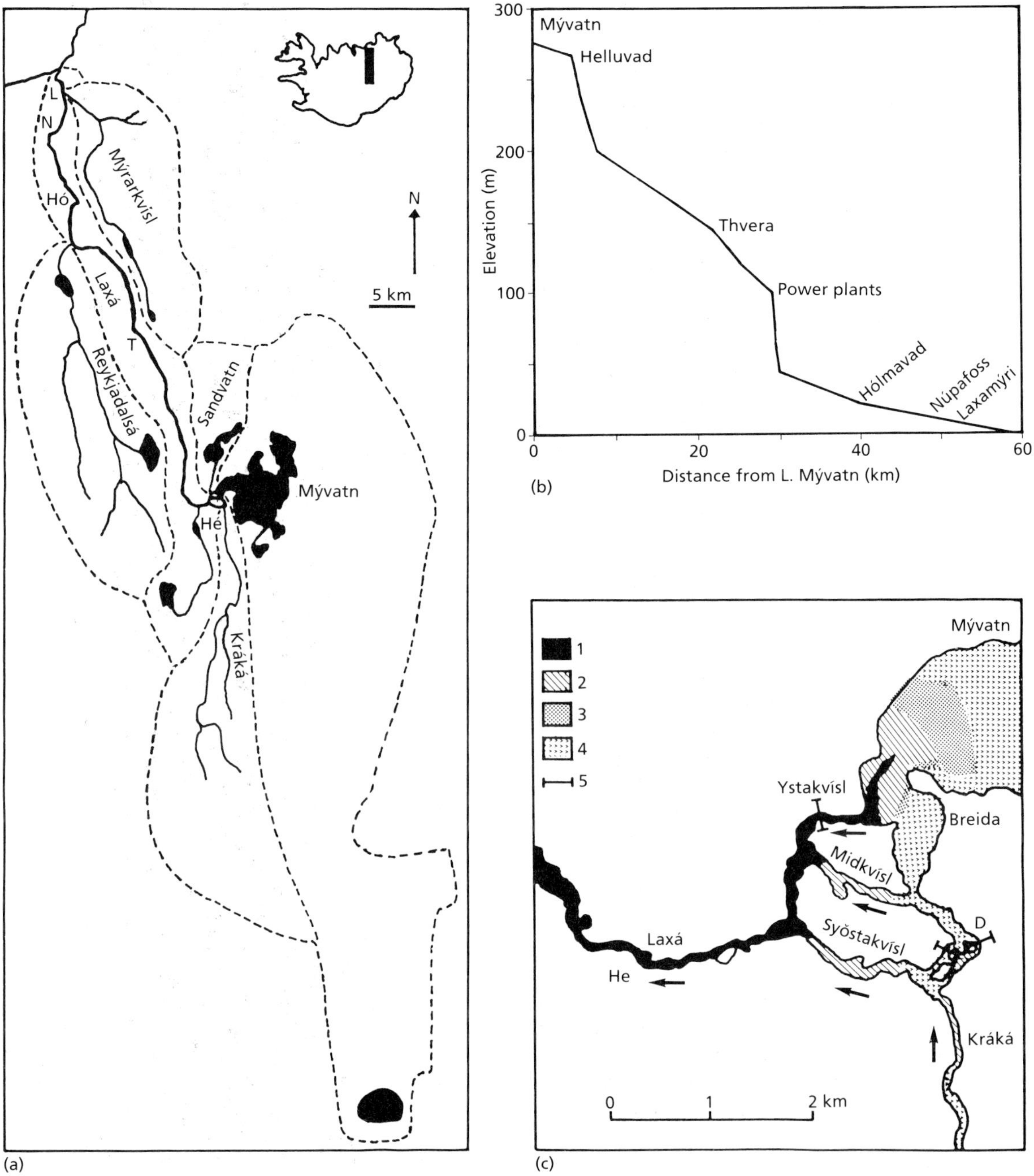

Fig. 24.1 (a) General location of Lake Mývatn and the River Laxá. Location of window traps: D = Dragsey (outlet), He = Helluvad, T = Thverá, Hó = Hólmavad, N = Núpafoss, L = Laxamýri. Catchment areas shown with broken lines. (b) The profile of the River Laxá, showing the locations of window traps and the power plants. (c) Ice conditions in the outlet of the River Laxá in winter. 1 = always open water; 2 = variable ice cover, most frequently free from ice, only covered with ice in periods of severe frosts; 3 = open water in mild winter periods; 4 = always ice covered in winter; 5 = dams; arrows show flow direction of water (from Rist 1979b).

24.2 THE RIVER LAXÁ

The River Laxá flows out of Lake Mývatn (65°35′N, 17°00′W) at 277 m a.s.l. (Fig. 24.1). Lake Mývatn is a shallow lake in the volcanic zone of northeast Iceland. The present lake was formed during volcanic activity 2300 years ago in a volcanic eruption east of the lake (Thórarinsson 1979). The lava flowed over the lake bottom to the sea along the River Laxá channel. For most of its length the present river flows in the lava channel on top of the lava (Thórarinsson 1951). Its maximum depth is 4.2 m and its surface area is 37 km^2. The water in Lake Mývatn originates from warm and cold springs on the eastern shores of the lake and in Lake Graenavatn, a short distance to the south. The Graenilaekur stream carries 5 $m^3 s^{-1}$ from Lake Graenavatn to Lake Mývatn. The theoretical retention time of Lake Mývatn is 27 days, with an outflow rate of 33 $m^3 s^{-1}$ (Ólafsson 1979a; Rist 1979a). The warm spring water is rich in nitrogen and the cold spring water is rich in phosphorus (Ólafsson 1979b). High solar radiation and nutrients in the spring water make Lake Mývatn one of the most productive lakes in northern Europe (Jónasson 1979). The Laxá River can be divided into two parts, the upper 33 km from Lake Mývatn to the Brúafossar waterfalls, and the lower 26 km from the waterfalls to the Arctic Ocean. The Brúafossar waterfalls in the Laxárgljúfur canyon are impassable to fish, even though a salmon ladder was constructed there in 1979.

Five spring- or lake-fed tributaries, with flow rates in the range 1–7 $m^3 s^{-1}$, enter the Laxá. The mean discharge rate of the Laxá is 55 m^3 s^{-1} when it enters the sea (Rist 1979a). The river floor consists of lava with boulders (diameter 20–50 cm), stones (diameter 10–20 cm), and gravel (<10 cm). At intervals the river widens. In the wider areas below the Kráká tributary the bottom consists of sand. The sand is brought in by the Kráká tributary, which originates in springs in sandy areas south of Lake Mývatn (36% vegetation cover in the catchment, Table 24.1). At the Laxá outlet, the mean water velocity is about 1 $m s^{-1}$. The mean water velocity in the upper part of the river is 1.3 $m s^{-1}$ (Rist 1979a). The lower part of the river, with an overall fall of 40 m, has a mean water velocity of 0.7 $m s^{-1}$ (Rist 1979a). The bottom is sandy, with riffles consisting of stones and boulders on the lava floor. At the lowest reach of the river, where the river widens, the mean velocity is about 0.4 $m s^{-1}$.

The catchment of Lake Mývatn (Fig. 24.1) is 1359 km^2, mostly a desert with only 17% of the area covered with vegetation (Table 24.1). The catchments of the Laxá and its tributaries have higher vegetation cover, ranging from 25% (Lake Sandvatn) to 88% (River Reykjadalsá), increasing with lower elevation. The total catchment of the Laxá is 2385 km^2, of which only 39% is covered with vegetation (based on Landsat satellite images: Landsat 1 MSS, 19 August 1973, and Landsat 5 TM 217/14, 11 August 1987). Agricultural use of land increases with lower elevation, with the shores of Lake Mývatn and the banks of the Laxá and its tributaries partly changed into cultivated land. The cover of cultivated land in the catchments varies between 0 and 4.8%. It is highest in the river valleys of Reykjadalsá and Laxá (Jónasson *et al* 1986).

The discharge of the Laxá River is stable (Ólafsson 1979a; Rist 1979a), and floods rarely occur. In the spring thaw in May, floods increase the discharge by about 35%, but occasionally by as much as 100% (three times in the last 20 years) (data from National Energy Authority, Hydrological Department). A small daily variation in discharge is caused by wind. River ice obstruction causes the largest variations in discharge (Rist 1979a). The chemical composition of the spring water is stable (Ólafsson 1979b), but changes in the chemical composition of the river water are caused by the lake's metabolism. The river is relatively warm compared to other rivers in Iceland. During winter (October–April) the monthly mean water temperature ranges from 0.3 to 2.3°C, compared to a range of 9.5–13.0°C in the summer (June to August) (Ólafsson 1979a; Gíslason & Jóhannsson 1991). Lake Mývatn primarily controls the thermal conditions in the river. The daily variations during the ice-free period indicate rapid heat exchange with the atmosphere. Small differences were found in the water temperature upstream and downstream in the river (Ólafsson 1979a).

Lake Mývatn is usually frozen over from

October or November, and normally thaws in May, or sometimes in early June (Rist 1979b). The mean (44 years' average) duration of ice cover on the lake is 189 days a year (Rist 1979b). The distance from Lake Mývatn to the power plants is 33 km; the water travel time is 6.0–7.5 h, with an average speed of 1.3 m s^{-1} (Rist 1979a). Where the Laxá flows out of Lake Mývatn, it passes through a 150-m wide passage into a wider slow-flowing part of the river, the so-called Breida (Fig. 24.1), at the outlet. From Breida, the river is divided into three main branches. Ystakvísl branch runs from the upper reaches; Midkvísl and Sydstakvísl branches flow out 700 m further below.

Ecology of the River Laxá

Fifty-two species of invertebrates have been identified from the river (Lindegaard 1979; Gíslason 1991). The blackfly *Simulium vittatum* Zett. (Simuliidae: Diptera) is the most abundant benthic invertebrate at the outlet, with midges (Chironomidae: Diptera) being the most abundant benthic invertebrates farther downstream (Fig.

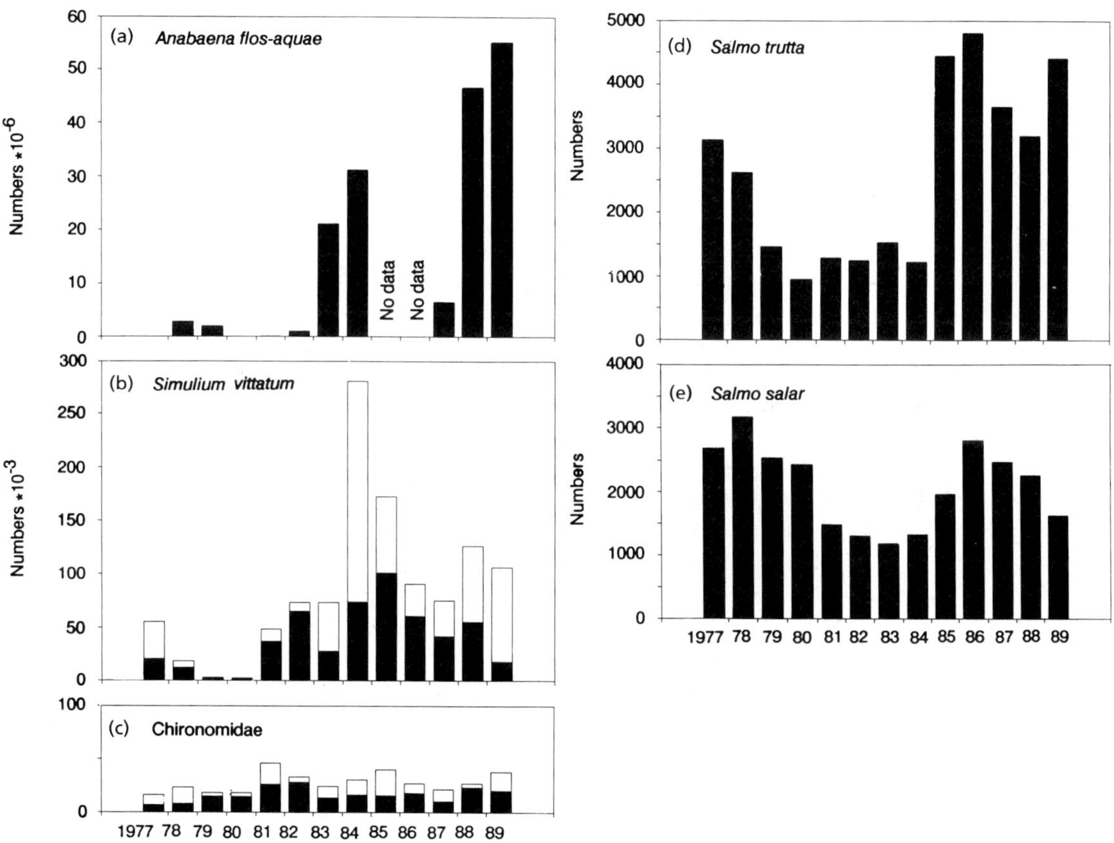

Fig. 24.2 Fluctuations in algal drift and animal populations in the River Laxá. (a) Mean number of cells l^{-1} of *Anabaena flos-aquae* from 15 June to 31 August; (b) catches of *Simulium vittatum* in window traps at the outlet, divided into early summer (filled) and late summer generations (open); (c) Chironomidae caught in window traps at the outlet, divided into catches before (filled) and after (open) 20 July; (d) catches of *Salmo trutta* in upper part of River Laxá and (e) catches of *S. salar* in lower part of River Laxá.

24.2). The Chironomidae account for 26 of the 52 benthic species.

The river is rich in seston (Gíslason 1985; Gíslason & Gardarsson 1988). It originates from the plankton and resuspended sediment of Lake Mývatn. The Kráká River adds sand to the seston. The quantity of fine particulate organic matter (FPOM) has fluctuated from year to year. At the outlet, it was about 3 mg l^{-1} from May to October in 1977–78, from 1979 to 82 it was 1–2 mg l^{-1}. It increased in 1982–85 to 3–4 mg l^{-1}, remained at similar levels between 1987 and 1989, but fell to 1–1.5 mg l^{-1} in 1990 (Gíslason & Gardarsson 1988; Gíslason unpublished results).

The flora of the Laxá has not received much attention (Jónasson 1979; Gíslason 1991). Several algal species dominate the flora. In summer, *Cladophora* sp. (Chlorophyta), which forms trailing threads up to 1 m long, attaches itself to stones in fast-flowing water, whereas *Spirogyra* sp. (Chlorophyta) grows in sheltered places. *Fragilaria construens* (Diatoma) is a common epilithic diatom. *Nostoc* cf. *opaca* (Cyanobacteria) forms pea-sized colonies on stones. The aquatic moss *Fontinalis antipyretica* is common where sand settles into crevices of stones. Vascular plants – *Potamogeton filiformis*, *Callitriche hermaphroditica* and *Carex rostrata* – occur in slow-flowing stretches of the river.

Studies on the production of benthic invertebrates in the river have shown that *Simulium vittatum* contributed about two-thirds of the total production in the upper reaches of the river (Gíslason 1985; Gíslason & Gardarsson 1988). In 1978–85, the *S. vittatum* production varied considerably. At the outlet it was 150 g ash free dry weight (AFDW) m^{-2} $year^{-1}$ (1977–78), 40–80 g AFDW m^{-2} $year^{-1}$ (1979–83) and increased to 900 (1984–85) g AFDW m^{-2} $year^{-1}$. At the sampling sites farther downstream (4.5 and 22 km) a similar fluctuation was observed, but the production values were lower (Gíslason 1985; Gíslason & Gardarsson 1988). Production of *S. vittatum* depends on the drifting FPOM from Lake Mývatn and on temperature (Gíslason & Gardarsson 1988). The number of life cycles per year also depends on temperature and drifting algae (Gíslason & Jóhannsson 1991). Production of Orthocladiinae chironomids fell in a similar manner as *S. vittatum* between 1977 and 1979 (Erlendsdóttir 1984). At the outlet it was about 20 g AFDW m^{-2} $year^{-1}$ in 1978 and 12 g AFDW m^{-2} $year^{-1}$ in 1979. At 4.5 and 22 km downstream, production was between 6 and 10 g AFDW m^{-2} $year^{-1}$ during the same years. These fluctuations were reflected in catches of adult flies in window traps at the river (Fig. 24.2) (Gíslason 1985; Gíslason & Gardarsson 1988). The food of *S. vittatum* was mainly detritus and drifting phytoplankton from the lake, and its composition was similar to the seston in the river (Gíslason & Jóhannsson 1985, 1991). *Eukiefferiella* spp. and *Orthocladius* spp. fed on detritus and epilithic diatoms (Gíslason, unpublished results). Presumably the detritus mostly originates in the lake, drifts down the river, and finally precipitates there. The numbers of *S. vittatum* and five species of Chironomidae caught in window traps declined downstream, i.e. *Eukiefferiella minor* (Edwards), *E. claripennis* (Lundbeck), *Orthocladius* (*Pogonocladius*) *consobrinus* (Holmgren), and *O.* (*O.*) *oblidens* (Walker), as did the quantity of FPOM (Fig. 24.3). This indicates that production of chironomids might be limited by availability of food drifting from the lake. However, certain species of Chironomidae, *Metriocnemus hygropetricus* (Kieffer), *Micropsectra atrofasciata* Kieffer and *Tanytarsus gracilentus* Holmgren, did not show this trend (Fig. 24.3) (Hrafnsdóttir 1991; Gíslason *et al* 1994).

There is a local stock of brown trout (*Salmo trutta* L.) in the upper part of the Laxá, and three-spined stickleback (*Gasterosteus aculeatus* L.) migrates from the lake into the river, whereas Atlantic salmon (*S. salar* L.), brown trout and arctic charr (*Salvelinus alpinus* (L.)) are found in the lower part. The fish in the lower part are anadromous. Catch statistics are available for brown trout in the upper part and for Atlantic salmon in the lower part of the river. The numbers of caught trout and salmon related to *S. vittatum* production (Fig. 24.2). The same applies for the duckling production of harlequin duck (*Histrionicus histrionicus* (L.)) (Gíslason & Gardarsson 1988; Gardarsson & Einarsson 1993) and the

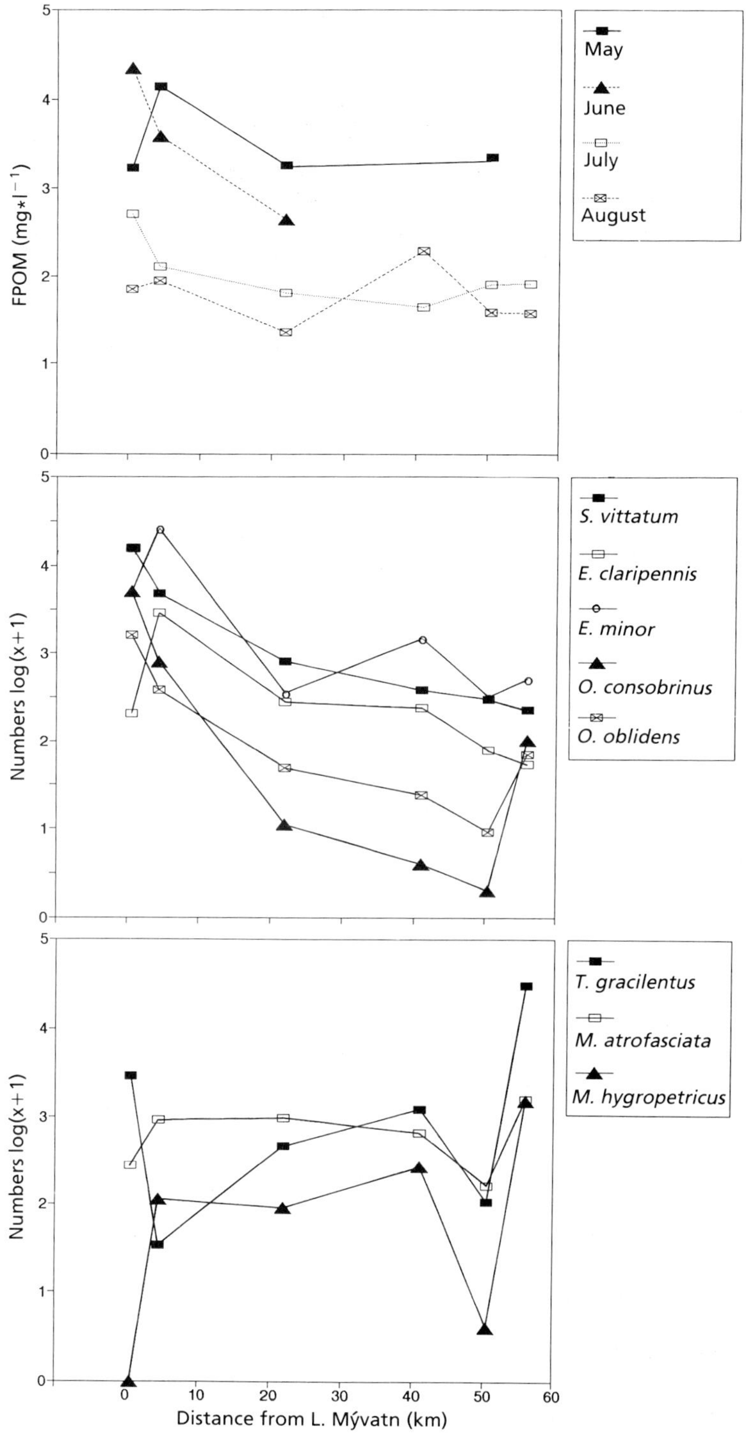

Fig. 24.3 Quantity of FPOM in River Laxá from 7 May to 10 August 1978 and numbers of blackflies and chironomid midges caught in window traps from May to October 1978 (based on Hrafnsdóttir 1991 and Gíslason *et al* 1994).

Barrow's goldeneye (*Bucephala islandica* (Gmelin)) (Gardarsson 1978).

24.3 MONITORING FAUNAL POPULATIONS AND ENVIRONMENTAL PARAMETERS

To understand an ecosystem, it is necessary to collect information on different parameters over a long time. Most limnological studies are conducted over relatively short periods, often for only 2–3 years. Although such studies may show considerable variation between years, the study period is too short to assess population fluctuations. Long-term population studies in lakes and rivers are surprisingly few (Gardarsson *et al* 1988), although large fluctuations on a timescale exceeding the approximate turnover time of the populations are commonplace in freshwater populations. Equilibrium, if it exists (see Connell & Sousa 1983), should perhaps be looked for on a larger time-scale (e.g. Elliott 1984; Gardarsson *et al* 1988). Monitoring has to meet two criteria. First it has to reflect changes in abiotic and biotic parameters or population sizes; secondly, it has to be simple and inexpensive to run.

Monitoring studies in the River Laxá were carried out on water temperature (see Ólafsson 1979a for methods), the quantity of seston, FPOM and number of cyanobacteria and other bacteria in the water (fortnightly samples from May to September from 1987, less frequent earlier) (Gíslason & Jóhannsson 1991), the population sizes of midges (Gíslason 1985; Gíslason & Gardarsson 1988; Jónsson *et al* 1986), number of ducks and duckling production (Gardarsson 1979a; Gardarsson & Einarsson 1993), and the catch of fish (Kristjánsson 1991; Gíslason, unpublished results). Temperature measurements together with seston and samples of algae were taken at the outlet, but other data were collected along the river.

Adult ducks on the River Laxá have been counted in the spring, and adults and ducklings in early August since 1975 (Gardarsson 1979a; Gardarsson *et al* 1988; Gardarsson & Einarsson 1993), thus giving population census and duckling production.

Statistics of brown trout caught in the upper part of the river have been available from 1973, since every rod fisherman has to return a report on his catch. These data are collected by the Institute of Freshwater Fisheries.

In 1977, when monitoring studies on invertebrate populations in the River Laxá and Lake Mývatn began, window traps were developed to catch flying insects and monitor populations of midges (Chironomidae) and blackflies (Simuliidae) (Jónsson *et al* 1986). Three traps were placed around Lake Mývatn, one trap at Lake Graenavatn and six traps were placed along the River Laxá. From 1978, the number of traps by the river was reduced to two, one at the outlet and another 4.5 km downstream (Gíslason 1985; Gíslason & Gardarsson 1988). By studying the production of *Simulium vittatum* from 1977 to 1985 at the outlet and 4.5 km downstream, it was possible to test the relationship between the window trap catches and *S. vittatum* production. A highly significant positive correlation was found between the window trap catches and the production of *S. vittatum* at the two sites ($r^2 = 0.590$, $p < 0.001$, df = 14) (Gíslason & Gardarsson 1988). Since then traps have only been complemented with sporadic sampling of benthic animals in the river (Fig. 24.2). The proportional numbers of different Chironomidae species found in traps coincided with their proportional numbers in benthic samples in Lake Mývatn in 1972–74 (Lindegaard & Jónasson 1979; Gardarsson 1988). The cost of analysing the trap catches is estimated to be less than 10% of the cost of sorting benthic samples for production studies.

These studies have increased the understanding of the relationships between different populations, and they are therefore of significance for management of the river. Results of monitoring highlighted the key role of *S. vittatum* in the aquatic ecosystem. Three important relationships are highlighted:

1 The number of *S. vittatum* in the summer generation was positively correlated with the *Anabaena flos-aquae* in the river ($r^2 = 0.630$, $0.01 > p > 0.001$, df = 8) (Figs 24.2 and 24.4), both measured at the outlet.

2 The number of trout caught in the upper part of the river was positively correlated with the window trap catches of *S. vittatum* at the outlet the previous year ($r^2 = 0.752$, $p = 0.001$, df = 11) (Figs 24.2 and 24.4). Multiple regression analysis of trout catches against *S. vittatum* and Chironomidae in the previous year did not yield higher positive significance ($R^2 = 0.643$, $0.01 > p > 0.001$, df = 10); the *S. vittatum* was significant ($t_{12} = 3.968$, $p = 0.002$), but the Chironomidae was not ($t_{12} = 0.202$, $p = 0.844$).

3 The production of Barrow's goldeneye ducklings was also found to be significantly positively correlated with window trap catches of *S. vittatum* ($r^2 = 0.455$, $0.05 > p > 0.01$, df = 10) (data on duckling numbers from Einarsson & Gardarsson 1990).

Four abundant Chironomidae species declined in number downstream (Fig. 24.3), but their numbers were relatively stable at each site from 1977 to 1989 (Gíslason *et al* 1994). Their food is to some extent provided by epilithic algae

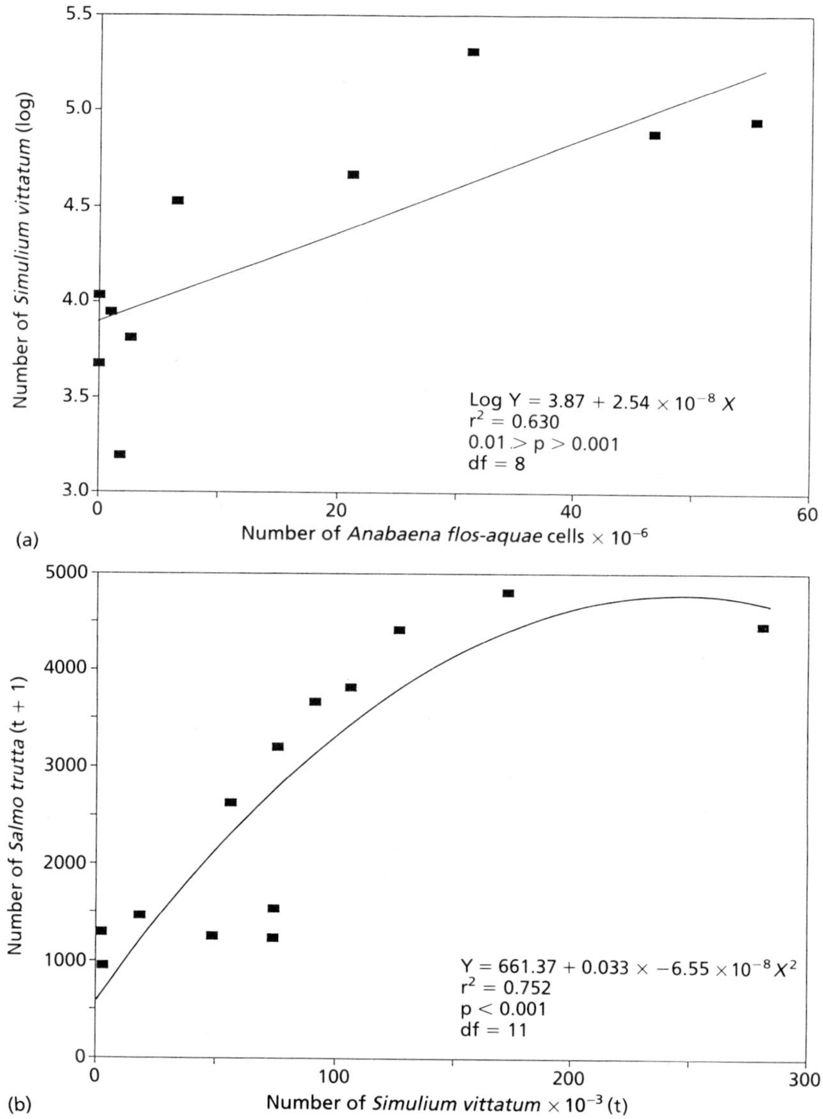

Fig. 24.4 (a) Relationship between numbers of *Anabaena flos-aquae* and *Simulium vittatum* from the summer generation. (b) Relationship between the numbers of *S. vittatum* caught in the outlet window trap and the number of *Salmo trutta* caught the following year. (A linear relationship is: $Y = 902.85 + 0.015X$, $r^2 = 0.642$, $p = 0.001$, df = 11).

on the river bottom; *Eukiefferiella* spp. and *Orthocladius consobrinus* (Holmgren) fed more on epilithic algae (*Synedra* sp. and *Epithemia* sp.) on the river bottom and less on drifting seston (Gíslason, unpublished results).

Possibly, long-term stability of the Chironomidae populations away from the outlet reflects the stability of the physical parameters of the river. If Chironomidae larvae rely on primary production of the river, their populations would be expected to fluctuate less than that of *S. vittatum*, which feeds mostly on drifting matter originating from Lake Mývatn, known for its large-scale fluctuations in phytoplankton production and animal populations (Gardarsson *et al* 1988). Fluctuations of *S. vittatum* are small compared with *Tanytarsus gracilentus* Holmgren of the lake, where a 3000-fold difference has been recorded between highest and lowest annual catches (Gardarsson *et al* 1988).

About 70% of the stomach content of brown trout in summertime in the Laxá is blackfly larvae, pupae and adults (Gíslason, unpublished results) and Barrow's goldeneye young feed almost entirely on blackfly larvae and pupae (Einarsson 1988). Over 96% of the stomach content of harlequin ducks on rivers in Iceland has been found to be blackfly larvae and pupae (Bengtson 1972) and production of harlequin ducklings fluctuated with *S. vittatum* production in the Laxá (Gíslason 1985; Gíslason & Gardarsson 1988; Gardarsson & Einarsson 1993).

24.4 DEVELOPMENT AND RIVER MANAGEMENT

Lake Mývatn, the Laxá River and its surroundings are considered one of the most scenic natural areas in Iceland. The landscape is shaped by volcanism and solfataras. The biota is renowned for its diversity and abundance and is considered unique for its latitude (65°35′N) (Jónasson 1979). Lake Mývatn and the Laxá are an important breeding area for large stocks of migratory waterfowl, including species not common elsewhere in Iceland (Gardarsson 1975). The unique geological setting, diverse biota and high productivity combined make the lake and river the most valuable area in Iceland for conservation and recreation (Nature Conservation Council 1987). This was recognized in 1974 by the Nature Conservation Act of Mývatn and Laxá, which covers almost the whole of the water catchment area (1760 km^2) (Skútustadir municipality and 200-m zone from the River Laxá on each side). In 1977 the area was designated according to the Convention of Wetlands of International Importance, especially as Waterfowl Habitats (Ramsar Convention). The scientific value of the lake and the river is also considerable, and increases with accumulated knowledge on their biota.

Since the time of settlement (late 9th century), the area has been inhabited by farmers. Agricultural land is either cut for hay or used as pasture, but the lifestock (especially sheep and horses) (Table 24.1) often range freely in summer. Cultivated land in the catchment is 37 km^2 of a total 2385 km^2 catchment area. However, only 936 km^2 of the catchment is covered with vegetation (Table 24.1). The farmers also have fished from Lake Mývatn and the River Laxá, and exploited the ducks by gathering eggs (Gudmundsson 1979).

The area has been a tourist attraction this century, but in the last 40 years tourism has increased substantially. Recreational fishing in the river is a 20th-century phenomenon, with increasing revenue for the landowners along the Laxá.

Since 1939, the river has been used for electricity production, with Lake Mývatn acting as a reservoir since 1945. A second power station was constructed in 1953 and a third in 1973. The total energy production is 20 MW. From 1967 sediment dredging from Lake Mývatn has been carried out for diatomite production (*c.* 25 000 tonnes per year). The diatomite production and increased tourism has led to urbanization at Lake Mývatn.

In 1970 a proposal to increase the electricity production caused a dispute over development of the River Laxá. This dispute was ended by an agreement in 1973 between the landowners and the National Power Company, which envisaged no water level changes of the river and the abandonment of further development after con struction of the third power station in 1973. The settlement was further strengthened by the Nature Conservation Act of Mývatn and Laxá in 1974. Since then, the Nature Conservation Council has had powers to restrict development in the area.

Management of river ice

Problems with river ice in arctic and subarctic rivers vary, depending on whether the local climate is continental or oceanic in character. In arctic and subarctic areas with a continental climate, river ice is formed at the onset of winter and remains on the river until the spring thaw (Rist 1962). The spring thaw is then followed by an ice-free summer season. In some areas, this cycle is so stable that the beginning and the end of its phases, or periods, may be predicted with an error of only a few days (Rist 1962). The thickness of the ice in such areas determines whether water can be extracted for use, e.g. for industry, homes, etc. (Mellor 1987), without harming the animal community in the rivers. Some of these rivers have been utilized for hydro-electric power production. Dams are constructed in such a way that when the upstream reservoirs freeze over, water can be taken from underneath. The ice cover formation is accelerated where dams and floating beams cause the water velocity to fall below a critical value of 0.5–0.6 m s^{-1} (Devik & Kanavin 1965; Rist 1979a).

In arctic and subarctic rivers with an oceanic climate, the problem is of another nature. Only rivers in the highlands of Iceland remain frozen over throughout most of the winter (Rist 1962, 1970) if they are under little or no influence of spring water. High winds are common, leading to rapid heat loss from the water surface. In the lowlands, ice cover is not continuous through the winter. Freezing and thawing occurs during winter due to changeable weather, resulting in drifting ice during thawing periods, and frazil ice and slush ice from river sections upstream during freezing periods.

As mentioned above, hydro-electric production in the Laxá started in 1939, and there are now three power plants (Gíslason 1991). They were constructed in the Brúafossar waterfalls, about 33 km below Lake Mývatn. Since Laxá has an even discharge, due to the spring-water that feeds Lake Mývatn, only small dams were constructed above the power plants. The first and third plants use the dam at the top of the Brúafossar waterfalls, but the second plant uses a low dam midway down the canyon.

Ice formerly blocked the water flow in the outlet of Laxá, about three or four times every winter, impeding power production for 2–5 days each time. The power plant management originally tried to manage ice jams by the use of dynamite (Rist 1950). Most of the ice jams were formed in the Breida, below the Ystakvísl branch outlet (Fig. 24.1c). The ice jams were of three types:

1 Slush ice and anchor ice, formed in northern high winds, severe frost and drift snow.
2 Anchor ice and ice broken-up from ice cover along the edges of the river following lowered water level and rapid evaporation from open water, formed in westerly winds and temperatures of below −6°C.
3 Broken-up ice cover in easterly winds, at temperatures above zero (Rist 1950, 1979b).

To control the flow of water, a dam with sluices was constructed on the Sydstakvísl branch in 1945. The Sydstakvísl branch was the second largest of the three branches, carrying 36% of the water volume. These sluice dams, however, did not solve the problem of ice jam in the river. It was found that ice formation occurred in the slow-flowing Breida region above the Sydstakvísl branch, sometimes above the Midkvísl branch (with 42% of river volume), and almost always below the smallest branch Ystakvísl (22% of the river volume) (Rist 1950, 1979b) (Fig. 24.1c). In 1959 the Midkvísl branch was completely dammed, with the exception of a fish passage. The Ystakvísl branch was then canalized, the canal being over 2.25 m deep and 8 m wide. A sluice dam was constructed to regulate the river flow. The bottom of the canal is even, to prevent frazil ice forming into anchor ice, and close to the bottom of the canal sluices are used to regulate the flow. The Ystakvísl branch has the capacity to pass 30–40 $m^3 s^{-1}$ of water from Lake Mývatn.

The dams can be used to raise the water level in Lake Mývatn but, according to an agreement between landowners, the National Power Company and the Nature Conservation Council, only 15-cm water level fluctuations are allowed in the lake (Adalsteinsson 1986). The dam in the Midkvísl branch was blown up by local farmers in 1970, because they claimed it was reducing brown trout migration between the Laxá and Lake Mývatn.

Ice blockage in the outlet region has not been a serious problem since these dams were constructed. However, after periods of severe frosts (temperatures below −10°C for a number of days), frazil ice and slush ice can form on the river below the outlet. Ice can then drift down and block the intake, or sometimes blocks of ice damage the turbines of the top power plant. Presently, only a low dam of the same height as the water intake (*c*. 2–3 m high) diverts the water flow into the intake above the uppermost power plant. A suggestion has been made to raise the water level of the river by 8 m and form a larger reservoir where ice can form above the power plants; the water intake will then be just above the bottom of the new reservoir and well below the influences of drifting ice. By this method, the power company hopes to overcome the problem of ice completely.

Fisheries management

In 1932, legislation was passed in order to direct a greater proportion of fishing income to locations upriver, i.e. to the farmers inhabiting the valleys. This law was repeatedly altered (1934, 1957 and 1970) in order to improve salmon management. Landowners along each salmon river are required to form a fishing association. These associations supervise fishing activities along their respective rivers (Gudjónsson 1978). Associations either lease fishing rights to angling clubs or sell angling privileges directly to individuals. The associations are also obliged to practise fish management, i.e. 'the protection of fish, improvement of its conditions for life, the importation of fish into fishing water, the easing of fish passage, fishing supervision, and anything else which may be conductive to increase or maintenance of the stock of fish' (The Salmon, Trout and Charr Fishing Act 1970).

The associations thus supervise fish management on a local basis and the expertise of the Institute of Freshwater Fisheries biologists is commonly sought in making management decisions. The duration of angling is limited to 3 months in the period between 20 May and 20 September, and angling is in its turn limited to 12 hours a day (commonly from 0700 to 1300 and 1600 to 2200 hours). The director of the Institute of Freshwater Fisheries, with the consent of the Freshwater Fisheries Council, determines the number of rods that may be fished at any one time in the rivers. In setting these rod limits, the director also seeks the views of the fishing association(s) of each respective river.

Because salmon fishing in Icelandic sea waters is prohibited, fishing is managed exclusively at the level of the individual river or tributary and, usually, the individual stock (Scarnecchia 1983). A comprehensive data collection system was established for angling catches on each river: an angling log is kept yearly and data on sex, length, weight, bait, site and date of capture of fish, and name of angler are recorded for each salmon caught. These records are returned to the Institute of Freshwater Fisheries each year. Site of capture along the river determines the allocation of fishing income among the members of the association. Landowners adjacent to more productive sections of the river for salmon receive proportionately more income from the fishing.

In the Laxá, brown trout, Atlantic salmon and Arctic charr are caught for local consumption and for recreational purposes. The rod-fishing season for Atlantic salmon in the lower part is from 10 June to 10 September and for brown trout from 1 June to 31 August. Prior to the rod-fishing season, anadromous charr and brown trout are fished with gill nets in the lower part of the river. No studies have been made on these populations. In the upper part of the river, less than 10% of the total catch of brown trout comes from netting, due to the prevalence of rod fishing. The purpose of fisheries management in the Laxá, as for all other salmon rivers in Iceland, is exclusively for 'high quality' sport angling (Gudjónsson 1978; Gudjónsson & Mills 1982). The prime task of salmon management has been not to maximize the yield, as in commercial fisheries, but to achieve the greatest satisfaction of the angling public (Mathisen & Gudjónsson 1978), which gives more revenue from the fishing than simply the value of the fish caught. Management of salmon fishing in the lower part of the Laxá is conducted in the same way as other salmon rivers in Iceland. In addition a small hatchery has been built close to the river mouth for rearing smolts and parr for release into the river.

The annual average angling catch of Atlantic salmon from 1971 to 1980 was 41 700 in all Icelandic rivers (Scarnecchia 1989). Commercial gill-net catches (exclusively from three glacial rivers) over the same period averaged 21 500 salmon. The average annual catch from the Laxa between 1977 and 1990 was 2050 salmon (range 1162–3161).

The management of the brown trout population in the upper river is similar. It is managed as a 'high quality' rod-fishing river and it is permitted to land only brown trout larger than 35 cm long; smaller brown trout are returned alive to the river. In 1979–81, the size limit was 40 cm. The maximum number of brown trout caught per rod is limited to 12 per day.

Between the years 1976 and 1983, and also in 1987, 12 000–31 000 brown trout parrs were released yearly into the upper river (Tómasson 1989). The angling association of the upper Laxá river has experimented with releasing salmon parrs and smolts in the upper Laxá river. From 1971 to 1980, 2000–6000 smolts were released yearly, and in 1971–75 and 1981–87 (except 1985) 11 000–75 000 parrs were released (Tómasson 1989). The young brown trout and salmon were reared from local brown trout and salmon stocks (Tómasson 1989). In spite of this and the construction of a fish ladder in the Laxárgljúfur canyon, salmon has only rarely been caught in the upper part of the river. Furthermore the number of salmon caught in the uppermost section of the lower part has remained between 60 and 120 per year, and cannot be related to the number of young salmon released above the hydro-electric power plants. The catch of salmon in the lower part of the river correlated positively with the catch of blackflies in window traps 3 years earlier ($r^2 = 0.364$, $0.05 > p > 0.01$, df = 9). It has been shown that the return of sea-going salmon to Icelandic rivers correlates significantly with temperature, and primary and secondary production in the adjacent seas in the year of smoltification (Scarnecchia 1984). The return of Atlantic salmon in the Laxá shows fluctuations similar to other rivers in northern Iceland.

The release of salmon parr and smolts near the river mouth has varied since 1967. A significant positive correlation has been found between the numbers of these releases with grilse (one sea-winter salmon) a year later ($p = 0.03$) in a multiple linear regression analysis, but not with two sea-winter salmon 2 years later (Thorvaldsson 1991). The return of grilse was related to the temperature of the Laxá between 4 May and 24 May, the sea temperature off north Iceland in May/June, zooplankton in the Arctic Ocean and smolt release the previous year ($R^2 = 0.943$, df = 15); the return of two sea-winter salmon was related to river discharge between 19 April and 26 May, zooplankton in the Arctic Ocean and sea temperature ($R^2 = 0.708$, df = 14).

There has been no increase in the number of permitted rods in the lower part of the river since 1970 and the number of permitted rods in the upper part remained at 32 from 1970 to 1987, but was lowered to 24 in 1988. From 1980 two rod permits at the outlet were sold outside the fishing association, which does not provide fishing statistics. A significant relationship has been observed between the production of Barrow's goldeneye ducklings and brown trout catch 1 year later (Einarsson & Gardarsson 1990).

The relationship between *Simulium vittatum* and brown trout (Fig. 24.4) indicates that following a year of large populations of *S. vittatum* the catch of brown trout reaches an asymptote. This might be due to the fishing quota (limited number of rods and catch limits) or that the growth of a large brown trout population is controlled by resources other than *S. vittatum*.

It may be concluded that the likelihood of return for Atlantic salmon depends on ocean conditions during its first spring in the sea. Brown trout catches seem to relate more strongly to events happening in the river, especially benthic production (*S. vittatum*).

Egg gathering

According to the Bird Protection Act (1966) all duck species are protected, except from hunting in autumn. However, farmers around Lake Mývatn and the Laxá River are allowed to gather eggs, since it is their traditional way of life. At the river, eggs from Barrow's goldeneye, tufted duck (*Aythya fuligula* (L.)), scaup (*A. marila* (L.)), wigeon (*Anas penelope* L.) and mallard (*A. platyrhynchos* L.) have been gathered for centuries for human consumption (Gudmundsson 1979). Eggs of Barrow's goldeneye and harlequin ducks have also been collected for zoos in Europe and elsewhere. Historically, farmers restricted egg gathering in the area by leaving four eggs (out of 9–10) for all species except Barrow's goldeneye, where five eggs are left intact in each nest (Gudmundsson 1979). This number is higher than the average number of 2–4 week-old ducklings per female, ranging from 1.05 for Barrow's goldeneye to 3.68 for mallard (Gardarsson 1991). No numbers of gathered eggs are available for the River Laxá area, but they could be estimated from the number of nests along the upper part of the river in May. The average number of drakes of Barrow's goldeneye, tufted duck, scaup, wigeon and mallard in May (1975–89) was 915, which indicates about 550 nests based on female/male ratio (Gardarsson 1991). It has been estimated that 2–3 eggs were removed from each nest on the average (Gardarsson 1979a). Thus between 1100 and 1650 eggs might be collected. Of those, eggs of Barrow's goldeneye are most numerous, both because it is the most abundant species on the Laxá and because farmers construct special nesting boxes for it to lay eggs in. There is no indication that this has affected the stocks of ducks (Gudmundsson 1979). At the river mouth, down is harvested from eider nests (*Somateria mollissima* (L.)).

Blackfly management

Many rivers harbour blackfly species (Simuliidae), which can be a pest in some areas. They constitute a serious menace to domestic animals, as well as to humans, in the temperate and subarctic regions of Canada and Russia (Saha 1973). Blackfly larvae have highest densities at lake outlets, where they feed on drifting organic matter from the source lake.

The blackfly, *Simulium vittatum*, requires blood before it can complete its second and subsequent gonotrophic cycles (Chutter 1970; Jónsson 1979; Jóhannsson 1988). It takes its blood meals mainly from domestic animals and humans. Although blackflies can sometimes occur in such numbers in the vicinity of mammals in the Mývatn district that their density obscures visibility (Peterson 1977), bites are not as common as might be expected. At time of emergence (early June and August) they cause a great annoyance, both to local people and to tourists that visit the area.

The blackflies from the River Laxá do not bite inside buildings or shelters, and to minimize the suffering of domestic animals (sheep, cattle and horses), the farmers living in the proximity of the outlet of the Laxá have made use of natural caves. These caves, in the lava fields that surround the lake and the river, have been made more accessible by widening of the cave mouths, or by building ramps from the cave mouths into the main cavities (Fig. 24.5). Animals enter these caves for protection when the blackflies are most aggressive.

In the tropics, where blackflies are vectors for onchocerciasis (Service 1980), and in some areas of the temperate region, blackfly control is common. Such control is not practised in the Arctic, because the region is so sparsely populated. In northern regions of Alberta and Saskatchewan, Canada, *Simulium* species, mainly *S. arcticum* Malloch, have caused a long-standing problem for livestock producers near the Athabasca and Saskatchewan Rivers. After DDT was virtually banned in Canada in 1970 (Wallace & Hynes 1981), methoxychlor has been used for blackfly control at a dosage of 9.6–10.7 ppb h^{-1} for a lethal dose of 90% of the larvae (LD_{90}), without seriously affecting other animals in the river (Charnetski & Haufe 1981; Freeden 1983). Other methods of keeping blackflies off livestock, such as repellents and construction of shelters, were not found to be economically feasible (Charnetski & Haufe 1981).

No attempts have been made to control the numbers of blackflies in the Laxá. Farmers along the river have traditionally had a considerable portion of their income from fisheries and to a lesser extent from egg gathering. They have long been aware of the importance of blackflies as a

Fig. 24.5 Photograph of a cave opening, which has been made accessible to domestic animals.

food for the fish and ducks. Were blackfly control applied, it could lead to a reduction of benthic production by two-thirds, with devastating effect on brown trout and Atlantic salmon fisheries and the number of ducks along the river.

Conservation management and threats to the River Laxá

The management of the various resources of the Laxá is administered by diverse interested parties: water and ice management is in the hands of the National Power Company; river fishing is managed by the two angling societies, one managing the Atlantic salmon in the lower part of the river, and the other the brown trout in the upper part. A third body, the Nature Conservation Council (NCC) by law (The Nature Conservation Act of Lake Mývatn and River Laxá 1974) makes conservation proposals to the government of Iceland. According to the act, the Mývatn Research Station was established as a scientific advisory body on conservation in the area. The board of the Research Station is appointed by the Minister for the Environment following nominations. The chairman and the vice-chairman are nominated by the NCC. Other members on the board are nominated by the Faculty of Science, University of Iceland, the Natural History Museum in Reykjavík, the Skútustadahreppur municipality (Mývatn district), and the Landowners Association of Lake Mývatn and the River Laxá.

Conflict may arise between the interested bodies, as well as with the developers at Lake Mývatn, the source of the Laxá. The plan for building a new 8-m-high dam at Laxárgljúfur canyon was proposed by the National Power Company in association with the angling society of the upper Laxá River because, according to the Nature Conservation Act of Mývatn and Laxá, any construction affecting water level changes has to be amenable to fish stock improvements or conservation of the river. The plan was to secure electricity production, but at the same time allow Atlantic salmon to spawn in the upper part of the river. Allegedly a high dam would divert smolts from the turbines of the hydro-electric power plants. By adding a new species, in this case salmon, increased competition may occur between the vertebrate populations, and this could lead to decreasing populations of Barrow's goldeneye, harlequin duck and brown trout. The Laxá and Lake Mývatn and a few nearby streams are the sole breeding grounds for Barrow's goldeneye in Europe (Gardarsson 1978; Einarsson 1988, 1990). Therefore the NCC rejected the plan.

All changes or modifications to the river or the river bottom have to be approved by the NCC. In addition, development in the Mývatn district may affect the river. Sediment dredging for diatom production has increased nutrient loading (phosphorus by 7% and nitrogen by 90%) (Ólafsson 1991), which may alter phytoplankton production and species composition of algae in the lake. Increased phosphorus loading is likely to increase algal production but increased nitrogen loading will probably reduce algal production by reducing the competitive advantage of the nitrogen-fixing *Anabaena flos-aquae* (Lúdvíksson 1991). Two special committees of experts, one operating from 1986 to 1991 and the second from 1992 to 1993, have been evaluating the effects of the sediment dredging. Increased urbanization and tourism will also increase nutrient loading. Failure of *A. flos-aquae* will reduce drift and reduce production of blackflies.

Erosion in the catchment area of the Laxá, where about 83% of the catchment of Lake Mývatn and 64% of the catchment of the River Kráká is sand deserts and lavas, will increase sand drift in the river, not only damaging turbines of hydro-electric power plants, but also making filter-feeding of blackfly larvae less efficient. The sand drift may also make river bottoms less favourable to benthic invertebrates and may decrease spawning sites of Salmonidae, by covering the bottom gravel with sand. Such threats will presumably become subjects for investigation in the future.

24.5 CONCLUSION

The human impact on rivers in cold regions is increasing. Rivers are particularly threatened by hydro-electric development, diversion, overfishing and pollution. Sound management of an ecosystem, its physical parameters (discharge, river ice), utilization of animal populations (brown trout, Atlantic salmon and duck eggs), pollution and for conservation purposes, must be based on the best available scientific knowledge of the stocks, their interrelationships and their relationships with physical parameters.

Long-term research has shown that sizes of natural biological populations of the Lake Mývatn and River Laxá ecosystems are highly variable. Fluctuations in phytoplankton production of Lake Mývatn and changes in resuspension of the lake's sediment leads to changes in seston and FPOM, fluctuations that were reflected in the food web of the animal populations in the river. By long-term studies, it was possible to relate these factors. However, it has not been possible to relate variable phytoplankton production of Lake Mývatn to known parameters. They are possibly related to different solar radiation from year to year, variable duration of ice cover on the lake, fluctuations in temperature and variable wind conditions in summer.

Even though the research in the area started in 1971 and monitoring studies on animal populations started in 1975–77, 15–17 years of monitoring is inadequate to analyse all relationships. Information is still emerging and a better picture of the ecosystem will be obtained in the future. If the experiences obtained from the research and management of the River Laxá can be extrapolated to other rivers in cold regions, then long-term studies will be necessary for proper management and to detect changes brought about by human involvement.

ACKNOWLEDGEMENTS

I should like to thank my friends and colleagues Professor Arnthór Gardarsson, Dr Árni Einarsson and Mrs Thóra Hrafnsdóttir, Institute of Biology, University of Iceland, for their co-operation through the study. I am also indebted to several people of the Institute of Biology, too numerous to name here, for various assistance during the research. Thanks are due to Mrs Halldóra Thorsteinsdóttir at the Library of the University of Iceland for searching for literature on arctic rivers, and to Mrs Erla Sigthórsdóttir and Mrs Ingibjörg Gísladóttir, librarians of the National Energy Authorities, who provided published and unpublished reports on water and river ice management. Mr Gudni Gudbergsson, the Institute of Freshwater Fisheries, provided data on Salmonidae catches in the River Laxá. Mr Hans H. Hansen, Iceland Geodetic Survey, Reykjavík, analysed the Landsat satellite images for vegetation cover. Logistic and financial support was received from the Science Fund of Iceland, The University of Iceland Research Fund, the Icelandic Aluminium Co. Ltd and The Nature Conservation Council. I thank Professor Arnthór Gardarsson, Dr Árni Einarsson, Dr Thóra E. Thórhallsdóttir (University of Iceland), Mr Gudni Gudbergsson (Institute of Freshwater Fisheries), Mr Árni Snorrason (National Energy Authorities), Dr Richard C. Russell, Messrs Peter Wells and John Clancy (University of Sydney Medical Entomology Department at Westmead Hospital, Sydney, Australia), and my wife Kristín Hafsteinsdóttir (Department of English, University of Sydney) for comments on the manuscript of this paper. Most of the writing was done when I was on a sabbatical leave at the Medical Entomology Department, University of Sydney at Westmead Hospital.

REFERENCES

Adalsteinsson H. (1986) Vatnsaflsvirkjanir og vötn (Water power plants and watersystems). *Náttúrufraedingurinn* **56**: 109–31. [In Icelandic, English summary] [24.4]

Bengtson S-A. (1972) Breeding ecology of the harlequin duck *Histrionicus histrionicus* (L.) in Iceland. *Ornis Scandinavica* **3**: 1–19. [24.3]

Charnetski WA, Haufe WO. (1981) Control of *Simulium arcticum* Malloch in Northern Alberta, Canada. In: Laird M (ed) *Blackflies. The Future for Biological Methods in Integrated Control*, pp. 117–32. Academic Press, New York. [24.4]

Chutter FM. (1970) A preliminary study of oocytes present in newly emerged blackflies (Diptera: Simuliidae) in Ontario. *Canadian Journal of Zoology* **48**: 1389–400. [24.4]

Connell JH, Sousa WP. (1983) On the evidence needed to judge ecological stability or persistence. *American Naturalist* **121**: 789–824. [24.3]

Devik O, Kanavin EV. (1965) *Final Report on Analysis and Consideration of the Ice Conditions in the Hvítá and Thjórsá River Systems, Southern Iceland*. State Electricity Authority, Reykjavik. [24.4]

Einarsson Á. (1988) Distribution and movements of Barrow's goldeneye *Bucephala islandica* young in relation to food. *Ibis* **130**: 153–63. [24.3, 24.4]

Einarsson Á (1990) *Lax í efri Laxá. Greinargerd um líkleg áhrif laxa á lífríki árinnar ofan Brúa* [Atlantic salmon in upper part of River Laxá. Report on the possible effects of Atlantic salmon on the biota in the upper part of River Laxá]. Mimeograph No. 22, 1–31. Nature Conservation Council, Reykjavik. [In Icelandic] [24.4]

Einarsson Á, Gardarsson A. (1990) *Efri Laxá: Húsandarungar og urridaveidi* [Upper part of River Laxá: Barrow's goldeneye ducklings and catch of brown trout]. Mimeograph No. 22, 32–6. Nature Conservation Council, Reykjavik. [In Icelandic] [24.3, 24.4]

Einarsson MÁ. (1979) Climatic conditions of the Lake Mývatn area. *Oikos* **32**: 29–37 [24.1]

Elliott JM. (1984) Growth, size, biomass and production of young migratory trout *Salmo trutta* in a Lake District stream, 1966–83. *Journal of Animal Ecology* **53**: 979–94. [24.3]

Erlendsdóttir A. (1984) *Framleidsla rykmýs í Laxá* (The production of Chironomidae in the River Laxá 1978–1979). Mimeograph No. 14, 73–6. Nature Conservation Council, Reykjavík. [In Icelandic, English summary] [24.2]

Eythórsson J, Sigtryggsson H. (1971) The climate and weather of Iceland. *The Zoology of Iceland* **1**(3): 1–62. [24.1]

Freeden FJH. (1983) Trends in numbers of aquatic invertebrates in a large Canadian river during four years of black fly larviciding with methoxychlor (Diptera: Simuliidae). *Questiones Entomologicae* **19**: 53–92. [24.4]

Gardarsson A. (ed) (1975) *Votlendi* (*Wetlands*). Rit Landverndar 4, Reykjavík. [In Icelandic, English summary] [24.4]

Gardarsson A. (1978) Íslenski húsandarstofninn (Distribution and numbers of Barrow's goldeneye (*Bucephala islandica*) in Iceland). *Náttúrufraedingurinn* **48**: 162–91. [In Icelandic, English summary] [24.2, 24.4]

Gardarsson A. (1979a) Waterfowl populations of Lake Mývatn and recent changes in numbers and food habits. *Oikos* **32**: 250–70. [24.3, 24.4]

Gardarsson A. (1979b) Vistfraedileg flokkun íslenskra vatna (A classification of Icelandic freshwaters). *Týli* **9**: 1–10. [In Icelandic, English summary] [24.1]

Gardarsson A. (1988) Stofnbreytingar í Mývatni og líklegar orsakir theirra (Population changes at Lake Mývatn and its causes). *Tímarit Háskóla Íslands* **3**: 55–66. [In Icelandic] [24.3]

Gardarsson A. (1991) Fuglalíf vid Mývatn og Laxá (The bird-life of Mývatn and Laxá). In: Gardarsson A, Einarsson Á (eds) *Náttúra Mývatns*, pp. 278–319. The Icelandic Natural History Society, Reykjavík. [In Icelandic] [24.4]

Gardarsson A, Einarsson Á. (1993) Responses of breeding duck populations to changes in food supply. *Hydrobiologia* (in press). [24.2, 24.3]

Gardarsson A, Gíslason GM, Einarsson Á. (1988) Long term changes in the Lake Mývatn ecosystem. *Aqua Fennica* **18**: 125–35. [24.1, 24.3]

Gíslason GM. (1985) The life cycle and production of *Simulium vittatum* Zett. in the River Laxá, North-East Iceland. *Verhandlungen der Internationalen Vereinigung für theoretische und angewandte Limnologie* **22**: 3281–7. [24.2, 24.3]

Gíslason GM. (1991) Lífid í Laxá (Life in River Laxá). In: Gardarsson A, Einarsson Á (eds) *Náttúra Mývatns*, pp. 218–35. The Icelandic Natural History Society, Reykjavik. [In Icelandic] [24.2, 24.4]

Gíslason GM, Gardarsson A. (1988) Long term studies on *Simulium vittatum* Zett. (Diptera: Simuliidae) in the River Laxá, North Iceland, with particular reference to different methods used in assessing population changes. *Verhandlungen der Internationalen Vereinigung für theoretische und angewandte Limnologie* **23**: 2179–88. [24.1, 24.2, 24.3]

Gíslason GM, Jóhannsson V. (1985) Bitmýid í Laxá (The biology of the blackfly *Simulium vittatum* Zett. (Diptera: Simuliidae) in the River Laxá, northern Iceland). *Náttúrufraedingurinn* **55**: 175–94. [In Icelandic, English summary] [24.2]

Gíslason GM, Jóhannsson V. (1991) Effects of food and temperature on the life cycle of *Simulium vittatum* Zett. (Diptera: Simuliidae) in the River Laxá, N-Iceland. *Verhandlungen der Internationalen Vereinigung für theoretische und angewandte Limnologie* **24**: 2912–16. [24.2, 24.3]

Gíslason GM, Hrafnsdóttir Th, Gardarsson A. (1994) Long term monitoring of numbers of Chironomidae and Simuliidae in the River Laxá, North Iceland. *Verhandlungen der Internationalen Vereinigung für theoretische und angewandte Limnologie* **25**: (in press). [24.2, 24.3]

Gudjónsson S. (1990a) *Classification of Icelandic watersheds and rivers to explain life history strategies of Atlantic salmon*. PhD Thesis, Oregon State University. [24.1]

Gudjónsson S. (1990b) Íslensk vötn og vistfraedileg flokkun theirra (Icelandic waters and their ecological classification). In: Sigbjarnarson G (ed) *Vatnid og landid*, pp. 219–23. National Energy Authority, Reykjavík. [In Icelandic] [24.1]

Gudjónsson Th. (1978) The Atlantic salmon in Iceland. *Journal of Agricultural Research in Iceland* **10**: 11–39. [24.4]

Gudjónsson Th, Mills D. (1982) *Salmon in Iceland*. The Atlantic Salmon Trust, Farnham. [24.4]

Gudmundsson F. (1979) The past status and exploitation of the Mývatn waterfowl populations. *Oikos* **32**: 232–49. [24.4]

Hrafnsdóttir Th. (1991) *My í Laxá í S-Thingeyjarsýsla: Dreifing tegunda frá upptökum til ósa* (Longitudinal distribution and phenology of midges (Chironomidae: Simuliidae) in the River Laxá, N. Iceland). Unpublished thesis, University of Iceland, Reykjavík. [In Icelandic, English summary] [24.2]

Jóhannsson V. (1988) The life cycles of *Simulium vittatum* Zett. in Icelandic lake-outlets. *Verhandlungen der Internationalen Vereinigung für theoretische und angewandte Limnologie* **23**: 2170–8. [24.4]

Jónasson H, Steingrímsdóttir JA, Arnórsson E. (1986) *Byggdir og bú Sudur-Thingeyinga 1985* (Settlements and farms in the county of South-Thingeyjarsýsla). Búndadarsamband Sudur-Thin-geyjarsýslu. Húsavík. [In Icelandic] [24.1, 24.2]

Jónasson PM (ed). (1979) Ecology of eutrophic, subarctic Lake Mývatn and the River Laxá. *Oikos* **32**: 1–308. [24.1, 24.2, 24.4]

Jónsson E. (1979) *Athugun á mýflugum í Mývatnssveit, 1977* (Studies on chironomids and blackflies in the Lake Mývatn area, summer 1977). Unpublished thesis, University of Iceland, Reykjavík. [In Icelandic, English summary] [24.4]

Jónsson E, Gardarsson A, Gíslason GM. (1986) A new window trap used in the assessment of the flight periods of Chironomidae and Simuliidae (Diptera). *Freshwater Biology* **16**: 711–19. [24.3]

Kjartansson G. (1945) Vatnfallategundir (Types of running waters). *Náttúrufraedingurinn* **15**: 113–26. [In Icelandic] [24.1]

Kristjánsson J. (1991) Fiskurinn í Mývatni og Laxá (The fish of Lake Mývatn and River Laxá). In: Gardarsson A, Einarsson Á (eds) *Náttúra Mývatns*, pp. 256–77. The Icelandic Natural History Society, Reykjavík. [In Icelandic] [24.3]

Lindegaard C. (1979) A survey of the macroinvertebrate fauna, with special reference to Chironomidae (Diptera) in the rivers Laxá and Kráká, northern Iceland. *Oikos* **32**: 281–8. [24.2]

Lindegaard C, Jónasson PM. (1979) Abundance, population dynamics and production of zoobenthos in Lake Mývatn, Iceland. *Oikos* **32**: 202–27. [24.3]

Longton RE. (1988) *The Biology of Polar Bryophytes and Lichens*. Cambridge University Press, Cambridge. [24.1]

Lúdvíksson V (ed). (1991) *Effects of the Operation of Kisilidjan Inc., on the Lake Mývatn Biota*. Final report of Committee of Experts for Lake Mývatn Research. Ministry for the Environment, Reykjavík. [24.4]

Mathisen OA, Gudjónsson Th. (1978) Salmon management and ocean ranching in Iceland. *Journal of Agricultural Research in Iceland* **10**: 156–74. [24.4]

Mellor JC. (1987) Arctic water availability: Data bases to meet present and future needs. In: Dworsky RF (ed.) *Water Resources Related to Mining and Energy – Preparing for the Future*, pp. 71–8. American Water Resources Association, Technical Publication Series TPS-87-4, Bethesda, Maryland. [24.1, 24.4]

Nature Conservation Council. (1987) *Mývatnssveit. Náttúruverndarkort* (*Mývatn district. Nature*

Conservation Map). Mimeograph No. 19. Nature Conservation Council, Reykjavik. [In Icelandic] [24.4]

Ólafsson J. (1979a) Physical characteristics of Lake Mývatn and River Laxá. *Oikos* **32**: 38–66. [24.2, 24.3]

Ólafsson J. (1979b) The chemistry of Lake Mývatn and River Laxá. *Oikos* **32**: 82–112. [24.1, 24.2]

Ólafsson J. (1991) Undirstödur lífríkis í Mývatni (The physical basis of life in Mývatn) In: Gardarsson A, Einarsson Á (eds) *Náttúra Mývatns*, pp. 140–65. The Icelandic Natural History Society, Reykjavík. [In Icelandic] [24.4]

Orians GH, Buckley J, Clark W, Gilpin ME, Jordan CF, Lehman JT, May RM, Robilliard GA, Simberloff DS. (1986) *Ecological Knowledge and Environmental Problem-Solving. Concepts and Case Studies*. National Academy Press, Washington, DC. [24.1]

Pálsson S. (1792) *Ferdabók* (*Travels in Iceland*). Eythórsson J (ed.) (1945). Reykjavík. [In Icelandic] [24.1]

Petersen RC, Gíslason GM, Petersen LB-M. (1994) Rivers of the Nordic countries. In: Cushing CE, Cummins KW, Minschall GW (eds) *River and Stream Ecosystems: Ecosystems of the World*, Vol. 22. Elsevier, Amsterdam (in press). [24.1]

Peterson BV. (1977) The black flies of Iceland (Diptera: Simuliidae). *Canadian Entomologist* **109**: 449–72. [24.4]

Reist JD, Bond WA. (1988) Life history characteristics of migratory coregonids of the lower Mackenzie River, Northwest Territories, Canada. *Finnish Fisheries Research* **9**:133–44. [24.1]

Rist S. (1950) *Ístruflanir vid Mývatnsósa. Skýrsla um athugun á rennsli úr Mývatni í janúar 1950* (Ice disturbances of the water flow at the outlet of Lake Mývatn in January 1950). Raforkumálastjóri-Vatnamaelingar, Mimeograph S2 Report 5, Reykjavík. [In Icelandic] [24.4]

Rist S. (1956) *Íslensk vötn* (Icelandic fresh waters). Hydrological Survey, State Electricity Authority, Reykjavík. [In Icelandic, English summary] [24.1]

Rist S. (1962) Thjórsárísar (Winter Ice of Thjórsá River System). *Jökull* **12**: 1–30. [In Icelandic, English summary] [24.4]

Rist S. (1970) Ice conditions in the Thjorsa River System. In: *International Association for Hydraulic Research (IAHR) Ice Symposium and its Action on Hydraulic Structures*. National Energy Authority, Reykjavík. [24.1, 24.4].

Rist S. (1979a) The hydrology of River Laxá. *Oikos* **32**: 271–80. [24.1, 24.2, 24.4]

Rist S. (1979b) Water level fluctuations and ice cover of Lake Mývatn. *Oikos* **32**: 67–81. [24.1, 24.2, 24.4]

Rosenberg DM, Barton DR. (1986) The MacKenzie River system. In: Davies BR, Walker KF (eds) *The Ecology of River Systems*, pp. 425–33. W. Junk Publishers, Dordrecht. [24.1]

Ryan RR. (1986) Soviets shelve plan on diverting rivers in Arctic regions. *Oceanus* **29**: 78–80. [24.1]

Saha JG. (1973) Insecticides used for the control of biting flies and their pollution potential. In: Hudson A (ed) *Biting Fly Control and Environmental Quality*. Proceedings of a Symposium held at the University of Alberta, Edmonton, pp. 19–34. Defence Research Board, Ottawa. [24.4]

Scarnecchia DL. (1983) Age at sexual maturity in Icelandic stocks of Atlantic salmon (*Salmo salar*). *Canadian Journal of Fisheries and Aquatic Sciences* **40**: 1456–68. [24.4]

Scarnecchia DL. (1984) Climatic and oceanic variations affecting yield of Icelandic stocks of Atlantic salmon (*Salmo salar*). *Canadian Journal of Fisheries and Aquatic Sciences* **41**: 917–35. [24.4]

Scarnecchia DL. (1989) The history and development of Atlantic salmon management in Iceland. *Fisheries* **14**: 14–21. [24.4]

Service MW. (1980) *A Guide to Medical Entomology*. Macmillan Press, London. [24.4].

Sjörs H. (1963) Amphi-Atlantic biota and their history. In: Löve A, Löve D (eds) *North Atlantic Biota and Their History*, pp. 109–25. Pergamon Press, Oxford. [24.1]

Slack KV, Nauman JW, Tilley LJ. (1977) Benthic invertebrates in an arctic mountain stream, Brooks Range, Alaska. *Journal of Research US Geological Survey* **5**: 519–27. [24.1]

Steindórsson S. (1964) *Gródur á Íslandi* (Vegetation in Iceland). A.B., Reykjavík. [In Icelandic]. [24.1]

Stewart KW, Hassage RL, Holder SJ, Oswood MW. (1990) Life cycles of six stonefly species (Plecoptera) in subarctic and arctic Alaska streams. *Annals of the Entomological Society of America* **83**: 207–14. [24.1]

Thórarinsson S. (1951) Laxárgljúfur and laxárhraun (A tephrological study). *Geografiska Annaler* **33**: 1–89. [24.2]

Thórarinsson S. (1979) The postglacial history of the Mývatn area. *Oikos* **32**: 17–28. [24.2]

Thorvaldsson JTh. (1991) *Spálíkan fyrir laxveidi í Laxá í Adaldal* (Model for forecasting catch of Atlantic salmon in the lower part of R. Laxá). Thesis, Department of Mechanical Engineering, University of Iceland. [In Icelandic] [24.4]

Tómasson T. (1989) *Laxaseidasleppingar í Laxá ofan Brúa* (Release of Atlantic salmon fry in upper part of River Laxá). Report to the Nature Conservation Council of Iceland. [In Icelandic] [24.4]

Ulfstrand S. (1968) Benthic animal communities in Lapland streams (Ephemeroptera, Plecoptera, Trichoptera and Diptera: Simuliidae). *Oikos* **10** (Suppl.): 1–20. [24.1]

Wallace RR, Hynes HBN. (1981) The effect of chemical treatments against blackfly larvae on the fauna of running waters. In: Laird M (ed) *Blackflies. The Future for Biological Methods in Integrated Control*, pp. 237–58. Academic Press, New York. [24.4]

25: Dryland Rivers: Their Ecology, Conservation and Management

B. R. DAVIES, M. C. THOMS, K. F. WALKER,
J. H. O'KEEFFE AND J. A. GORE

25.1 INTRODUCTION

Many of the world's people live in arid and semi-arid areas but most river ecologists live in the humid, temperate areas of Europe and North America. As a direct result of this geographical bias, both the theoretical approaches to, and the popular ideas about, river ecology tend to overlook the peculiar biophysical features of dryland rivers. Such features are readily seen from the viewpoints of hydrology and geomorphology (Alexander 1985; Graf 1987; Finlayson & McMahon 1988; Puckridge *et al* 1993), and ecological functioning (e.g. Vannote *et al* 1980; see Davies *et al* 1993b). Indeed, it is doubtful if many of the ideas developed in temperate systems can usefully be applied to the rivers of dry regions (cf. Williams 1988). Conventional management practices established in temperate regions may also be inappropriate, given that the patterns of water use in dry regions are quite unlike those in humid areas.

Such issues are *critical* because all countries with extensive dry regions, and increasing human populations, face rapidly growing crises of water supply and demand (e.g. Davies & Day 1986). In South Africa, for example, all available water resources will be committed within the next 10 years, despite conservative use, increased prices and the inclusion of ground-water reserves (but without regard for desalination and recycling). Figure 25.1 compares estimates of South African water demand and population growth projected to the year 2050. The curves illustrate 'best' and 'worst' scenarios relative both to water use and to the total exploitable surface and groundwater resources ($4959 \times 10^9 \, m^3 \, yr^{-1}$), which together suggest that a national water deficit will develop between 2003 and 2017. It is ominous that the demographic trends were underestimated in 1986, when the curves were drawn, and that the population now is more than 32 million and growing at *c.* 2.3% yr^{-1}. Most of the demand (*c.* 50%) is from the agricultural sector (Table 25.1). Moreover, the crisis is not simply one of water quantity, but includes water quality, as most South African rivers have been profoundly changed by impoundment and interbasin transfers (IBTs; cf. Davies *et al* 1992), and by pollution and salinization. The need for conservation is manifestly obvious, but it is difficult to persuade the poor rural and urban dwellers of South Africa, let alone those of other dryland countries, that short-term economies of water use contribute to greater security of supplies in the long term (e.g. O'Keeffe & Davies 1991).

As the rivers of dryland regions frequently form the *only* exploitable water resources (e.g. Davies & Day 1986), they present critical problems for water conservation and management for which all sectors of the 'water industry', including researchers, are largely unprepared. Almost by definition, the resource is often placed under far greater stresses than more humid-zone systems, so much so that even in moderately populated dryland regions: 'There are few rivers [in such zones] which have not been regulated by impoundment/s, and ... many that were once perennial ... now flow only ["seasonally"]' (Davies *et al* 1993b).

To give one pertinent example, IBTs, common solutions to water-supply problems of dryland regions, cause the breakdown of biogeographical barriers, the loss of basin integrity, the spread

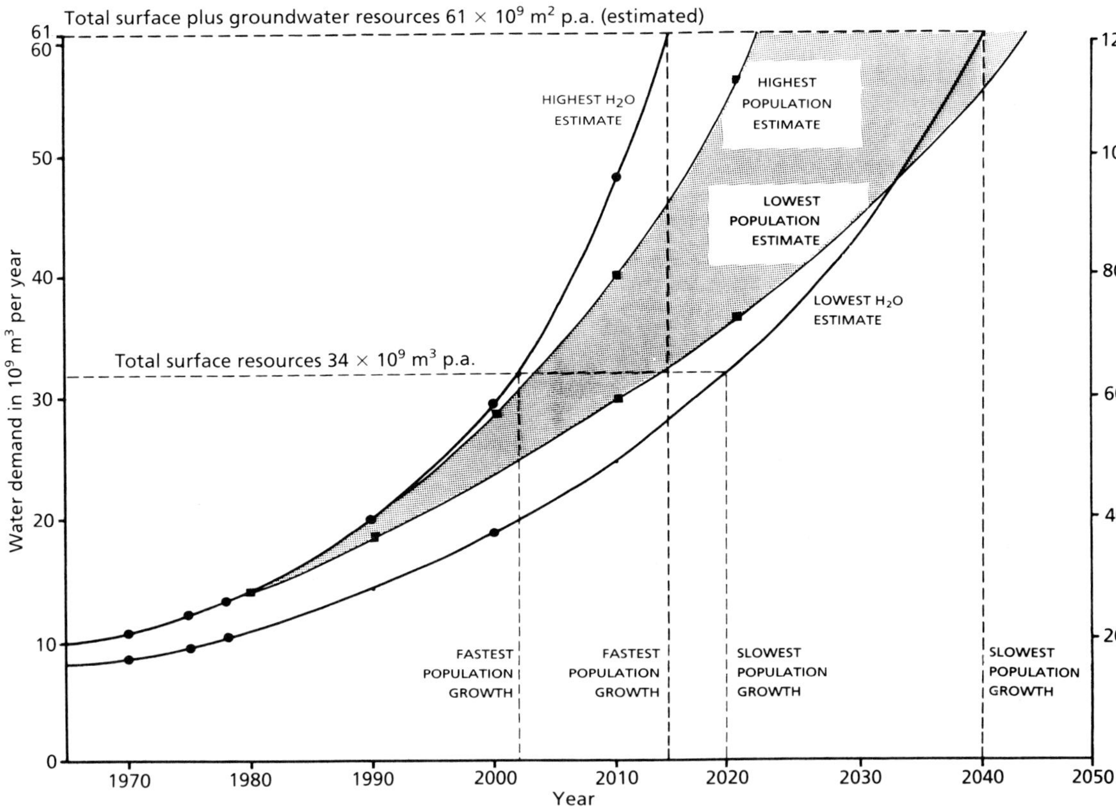

Fig. 25.1 Water demand against population growth over time for South Africa, projected to the year 2050 (after Davies & Day 1986).

Table 25.1 Water demand by sector of the South African economy – projection for 2010. Note that agricultural demand will not increase as fast as domestic consumption, which will practically treble between 1980 and 2010 (from South African Department of Water Affairs 1986)

	1980		1990		2000		2010	
Demand sector	(million $m^3 a^{-1}$)	(%)	(million $m^3 a^{-1}$)	(%)	(million $m^3 a^{-1}$)	(%)	(million $m^3 a^{-1}$)	(%)
Direct use								
Municipal and domestic	1516	9.3	2281	12.0	3220	14.4	4477	17.3
Industrial	1031	6.3	1448	7.6	2043	9.1	2961	11.4
Mining	466	2.9	511	2.7	582	2.6	649	2.5
Power generation	282	1.7	444	2.3	779	3.5	900	3.5
Irrigation	8504	52.2	9695	50.9	10974	48.9	11885	45.9
Stock-watering	262	1.6	288	1.5	316	1.4	358	1.4
Nature conservation	178	1.1	182	1.0	187	0.8	191	0.7
Indirect use								
Forestry runoff reduction	1284	7.9	1427	7.5	1570	7.0	1700	6.6
Ecological use, estuaries and lakes	2768	17.0	2767	14.5	2767	12.3	2767	10.7
Total	16291	100.0	19043	100.0	22438	100.0	25888	100.0
% Possible expected runoff				49		58		67

of alien and invasive species, the swamping of indigenous biotas by new species/populations, and the spread of water-borne human diseases, but such problems generally have been overlooked (e.g. Petitjean & Davies 1988a,b; Davies *et al* 1992). In most instances, no pre-transfer data are available, and dryland river ecologists are now at a great disadvantage in terms of understanding the original condition of many of their rivers.

This chapter elaborates these arguments by outlining some key features of dryland rivers, discussing the applicability of current ecological concepts and by highlighting issues of conservation and management. In conclusion, a partial agenda is proposed for research in support both of management and conservation of dryland rivers.

25.2 THE NATURE AND DISTRIBUTION OF DRYLAND RIVERS

Dry regions occur where climatic, topographic or oceanographic factors bar the passage of moisture-bearing weather systems (Thomas 1989a,b). Parts of central and eastern Asia, and the western regions of North and South America, are in the shadow of mountain ranges that block the prevailing winds. On the other hand, the dry areas of Australia, southwestern and northern Africa, India and southwestern Asia are governed by descending, stable air masses associated with high-pressure weather systems that circle the globe between 30°N and 30°S. Furthermore, the climates of dryland regions are frequently unstable and are driven by events such as the El Nino-Southern Oscillation (cf. Molles *et al* 1992), and the boundaries of dryland regions are dynamic rather than static, reflecting variations in the patterns of atmospheric circulation.

A logical way to comprehend the diversity of dryland river environments would be to formulate a classification with regard to flow patterns or some other distinguishing feature. However, virtually every feature of dryland rivers shows extreme variability, both in space and time, and any attempt at a rigid classification is self-defeating. Such a classification would need to accommodate rivers whose flows are periodic (e.g. seasonal) as well as those whose flows are aperiodic and which, perhaps, would better be described in probabilistic terms. It would also need to cover systems that originate within arid or semi-arid zones, and those that traverse these zones but which rise in better-watered areas (allogenic) – a challenge in no uncertain terms. Examples of the former are Cooper Creek in central Australia, the Gila River in New Mexico, the Kuiseb River in Namibia, and the Molopo River in southern Africa, while the latter would include many large systems like the Colorado and Columbia (North America), the Murray–Darling (Australia) and the Orange–Vaal (South Africa). The sheer scale of climatic, hydrological and other environmental patterns in large river systems also militates against a simple classification, and it may be more useful simply to group rivers in their broad climatic associations.

Climatic definitions of aridity commonly use analyses of precipitation and temperature or some derivative of the two. Meigs (1953) defined semi-arid, arid and hyper-arid climates using Thornthwaite's (1948) Index of Moisture Availability (*Im*), calculated from the annual moisture surplus and deficit, and from potential evapotranspiration. Meigs' categories correspond broadly to ranges of mean annual precipitation (Grove 1977):

Semi-arid	200–500 mm
Arid	25–200 mm
Hyper-arid	<25 mm

Figure 25.2 shows the global distribution of Meigs' categories (this differs somewhat from the distribution illustrated in Vol. 1, Chapter 1). Data from Thomas (1989a) suggest that 43.3% of the world's land area is semi-arid, 18.3% is arid and 6.3% is hyper-arid, adding strength to Williams' (1988) viewpoint on aquatic ecosystem theory and application (cf. Section 25.5). Although Africa contains the most extensive dry areas (including hyper-arid regions, e.g. the Namib Desert and its main river, the Kuiseb), Australia is proportionately the driest continent (75% is semi-arid or arid, compared to *c*. 60% of Africa). Some authors have argued, somewhat pedantically, that Antarctica is the driest continent with regard to rainfall, and that Australia should be regarded as the driest *inhabited* continent.

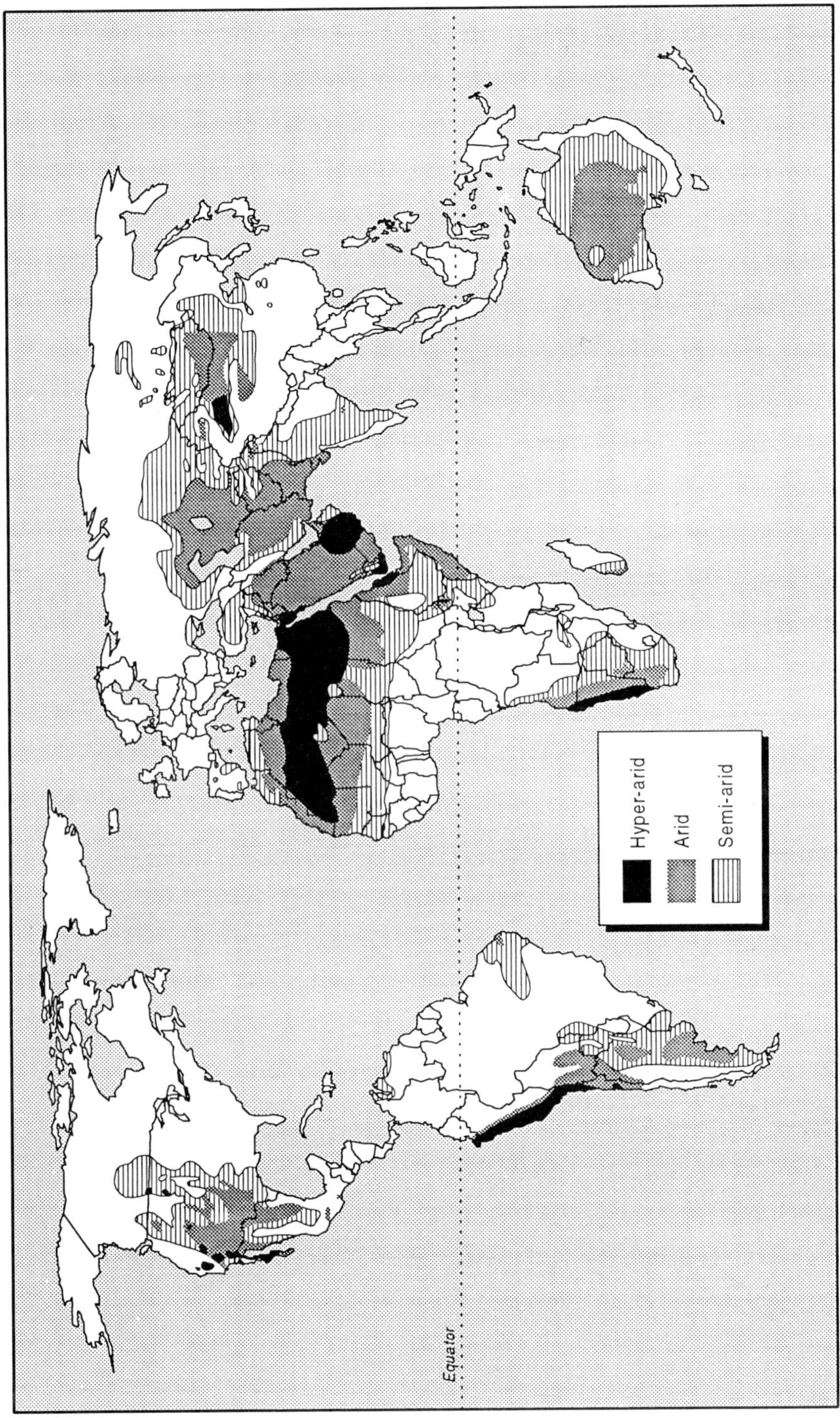

Fig. 25.2 The global distribution of dry regions (after Meigs 1953).

With regard to temperature, it is often supposed that dry regions occur in hot climates, but the only common element of drylands is a wide temperature range (Thomas 1989b). Dry regions include some that are hot year-round and others with mild, cool or cold winters. In the hot deserts of central Australia, temperatures often exceed 35°C, and rainfall may occur during any month of the year (e.g. Table 25.2), whereas in the cold deserts of western China, average monthly temperatures in winter drop to between −10°C and −20°C, depending on altitude; winter snowfall is a significant part of the annual precipitation.

Hydrology

A salient feature of dryland rivers is extreme variability in discharge, reflecting variable rainfall (see Molles *et al* 1992; Table 25.2; see also Vol. 1, Chapter 2). For example, Graf (1988) showed that in arid areas of the western USA, annual precipitation over a 20-year period fluctuated up to 200%, and that the centres of greatest variability were changeable. In the Murray–Darling river system, draining about one-seventh ($1.073 \times 10^6\,km^2$) of continental Australia, annual discharge in the Darling has varied from 0% to 911% of the long-term average, and discharge in the Murray has varied from 6% to 491% of the average (Walker 1986; see also Vol. 1, Chapter 22). The effects of rainfall variability are still more dramatic in small catchments (see Vol. 1, Chapter 1). In the Hartbees River (catchment $4112\,km^2$), a tributary of the Orange–Vaal system in South Africa, the runoff has varied from 9% to 1700% of the long-term average, and in 1961 alone was more than the total from the previous 27 years of record (Noble & Hemens 1978; Cambray *et al* 1986).

Mean annual runoff values in semi-arid regions generally are low (<75 mm) (McMahon 1979), and the efficiencies of conversion to runoff are so low that flowing water is a very limited resource. The ratio of mean annual precipitation (MAP) and mean annual runoff (MAR) is low for regions with a high proportion of 'hot-arid' and 'mild-winter-arid' environments, mainly because of high evaporation. For example, Australia and southern Africa have ratios respectively of 9.8% and 8.6%, compared to a world average of 48% (cf. Alexander 1985). In comparison, 'cool winter arid' and 'cold winter arid' regions have higher ratios, approaching 30% in parts of Canada.

McMahon (1979) and Finlayson and McMahon (1988) used a global database of annual flow records and peak discharge series to demonstrate that dryland rivers are more variable than those in humid regions (see also Alexander 1985; Braune 1985; Allanson *et al* 1990; Davies *et al* 1993b). It may be inferred that the extrapolation of the findings for humid regions to drylands may tend to overestimate the water yields of the latter (Fig. 25.3). The variability of dryland rivers is amply demonstrated in Table 25.3. This data set was developed for some South African rivers and exists only as an unpublished manuscript by Van Biljon and Visser, where the mean coefficient of variation (C_v) for 83 rivers was calculated at 0.89. King and Tharme (1993) have generated more up-to-date data (showing even greater variability) on the C_v for both MAP and MAR of major South African catchments (these data are still in report form, and as such are not available for publication here). Generally, annual flows in dry regions have a mean C_v – the method of calculation is crucial and there are varying figures in the literature – of 0.99, higher than humid regions in North America (0.3), Europe (0.2) and Asia (0.2). Maxi-

Table 25.2 Variability of rainfall seasons in tertiary catchment (Q_{93}) of the Great Fish River (eastern Cape, South Africa), over 19 years (1958–76) (data after Middleton *et al* 1981)

Year	Oct	Nov	Dec	Jan	Feb	Mar	Apr	May	Jun	Jul	Aug	Sep
Mean	11	11	9	9	10	13	7	7	5	5	6	9
Maximum	21	23	30	21	18	42	16	23	17	15	33	36
Minimum	3	2	1	2	2	4	3	0	0	0	0	1
Number of years for which wettest month	1	0	0	1	2	6	0	2	1	1	2	3

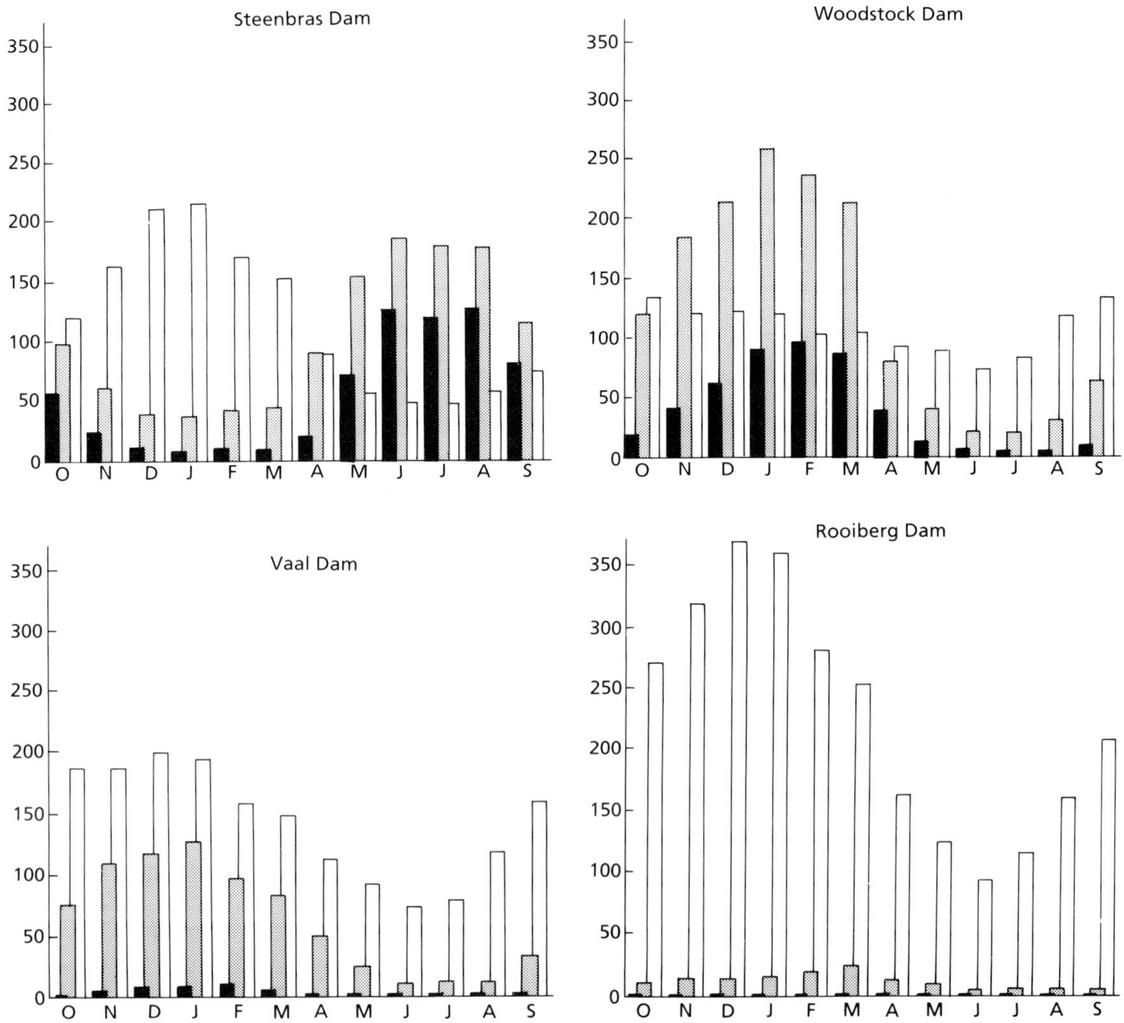

Fig. 25.3 The relationship between mean monthly precipitation (partial shading), potential open-water surface evaporation (unshaded), and runoff (solid shading) (all in mm), for four different impoundements in South Africa (after Alexander 1985).

mum C_v values are 1.27 for Australia, 1.25 for the Mediterranean region and 1.14 for southern Africa (again depending upon how they are calculated; see Table 25.3; see also Vol. 1, Chapter 2). Finlayson and McMahon (1988) have suggested that regions of low precipitation have coefficients twice those of humid regions, irrespective of the size of the drainage basins.

Flood flows in dry regions are also more highly variable than those in other climatic regions (cf. Vol. 1, Chapter 2). Thus, average values of the ratio Q_{100}/Q (the 100-year recurrence interval flood divided by the mean annual flood) are >1.5 times higher in dry regions than in other climatic zones, and values for Australia and southern Africa may exceed 12 (Finlayson & McMahon 1988).

Hydrological characteristics determined from data for humid regions cannot safely be extrapolated to dry regions, as these tend to overestimate

Table 25.3 Characteristics of representative South African rivers (a) and mean variability for rivers in other parts of the world as a comparison (b) using the coefficient of variation (C_v) of MAR* (modified from Van Biljon & Visser, in manuscript; see text)

(a)

River	Locality	Catchment area (km^2)	Mean precipitation (mm)	Runoff	
				MAR (m^3 × 10^6 yr^{-1})	C_v (%)
Dorpspruit	Victoria West	280	216	1	162
Hartz	Taung	10 990	546	52	158
Sundays	Vanrynevelds Pass	3 681	337	35	154
Hennops	Zwartkops	808	709	13	148
Great Marico	Marico-Bosveld	1 219	613	32	118
Kammanassie	Kammanassie	1 505	637	39	110
Luvuvhu	Albasini	509	706	24	108
Brak	Bellair	558	288	2	107
Olifants	Kromdraai	3 989	693	139	99
Koonap	Adelaide	1 245	872	39	93
Vaal	Standerton	8 193	726	550	82
Kubusi	Thornhill	491	786	54	79
Crocodile	Hartbeespoort	4 112	688	154	78
Doring	Melkboom	24 044	214	1	73
Berg	Wellington	713	1 034	547	64
Komati	Hooggenoeg	5 499	670	528	62
Pongolo	Intulembi	7 081	787	1 059	59
Orange	Aliwal North	37 075	725	5 131	56
Olifants	Clanwilliam	2 033	396	394	53
Slang	Vlakdrift	676	786	131	53
Tugela	Colenso	4 176	889	944	47
Noorkaap	Bellevue	126	1 123	43	42
Sabie	Sabie	174	1 268	67	41
Bonnie Brook	Broadholms	119	844	13	39
Hermitage	Swellendam	9	773	8	35
Steenbras	Steenbras	67	1 076	45	33
Mzimkulu	F.P. 160	545	1 050	225	30

Mean coefficient of variation in discharge for 83 South African rivers (including those listed above): 89%.

(b)

Country/Region	Number of rivers in the analysis	Coefficient of variation (%)
South Africa†	>83	117
Victoria State (Australia)	10	53
Australia + New Zealand†	–	50
USA	72	38
Asia†	–	27
Africa†	–	23
Europe	37	22
Canada	13	20

* MAR = mean annual runoff.
† Data generated by C. Stewart.

water yields and certain flood characteristics (McMahon 1979; Molles *et al* 1992). Nevertheless, some basic comparisons can be made. For instance, the greater flow variability in dryland rivers means that larger reservoir capacities are required, compared with rivers in humid regions (see e.g. Fig. 25.3), to ensure a stable annual flow pattern. McMahon and Mein (1986) suggest that reservoir storage capacity is approximately proportional to the square of the annual C_v (see also Lof & Hardison 1966; McMahon 1978; Alexander 1985). Finlayson and McMahon (1988) suggest that the potential maximum flow regulation in arid regions is about 10% of the mean annual flow, except in North America, where up to 40% regulation is possible – already many dryland rivers are well over the suggested limits (e.g. Davies *et al* 1993b)!

Perennial, 'temporary' and ephemeral rivers

Many rivers in arid and semi-arid regions flow perennially, and others may cease flow for indefinite periods. For example, of some 65 000 km of river channels in South Africa, perhaps 40% (26 000 km) are subject to interruptions of flow (O'Keeffe *et al* 1992; Davies *et al* 1993b). Boulton (1988, 1989) distinguished 'temporary' or 'intermittent' rivers, with more-or-less regular, seasonally intermittent discharge, and 'ephemeral' or 'episodic' rivers, which flow only after unpredictable rainfall. Matthews (1988) regarded intermittent streams as those which flow 20–80% of the time, and ephemeral ('interrupted') streams as those which flow <20% of the time (and so are dry, or exist as a series of pools, for much of the time). Poff and Ward (1990) defined six categories: intermittent flashy, harsh intermittent, mesic groundwater, perennial flashy, snowmelt and winter rain.

The simple distinction between 'temporary' and 'ephemeral' rivers will suffice here, as the underlying criterion has special biological significance. When flow ceases and surface water is reduced to standing pools, species dependent on currents to supply food or oxygen, or to maintain a suitable substratum or thermal environment, will be eliminated. When all surface water disappears, the survivors will be fewer still (and will depend upon the rates at which the system dries; cf. Stanley & Fisher 1992), and will include species able to disperse elsewhere (e.g. flying insects), those able to tolerate or to avoid desiccation (e.g. certain molluscs and fish), and invertebrate species able to exploit hyporheic refuges (see below, Section 25.4, and e.g. Stanford & Ward 1988; Boulton *et al* 1992).

The connectivity of rivers, or reaches within rivers, also influences the composition of the biota, although few studies have been made of the processes involved (Boulton & Suter 1986; Matthews 1988); in the last case, for example, Matthews (1988) noted that: 'Especially at lower trophic levels, the biota of prairie streams is poorly known', while Boulton and Suter (1986) considered that: 'Worldwide, comparatively few studies of the general ecology of temporary streams have been published' (see also Boulton *et al* 1992). This lack of research on temporary streams is both a reflection of the dominance of research in north-temperate regions (cf. Williams 1988; see above), and of the difficulty in predicting hydrological events, which makes research planning difficult, if not impossible at times. The paucity of information is one of the major deficiencies in river ecology since, as outlined in the introductory section of this chapter, temporary streams form a major proportion of the water resources of the planet.

There appears to be a correlation between the species richness of temporary-stream communities and the permanence of water (Boulton & Lake 1988). Temporary streams in Australia appear to have a comparatively rich invertebrate fauna (Boulton & Lake 1988; Boulton & Lloyd 1992) (the converse appears to be true for some South African systems; cf. Agnew 1986), although this could reflect the intensity of study. Aquatic insects predominate, particularly Diptera, Coleoptera and Trichoptera, and may comprise 75% of the species present (Boulton & Lake 1988).

In Australia, a high proportion of the invertebrate fauna of perennial streams occurs in temporary streams (Boulton & Lake 1988) (preliminary data from the Sabie and Sand systems in the eastern Transvaal of South Africa appear to confirm this; O'Keeffe & Davies unpublished; see Section 25.7 below), in contrast to North America where there appears to be little faunal overlap (Williams & Hynes 1977). This may mean that differences between permanent and temporary stream environments have been less well-sustained in Australia, reflecting the effects of a harsh, changeable climate over a long period of geological isolation (see also Agnew 1986).

Environmental extremes in temporary streams tend to exceed those in permanent streams, particularly after prolonged dry periods, when only isolated pools remain (e.g. Boulton *et al* 1992; Stanley & Fisher 1992). Perhaps as a result, the life-history patterns of temporary-stream animals are variable, and many differ considerably from year to year (e.g. Boulton & Suter 1986). The characteristics of temporary stream invertebrates seem to be: high mobility; continuous, opportunistic reproduction; rapid development; tolerance of extreme conditions coupled with desiccation tolerance and evasion behaviours. Surprisingly few have dormant life stages (e.g. Gray 1981).

Sediment transport

Sediment transport is no less an important feature of the physical environment than transport of water (cf. Vol. 1, Chapter 5), influencing channel morphology, substratum composition and other habitat features, and hence biotic composition and diversity (see Chapter 4). The quantity of sediment produced within a catchment (the sediment yield) is determined by the climate, particularly precipitation, subject to the effects of vegetation and other factors (e.g. Vol. 1, Chapter 3). The relationship between precipitation and sediment yield has been defined in terms of the annual precipitation required to generate the observed annual runoff at a standardized annual mean temperature (Langbein & Schumm, 1958). This shows that the greatest yields occur at an annual precipitation of 400–500 mm (i.e. typical of semi-arid regions). Lower sediment yields are seen to occur in low-precipitation areas (<200 mm), because there is not enough water (energy) to initiate erosion, and in high-precipitation areas (>800 mm), because vegetation retards erosion. Walling and Webb (1983), using a data set from 1246 drainage basins from all continents, confirmed that peak sediment yields occur in semi-arid regions.

Rivers draining semi-arid areas generally have high annual sediment loads. The greatest load is transported by the Yellow River (Huanghe) in China (1890 t yr^{-1}, or 2640 t km^{-2}), which drains semi-arid loess terrain (see review by Holeman 1968; see also Vol. 1, Chapter 3). Comparisons of suspended sediment concentration and discharge in rivers from different regions suggest that regression equations for semi-arid rivers are distinguishable from those for rivers in humid regions by their higher intercepts and lower exponents (Thoms & Walker 1992).

Much of the sediment load transported by dryland rivers is carried by discharges of high magnitude and low frequency (e.g. Baker 1977; Bremner *et al* 1990; see also Vol. 1, Chapter 3), so that semi-arid rivers are characterized by relatively high 'dominant discharges' (Andrews 1980; Graf 1985). Up to 60% of the total load may be transported by flow events with return intervals >10 years, compared with only 10% for rivers in more humid regions (Neff 1967).

Channel morphology

The variable hydrological and sediment transport regimes of semi-arid rivers have implications for channel morphology and stability. In the rivers of humid regions, equilibrium channel conditions are maintained by continuous transfers of energy and mass (see Vol. 1, Chapter 6), but the more extreme, rapid and short-lived changes in water and sediment transport in dryland rivers suggest that they do not tend towards long-term equilibria (Stevens *et al* 1975). Dryland rivers are subject to more variable energy-mass transfers and usually have compound channels (Graf 1987), with a low-water configuration to accommodate near-continuous low flows and a wider, often braided, high-water configuration to accommodate rarer high flows. Compound channels can have variable forms. For example, the Gila River, Arizona, has a well-defined meandering inner channel and an outer, braided channel (Graf 1988). Rivers of the Murray–Darling system in Australia lack distinct low- and high-flow channels, but have in-channel benches representing adjustments to variable discharge (Woodyer 1968; Thoms & Walker 1993).

The braided and compound channels of many rivers of India and the central USA have been reported to have high rates of lateral movement (cf. Graf 1988). In a study of the channel and sedimentation processes in the Brahmaputra River, India, Coleman (1969) reported rates of movement as high as 792 m yr^{-1}, and attributed these to high discharges and sediment transport rates as well as the sandy incohesive nature of the bank material. However, single-channel rivers, with highly cohesive banks, experiencing low stream powers (e.g. the lower Murray and Darling rivers of Australia), are reported to be laterally stable (cf. Thoms & Walker 1992).

Physical responses to regulation

The water resources of dryland regions are often heavily exploited, and flows are subject to intensive regulation through dams, weirs, levees, canals, pipes and other structures. The role of minor regulating structures is often assumed to be small, but they may have profound effects overall, by virtue of their numbers and ubiquity (e.g. Adams 1991; Walker & Thoms 1993). The physical and biological effects of regulation on fluvial systems generally are well documented (e.g. Petts 1984), but most examples are from humid regions, and the responses of dryland rivers may be different, particularly in regard to scales of time and space (cf. Walker & Thoms 1993; Thoms & Walker 1993; Davies *et al* 1993b). Indeed, the environmental impact of regulation in dryland rivers may be greatly enhanced because the regulated regime is so different from the natural flow pattern. Figure 25.4 shows this situation for the Orange River, South Africa; pre-regulation variability has been completely lost, and periods of high- and low- or no-flow have been replaced by continuous flows (see Cambray *et al* 1986).

The simple concept of 'dominant discharge' (the discharge interval that transports the largest part of the annual sediment load) may be used to illustrate the impact of regulation (Thoms & Walker 1993). In contrast to humid-zone rivers, dryland rivers have higher dominant discharges (Graf 1985), meaning that the transport of water and sediment in these systems is dominated by events of low frequency and high magnitude. Regulation reduces the dominant discharge, imposing a fluvial regime more like that of a (regulated) river in a humid region. Figure 25.5 outlines a simple model of the impacts of flow regulation on sediment transport; the impacts, and hence

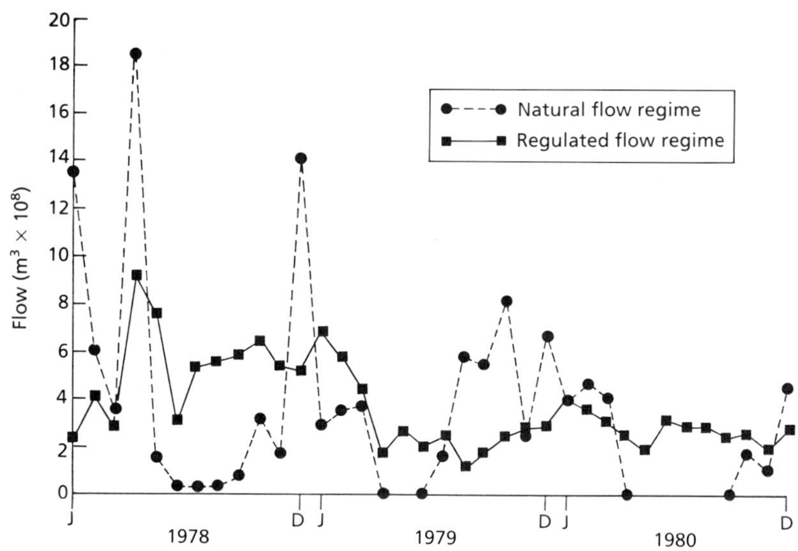

Fig. 25.4 Natural (dashed line) and regulated (solid line) flow regimes of the Orange River, South Africa (after Cambray *et al* 1986).

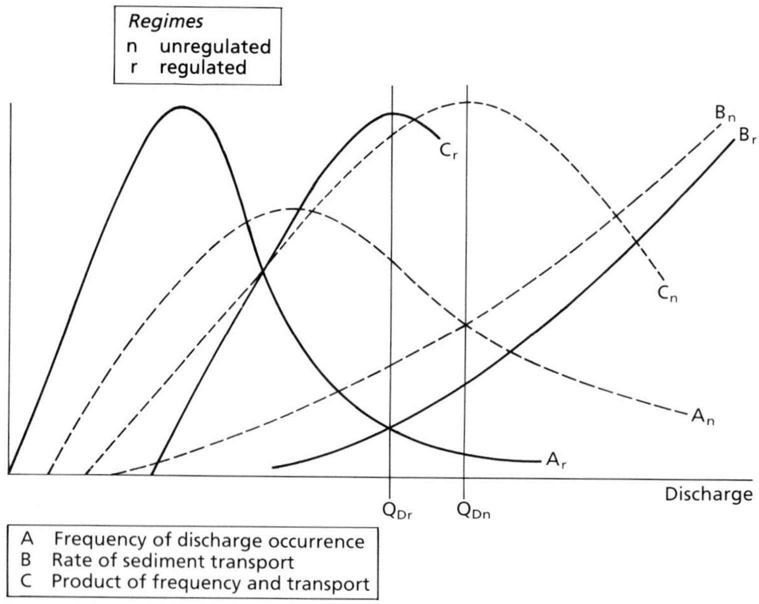

Fig. 25.5 A conceptual model of the impact of flow regulation on water and sediment regimes (see text for explanation) (after Thoms & Walker 1992).

the parameters of the model, depend on three factors:

1 *Changes in the flow regime.* Curves A_n and A_r represent the frequencies of discharge under natural (n) and regulated (r) conditions, respectively. As reservoir operations decrease the magnitude of downstream flows, A_r has a more pronounced peak and is skewed left of A_n.

2 *Sediment transport.* Curves B_n and B_r represent sediment transport rates under natural and regulated conditions. The positions and slopes of the curves reflect the availability of sediment. Regulated flows decrease the river's capacity to transport sediment, particularly coarse sediment (cf. Petts & Thoms 1986), and limit entrainment from floodplain sources. However, more importantly here, upstream sediment source areas are isolated by reservoirs, restricting the supply to downstream areas, so that in relative terms the transport of finer sediment is increased.

3 *Interaction of the water and sediment regimes.* The change in dominant discharge (Q_d in Fig. 25.5), representing changes in the interaction of the water and sediment regimes, is represented by the curves C_n and C_r. Each curve is peaked, with the apex representing the dominant discharge. Curve C_r is to the left of curve C_n, showing that discharges which occur most frequently dominate the sediment transport regime under regulated conditions.

This model serves to describe the impact of flow regulation on dryland rivers in terms of reductions in the dominant discharge and the overall variability of the sediment transport regime. It may provide a useful framework for comparing the physical impacts of catchment disturbances, as well as of flow regulation.

Ecological responses to flow regulation

The effects of flow regulation can be ecologically catastrophic for dryland river ecosystems, for they are adapted to unregulated and highly variable flow, sediment and chemical regimes. The spatial and temporal variability of dryland river ecosystems suggests that the time required to respond to, and to recover from, environmental change may be more prolonged than for rivers in more stable environments (cf. Walker & Thoms 1993).

Flow variability strongly influences the evolutionary character of the biota (e.g. Boulton *et al* 1992). Thus, many of the species native to dryland rivers are tolerant of environmental extremes and are possibly far more opportunistic, both in their feeding and reproductive adaptations, than their counterparts in humid-zone (and, therefore, more predictable) systems (see above). In terms of flow reduction relative to the natural flow regime, stream regulation by impoundments (*sensu* Ward & Stanford 1979), over-exploitation of water resources and associated 'instream flow requirements', including the concomitant appearance of pest species, the desynchronization of life cycles and ultimately, the elimination of part of the natural biota of a system, are undoubtedly of greatest immediate concern to dryland lotic ecologists; see e.g. Davies (1979, 1986) and Cambray *et al* (1986) for African examples, Walker (1979, 1986) and Cadwallader (1986) for Australia, and Stanford and Ward (1986a,b) for the USA.

River regulation generally means loss of water to abstraction, and the gross alteration of natural regime, changing temporal patterns of flow as well as the duration, timing, frequency, magnitude, and rate of recession of floods. Regulating structures may also impede upstream migrations of fish and of other animals and, depending on their volume, location on the river and the nature of their outlets to the river, may cause environmental changes which require long 'recovery distances' (*sensu* O'Keeffe *et al* 1990) downstream (of the many hundreds of dams constructed in South Africa only two were designed with fish passes; one is now buried under silt, while the other is ineffective – see Davies *et al* 1993b). Impoundments also may permanently flood areas that once were productive wetlands or floodplains, subject to cycles of wetting and drying. To further complicate the impacts and exploitation of dryland rivers, many systems now have flows augmented by IBTs, with concomitant breakdown of biogeographical barriers, spread of alien biotas and disease vectors, and mixing of population gene pools, amongst many other impacts. These issues have recently been raised by Davies *et al* (1992) who, using a combination of Northern Hemisphere river theories (see Section 25.5, below), pointed out that dryland systems are now subjected to over 60 different types of manipulation from IBTs alone (Fig. 25.6), irrespective of regulation by other structures. The ecological impacts of such large-scale manipulation have been virtually ignored on a global basis (Davies *et al* 1992).

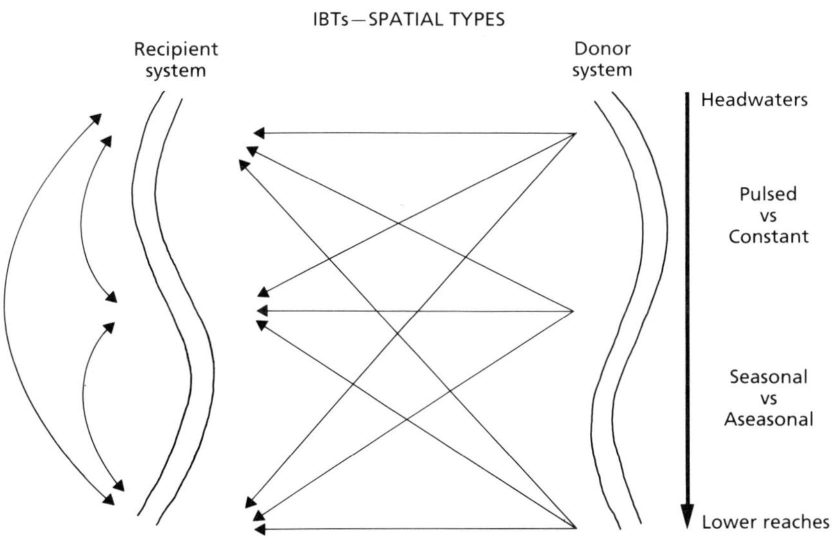

Fig. 25.6 Diagrammatic representation of the major types of IBTs that are possible, using the RCC of Vanotte *et al* (1980) as a basis. Some 60 or more possible permutations can be derived from this scheme (after Davies *et al* 1992).

25.3 INTERACTION OF HYDROLOGICAL AND HYDRAULIC FACTORS

Significance of hydraulic factors

Lotic animals are limited to particular flow environments by hydrodynamic factors against which they must expend energy to forage, to grow and to reproduce (see Chapters 1, 3 and 4 for discussion). Such observations are significant because the influence of hydraulic conditions is related to the harshness of the environment. Connell's (1975) 'harsh-benign' hypothesis suggests that, as environmental stress declines, biological interactions become more significant in the dynamics of communities (cf. Peckarsky *et al* 1990). In the special case of dryland rivers, where environmental stresses may be extreme, this seems to be the case. For example, as systems dry out, competition for ever-dwindling resources (suitable habitat/food) must markedly increase.

As shown earlier, the hydrological conditions that dominate dryland rivers are considerably more variable than those of humid-zone systems. To survive in such changeable ('flashy') environments, organisms need to be highly flow-adapted (widely tolerant), particularly if critical phases of the life cycle are likely to occur during times when spates are most probable (cf. King *et al* 1987a,b, 1988, 1991; Davies *et al* 1993a,b).

As hydraulic stresses decline, they are less likely to influence the biological interactions of arid-zone lotic biota. Matthews (1988) suggested that in intermittent or ephemeral dryland streams (flowing >20% and <20% of the year, respectively), hydraulic patterns are not critical for such short periods of the year, and other factors are more likely to influence community structure. For example, it is more likely that the nature and availability of refugia will influence the survival of lotic species (e.g. Boulton *et al* 1992). The declining volume and wetted area of pools in intermittent lotic systems, together with concomitant increases in temperatures and changing chemical characteristics, are more likely to determine the success of populations occupying refugia until recharge and reconnection occurs during the next runoff event.

Instream flow management

Over the past two decades, management to conserve fish and other riverine biota has been directed towards modelling techniques that combine descriptions of the physical habitat (flow characteristics) and stage-discharge relationships that predict 'minimum flow' conditions for target species (see review by Nestler *et al* 1989). The widely-used Instream Flow Incremental Methodology (IFIM: Bovee 1982; see also Chapters 14 and 15) has a number of useful capabilities, but may have only limited applications away from the perennial, snowmelt-dominated, temperate-zone streams for which it was developed (cf. Gan & McMahon 1990; King & Tharme 1993). IFIM may be useful for the management of some semi-arid prairie rivers (Gore 1989), but these are perennial and most unlike the 'flashy' streams of dry regions elsewhere. Furthermore, some of the smaller dryland rivers are unlikely to have economically-important species that can be targeted for by IFIM and related procedures. In these cases, the methodology needs to be modified to become more community- or ecosystem-oriented, with concomitant increases in labour, logistical problems and data requirements (King & Tharme 1993). In addition, the problem of assessing the instream flow requirements of dryland rivers depends greatly upon the availability of long-term hydrological and climatic data, which are often lacking (King & Tharme 1993).

Gore *et al* (1991, 1992) made tentative minimum-flow recommendations for the semi-arid zone rivers of southern Africa, using flow targets over short intervals, such as 4 weeks, to compensate for the flashiness of the natural hydrographs. These evaluations require intensive development of 'habitat suitability' criteria in order to differentiate between flow-related conditions that can be predicted by existing instream-flow models, and the environmental conditions that cannot be used to predict flow reservations with current models. In keeping with Connell's 'harsh–benign' hypothesis, competitive interactions and predator–prey relationships will become more significant as discharge and the availability of space, food and other resources decline. Gore *et al* (1992) suggested that the biotic diversity of flashy systems is related to the heterogeneity of the flow environment rather than to complex hydraulics, hydrographic conditions, and habitat suitability criteria.

25.4 SEASONALITY, THE HYPORHEOS AND THE ROLE OF PARTICULATE ORGANIC MATTER

Seasonality

In temperate streams, seasonality plays a major role in community composition, particularly of the benthos (Hynes 1970). For example, thermal cues in spring may determine the emergence of an over-wintering univoltine insect assemblage, to be replaced by multivoltine species with several emergences during summer. Seasonal synchrony is less certain in the semi-arid prairie streams of central North America. A winter–summer community shift occurs in northerly latitudes, where the prairie rivers are influenced by snowmelt from the Rocky Mountains, but not in southerly latitudes (Gore 1977).

Benthic invertebrates in the dryland streams of southern Africa (Hart 1985) and Australia (Boulton & Suter 1986) also appear to be more opportunistic and less constrained by seasonal changes. For example, in Cibolo Creek, south Texas, there are no apparent seasonal changes in benthic communities, and alterations in abundance and distribution are more influenced by the frequency of spates and the degree of intermittency (Davis 1982). In Aquila Creek, a Texan 'Hill Country' stream, the degree of intermittency again is more important than seasonality in governing the diversity and productivity of benthic communities (Campbell & Clark 1982). The contributions of perennial tributaries determine the persistence of main-channel pools, which are refugia for resident species, and the abundance and diversity of species in the pools are related to depth, the degree of isolation from other pools, and substratum composition (Richardson 1990). In these warmer climates, thermal conditions in the isolated pools, combined with flooding frequency, also influenced primary production and zooplankton abundance; zooplankton species were restored to isolated pools by drift during flood events (Richardson 1990).

The importance of the hyporheos

The importance of the hyporheos as a refugium and, indeed, as a major lotic ecosystem habitat has frequently been overlooked (e.g. Stanford & Ward 1988) and in seasonal dryland streams the hyporheos assumes an extremely important role in maintaining invertebrate benthic populations. For example, Boulton *et al* (1992) recorded 23 invertebrate taxa in the hyporheic zone of two Australian intermittent streams during dry periods. Surface-water populations may be more diverse than their hyporheic counterparts (McElravy & Resh 1991), but they are subject to unstable environmental conditions during the wet season. High silt loads in some streams may prevent early instars from entering interstices in the substratum, but hyporheic populations are less affected.

The hyporheos may play a key role by providing points (nodes) where nutrient cycling may take place, particularly in times of environmental stress (Davies *et al* 1993a,b). 'Fixed' nutrient-processing loops in the subsurface environment could allow nutrient cycling rather than nutrient spiralling (cf. Webster 1975; Newbold *et al* 1982; Newbold 1987; see also Vol. 1, Chapter 18), and promote the survival of communities during low-flow conditions. These cycles would be 'reset' during spates, and new food resources also would be entrained at such times (see Davies *et al* 1993a,b; Koch *et al* 1993; Ractliffe 1991). In some blackwater, low-order streams of the western Cape Province, southern Africa, spate-generated foams formed from material accumulated in the substratum during low flows have yielded high concentrations of sugars and amino acids, and up to 20% dry mass nitrogen (Koch *et al* 1993).

Particulate organic matter (POM) in dryland systems

Particulate organic matter (POM) affects the distribution and productivity of lotic communities, as POM is a link in nutrient cycling (see Vol. 1, Chapters 11 and 15). Seston transport, the availability of POM and the ability of the stream to retain POM all appear to be correlated with hydrological variability (e.g. Sedell *et al* 1978; Snaddon *et al* 1992). The prairie streams of North America experience prolonged periods of desiccation, and should be less retentive of POM than their perennial, temperate-zone counterparts. The flashy nature of these streams may suggest that there should be a low correlation between POM concentration and discharge (Gurtz *et al* 1982), but in the Trinity River, Texas, a typical prairie stream, Hill *et al* (1992) reported that POM availability was predictable, largely as a consequence of stable POM concentrations.

The stability of POM levels may be associated with retention in isolated pools during dry periods. The mode of processing POM in dryland rivers also may be a factor. Whilst shredders play a significant role in the transformation of coarse POM (CPOM) to fine POM (FPOM), shredder densities in dryland rivers are frequently low (e.g. King *et al* 1988). In the absence of shredders, the mechanisms of POM processing, and particularly CPOM processing, are little understood. Leaching, microbial respiration and physical breakdown may be most important (cf. Stewart & Davies 1989; see Vol. 1, Chapter 15). If so, the rate of processing must be related to temperature, which influences microbial and chemical processes, as well as to the variable flows that characterize such systems. Intermittent flows in warm, dry seasons, with inputs from thermal springs and groundwater, may be the key environmental factors regulating the rate of processing of POM. These factors are related to hydrological behaviour rather than to biological events, and again conform with Connell's (1975) 'harsh-benign' concept.

25.5 APPLICABILITY OF BASIC CONCEPTS

As longitudinal ecosystems, with lateral, vertical and temporal components, rivers may be viewed as exceedingly complex four-dimensional systems (Ward 1989; see also Chapter 1) within which communities must exist and respond to environmental variables. At the same time, all rivers traverse multiple geographical boundaries (e.g. Davies & Walker 1986), introducing numerous factors such as altitude, climate, topography, geochemistry, hydrology and catchment land use within the same basin (as well as political and ownership/management divisions), and these all in turn influence the distribution of species, communities and habitats. Management has been at least partially assisted by the relatively recent development of a series of ecological concepts.

The *River Continuum Concept* (RCC) of Vannote *et al* (1980) (see also Vol. 1, Chapter 11) portrays a river as an environmental gradient dependent upon inputs of organic matter from its headwaters. There has been much debate over the validity of the concept in areas other than the deciduous forests of eastern North America, where it was developed (e.g. Barmuta & Lake 1982; Lake *et al* 1986). In addition, the concept fails to provide an adequate model for lowland rivers, where there is greater emphasis upon lateral exchanges than on longitudinal transport (e.g. Junk *et al* 1989).

The concept of *disturbance* (see Chapters 1, 5 and 6) has become very confused, through a proliferation of conflicting definitions (see also Stanley & Fisher 1992). For investigations of dryland rivers floods should not be viewed as disturbances in the sense of aberrant phenomena; they are rather important cues for the dynamic changes in the ecosystem (Puckridge *et al* 1993). In this context the *Intermediate Disturbance Hypothesis* (IDH) of Connell (1978) predicts that biotic diversity is greatest in communities subject to moderate levels of disturbance.

The ecological significance of exchanges between rivers and their floodplains is encapsulated by the *Flood Pulse Concept* (Junk *et al* (1989), which suggests that flooding is the major 'driving variable' in the dynamics of lowland river ecosystems (see Chapter 15). The emphasis on lateral exchanges complements the RCC, with its emphasis on longitudinal transport. Although the Flood Pulse Concept was first proposed for rivers in humid regions where the flood pulse is regular, it is no less applicable to dryland rivers like the Murray–Darling and Orange–Vaal, where the pulse is more erratic (e.g. Puckridge *et al* 1993). The characteristics of the pulse, including its frequency, magnitude and rates of rise and fall, are encoded in the ecological character of riverine plants and animals; those in the Murray system, for example, tend to have opportunistic life histories and wide tolerances to environmental change.

The *Serial Discontinuity Concept* (SDC) (Ward & Stanford 1983) refers to the effects of impoundments, which create discontinuities within the river continuum, and require certain *recovery distances* (*sensu* O'Keeffe *et al* 1990). Recovery is system and site (position of impoundment along the continuum) dependent, and also varies according to the operational criteria pertaining to each impoundment. Regulation, either by impoundment or by IBTs, is assuredly one of the greatest threats to river conservation and management in drylands (see Section 25.7), and as a result, the SDC has received considerable attention in a number of dryland systems (see e.g. Byren & Davies 1989; Palmer & O'Keeffe 1990a,b; O'Keeffe *et al* 1990).

Although such ideas are useful, because many dryland rivers occur in 'Third-World' countries, with their attendant problems of high human population growth rates and a general lack of skilled 'manpower', managers often rely heavily on these imported concepts, often at the expense of locally developed approaches. For example, the testing of Instream Flow Incremental Methodology (IFIM: e.g. Bovee 1982; Gore & Nestler 1988; Gore & Petts 1989; Milhous *et al* 1989; review by King & Tharme 1993) is currently consuming a large proportion of the manpower and funds available for lotic system research in South Africa; the labour-intensive nature of IFIM runs contrary to the available resources (e.g. O'Keeffe & Davies 1991; King & Tharme 1993).

25.6 FLOODPLAINS IN DRYLAND REGIONS

Perceptions of floodplains

River alluvium typically comprises silts and clays, and forms broad plains through which the channel meanders (see Vol. 1, Chapter 7). Floodplain development depends primarily upon the width of the meander train, and its depth may extend to bedrock. This pattern is typical of so-called 'fringing' floodplains (Welcomme 1979), but does not apply to those associated with internal or coastal deltas (e.g. Davies & Walker 1986).

In describing the Flood Pulse Concept, Junk *et al* (1989) defined a floodplain as an 'aquatic–terrestrial transition zone' (ATTZ), highlighting the importance of wetting and drying cycles in floodplain dynamics (e.g. Heeg & Breen 1982). But, as pointed out by Finlayson and McMahon (1988) and Puckridge *et al* (1993), the 'aquatic' phase is quite different from permanent aquatic environments, and the 'terrestrial' phase also is different from terrestrial habitats that are not flooded (see e.g. McLachlan 1974). For these reasons, the floodplain has been viewed as an ecotone (Naiman & Décamps 1990; Puckridge *et al* 1993) requiring special consideration, rather than as an ecosystem that alternates between the two phases (see also Ferrington & Goldhammer 1992) for an interesting variation on this theme in terms of human impact upon dryland rivers, and indeed, upon rivers in general – namely the Capillary Fringe Habitat).

Puckridge *et al* (1993) suggest that in an ecological context, the spatial limits of the floodplain are marked by the limits of inundation in the lifetime of the longest-lived flood-dependent species. For many rivers this would implicate species of trees which, like the river red gums (*Eucalyptus camaldulensis*) of the Murray floodplain (Australia) and the sycamore figs (*Ficus sycamorus*) of the Luvuvhu floodplain (southeast Africa), may live for hundreds of years.

Where linkages between river and floodplain are broken, perhaps by regulation or by IBTs, both river and floodplain environments will be changed (e.g. Davies *et al* 1992; see also Chapter 21). In this respect, large floodplain rivers like the Murray, Pongolo, Zambezi and Colorado provide clear illustrations of situations repeated in other dryland regions of the world (e.g. Heeg & Breen 1982; Davies 1986; Carlson & Muth 1989; Walker & Thoms 1993; see also Vol. 1, Chapter 22).

Responses to the flood pulse

The biomass and productivity of floodplain communities depend directly upon the flood pulse, but this has a complex character. It may be regular, as in large tropical rivers, or less predictable, as in smaller, temperate-zone rivers. Flood pulses in dryland rivers are likely to be still less predictable (e.g. Puckridge *et al* 1993; Davies *et al* 1993a,b). Under these circumstances it is relevant to consider the responses of the ecosystem to drought as well as to flood.

Flow variability is pertinent because one of the intriguing implications of the Flood Pulse Concept is that ecological responses to flooding may be related, both qualitatively and quantitatively, to features of the flood. A starting-point would be to attempt to relate the biological responses to flooding and drying to elements of the simple flood-stage-hydrograph (cf. Vol. 1, Chapter 1). Following Puckridge *et al* (1993), the hydrograph could be elaborated in terms of distinctions between a *particular flood* (an increase in river stage), a *flood pulse* (an alternation of stage), the *flood history* (antecedent pulses at one location) and the *flood regime* (a statistical description of the system's behaviour over time and space).

Some biologically meaningful flood-pulse parameters include the duration, frequency, timing and magnitude of floods, and the rates of rise and fall of flood waters. For example, the rate of flood recession in the Murray has increased sharply over the past century, apparently as a result of flow regulation (Walker & Thoms 1993), and this probably has limited recruitment to fish populations dependent on wetlands for spawning and larval growth. The duration of inundation, regardless of the rate of rise or fall, will generally affect decomposition processes, nutrient fluxes and the biomass and productivity of wetland plants and animals (e.g. for the Pongolo Floodplain see Furness & Breen 1980; Furness 1981; Heeg & Breen 1982; Rogers & Breen 1982). Flooding frequency is significant because responses may be tempered by the recent flood history – a succession of big floods is likely to produce diminishing returns, whereas there may be a spectacular response to a flood that follows a prolonged drought (cf. Finlayson & McMahon 1988; Puckridge *et al* 1993).

The magnitude of flooding is also important. For species whose recruitment is enhanced by flooding, minor floods may sustain relatively low 'bridging' levels of recruitment between major floods (e.g. Heeg & Breen 1982). For example, the decline of certain native fish populations in the Murray may be related to changes in the frequency of minor (seasonal) flows, induced by regulation, while the frequency of major floods is little changed. Major floods flush accumulated debris from the floodplain, including wetlands, and may introduce large numbers of 'snags' (fallen tree-trunks and branches) which, in the main channel, are a vital refuge for riverine plants and animals.

Finally, variations in both the frequency and magnitude of floods may maintain river and floodplain communities in an unstable, but highly productive, state of early succession; detrital decomposition and nutrient cycling is fastest in flooded, aerobic sediments and, given that wetland sediments rapidly become anaerobic, efficient decomposition depends upon a fine balance between submergence and exposure (e.g. Howard-Williams 1985). Temporary wetlands support diverse communities of plants and animals, and in some circumstances may be more productive than those subjected either to prolonged drying or to prolonged inundation (Ward 1989).

25.7 CONSERVATION AND MANAGEMENT ISSUES

Rivers are often seen as 'lifelines' for people in arid and semi-arid regions (e.g. Carlson & Muth 1989), and water flowing to the sea is often perceived as 'wastage' of an essential resource that should be stored and manipulated for the benefit of people. This underscores the importance of dryland rivers as resources, and implies that careful management is needed to maintain them.

Management of dryland river systems is an inordinately difficult task, and the nature of human demands for water differs greatly from the patterns of water availability imposed by the 'natural' environment. On the one hand, pressures on dryland river systems suggest that prospects for their conservation as natural, self-regulating systems are poor. On the other hand, present rates of change and degradation cannot be allowed to continue if these rivers are to be maintained as renewable resources. Some form of compromise is indicated, perhaps not based on the presumption that we should strive to manage rivers, but that we should endeavour to manage our relationship with them (Regier *et al* 1989).

Virtually all major dryland rivers of the world have been modified by flow regulation, mainly to serve irrigated agriculture (e.g. Cambray *et al* 1986; Dumont 1986; Stanford & Ward 1986a; Carlson & Muth 1989; Ebel *et al* 1989; Walker & Thoms 1993; Davies *et al* 1992, 1993b; see also Vol. 1, Chapter 22). The environmental effects of irrigation development are a suite of long-lasting, predictable and reasonably well-understood changes that, together, represent a massive ecological disturbance. The effects of flow regulation are most pervasive (see reviews by Petts 1984; Graf 1987). However, the 'irrigation syndrome' also includes: desertification; salinization; erosion and siltation; loss of native vegetation and animals; disruption of food chains; loss of habitat; barriers to migratory routes; breakdown of biogeographical barriers; spread of water-borne diseases and their vectors; alienation of river and associated wetlands and floodplains, and their ultimate loss; invasion by weeds and other exotic plants and animals; and pollution of land and water, including groundwater, by fertilizers and the chemicals used in pest control. For the second and last points in the preceding catalogue, the ultimate irony is that there usually is an insufficiency of water to 'dilute' effluents anyway, thereby exacerbating the problems facing dryland systems.

Irrigation has had adverse effects on dryland regions from the beginnings of western civilization in Mesopotamia and Egypt more than 5000 BP. In the modern era, developments have expanded greatly to meet the needs of burgeoning human populations, and the extent and severity of environmental impacts have increased accordingly. Although technology exists to limit salinization, to minimize the volume of water used and to maximize the efficiency of its use, it is not implemented on a scale sufficient to arrest the global degradation of drylands. For reasons of politics, economics, sociology and other factors, irrigation technology in many parts of the world differs little from that used thousands of years ago, and even in so-called 'developed' countries, applications of new technologies are often constrained by economic factors. The essence of human conflicts with dryland river resources therefore is not irrigation *per se*, but a failure to effectively manage highly complex social and ecological systems that become still less tractable as the pressure of human population growth increases. Although further analysis is beyond the scope of this chapter, it makes little sense to discuss the conservation of dryland rivers outside this perspective. Ultimately, of course, we may hope that a philosophy of conservation will become the arbiter of management in both theory and practice.

Case study

Background

For our purposes we have selected a South African system, the Sabie–Sand River, because its problems are so typical of all dryland rivers. The reader is referred to additional case studies of dryland systems incorporated in Davies and Walker (1986) (namely, the Colorado, Orange–Vaal and the Murray–Darling). Other studies may be found in Davies *et al* (1992) – one Namibian and one Lesotho-South African, two Australian and two American – in Dodge (1989), in Robarts and Bothwell (1992) – ten studies, including system studies of the Australian Murray and the Canadian Thompson rivers, together with eight community or species-specific works – and in Davies *et al* (1993b) – five additional South African studies.

The Sabie – Sand River, Kruger National Park, South Africa

In the case of southern Africa, river conservation traditionally has been the poor relation of terrestrial conservation, chiefly because of a preoccupation with the spectacular large mammals that inhabit the subcontinent, instead of the ecosystems that support them, and because rivers are difficult to manage. However, they have recently received increasing attention due to the stresses placed on them (e.g. O'Keeffe 1986; Ferrar *et al* 1988; Ferrar 1989; O'Keeffe *et al* 1987,

1989a,b, 1992; Allanson *et al* 1990; King *et al* 1991; O'Keeffe & Davies 1991), such that water allocations for primary drainage regions have now included nature conservation as a legitimate water user (see Table 25.1), albeit at disturbingly low levels – an allocation of only about 2% in the *best* case (Walmsley & Davies 1991). Ironically, the first concerns for the conservation of rivers in southern Africa were associated with the introduction of alien species (namely trout, *Salmo gairdneri, S. trutta*; De Moor & Bruton 1988)! Day (1932) made the now extraordinary remark concerning the 'threat' to trout posed by the indigenous *Barbus* species which 'eat trout ova', while Bush (1933) identified similar problems for the establishment of trout in Natal rivers and recommended that suitable insect species be imported from Europe as trout food!

In the following abbreviated study, the 'River Conservation System' (RCS) of O'Keeffe *et al* (1987), developed as a tool for understanding the conservation status of South African rivers, has been used. This computer-based expert system was designed to assess river conservation status, defined as 'the extent to which any river has been changed from its natural state, and from its original importance for conservation, and which gives a numerical assessment of the worth of the river for conservation' (O'Keeffe *et al* 1987). It asks up to 58 questions regarding river size, regulation, pollution, water use, catchment land use, vegetation condition, degree of erosion, indigenous, endemic and introduced biota, and recreation uses. The responses are weighted to indicate their relative importance and the computer programme then provides both an overall percentage maximum and minimum score for the river in question, together with individual scores for river channel condition, and condition of the catchment, riverine biota, etc. The differences between the minima and maxima indicate the state of knowledge for each component and system, thereby identifying conservation attributes for which more information is required.

Of the six major rivers flowing west–east through the Kruger National Park (KNP), one was naturally seasonal and five were perennial. Of the perennial rivers, the Luvuvhu and Letaba now flow only for a few weeks annually, the Sabie stopped flowing for the first time in 1989, and again in 1992, the Olifants is heavily polluted with mining effluent and has lower than natural flows, and the Crocodile is constantly regulated at $5\,m^3\,s^{-1}$. A mutli-disciplinary and multi-institutional research programme commenced 3 years ago to address these problems (O'Keeffe & Davies 1991; Davies *et al* 1993b). The Sabie–Sand system is of special interest because it forms the major 'lifeline' for the premier conservation area in South Africa (KNP); it is now facing rapidly escalating threats from overutilization by irrigation agriculture and forestry, and regulation by a suite of proposed impoundments (e.g. O'Keeffe *et al* 1989a; O'Keeffe & Davies 1991). Because of the seriousness of the situation, a great deal of research effort has been invested in the system and an excellent database for conservation and management purposes now exists.

Rising at more than 2000 m on the eastern Transvaal escarpment, the Sabie flows through an afforested upper catchment (introduced pine and eucalypts – here, water demand has reduced MAR by 20.5%), supplies water to rural communities and some irrigation farming in its middle reaches, and flows through the Kruger National Park (here it supplies water for conservation purposes and for human consumption – tourism) into Mozambique, where it joins the Komati River (see O'Keeffe & Davies 1991). The river may be biologically the most diverse in South Africa, in terms of indigenous fish (47) (Russell & Rogers 1989), and invertebrates, notably Ephemeroptera and Trichoptera (Moore & Chutter 1988; Moore 1991; O'Keeffe *et al* 1989a). Interestingly, nearly all the fish and invertebrate species in the perennial Sabie mainstem occur in the tributary Sand River, a temporary stream, presumably because of recolonization during wet periods (see Section 25.3). A graphic illustration of the problems facing the two rivers has been highlighted during the drought of 1990–92. For example, the pre-drought fish communities of the Sand River were dominated by cyprinid minnows (74% of the assemblage in May 1990); this situation has now changed to favour more robust cichlid species (72% in May 1992) (Desmond Weeks, Institute for Water Research, Rhodes University, Grahamstown, personal communication). At the same time, the entire system changed from strong flow to isolated and dwindling pools, and *Oreochromis mossambicus*, the most abundant cichlid, bred continuously (cf. Gray 1981).

The main conservation objective has been to estimate the instream-flow needs of the river. Methods such as IFIM (see Section 25.3) and Physical Habitat Simulation (PHABSIM II: Milhous *et al* 1989) are being assessed, with special regard for the benthic macroinvertebrate and fish populations (e.g. King & Tharme 1993). The labour-intensive nature of PHABSIM, and its concentration on target species, rather than habitats or communities, makes it unlikely that this approach will be widely applied in South Africa, and it has now become essential that other, more holistic, and less intensive methods for instream-flow assessment be developed (King & Tharme 1993).

Carter and Rogers (1989) have approached the problem of instream-flow assessment from a very different angle, involving Markov models to investigate the effects of reduced flow on macrophytes in the Sabie and other rivers of the KNP. They recorded a complex series of temporal and spatial changes, governed by exposure of sediment due to flow decrease, sediment deposition and vegetation removal by scouring, colonization by reeds (*Phragmites mauritianus*) and reed-bed colonization by woody riparian vegetation. They concluded that flow reductions are responsible for marked changes in channel form and processes, and increasingly prevalent transitions from water to sand to reeds to woody vegetation. Furthermore, and of considerable concern, they also noted a steadily increasing tendency towards resistance to changes that will allow reversal of the sequence.

In a third alternative approach, O'Keeffe and Davies (1991) obtained information about past and present flows, and flows that were predicted under various management options, and applied the RCS model to assess changes in the conservation status of the river. They assumed that the conservation status of the pristine Sabie was 100% and that, if all flows were impounded, its conservation status would be zero. The model was run for present conditions, and for management regimes involving the construction of two dams (Fig. 25.7).

The present conservation status of the Sabie is deemed very high (84%), reflecting the absence of regulation, low pollution and a diverse indigenous fauna and flora. If the proposed Injaka Dam on the tributary Marite River is maintained at maximal storage capacity, the conservation status of the Sabie would fall to 69% due to loss of water from the Marite during the low-flow period in the Sabie. If, however, the dam operations are eased so that water is stored at high and at intermediate flows, and compensation flows are provided in the dry months, the loss of storage would be only 2% of MAR, but the conservation status of the river would be little affected (81% – a decrease of only 3%; Fig. 25.7). On the other hand, the proposed Madras Dam on the Sabie mainstem would have severe impacts: water chemistry, temperature and sediment loads would change significantly, species would be lost (Russell & Rogers 1989), and the conservation status would fall to 43%, converting the Sabie to yet one more degraded regulated river in southern Africa.

The Sabie River will almost certainly be impounded, if only to ensure that water can be provided to the downstream reaches during the dry season. The encouraging aspect of this case study, however, is that the ecosystem requirements of the Sabie are seriously being assessed, and ecologically sound operation criteria for impoundments are being designed, together with the development of sensible compromises and water allocations for a wide variety of uses and users.

25.8 PRIORITIES FOR RESEARCH AND MANAGEMENT

Generalizations

The foregoing comments and discussion have outlined many of the special attributes of dryland rivers and have drawn attention to the urgency of measures needed for their conservation. Although research and management are progressing – with some outstanding examples of collaborative progress – the pressures of human population growth and the escalating demand for water make it imperative that researchers and managers develop a much more cohesive approach. In this final section, an incomplete agenda of problems requiring attention is listed in no particular order.

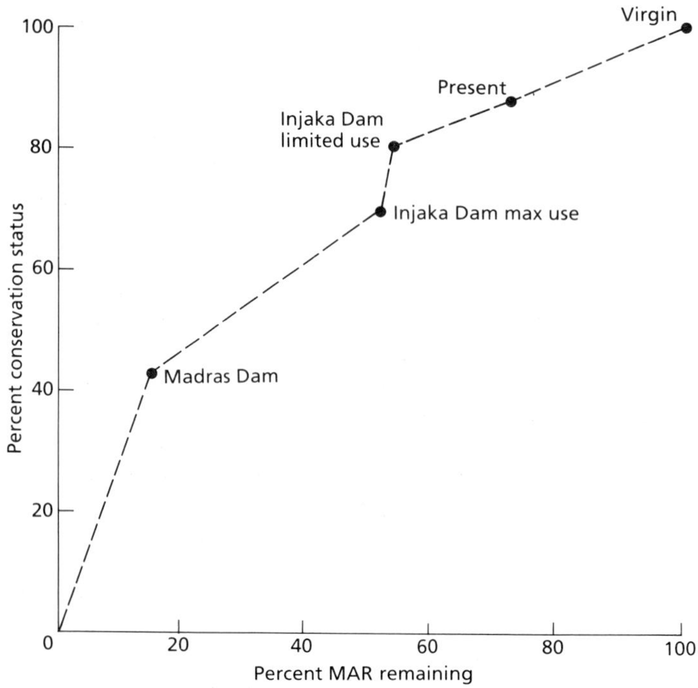

Fig. 25.7 Application of the RCS (O'Keeffe *et al* 1987) to the Sabie River, eastern Transvaal (see text for explanation). The present conservation value of the Sabie is rated at 84%. If the Injaka Dam on the Marite tributary is used to maximum effect, the conservation value would fall to 69%, but with controls, the status would fall only marginally to 81%, with little loss in water yield (3%). The Madras Dam option on the Sabie mainstem would have a much more deleterious effect on the river (after O'Keeffe & Davies 1991).

The list is prefaced by a consolidated statement of the ecological nature of dryland rivers, in order to reinforce the view that such river systems cannot be seen merely as special cases of well-studied rivers in the more humid regions of the world.

Dryland rivers occur in arid and semi-arid regions (mean annual precipitation <500 mm yr^{-1}), in both hot and cold climates, and provide vital water resources for most of the world's people. Their salient feature is hydrological variability, and a tendency towards 'flash' flows, reflecting erratic rainfall patterns and a generally low conversion of MAP : MAR. Channels typically are complex, reflecting the influence of changeable discharges, while sediment loads are high and are augmented by erosion of river channels and catchment soils. Native plants and animals tend to be highly mobile or tolerant of environmental extremes, and often are opportunistic breeders cued especially by flooding. Refugia, such as river-bed pools and subsurface water in the hyporheos, are essential for the survival of the biota during dry periods, and the maintenance of natural communities is dependent upon river-floodplain linkages. Flow regulation, mainly for irrigated agriculture, is a major disturbance because it alters the flow regime as well as the hydrological and geomorphological environment, creating further disadvantages for native biotas (note: we do not regard drying as a disturbance in the frequently used sense of the word – it is a natural and predictable feature of dryland systems).

Management and research priorities

There are many technical and scientific needs for improving river management in drylands. In this context, 15 needs for river management in drylands can be identified (Table 25.4). Six priority needs are emphasized here – they are not listed in any particular order.

The primary threat/disturbance to dryland rivers appears to be flow regulation, and irrigation in particular. The strategy for conservation and management requires a more appropriate balance between the water needs of the environment and those of irrigation. In drylands the balance could be improved by accepting that the margin for

Table 25.4 Priority needs for river management in drylands

Initiate holistic, long-term ecological studies of dryland river ecosystems
Establish framework for collaborative investigations by managers, engineers, geomorphologists and ecologists
Develop research programmes on the impacts of inter-basin transfers
Conserve functional systems
Manage for variability
Maintain habitat heterogeneity
Preserve functional ecotones
Identify and protect native biota
Investigate the population dynamics and life histories of the biota
Initiate studies of flood-pulse dynamics
Determine the role of refugia in recolonization, and their role in conservation
Study hydrological source areas
Protect natural flow regimes
Investigate ecosystem response to low flows
Develop new, less labour-intensive and technology-intensive methods for instream flow assessment and environmental flow allocations

exploitation by irrigators should rise and fall with variations in rainfall and flow. The philosophy of management needs to be fundamentally different from that applied in more stable, seasonal climates, in that patterns of resource utilization require adjustment according to environmental fluctuations. However difficult this may be, *vis-à-vis* traditional economics, it offers the only logical means for conservation.

Piecemeal conservation does little to promote a sustained balance. It is axiomatic that all parts of an ecosystem are interdependent, and it follows that conservation should target functional 'subsets' of dryland river communites. For example, a river channel cannot properly be conserved in isolation from the floodplain, and individual wetlands cannot be conserved without regard for other parts of the floodplain community, or the river itself.

Effective conservation of river-floodplain systems requires that the river corridor be maintained intact, with a representative subset, at least, of each of its key elements. A determination of what is representative must take account of environmental patchiness (e.g. minor topographic relief: Buchholz 1981). The heterogeneity of wetland environments in time and space may be largely responsible for greater biodiversity compared to other floodplain environments, including the river channel (Ward 1989). Indeed, patchiness is undoubtedly responsible, in part, for the comparatively high biodiversity associated with ecotone environments. Both spatial and temporal patchiness are significant at all levels of scale, and are vital considerations for management.

The preservation of functional ecotones is frequently overlooked. For example, the riverine-littoral zone in the lower Murray is constrained particularly by high turbidities and water-level fluctuations associated with flow regulation, including the operation of weirs (e.g. Walker *et al* 1992). This zone supports a comparatively high diversity of plants and animals, including aquatic, terrestrial and amphibious species, and could be considered to be a 'corridor within a corridor'. Protection of the river-edge zone is crucial because it is an ecotone linking the river channel and floodplain environments. All semi-arid ecotones warrant special protection; they provide refuges for comparatively high diversities of plants and animals, and play a crucial role in maintaining the structure and functioning of the river ecosystem (e.g. Naiman & Décamps 1990; Ferrington & Goldhammer 1992).

Walmsley and Davies (1991), reporting on South African systems, have noted that the allocation of water to maintain ecological functioning has been neglected and have urged a change in national environmental ethics and integrated catchment planning and management, in order to take *all* aspects of water use into consideration. This followed Roberts' (1983) suggestion that *c.* 11% of the MAR (13% of total water demand) be allocated for the environmental needs of the rivers of the region. The response by the South African Department of Water Affairs (see Table 25.1) has been to allocate to each user a proportion of the MAR. As Walmsley and Davies (1991) point out, these allocations are still grossly skewed away from conservation needs and the allocation of research funds and manpower resources required to address the environmental water allocation issue properly has, until very recently, been totally inappropriate. The fact that an allocation has been made at all may be viewed, at least, as a small crumb, but system-specific requirements are still neglected (e.g. Davies *et al* 1993b).

In many dryland rivers, most of the discharge originates in a relatively small part of the catchment. These 'source areas', usually in upland regions, are disproportionately important because of their influence on the quality and quantity of water in the system, and warrant special protection and management.

ACKNOWLEDGEMENTS

The funds for the production of this chapter came from the South African Special Programme on Rivers Research, a grant held by the senior author from the South African Foundation for Research Development, and forms part of the 'technology transfer' component of the grant. Part of this chapter draws extensively from a discussion paper prepared by KF Walker and F Sheldon for a UNESCO Regional Ecotones Workshop held at Barmera, South Australia, in September 1992.

REFERENCES

Adams JR. (1991) The impact of stream flow storage in farm dams on the Upper Klein Berg River catchment. Unpublished BSc (Hons) thesis, Department of Environmental and Geographical Science, University of Cape Town. [25.2]

Agnew JD. (1986) Invertebrates of the Orange–Vaal system, with emphasis on the Ephemeroptera. In: Davies BR, Walker KF (eds) *The Ecology of River Systems, Monographiae Biologicae, 60*, pp. 123–34. Dr W Junk, Dordrecht. [25.2]

Alexander WJR. (1985) Hydrology of low-latitude Southern Hemisphere land masses. In: Davies BR, Walmsley RD (eds) *Perspectives in Southern Hemisphere Limnology, Developments in Hydrobiology, 28*, pp. 75–83. Dr W Junk, Dordrecht. [25.1, 25.2]

Allanson BR, Hart RC, O'Keeffe JH, Robarts RD. (1990) *Inland Waters of Southern Africa: An Ecological Perspective. Monographiae Biologicae, 64.* Kluwer Academic Publishers, Dordrecht. [25.2, 25.7]

Andrews ED. (1980) Effective and bankfull discharge of streams in the Yampa River Basin, Colorado and Wyoming. *Journal of Hydrology* **46**: 311–30. [25.2]

Baker VR. (1977) Stream-channel response to floods, with examples from central Texas. *Bulletin of the Geological Society of America* **88**: 1057–77. [25.2]

Barmuta LA, Lake PS. (1982) On the value of the River Continuum Concept. *New Zealand Journal of Marine and Freshwater Research* **16**: 227–31. [25.5]

Boulton AJ. (1988) Composition and dynamics of macroinvertebrate communities in two intermittent streams. Unpublished PhD thesis, Monash University, Clayton, Victoria. [25.2]

Boulton AJ. (1989) Over-summering refuges of aquatic macroinvertebrates in two intermittent streams in Victoria. *Transactions of the Royal Society of South Australia* **113**: 23–34. [25.2]

Boulton AJ, Lake PS. (1988) Australian temporary streams – some ecological characteristics. *Verhandlungen Internationale Vereinigung für theoretische und angewandte Limnologie* **23**: 1380–3. [25.2]

Boulton AJ, Lloyd LN. (1992) Flooding frequency and invertebrate emergence from dry floodplain sediments of the River Murray, Australia. *Regulated Rivers: Research and Management* **7**: 137–51. [25.2]

Boulton AJ, Suter PJ. (1986) Ecology of temporary streams – an Australian perspective. In: De Deckker P, Williams WD (eds) *Limnology in Australia*, pp. 486–96. Council for Scientific and Industrial Research, Dr W Junk, Melbourne. [25.2, 25.4]

Boulton AJ, Stanley EH, Fisher SG, Lake PS. (1992) Over-summering strategies of macroinvertebrates in intermittent streams in Australia and Arizona. In: Robarts RD, Bothwell ML (eds) *Aquatic Ecosystems in Semi-arid Regions: Implications for Resource Management*, pp. 227–37. Environment Canada, National Hydrology Research Institute Symposium Series 7, Saskatoon, Saskatchewan. [25.2, 25.3, 25.4]

Bovee KD. (1982) A guide to stream habitat analysis using the instream flow incremental methodology. *Instream Flow Information Paper Number 12.* US Fish and Wildlife Service, FWS/OBS-82/26, Fort Collins, Colorado. [25.3, 25.5]

Braune E. (1985) Aridity and hydrological characteristics: Chairman's summary. In: Davies BR, Walmsley RD (eds) *Perspectives in Southern Hemisphere Limnology, Developments in Hydrobiology, 28*, pp. 131–6. W. Junk, Dordrecht. [25.2]

Bremner JM, Rogers J, Willis JP. (1990) Sedimentological aspects of the 1988 Orange River floods. *Transactions of the Royal Society of South Africa* **47**: 247–94. [25.2]

Buchholz K. (1981) Effects of minor drainages on woody species distributions in a successional floodplain forest. *Canadian Journal of Forest Research* **11**: 671–6. [25.8]

Bush SF. (1933) The establishment and well-being of trout in Natal waters: An ecological problem. *South African Journal of Science* **30**: 388–93. [25.7]

Byren BA, Davies BR. (1989) The effect of stream regulation on the physico-chemical properties of the

Palmiet River, South Africa. *Regulated Rivers: Research and Management* **3**: 107–21. [25.5]

Cadwallader PL. (1986) Fish of the Murray–Darling system. In: Davies BR, Walker KF (eds) *The Ecology of River Systems, Monographiae Biologicae, 60*, pp. 679–94. Dr W. Junk, Dordrecht. [25.2]

Cambray JA, Davies BR, Ashton PJ. (1986) The Orange–Vaal River system. In: Davies BR, Walker KF (eds) *The Ecology of River Systems, Monographiae Biologicae, 60*, pp. 89–122. Dr W. Junk, Dordrecht. [25.2, 25.7]

Campbell JM, Clark WJ. (1982) Spatial variability of benthic macroinvertebrate populations inhabiting the impact zone of Aquilla Creek, Hill Country, Texas, prior to its impoundment. In: Davis JR (ed) *Proceedings of the Symposium on Recent Benthological Investigations in Texas and Adjacent States*, pp. 123–38. Texas Academy of Science, Austin, Texas. [25.4]

Carlson CA, Muth RT. (1989) The Colorado River: lifeline of the American southwest. In: Dodge DP (ed) *Proceedings of the International Large Rivers Symposium, Canadian Special Publication of Fisheries and Aquatic Sciences* **106**: 220–39. [25.6, 25.7]

Carter AJ, Rogers AH. (1989) *Phragmites* reedbeds in the Kruger National Park: The complexity of change in riverbed state. In: Kienzle S, Maaren H (eds) *Proceedings of the Fourth South African National Hydrological Symposium*, pp. 339–46. University of Pretoria. [25.7]

Coleman JM. (1969) Brahmaputra River: Sedimentary processes and sedimentation. *Sedimentary Geology* **3**: 129–39. [25.2]

Connell JH. (1975) Some mechanisms producing structure in natural communities: a model and evidence from field experiments. In: Cody ML, Diamond JM (eds) *Ecology and Evolution of Communities*, pp. 460–90. Belknap Press, Harvard, Cambridge, MA. [25.3, 25.4]

Connell JH. (1978) Diversity in tropical rain forests and coral reefs. *Science* **199**: 1302–10. [25.5]

Davies BR. (1979) Stream regulation in Africa: a review. In: Ward JV, Stanford JA (eds) *The Ecology of Regulated Streams*, pp. 113–42. Plenum Press, New York. [25.2]

Davies BR. (1986) The Zambezi River system. In: Davies BR, Walker KF (eds) *The Ecology of River Systems, Monographiae Biologicae, 60*, pp. 225–67. W. Junk, Dordrecht. [25.2, 25.6]

Davies BR, Day JA. (1986) *The Biology and Conservation of South Africa's Vanishing Waters*. Department of Extra-Mural Studies, University of Cape Town, and the Wildlife Society of Southern Africa, Western Cape Branch, Cape Town. [25.1]

Davies BR, Walker KF. (eds) (1986) *The Ecology of River Systems, Monographiae Biologicae, 60*. W. Junk, Dordrecht. [25.5, 25.6, 25.7]

Davies BR, Thoms MC, Meador M. (1992) Viewpoint: An assessment of the ecological impacts of inter-basin water transfers, and their threats to river basin integrity and conservation. *Aquatic Conservation: Marine and Freshwater Ecosystems* **2**: 325–49. [25.1, 25.2, 25.6, 25.7]

Davies BR, O'Keeffe JH, Snaddon CD. (1993a) Southern Africa. In: Cushing CE, Cummins KW, Minshall GW (eds) *Lotic Ecosystem Paradigms*. Elsevier, Amsterdam (in press). [25.3, 25.4, 25.6]

Davies BR, O'Keeffe JH, Snaddon CD. (1993b) *A Synthesis of the Ecological Functioning, Conservation and Management of Southern African River Ecosystems*. Special Publication of the Water Research Commission, Pretoria TT62/93. [25.1, 25.2, 25.3, 25.4, 25.6, 25.7, 25.8]

Davis JR. (1982) Macrobenthic diversity of Cibolo Creek, an interrupted, south-central Texas stream. In: Davis JR (ed) *Proceedings of the Symposium on Recent Benthological Investigations in Texas and Adjacent States*, pp. 79–94. Texas Academy of Science, Austin, Texas. [25.4]

Day LA. (1932) The introduction of trout in Natal. *Salmon and Trout Magazine* **69**: 345–52. [25.7]

De Moor IJ, Bruton MN. (1988) *Atlas of Alien and Translocated Indigenous Aquatic Animals in Southern Africa*. South African National Scientific Programmes, Report 144: Council for Scientific and Industrial Research, Pretoria. [25.7]

Dodge DP. (ed) (1989) *Proceedings of the International Large River Symposium (LARS)*. Canadian Special Publication of Fisheries and Aquatic Sciences 106. [25.7]

Dumont HJ. (1986) The Nile River system. In: Davies BR, Walker KF (eds) *The Ecology of River Systems*, pp. 61–74. W. Junk, Dordrecht. [25.7]

Ebel WJ, Becker CD, Mullan JW, Raymond HL. (1989) The Columbia River: towards a holistic understanding. In: Dodge DP (ed) *Proceedings of the International Large Rivers Symposium (LARS)*, pp. 205–19. Canadian Special Publication of Fisheries and Aquatic Sciences 106. [25.7]

Ferrar AA. (ed) (1989) *Ecological Flow Requirements for South African Rivers*. South African National Scientific Programmes Report 162. Council for Scientific and Industrial Research, Pretoria. [25.7]

Ferrar AA, O'Keeffe JH, Davies BR. (1988) *The River Research Programme*. South African National Scientific Programmes Report 146. Council for Scientific and Industrial Research, Pretoria. [25.7]

Ferrington LC, Goldhammer DS. (1992) The biological significance of Capillary Fringe Habitat in large rivers of the Central High Plains of the United States. In: Robarts RD, Bothwell ML (eds) *Aquatic Ecosystems in Semi-arid Regions: Implications for Resource*

Management, pp. 203–25. Environment Canada, National Hydrology Research Institute Symposium Series 7, Saskatoon, Saskatchewan. [25.6, 25.8]

Finlayson BL, McMahon TA. (1988) Australia *vs.* the World: A comparative analysis of stream flow characteristics. In: Warner RF (ed) *Fluvial Geomorphology of Australia*, pp. 17–40. Academic Press, Sydney. [25.1, 25.2, 25.6]

Furness HD. (1981) *The plant ecology of seasonally flooded areas of the Pongolo River floodplain, with particular reference to* Cynodon dactylon *(L.) Pers.* Unpublished PhD thesis, University of Natal, Pietermaritzburg. [25.6]

Furness HD, Breen CM. (1980) The vegetation of seasonally flooded areas of the Pongolo River floodplain. *Bothalia* **13**: 113–18. [25.6]

Gan K, McMahon TA. (1990) Variability of results from the use of PHABSIM in estimating habitat area. *Regulated Rivers: Research and Management* **5**: 233–9. [25.3]

Gore JA. (1977) Reservoir manipulations and benthic macroinvertebrates in a prairie river. *Hydrobiologia* **55**: 113–23. [25.4]

Gore JA. (1989) Case histories of instream flow analyses for permitting environmental impact assessments in the United States. *Southern African Journal of Aquatic Science* **16**: 194–208. [25.3]

Gore JA, Nestler JM. (1988) Instream flow studies in perspective. *Regulated Rivers: Research and Management* **2**: 93–101. [25.5]

Gore JA, Petts GE. (eds) (1989) *Alternatives in Regulated River Management*. CRC Press, Boca Raton, Florida [25.5]

Gore JA, King JM, Hamman KCD. (1991) Application of the instream flow incremental methodology to Southern African rivers: protecting endemic fish of the Olifants River. *Water South Africa* **17**: 225–36. [25.3]

Gore JA, Layzer JB, Russell IA. (1992) Non-traditional applications of instream flow techniques for conserving habitat of biota in the Sabie River of Southern Africa. In: Boon PJ, Calow P, Petts GE (eds) *River Conservation and Management*, pp. 161–77. John Wiley, Chichester. [25.3]

Graf WL. (1985) Magnitude and frequency effects. In: Goudie A (ed) *The Encyclopaedic Dictionary of Physical Geography*, pp. 272–4. Blackwell Scientific Publications, Oxford. [25.2]

Graf WL. (1987) *Fluvial Processes in Dryland Rivers*. Springer-Verlag, Berlin. [25.1, 25.2, 25.7]

Graf WL. (1988) Definition of flood plains along arid-region rivers. In: Baker VR, Kochel RC, Patton PC (eds) *Flood Geomorphology*, pp. 231–42. John Wiley, New York. [25.2]

Gray LJ. (1981) Species composition and life histories of aquatic insects in a lowland Sonoran desert stream. *American Midland Naturalist* **106**: 229–42. [25.2, 25.7]

Grove AT. (1977) The geography of semi-arid lands. *Philosophical Transactions of the Royal Society of London, Series B* **278**: 457–75. [25.2]

Gurtz ME, Marzolf GR, Killingbeck KT, Smith DL, McArthur JV. (1982) *Organic Matter Loading and Processing in a Pristine Stream Draining a Tallgrass Prairie/Riparian Forest Watershed.* Report to Kansas Water Resources Research Institute, Manhattan, KS. [25.4]

Hart RC. (1985) Seasonality of aquatic invertebrates in low-latitude and Southern Hemisphere inland waters. In: Davies BR, Walmsley RD (eds) *Perspectives in Southern Hemisphere Limnology, Developments in Hydrobiology* **28**, pp. 151–78. Dr W. Junk, Dordrecht. [25.4]

Heeg J, Breen CM. (1982) *Man and the Pongolo Floodplain.* South African National Scientific Programmes Report, 56. Council for Scientific and Industrial Research, Pretoria. [25.6]

Hill BH, Gardner TJ, Ekisola OF. (1992) Predictability of streamflow and particulate organic matter concentration as indicators of stability in prairie streams. *Hydrobiologia* **242**: 7–18. [25.4]

Holeman JN. (1968) The sediment yield of major rivers of the world. *Water Resources Research* **4**: 737–47. [25.2]

Howard-Williams C. (1985) Cycling and retention of nitrogen and phosphorus in wetlands: a theoretical and applied perspective. *Freshwater Biology* **15**: 391–431. [25.6]

Hynes HBN. (1970) *The Ecology of Running Waters.* University of Toronto Press, Toronto. [25.4]

Junk WJ, Bayley PB, Sparks RE. (1989) The Flood Pulse Concept in river-floodplain systems. In: Dodge DP (ed) *Proceedings of the International Large Rivers Symposium (LARS), Canadian Special Publication of Fisheries and Aquatic Sciences* **106**: 110–27. [25.5, 25.6]

King JM, Tharme RE. (1993) *Assessment of the Instream Flow Requirements of Rivers.* Final Report to the Water Research Commission, Pretoria. Contract K5/295. [25.2, 25.3, 25.5, 25.7]

King JM, Day JA, Davies BR, Henshall-Howard M-P. (1987a) Particulate organic matter in a mountain stream in the south-western Cape, South Africa. *Hydrobiologia* **154**: 165–87. [25.3]

King JM, Henshall-Howard M-P, Day JA, Davies BR. (1987b) Leaf-pack dynamics in a Southern African mountain stream. *Freshwater Biology* **18**: 325–40. [25.3]

King JM, Day JA, Hurley PR, Henshall-Howard M-P, Davies BR. (1988) Macroinvertebrate communities and environment in a southern African mountain stream. *Canadian Journal of Fisheries and Aquatic*

Sciences **45**: 2168–81. [25.3, 25.4]
King JM, De Moor FC, Chutter FM. (1991) Alternative ways of classifying rivers in southern Africa. In: Boon PJ, Callow P, Petts GE (eds) *River Conservation and Management*, pp. 213–28. John Wiley, Chichester. [25.3, 25.7]
Kirby *et al.* (1983) [25.4]
Koch KR, Churms SC, Spit AT, Davies BR. (1993) A preliminary assessment of the distribution, chemical composition and variability of potential nutrients in spate-generated foam from western Cape black-water rivers. *South African Journal of Science* (in press). [25.4]
Lake PS, Barmuta LA, Boulton AJ, Campbell S, St Claire RM. (1986) Australian streams and Northern Hemisphere stream ecology: Comparisons and problems. *Proceedings of the Ecological Society of Australia* **14**: 61–82. [25.5]
Langbein WB, Schumm SA. (1958) Yield of sediment in relation to mean annual precipitation. *Transactions of the American Geophysical Union* **39**: 1076–84. [25.2]
Lof GOG, Hardison CH. (1966) Storage requirements for water in the United States. *Water Resources Research* **2**: 323–54. [25.2]
McElravy EP, Resh VH. (1991) Distribution and seasonal occurrence of the hyporheic fauna in a northern California stream. *Hydrobiologia* **220**: 233–46. [25.4]
McLachlan AJ. (1974) Development of some lake ecosystems in tropical Africa with special reference to the invertebrates. *Biological Reviews* **49**: 365–97. [25.6]
McMahon TA. (1978) Australia's surface water resources: potential development based on hydrologic factors. *Institution of Engineers Australia* **20**: 155–64. [25.2]
McMahon TA. (1979) Hydrological characteristics of arid zones. In: *The Hydrology of Areas of Low Precipitation*, pp. 105–23. IAHS-AISH Publication 128, Canberra. [25.2]
McMahon TA, Mein RG. (1986) *River and Reservoir Yield*. Water Resources Publications, Colorado. [25.2]
Matthews WJ. (1988) North American prairie streams as systems for ecological study. *Journal of the North American Benthological Society* **7**: 387–409. [25.2, 25.3]
Meigs P. (1953) World distribution of arid and semi-arid homoclimates. In: *Arid Zone Hydrology, United Nations Educational, Scientific and Cultural Organisation, Arid Zone Research Series* **1**: 203–9. [25.2]
Middleton BJ, Lorenz SA, Pitman WV, Midgley DC. (1981) *Surface Water Resources of South Africa, Vol. V: Drainage Regions MNPQRST, Eastern Cape*. Hydrological Research Unit, Report 12/81. University of the Witwatersrand, Johannesburg. [25.2]
Milhous RT, Updike MA, Schneider DM. (1989) *Physical Habitat Simulation System Manual – Version 2*. Instream Flow Information Paper 26, USDI, Fish and Wildlife Service, Biology Report 89(16), Fort Collins, Colorado. [25.5, 25.7]
Molles MC, Dahm CN, Crocker MT. (1992) Climatic variability and streams and rivers in semi-arid regions. In: Robarts RD, Bothwell ML (eds) *Aquatic Ecosystems in Semi-arid Regions: Implications for Resource Management*, pp. 197–202. Environment Canada, National Hydrology Research Institute Symposium Series 7, Saskatoon, Saskatchewan. [25.2]
Moore CA. (1991) A survey of the conservation status and benthic biota of the six major rivers of the Kruger National Park. Unpublished MSc thesis, Zoology Department, University of Pretoria. [25.7]
Moore CA, Chutter FM. (1988) *A Survey of the Conservation Status and Benthic Biota of the Six Major Rivers of the Kruger National Park*. Contract Report, National Institute for Water Research, Council for Scientific and Industrial Research, Pretoria. [25.7]
Naiman RJ, Décamps H (eds). (1990) *The Ecology and Management of Aquatic–Terrestrial Ecotones*. United Nations Educational, Scientific and Cultural Organisation, Paris; Parthenon, Carnforth, England. [25.6, 25.8]
Neff EL. (1967) *Discharge Frequency Compared to Long Term Sediment Yields*. IAHS Publication 75. [25.2]
Nestler JM, Milhous RT, Layzer JB. (1989) Instream habitat modeling techniques. In: Gore JA, Petts GE (eds) *Alternatives in Regulated River Management*, pp. 295–315. CRC Press, Boca Raton, Florida. [25.3]
Newbold JD. (1987) Phosphorus spiralling in rivers and river–reservoir systems: implications of a model. In: Craig JF, Kemper JB (eds) *Regulated Streams: Advances in Ecology*, pp. 303–27. Plenum Press, New York. [25.4]
Newbold JD, O'Neill RV, Elwood JW, Van Winkel W. (1982) Nutrient spiralling in streams. Implications for nutrient limitation in invertebrate activity. *American Naturalist* **120**: 628–52. [25.4]
Noble RG, Hemens J. (1978) *Inland Water Ecosystems in South African – A Review of Research Needs*. South African National Scientific Programmes Report, 34. Council for Scientific and Industrial Research, Pretoria. [25.2]
O'Keeffe JH. (ed) (1986) *The Conservation of South African Rivers*. South African National Scientific Programmes Report 131. Council for Scientific and Industrial Research, Pretoria. [25.7]
O'Keeffe JH, Davies BR. (1991) Conservation and management of the rivers of the Kruger National Park: suggested methods for calculating instream flow needs. *Aquatic Conservation: Marine and Freshwater Ecosystems* **1**: 55–71. [25.1, 25.5, 25.7]
O'Keeffe JH, Danilewitz DB, Bradshaw JA. (1987) An expert system approach to the assessment of the conservation status of rivers. *Biological Conservation*

40: 69–84. [25.7]

O'Keeffe JH, Davies BR, King JM, Skelton PH. (1989a) The conservation status of Southern African rivers. In: Huntley BJ (ed) *Biotic Diversity in Southern Africa: Concepts and Conservation*, pp. 276–99. Oxford University Press, Cape Town. [25.7]

O'Keeffe JH, King JM, Day JA, Roussouw D, Van Der Zel DW, Skorozewski R. (1989b) General concepts and approaches to Instream Flow Assessment. In: Ferrar AA (ed) *Ecological Flow Requirements for South African Rivers*, pp. 41–5. South African National Scientific Programmes Report 162. Council for Scientific and Industrial Research, Pretoria. [25.7]

O'Keeffe JH, Byren BA, Davies BR, Palmer RW. (1990) The effects of impoundment on the physico-chemistry of two southern African river systems. *Regulated Rivers: Research and Management* **5**: 97–110. [25.2, 25.5]

O'Keeffe JH, Uys M, Bruton MN. (1992) Freshwater Systems. In: Fuggle RF, Rabie MA (eds) *Environmental Management in South Africa*, pp. 277–315. Juta & Company, Cape Town. [25.2, 25.7]

Palmer RW, O'Keeffe JH. (1990a) Downstream effects of impoundments on the water chemistry of the Buffalo River (Eastern Cape), South Africa. *Hydrobiologia* **202**: 71–83. [25.5]

Palmer RW, O'Keeffe JH. (1990b) Downstream effects of an impoundment on a turbid river. *Archiv für Hydrobiologie* **119**: 457–73. [25.5]

Peckarsky BL, Horn SC, Statzner B. (1990) Stonefly predation along a hydraulic gradient: a field test of the harsh-benign hypothesis. *Freshwater Biology* **24**: 181–91. [25.3]

Petitjean MOG, Davies BR. (1988a) *Ecological and Environmental Impacts of Inter-Basin Water Transfer Schemes in Southern Africa – A Review*. Foundation for Research Development, Council for Scientific and Industrial Research, Pretoria. [25.1]

Petitjean MOG, Davies BR. (1988b) Ecological impacts of inter-basin water transfers: some case studies, research requirements and assessment procedures in Southern Africa. *South African Journal of Science* **84**: 819–28. [25.1]

Petts GE. (1984) *Impounded Rivers: Perspectives for Ecological Management*. John Wiley, Chichester. [25.2, 25.7]

Petts GE, Thoms MC. (1986) Channel aggradation below Chew Valley Lake, Somerset, England. *Catena* **13**: 305–20. [25.2]

Poff L, Ward JV. (1990) Physical habitat template of lotic systems: recovery in the context of historical pattern of spatio-temporal heterogeneity. *Environmental Management* **14**: 629–45. [25.2]

Puckridge JT, Sheldon F, Boulton AJ, Walker KF. (1993) The Flood Pulse Concept applied to rivers with variable flow regimes (in manuscript). [25.1, 25.5, 25.6]

Ractliffe G. (1991) The influence of discharge on bank litter distribution in Langrivier, a south-western Cape stream. Unpublished BSc (Hons) thesis, Freshwater Research Unit, University of Cape Town. [25.4]

Regier HA, Welcomme RL, Steedman RJ, Henderson HF. (1989) Rehabilitation of degraded river ecosystems. In: Dodge DP (ed) *Proceedings of the International Large Rivers Symposium (LARS), Canadian Special Publication of Fisheries and Aquatic Sciences* **106**: 86–97. [25.7]

Richardson WB. (1990) Seasonal dynamics, benthic habitat use, and drift of zooplankton in a small stream in southern Oklahoma, U.S.A. *Canadian Journal of Zoology* **69**: 748–56. [25.4]

Robarts RD, Bothwell ML (eds). (1992) *Aquatic Ecosystems in Semi-arid Regions: Implications for Resource Management*. Environment Canada, National Hydrology Research Institute Symposium Series 7, Saskatoon, Saskatchewan. [25.7]

Roberts CPR. (1983) Environmental constraints on water resources development. *Proceedings of the South African Institute of Civil Engineers* **1**: 16–23. [25.8]

Rogers KH, Breen CM. (1982) Decomposition of *Potamogeton crispus* L.: the effects of drying on the pattern of mass and nutrient loss. *Aquatic Botany* **12**: 1–12. [25.6]

Russell IA, Rogers KH. (1989) The distribution and composition of fish communities in the major rivers of the Kruger National Park. In: Kienzle S, Maaren H (eds) *Proceedings of the Fourth South African National Hydrological Symposium*, pp. 281–8. University of Pretoria. [25.7]

Sedell JR, Naiman RJ, Cummins KW, Minshall GW, Vannote RL. (1978) Transport of particulate organic material in streams as a function of physical processes. *Verhandlungen Internationale Vereinigung für theoretische und angewandte Limnologie* **10**: 1366–75. [25.4]

Snaddon CD, Stewart BA, Davies BR. (1992) The effect of discharge on retention in two south-western Cape streams. *Archiv für Hydrobiologie* **125**: 109–20. [25.4]

South African Department of Water Affairs. (1986) *Management of the Water Resources of the Republic of South Africa*. Department of Water Affairs, Pretoria. [25.1, 25.8]

Stanford JA, Ward JV. (1986a) The Colorado River System. In: Davies BR, Walker KF (eds) *The Ecology of River Systems, Monographiae Biologicae 60*, pp. 353–74. W. Junk, Dordrecht. [25.2, 25.7]

Stanford JA, Ward JV. (1986b) Fish of the Colorado system. In: Davies BR, Walker KF (eds) *The Ecology of River Systems, Monographiae Biologicae 60*, pp. 385–402. W. Junk, Dordrecht. [25.2]

Stanford JA, Ward JV. (1988) The hyporheic habitat of

river ecosystems. *Nature [London]*, **335**: 64–66. [25.2, 25.4]

Stanley EH, Fisher SG. (1992) Intermittency, disturbance, and stability in stream ecosystems. In: Robarts RD, Bothwell ML (eds) *Aquatic Ecosystems in Semi-arid Regions: Implications for Resource Management*, pp. 271–80. Environment Canada, National Hydrology Research Institute Symposium Series 7, Saskatoon, Saskatchewan. [25.2, 25.5]

Stevens MA, Simons DB, Richardson EV. (1975) Non-equilibrium river form. *Journal of the Hydraulic Division of the Australian Society of Civil Engineers* **101**: 557–66. [25.2]

Stewart BA, Davies BR. (1989) The influence of different litter bag designs on the break-down of leaf material in a small mountain stream. *Hydrobiologia* **183**: 173–7. [25.4]

Thomas DSG. (1989a) The nature of arid environments. In: Thomas DSG (ed) *Arid Zone Geomorphology*, pp. 1–10. Belhaven Press, London; Halsted Press, New York. [25.2]

Thomas DSG. (ed) (1989b) *Arid Zone Geomorphology*. Bellhaven Press, London; Halsted Press, New York. [25.2]

Thoms MC, Walker KF. (1992) Sediment transport in a regulated semi-arid river: the River Murray, Australia. In: Robarts RD, Bothwell ML (eds) *Aquatic Ecosystems in Semi-Arid Regions: Implications for Resource Management*, pp. 239–50. Environment Canada, National Hydrology Research Institute Symposium Series 7, Saskatoon, Saskatchewan. [25.2]

Thoms MC, Walker KF. (1993) Channel change associated with two adjacent weirs on a regulated lowland alluvial river. *Regulated Rivers: Research and Management* **8**(3): 271–84. [25.2]

Thornthwaite CW. (1948) An approach towards a rational classification of climate. *Geographical Review* **38**: 55–94. [25.2]

Van Biljon S, Visser AA. (in manuscript) The effect of dam construction on downstream flow. Division of Hydrology, Pretoria. [25.2]

Vannote RL, Minshall GW, Cummins KW, Sedell JR, Cushing CE. (1980) The River Continuum Concept. *Canadian Journal of Fisheries and Aquatic Sciences* **37**: 130–7. [25.1, 25.5]

Walker KF. (1979) Regulated streams in Australia: The Murray–Darling River System. In: Ward JV, Stanford JA (eds) *The Ecology of Regulated Streams*, pp. 143–63. Plenum Press, New York. [25.2]

Walker KF. (1986) The Murray–Darling River System. In: Davies BR, Walker KF (eds) *The Ecology of River Systems, Monographiae Biologicae 60*, pp. 631–59. Dr W. Junk, Dordrecht. [25.2]

Walker KF, Thoms MC. (1993) Environmental effects of flow regulation on the lower River Murray, Australia. *Regulated Rivers: Research and Management* **8**: 103–19. [25.2, 25.6, 25.7]

Walker KF, Thoms MC, Sheldon F. (1992) Effects of weirs on the littoral environment of the River Murray, South Australia. In Boon PJ, Calow P, Petts GE (eds) *River Conservation and Management*, pp. 271–92. John Wiley, Chichester. [25.8]

Walling DE, Webb BW. (1983) Patterns of sediment yield. In: Gregory KJ (ed) *Background to Palaeohydrology: A Perspective*, pp. 69–100. John Wiley, Chichester. [25.2]

Walmsley RD, Davies BR. (1991) Water for environmental management: an overview. *Water South Africa* **17**: 67–76. [25.7, 25.8]

Ward JV. (1989) The four-dimensional nature of lotic ecosystems. *Journal of the North American Benthological Society* **8**: 2–8. [25.5, 25.6, 25.8]

Ward JV, Stanford JA. (eds) (1979) *The Ecology of Regulated Streams*. Plenum Press, New York. [25.2]

Ward JV, Stanford JA. (1983) The serial discontinuity concept of lotic ecosystems. In: Fontain TD, Bartell SM (eds) *Dynamics of Lotic Ecosystems*, pp. 29–42. Ann Arbor Science Publishers. Michigan. [25.5]

Webster JR. (1975) *Analysis of potassium and calcium dynamics in stream ecosystems on three southern Appalachian watersheds of contrasting vegetation*. Unpublished PhD thesis, University of Georgia, Athens, USA. [25.4]

Welcomme RL. (1979) *Fisheries Ecology of Floodplain Rivers*. Longman, London. [25.6]

Williams DD, Hynes HBN. (1977) The ecology of temporary streams. II. General remarks on temporary streams. *Internationale Revue der gesampten Hydrobiologie* **62**: 53–61. [25.2]

Williams WD. (1988) Limnological imbalances: an antipodean viewpoint. *Freshwater Biology* **20**: 407–20. [25.1, 25.2]

Woodyer KD. (1968) Bankfull frequency in rivers. *Journal of Hydrology* **6**: 114–42. [25.2]

Index

Page numbers in *italic* refer to pages with figures and/or tables.